D1278663

Introduction to Medical Immunology

fourth edition

Introduction to Medical Immunology

fourth edition

EDITED BY

GABRIEL VIRELLA

Medical University of South Carolina
Charleston, South Carolina

MARCEL DEKKER, INC.   NEW YORK • BASEL • HONG KONG

Library of Congress Cataloging-in-Publication Data

Introduction to medical immunology / edited by Gabriel Virella. — 4th ed.
p. cm.
Includes bibliographical references and index.
ISBN 0-8247-9897-X (hardcover : alk. paper)
1. Clinical immunology. 2. Immunology. I. Virella, Gabriel.
[DNLM: 1. Immunity. 2. Immunologic Diseases. QW 504 I6286 1997]
RC582.I59 1997
616.07'9—dc21
DNLM/DLC
for Library of Congress

97-22373
CIP

The publisher offers discounts on this book when ordered in bulk quantities. For more information, write to Special Sales/Professional Marketing at the address below.

This book is printed on acid-free paper.

MARCEL DEKKER, INC.
270 Madison Avenue, New York, New York 10016
http://www.dekker.com

Current printing (last digit):
10 9 8 7 6 5 4 3 2 1

PRINTED IN THE UNITED STATES OF AMERICA

Preface

Ten years after the publication of the first edition of *Introduction to Medical Immunology*, the ideal immunology textbook continues to be a very elusive target. The discipline continues to grow at a brisk pace, and the concepts tend to become obsolete as quickly as we put them in writing. It is very true of immunology that the more we know, the greater our ignorance. This represents the challenge that makes teaching immunology so exceptional and writing immunology textbooks such a daunting task.

The fourth edition of *Introduction to Medical Immunology* retains the features that make this textbook unique—particularly, its emphasis on the clinical application of immunology—but represents a significant departure from the earlier editions. Most changes have resulted from our strong conviction that this textbook is not written to impress our peers with extraordinary insights or revolutionary knowledge, but rather to be helpful to medical students and young professionals who need an introduction to the field. The requirements that we tried to fulfill were sometimes difficult to conciliate. The text needs to be updated and relatively complete, but not overwhelming. The scientific basis of immunology needs to be clearly conveyed without allowing the detail to obscure the concept. The application to medicine needs to be transparently obvious, but without unnecessary exaggeration. The text must present a reasonably general and succinct overview, while covering areas that appear likely to have a strong impact in the foreseeable future. The book needs to stimulate students to seek more information and to develop their own "thinking" without being merely a model of theoretical dreams (and nightmares).

In what is probably not an entirely successful attempt to fulfill some of these goals, we have extensively revised the book, added significant new concepts, and deleted areas that were clearly obsolete. The clinical sections are peppered with cases in order to provide a solid link between the discussion of concrete problems presented by patients with diseases of immunological basis and the relevant immunological principles. More significantly, the book has been rewritten in an outline format. This format allows us to keep the conceptual approach while facilitating the understanding of a reader facing the complexities of immunology with very little background. Of necessity, the book emphasizes that which is well understood, as clearly as we can present it, and we try to promote a general understanding of the discipline at the end of the twentieth century. It is not, and never will be, a finished work. We are certain that we will always wish we could add here and revise there. But we hope that this new edition will be even more successful in focusing the attention of our readers toward an intrinsically fascinating discipline that seeks understanding of fundamental biological knowledge that has direct impact on the diagnosis and treatment of a variety of conditions in which the immune system plays a key role.

Gabriel Virella, M.D., Ph.D.

Contents

Contributors

Barbara E. Bierer, M.D. Associate Professor of Medicine, Department of Pediatric Oncology, Dana Farber Cancer Institute, Boston, Massachusetts

Robert J. Boackle, Ph.D. Professor and Director of Oral Biology and Professor of Immunology, Division of Stomatology, Medical University of South Carolina, Charleston, South Carolina

Jonathan S. Bromberg, M.D., Ph.D. Associate Professor of Surgery, Microbiology, and Immunology, Department of General Surgery, Transplant Division, University of Michigan Hospitals, Ann Arbor, Michigan

Albert F. Finn, Jr., M.D. Clinical Assistant Professor, Departments of Medicine, Microbiology and Immunology, Medical University of South Carolina, Charleston, South Carolina

Jean-Michel Goust, M.D. Professor of Immunology, Department of Microbiology and Immunology, Medical University of South Carolina, Charleston, South Carolina

Anne L. Jackson, Ph.D. Consultant, Ridgefield, Washington

Janardan P. Pandey, Ph.D. Professor, Department of Microbiology and Immunology, Medical University of South Carolina, Charleston, South Carolina

Christian C. Patrick, M.D., Ph.D. Director of Academic Programs and Associate Member, Department of Infectious Diseases and Pathology and Laboratory Medicine, St. Jude Children's Research Hospital, Memphis, Tennessee

Mary Ann Spivey, M.H.S., M.T. (A.S.C.P.), S.B.B. Department of Pathology–Laboratory Medicine, Transfusion Medicine Section, Medical University of South Carolina, Charleston, South Carolina

Henry C. Stevenson-Perez, M.D. Senior Investigator, Biologics Evaluation Section, Investigational Drug Branch, National Cancer Institute, National Institutes of Health, Rockville, Maryland

Kwong-Y. Tsang, Ph.D. Senior Scientist, Laboratory of Tumor Immunology and Biology, National Cancer Institute, National Institutes of Health, Bethesda, Maryland

George C. Tsokos, M.D. Professor, Department of Medicine, Uniformed Services University of Health Sciences, Bethesda, Maryland, and Department of Clinical Investigations, Walter Reed Army Medical Center, Washington, D.C.

Gabriel Virella, M.D., Ph.D. Professor and Vice Chairman of Education, Department of Microbiology and Immunology, Medical University of South Carolina, Charleston, South Carolina.

An-Chuan Wang, Ph.D. Professor, Department of Microbiology and Immunology, Medical University of South Carolina, Charleston, South Carolina

1
Introduction

Gabriel Virella

I. INTRODUCTION

A. The fundamental observation that led to the development of immunology as a scientific discipline was that an individual can become resistant for life to a certain disease after having contracted it only once. The term immunity, derived from the Latin "immunis" (exempt), was adopted to designate this naturally acquired protection against diseases such as measles or smallpox.

B. The emergence of immunology as a discipline was closely tied to the development of microbiology. The work of Pasteur, Koch, Metchnikoff, and many other pioneers of the golden age of microbiology resulted in the rapid identification of new infectious agents, closely followed by the discovery that infectious diseases could be prevented by exposure to killed or attenuated organisms, or to compounds extracted from the infectious agents. The impact of immunization against infectious diseases such as tetanus, pertussis, diphtheria, and smallpox, to name just a few examples, can be grasped when we reflect on the fact that these diseases, which were significant causes of mortality and morbidity, are now either extinct or very rarely seen. Indeed, it is fair to state that the impact of vaccination and sanitation on the welfare and life expectancy of humans has had no parallel in any other developments of medical science.

C. In the second part of this century, immunology started to transcend its early boundaries and become a more general biomedical discipline. Today, the study of immunological defense mechanisms is still an important area of research, but immunologists are involved in a much wider array of problems, such as self–nonself discrimination, control of cell and tissue differentiation, transplantation, cancer immunotherapy, etc. The focus of interest has shifted toward the basic understanding of how the immune system works in the hope that this insight will allow novel approaches to its manipulation.

II. GENERAL CONCEPTS

A. **Specific and Nonspecific Defenses.** The protection of the organism against infectious agents involves many different mechanisms, some nonspecific (i.e., generically applicable to many different pathogenic organisms), and others specific (i.e., their protective effect is directed to one single organism).

1. **Nonspecific defenses**, which as a rule are innate (i.e., all normal individuals are born with it), include:
    a. Mechanical barriers, such as the integrity of the epidermis and mucosal membranes.
    b. Physicochemical barriers, such as the acidity of the stomach fluid.
    c. Antibacterial substances (e.g., lysozyme) present in external secretions.
    d. Normal intestinal transit and normal flow of bronchial secretions and urine, which eliminate infectious agents from the respective systems.
    e. Ingestion and elimination of bacteria and particulate matter by **granulocytes**, which is independent of the immune response.
2. **Specific defenses**, as a rule, are induced during the life of the individual as part of the complex sequence of events designated as the **immune response**.

B. Unique Characteristics of the Immune Response. The immune response has two unique characteristics:

1. **Specificity** for the eliciting antigen. For example, immunization with poliovirus only protects against poliomyelitis, not against the flu. The specificity of the immune response is due to the existence of exquisitely discriminative antigen receptors on lymphocytes. Only a single or a very limited number of similar structures can be accommodated by the receptors of any given lymphocyte. When those receptors are occupied, an activating signal is delivered to the lymphocytes. Therefore, only those lymphocytes with specific receptors for the antigen in question will be activated.
2. **Memory**, meaning that repeated exposures to a given antigen elicit progressively more intense specific responses. Most immunizations involve repeated administration of the immunizing compound, with the goal of establishing a long-lasting, protective response. The increase in the magnitude and duration of the immune response with repeated exposure to the same antigen is due to the proliferation of antigen-specific lymphocytes after each exposure. The numbers of responding cells will remain increased even after the immune response subsides. Therefore, whenever the organism is exposed again to that particular antigen, there is an expanded population of specific lymphocytes available for activation and, as a consequence, the time needed to mount a response is shorter and the magnitude of the response is higher.

C. Stages of the Immune Response. To better understand how the immune response is generated, it is useful to consider it as divided into separate sequential stages (Table 1.1). The first stage (induction) involves a small lymphocyte population with specific receptors able to recognize an antigen or a fragment generated by specialized cells known as antigen-presenting cells (APC). The proliferation and differentiation of antigen-responding lymphocytes is usually enhanced by amplification systems involving APC and specialized T-cell subpopulations (T helper cells, defined below) and is followed by the production of effector molecules (antibodies) or by the differentiation of effector cells (cells which directly or indirectly mediate the elimination of undesirable elements). The final outcome, therefore, is the elimination of the microbe or compound that triggered the

Table 1.1 A Simplified Overview of the Three Main Stages of the Immune Response

Stage of the immune response	Induction	Amplification	Effector
Cells/molecules involved	Antigen-presenting cells; lymphocytes	Antigen presenting cells; helper T lymphocytes	Antibodies (+ complement or cytotoxic cells); cytotoxic T lymphocytes; macrophages
Mechanisms	Processing and/or presentation of antigen; recognition by specific receptors on lymphocytes	Release of cytokines; signals mediated by interaction between cell membrane molecules	Complement-mediated lysis; opsonization and phagocytosis; cytotoxicity
Consequences	Activation of T and B lymphocytes	Proliferation and differentiation of T and B lymphocytes	Elimination of non-self; neutralization of toxins and viruses

reaction by means of activated immune cells or by reactions triggered by mediators released by the immune system.

III. THE CELLS OF THE IMMUNE SYSTEM

A. **Lymphocytes and Lymphocyte Subpopulations.** The peripheral blood contains two large populations of cells: the red cells, whose main physiological role is to carry oxygen to tissues, and the white blood cells, which have as their main physiological role the elimination of potentially harmful organisms or compounds. Among the white blood cells, lymphocytes are particularly important because of their primordial role in the immune response. Several subpopulations of lymphocytes have been defined:

1. **B lymphocytes**, which are the precursors of antibody-producing cells, known as plasma cells.
2. **T lymphocytes**, or T-cells, which are further divided into several subpopulations:
   a. **Helper T lymphocytes (TH)**, which play a very significant amplification role in the immune responses. Two functionally distinct subpopulations of T helper lymphocytes have been well defined in mice.
      i. TH1 lymphocytes, which assist the differentiation of cytotoxic cells and also activate macrophages, which after activation play a role as effectors of the immune response.
      ii. TH2 lymphocytes, which are mainly involved in the amplification of B lymphocyte responses.

   These amplifying effects of helper T lymphocytes are mediated in part by soluble mediators—**interleukins**—and in part by signals delivered as a consequence of cell–cell contact.

b. **Cytotoxic T lymphocytes**, which are the main immunological effector mechanisms involved in the elimination of non-self or infected cells.

3. **Antigen-presenting cells**, such as the macrophages and macrophage-related cells, play a very significant role in the induction stages of the immune response by trapping and presenting both native antigens and antigen fragments in a most favorable way for the recognition by lymphocytes. In addition, these cells also deliver activating signals to lymphocytes engaged in antigen recognition, both in the form of soluble mediators (interleukins such as IL-12 and IL-1) and in the form of signals delivered by cell–cell contact.
4. **Phagocytic and cytotoxic cells**, such as monocytes, macrophages, and granulocytes, also play significant roles as effectors of the immune response. Once antibody has been secreted by plasma cells and is bound by the microbes, cells, or compounds that triggered the immune response, it is able to induce their ingestion by phagocytic cells. If bound to live cells, antibody may induce the attachment of cytotoxic cells that cause the death of the antibody-coated cell (**antibody-dependent cellular cytotoxicity; ADCC**). The ingestion of microorganisms or particles coated with antibody is enhanced when an amplification effector system known as **complement** is activated.
5. **Natural killer (NK) cells** play a dual role in the elimination of infected and malignant cells. These cells are unique in that they have two different mechanisms of recognition: they can identify directly virus-infected and malignant cells and cause their destruction, and they can participate in the elimination of antibody-coated cells by ADCC.

IV. ANTIGENS AND ANTIBODIES

A. **Antigens** are non-self substances (cells, proteins, polysaccharides) that are recognized by receptors on lymphocytes, thereby eliciting the immune response. The receptor molecules located on the membrane of lymphocytes interact with small portions of those foreign cells or proteins, designated as **antigenic determinants** or **epitopes**. An adult human being has the capability of recognizing millions of different antigens, some of microbial origin, others present in the environment, and even some artificially synthesized.

B. **Antibodies** are proteins that appear in circulation after immunization and that have the ability to react specifically with the antigen used to immunize. Because antibodies are soluble and are present in virtually all body fluids ("humors"), the term **humoral immunity** was introduced to designate the immune responses in which antibodies play the principal role as effector mechanisms. Antibodies are also generically designated as **immunoglobulins**. This term derives from the fact that antibody molecules structurally belong to the family of proteins known as globulins (globular proteins) and from their involvement in immunity.

C. **Antigen–Antibody Reactions, Complement, and Phagocytosis**. The knowledge that the serum of an immunized animal contained protein molecules able to bind specifically to the antigen led to exhaustive investigations of the characteristics and consequences of the **antigen–antibody reactions**.

1. If the antigen is soluble, the reaction with specific antibody under appropriate conditions results in **precipitation** of large antigen–antibody aggregates.
2. If the antigen is expressed on a cell membrane, the cell will be cross-linked by antibody and form visible clumps (**agglutination**).
3. Viruses and soluble toxins released by bacteria lose their infectivity or pathogenic properties after reaction with the corresponding antibodies (**neutralization**).
4. Antibodies complexed with antigens can activate the **complement system**. This system is composed of nine major proteins or components which are sequentially activated. Some of the complement components are able to promote ingestion of microorganisms by phagocytic cells (**phagocytosis**), while others are inserted into cytoplasmic membranes and cause their disruption, leading to cell death.
5. Antibodies can cause the destruction of microorganisms by promoting their ingestion by **phagocytic cells** or their destruction by **cytotoxic cells**. Phagocytosis is particularly important for the elimination of bacteria and involves the binding of antibodies and complement components to the outer surface of the infectious agent (**opsonization**) and recognition of the bound antibody and/or complement components as a signal for ingestion by the phagocytic cell.
6. Antigen–antibody reactions are the basis of certain pathological conditions, such as allergic reactions. Antibody-mediated allergic reactions have a very rapid onset, in a matter of minutes, and are known as **immediate hypersensitivity reactions**.

V. LYMPHOCYTES AND CELL-MEDIATED IMMUNITY

A. Lymphocytes as Effector Cells. Lymphocytes play a significant role as effector cells in two types of situations:

1. **Immune destruction of infected cells**, which are not amenable to destruction by phagocytosis or complement-mediated lysis. The study of how the immune system recognizes and eliminates infected cells resulted in the definition of the **biological role of the histocompatibility antigens** that had been described as responsible for graft rejection (see below).
   a. Intracellular organisms, such as viruses, need to replicate. During the replication cycle of a virus, for example, the infected cells will synthesize viral proteins and viral nucleic acids.
   b. Some of the synthesized viral proteins are cleaved by proteolytic enzymes, and the small peptides resulting from this process become associated with histocompatibility antigens, at which point the complex is transported to the membrane and presented to the immune system.
   c. The immune system does not respond (i.e., is tolerant) to self antigens, including antigens of the **major histocompatibility complex** (**MHC**), an extremely polymorphic system with hundreds of alleles, which is responsible for the rejection of tissues and organ grafts (see below). However, the complex formed by an autologous MHC antigen and a

non-self viral peptide is recognized by the immune system and an immune response is mounted against cells expressing these complexes.
   d. The same general process is involved in the elimination of cells infected by bacteria, parasites, or fungi.
2. The fight against intracellular infections involves several effector mechanisms.
   a. Specific **cytotoxic T lymphocytes** are able to destroy infected cells expressing complexes of MHC molecules and microbial-derived peptides on their membrane.
   b. **TH1 lymphocytes** can also recognize microbial peptides expressed on the membrane of infected cells, particularly of macrophages and related APC. The responding TH1 cells, in turn, release cytokines, such as **interferon-γ**, which activate macrophages and increase their ability to destroy the intracellular infecting agents.
3. Because of the primary involvement of lymphocytes in these reactions, they fall under the category of **cell-mediated immunity**. The elimination of intracellular infectious agents can be considered as the main physiological role of cell-mediated immunity. Other immune reactions mediated by cells are responsible for pathological conditions, as described below.
4. Some inflammatory processes, particularly skin reactions known as **cutaneous hypersensitivity**, which are induced by direct skin contact or by intradermal injection of antigenic substances, are also mediated by T lymphocytes. These reactions express themselves 24 to 48 hours after exposure to an antigen to which the patient had been previously sensitized. For this reason, these reactions are designated as **delayed hypersensitivity** and are a pathological manifestation of **cell-mediated immunity**.
5. Transplantation of tissues among genetically different individuals of the same species or across species is followed by rejection of the grafted organs or tissues (**graft rejection**). Cell-mediated immunity triggered by differences in **transplantation** or **histocompatibility antigens**, which are generically grouped as the **MHC**, is responsible for graft rejection.

VI. SELF VERSUS NON-SELF DISCRIMINATION

The immune response is triggered by the interaction of an antigenic determinant with specific receptors on lymphocytes. It is calculated that there are several million different receptors in lymphocytes necessary to respond to the wide diversity of epitopes presented by microbial agents and exogenous particles that stimulate immune responses. The basis for such discrimination between self and non-self is the array of structural differences between self and non-self:

A. Infectious agents have marked differences in their chemical structure, easily recognizable by the immune system.

B. Cells, proteins, and polysaccharides from animals of different species have differences in chemical constitution which, as a rule, are directly related to the degree of phylogenetic divergence between species. Those also elicit potent immune responses.

C. Many polysaccharides and proteins from individuals of the same species show antigenic heterogeneity, reflecting the genetic diversity of individuals within a species. Those differences are usually minor (relative to differences between species), but can still be recognized by the immune system. Transfusion reactions, graft rejection, and hypersensitivity reactions to exogenous human proteins are clinical expressions of the recognition of this type of difference between individuals.

D. An important corollary of the exquisite ability of the immune system to recognize differences in chemical structure between self tissues and foreign cells or substances is the need for the immune system not to respond to self, in spite of having the potential to generate lymphocytes with receptors able to interact with epitopes expressed by self-antigens. During embryonic differentiation the immune system eliminates or turns off auto-reactive lymphocytes. The state of tolerance is maintained during the lifetime of healthy individuals by mechanisms not fully understood.

VII. GENERAL OVERVIEW

One of the most difficult intellectual exercises in immunology is to try to understand the organization and control of the immune system. Its extreme complexity and the wide array of regulatory circuits involved in fine-tuning the immune response pose a formidable obstacle to our understanding. A concept map depicting a simplified view of the immune system is reproduced in Figure 1.1.

A. If we use as an example the activation of the immune system by an infectious agent that has managed to overcome the innate anti-infectious defenses, the first step must be the uptake of the infectious agent by an **antigen-presenting cell**, such as a tissue **macrophage**. Such uptake will most likely be productive in terms of the activation of an immune response when it takes place in a lymphoid organ (lymph node, spleen), where there is ample opportunity for interactions with the other cellular elements of the immune system.

B. The antigen-presenting cells will adsorb the infectious agent to their surface, ingest some of the absorbed microorganism, and process it into small antigenic subunits. These subunits become intracellularly associated with histocompatibility antigens, and the resulting complex is transported to the cytoplasmic membrane, allowing stimulation of **helper T lymphocytes**.

C. The interaction between surface proteins expressed by antigen-presenting cells and T lymphocytes as well as interleukins released by the antigen-presenting cells act as co-stimulants of the helper T cells.

D. Once stimulated to proliferate and differentiate, helper T cells become able to assist the differentiation of **effector cells**.

1. Activated **TH1 helper lymphocytes** secrete interleukins that will act on a variety of cells, including **macrophages** (further increasing their level of activation and enhancing their ability to eliminate infectious agents that may be surviving intracellularly), and **cytotoxic T cells**, which are very efficient in the elimination of virus-infected cells.
2. Activated **TH2 helper lymphocytes** secrete a different set of cytokines

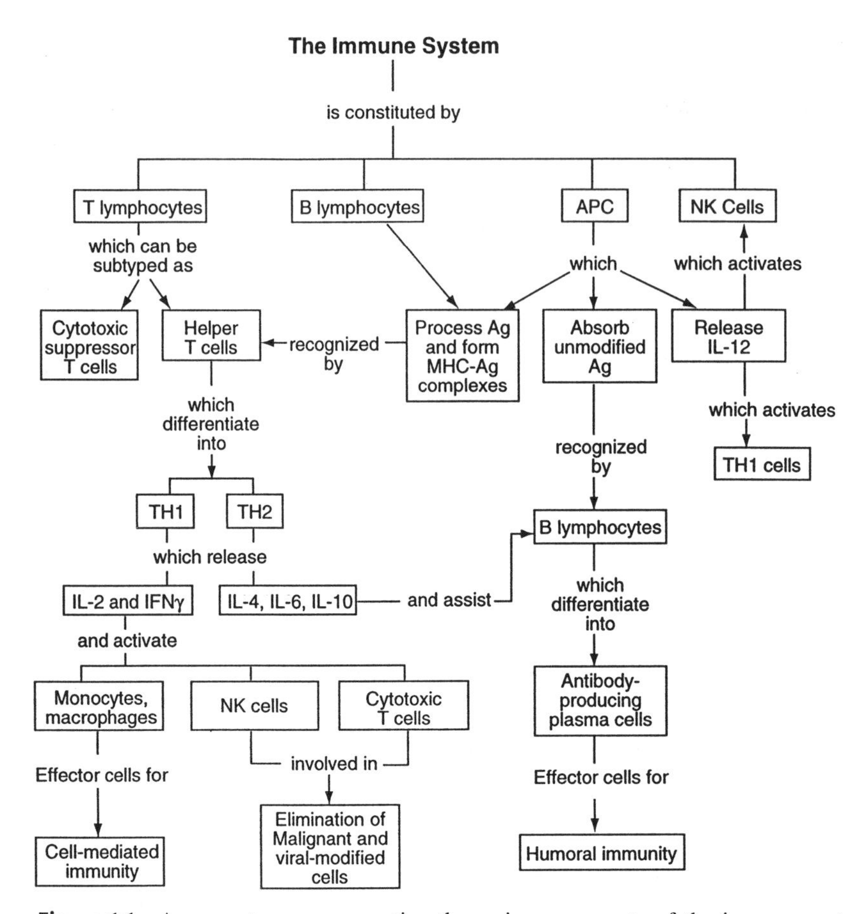

Figure 1.1 A concept map representing the main components of the immune system and their interactions.

that will assist the proliferation and differentiation of antigen-stimulated B lymphocytes, which then differentiate into **plasma cells**. The plasma cells are engaged in the synthesis of large amounts of antibody.

E. Specific antibody will bind to the microorganism and promote its elimination, by one or several of three major mechanisms:
1. Complement-mediated lysis
2. Phagocytosis
3. ADCC

F. Once the microorganism is removed, negative feedback mechanisms become predominant, turning off the immune response. The down-regulation of the immune response appears to result from the combination of several factors, such as the elimination of the positive stimulus that the microorganism represented, and the activation of lymphocytes with suppressor activity, known as **suppressor T cells**.

G. At the end of the immune response, a residual population of long-lived lymphocytes specific for the offending antigen will remain. This is the population of

memory cells that is responsible for protection after natural exposure or immunization. It is also the same generic cell subpopulation that may cause accelerated graft rejections in recipients of multiple grafts. As discussed in greater detail below, the same immune system that protects us can be responsible for a variety of pathological conditions.

VIII. IMMUNOLOGY AND MEDICINE

Immunological concepts have found ample application in medicine, in areas related to diagnosis, treatment, prevention, and pathogenesis.

A. The exquisite specificity of the antigen-antibody reaction has been extensively applied to the development of **diagnostic assays** for a variety of substances.

B. Also, experiments with malignant plasma cell lines obtained from mice with plasma cell tumors culminated serendipitously in the discovery of the technique of hybridoma production, the basis for the production of **monoclonal antibodies**, which have had an enormous impact in the fields of diagnosis and immunotherapy.

C. **Immunotherapy**, once derided as little more than wishful thinking, is coming of age. The therapeutic use of interleukins, activated cytotoxic cells, monoclonal antibodies, anti-idiotypic antibodies, and immunotoxins are being extensively investigated, particularly in oncology and transplantation.

D. The study of children with deficient development of their immune systems (**immunodeficiency diseases**) has provided the best tools for the study of the immune system in humans, while at the same time giving us ample opportunity to devise corrective therapies. The emergence of the acquired immunodeficiency syndrome (AIDS) underscores the delicate balance that is maintained between the immune system and infectious agents in the healthy individual.

E. The importance of maintaining self tolerance in adult life is obvious when we consider the consequences of the loss of tolerance. Several diseases, some affecting single organs, others of a systemic nature, have been classified as **autoimmune diseases**. In those diseases, the immune system reacts against cells and tissues and this reactivity can either be the primary insult leading to the disease, or may represent a factor contributing to the evolution and increasing severity of the disease. New knowledge of how to induce a state of unresponsiveness in adult life through oral ingestion of antigens has raised hopes for the rational treatment of autoimmune conditions.

F. Not all reactions against non-self are beneficial. If and when the delicate balance that keeps the immune system from overreacting is broken, **hypersensitivity diseases** may become manifest. The common allergies, such as asthma and hay fever, are prominent examples of diseases caused by hypersensitivity reactions. The manipulation of the immune response to induce a protective rather than a harmful immunity was first attempted with success in this type of disease.

G. Research into the mechanisms underlying the normal state of tolerance against non-self attained during normal **pregnancy** continues to be intensive, since this knowledge could be the basis for more effective manipulations of the immune response in patients needing organ transplants and for the treatment or prevention of infertility.

H. The concept that malignant mutant cells are constantly being eliminated by the immune system (**immune surveillance**) and that malignancies develop when the mutant cells escape the protective effects of the immune system has been extensively debated, but not quite proven. However, anticancer therapies directed at the enhancement of **antitumoral responses** continue to be evaluated.

In the following chapters of this book, we will illustrate abundantly the productive interaction that has always existed in immunology between basic concepts and clinical applications. In fact, no other biological discipline illustrates better the importance of the interplay between basic and clinical scientists; in this probably lies the main reason for the prominence of immunology as a biomedical discipline.

2

Cells and Tissues Involved in the Immune Response

Gabriel Virella and Jean-Michel Goust

I. INTRODUCTION

The fully developed immune system of humans and most mammals is constituted by a variety of cells and tissues whose different functions are remarkably well integrated. Among the cells, the lymphocytes play the key roles in the control and regulation of immune responses as well as in the recognition of infected or heterologous cells, which the lymphocytes can recognize as undesirable and promptly eliminate. Among the tissues, the thymus is the site of differentiation for T lymphocytes during embryonic differentiation and, as such, is directly involved in critical steps in the differentiation of the immune system.

II. CELLS OF THE IMMUNE SYSTEM

A. **Lymphocytes.** The lymphocytes (Fig. 2.1A) occupy a very special place among the leukocytes that participate in one way or another in immune reactions due to their ability to interact specifically with antigenic substances and to react to non-self antigenic determinants. Lymphocytes differentiate from stem cells in the fetal liver, bone marrow, and thymus into two main functional classes. They are found in the peripheral blood and in all lymphoid tissues.

1. **B lymphocytes** or **B cells** are so designated because the **Bursa of Fabricius**, a lymphoid organ located close to the caudal end of the gut in birds, plays a key role in their differentiation. Removal of this organ, at or shortly before hatching, is associated with lack of differentiation, maturation of B lymphocytes, and the inability to produce antibodies. A mammalian counterpart to the avian bursa has not yet been found. Some investigators believe that the bone marrow is the most likely organ for B lymphocyte differentiation, while others propose that the peri-intestinal lymphoid tissues play this role.
   a. B lymphocytes carry **immunoglobulins** on their cell membranes, which function as antigen receptors. After proper stimulation, B cells differentiate into antibody-producing cells (plasma cells).

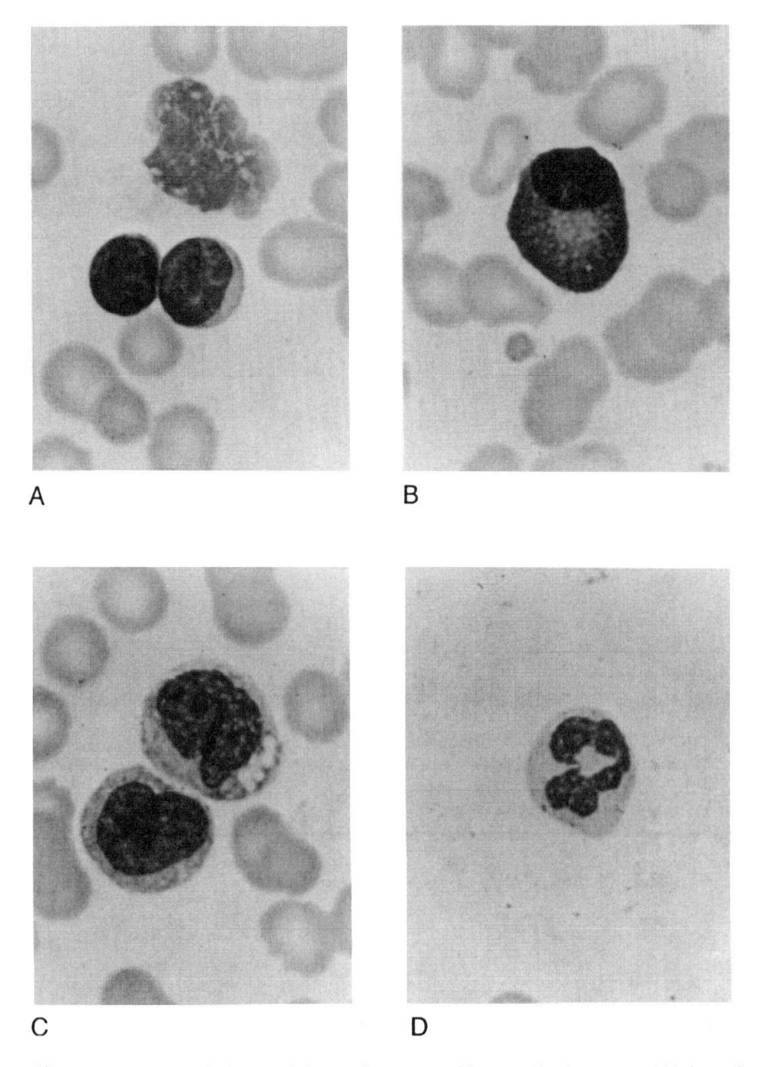

Figure 2.1 Morphology of the main types of human leukocytes. (A) lymphocyte; (B) plasma cell; (C) monocyte; (D) granulocyte. (Reproduced with permission from Reich, P.R., *Manual of Hematology*. Upjohn, Kalamazoo, MI, 1976.)

b. B lymphocytes can also play the role of **antigen-presenting cells** (**APC**), which is usually attributed to cells of monocyte/macrophage lineage.

2. **T lymphocytes** or **T cells** are so designated because the **thymus** plays a key role in their differentiation.

   a. The functions of the T lymphocytes include the **regulation of immune responses**, and various effector functions (cytotoxicity and lymphokine production being the main ones) that are the basis of **cell-mediated immunity** (**CMI**).

b. T lymphocytes also carry an antigen-recognition unit on their membranes, known as **T-cell receptor**. T-cell receptors and immunoglobulin molecules are structurally unrelated.
c. Several subpopulations of T lymphocytes with separate functions have been recognized:
    i. **Helper T lymphocytes** are involved in the induction and regulation of immune responses
    ii. **Cytotoxic T lymphocytes** are involved in the destruction of infected cells.
    iii. It is also known that at specific stages of the immune response T lymphocytes can have **suppressor** functions.
d. To date, there are no known markers that perfectly differentiate T lymphocytes with different functions, although it is possible to differentiate cells with predominant helper function from those with predominant cytotoxic function.
e. **T-cell mediated cytotoxicity is an apoptotic process** that appears to be mediated by two separate pathways. One involves the release of proteins known as **perforins**, which insert themselves in the target cell membranes forming channels. These channels allow the diffusion of enzymes (**granzymes**, which are serine esterases) into the cytoplasm. The exact way in which granzymes induce apoptosis has not been established, but granzyme-induced apoptosis is Ca^{2+}-dependent. The other pathway, which can be easily demonstrated in knock-out laboratory animals in whom the perforin gene is inactivated or by carrying out killing experiments in buffers without Ca^{2+}, depends on signals delivered by the cytotoxic cell to the target cell which require **cell–cell contact** (see Chapter 11).
f. T lymphocytes have a longer life span than B lymphocytes. Long-lasting lymphocytes are particularly important because of their involvement in immunological **memory**.

3. Upon recognizing an antigen and receiving additional signals from auxiliary cells, a small, resting T lymphocyte rapidly undergoes **blastogenic transformation** into a large lymphocyte (13–15 μm). This large lymphocyte (lymphoblast) then subdivides to produce an expanded population of medium (9–12 μm) and small (5–8 μm) lymphocytes with the same antigenic specificity.
    a. Activated and differentiated T lymphocytes are morphologically indistinguishable from small, resting lymphocytes.
    b. Activated B lymphocytes differentiate into plasma cells, easy to distinguish morphologically from resting B lymphocytes.

B. Plasma Cells are morphologically characterized by their eccentric nuclei with clumped chromatin, and a large cytoplasm with abundant rough endoplasmic reticulum (Fig. 2.1B). Plasma cells produce and secrete large amounts of immunoglobulins, but do not express membrane immunoglobulins. Plasma cells divide very poorly, if at all. Plasma cells are usually found in the bone marrow and in the perimucosal lymphoid tissues.

C. Natural Killer (NK) Cells are morphologically described as **large granular lymphocytes**. These cells do not carry antigen receptors of any kind, but can recognize antibody molecules bound to target cells and destroy those cells using the same general mechanisms involved on T-lymphocyte cytotoxicity (**antibody-dependent cellular cytotoxicity**). They also have a recognition mechanism that allows them to destroy tumor cells and virus-infected cells.

D. **Monocytes and Macrophages**

1. The **monocyte** (Fig. 2.1C) is considered a leukocyte in transit through the blood which will become a **macrophage** when fixed in a tissue.
2. Monocytes and macrophages, as well as granulocytes (see below), are able to ingest particulate matter (microorganisms, cells, inert particles) and for this reason are said to have phagocytic functions. The phagocytic activity is greater in macrophages (particularly after activation by soluble mediators released during immune responses) than in monocytes.
3. Macrophages, monocytes, and related cells play an important role in the inductive stages of the immune response by processing complex antigens and concentrating antigen fragments on the cell membrane. In this form, the antigen is recognized by helper T lymphocytes, as discussed in detail in Chapters 3 and 11. For this reason, these cells are known as antigen-presenting cells.
4. APC include other cells sharing certain functional properties with monocytes and macrophages are present in skin (**Langerhans cells**), kidney, brain (microglia), capillary walls, and lymphoid tissues. The Langerhans cells can migrate to the lymph nodes, where they interact with T lymphocytes and assume the morphological characteristics of **interdigitating cells** (see below).
5. One type of monocyte-derived cell, the **dendritic cell** (Fig. 2.2), is present in the spleen and lymph nodes, particularly in follicles and germinal centers. This cell, apparently of monocytic lineage, is not phagocytic, but appears particularly suited to carry out the antigen-presenting function by concentrating antigen on its membrane and keeping it there for relatively long periods of time, a factor that may be crucial for a sustained immune response. The dendritic cells form a network in the germinal centers, known as the **antigen-retaining reticulum**.
6. All antigen-presenting cells express one special class of histocompatibility antigens, designated as **class-II MHC** or **Ia** (I region-associated) antigens (see Chapter 3). The expression of MHC-II molecules is essential for the interaction with helper T lymphocytes.
7. Antigen-presenting cells also release cytokines, which assist the proliferation of antigen-stimulated lymphocytes, including interleukins 1, 6, and 12.

E. **Granulocytes** are a collection of white blood cells with segmented or lobulated nuclei and granules in their cytoplasm which are visible with special stains.

1. Because of their segmented nuclei, which assume variable sizes and shapes, these cells are generically designated as **polymorphonuclear (PMN)** leukocytes (Fig. 2.1D).
2. Different subpopulations of granulocytes (**neutrophils**, **eosinophils**, and **basophils**) can be distinguished by differential staining of the cytoplasmic granules, which reflect their different chemical constitution.
3. **Neutrophils** are the largest subpopulation of white blood cells and have two types of cytoplasmic granules containing compounds with bactericidal activity.

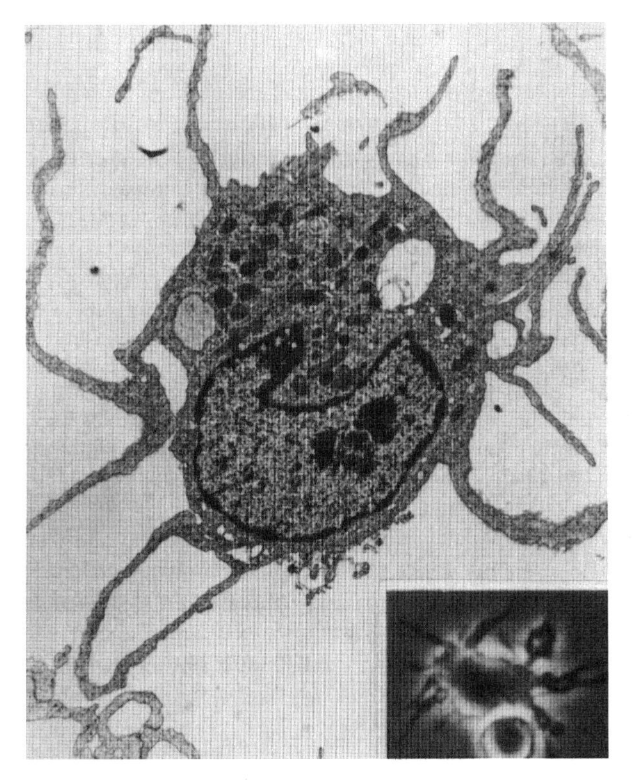

Figure 2.2 Electron microphotograph of a dendritic cell isolated from a rat lymph node (×5000). The inset illustrates the in vitro interaction between a dendritic cell and a lymphocyte as seen in phase contrast microscopy (×300). (Reproduced with permission from Klinkert, W.E.F., Labadie, J.H., O'Brien, J.P., Beyer, L.F., and Bowers, W.E., *Proc. Natl. Acad. Sci. USA*, *77*:5414, 1980.)

i. Neutrophils are phagocytic cells. As with most other phagocytic cells, they ingest with greatest efficiency microorganisms and particulate matter coated by antibody and complement (see Chapter 9). However, nonimmunological mechanisms have also been shown to lead to phagocytosis by neutrophils, perhaps reflecting phylogenetically more primitive mechanisms of recognition.
ii. Neutrophils are attracted by chemotactic factors to areas of inflammation. Those factors may be released by microbes (particularly bacteria) or may be generated during complement activation as a consequence of an antigen–antibody reaction.
iii. The attraction of neutrophils is specially intense in bacterial infections. Great numbers of neutrophils may die trying to eliminate the invading bacteria. Dead PMN and their debris become the primary component of **pus**, characteristic of many bacterial infections. Bacterial infections associated with the formation of pus are designated as purulent.

4. **Eosinophils** are PMN leukocytes with granules that stain orange–red with cytological stains containing eosin. These cells are found in high concentra-

tions in allergic reactions and during parasitic infections, and their roles in both areas will be discussed in later chapters.

5. **Basophils** have granules that stain metachromatically due to their contents of histamine and heparin. The tissue-fixed **mast cells** are very similar to basophils, even though they appear to evolve from different precursor cells. Both basophils and mast cells are involved in antiparasitic immune mechanisms and play a key pathogenic role in allergic reactions.

III. LYMPHOID TISSUES AND ORGANS

The immune system is organized on several special tissues, collectively designated as lymphoid or immune tissues. These tissues, as shown in Figure 2.3, are distributed throughout the entire body. Some lymphoid tissues achieve a remarkable degree of organization and can be designated as **lymphoid organs**. The most ubiquitous of the lymphoid organs are the **lymph nodes** which are located in groups along major blood vessels and loose connective tissues. Other mammalian lymphoid organs are the **thymus** and the **spleen** (white pulp). Lymphoid tissues include the **gut-associated lymphoid tissues (GALT)**—tonsils, Peyer's patches, and appendix—as well as aggregates of lymphoid tissue in the submucosal spaces of the respiratory and genitourinary tracts.

A. **Primary and Secondary Lymphoid Tissues**. Lymphoid tissues can be subdivided into primary and secondary lymphoid tissues based on the ability to produce progenitor cells of the lymphocytic lineage, which is characteristic of primary lymphoid tissues.

B. **Distribution of T and B Lymphocytes on Lymphoid Organs and Tissues**. Table 2.1 shows the relative percentages of T and B lymphocytes within human immune tissues. T lymphocytes predominate in the lymph, peripheral blood, and, above all, in the thymus. B lymphocytes predominate in the bone marrow and perimucosal lymphoid tissues.

C. **Lymph Nodes**. The lymph nodes are extremely numerous and disseminated all over the body. They measure 1 to 25 mm in diameter and play a very important and dynamic role in the initial or inductive states of the immune response.

1. **Anatomical organization**. The lymph nodes are circumscribed by a connective tissue capsule. Afferent lymphatics draining peripheral interstitial spaces enter the capsule of the node and open into the subcapsular sinus. The lymph node also receives blood from the systemic circulation through the hilar arteriole. Two main regions can be distinguished in a lymph node: the cortex and the medulla.

   a. The **cortex** and the **deep cortex** (also known as **paracortical area**) are densely populated by lymphocytes, in constant traffic between the lymphatic and systemic circulation. In the cortex, at low magnification, one can distinguish roughly spherical areas containing densely packed lymphocytes, termed **follicles** or **nodules** (Fig. 2.4).

   b. T and B lymphocytes occupy different areas in the cortex. B lymphocytes predominate in the follicles (hence, the follicles are designated as **T-independent areas**), which also contain macrophages, dendritic cells, and some T lymphocytes. The follicles can assume two different morphologies:

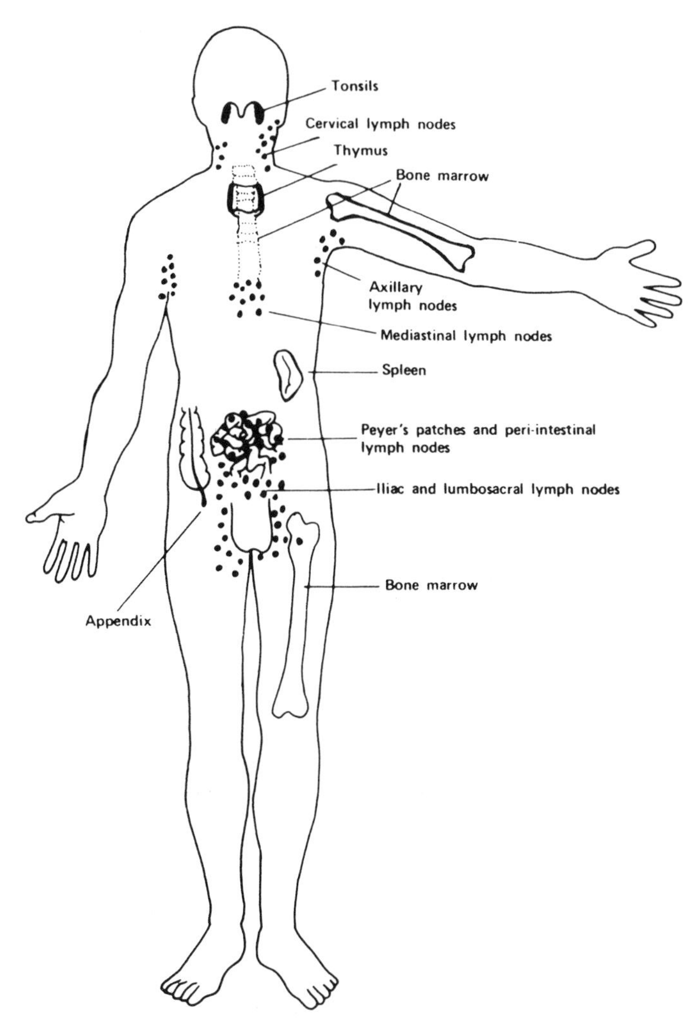

Figure 2.3 Diagrammatic representation of the distribution of lymphoid tissues in humans. (Modified from Mayerson, H.S., *Sci. Am.*, *208*:80, 1963.)

Table 2.1 Distribution of T and B Lymphocytes in Humans

	Lymphocyte distribution (%)[a]	
Immune tissue	T lymphocyte	B lymphocyte
Peripheral blood	80	10[b]
Thoracic duct	90	10
Lymph node	75	25
Spleen	50	50
Thymus	100	<5
Bone marrow	<25	>75
Peyer's patch	10–20	70

[a]Approximate values.
[b]The remaining 10% would correspond to non-T, non-B lymphocytes.

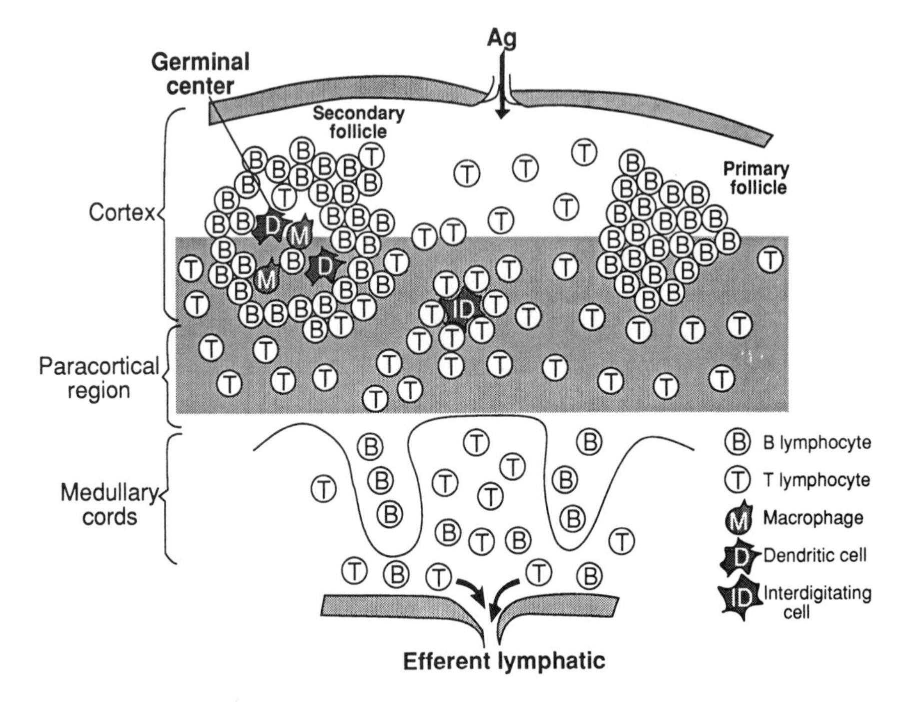

Figure 2.4 Diagrammatic representation of the lymph node structure. B lymphocytes are predominantly located on the lymphoid follicles and medullary cords (B-dependent areas), while T lymphocytes are mostly found in the paracortical area (T-dependent area).

i. The **primary follicles** are very densely packed with small lymphocytes in lymph nodes not actively involved in an immune response.

ii. In a lymph node draining in an area in which an infection has taken place, one will find larger, less dense follicles, termed **secondary follicles**, containing clear **germinal centers** where B lymphocytes are actively dividing as a result of antigenic stimulation.

c. In the deep cortex or paracortical area, which is not as densely populated as the follicles, T lymphocytes are the predominant cell population and, for this reason, the paracortical area is designated as **T-dependent**. Interdigitating cells are also present in this area, where they present antigen to T lymphocytes.

d. The medulla, less densely populated, is organized into medullary cords draining into the hilar efferent lymphatic vessels. Plasma cells can be identified in the medullary cords.

2. **Physiological role**. The lymph nodes can be compared to a network of filtration and communication stations where antigens are trapped and messages are interchanged between the different cells involved in the immune response.

a. **The dual circulation in the lymph nodes**. Lymph nodes receive both lymph and arterial blood flow. The afferent lymph, with its cellular elements, percolates from the subcapsular sinus to the efferent lymphatics via cortical and medullary sinuses, and the cellular elements of the lymph have ample opportunity to migrate into the lymphocyte-rich cortical structures during their transit through the nodes. The artery that penetrates through the hilus brings peripheral blood lymphocytes into the lymph node; these lymphocytes can leave the vascular bed at the level of the **high endothelial venules** located in the paracortical area.

b. **Lymph nodes as the anatomical fulcrum of the immune response**.

i. Soluble or particulate antigens reach the lymph nodes primarily through the lymphatic circulation. Once in the lymph nodes, antigen is concentrated on a network formed by the dendritic cells, designated as **antigen-retaining reticulum**. The antigen is retained by these cells in its unprocessed form, often associated with antibody (particularly during secondary immune responses), and is efficiently presented to B lymphocytes. The B lymphocytes recognize specific epitopes, but are also able to internalize and process the antigen, presenting antigen-derived peptides associated to MHC-II molecules to helper T lymphocytes, whose "help" is essential for the proper activation and differentiation of the B cells presenting the antigen (see Chapter 3).

ii. Antigens can also reach the lymph nodes in association with trafficking cells, particularly the **Langerhans cells** of the dermis. Those cells express MHC-II molecules, and therefore can function as APC. From the dermis they migrate to the paracortical areas, where they assume the morphology of **interdigitating cells** and interact with the T lymphocytes that abound in that region. The close contact between the interdigitating cells presenting antigen-derived peptides on their MHC-II molecules and helper T lymphocytes able to specifically recognize those MHC-associated peptides is essential for proper initiation of the immune response (see Chapters 3 and 11).

D. **Spleen**

1. **Anatomical organization**. Surrounded by a connective tissue capsule, the parenchyma of this organ is heterogeneous, constituted by the white and the red pulp.

a. **White pulp**. The spleen receives blood from the splenic artery. The narrow central arterioles, derived from the splenic artery after multiple branchings, are surrounded by lymphoid tissue (**periarteriolar lymphatic sheath**). In the white pulp, T lymphocytes are in close proximity to the arteriole, whereas B lymphocytes are concentrated in follicles, which lie more peripherally relative to the arterioles (Fig. 2.5), and which may or may not show germinal centers depending on the state of activation of the resident cells.

b. The **red pulp** surrounds the white pulp. Blood leaving the white pulp through the central arterioles flows into the penicillar arteries and from there flows directly into the venous sinuses. The red pulp is formed by these venous sinuses which are bordered by the splenic cords (cords of Billroth) and venous sinuses, where macrophages abound. From the sinuses, blood reenters the systemic circulation through the splenic vein.

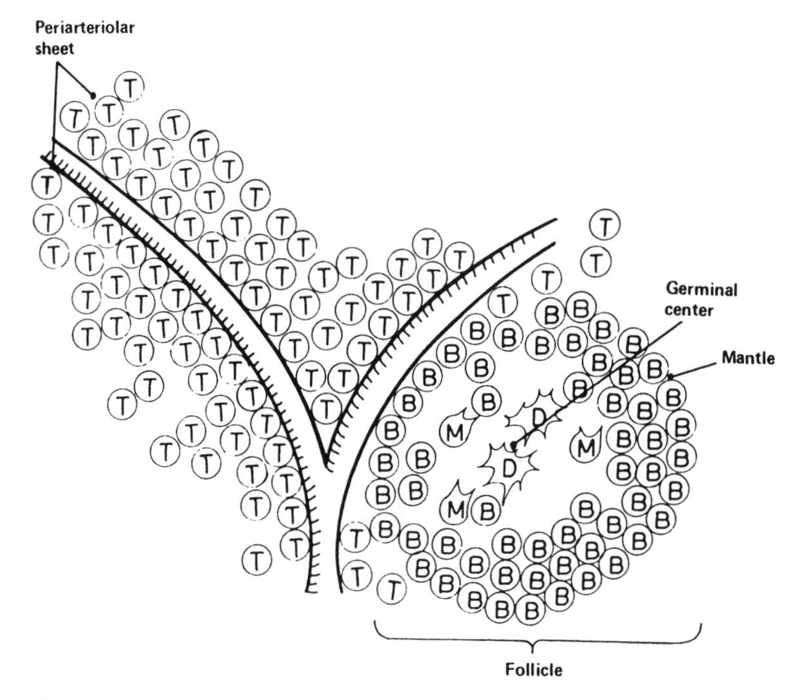

Figure 2.5 Diagrammatic representation of the topography of the splenic lymphoid tissue. The lymphocytic periarteriolar sheet is a T-dependent area, while the B lymphocytes are localized on lymphoid follicles (B-dependent areas). Same key as in Figure 2.4.

c. Between the white and the red pulp lies an area known as the **marginal zone**, more sparsely cellular than the white pulp, but very rich in macrophages and B lymphocytes.

2. **Physiological role**. The spleen is the lymphoid organ associated with filtering or clearing of particulate matter, infectious organisms, and aged or defectively formed elements (e.g., spherocytes, ovalocytes) from the peripheral blood. The main filtering function is performed by the macrophages lining up the splenic cords. In the marginal zone, circulating antigens are trapped by the macrophages which will then be able to process the antigen, migrate deeper into the white pulp, and initiate the immune response by interacting with T and B lymphocytes.

E. **Thymus**. The thymus is the only clearly individualized primary lymphoid organ in mammals. It is believed to play a key role in determining the differentiation of T lymphocytes.

1. **Anatomical organization**. The thymus, whose structure is diagrammatically illustrated in Fig. 2.6, is located in the superior mediastinum, anterior to the great vessels. It has a connective tissue capsule from which emerge the trabeculae, which divide the organ into lobules. Each lobule has a **cortex** and **medulla**, and the trabeculae are coated with epithelial cells.

a. **Cortex**. Lymphocyte aggregates, composed mainly of immunologically immature T lymphocytes, are located in the cortex, an area of intense cell proliferation. A small number of macrophages and plasma cells are also present. In addition, the cortex contains two subpopulations of epithelial cells, the epithelial nurse cells and the cortical epithelial cells which form a network within the cortex.

b. **Medulla**. Not as densely populated as the cortex, the medulla contains predominantly mature T lymphocytes, and has a larger epithelial cell-to-lymphocyte ratio than the cortex. Unique to the medulla are concentric rings of squamous epithelial cells known as **Hassall's corpuscles**.

2. **Physiological role**.

   a. **T-lymphocyte differentiation**. The thymus is believed to be the organ where T lymphocytes differentiate during embryonic life. The thymic cortex is an area of intense cell proliferation and death (only 1% of the

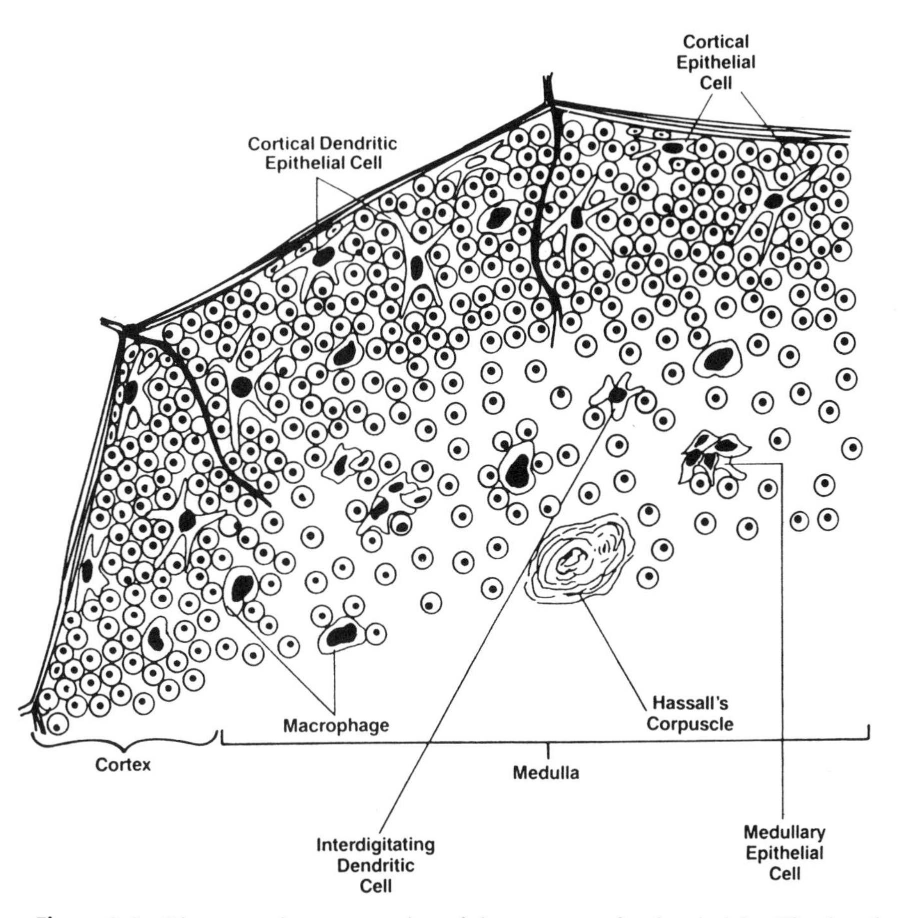

Figure 2.6 Diagrammatic representation of the structure of a thymic lobe. The densely packed cortex is mostly populated by T lymphocytes and by some cortical dendritic epithelial cells and cortical epithelial cells. The more sparsely populated medulla contains epithelial and dendritic cells, macrophages, T lymphocytes, and Hassall's corpuscles. (Adapted from Butcher, E.C., and Weissman, I.L. Lymphoid tissues and organs. In *Fundamental Immunology*, W.E. Paul, ed. Raven Press, New York, 1984.)

cells generated in the thymus eventually mature and migrate to the peripheral tissues). The mechanism whereby the thymus determines T-lymphocyte differentiation is believed to involve the interaction of T-lymphocyte precursors with thymic epithelial cells. These interactions result in the elimination or inactivation of self-reactive T-cell clones and in the differentiation of two separate lymphocyte subpopulations with different membrane antigens and different functions. Most T-lymphocyte precursors appear to reach full maturity in the medulla.

b. **Hormone synthesis**. The thymic epithelial cells are believed to produce hormonal factors (e.g., **thymosin** and **thymopoietin**), which may play an important role in the differentiation of T lymphocytes.

F. **Mucosa-Associated Lymphoid Tissues** (MALT) encompass the lymphoid tissues of the intestinal tract, genitourinary tract, tracheobronchial tree, and mammary glands. All of the mucosa-associated lymphoid tissues are unencapsulated and contain both T and B lymphocytes, the latter predominating.

G. **Gut-Associated Lymphoid Tissue** is the designation proposed for all lymphatic tissues found along the digestive tract. Three major areas of GALT that can be identified are the tonsils, the Peyer's patches, located on the submucosa of the small intestine, and the appendix. In addition, scanty lymphoid tissue is present in the lamina propria of the gastrointestinal tract.

1. **Tonsils**, located in the oropharynx, are predominantly populated by B lymphocytes and are the site of intense antigenic stimulation, as reflected by the presence of numerous secondary follicles with germinal centers in the tonsilar crypts (Fig. 2.7).
2. **Peyer's patches** are lymphoid structures disseminated through the submucosal space of the small intestine (Fig. 2.8).
   a. The follicles of the intestinal Peyer's patches are extremely rich in B cells, which differentiate into IgA-producing plasma cells.
   b. Specialized epithelial cells, known as **M cells** abound in the dome epithelia of Peyer's patches, particularly at the ileum. These cells take up small particles, virus, bacteria, etc., and deliver them to submucosal macrophages, where the engulfed material will be processed and presented to T and B lymphocytes.
   c. T lymphocytes are also present in the intestinal mucosa, the most abundant of them expressing membrane markers that are considered typical of memory helper T cells. This population appears to be critically involved in the induction of humoral immune responses.
   d. A special subset of T cells, with a different type of T-cell receptor (γ/δ **T lymphocytes**) is well represented on the small intestine mucosa. These lymphocytes appear to recognize and destroy infected epithelial cells by a nonimmunological mechanism (i.e., not involving the T-cell receptors).

IV. LYMPHOCYTE TRAFFIC

A. General considerations. The lymphatic and circulatory systems are intimately related (Fig. 2.9) and there is a constant traffic of lymphocytes throughout the body, moving from one system to another.

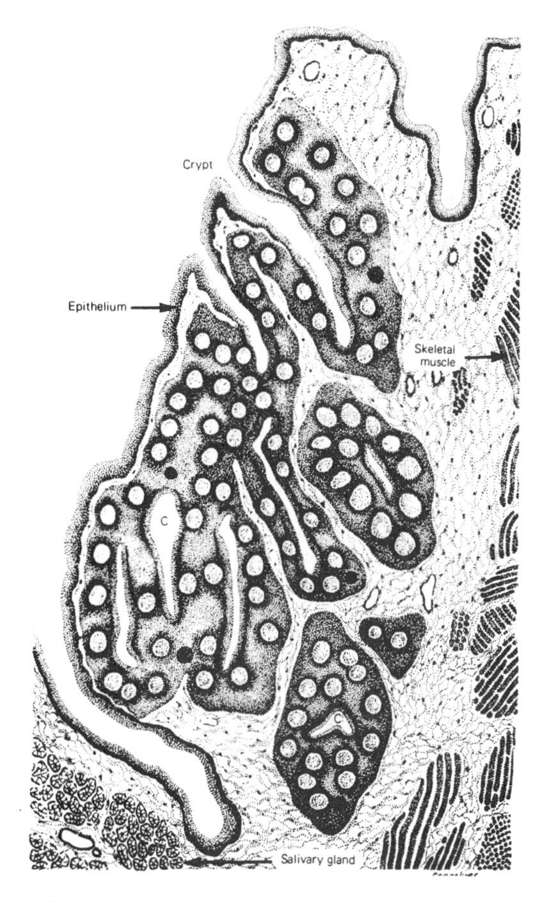

Figure 2.7 Diagrammatic representation of the histological structure of the tonsils. (Reproduced with permission from Junqueira, L.C., Carneiro, J., and Contopoulas, J., *Basic Histology*, 2nd ed. Lang, Los Altos, CA, 1971.)

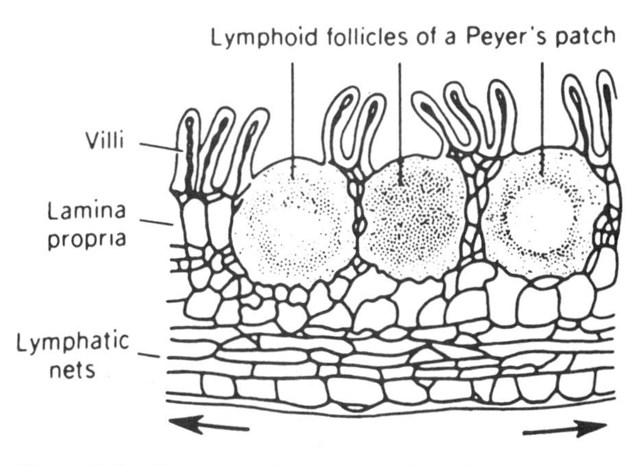

Figure 2.8 Diagrammatic representation of the topography of the lymphoid follicles of a Peyer's patch. (Reproduced with permission from Kampmeier, O.F., *Evolution and Comparative Morphology of the Lymphatic System*. Charles C Thomas, Springfield, IL, 1969.)

1. **Lymphatic circulation**. Afferent lymphatics from interstitial spaces drain into lymph nodes that "filter" these fluids, removing foreign substances. "Cleared" lymph from below the diaphragm and the upper left half of the body drains via efferent lymphatics, emptying into the thoracic duct for subsequent drainage into the left innominate vein. "Cleared" lymph from the right side above the diaphragm drains into the right lymphatic duct with

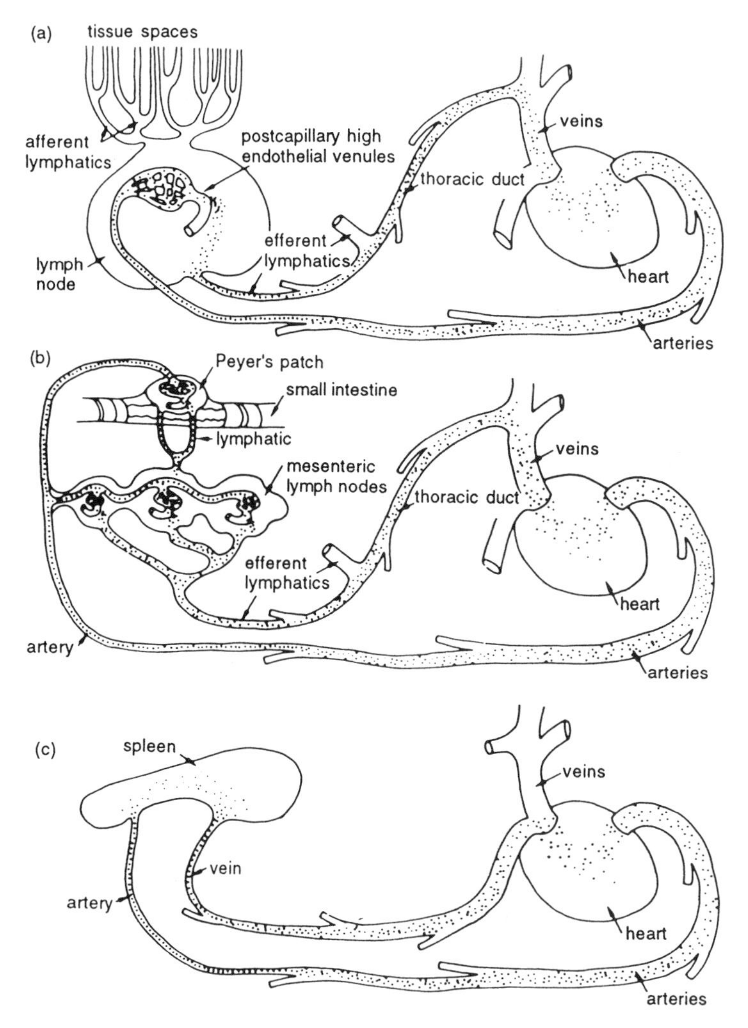

Figure 2.9 Pathways of lymphocyte circulation: (a) blood lymphocytes enter lymph nodes, adhere to the walls of specialized postcapillary venules, and migrate to the lymph node cortex. Lymphocytes then percolate through lymphoid fields to medullary lymphatic sinuses and on to efferent lymphatics, which in turn collect in major lymphatic ducts in the thorax, which empty into the superior vena cava; (b) the gut-associated lymphoid tissues (Peyer's patches and mesenteric lymph nodes) drain into the thoracic duct, which also empties into the superior vena cava; (c) the spleen receives lymphocytes and disburses them mainly via the blood vascular system (inferior vena cava). (Reproduced with permission from Hood, L.E., Weissman, I.L., Wood, W.B., and Wilson, J.H., *Immunology*, 2nd ed. Benjamin/Cummings, Menlo Park, CA, 1984.)

subsequent drainage into the origin of the right innominate vein. The same routes are traveled by lymphocytes stimulated in the lymph nodes or peripheral lymphoid tissues, which eventually will reach the systemic circulation.

2. **Systemic circulation**. Peripheral blood, in turn, is "filtered" by the spleen and liver, the spleen having organized lymphoid areas while the liver is rich in Kupffer's cells, which are macrophage-derived phagocytes. Organisms and antigens that enter directly into the systemic circulation will be trapped in these two organs, of which the spleen plays the most important role as a lymphoid organ.
3. **Lymphocyte recirculation**.
   a. Lymphocytes circulating in the systemic circulation eventually enter a lymph node, exit the systemic circulation at the level of the **high endothelial venules**, leave the lymph node with the efferent lymph, and eventually reenter the systemic circulation.
   b. B lymphocytes circulate between different segments of the mucosal-associated lymphoid tissues, including the GALT, the mammary gland-associated lymphoid tissue, and the lymphoid tissues associated with the respiratory tree and urinary tract.
   c. The crucial step in the traffic of lymphocytes from the systemic circulation to a lymphoid tissue is the crossing of the endothelial barrier by diapedesis at specific locations. Under physiological conditions, this seems to take place predominantly at the level of the high endothelial venules. These specialized endothelial cells express surface molecules—**cell adhesion molecules (CAMs)**—which interact with ligands, including other cell adhesion molecules, expressed on the membrane of T and B lymphocytes. The interplay between endothelial and lymphocyte CAMs determines the traffic and homing of lymphocytes.
4. **Cell adhesion molecules**. Three main families of cell adhesion molecules have been defined (Table 2.2). The **addressins** or **selectins** are expressed on endothelial cells and leukocytes and mediate leukocyte adherence to the endothelium. The **immunoglobulin superfamily of CAMs** includes a variety of molecules expressed by leukocytes, endothelial cells, and other cells. The **integrins** are defined as molecules that interact with the cytoskeleton and tissue matrix compounds. The following CAMs have been reported to be involved in lymphocyte traffic and homing.
   a. **LAM-1**, **ICAM-1**, and **CD44** are primarily involved in controlling lymphocyte traffic and homing in peripheral lymphoid tissues.
   b. **MadCAM-1** is believed to control lymphocyte homing to the mucosal lymphoid tissues.

   The interaction between adhesion molecules and their ligands takes place in several stages. First, the cells adhere to endothelial cells at the level of the high endothelium venules (HEV), and the adhering lymphocyte is then able to migrate through endothelial slits into the lymphoid organ parenchyma. Different CAMs and ligands are involved in this sequence of events.
5. **Regulation of lymphocyte traffic and homing**. The way in which cell adhesion molecules regulate lymphocyte traffic and homing seems to be a result both of differences in the level of their expression and of differences in the nature of the CAMs expressed in different segments of the microcirculation.

Table 2.2 Main Adhesion Molecules, Their Families, Ligands, and Functions

Family	Members	Ligand	Function
Selectins			
	Endothelial-leukocyte adhesion molecule (ELAM-1, E-selectin)	Sialylated/fucosylated molecules	Mediates leukocyte adherence to endothelial cells in inflammatory reactions
	Leukocyte adhesion molecule-1 (LAM-1, L-selectin)	Immunoglobulin superfamily CAMs; mucins and sialomucins	Interaction with HEV (lymphocyte homing); leukocyte adherence to endothelial cells in inflammatory reactions
Immunoglobulin Superfamily CAMs			
	Intercellular adhesion molecule-1 (ICAM-1)	LFA-1 (CD11a/CD18), Mac-1 (CD11b)	Expressed by leukocytes, endothelial cells, dendritic cells, etc.; mediates leukocyte adherence to endothelial cells in inflammatory reactions
	ICAM-2	LFA-1	Expressed by leukocytes, endothelial cells, and dendritic cells; involved in control of lymphocyte recirculation and traffic
	Vascular CAM-1 (VCAM-1)	VLA-4	Expressed primarily by endothelial cells; mediates leukocyte adherence to activated endothelial cells in inflammatory reactions
	Mucosal addressin CAM-1 (MadCAM-1)	β7, α4, L-Selectin	Expressed by mucosal lymphoid HEV; mediates lymphocyte homing to mucosal lymphoid tissues
	Platelet/endothelial CAM-1 (PECAM-1)	PECAM-1	Expressed by platelets, leukocytes, and endothelial cells; involved in leukocyte transmigration across the endothelium in inflammation
Integrins			
VLA family	VLA1 to 6	Fibronectin, laminin, collagen	Ligands mediating cell–cell and cell–substrate interaction
LEUCAM family	LFA-1	ICAM-1, ICAM-2, ICAM-3	Ligands mediating cell–cell and cell–substrate interaction
	Mac-1	ICAM-1, fibrinogen, C3bi	
Other			
	CD44	Hyaluronate, collagen, fibronectin	Expressed on leukocytes; mediates cell–cell and cell–matrix interactions; involved in lymphocyte homing

a. The involvement of **HEV** as the primary site for lymphocyte egress from the systemic circulation is a consequence of the high density of selectins in HEV cells. Thus, the opportunity for cell adhesion and extravascular migration is considerably higher in HEV than on segments covered by flat endothelium.

b. **Inflammatory and immune reactions** often lead to the release of mediators which up-regulate the expression of CAM in venules or in other segments of the microvasculature near the area where the reaction is taking place. This results in a sequence of events that is mediated by different sets of CAMs and respective ligands:
   i. First the leukocytes slow down and start rolling along the endothelial surface. This stage is mediated primarily by selectins.
   ii. Next, leukocytes adhere to endothelial cells expressing integrins such as VLA and CAMs of the immunoglobulin superfamily, such as ICAM and VCAM.
   iii. Finally, the adherent leukocytes squeeze between two adjoining endothelial cells and move to the extravascular space.

   The end result of this process is an increase in leukocyte migration to specific areas where those cells are needed to eliminate some type of noxious stimulus or to initiate an immune response. As a corollary, there is great interest in developing compounds able to block up-regulated CAMs to be used as anti-inflammatory agents.

c. It is known that the lymphocyte constitution of lymphoid organs is variable (Table 2.1). T lymphocytes predominate in the lymph nodes but B lymphocytes and IgA-producing plasma cells predominate in the Peyer's patches and the GALT in general. This **differential homing** is believed to be the result of the expression of specific addressins such as MadCAM-1 on the HEV of the perimucosal lymphoid tissues, which are specifically recognized by the B cells and plasma cells residing in those tissues. Most B lymphocytes recognize specifically the GALT-associated HEV and do not interact with the lymph node-associated HEV, while most naive T lymphocytes recognize both the lymph node-associated HEV and the GALT-associated HEV.

d. The differentiation of T-dependent and B-dependent areas in lymphoid tissues is a poorly understood aspect of lymphocyte "homing." It appears likely that the distribution of T and B lymphocytes is determined by their interaction with nonlymphoid cells. For example, the interaction between interdigitating cells and T lymphocytes may determine the predominant location of T lymphocytes in the lymph node paracortical areas and periarteriolar sheets of the spleen, while the interaction of B lymphocytes with follicular dendritic cells may determine the organization of lymphoid follicles in the lymph nodes, spleen, and GALT.

e. The modulation of CAM at different states of cell activation explains changing **patterns in lymphocyte recirculation** seen during immune responses.
   i. Immediately after antigen stimulation, the recirculating lymphocyte appears to transiently lose its capacity to recirculate. This loss of recirculating ability is associated with a tendency to self-

aggregate (perhaps explaining why antigen-stimulated lymphocytes are trapped at the site of maximal antigen density), due to the up-regulation of CAMs involved in lymphocyte–lymphocyte and lymphocyte–accessory cell interactions.

ii. After the antigenic stimulus ceases, a population of **memory T lymphocytes** carrying distinctive membrane proteins can be identified. This population seems to have a different recirculation pattern than that of the naive T lymphocyte, leaving the intravascular compartment at sites other than the HEV and reaching the lymph nodes via the lymphatic circulation. This difference in migration seems to result from the down-regulation of the CAMs, which mediate the interaction with HEV selectins and upregulation of other CAMs, which interact with selectins located in other areas of the vascular tree.

iii. **B lymphocytes** also change their recirculation patterns after antigenic stimulation. Most B cells will differentiate into plasma cells after stimulation, and this differentiation is associated with marked changes in the antigenic composition of the cell membrane. Consequently, the plasma cells exit the germinal centers, move into the medullary cords, and, eventually, into the bone marrow, where most of the antibody production in humans takes place.

f. Memory lymphocytes appear to home preferentially in the type of lymphoid tissue where the original antigen encounter took place (i.e., a lymphocyte that recognized an antigen in a peripheral lymph node will recirculate to another peripheral lymph node, while a lymphocyte that was stimulated at the GALT level will recirculate to the GALT). Memory B lymphocytes remain in the germinal centers while memory T lymphocytes "home" in T-cell areas.

SELF-EVALUATION

Questions

Choose the ONE *best* answer.

2.1 A patient born without the human bursa-equivalent would be expected to have normal:
- A. Cellularity in the paracortical areas of the lymph nodes
- B. Differentiation of germinal centers in the lymph nodes
- C. Numbers of circulating lymphocytes bearing surface immunoglobulins
- D. Numbers of plasma cells in the bone marrow
- E. Tonsils

2.2 Which one of the following anatomical regions is most likely to show a predominance of T lymphocytes?
- A. A periarteriolar sheet in the spleen
- B. A Peyer's patch in the small intestine
- C. A tonsilar follicle
- D. The bone marrow
- E. The germinal center of a lymph node follicle

2.3 The role of selectins in the microvasculature is to:
   A. Attract lymphocytes to the extravascular compartment in specific tissues
   B. Mediate the adhesion of leukocytes to endothelial cells
   C. Promote cell–cell interaction in the lymphoid tissues
   D. Promote trapping of antigen in the antigen-retaining reticulum
   E. Regulate blood flow in or out of specific areas of the organism

Questions 2.4–2.10

Match the listed properties and characteristics with the right type of lymphocytes.
   A. B lymphocytes
   B. T lymphocytes
   C. Follicular dendritic cells
   D. Neutrophils
   E. Plasma cells

2.4 Are present in large numbers in the follicles and germinal centers of the lymph nodes
2.5 Release perforins and granzymes
2.6 Recirculate between different segments of the GALT
2.7 Found in the spleen follicles
2.8 Concentrate antigen on their surface
2.9 Migrate to the bone marrow after antigenic stimulation
2.10 Produce and secrete large amounts of immunoglobulins

Answers

2.1 (A) The lack of a bursal equivalent would result in virtually no differentiation of B lymphocytes and plasma cells and this would be reflected in the peripheral blood and B-cell rich lymphoid tissues. However, the paracortical areas of the lymph nodes are mostly populated by T cells and, as such, would not be affected.

2.2 (A)

2.3 (A) Selectins are surface receptors expressed in endothelial cells, which are recognized by specific ligands on leukocytes. Their physiological function is to promote adhesion of circulating leukocytes to the endothelial cells, initiating a sequence of interactions that eventually results in the "homing" of the circulating cell into a given lymphatic tissue. The actual migration of lymphocytes out of the vessel wall requires firm attachment mediated by additional cell adhesion molecules and the release of chemoattractant cytokines in the extravascular compartment.

2.4 (A)

2.5 (B) Cytotoxic T lymphocytes mediate their function through several mechanisms, one of which includes the release of perforins and granzymes.

2.6 (A)

2.7 (A) The white pulp of the spleen is organized in two different areas: the periarteriolar sheet (where T lymphocytes predominate) and the follicles (where B lymphocytes predominate).

2.8 (C)

2.9 (E)

2.10 (E) Immunoglobulin secretion is a property of the plasma cell which, although derived from B lymphocytes, has unique functions and membrane markers.

BIBLIOGRAPHY

Bevilaqua, M.P. Endothelial-leucocyte adhesion molecules. *Annu. Rev. Immunol., 11*:767, 1993.

Berke, G. The CTL's kiss of death. *Cell, 81*:9, 1995.

Blumberg, R.S., Yockey, C.E., Gross, G.C., et al. Human intestinal intraepithelial lymphocytes are derived from a limited number of T cell clones that utilize multiple V beta T cell receptor genes. *J. Immunol., 150*:5144, 1993.

Camerini, V., Panwala, C., and Kronenberg, M. Regional specialization of the mucosal immune system. Intraepithelial lymphocytes of the large intestine have a different phenotype and function than those of the small intestine. *J. Immunol., 151*:1765, 1993.

Collins, T. Adhesion molecules and leukocyte emigration. *Sci. Med., 2(6)*:28–37, 1995.

Dustin, M.L., and Springer, T.A. Role of lymphocyte adhesion receptors in transient interactions and cell locomotion. *Annu. Rev. Immunol., 9*:27, 1991.

Fujihashi, K., Yamamoto, M., McGhee, J.R., and Kiyono, H. Function of α/β TCR^+ and γ/δ TCR^+ IELs for the gastrointestinal immune response. *Int. Rev. Immunol., 11*:1, 1994.

JunMichl, J., Qiu, Q.Y., and Kuerer, H.M. Homing receptors and addressins. *Curr. Opin. Immunol., 3*:373, 1991.

MacLennan, I.C.M. Germinal centers. *Annu. Rev. Immunol., 12*:117, 1994.

Pellas, T.C., and Weiss, L. Migration pathways of recirculating murine B cells and $CD4^+$ and $CD8^+$ T lymphocytes. *Am. J. Anat., 187*:355, 1990.

Picker, L.J. and Butcher, E.C. Physiological and molecular mechanisms of lymphocyte homing. *Annu. Rev. Immunol., 10*:561, 1992.

Szakal, A.K., Kosco, M.H., and Tew, J.G. Microanatomy of lymphoid tissue during humoral immune responses: Structure function relationships. *Annu. Rev. Immunol. 7*:91, 1989.

3

Major Histocompatibility Complex

Jean-Michel Goust

I. INTRODUCTION: GRAFT REJECTION

Grafting of tissues or organs between genetically unrelated individuals is inevitably followed by rejection of the grafted tissue or organ. On the other hand, if tissues or organs are transplanted between genetically identical individuals, rejection does not take place.

A. **Graft Rejection Is Under Genetic Control**. The understanding of the factors controlling rejection became possible after inbred strains of mice became available. Inbred strains are obtained after 20 or more generations of brother/sister mating, and for all purposes are constituted genetically identical animals.
   1. When skin is transplanted among laboratory animals of different inbred strains, the recipient animal will reject the graft, after varying amounts of time depending on the degree of genetic relatedness between the strains used in the experiment.
   2. When skin is grafted among animals of the same inbred strain, no rejection is observed.
   3. First generation hybrids (Fl hybrids), produced by mating animals of two genetically different strains, do not reject tissue from either parent, while the parents reject skin from the Fl hybrids.

   These observations showed that the genetic differences explaining transplant rejection are due to co-dominant histocompatibility determinants. Further studies, diagrammatically summarized in Figure 3.1, showed that graft rejection shares two important characteristics with classic immune responses: **specificity** and **memory**.

B. **The Antigens Responsible for Graft Rejection are Expressed in Most Cells and Tissues**. Laboratory animals receiving skin grafts from animals of the same species but of different strains developed antibodies that reacted specifically with skin and peripheral blood lymphocytes of the donor strain. These findings pointed to the sharing of antigens by different tissues of the donor animal. This possibility was confirmed through reverse immunizations in which mice preinjected with lymphocytes obtained from a different strain showed accelerated rejection of a skin graft taken from the animals of the same strain from which the lymphocytes were obtained. It is now well established that most nucleated cells of the organism express the antigens responsible for rejection, which are designated as **histocompatibility antigens**.

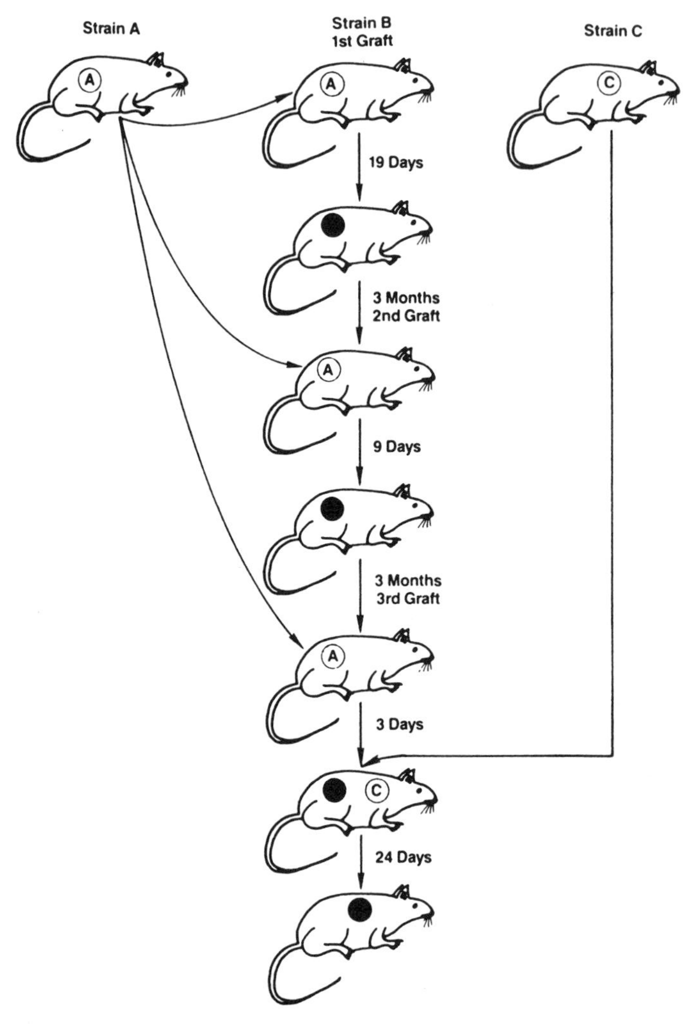

Figure 3.1 Diagrammatic representation of an experiment designed to demonstrate the memory and specificity of graft rejection. Memory is demonstrated by the progressive shortening of the time it takes a mouse of strain B to reject consecutive skin grafts from strain A. Specificity is demonstrated by the fact that the mouse of strain B is already able to reject a graft from strain A in an accelerated fashion, and if given a graft from a third, unrelated strain (C), rejection will take as long as the rejection of the first graft from strain A. In other words, sensation of mouse B to strain A was strain-specific and did not extend to unrelated strains.

II. THE MAJOR HISTOCOMPATIBILITY COMPLEX

A. **General Concepts**. After many years of detailed genetic analysis, it became clear that the system that determines the outcome of a transplant is complex and highly polymorphic. It was also determined that this system contained antigens of variable strength. The **major antigens** are considered as responsible for most graft rejection responses and trigger a stronger immune response than the others, which are designated as **minor**. The aggregate of major histocompability antigens is known as **major histocompatibility complex** (**MHC**). All mammalian species express MHC antigens on their nucleated cells.

B. **The Human MHC: Human Leukocyte Antigens (HLA)**. Historically, the human histocompatibility antigens were defined after investigators observed that the serum of multiparous women contained antibodies agglutinating their husband's lymphocytes. These *leukoagglutinins* were also found in the serum of multitransfused individuals, even when the donors were compatible with the transfused individual for all the known blood groups. The antigens responsible for the appearance of these antibodies were thus present on leukocytes and received the designation of **Human Leukocyte Antigens** (**HLA**).
   1. **HLA and transplantation**. It soon became apparent that the immune response of an individual to the HLA antigens of another individual was responsible for the rejection of tissues grafted between genetically unrelated individuals (see Chapter 27). The study of HLA antigens received its initial impetus from the desire to transplant tissues with minimal risk of rejection and from their interest to geneticists as one of the most polymorphic antigenic systems in humans.
   2. **The biological role of HLA (MHC) molecules**. It took several decades for a wider picture of the biological significance of HLA antigens to become obvious. Today, we know that these molecules are at the very core of the immune response and at the basis of the establishment of tolerance (lack of response) to self antigens, as discussed later in this chapter, as well as in Chapters 4, 10, and 22.

III. CLASSIFICATION, STRUCTURE, AND DETECTION OF HLA GENES AND GENE PRODUCTS

A. **MHC and HLA Classes**. Six different loci of the HLA system have been identified and are divided into two classes (Table 3.1). Homologous MHC classes have been defined in other mammalian species.
   1. **Class I** includes three major loci (HLA-A, HLA-B, HLA-C) and four minor loci (HLA-E, F, G, and H). Each locus has multiple alleles, ranging from 6 in the case of HLA-C to more than 50 in the case of HLA-B. Each allele is designated by a number (e.g., HLA-B27).
   2. **Class II antigens** are the most polymorphic and include three main loci

Table 3.1 The Main Characteristics of MHC Antigen Classes

Characteristics	Class I	Class II
Major loci (mouse)	K, D, L	I (-A,-E)
Major loci (man)	A, B, C	DP, DQ, DR
Alleles (No.)	>100	>20
Specificities (No.)	>100	>100
Distribution of gene products	All nucleated cells	Monocytes, macrophages, B lymphocytes

(HLA-DP, HLA-DQ, and HLA-DR) and less well-defined loci (HLA-DM, -DN, and -DO).

B. **Structure of the MHC Antigens**

1. **Class I MHC molecules**. The HLA or H2 molecules coded by MHC class I genes have molecular weights that may vary from 43,000 to 48,000. They are formed by two nonidentical polypeptide chains.
   a. The major chain (**α chain**) has a long extracellular region folded in three domains, named α_1, α_2, and α_3. The extracellular domains are attached to a short transmembrane, hydrophobic region of 24 amino acids and an intracytoplasmic "tail" composed of about 30–35 amino acids, which includes the carboxyl terminus attached to cytoskeletal structures.
   b. **β_2-microglobulin**, a 12,000 dalton protein coded by a gene located on **chromosome 15**, is postsynthetically and noncovalently associated with the major polypeptide chain.
   c. The comparison of amino acid sequences and nucleotide sequences of the various domains of class I MHC shows that the α_1 and α_2 domains are highly variable, and that most of the amino acid and nucleotide changes responsible for the differences between alleles occur in these domains. It also shows areas in these domains that are relatively constant and closely related in different alleles. This explains why polyclonal antibodies raised against MHC molecules can recognize several of them, an occurrence designated as the existence of **"public" specificities**, as opposed to the **"private" specificities** unique to each different allele.
   d. X-ray crystallography studies have determined the **tridimensional structure of the HLA class I molecules** (Fig. 3.2) and illuminated the relation between the structure and the function of this molecule. The most polymorphic areas of the molecule are located within and on the edges of a groove formed at the junction of the helical α_1 and α_2 domains. This groove is usually occupied by a short peptide (10–11 residues) of endogenous or exogenous origin. The α_3 domain shows much less genetic polymorphism and, together with β_2-microglobulin, is like a frame supporting the external deployment of the more polymorphic α_1 and α_2 domains. In addition, the α_3 domain has a binding site for the CD8 molecule characteristic of cytotoxic T cells (see Chapters 4, 10, and 11).
2. **Class II MHC molecules**. Although a remarkable degree of tertiary structure homology seems to exist between class I and class II gene products (Fig. 3.3), there are important differences in their primary structure.
   a. First, class II gene products are not associated with β_2-microglobulin. The MHC-II molecules consist of two distinct polypeptide chains, a **β chain** (M.W. 28,000) which expresses the greatest degree of genetic polymorphism, and a less polymorphic, heavier chain (**α chain**, M.W. 33,000).
   b. Each polypeptide chain has two extracellular domains (α_1 and α_2; β_1 and β_2), a short transmembrane domain, and an intracytoplasmic tail. The NH2 terminal ends of the terminal α_1 and β_1 domains contain **hypervariable regions**.

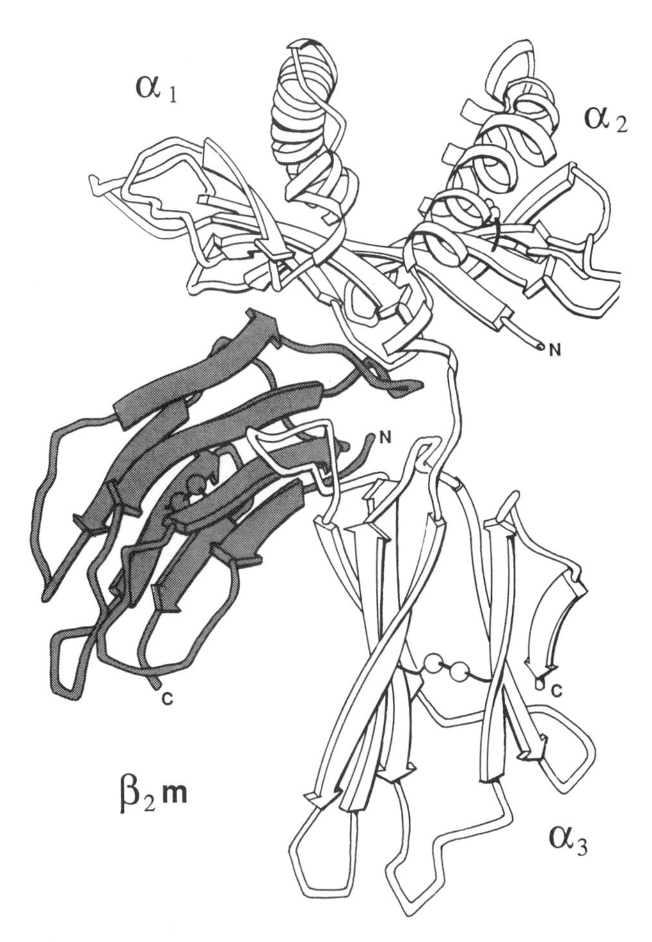

Figure 3.2 Schematic representation of the spatial configuration of the HLA-A2 molecule, based on x-ray crystallography data. The diagram shows the immunoglobulinlike domains (α_3, β_2m) at the bottom and the polymorphic domains (α_1, a_2) at the top. The indicated C terminus corresponds to the site of papain cleavage; the native molecule has additional intramembrane and intracellular segments. The α_1 and α_2 domains form a deep groove that is identified as the antigen recognition site. (Modified from Bjorkman, P.J., et al., *Nature*, *329*:506, 1987.)

c. The **three-dimensional structure of class II antigens** has been established. The β_1 domains of class II MHC antigens resemble the α_2 domain of their class I counterparts. The junction of α_1 and β_1 domains forms a groove similar to the one formed by the α_1 and α_2 domains of class I MHC antigens, which also binds antigen-derived oligopeptides.

d. The β1 domain also contains two important sites located below the antigen-binding site. The first acts as a receptor for the CD4 molecule of helper T lymphocytes (see Chapters 4, 10, and 11). The second site, which overlaps the first, is a receptor for the envelope glycoprotein (gp 120) of the human immunodeficiency virus (HIV).

C. Identification of HLA Antigens. The different antigenic specificities of each loci are recognized by two major techniques.

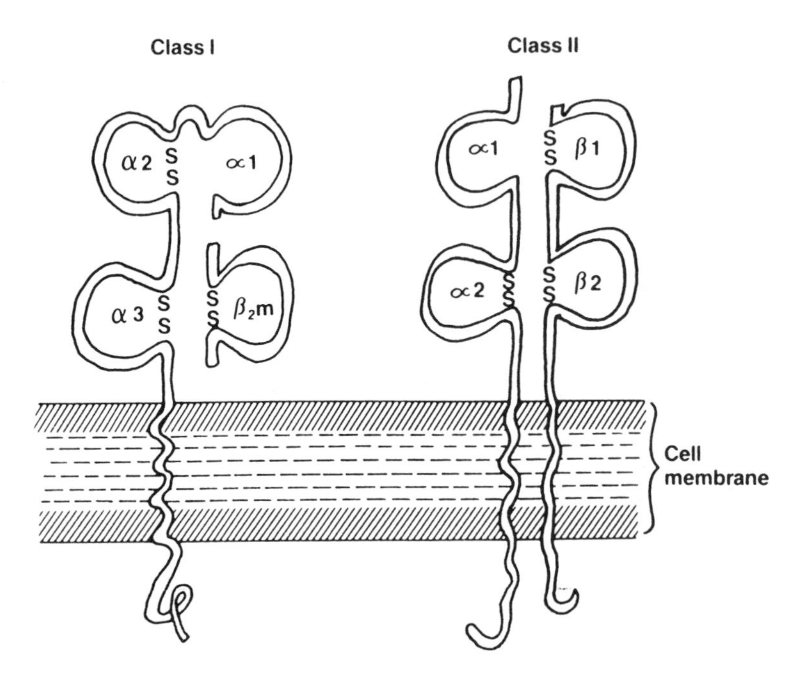

Figure 3.3 Diagrammatic representation of the structure of human class I and class II histocompatibility antigens. (Modified from Hood, L.E., et al., *Immunology*, 2nd ed. Benjamin/Cummings, Menlo Park, CA, 1984.)

1. The **serological technique**, which is the oldest and most widely used, is based on the lymphocytotoxicity of anti-HLA antibodies of known specificity in the presence of complement. The antibodies used for HLA typing were initially obtained from multiparous females or from recipients of multiple transfusions. Such antibodies are still in use, but monoclonal antibodies (see Chapter 10) have also been raised against HLA specificities. These antibodies identify several groups of HLA that are therefore designated as **serologically defined**.
2. Hybridization with **sequence-specific oligonucleotides** is particularly useful for typing MHC-II specificities. A very large number of complete DNA sequences of class-I and class-II alleles has been completed. All HLA-DR molecules share the same α chain, but their β-chain genes are extremely polymorphic. In contrast, HLA-DP and DQ molecules have polymorphic α and β chains, and thus are much more diverse than the DR molecules. As the sequences of different HLA genes became known it became possible to produce specific probes for different specificities. Typing usually involves DNA extraction, fragmentation with restriction enzymes, amplification by PCR, and finally hybridization with labeled cDNA probes specific for different alleles of the corresponding genes.
3. **Not all HLA specificities have been defined**. Some individuals express unknown specificities at some loci (usually class II) which the typing laboratory reports as "blank." Investigation of these "blank" specificities often leads to the discovery of new HLA antigens. To avoid unnecessary

confusion, they are assigned a numerical designation by regularly held workshops of the World Health Organization. At first, the designation is preceded by a **w**, indicating a provisional assignment. For example, DQw3 designates an antigenic specificity of the DQ locus that has been tentatively designated as w3 by a workshop. When worldwide agreement is reached about the fact that this is a new specificity, the **w** is dropped.

IV. CELLULAR DISTRIBUTION OF THE MHC ANTIGENS

A. **MHC Class I.** Class I molecules (HLA-A, B, and C alleles in humans and H2-K, D, and L alleles in mice) are expressed on all nucleated cells with only two exceptions: neurons and striated muscle cells. They are particularly abundant on the surface of lymphocytes (1,000 to 10,000 molecules/cell).

B. **MHC Class II.** The **I-A** and **I-E** alleles of the mouse H2 complex and the **DP**, **DQ**, and **DR** alleles of the human HLA system are exclusively expressed in two groups of leukocytes: B lymphocytes and cells of the monocyte-macrophage family, which includes all **antigen-presenting cells** (Langerhans cells in the skin, Kupffer cells in the liver, microglial cells in the central nervous system, and interdigitating cells in the spleen and lymph nodes). While resting T lymphocytes do not express MHC-II molecules, these antigens can be detected after cell activation. It must be stressed that all cells expressing class II MHC simultaneously express class I MHC.

V. CHROMOSOMAL LOCALIZATION AND ARRANGEMENT OF THE MHC GENES

A. **Mapping of the MHC region** has been established based on the study of crossover gene products and on in situ hybridization studies with DNA probes. The MHC genes are located on **chromosome 6** of humans and on **chromosome 17** of the mouse. In both cases, the MHC genes are located between the centromere and the telomere of the short arm of the respective chromosomes. A simplified map of human chromosome 6 is shown in Figure 3.4.

B. The MHC genes can be grouped in the same classes as the antigens detected in cell membranes (i.e., MHC class I and class II genes). The MHC region occupies 0.5 (mice) and 1.8 centimorgans (humans) of their respective chromosomes. The larger size of the human HLA region suggests that it includes more genes and is more polymorphic than the murine H2.

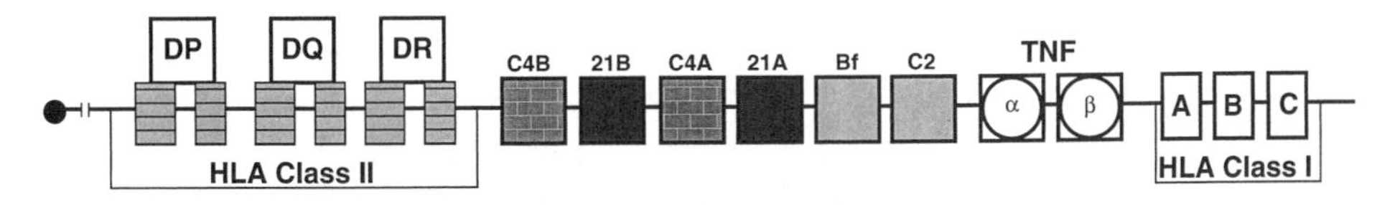

Figure 3.4 Simplified map of the region of human chromosome 6 where the HLA locus is located. The genes for 21 α and 21 β hydroxylase and for tumor necrosis factors (TNF) α and β have also been located to chromosome 6 but are not considered as part of the HLA complex.

C. In mice, the H2-K locus and the H2-L/H2-D loci (class I genes) are separated by the I region which includes two loci: I-A and I-E (class II genes). One related locus (locus S) codes for the C4 molecule of the complement system and is located between the H-2K and H-2D loci.

D. The organization of the **HLA gene complex in humans** is similar. However, the class II genes are closer to the GLO1 locus (coding for one isoenzyme of glioxylase), and followed by several loci coding for proteins related to the complement cascade such as Bf, C2, and C4, and by the HLA, B, and C loci (class I genes). The human MHC region includes other non-MHC genes such as those coding for tumor necrosis factors α and β (TNF-α, TNF-β; see Chapter 11) which are located near the C4 genes. The two C4 alleles are separated from each other by the genes coding for the enzyme 21 α hydroxylase (Fig. 3.4). In addition, the MHC-II region includes genes that code for proteins involved in loading peptides into the MHC molecules (see Chapter 4).

VI. GENETICS OF THE MHC

A. **The MHC as an Alloantigenic System**. As a consequence of the multiplicity of alleles for the different MHC antigens, there is only an extremely remote chance that two unrelated humans will be found who share an individual set of MHC antigens. Therefore, the MHC antigens are **alloantigens** (from the Greek "allos," different) distinguishing individuals within a given species. This is the basis for their use as genetic markers, which finds a major practical application in paternity studies.

B. **Haplotypes**. Each antigenic specificity of any given MHC locus is determined by one structural gene. For each MHC locus, a given individual carries two structural genes, one inherited with the paternal chromosome, and the other inherited with the maternal chromosome. Each chromosome, on the other hand, contains one set of structural genes coding for all the possible MHC molecules. The set of alleles that an individual carries at each locus on a single chromosome forms the **haplotype**, transmitted as a single unit except in very rare cases of recombination within the complex. Haplotypes can only be determined by family studies, which establish which MHC specificities are "linked" (i.e., transmitted as a bloc in a single chromosome).

C. **Co-Dominance of the MHC Alleles**. For each MHC locus, a given individual may be homozygous or heterozygous. In homozygous individuals, both chromosomes carry the same structural gene for that locus, and the cells of the individual express one single antigenic specificity (e.g., an individual homozygous for HLA-B27 carries two genes for the B27 specificity, one in each chromosome). Most individuals are heterozygous for any given locus and will express the two specificities inherited from each parent that are coded by the two DNA strands of the same chromosome (e.g., a heterozygous B8/B27 will have a gene coding for B8 in one chromosome and a gene coding for a B27 in the other). Both specificities for each locus will be expressed by every individual cell of this heterozygous individual. Therefore, the MHC genes are **co-dominant** at the cellular level, and there is no allelic exclusion in their expression, as observed in the case of immunoglobulin genes (see Chapter 7).

D. Tolerance to MHC Antigens. During embryonic differentiation, all mammals develop tolerance to the MHC specificities that they express. As a corollary, all adult mammals are able to respond to an MHC alloantigen that they do not express. Because of the rules of inheritance of the MHC, F1 hybrids of two inbred strains of mice will accept tissues from both parents (since inbred animals are homozygous the F1 hybrids are tolerant to all paternal and maternal specificities), but the animals of the parental strains will reject tissues from the hybrids, which will express MHC antigens of the other nonidentical strain.

E. MHC Phenotypes. Since all the alleles of any individual are co-dominant, it follows that both haplotypes that form an individual genotype will be expressed in the cells of that individual. The sum of all the specificities coded by the genome of the individual is known as that individual's **phenotype**. An example of the notation of a given individual's phenotype is as follows: HLA-A1,2; B8,27; Dw3,-; DR23,-. The hyphen indicates that only one antigen of a particular locus can be typed; this can signify that the individual is homozygous, or that he or she possesses an antigen that cannot be typed because no appropriate reagents are available. Family studies are the only way to distinguish between these two possibilities. Figure 3.5 shows an example of haplotype inheritance within the

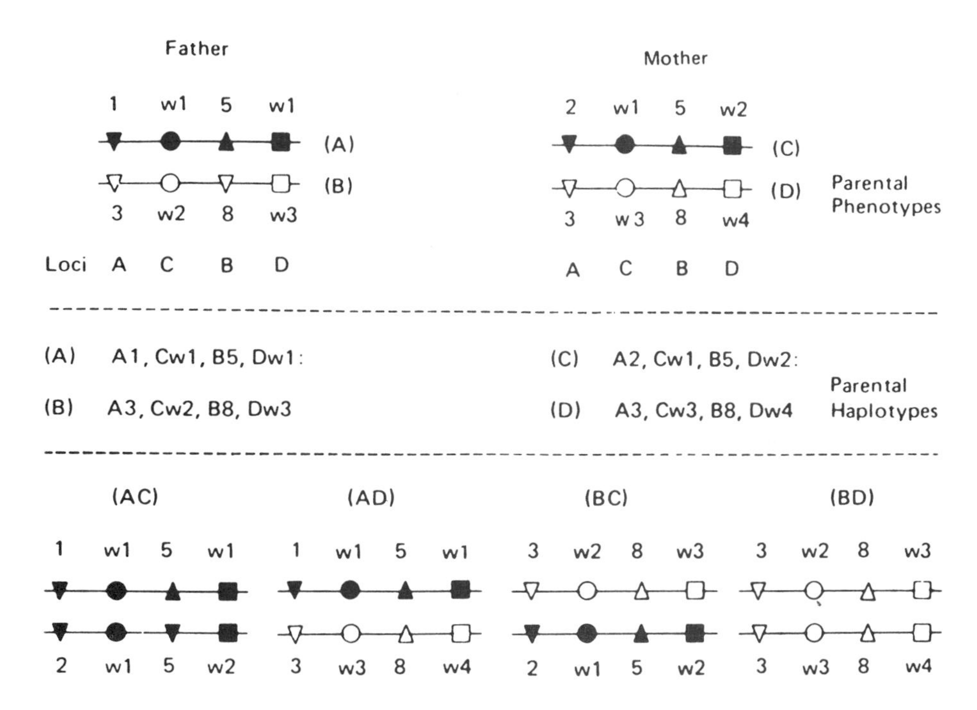

Figure 3.5 Diagrammatic representation of the genetic transmission of HLA haplotypes. Each parent has two haplotypes (one in each chromosome). Paternal haplotypes are designated A and B and maternal haplotypes C and D. Each offspring has to receive one paternal haplotype and one maternal haplotype. In a large family, 25% of the children share both haplotypes, 50% share one haplotype, and 25% have no haplotype in common. (Reproduced with permission from Hokama, Y., and Nakamura, R.M., *Immunology and Immunopathology*, Little Brown, Boston, 1982.)

HLA complex in humans, illustrating the fact that the haplotypes are transmitted, as a rule, as single units following the rules of simple Mendelian heredity. This rule is broken in cases of recombination between the parental and maternal chromosomes.

F. **Linkage Disequilibrium**. In an outbred population in which mating takes place at random, the frequency of finding a given allele at one HLA locus associated with a given allele at a second HLA locus should simply be the product of the frequencies of each individual allele in the population. However, certain combinations of alleles (i.e., certain haplotypes) occur with a higher frequency than expected. Thus, many HLA antigens occur together on the same chromosome more often than is expected by chance. This phenomenon is termed "**linkage disequilibrium**." As an example, the HLA-A1 allele is found in the Caucasian population with a frequency of 0.158, and the HLA-B8 allele is found with a frequency of 0.092. The A1, B8 haplotype should therefore be found with a frequency of $0.158 \times 0.092 = 0.015$. In reality, it is found with a frequency of 0.072. The linkage disequilibrium is expressed as the difference (Δ) between the observed and expected frequencies of the alleles (i.e., $0.072 - 0.015 = 0.057$).

VII. THE MHC COMPLEX AND THE IMMUNE RESPONSE

A. MHC-I and the Cytotoxic Response

1. **Modified self and intracellular infections**. The major function of the immune system is to protect the organism from non-self material, such as infectious agents. This function requires the recognition of non-self antigenic structures and the induction of an immune response through a set of cell–cell interactions involving macrophages, T lymphocytes, and B lymphocytes. Viruses and other intracellular parasites present a special problem to the immune system due to their shielding from contact with immunocompetent cells. However, the immune system can recognize infected cells and destroy them. Thus, antigens specific for the infectious agent must be presented to the immune system and trigger a response. As discussed in greater detail in Chapter 4, this is explained by the fact that oligopeptides derived from non-self proteins coded by intracellular infectious agents become associated with MHC molecules, and are transported to the membrane. The immune system recognizes the MHC-oligopeptide complexes as non-self ("**modified**" self in the sense that part of the antigen is self, and part is not).
2. **MHC restriction of cytotoxic T lymphocytes**. It became evident from detailed studies of the cell-mediated immune responses to viruses that the immune system relies on cytotoxic T lymphocytes to destroy the infected cells. An unexpected observation was that $CD8^+$ cytotoxic T lymphocytes would only kill viral-infected cells if both shared identical class I MHC antigens. The specificity of the cytotoxic reaction is directed to a specific configurational determinant formed by the association of a viral oligopeptide with an autologous MHC-I molecule. Cells infected by the same virus, but expressing different MHC-I molecules, will present a different MHC-I oligopeptide complex, not recognized by a cytotoxic T cell from an animal

expressing different MHC-1 molecules. Thus, the MHC-I cytotoxicity mediated by T lymphocytes is MHC restricted and the MHC-I molecule is the "restricting element."

3. **The biological role of MHC-I molecules** is to capture endogenous peptides and present them to the immune system. To reach the membrane in a stable conformation, the MHC molecule must always be loaded with a peptide. In the absence of an intracellular infection, peptides derived from autologous proteins occupy the peptide-binding groove. During an infection, proteins synthesized by a replicating intracellular infectious agent are presented by MHC-I molecules, replacing the endogenous peptides. The immune system learns to tolerate self MHC-I/self peptide complexes during embryonic differentiation, but will respond to a self MHC–non-self peptide complex (see Chapters 4 and 10).

B. MHC-II and the Magnitude of the Immune Response

1. **High and low responders**. Experimental studies carried out in inbred strains of mice and guinea pigs showed that the amplitude of the response to a given antigen is genetically controlled; some strains are **high** responders while others are **low** responders to specific antigens. Of great interest to immunologists was the fact that these quantitative differences in the amplitude of the response appeared to be under the control of genes closely linked or identical to MHC class II genes.
2. **MHC-II and antigen presentation to helper T lymphocytes**. T lymphocytes cannot respond to unmodified antigens. Their activation usually requires endocytosis and processing of the antigen by a specialized **antigen-presenting cell**. During "processing," soluble antigens are broken down into peptides of 12 to 23 amino acids, which become associated in the cytoplasm with newly synthesized MHC-II molecules (the groove of the HLA-class-II dimer is longer than the groove of MHC-I molecules and accommodates slightly larger oligopeptides). The peptide-MHC-II complex is then transported across the cytoplasm and inserted in the cell membrane. The MHC-II-associated peptides can be recognized by $CD4^+$ helper T lymphocytes carrying a peptide-specific T-cell receptor (TcR), but are not recognized by TCR on $CD8^+$ lymphocytes (see Chapters 4, 10, and 11). **All cells able to synthesize and express MHC-II molecules can function as antigen-presenting cells** (APC).
3. **Restrictive role of MHC-II molecules**. The interaction between MHC-expressing cells and the immune system is subjected to two levels of MHC restriction:
   a. In the first level, peptides bind to MHC molecules of one class or another.
   b. The CD8-binding site on class-I, and the CD4-binding site on class-II, determine the second level in which (CD4 or CD8) T-lymphocyte subsets will be able to interact with the MHC-bound peptide.

   Detailed discussions of the role of MHC-restriction in the immune response (Chapters 4 and 11), T-lymphocyte ontogeny (Chapter 10), cell-mediated immunity (Chapter 11), and tolerance (Chapter 22) are included throughout this book in the chapters indicated.

4. **MHC-II diversity**. A theoretical stumbling block is whether or not the limited number of MHC-II antigens is sufficient to bind a vast repertoire of peptides (on the order of 10^8–10^{10}). The peptides presented to T lymphocytes are helical structures with two parts. The part of the peptide that protrudes above the surface of the groove and is accessible to the T lymphocyte is known as the "**epitope**." The rest of the peptide interacts with the groove of the MHC molecule. This interaction is mediated by "**anchoring residues**," shared by many peptides. Thus a given peptide can bind to many different class I or class II HLA molecules and a limited repertoire of these molecules can accommodate a wide diversity of peptides.
5. **MHC binding and the immune response**. The anchoring residues determine the binding affinity of a given oligopeptide to specific MHC-II alleles, which varies over 2 or 3 orders of magnitude between different alleles. These differences in binding affinity are believed to determine the strength of the response. A peptide bound with high affinity will be presented to the T cells in optimal conformation determining a high response to this epitope. In contrast, if the binding is of low affinity, the individual will be a low responder or a nonresponder. Thus, the magnitude of the immune response is determined by the close fit between peptides and MHC molecules.
6. **Importance of antigen complexity and MHC heterozygosity**. Even with a limited MHC-II repertoire, the probabilities of mounting a good response to a complex antigen are high, since such antigen is likely to generate many different oligopeptides during processing and the odds will be favorable to generate some peptide(s) able to bind to the MHC-II molecules of a particular individual. These odds will increase if the individual is heterozygous, as happens in over 95% of humans. The expression of two specificities per locus (one paternal and the other maternal) doubles their chance of finding a good fit between peptides and MHC-II molecules, and therefore of mounting an adequate immune response to that particular antigen.
7. **MHC and malaria**. The association of particular MHC phenotypes with stronger responses to a given infectious agent was recently demonstrated in the case of malaria. In sub-Saharan Africa, malaria is responsible for the death of up to 2 million individuals/year. Researchers very recently found that individuals who do **not** express the class I HLA-Bw53 and the class II DQw5 alleles are at very high risk for the most severe and lethal forms of malaria. Reciprocally, since carrying the better haplotype improves the chances of survival, its frequency in the population increased by natural selection. Nowadays, 40% of the natives in sub-Sarahan Africa are positive for HLA-Bw53 and/or DQw5.

VIII. MHC-DISEASE ASSOCIATIONS

A. **General Considerations**. As discussed, while carrying some HLA alleles may protect an individual against specific diseases, it has also been found that carrying some other HLA alleles is associated with unusual frequency of diseases generally classified as "autoimmune." A list of the most important associations between HLA antigens and specific diseases is given in Table 3.2.

Table 3.2 Examples of Associations Between Particular Diseases and MHC in Humans[a]

Disease	Linked HLA region determinant	Relative risk of developing the disease[b]	Description of the disease
Inflammatory diseases			
Ankylosing spondylitis	B27	90–100	Inflammation of the spine, leading to stiffening of vertebral joints
Reiter's syndrome	B27	30–40	Inflammation of the spine, prostate, and parts of eye (conjunctiva, uvea)
Juvenile rheumatoid arthritis	B27	4–5	A multisystem inflammatory disease of children characterized by rapid onset of joint lesions and fever
Adult rheumatoid arthritis	DR4	6–12	Autoimmune inflammatory disease of the joints often associated with vasculitis
Psoriasis	B13, 17, 37	4–7	An acute, recurrent, localized inflammatory disease of the skin (usually scalp, elbows), associated with arthritis
Celiac disease	B8 Dw3	8–11	A chronic inflammatory disease of the small intestine; probably a food allergy to a protein in grains (gluten)
Multiple sclerosis	DR2	5	A progressive chronic inflammatory disease of brain and spinal cord that destroys the myelin sheath
Endocrine diseases			
Addisons's disease	DR3	4–10	A deficiency in production of adrenal gland cortical hormones
Diabetes mellitus	DQ alleles DR3, DR4	⩾100?[c] 2–5	A deficiency of insulin production; pancreatic islet cells usually absent or damaged
Miscellaneous diseases			
Narcolepsy	DR2	100	A condition characterized by the tendency to fall asleep unexpectedly

[a]Modified from Hood, L.E., Weissman, I.L., Wood, W.B., and Wilson, J.H., *Immunology*, 2nd ed. Benjamin/Cummings, Menlo Park, CA, 1984.

[b]Ratio of incidence rate of a disease in individuals expressing a given genetic marker relative to the incidence of disease in a group not expressing the marker, determined by the following formula:

$$\text{relative risk} = \frac{\text{no. of patients with the marker/total no. of patients and controls with the marker}}{\text{no. of patients without the marker/total no. of patients and controls without the marker}}$$

[c]Some alleles are associated with strong predisposition and others with strong resistance, but the precise relative risks for each allele have not yet been determined.

B. **Mechanisms**. Several mechanisms have been postulated to be on the basis of the associations between MHC antigens and specific diseases.

1. **MHC binding of exogenous peptides structurally similar to endogenous peptides**. As mentioned, MHC genes control the immune response by acting as binding sites for peptides derived from the processing of antigens from common pathogens. If the peptides in question are structurally similar to those derived from an endogenous protein and are bound with relatively high affinity to the MHC molecule, an immune response that eventually affects cells expressing the autologous peptide may ensue.
   a. Most human diseases in which autoimmune phenomena play a pathogenic role are strongly linked with **class II HLA genes**, particularly with the genes of the **DQ locus**. This probably reflects the fact that DQ molecules are involved in the interactions of APCs with helper/inducer CD4 T lymphocytes.
   b. It must be stressed that carrying the genes associated with any given autoimmune disease implies only an increased susceptibility to the disease. The individual may remain asymptomatic for life, but a chance encounter with a pathogen can trigger an autoimmune response.
   c. One well-studied example of the association between DQ molecules and autoimmunity is **diabetes mellitus** (DM). The susceptibility or lack of susceptibility of humans to the autoimmune response against pancreatic islet cells that results in insulin-dependent diabetes mellitus (IDDM) is determined by allelic polymorphisms of the β1 chain of the DQ antigen. More specifically, the charge of the residue at position 57 of that chain seems to determine whether or not the immune system will be presented with peptides able to trigger the autoimmune response. Most IDDM patients have a strongly charged valine in that position, which is usually occupied by an aspartate in nondiabetics. The binding of the anchoring residues of a diabetes-inducing peptide may be prevented by the presence of a charged aspartate residue in the groove, but may take place with high affinity binding when the 57 residue is uncharged (i.e., valine). The source of the diabetes-inducing peptide remains in question, but the fact that diabetes has been observed to follow some viral infections has led several authors to postulate that a viral protein-derived peptide may play that role.
   d. MHC-I antigens can also present antigen-derived peptides (particularly from intracellular pathogens) to CD8+ cells and this could be the basis for an autoimmune response in the case of ankylosing spondylitis and other reactive arthropathies, in which an infectious peptide presented by HLA-B27 could be cross-reactive with an endogenous collagen-derived peptide, equally associated with HLA-B27.
2. **MHC molecules may act as receptors for intracellular pathogens**. Such pathogens would interact with specific HLA antigens in the cell membrane and, as a result, infect the cells carrying those antigens. The infected cell would undergo long-lasting changes in cell functions, which would eventually result in disease. This could be the case in **ankylosing**

spondylitis and related disorders (**acute anterior uveitis**, **Reiter's syndrome**). Over 90% of the individuals with ankylosing spondylitis are HLA-B27 positive and about 75% of the patients developing Reiter's syndrome are HLA-B27 positive. Reiter's syndrome frequently follows an infection with *Chlamydia trachomatis* and some evidence for persistent infection with this intracellular organism has been obtained, but this still remains controversial.

3. **Molecular mimicry between antigenic determinants in infectious agents and HLA antigens**. This mechanism has been postulated to explain the relationship between *Yersinia pseudotuberculosis* and ankylosing spondylitis. This bacterium has been shown to contain epitopes cross-reactive with HLA-B27. Therefore, it could be speculated that an immune response directed against *Y. pseudotuberculosis* could lead to an autoimmune reaction. However, why this reaction would affect specific joints remains to be explained.
4. **Linkage disequilibrium between HLA genes and disease-causing genes**. A very strong association of various forms of **21 hydroxylase deficiency** (the molecular basis of a disease known as congenital adrenal hyperplasia) with various HLA haplotypes (HLA-Bw47, DR7) suggested a possible link of this disease with the MHC genes. However, data obtained in classic genetic studies suggested that this disorder was determined by alleles of a single locus. These apparent contradictions were resolved when it was demonstrated that the genes coding for 21-hydroxylase are located on chromosome 6, in the segment flanked by MHC-I and MHC-II genes (Fig. 3.3). Given this physical proximity between MHC genes and the 21-hydroxylase genes, a strong linkage disequilibrium between them is not surprising. A similar explanation may account for the very strong associations between **HLA-DR2** and **narcolepsy, HLA-B27** and **ankylosing spondylitis**, and **HLA-Bw35** and hemochromatosis.

SELF-EVALUATION

Questions

Please choose the ONE *best* answer.

3.1 The molecular basis for the control that MHC-II genes have over the immune response is best explained by the:
  A. Ability of MHC-II antigen complexes to be released from antigen-processing cells and activate helper T cells
  B. Involvement of MHC-II molecules in targeting reactions mediated by cytotoxic T lymphocytes
  C. Existence of genes controlling the immune response in linkage disequilibrium with MHC-II genes
  D. Need for antigen-derived peptides to bind to an MHC-II molecule for proper presentation to the TCR of a helper T lymphocyte
  E. Special affinity of unprocessed antigens for MHC-II molecules

3.2 Which of the following cellular antigens is(are) not coded by genes in the MHC region?
A. β_2-Microglobulin
B. Class I HLA antigens
C. Class II HLA antigens
D. C4
E. Tumor necrosis factor α

3.3 The HLA phenotypes of a married couple are:
Father A1,3; Cw1,w2; B5,8; Dw1,3
Mother A2,3; Cw1,3; B5,8; Dw2,4
Which of the following phenotypes would definitely *not* be possible in an offspring of that couple?
A. A1,2; Cw1-; B5-; Dw1,w2
B. A2,3; Cw1,w2; B5,8; Dw2,3
C. A1,3; Cw1,w3; B5,8; Dw1,w4
D. A1,-; Cw1,w2; B5-; Dw1,w3
E. A3-; Cw2,w3; B8,-; Dw3,w4

3.4 Which of the following is a unique characteristic of class I MHC proteins?
A. Expression of the membrane of activated T lymphocytes
B. Inclusion of β_2 microglobulin as one of their constituent chains
C. Interaction with the CD4 molecule
D. Limited serological diversity
E. Presentation of immunogenic peptides to helper T lymphocytes

3.5 Which of the following groups of MHC genes include alleles closely associated with the susceptibility or resistance to develop diabetes mellitus?
A. HLA-A
B. HLA-B
C. HLA-DP
D. HLA-DQ
E. HLA-DR

3.6 Over 80% of the patients with ankylosing spondylitis are positive for:
A. HLA-DW3
B. HLA-B8
C. HLA-B27
D. HLA-B7
E. HLA-A5

3.7 Cells not expressing class I HLA antigenic products include:
A. Monocyte
B. B lymphocytes
C. Skin cells
D. T lymphocytes
E. Striated muscle cells

3.8 The frequency of the HLA-A1 and HLA-B8 alleles in the general population is 0.158 and 0.092, respectively. Assuming that these alleles are transmitted independently (without linkage disequilibrium), the expected frequency of the A1, B8 haplotype would be:

A. 0.072
B. 0.057
C. 0.030
D. 0.015
E. 0.003

3.9 A relative risk of 5 for the association between DR4 and rheumatoid arthritis means that:
A. The ratio of expression of DR4 is five times higher among patients with rheumatoid arthritis than in healthy controls
B. $DR4^+$ individuals are five times more likely to develop rheumatoid arthritis than $DR4^-$ individuals
C. Five percent of $DR4^+$ individuals will develop rheumatoid arthritis
D. The autoimmune reaction that causes rheumatoid arthritis is directed against DR4 epitopes in 5 of every 10 patients.
E. The frequency of rheumatoid arthritis is fivefold higher in $DR4^+$ individuals than in $DR4^-$ individuals

3.10 The stimulation of cytotoxic ($CD8^+$) T lymphocytes requires the association of an antigen-derived peptide with a(n):
A. MHC-I molecule
B. MHC-I molecule identical to one of those expressed by the cytotoxic T lymphocyte
C. MHC-II molecule
D. MHC-II molecule identical to one of those expressed by the cytotoxic T lymphocyte
E. Newly synthesized MHC-II molecule in the endoplasmic reticulum

Answers

3.1 (D) The need for antigen-derived peptides to be presented in association with MHC-II molecules for proper stimulation of helper T cells is the limiting factor controlling the immune response. Peptides that bind with high affinity to MHC-II molecules will elicit strong responses, while no response will be elicited when the peptides cannot bind (or bind very weakly) to the available MHC-II molecules.

3.2 (A) β_2 microglobulin is coded by a gene in chromosome 15 and becomes associated to the heavy chain of HLA-I antigens postsynthetically. The gene coding for tumor necrosis factor α is located on chromosome 6, between the MHC-I and MHC-II loci, in close proximity to the genes that code for several components of the complement system.

3.3 (D) The phenotype should include specificities 2 or 3 for locus A and W2 or W4 for locus D (one specificity being of paternal origin, and the other of maternal origin).

3.4 (B) Class I MHC antigens are highly polymorphic and have hundreds of different serological specificities. Structurally, they are constituted by one MHC-coded polypeptide and β_2-microglobulin and are expressed in the membranes of almost all nucleated cells. They interact with the CD8 molecule and their expression in lymphocytes is independent on the state of activation of the cell.

3.5 (D) Variations in the structure of the DQ β chain, at the point critically associated with peptide binding, seem to be most strongly associated with protection or sensitivity for diabetes; the structural variations are detected as DQ alleles by serological techniques and by the use of DNA probes.

3.6 (C)

3.7 (E) Class I antigens are expressed in most nucleated cells, except nervous tissue cells and striated muscle cells.

3.8 (D) In the absence of linkage disequilibrium, the frequency of the A1, B8 haplotype should equal the product of the frequencies of each antigen in the general population.

3.9 (A) Relative risk is a measurement of the frequency of detection of a given genetic marker in a given patient population, relative to the frequency of the same marker in individuals not affected by the disease. The presence of the marker is not diagnostic by itself, and does not necessarily imply that all positive individuals will develop the disease.

3.10 (B) Cytotoxic T cells recognize peptides associated with self MHC-I molecules.

BIBLIOGRAPHY

Bjorkman, P.J., Saper, M.A., Samroui, B., Bennett, W.S., Strominger, J.L., and Wiley, D.C. Structure of the human class I histocompatibility antigen, HLA-A2. *Nature*, *329*:506, 1987.

Careless, D.J., and Inman, R.D. Etiopathogenesis of reactive arthritis and ankylosing spondylitis. *Curr. Opin. Rheumatol.*, *7*:290, 1995.

Fremont, D.H., Hendrickson, W.A., Marrack, P., Kapler, J. Structures of an MHC Class II molecule with covalently bound single peptides. *Science*, *272*:1001, 1996.

Gladman, D.D., and Farewell, V.T. The role of HLA antigens as indicators of disease progression in psoriatic arthritis. Multivariate relative risk model. *Arthritis Rheum.*, *38*:845, 1995.

Hill, A.V.S., Allsopp, C.E.M., Kwiatkowski, D., Anstey, N.M., Twumasi, P., Rowe, P.A., Bennett, S., Brewster, D., McMichael, A.J., and Greenwood, B.M. Common West African HLA antigens are associated with protection from severe malaria. *Nature*, *342*:595, 1991.

Howard, J.C. Disease and evolution. *Nature*, *352*:565, 1991.

Klein, J. *Immunology*. Blackwell Scientific Publications, Oxford/Brookline, MA, 1990.

Lopez de Castro, J.A. Structural polymorphism and function of HLA-B27. *Curr. Opin. Rheumatol.*, *7*:270, 1995.

Parham, P. Antigen processing: Transporters of delight. *Nature*, *348*:674, 1990.

Pugliese, A., Gianani, R., Eisenbarth, G.S. et al. HLA-DQB1*0602 is associated with dominant protection from diabetes even among islet cell antibody-positive first-degree relatives of patients with IDDM. *Diabetes*, *44*:608, 1995.

Weyand, C.M., and Goronzy, J.J. Functional domains on HLA-DR molecules: Implications for the linkage of HLA-DR genes to different autoimmune diseases. *Clin. Immunol. Immunopathol.*, *70*:91, 1994.

4
The Induction of an Immune Response: Antigens, Lymphocytes, and Accessory Cells

Gabriel Virella and Barbara E. Bierer

I. INTRODUCTION

The mechanisms involved in recognizing a given substance as "foreign," the sequence of events that follow such recognition and lead to the stimulation of immune competent cells, and the induction of specific effector mechanisms that result in its elimination are beginning to be elucidated. In spite of the limitations of in vitro models available for analysis, our knowledge of the steps involved in the generation of an immune response is rapidly increasing.

II. ANTIGENICITY AND IMMUNOGENICITY

A. **Antigenicity** is defined as the property of a substance (**antigen**) that allows it to react with the products of a specific immune response (**antibody** or **T-cell receptor**).

B. **Immunogenicity** is defined as the property of a substance (**immunogen**) that endows it with the capacity to provoke a specific immune response.

C. **All immunogens are antigens, but the reverse is not true.**

1. Immunogens are usually complex, large molecules, that are able to induce an immune response by themselves. Immunogens are also antigens in that specific antibodies or specific T cells recognizing them will be formed as a consequence of the immune response they elicit.
2. Certain low molecular weight substances, known as **haptens**, are unable to induce an immune response by themselves. However, if haptens are coupled to an immunogenic **carrier** molecule, the immune system will recognize them as separate epitopes and produce antihapten antibodies that react with soluble hapten molecules, free of carrier protein. Thus, a hapten is an antigen, but not an immunogen.
3. Unfortunately, the terms antigen and immunogen are often interchanged.

III. ANTIGENIC DETERMINANTS

A. Most antigens are complex molecules (mostly proteins and polysaccharides). Only a restricted portion of the antigen molecule—known as an **antigenic determinant** or **epitope**—is able to react with the specific binding site of a B-lymphocyte membrane immunoglobulin, a soluble antibody, or a T-lymphocyte antigen receptor.

B. While B lymphocytes recognize epitopes expressed by native antigens, T lymphocytes recognize small peptides generated during antigen processing or derived from newly synthesized proteins.

C. Studies with x-ray crystallography and two-dimensional nuclear magnetic resonance imaging have resulted in the detailed characterization of epitopes presented by some small proteins, such as lysozyme, in their native configuration. From such studies the following rules have been derived for antibody–antigen recognition:

1. Most epitopes are composed of a series of 15 to 22 amino acids located on discontinuous segments of the polypeptide chain, forming a roughly flat area with peaks and valleys that establish contact with the folded hypervariable regions of the antibody heavy and light chains.
2. Specific regions of the epitope constituted by a few amino acids bind with greater affinity to specific areas of the antibody binding site, and thus are primarily responsible for the specificity of antigen–antibody interaction. On the other hand, the antibody binding site has some degree of flexibility, which contributes to the good fit with the corresponding epitope.
3. A polypeptide with 100 amino acids may have as many as 14 to 20 non-overlapping determinants. However, a typical 100 amino acid globular protein is folded over itself, and most of its structure is hidden from the

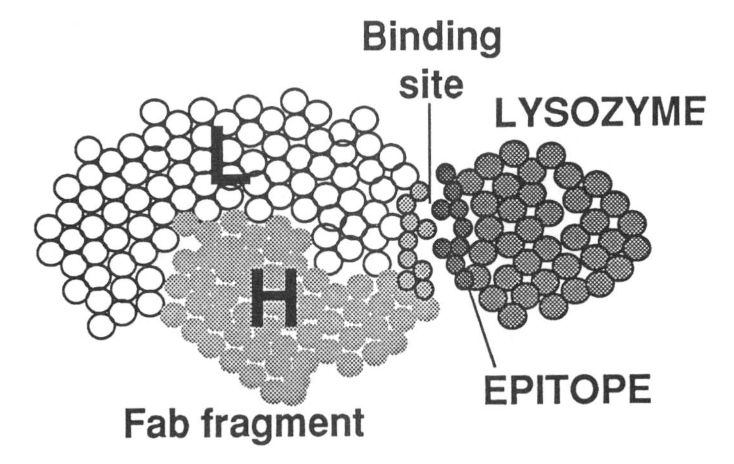

Figure 4.1 Diagrammatic representation of a space-filling model showing the fit between an epitope of lysozyme and the antigen-binding site on a Fab fragment obtained from an antilysozyme antibody. (Adapted from Amit, A.G., Mariuzza, R.A., Phillips, S.E.V., and Poljack, R.J. Three-dimensional structure of an antigen–antibody complex at 2.8 Å resolution. *Science, 233*:747, 1986.)

outside. Only surface determinants on molecules will usually be accessible for recognition by B lymphocytes and for interactions with antibodies (Fig. 4.1).

IV. CHARACTERISTICS OF IMMUNOGENICITY

Many different substances can induce immune responses. The following characteristics have an important influence in the ability that a substance has to behave as an immunogen.

A. **Foreigness**. As a rule, only substances recognized as "non-self" will trigger the immune response. Microbial antigens and heterologous proteins are obviously "non-self" and are strongly immunogenic.

B. **Molecular Size**. The most potent immunogens are macromolecular proteins [molecular weight (M.W.) > 100,000]. Molecules smaller than 10,000 daltons are weakly immunogenic.

C. **Chemical Structure and Complexity**. Proteins and polysaccharides are among the most potent immunogens, although relatively small polypeptide chains, nucleic acids, and even lipids can, given the right circumstances, be immunogenic.

1. **Proteins**. Large heterologous proteins expressing a wide diversity of antigenic determinants are potent immunogens.
    a. The immunogenicity of a protein is strongly influenced by its chemical composition. Positively charged (basic) amino acids, such as lysine, arginine, and histidine are repeatedly present in the antigenic sites of lysozyme and myoglobin, while aromatic amino acids (such as tyrosine) are found in two of albumin's six antigenic sites. Therefore, it appears that basic and aromatic amino acids may contribute more strongly to immunogenicity than other amino acids; basic proteins with clusters of positively charged amino acids are strong immunogens.
    b. There appears to be a direct relationship between antigenicity and chemical complexity: aggregated or chemically polymerized proteins are much stronger immunogens than their soluble monomeric counterparts.
2. **Polysaccharides**. Polysaccharides are among the most important natural antigens, since either pure polysaccharides or the sugar moieties of glycoproteins, lipopolysaccharides, glycolipid-protein complexes, etc., are immunogenic. Many microorganisms have polysaccharide-rich capsules or cell walls, and a variety of mammalian antigens, such as the erythrocyte antigens (A, B, Le, H), are short-chain polysaccharides (oligosaccharides).
3. **Nucleic acids**. Nucleic acids usually are not immunogenic, but can induce antibody formation if coupled to a protein to form a nucleoprotein. The autoimmune responses characteristic of some of the so-called **autoimmune diseases** (e.g., systemic lupus erythematosus) are often directed to DNA and RNA that may have stimulated the immune system as nucleoproteins.
4. **Polypeptides**. Hormones such as insulin and other polypeptides, although relatively small in size (M.W. 1500), are usually immunogenic when isolated from one species and administered over long periods of time to an individual of a different species.

V. HAPTENS AND CARRIERS

A. **Landsteiner**, **Pauling**, and others discovered in the 1930s and 1940s that small aromatic groups, such as amino-benzene sulphonate, amino-benzene arsenate, and amino-benzene carboxylate could be chemically coupled to immunogenic proteins (**carriers**) and induce antihapten antibodies in this form. These authors used hapten-carrier conjugates to study the specificity of the immune response (see Chapter 8). Later in this chapter we will discuss the use of haptens and carriers to define T-B lymphocyte cooperation.

B. In the last two decades the hapten-carrier concept has found significant applications in medicine, particularly to enhance immunization protocols, and has also been demonstrated to be the pathological basis for some abnormal immune reactions:
   1. Poorly immunogenic polysaccharides have been shown to induce strong immune responses when conjugated to immunogenic proteins, and this observation has resulted in vaccine development for immunoprophylaxis of infectious diseases.
   2. Using drug-protein conjugates it has been possible to produce antibodies to a wide variety of drugs, which are used in numerous drug immunoassays (e.g., plasma digoxin levels).
   3. Some hypersensitivity reactions to some drugs, chemicals, and metals is believed to result from spontaneous coupling of these compounds to endogenous proteins, creating hapten-carrier combinations. One example of this mechanism is the spontaneous coupling of the penicilloyl derivative of penicillin to a host protein, believed to be the first step toward developing hypersensitivity to penicillin.

VI. FACTORS ASSOCIATED WITH THE INDUCTION OF AN IMMUNE RESPONSE

Besides the chemical nature of the immunogen, other factors strongly influence the development of an immune response.

A. Genetic Constitution of the Animal. Different animal species or different strains of one given species show different degrees of responsiveness to a given antigen. In humans, different individuals can behave as "high responders" or "low responders" to any given antigen. The genetic control of the immune response seems mainly related to the repertoire of MHC molecules, which bind antigen fragments and present them to the immune system, as we will discuss later. The animal will respond well to those antigens that are processed into peptides or oligopeptides with high affinity for the binding sites of the MHC molecules (see Chapter 3).

B. Method of Antigen Administration. A given dose of antigen may elicit no detectable response when injected intravenously, but may elicit a strong immune response if injected intradermally. This last route of administration results in slow removal from the site of injection and prolonged antigenic stimulation. The presence of dendritic cells in the dermis (where they are known as Langerhans cells) is believed to be a significant factor determining the vigorous immune responses obtained when antigens are injected intradermally. The injected antigen is trapped on the surface of those dendritic cells, which migrate to the lymph node follicles, where the initial stages of the immune response take place.

C. **Use of Adjuvants**. Adjuvants are agents that, when administered along with an antigen, enhance the specific response. Several factors seem to contribute to this enhancement, including delayed release of antigen, nonspecific inflammatory effects, and the activation of monocytes and macrophages.

1. One of the most effective adjuvants is **Complete Freund's Adjuvant** (CFA), a water-in-oil emulsion with killed mycobacteria in the oil phase. **Bacillus Calmette-Guérin (BCG)** is an attenuated strain of *Mycobacterium bovis* used as vaccine against tuberculosis and is also an effective adjuvant. **Muramyl-dipeptide (MDP)**, the active moiety of *Mycobacterium tuberculosis* and of BCG, also has adjuvant properties.
2. Several other **microbial** and **inorganic compounds** have been used as adjuvants. Some of these adjuvants have been used therapeutically with the aim of boosting the immune response of compromised patients. Many of these compounds cause intense inflammatory reactions and discomfort.
3. **Aluminum hydroxide**, an inert compound that absorbs the immunogen, stimulates phagocytosis, and delays removal from the inoculation site, is the adjuvant most frequently used with human vaccines. Aluminum hydroxide is not as effective as most of the adjuvants listed above, but is considerably less toxic.

VII. EXOGENOUS AND ENDOGENOUS ANTIGENS

A. Most of the antigens to which we react are of **exogenous** origin, and include microbial antigens, environmental antigens (such as pollens and pollutants), and medications. The objective of the immune response is the elimination of foreign antigens, but, in some instances, the immune response itself may have a deleterious effect (**hypersensitivity states**).

B. "**Alloantigens**," i.e. antigenic determinants that distinguish one individual from another within the same species, are unique exogenous antigens. These are alleles of highly polymorphic systems, which define the antigenic makeup of the cells and tissues of an individual. A classic example is the A, B, O blood group antigens: some individuals carry the A specificity, some are B positive, some are AB positive, and some express neither A nor B (O). Other alloantigenic systems are the histocompatibility (MHC or in the human, HLA) antigens of nucleated cells and tissues, the platelet (Pl) antigens, and the immunoglobulin allotypes. Examples of sensitization to exogenous alloantigens include:

1. Women who may become sensitized to fetal red cell antigens or immunoglobulin alloantigens during pregnancy.
2. Repeated blood transfusions that can induce sensitization against cellular or immunoglobulin alloantigens from the donor(s).
3. Organ transplantation that usually results in sensitization against histocompatibility alloantigens expressed in the transplanted organ.

These situations are discussed in greater detail in Chapters 24 and 27.

C. **Endogenous** antigens, by definition, are part of self, and the immune system should not react against them. The response to self antigens can be the cause of severe pathological situations (autoimmune diseases), although it may also have an important role in normal catabolic processes (i.e., antibodies to denatured IgG may help in eliminating antigen–antibody complexes from circulation; antibodies to oxidized LDL may help in eliminating a potentially toxic lipid).

VIII. T-DEPENDENT AND T-INDEPENDENT ANTIGENS

A. **Functional Definition**. Early studies on the physiological role of T lymphocytes included experiments in which inbred rodents were sublethally irradiated to render them immunoincompetent, and their immune systems were then reconstituted with T lymphocytes, B lymphocytes, or mixtures of T and B lymphocytes obtained from normal animals of the same strain. After reconstitution of the immune system, the animals were challenged with a variety of antigens, and their antibody responses were measured.

1. For most antigens, including complex proteins, heterologous cells, and viruses, a measurable antibody response was only observed in animals reconstituted with mixtures of T and B lymphocytes. Since T lymphocytes do not synthesize antibody, their role must be one of assisting the proliferation and/or differentiation of B lymphocytes. The antigens that can elicit antibody responses exclusively in animals reconstituted with both T and B cells are known as **T-dependent** antigens. Structurally, T-dependent antigens are usually complex proteins with large numbers of different antigenic determinants with little repetition among themselves.
2. Other antigens, particularly polysaccharides, can induce antibody synthesis in animals depleted of T lymphocytes, and are known as **T-independent** antigens.
3. It should be noted that there appears to be a continuous gradation from T-dependence to T-independence, rather than two discrete groups of antigens in many species. However, this differentiation is useful as a "working classification" of antigens.

B. **Biological Basis of T-Independence**. Two theories concerning the signaling by T-independent antigens have been proposed:

1. Some T-independent antigens, such as bacterial lipopolysaccharides (LPS), have mitogenic properties and can deliver dual signals to B cells, one by occupying the antigen-specific receptor (membrane immunoglobulin), and the other by a nonspecific mitogenic signal involving the lipid moiety of LPS, whose nature has not been well characterized. The association of these two signals would be sufficient to stimulate B cells and promote their differentiation into antibody-producing cells.
2. Other T-independent antigens (such as polysaccharides) do not have mitogenic properties, but are constituted by multiple repeats of a limited number of sugar molecules, allowing extensive cross-linking of membrane immunoglobulins. Receptor cross-linking delivers strong activating signals that apparently override the need for co-stimulatory signals.

C. **Special Characteristics of the Immune Response to T-Independent Antigens**. The antibody produced in response to stimulation with T-independent antigens is predominantly IgM. The switch to IgG production requires T-cell help or cytokine production; this switch does not occur, therefore, with T-independent antigens. Immunological memory for T-independent antigens is either not evident or very weak.

IX. THE HAPTEN-CARRIER EFFECT

Experiments carried out with hapten-carrier complexes have contributed significantly to our understanding of T-B lymphocyte cooperation.

A. **Experiments with Immunocompetent Animals**

1. If an animal is primed with a hapten-carrier conjugate prepared by chemically coupling the hapten 2-dinitrophenyl (2-DNP) radical to the carrier egg albumin (ovalbumin, OVA), the animal will produce antibodies both to DNP and OVA. A recall challenge with DNP-OVA will trigger a secondary response of higher magnitude against both hapten and carrier. In contrast, if the DNP-OVA primed animal is challenged with the same hapten coupled to a different carrier, such as bovine gamma globulin (DNP-BGG), the ensuing response to DNP is of identical magnitude to that observed after the first immunization with DNP-OVA (Fig. 4.2).
2. If the animal is primed with ovalbumin alone, and then challenged with DNP-OVA, the response to DNP is as high as that observed when the animal is primed with DNP-OVA (Fig. 4.3). If DNP and OVA are administered as a

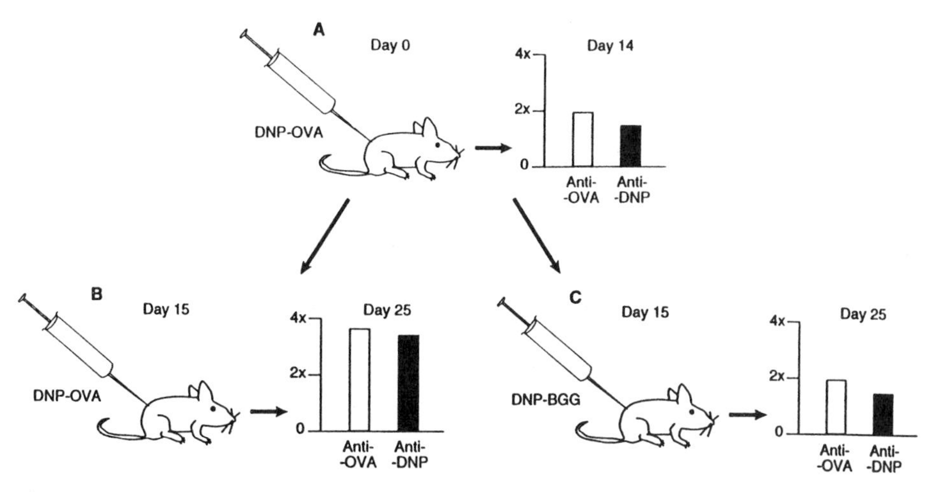

Figure 4.2 The hapten-carrier effect: To obtain a secondary immune response to the hapten (DNP), the animal needs to be (A) immunized and (B) challenged with the same DNP-carrier combination; (C) boosting with a different DNP-carrier conjugate will result in an anti-DNP response of identical magnitude to that obtained after the initial immunization. The memory response, therefore, appears to be carrier-dependent.

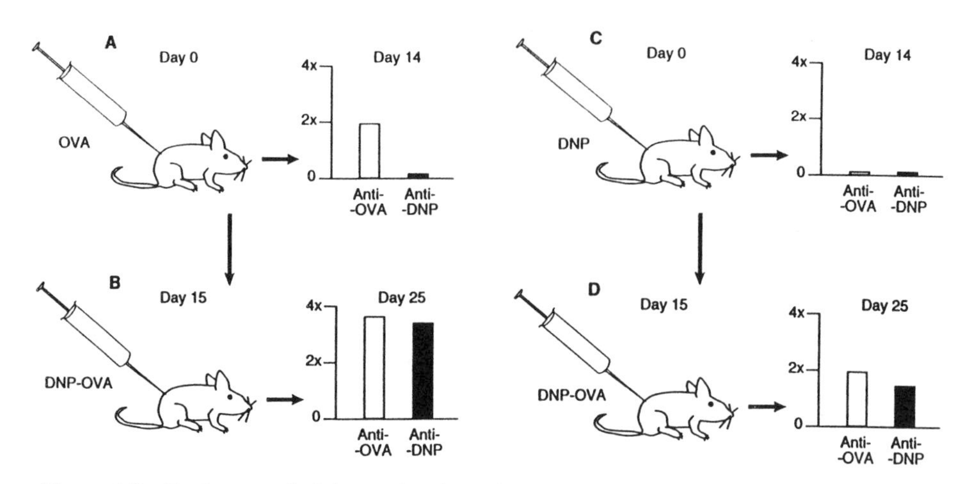

Figure 4.3 Further proof of the carrier dependency of the memory response to a hapten-carrier conjugate was obtained by studying the effects of primary immunization with carrier (e.g., OVA) or hapten (e.g., DNP) alone on a booster response with the hapten-carrier conjugate. A primary immunization with OVA (A) was followed by a "secondary" response to both hapten and carrier when the animals were challenged with OVA-DNP (B). A primary immunization with DNP (C) did not induce anti-DNP antibodies and the animal reacted to a challenge with OVA-DNP (D) as if it was a primary immunization to either carrier or hapten.

mixture of unconjugated molecules to an OVA-primed animal, no response to DNP will be observed.

3. From these experiments the following conclusions can be drawn:
   a. While the response to a hapten can only be elicited if the hapten is chemically conjugated to the carrier, carrier and hapten are recognized independently.
   b. Immunological memory depended upon the carrier moiety.
   c. Once a "memory" response is elicited by the carrier, the antihapten response will be of the high magnitude characteristic of secondary responses; therefore, whatever expands the response to the carrier is equally able to influence the response to the hapten.
   d. The same way that a hapten by itself cannot elicit an immune response, administration of a mixture of unconjugated DNP and OVA to an animal primed with DNP-OVA will elicit a secondary response to OVA but not to DNP. Thus, the influence of OVA memory into the DNP response cannot be seen unless the hapten is administered in an immunogenic form.

B. Experiments with Irradiated Animals. Groups of inbred mice were immunized with DNP or with two different immunogenic proteins, such as keyhole limpet hemocyanin (KLH) and ovalbumin (OVA), and a control group was injected with saline (see Fig. 4.4). T and B lymphocytes were purified from the immunized animals and transferred to sublethally irradiated mice of the same strain.

1. If both DNP-primed B lymphocytes and OVA-primed T lymphocytes were

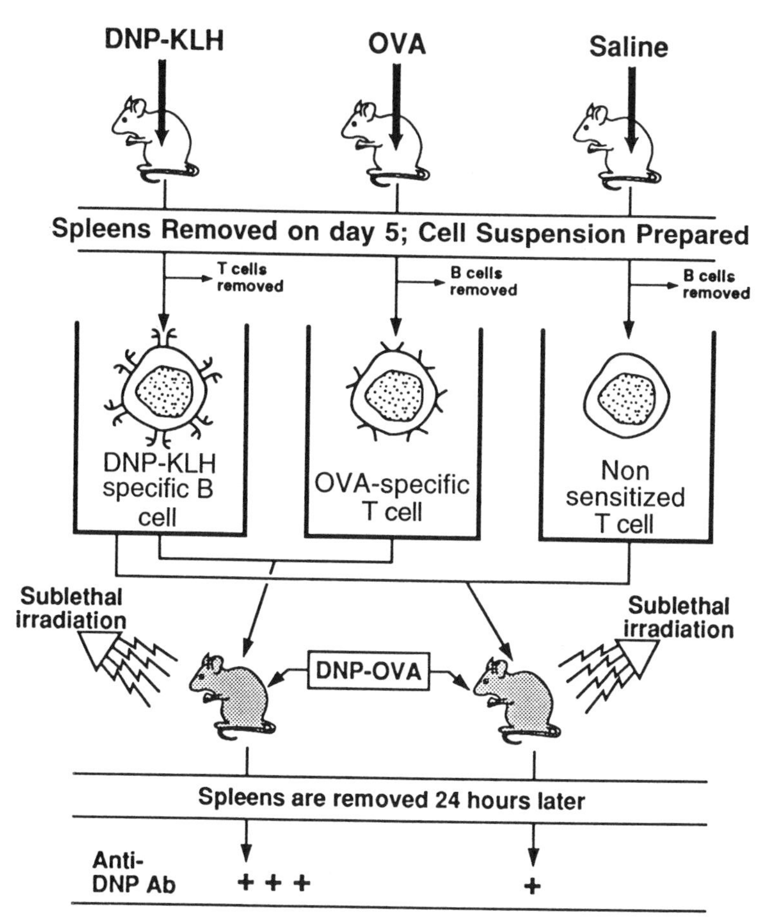

Figure 4.4 Diagrammatic representation of an experiment designed to determine the nature of T-cell help in a classic hapten-carrier response. Sublethally irradiated mice were reconstituted with different combinations of T and B cells obtained from nonimmune mice or from mice immunized with DNP-OVA. T cells from mice preimmunized with OVA "helped" B cells from animals preimmunized with DNP-KLH to produce large amounts of anti-DNP antibodies, but the same B cells did not receive noticeable help from T cells separated from nonimmune mice. Thus, B cells are most efficiently helped by T cells with "carrier" memory.

transferred to sublethally irradiated recipients, animals produced relatively large amounts of anti-DNP antibody upon challenge with DNP-OVA.

2. If nonimmune or KLH-primed T lymphocytes were co-transferred with DNP primed B lymphocytes, only a meager anti-DNP antibody response was obtained after challenge with DNP-OVA.
3. An irradiated mouse reconstituted with DNP-specific B lymphocytes and ovalbumin (OVA)-specific T lymphocytes will make an excellent anti-DNP antibody response upon challenge with DNP-OVA, but will show a suboptimal response upon challenge with DNP-KLH, and no anti-DNP antibody response when challenged with an unconjugated mixture of OVA and DNP.

4. The following conclusions can be drawn from these experiments:
   a. The amplification of the B-cell response requires T cells.
   b. Both carrier and hapten-specific antibody-producing cells are "helped" by carrier-specific T lymphocytes. In other words, T-lymphocyte "help" is not antigen-specific, since the T and B lymphocytes collaborating in the immune response may recognize antigenic determinants from totally unrelated compounds (hapten and carrier).
   c. The efficient collaboration between T and B lymphocytes requires that the antigenic determinants for which each cell type is specific must be on the same molecule. This suggests that the helper effect is most efficient if the collaborating B and T lymphocytes are brought into intimate, cell-to-cell contact, each cell reacting with distinct determinants on the same molecule.

X. INDUCTION OF THE IMMUNE RESPONSE

A. **Immune Recognition: Clonal Restriction and Expansion**. For an immune response to be initiated, the antigen or a peptide associated with an MHC molecule must be recognized as "non-self" by the immunocompetent cells. This phenomenon is designated **immune recognition**.
   1. It is calculated that as many as 10^6–10^8 different antigenic specificities can be recognized by the immune system of a normal individual, and it is believed that an equal number of different small families (clones) of immunocytes, bearing receptors for those different antigens, constitutes the normal repertoire of the immune system.
   2. Each immunocompetent cell expresses on its membrane many identical copies of a receptor for one single antigen. Thus, a major characteristic of the immune response is its **clonal restriction** (i.e., one given epitope will be recognized by a single family of cells with identical antigen receptors, known as a **clone**). When stimulated, each cell will proliferate, and the clone of reactive cells will become more numerous (**clonal expansion**).
   3. Since immunogens present many different epitopes to the immune system, the normal immune responses are **polyclonal** (i.e., multiple clones of immunocompetent cells, each one of them specific for one unique epitope, are stimulated by any complex immunogen).

B. **The Antigen Receptor on T and B Lymphocytes**
   1. In **B lymphocytes**, the antigen receptors are membrane-inserted immunoglobulins, particularly IgD and monomeric IgM molecules (see Chapter 5).
   2. In **T lymphocytes**, the antigen receptors are known as **T-cell receptors (TcR)**. As discussed in Chapters 10 and 11, two types of TcR have been identified, depending on the polypeptide chains that constitute them.
      a. In differentiated T cells, the TcR most frequently found is composed of two polypeptide chains, designated as α and β (α/β TcR), with similar molecular weight (40–45,000).
         i. The two chains each have extracellular segments with variable and constant domains, a short cytoplasmic domain, and a transmembrane segment (Fig. 4.5). They are joined by a disulfide bridge just outside the transmembrane segment.

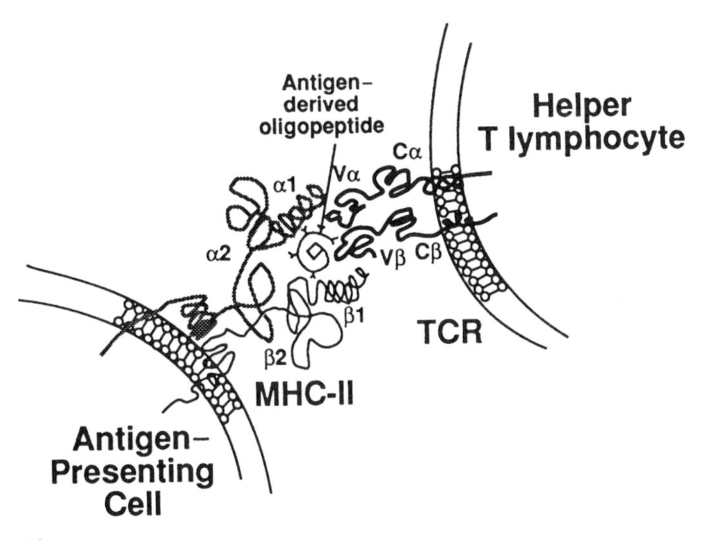

Figure 4.5 Schematic representation of the T-lymphocyte receptor and of its interaction with an MHC-II-associated peptide. (Redrawn from Sinha, A.A., Lopez, M.T., and McDevitt, H. Autoimmune disease: The failure of self tolerance. *Science*, *248*:1380, 1990.)

ii. Considerably more diversity exists among β chains than among α chains. The β chains are encoded by a multigene family that includes genes for regions homologous to the V, C, D, and J regions of human immunoglobulins. The α chains are encoded by a more limited multigene family, with genes for regions homologous to the V, C, and J regions of human immunoglobulins (see Chapter 7).

b. A second type of TcR, composed of two different polypeptide chains known as γ and δ (γ/δ TcR), is predominantly found in the perimucosal lymphoid tissues.

C. **Antigen Processing and Presentation**. Most immune responses to complex, T-dependent antigens (such as heterologous proteins) require the participation of several cell types, some directly involved in the generation of effector mechanisms (e.g., T and B lymphocytes), others assisting the inductive stages of the immune response (**accessory cells; AC**).

1. One special type of accessory cell that plays a very important role in the early stages of the immune response are **antigen-presenting cells** (i.e., cells that carry MHC-II molecules on their membrane where antigen fragments can be bound and "presented" to lymphocytes).
2. Tissue **macrophages** and related cells are ideally suited for antigen presentation, since they have phagocytic properties, can process the antigen, and express MHC-II molecules that are able to bind antigen-derived peptides and present them to helper T cells. However, the role of macrophages in the induction of a primary immune response has to be considered with some reservation, since these cells are most efficient ingesting antigens coated with antibody and complement, and antibody will not be available before the immune system is activated. Other types of cells, such as activated B lym-

phocytes or dendritic cells, appear to serve as APC in an immunologically naive individual (see later in this chapter).

3. **Antigen processing** is a complex sequence of events that involves endocytosis of membrane patches with attached organisms or proteins, and transport to an acidic compartment (lysosome) within the cell which allows for the breakdown of the engulfed material into small fragments. In the case of a microorganism, processing involves the breakdown of the infectious agent and the generation of immunogenic fragments. In the case of complex proteins, processing involves unfolding and breakdown into small peptides.
4. **Antigen presentation** to T-lymphocytes requires the assembly on MHC-II-peptide complexes and their transport to the cell membrane. As complex immunogens are broken down, vesicles coated with newly synthesized HLA-II molecules fuse with the lysosome. Some of the peptides generated during processing have affinity for the binding site located within the MHC-II α/β heterodimer. Once bound, these peptides seem protected against further degradation and the MHC-II-peptide complexes are transported to the cell membrane (Fig. 4.6).

D. **Activation of Helper T Lymphocytes**. The activation of resting T helper cells requires a complex sequence of signals. Of all the signals involved, the only antigen-specific signal is the recognition by the T-cell receptor (TcR) of the complex formed by an antigen-derived peptide and an MHC-II molecule expressed on the membrane of an APC.

1. The role of APC goes well beyond that of a site for generation and expression of antigen fragments of adequate size. The interaction between APC and helper T lymphocytes is essential for T-cell stimulation, because the binding of the antigen-derived peptide to the binding site of the TcR is of low affinity, and other receptor–ligand interactions are required to maintain T-lymphocyte adhesion to APC and for the delivery of required co-stimulatory signals.
2. The TCR on a helper T lymphocyte interacts with both the antigen-derived peptide and the MHC-II molecule. This selectivity of the TcR from helper T lymphocytes to interact with MHC-II molecules results from the fact that, during ontogeny, the differentiation of helper and cytotoxic T lymphocytes is based on the ability of their TcR to interact, respectively, with MHC-II molecules (helper T lymphocytes) or with MHC-I molecules (cytotoxic T lymphocytes) (see Chapter 10). The interactions between T lymphocytes and MHC-expressing cells are strengthened by special molecules on the lymphocyte membrane which also interact with MHC molecules: the **CD4 molecule** on helper T cells interacts with MHC-II molecules, and the **CD8 molecule** on cytotoxic lymphocytes interacts with MHC-I molecules.
3. Several other **cell adhesion molecules** (CAM) can mediate lymphocyte-APC interactions, including lymphocyte function-associated antigen (LFA)-1 interacting with the intercellular adhesion molecules (ICAM)-1, -2 and -3, and CD2 interacting with CD58 (LFA-3). All interactions other than the one between the MHC-associated peptide and the TcR are not antigen specific (i.e., they mediate adhesion between T lymphocytes and APC).
4. Accessory cells participate in the activation of helper T lymphocytes through the delivery of signals involving cell–cell contact as well as by the release of soluble factors, such as interleukin-1 and interleukin-12.

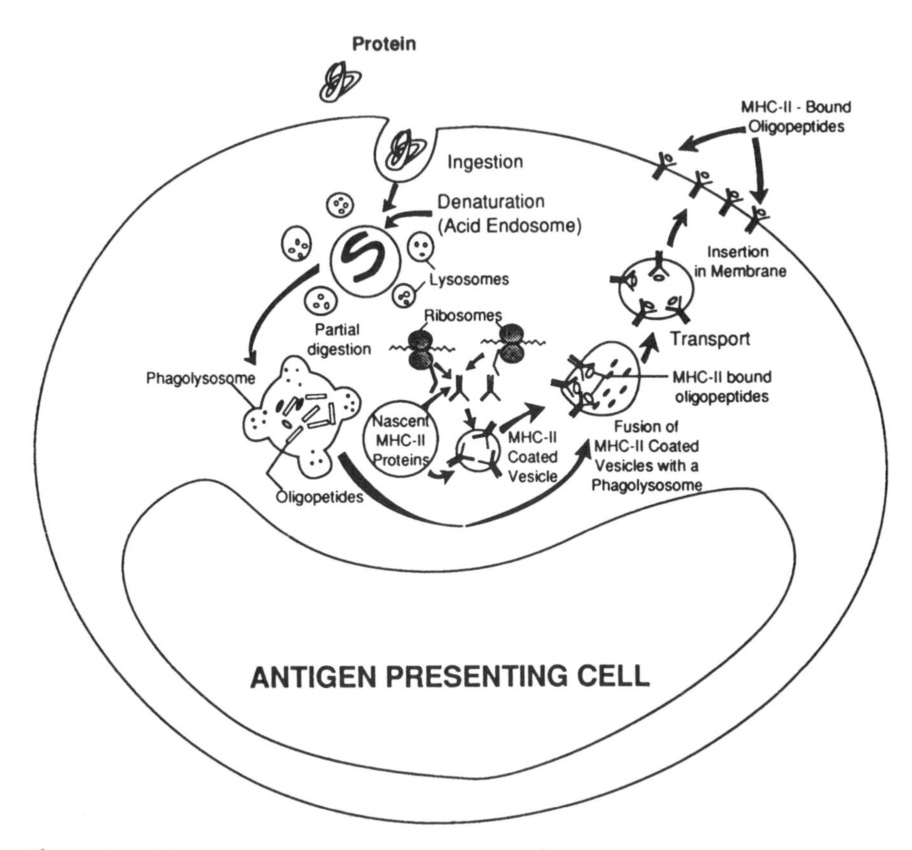

Figure 4.6 Diagrammatic representation of the general steps in antigen processing. The antigen is ingested, partially degraded, and, after vesicles coated with nascent MHC-II proteins fuse with the phagolysosomes, antigen-derived polypeptides bind to the MHC-II molecule. In this bound form, the oligopeptides seem protected against further denaturation and are transported together with the MHC-II molecule to the cell membrane, where they will be presented to CD4+ T lymphocytes in traffic through the tissue where the APC are located.

5. Given the predominance of nonspecific signals, what ensures that the activated lymphocytes are predominantly those involved in an antigen-specific response?
   a. An essential and first activation signal is delivered through the antigen-specific TcR. The signal is dependent upon appropriate binding of the TcR-bearing helper T lymphocyte and an APC presenting an antigen-derived peptide properly associated to an MHC-II molecule.
   b. One consequence of the activating signal is the up-regulation and modification of several membrane proteins on the T-cell membrane, such as CD2, CD28, CD40 ligand (CD40L, CD154, gp39), LFA-1 and ICAM-1. These molecules have counterparts on the APC: CD58 (LFA-3), CD80/86, CD40, ICAM-1, and LFA-1 (ICAM-1 and LFA-1 are expressed in both cell populations and interact with each other). The interactions involving these molecules contribute both to establishing intimate cell–cell contact and to delivering additional activating signals to T cells.

   c. In addition, APC produce cytokines such as **interleukin-1** (IL-1), which promotes growth and differentiation of many cell types, including T and B lymphocytes. Both membrane-bound and soluble IL-1 have been shown to be important in activating T lymphocytes in vitro. Membrane-bound IL-1 can only activate T lymphocytes in close contact with the APC.
   d. Thus, only helper T lymphocytes specifically recognizing the MHC-II-associated peptide undergo the changes that facilitate cell–cell contact and signaling, and that ensures their specific proliferation and differentiation.
6. The precise sequence of intracellular events resulting in T-cell proliferation and differentiation will be discussed in greater detail in Chapter 11. The following are the major steps in the activation sequence:
   a. The occupancy of the TcR signals the cell through a closely associated complex of molecules, known as CD3, which has signal-transducing properties.
   b. Co-stimulatory signals are delivered by CD4, as a consequence of the interaction with MHC-II, and by CD45, a tyrosine phosphatase, whose mechanism of activation has not yet been defined.
   c. The activation of CD45 initiates the sequential activation of several protein kinases closely associated with CD3 and CD4. The activation of the kinase cascade has several effects, namely:
      i. Increased expression of cell adhesion molecules, allowing additional signaling of the T lymphocyte.
      ii. Phospholipase C activation, leading to the mobilization of Ca^{2+}-dependent second messenger systems, such as the one involving **inositol triphosphate (IP_3)**, which promotes an increase in intracellular free Ca^{2+} released from intracellular organelles and taken up through the cell membrane. The increase in intracellular free calcium results in activation of a serine threonine phosphatase known as **calcineurin**.
      iii. Diacylglycerol (DAG), another product released by phospholipase C, activates **protein kinase C (PKC)**, and, consequently, other enzymes are activated in a cascading sequence.
      iv. The activation of second messenger systems results in the activation and translocation of nuclear binding proteins, such as the **nuclear factor-kappa B (NF-κB)** and the **nuclear factor of activated T cells (NF-AT)**. Once translocated to the nucleus, these factors induce genes controlling T-cell proliferation, such as those encoding **interleukin-2 (IL-2)**, the **IL-2 receptor gene**, and **c-myc**.
   d. The binding of IL-2 to its receptor triggers an additional activation pathway involving nuclear binding proteins that promote the entry of the cell into a division cycle. This activation pathway seems to promote primarily the proliferation of helper T cells, which assist the differentiation of B cells and of cytotoxic T cells.

E. Antigen Presentation and Activation of Cytotoxic T Lymphocytes. As mentioned in Chapter 3, cytotoxic T lymphocytes can be sensitized to react against virus-derived peptides embedded in self MHC-1 molecules expressed by virus-infected cells. The infected cell acts as an APC by expressing viral peptides complexed with their own MHC-I molecules. The way in which MHC-I molecules and viral peptides become associated has been recently elucidated (Fig. 4.7).

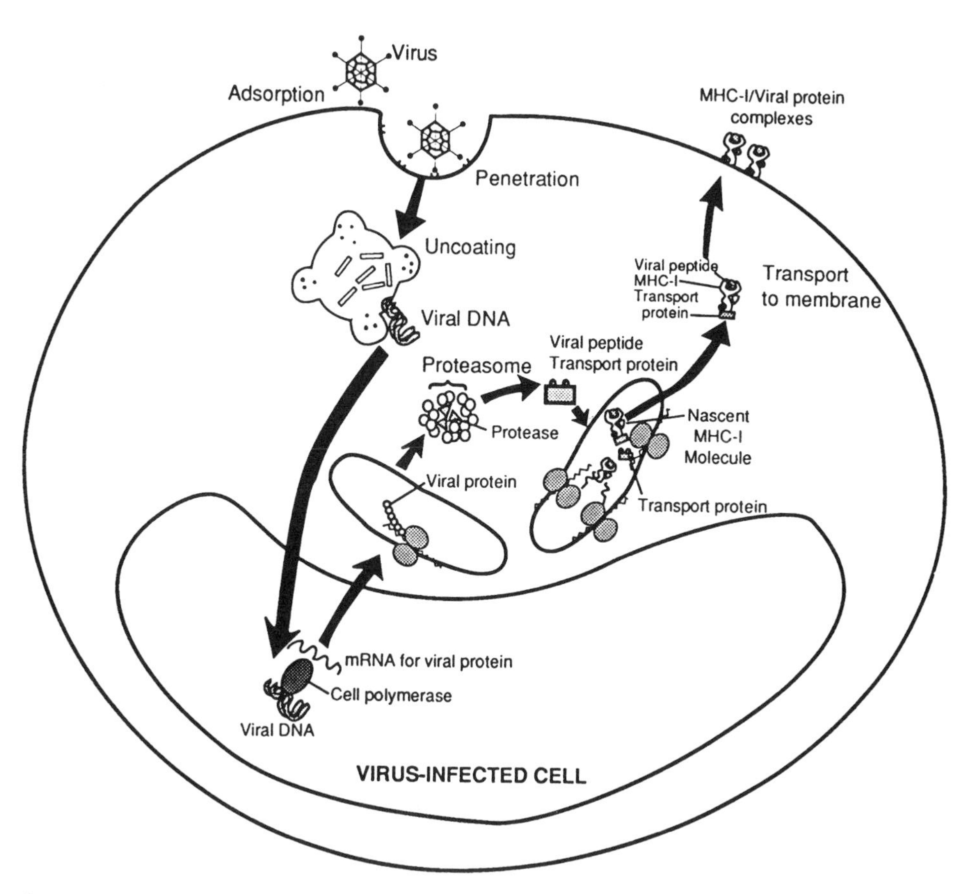

Figure 4.7 Diagrammatic representation of the general steps involved in the presentation of virus-derived peptides on the membrane of virus-infected cells. The virus binds to membrane receptors and is endocytosed, its outer coats are digested, and the viral genome (in this case DNA) is released into the cytoplasm. Once released, the viral DNA diffuses back into the nucleus where it is initially transcribed into mRNA by the cell's polymerases. The viral mRNA is translated into proteins that diffuse into the cytoplasm, where some will be broken down into oligopeptides. These small peptides are transported back into the endoplasmic reticulum where they associate with newly synthesized MHC-I molecules. The MHC-I/oligopeptide complex becomes associated to a second transport protein and is eventually inserted into the cell membrane. In the cell membrane, it can be presented to CD8+ T lymphocytes in traffic through the tissue where the virus-infected cell is located. A similar mechanism would allow an MHC-II synthesizing cell to present MHC-II/oligopeptide complexes to CD4+ lymphocytes.

1. When a virus infects a cell, it takes over the cell's synthetic machinery to produce its own proteins. In the early stages of infection, the cell will produce both is own proteins and viral proteins. Some of the nascent viral proteins diffuse into the cytoplasm where they become associated with degradative enzymes forming a peptide-enzyme complex (**proteasome**). In these complexes, the viral protein is partially digested, and the resulting peptides bind to transport proteins (**TAP**, transport-associated proteins), which deliver them to the endoplasmic reticulum, where MHC-I molecules are being synthesized.
2. In the endoplasmic reticulum, the viral peptides bind to newly synthesized MHC-class I molecules, and the resulting MHC-viral peptide complex is transported to the infected cell's membrane.
3. Among resting, circulating cytotoxic T lymphocytes, some carry antigen receptors able to recognize associations of MHC-I and non-self peptides; occupancy of the binding site on the TcR by MHC-I-associated peptide provides the antigen-specific signal that drives cytotoxic T cells.
4. Cytotoxic T lymphocytes also differentiate and proliferate when mixed with T lymphocytes from a different individual in vitro (mixed lymphocyte reaction) or when encountering cells from an individual of the same species but from a different genetic background, as a consequence of tissue or organ transplantation.
5. Similar to helper T cells, the stimulation of cytotoxic T cells also requires additional signals and interactions, some of which depend upon cell–cell contact, such as those mediated by the interaction of CD8 with MHC-I, CD2 with CD58 (LFA-3), LFA-1 with ICAM family members, and CD28 with CD80 and CD86, to name a few.
6. The expansion of antigen-activated cytotoxic T lymphocytes requires the secretion of **IL-2**. Rarely, activated cytotoxic T lymphocytes can secrete sufficient quantities of IL-2 to support their proliferation and differentiation, and thus proceed without help from other T-cell subpopulations.
7. Activated helper T lymphocytes may also provide the IL-2 necessary for cytotoxic T-lymphocyte differentiation, but their activation requires the presentation of antigen-derived peptides in association with MHC-II molecules.
   a. In the case of **antiviral responses**, virus-infected macrophages are likely to express viral peptide–MHC-II complexes on their membrane; these complexes are able to activate CD4+ helper T cells.
   b. In the case of **mixed lymphocyte reactions**, T cells recognize non-self peptides bound to MHC-II molecules, which are either shared between the two cell populations, or sufficiently alike to allow the stimulatory interaction. MHC-II-expressing cells have to be present for the reaction to take place.
      i. Naive helper T lymphocytes interact with non-self peptide–MHC-II complexes, while cytotoxic T lymphocytes are activated through the recognition of non-self peptide–MHC-I complexes.
      ii. The absolute requirement for MHC-II-expressing cells suggests that activation of helper T cells is essential for the differentiation of cytotoxic CD8+ cells. This reflects the requirement for helper T

cells to provide cytokines and probably other co-stimulatory signals essential for cytotoxic T-cell growth and differentiation.

F. **Antigen Presentation and Activation of B Lymphocytes**. In contrast to T lymphocytes, B lymphocytes recognize external epitopes of unprocessed antigens, which do not have to be associated to MHC molecules.

1. Some special types of APC, such as the **Langerhans cells** of the epidermis and the **dendritic cells** of the germinal centers, appear to adsorb complex antigens to their membranes, may be able to maintain them in that form for long periods of time, and may be able to present the antigen to B lymphocytes for as long as it remains adsorbed.
2. Additional signals necessary for B-cell activation, proliferation, and differentiation are provided by accessory cells and helper T lymphocytes. A major role is believed to be played by a complex of four proteins associated noncovalently with the membrane immunoglobulin, including CD19 and CD21. These proteins seem to play a role similar to CD4 or CD8 in T lymphocytes, potentiating the signal delivered through occupancy of the binding site on membrane immunoglobulin.
3. Similar to the TcR, membrane immunoglobulins have short intracytoplasmic domains, which do not appear to be involved in signal transmission. At least two heterodimers composed of two different polypeptide chains, termed Igα and Igβ, with long intracytoplasmic segments are associated to each membrane immunoglobulin. These heterodimers seem to have a dual function:
   a. They act as transport proteins, capturing nascent immunoglobulin molecules in the endoplasmic reticulum and transporting them to the cell membrane.
   b. They are believed to be the "docking sites" for a family of protein kinases related to the src gene product, including $p56^{lck}$ and $p59^{fyn}$, which also play a role in T-cell activation. Another parallel with T-cell activation lies in the essential role of the phosphatase CD45 for $p56^{lck}$ activation, thus initiating a cascade of tyrosine kinase activation. Specific to B-cell activation is the involvement of a specific protein kinase, known as Bruton's tyrosine kinase (Btk) in the activation cascade. The critical role of this kinase was revealed when its deficiency was found to be associated with infantile agammaglobulinemia (Bruton's disease).
   c. The subsequent sequence of events seems to have remarkable similarities with the activation cascade of T lymphocytes. Activation and translocation of common transcription factors (e.g., NF-AT, NF-κB) induce overlapping, but distinct, genetic programs. For instance, in B cells, NF-κB activates the expression of genes coding for immunoglobulin polypeptide chains. (**NF-κB** received its designation when it was originally described as a transcription factor that binds to the enhancer region controlling the gene coding for kappa-type immunoglobulin light chains.)
4. Additional signals necessary for B-cell proliferation and differentiation depend both on soluble molecules (interleukins-2,4,5, and 6) and cell–cell contact (see below).

XI. STIMULATION OF A B-LYMPHOCYTE RESPONSE BY A T-DEPENDENT ANTIGEN

The stimulation of a B-cell response with a T-dependent antigen involves several cell populations cooperating with each other in the activation, proliferation, and differentiation processes. T-cell help is mediated both by soluble factors (**cytokines**) and by interactions between complementary ligands (**co-stimulatory molecules**) expressed by T cells and B cells.

A. The naive B cell is initially stimulated by recognition of an epitope of the immunogen through the membrane immunoglobulin. Two other sets of membrane molecules are involved in this initial activation, the CD45 molecule and the CD19/CD21/CD81 complex. Whether the activation of CD45 involves interaction with a specific ligand on the accessory cell remains to be determined. In the CD19/CD21/CD81 complex, the only protein with a known ligand is CD21, a receptor for C3d (a fragment of the complement component 3, C3). It is possible that B cells interacting with bacteria coated with C3 and C3 fragments may receive a co-stimulatory signal through the CD19/CD21/CD81 complex.

B. In the same microenvironment where B lymphocytes are being activated, helper T lymphocytes are also activated. Two possible mechanisms could account for this simultaneous activation:

1. The same accessory cell (i.e., a macrophage) may present not only membrane-absorbed, unprocessed molecules with epitopes reflective of the native configuration of the immunogenic molecule to B lymphocytes, but also MHC-II-associated peptides derived from processed antigen to the helper T lymphocytes.
2. The activated B cell may internalize the immunoglobulin–antigen complex, process the antigen, and present MHC-II-associated peptides to the helper T cells.

C. The proper progression of the immune response will require that accessory cells (macrophages or B cells), helper T lymphocytes, and B lymphocytes interact in a circuit of mutual activation (Figures 4.8 and 4.9):

1. The T lymphocyte receives the following activation signals from accessory cells.
   a. Recognition of the MHC-II-associated peptide by the TcR.
   b. Signals mediated by CD4–MHC-II interactions.
   c. Signals mediated by the cell–cell interactions, which are facilitated by the up-regulation of some of the interacting molecules after initial activation, including:
      i. CD2 (T cell): CD58 (APC).
      ii. LFA-1 (T cell): ICAM-1, ICAM-2, ICAM-3 (APC).
      iii. CD40L (T cell): CD40 (APC).
      iv. CD28 (T cells): CD80, CD86 (APC).
   d. Signals mediated by interleukins, particularly IL-1 and IL-12.
2. The activated helper T cell, in turn, delivers activating signals to APC and B cells (Figure 4.9).
   a. Signals mediated by interleukins and cytokines.
      i. IL-2 and IL-4 which stimulate B-cell proliferation and differentiation.
      ii. Interferon-γ, which stimulates APC, particularly macrophages.

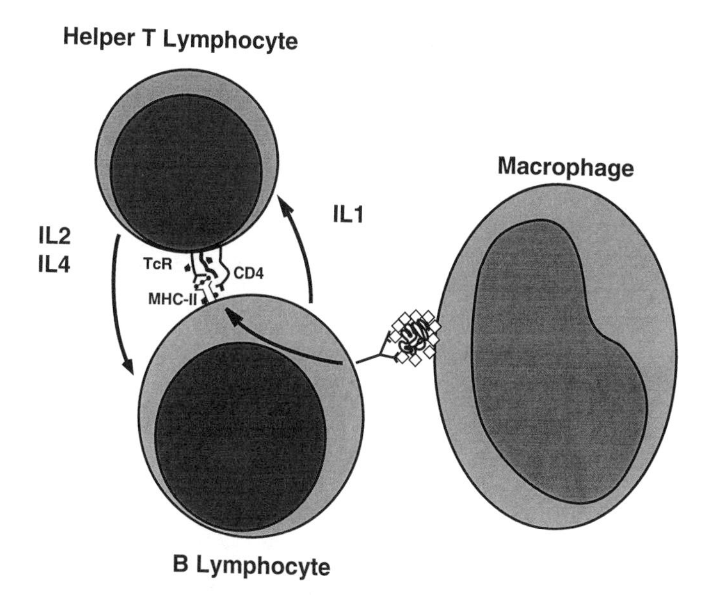

Figure 4.8 A diagrammatic representation of the induction of a T-dependent response to a hapten-carrier conjugate. In this diagram the professional APC (macrophage) is responsible for adsorbing and presenting the hapten-carrier conjugate to a B lymphocyte. The B lymphocyte depicted in the diagram recognizes the hapten, internalizes the hapten-carrier conjugate, processes the carrier, and presents a carrier-derived peptide to a helper T lymphocyte. This will result in the initial steps of cross-activation between T helper and B lymphocytes.

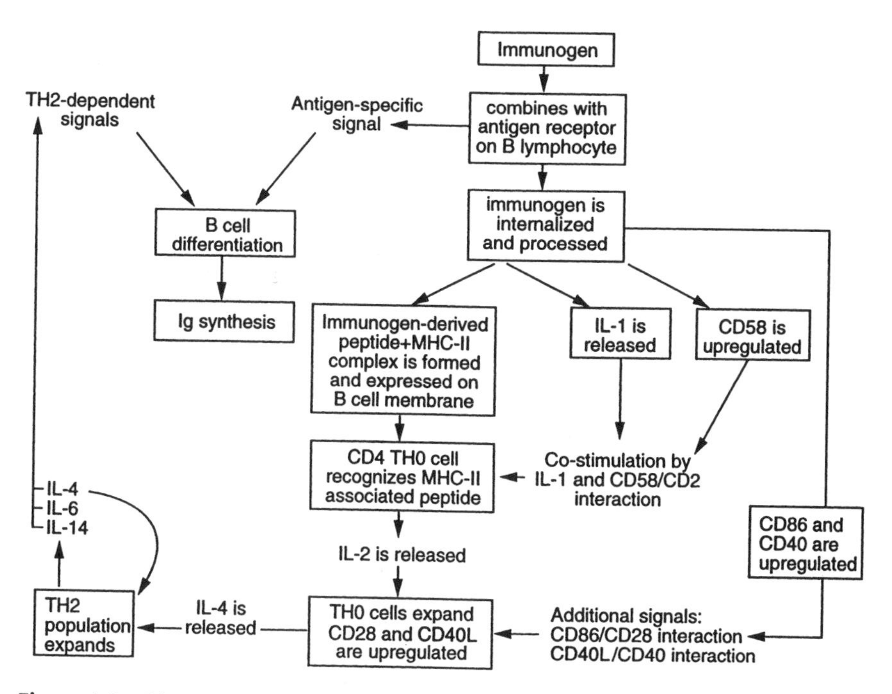

Figure 4.9 Diagrammatic representation of the sequence of events leading to the stimulation of a T-dependent B-cell response.

b. Signals mediated by cell–cell interactions, involving CD40L (gp39, on T cells) and CD40 (on B cells).

3. At the same time, the helper T cells continue to proliferate and differentiate.
   a. The IL-2 receptor is up-regulated and increases its affinity for IL-2; consequently, IL-2 participates in autocrine and paracrine signaling, which results in T-lymphocyte proliferation.
   b. As a consequence of signaling through the CD40 molecule, B cells express CD80 and CD86, which deliver differentiation signals to T cells through the CD28 family of molecules. Additional activation signals are delivered as a consequence of interactions involving other sets of membrane molecules.

D. In this type of response, a functional subpopulation of helper T lymphocytes specialized in assisting B-cell activation and differentiation emerges (**TH2 lymphocytes**). Another functional subpopulation, **TH1 lymphocytes**, assists the differentiation of cytotoxic T lymphocytes and NK cells as well as the activation of macrophages. Several factors appear to control the differentiation of TH2 cells as opposed to TH1 cells, including the affinity of the interaction of the TcR with the MHC-II-associated peptide, the concentration of MHC-associated peptide, cytokines, and signals dependent on cell–cell interactions (Table 4.1). The two subpopulations of helper T lymphocytes differ in the repertoire of cytokines they release (Table 4.2).

E. Cell–cell contact phenomena play at least an equally significant role as lymphokines in promoting B-cell activation, either by delivering co-stimulatory signals to the B cell, or by allowing direct traffic of unknown factors from helper T lymphocytes to B lymphocytes.
   a. Transient conjugation between T and B lymphocytes seems to occur constantly, due to the expression of complementary CAM on their membranes (for example, T cells express CD2 and CD4, and B cells express the respective ligands, CD58 (LFA-3) and MHC-II; both T and B lymphocytes

Table 4.1 Signals Involved in the Control of Differentiation of TH1 and TH2 Subpopulations of Helper T Lymphocytes

Signals favoring TH1 differentiation	Signals favoring TH2 differentiation
Interleukin-12	Interleukin-4[a]
Interferon γ[b]	Interleukin-1
Interferon-α	Interleukin-10
CD28/CD80 interaction	CD28/CD86 interaction
High density of MHC-II—peptide complexes on the APC membrane	Low density of MHC-II—peptide complexes on the APC membrane
High-affinity interaction between TcR and the MHC-II/peptide complex	Low-affinity interaction between TcR and the MHC-II/peptide complex

[a]Released initially by undifferentiated TH cells (also known as TH0) after stimulation in the absence of significant IL-12 release from APC; IL-4 becomes involved in an autocrine regulatory circuit that results in differentiation of TH2 cells and in paracrine regulation of B-cell differentiation.

[b]Interferon gamma does not act directly on TH1 cells, but enhances the release of IL-12 by APC, and as such has an indirect positive effect on TH1 differentiation.

Table 4.2 Interleukins and Cytokines Released by TH1 and TH2 Helper T Lymphocytes

TH1 interleukins/cytokines	Target cell/effect
Interleukin-2	TH1 and TH2 cells/expansion; B cells/expansion, differentiation
Interferon-γ	Macrophages/activation; TH1 cells/differentiation; TH2 cells/downregulation
TNF-β (lymphotoxin)	TH1 cells/expansion; B cells/homing
TH2 interleukins/cytokines	
Interleukin-4	TH2 cells/expansion; B cells/differentiation; APC/activation
Interleukin-5	Eosinophils/growth and differentiation
Interleukin-6	B cells/differentiation; plasma cells/proliferation; TH1, TH2 cells/activation; CD8+ T cells/differentiation, proliferation
Interleukin-10	TH1, TH2 cells/down-regulation; B cells/differentiation
Interleukin-13	Monocytes, macrophages/down-regulation; B cells/activation, differentiation
TH1-TH2 interleukins/cytokines	
Interleukin-3	B cell/differentiation; macrophage/activation
TNF-α	B cell/activation, differentiation
Granulocyte/monocyte CSF	B cell/differentiation

express ICAM-1 and LFA-1 and can engage in homotypic interactions through these molecules).

b. When the B lymphocyte presents an antigen-derived peptide on its MHC-II, which is specifically recognized by the T lymphocyte, the two cells modulate the expression of membrane molecules and cell–cell conjugates becomes more stable.

c. In stable conjugates of cooperating T and B cells, a reorganization of the cytoskeleton is seen on the T lymphocyte. The microtubule organizing center and the Golgi apparatus move to the pole of the T cell closer to the point of adhesion with the B cell. This reorganization implies unidirectional transport of proteins from the T lymphocyte to the B lymphocyte, and the membranes of the two interacting cells fuse for a brief period of time. Thus, it seems very likely that important signals are transmitted from cell to cell during conjugation. It is not known whether or not those signals are delivered in the form of interleukins or as of yet uncharacterized molecules.

F. The continuing proliferation and differentiation of B cells into plasma cells is assisted by several soluble factors, including **IL-4**, released by TH2 cells, **IL-6**, and **IL-14** (previously known as high molecular weight B-cell growth factor), released by T lymphocytes and accessory cells.

G. At the end of an immune response, the total number of antigen-specific T- and B-lymphocyte clones will remain the same, but the number of cells in those clones will remain increased severalfold. The increased residual population of antigen-specific T cells is long-lived, and is believed to be responsible for the phenomenon known as **immunological memory**.

SELF-EVALUATION

Questions

Choose the ONE *best* answer.

4.1 Which one of the following cytokines is believed to mediate the role of accessory cells in determining the differentiation of TH1 cells?
  A. Interferon γ
  B. Interleukin 1
  C. Interleukin 4
  D. Interleukin 12
  E. Tumor necrosis factor α

4.2 Which of the following concepts is an essential element in our understanding of how a humoral immune response to a T-dependent antigen can be elicited in the absence of monocytes or macrophages?
  A. Activated B lymphocytes must release cytokines that activate helper T cells without requirement for TcR occupancy
  B. B lymphocytes can process and present MHC-II-associated peptides to helper T lymphocytes
  C. Macrophages do not play a significant role in helper T lymphocyte activation
  D. Some TcR must recognize epitopes in unprocessed antigens
  E. Undifferentiated helper T cells can provide the necessary help for antigen-stimulated B cells to differentiate into antibody-secreting cells.

4.3 Which of the following steps of the immune response is likely to be impaired by a deficiency of cytoplasmic transport-associated proteins (TAP-1, TAP-2)?
  A. Assembly of a functional B-cell receptor
  B. Expression of the CD3-TcR complex
  C. Formation of stable complexes of viral derived peptides with MHC-I proteins
  D. Translocation of nuclear binding proteins from the cytoplasm to the nucleus
  E. Transport of MHC-II peptide complexes to the cell membrane

4.4 A significant number of individuals (as high as 1 in 100) fails to develop antibodies after immunization with tetanus toxoid, while other immune responses are perfectly normal. The most likely explanation for this observation is that:
  A. The accessory cells of those individuals are unable to process bacterial proteins
  B. The MHC-II proteins expressed on those individuals' accessory cells do not accommodate the peptides derived from processing of tetanus toxoid
  C. The repertoire of membrane immunoglobulins lacks variable regions able to accommodate the dominant epitopes of tetanus toxoid
  D. Those individuals lack a critical gene that determines the ability to respond to toxoids
  E. Those individuals lack TcR specific for tetanus toxoid-derived peptides

4.5 A rabbit has been immunized with DNP-BSA. Three weeks later you want to induce an anamnestic response to DNP-BGG. This can be accomplished by:
  A. Boosting with DNP 1 week before immunization with DNP-BGG
  B. Immunizing with BGG 1 week after the initial immunization with DNP-BSA
  C. Passively administering anti-BGG antibodies before immunization with BGG
  D. Passively administering anti-DNP antibodies prior to challenging with DNP-BGG
  E. Transfusing purified lymphocytes from a rabbit primed with DNP at least 2 days before challenging with DNP-BGG

4.6 What do you expect when you immunize a congenitally athymic (nude) mouse with type III pneumococcal polysaccharide?
A. Development of an overwhelming pneumococcal infection
B. No evidence of specific antibody synthesis
C. No immune response, either cellular or humoral
D. Production of significant amounts of IgG antibodies
E. Production of significant amounts of IgM antibodies

4.7 Alloantigens are best defined as antigens:
A. Identically distributed in *all* individuals of the same species
B. That define protein isotypes
C. That differ in distribution in individuals of the same species
D. Unique to human immunoglobulin G (IgG)
E. That do not induce an immune response in animals of the same species

4.8 Which of the following sets of characteristics is most closely associated with haptens?
A. Constituted by repeating units, are able to induce responses in sublethally irradiated mice reconstituted with B cells only
B. Do not induce an immune response by themselves, but induce antibody formation when coupled to an immunogenic molecule
C. Induce cellular immune responses but not antibody synthesis
D. Induce tolerance when injected intravenously in soluble form and induce an immune response when injected intradermally
E. Simple compounds able to interact directly with MHC molecules

4.9 Cells obtained from the tissues of an animal infected with *Leishmania major*, an intracellular parasite, show increased transcription of mRNA for IL-4 and IL-10. The synthesis of these two cytokines can be interpreted as meaning that:
A. TH1 cells are actively engaged in the immune response against the parasite
B. Antibody levels to *L. major* are likely to be elevated
C. IL-12 mRNA is also likely to be overexpressed in the same tissues
D. The ability of infected macrophages to eliminate *L. major* is enhanced
E. The mice carry an expanded population of cytotoxic T cells able to destroy *L. major*-infected cells

4.10 Which of the following procedures is *less* likely to enhance antigenicity?
A. Chemical polymerization of the antigen
B. High-speed centrifugation to eliminate aggregates
C. Immunization on antigen obtained from a phylogenetically distant species to that of the animal immunized
D. Injection of an antigen-adjuvant emulsion
E. Intradermal injection

Answers

4.1 (D) Interleukin-12 is released by accessory cells (particularly monocytes and macrophages) and is believed to be one of the primary determinants of the differentiation of TH1 lymphocytes. Interleukin-4, released by activated T cells in the absence of a co-stimulatory signal from IL-12, plays a similar role in the differentiation of the TH2 population.

4.2 (B) B lymphocytes express MHC-II molecules, and although they are not phagocytic cells, there is evidence suggesting that the mIg-antigen complex is internalized, the antigen is broken down, and peptides derived from it are coupled with MHC-II molecules and presented to helper T cells.

4.3 (C) The TAP proteins transport peptides derived from newly synthesized proteins (endogenous or viral) into the endoplasmic reticulum, where the peptides form complexes with newly synthesized MHC-I molecules. These complexes are then transported and expressed on the cell membrane.

4.4 (B) The differences in the level of immune response seen among different individuals of the same species are believed to depend on the repertoire of MHC-II molecules and their relative affinity toward the small peptides derived from the processing of the antigen in question. The existence of immune response genes transmitted in linkage disequilibrium with the MHC-II genes is an older theory now abandoned. The genes controlling immunoglobulin synthesis have a significant impact on the total repertoire of B-cell membrane immunoglobulins, but the lack of a given antigen-binding site is not as likely to result in a low response to a complex antigen, which presents many different epitopes to the immune system. The lack of helper T cells with specific receptors for toxoid-derived peptides is also unlikely, given the great diversity of TcR which exists in a normal individual. A general deficiency in processing would cause a general lack of responsiveness, not a specific inability to respond to one given immunogen.

4.5 (B) The development of a "memory" response (quantitatively amplified relative to the primary response) requires preimmunization with the carrier. Hence, the animal needs to be previously immunized either with the same hapten-carrier conjugate used to induce the secondary immune response or to the carrier alone.

4.6 (E) Polysaccharides are T-independent antigens and induce responses of the IgM type in mice, even if these mice lack T cells. Athymic mice obviously will lack T cells, because this population differentiates in the thymus. Infection will not occur as a consequence of injecting the isolated capsular polysaccharide of any bacteria.

4.7 (C) As do the A, B, O antigens or the immunoglobulin allotypes, alloantigens can induce strong immune responses in individuals of a different genetic makeup.

4.8 (B)

4.9 (B) IL-4 and IL-10 synthesis are characteristic of a TH2 response, associated with B-cell activation but with lack of differentiation of cytotoxic T cells and NK cells. IL-12 synthesis would induce a TH1 response, with overproduction of IL-2 and TNFα. The infected animals are not likely to eliminate the infection, since antibodies are not effective against intracellular organisms.

4.10 (B) Soluble proteins are *less* immunogenic than aggregated or polymerized proteins.

BIBLIOGRAPHY

Bierer, B.E., Sleckman, B.P., Ratnofsky, S.E., and Burakoff, S.J. The biologic roles of CD2, CD4, and CD8 in T cell activation. *Annu. Rev. Immunol.*, *7*:579, 1989.

Brunn, G.J., Falls, E.L., Nilson, A.E., and Abraham, R.T. Protein tyrosine kinase-dependent activation of STAT transcription factors in interleukin-2 or interleukin-4 stimulated T lymphocytes. *J. Biol. Chem.*, *270*:11628, 1995.

Constant, S., Pfeiffer, C., Woodard, A., Pasqualini, T., and Bottomly, K. Extent of T cell receptor ligation can determine functional differentiation of naive CD4+ T cells. *J. Exp. Med.*, *182*:1591, 1995.

Gold, M.R., and Matsuuchi, L. Signal transduction by the antigen receptors of B and T lymphocytes. *Int. Rev. Cytol.*, *157*:181, 1995.

June, C.H., Bluestone, J.A., Nadler, L.M., and Thompson, C.B. The B7 and CD28 receptor families. *Immunol. Today*, *15*:321, 1994.

Justement, L.B., Brown, V.K., and Lin, J. Regulation of B-cell activation by CD45: a question of mechanism. *Immunol. Today*, *15*:399, 1994.

Mond, J.J., Lees, A., and Snapper, C.M. T cell-independent antigens type-2. *Annu. Rev. Immunol.*, *13*:655, 1995.

Noelle, R.J. The role of gp39 (CD40L) in immunity. *Clin. Immunol. Immunopathol.*, *76*:S203, 1995.

Pleiman, C.M., D'Ambrosio, D., and Cambier, J.C. The B cell antigen receptor complex: structure and signal transduction. *Immunol. Today*, *15*:393, 1994.

Rothbard, J.B., and Gefter, M. Interactions between immunogenic peptides and MHC proteins. *Annu. Rev. Immunol.*, *9*:527, 1991.

Saouaf, S., Burkardt, A., and Bolen, J.B. Nonreceptor protein tyrosin kinase involvement in signal transduction in immunodeficiency disease. *Clin. Immunol. Immunopathol.*, *76*:S151, 1995.

Weiss, A., and Littman, D.R. Signal transduction by lymphocyte antigen receptors. *Cell*, *76*:262, 1994.

5 Immunoglobulin Structure

Gabriel Virella and An-Chuan Wang

I. GENERAL STRUCTURE OF IMMUNOGLOBULINS

A. Information concerning the precise structure of the antibody molecule started to accumulate as technological developments were applied to the study of the general characteristics of antibodies. By the early 1940s antibodies had been characterized electrophoretically as **gamma globulins** (Fig. 5.1) and also classified into large families by their **sedimentation coefficient** determined by analytical ultracentrifugation (**7S** and **19S** antibodies). It also became evident that **plasma cells** were responsible for immunoglobulin synthesis and that a malignancy known as **multiple myeloma** was a malignancy of immunoglobulin-producing plasma cells.

B. As protein fractionation techniques became available, complete immunoglobulins and their fragments were isolated in large amounts, particularly from the serum and urine of patients with multiple myeloma. These proteins were used both for studies of chemical structure and for immunological studies that led to the definition of antigenic differences between proteins from different patients; this was the basis for the initial identification of the different classes and subclasses of immunoglobulins and the different types of light chains.

II. IMMUNOGLOBULIN G (IgG): THE PROTOTYPE IMMUNOGLOBULIN MOLECULE

A. General Considerations. IgG, a 7S immunoglobulin, is the most abundant immunoglobulin in human serum and in the serum of most mammalian species. It is also the immunoglobulin most frequently detected in large concentrations in multiple myeloma patients. For this reason, it was the first immunoglobulin to be purified in large quantities and to be extensively studied from the structural point of view. The basic knowledge about the structure of the IgG molecule was obtained from two types of experiments:

1. **Proteolytic digestion**. The incubation of purified IgG with **papain**, a proteolytic enzyme extracted from the latex of *Carica papaya*, results in the splitting of the molecule into two fragments that differ both in charge and antigenicity. These fragments can be easily demonstrated by immunoelectro-

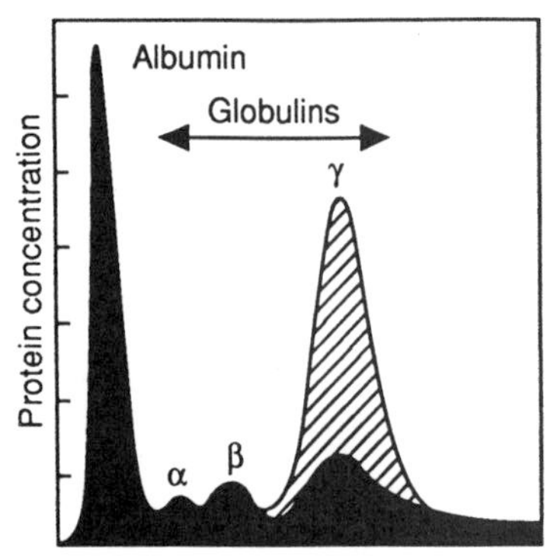

Figure 5.1 Demonstration of the gamma globulin mobility of circulating antibodies. The serum from a rabbit hyperimmunized with ovalbumin showed a very large gamma globulin fraction (shaded area), which disappeared when the same serum was electrophoretically separated after removal of antibody molecules by specific precipitation with ovalbumin. In contrast, serum albumin and the remaining globulin fractions were not affected by the precipitation step. (Redrawn after Tiselius, A., and Kabat, E.A., *J. Exp. Med.*, *69*:119, 1939.)

phoresis (Fig. 5.2), a technique that separates proteins by charge in a first step, allowing their antigenic characterization in a second step (as explained in greater detail in Chapter 14).

2. **Reduction of disulfide bonds**. If the IgG molecule is incubated with a reducing agent containing free SH groups and fractionated by gel filtration (a technique that separates proteins by size) in conditions able to dissociate noncovalent interactions, two fractions are obtained. The first fraction corresponds to polypeptide chains of M.W. 55,00 (**heavy chains**); the second corresponds to polypeptide chains of M.W. 23,000 (**light chains**) (Fig. 5.3).

B. **The IgG Structural Model**. The sum of data obtained by proteolysis and reduction experiments resulted in the conception of a diagrammatic two-dimensional model for the IgG molecule (Fig. 5.4).

C. **Proteolytic Fragments and Functional Topography of the Molecule**

1. **Papain digestion**, splitting the heavy chains in the hinge region (so designated because this region of the molecule appears to be stereoflexible) results in the separation of two **Fab** fragments and one **Fc** fragment per IgG molecule (Fig. 5.5).

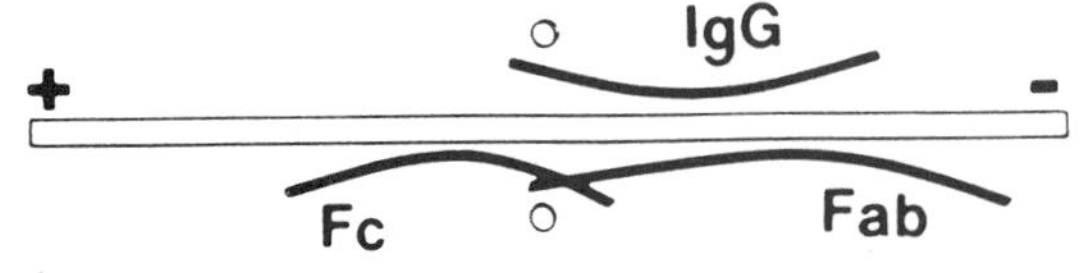

Figure 5.2 Immunoelectrophoretic separation of the fragments resulting from papain digestion of IgG. A papain digest of IgG was first separated by electrophoresis and the two fragments were revealed with an antiserum containing antibodies that react with different portions of the IgG molecule.

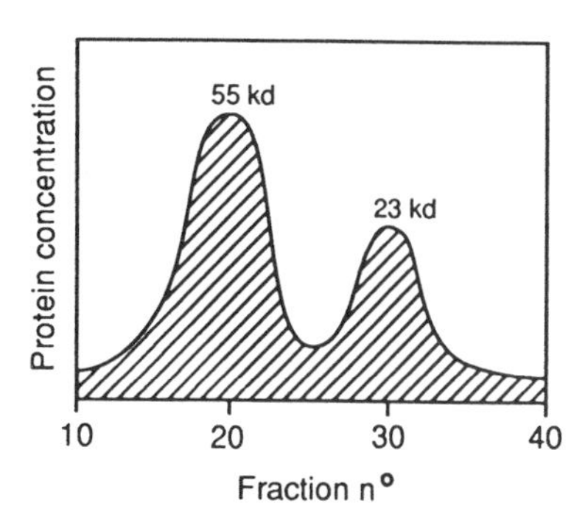

Figure 5.3 Gel filtration of reduced and alkylated IgG (M.W. 150,000) on a dissociating medium. Two protein peaks are eluted, the first corresponding to a M.W. 55,000 and the second corresponding to a M.W. 23,000. The 2:1 ratio of protein content between the high M.W. and low M.W. peaks is compatible with the presence of identical numbers of two polypeptide chains, one of which is about twice as large as the other.

a. The **Fab fragments** are so designated because they contain the **antigen binding site**.
b. The **Fc fragment** is so designated because it can be easily crystallized.
c. If the disulfide bond joining heavy and light chains in the Fab fragments is split, a complete light chain can be separated from a fragment that comprises about half of one of the heavy chains, the NH_2 terminal half. This portion of the heavy chain contained in each Fab fragment has been designated the **Fd fragment**.

2. A second proteolytic enzyme, **pepsin**, splits the heavy chains at the carboxyl

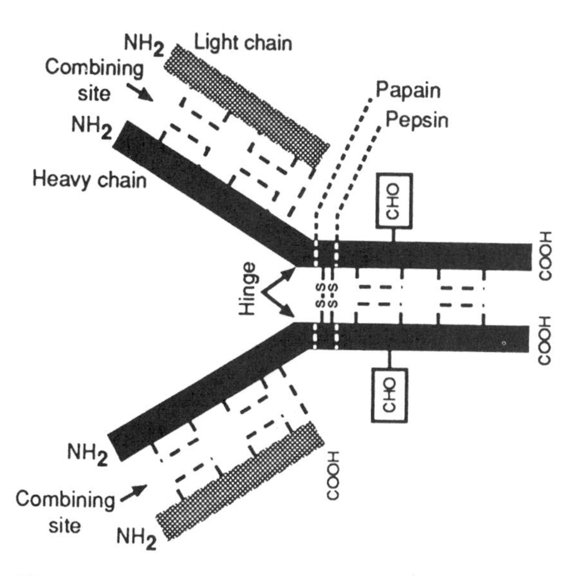

Figure 5.4 Diagrammatic representation of the IgG molecule. (Modified from Klein, J., *Immunology*. Blackwell Scientific Publishers, Boston/Oxford, 1990.)

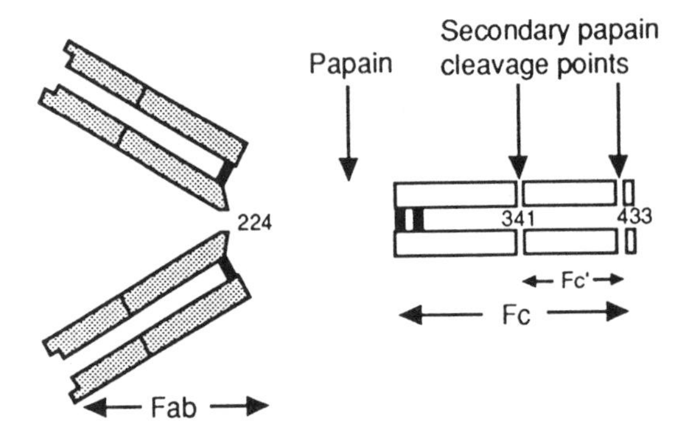

Figure 5.5 The fragments obtained by papain digestion of the IgG molecule (Modified from Klein, J., *Immunology*. Blackwell Scientific Publishers, Boston/Oxford, 1990.)

side of the disulfide bonds that join them at the hinge region, producing a double Fab fragment or **F(ab′)$_2$** (Fig. 5.6), while the Fc portion of the molecule is digested into peptides.

3. The comparison of Fc, Fab, F(ab′)$_2$, and whole IgG molecules shows both important similarities and differences between the whole molecule and its fragments.
   a. Both Fab and F(ab′)$_2$ contain antibody binding sites, but while the intact **IgG** molecule and the **F(ab′)$_2$** are **bivalent**, the **Fab** fragment is **monovalent**. Therefore, a Fab fragment can bind to an antigen, but cannot cross-link two antigen molecules.
   b. An antiserum raised against the Fab fragment reacts mostly against light-chain determinants; the immunodominant antigenic markers for the heavy chain are located in the Fc fragment.
   c. The F(ab′)$_2$ fragment is identical to the intact molecule as far as antigen-binding properties, but lacks the ability to fix complement, bind to cell membranes, etc., which are determined by the Fc region of the molecule.

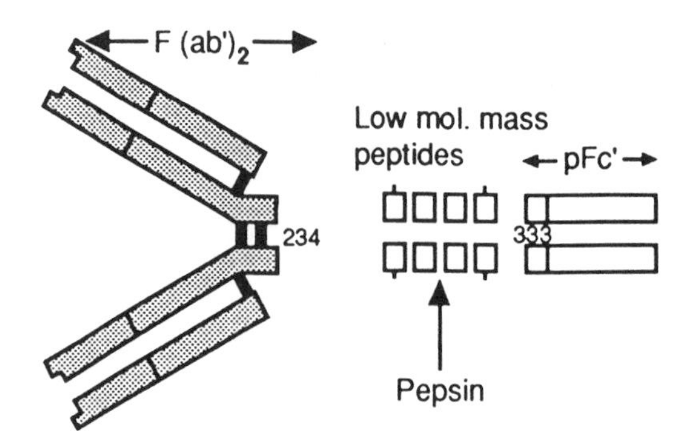

Figure 5.6 The fragments obtained by pepsin digestion of the IgG molecule. (Modified from Klein, J., *Immunology*. Blackwell Scientific Publishers, Boston/Oxford, 1990.)

III. THE STRUCTURAL AND ANTIGENIC HETEROGENEITY OF HEAVY AND LIGHT CHAINS

A. **Immunoglobulin Classes**. Five **classes** of immunoglobulins were identified due to antigenic differences of the heavy chains and designated as **IgG** (the classic 7S immunoglobulin), **IgA**, **IgM** (the classic 19S immunoglobulin), **IgD**, and **IgE**. IgG, IgA, and IgM together constitute over 95% of the whole immunoglobulin pool in a normal human being and are designated as **major immunoglobulin classes**. Because they are common to all humans, the immunoglobulin classes can also be designated as **isotypes**. The major characteristics of the five immunoglobulin classes are summarized in Table 5.1.

B. **Light-Chain Isotypes**. The light chains also proved to be antigenically heterogeneous and two main isotypes were defined: **kappa** and **lambda**. Each immunoglobulin molecule is constituted by a pair of identical heavy chains and a pair of identical light chains; hence, a given immunoglobulin molecule can have either kappa or lambda chains but not both. A normal individual will have a mixture of immunoglobulin molecules in his serum, some with kappa chains (e.g., IgGκ), and others with lambda chains (e.g., IgGλ). Normal serum IgG has a 2:1 ratio of kappa chain– over lambda chain–bearing IgG molecules. In contrast, **monoclonal immunoglobulins**, the results of the synthetic activity of malignant proliferations of plasma cells, such as multiple myeloma, have one single heavy-chain isotype and one single light-chain isotype, since they are the product of large number of cells all derived from a single mutant, constituting one large clone of identical cells.

C. **Immunoglobulin Subclasses**. Antigenic differences between the heavy chains of IgG and IgA exist and define **subclasses** of those immunoglobulins. The most important structural and biological characteristics of IgG and IgA subclasses are listed in Tables 5.2 and 5.3.

1. **IgG subclasses**. Some interesting biological and structural differences have been demonstrated for IgG proteins of different subclasses.
   a. From the functional point of view IgG1 and IgG3 are more efficient in terms of complement fixation and have greater affinity for monocyte receptors. Those properties can be correlated with a greater degree of biological activity, both in normal antimicrobial responses, in which these properties have direct consequences in opsonization and bacterial

Table 5.1 Major Characteristics of Human Immunoglobulins

	IgG	IgA	IgM	IgD	IgE
Heavy-chain class	γ	α	μ	δ	ε
H-chain subclasses	γ 1,2,3,4	α 1,2	—	—	—
L-chain type	κ and λ	κ and λ	κ and λ	κ and λ	κ and λ
Sedimentation coefficient	7S	7S, 9S, 11S	19S	7–8S	8S
Polymeric forms	no	dimers, trimers	pentamers	no	no
Molecular weight	150,000	(160,000)n	900,000	180,000	190,000
Serum concentration (mg/dl)	600–1300	60–300	30–150	3	0.03
Intravascular distribution	45%	42%	80%	75%	51%

Table 5.2 IgG Subclasses

	IgG1	IgG2	IgG3	IgG4
% of total IgG in normal serum	60%	30%	7%	3%
Half-life (days)	21	21	7	21
Complement fixation[a]	++	+	+++	−
Segmental flexibility	+++	+	++++	++
Affinity for monocyte and PMN receptors	+++	+	++++	+
Binding to protein A[b]	++	++	−	++
Binding to protein G[c]	+++	+++	+++	+++

[a]By the classical pathway.
[b]A protein isolated from select strains of *Staphylococcus aureus*, which has the ability to bind IgG of different species, including human.
[c]A protein similar to protein A, but isolated from Group G *Streptococci*, which also binds IgG proteins of different species.

killing, and in pathological conditions, in which the formation of immune complexes containing IgG1 and IgG3 antibodies is more likely to have pathogenic consequences.

b. From the structural point of view the **IgG3** subclass has the greatest number of structural and biological differences relative to the remaining IgG subclasses. Most differences appear to result from the existence of an extended hinge region (which accounts for the greater M.W.), with a large number of disulfide bonds linking the heavy chains together (estimates of their number vary between 5 and 15). This extended hinge region seems to be easily accessible to proteolytic enzymes, and this lability of the molecule is likely to account for its considerably shorter half-life.

2. **IgA subclasses**. Of the two subclasses known, it is interesting to note that a subpopulation of IgA2 molecules carrying the A2m(1) allotype is the only example of a human immunoglobulin molecule lacking the disulfide bond joining heavy and light chains. The IgA2 A2m(1) molecule is held together through noncovalent interactions between heavy and light chains.

IV. IMMUNOGLOBULIN REGIONS AND DOMAINS

A. **Constant and Variable Regions**. The light chains of human immunoglobulins are composed of 211 to 217 amino acids. As mentioned above, there are two

Table 5.3 IgA Subclasses

	IgA1	IgA2
Distribution	Predominates in serum	Predominates in secretions
Proportions in serum	85%	15%
Allotypes	?	A2m(1) and A2m(2)
H-s-s-L	+	− in A2m(1); + in A2m(2)

major antigenic types of light chains (κ and λ); when the amino acid sequences of light chains of the same type were compared, it became evident that two regions could be distinguished in the light-chain molecules: a **variable region**, comprising the portion between the amino terminal end of the chain and residues of 107 to 115, and a **constant region**, extending from the end of the variable region to the carboxyl terminus (Fig. 5.7).

1. The **light-chain constant regions** were found to be almost identical in light chains of the same type, but differ markedly in κ and λ chains. It is assumed that the difference in antigenicity between the two types of light chains is directly correlated with the structural differences in constant regions.
2. The amino acid sequence of the **light-chain variable regions** is different even in proteins of the same antigenic type, and early workers thought that this sequence would be totally individual to any single protein. With increasing data, it became evident that some proteins shared similarities in their variable regions, and it has been possible to classify variable regions into three groups, Vκ, Vλ, and VH. Each group has been further subdivided into several subgroups.
3. The light-chain **V-region subgroups** (Vκ, Vλ) are "type" specific (i.e., Vκ subgroups are only found in κ proteins and Vλ subgroups are always associated with λ chains). In contrast, the heavy-chain V-region subgroups

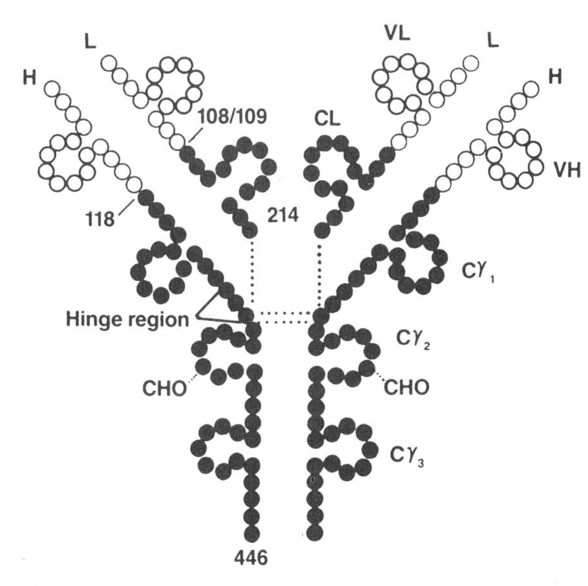

Figure 5.7 Schematic representation of the primary and secondary structure of a human IgG. The light chains are constituted by about 214 amino acids and two regions, variable (first 108 amino acids, white beads in the diagram) and constant (remaining amino acids, black beads in the diagram). Each of these regions contains a loop formed by intrachain disulfide bonds containing about 60 amino acids, which are designated as variable domain and constant domain (VL and CL in the diagram). The heavy chains have slightly longer variable regions (first 118 amino acids, white beads in the diagram), with one domain (VH) and a constant region that contains three loops or domains (Cγ1, Cγ2, and Cγ3), numbered from the NH2 terminus to the COOH terminus.

(VH) are not "class" specific. Thus, any given VH subgroup can be found in association with the heavy chains of any of the known immunoglobulin classes and subclasses.

4. The heavy chain of IgG is about twice as large as a light chain; it is composed of approximately 450 amino acids, and a **variable** and a **constant** region can also be identified. The **variable** region is composed of the first 113 to 121 amino acids (counted from the amino terminal end), and subgroups of these regions can also be identified. The **constant** region is almost three times larger; for most of the heavy chains, it starts at residue 116 and ends at the carboxyl terminus (Fig. 5.7). The maximal degree of homology is found between constant regions of IgG proteins of the same subclass.

B. **Immunoglobulin Domains**. The immunoglobulin molecule contains several disulfide bonds formed between contiguous cysteine residues. Some of them join two different polypeptide chains (**interchain disulfide bonds**), keeping the molecule together. Others (**intrachain bonds**) join different areas of the same polypeptide chain, leading to the formation of "loops." These "loops" and adjacent amino acids constitute the **immunoglobulin domains**, which are folded in a characteristic β-pleated sheet structure (Fig. 5.8).

1. Variable regions of both heavy and light chains have a single domain, which is involved in **antigen binding**.
2. Light chains have one single constant region domain (CL), while heavy chains have several constant region domains (three in the case of IgG, IgA, and IgD; four in the case of IgM and IgE). The constant region domains are generically designated as CH1, CH2, and CH3, or, if one wishes to be more specific, they can be identified by the class of immunoglobulins to which they belong by adding the symbol for each heavy chain class (γ, α, μ, δ, ε). For example, the constant region domains of the IgG molecule can be designated as Cγ1, Cγ2, and Cγ3.

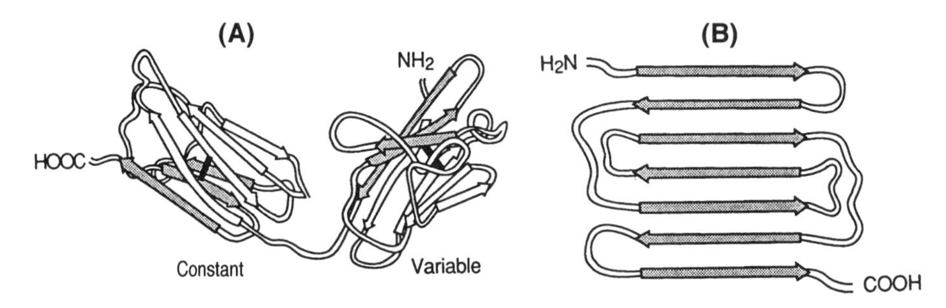

Figure 5.8 Model for the V and C domains of a human immunoglobulin light chain. Each domain has two β-pleated sheets consisting of several antiparallel b strands of 5 to 10 amino acids. The interior of each domain is formed between the two β sheets by in-pointing amino acid residues, which alternate with out-pointing hydrophilic residues, as shown in (A). The antiparallelism of the β-strands is diagrammatically illustrated in (B). This β-sheet structure is believed to be the hallmark of the extracellular domains of all proteins in the immunoglobulin superfamily. [(A) Modified from Edmundson, A.B., Ely, K.R., Abola, E.E., Schiffer, M., and Panagiotopoulos, N. *Biochemistry*, *14*:3953, 1975; (B) Modified from Amzel, L.M. and Poljak, R.J., *Annu. Rev. Biochem.*, *48*:961, 1979.]

3. Different functions have been assigned to the different domains and regions of the heavy chains. For instance, Cγ2 is the domain involved in complement fixation, while both Cγ2 and Cγ3 are believed to be involved in the binding to phagocytic cell membranes.
4. The **"hinge region"** is located between CH1 and CH2, and its name is derived from the fact that studies by a variety of techniques, including fluorescence polarization, spin-labeling, electron microscopy, and x-ray crystallography, have shown that the Fab fragments can rotate and waggle, coming together or moving apart. As a consequence, IgG molecules can change their shape from a "Y" to a "T" and vice versa, using the region intercalated between Cγ1 and Cγ2 as a "hinge." The length and primary sequence of the hinge regions play an important role in determining the segmental flexibility of IgG molecules. For example, IgG3 has a 12 amino acid hinge amino terminal segment and has the highest segmental flexibility. The hinge region is also the most frequent point of attack by proteolytic enzymes. In general, the resistance to proteolysis of the different IgG subclasses is inversely related to the length of the hinge amino terminal segments—IgG3 proteins are the most easily digestible, while IgG2 proteins, with the shortest hinge region, are the most resistant to proteolytic enzymes.

V. THE IMMUNOGLOBULIN SUPERFAMILY OF PROTEINS

The existence of globular "domains" (Fig. 5.8) is considered as the structural hallmark of immunoglobulin structure. A variety of other proteins that exhibit amino acid sequence homology with immunoglobulins also contain Ig-like domains (Fig. 5.9). Such proteins are considered as members of the immunoglobulin superfamily, based on the assumption that the genes which encode them must have evolved from a common ancestor gene coding for a single domain, much likely the gene coding for the Thy-1 molecule found on murine lymphocytes and brain cells.

1. The T-cell antigen receptor molecule, the major histocompatibility antigens, the polyimmunoglobulin receptor on mucosal cells (see below), and the CD2 molecule on T lymphocytes (see Chapters 10 and 11) are some examples of proteins included in the immunoglobulin superfamily.
2. The majority of the membrane proteins of the immunoglobulin superfamily seem to be functionally involved in recognition of specific ligands that may determine cell–cell contact phenomena and/or cell activation.

VI. THE ANTIBODY COMBINING SITE

The binding of antigens by antibody molecules takes place in the Fab region, and is basically a noncovalent interaction that requires a good fit between the antigenic determinant and the antigen binding site on the immunoglobulin molecule. The antigen binding site appears to be formed by the variable regions of both heavy and light chains folded in close proximity, forming a pouch where an antigenic determinant or epitope will fit (Fig. 5.10).

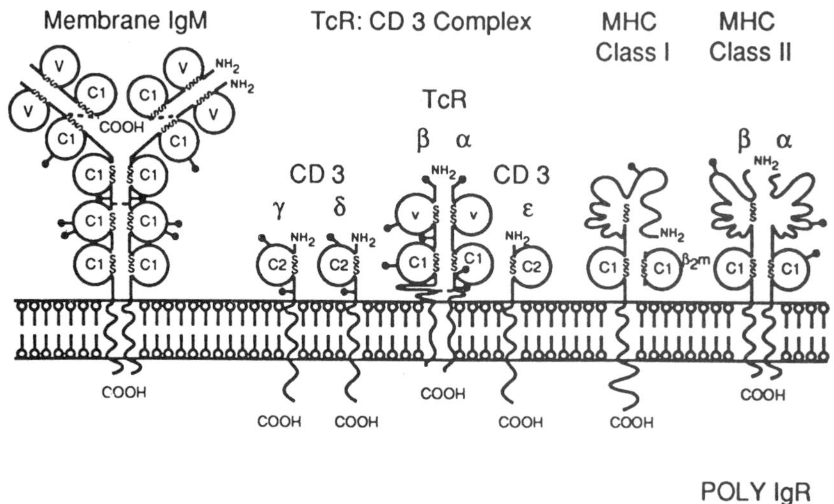

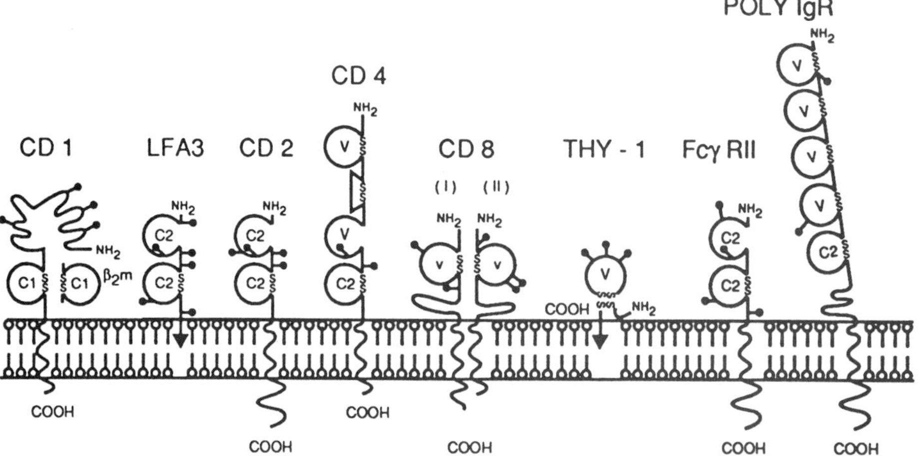

Figure 5.9 Model representations for the different proteins included in the immunoglobulin superfamily. (Modified from Williams, A.F., and Barclay, A.N. The immunoglobulin superfamily—domains for cell surface recognition. *Annu. Rev. Immunol.*, *6*:381, 1988.)

A. **Hypervariable Regions**. Certain sequence stretches of the variable regions vary widely from protein to protein, even among proteins sharing the same type of variable regions. For this reason, these highly variable stretches have been designated as **hypervariable regions**.

B. The structure of hypervariable regions is believed to play a critical role in determining **antibody specificity** since these regions are believed to be folded in such a way that they form a "pouch" where a given epitope of an antigen will fit. In other words, the hypervariable regions will interact to create a binding site whose configuration is *complementary* to that of a given epitope. Thus, these regions can be also designated as **complementarity-determining regions**.

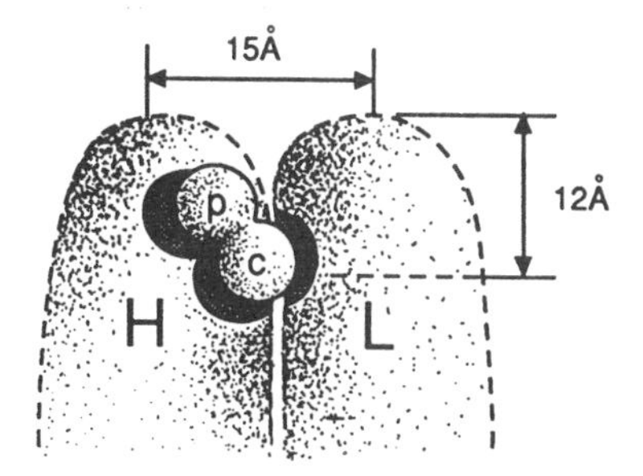

Figure 5.10 Diagrammatic representation of the hypothetical structure of an antigen binding site. The variable regions of the light and heavy chains of a mouse myeloma protein that binds specifically the phosphorylcholine hapten form a pouch in which the hapten fits. In this particular example the specificity of the binding reaction depends mostly on the structure of the heavy chain V region. (Modified from Padlan E., et al., *The Immune System: Genes, Receptors, Signals*. E. Secarz, A. Williamson, and C. Fox, eds., Academic Press, New York, 1975, p. 7)

VII. IMMUNOGLOBULIN M: A POLYMERIC MOLECULE

A. **The Pentameric Nature of IgM**. Serum **IgM** is basically composed of five subunits (monomeric subunits, IgMs), each one of them composed of two light chains (κ or λ) and two heavy chains (μ). The heavy chains are larger than those of IgG by about 20,000 daltons, corresponding to an extra domain on the constant region (Cμ4).

B. The **J chain**, a third polypeptide chain, can be revealed by adequate methodology in IgM molecules. This is a small polypeptide chain of 15,000 daltons, also found in polymeric IgA molecules. One single J chain is found in any polymeric IgM or IgA molecule, regardless of how many monomeric subunits are involved in the polymerization. It has been postulated that this chain plays some role in the polymerization process.

VIII. IMMUNOGLOBULIN A: A MOLECULARLY HETEROGENEOUS IMMUNOGLOBULIN

A. Serum **IgA** is molecularly heterogeneous, composed of a mixture of monomeric, dimeric, and larger polymeric molecules. In a normal individual, over 70–90% of serum IgA is monomeric. Monomeric IgA is similar to IgG, and composed of two heavy chains (α) and two light chains (κ or λ). The dimeric and polymeric forms of IgA found in circulation are covalently bonded synthetic products containing J chains.

B. IgA is the predominant immunoglobulin in secretions. **Secretory IgA** molecules are most frequently dimeric, contain J chains as do all polymeric immunoglobulin molecules, and, in addition, contain a unique polypeptide chain, designated as **secretory component** (SC) (Fig. 5.11).

C. **Secretory component** is constituted by a single polypeptide chain of approximately 70,000 daltons with five homologous immunoglobulin-like domains. It is synthesized by epithelial cells in the mucosa and by hepatocytes, initially

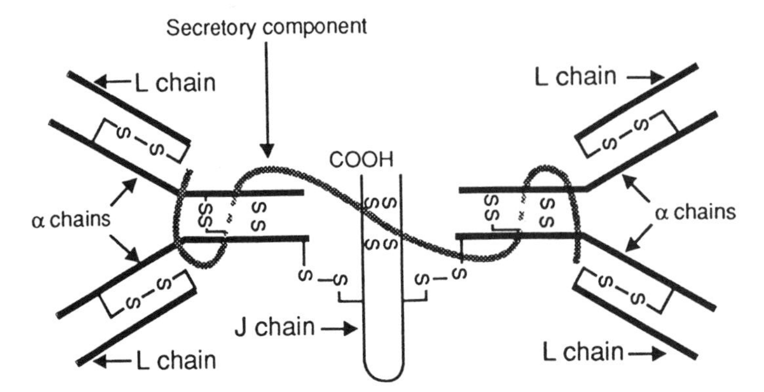

Figure 5.11 Structural model of the secretory IgA molecule. [Modified from Turner, M.W. In *Immunochemistry: An Advanced Textbook* (Glynn, L.E., and Steward, M.W., eds.), Wiley & Sons, New York, 1977.]

as a larger membrane molecule known as a **polyimmunoglobulin receptor**, from which SC is derived by proteolytic cleavage separating SC from the intramembrane and cytoplasmic segments of its membrane form (see Chapter 6).

IX. THE MINOR IMMUNOGLOBULIN CLASSES: IgD AND IgE

A. **General Concepts. IgD** and **IgE** were the last immunoglobulins to be identified due to their low concentrations in serum. Both are monomeric immunoglobulins, similar to IgG, but their heavy chains are larger than γ chains. IgE has five domains in the heavy chain (one variable and four constant); IgD has four heavy-chain domains (as most other monomeric immunoglobulins).

B. **IgD and IgM** are the predominant immunoglobulin classes in the B lymphocyte membrane, where they are the antigen-binding molecules in the antigen-receptor complex. The B-cell antigen complex is composed of membrane Ig and several other membrane proteins including Igα and Igβ, which have sequence motifs in their cytoplasmic portions that are required for signal transduction.
   1. Membrane IgD and IgM are monomeric. The heavy chains of membrane IgD and IgM (δm, μm) differ from that of the secreted forms at their carboxyl termini, where the membrane forms have a hydrophobic transmembrane section and a short cytoplasmic tail that are lacking in the secreted forms. In contrast, a hydrophilic section is found at the carboxyl termini of heavy chains of secreted Igs.
   2. The biological role of circulating IgD is not clear.

C. **IgE** is an extremely important immunoglobulin because of its biological properties. Its biological role appears to be predominantly related to antiparasitic responses, but its main clinical relevance is related to allergic reactions.
   1. IgE has the unique property of binding to Fcε receptors on the membranes of **mast cells** and **basophils**. The binding of IgE to those receptors has an extremely high affinity (7.7×10^9 $1/M^{-1}$), about 100-fold greater than the affinity of IgG binding to monocyte receptors.

2. The high affinity binding of IgE to basophil membrane receptors is the basis for its designation as **homocytotropic** antibody, and is responsible for its role in allergic reactions. In allergic individuals, if those IgE molecules have a given antibody specificity and react with the antigen while attached to the basophil or mast cell membranes, they will trigger the release of histamine and other substances that cause the symptoms of allergic reactions.
3. IgE is the most thermolabile immunoglobulin and loses biological activity (i.e., the ability to bind to high-affinity Fcϵ receptors) after heating at 56°C for 30 minutes. This binding depends on the tertiary structure of the C-terminal portion of Cϵ2 and the N-terminal portion of the Cϵ3 domain. Circular dichroism studies demonstrated that heating changes the configuration of Cϵ3 and Cϵ4 domains, and the changes in configuration of Cϵ3 are likely to be critical in preventing proper binding to the receptor.

SELF-EVALUATION

Questions

Choose the ONE *best* answer.

5.1 Which of the following antibodies would be most useful to assay human secretory IgA in secretions?
  A. Anti-IgA1 antibodies
  B. Anti-IgA2 antiserum
  C. Anti-J chain antibody
  D. Anti-kappa light chains
  E. Anti-secretory component

5.2 Which of the following is a likely event resulting from the mixture of an $F(ab')_2$ fragment of a given antibody with the corresponding antigen?
  A. Formation of a precipitate if the antigen is multivalent
  B. Formation of a precipitate with both univalent and monovalent antigens
  C. Formation of soluble complexes containing one single molecule of $F(ab')_2$ and one single molecule of divalent antigen
  D. Inhibition of the ability of a multivalent antigen to react with a complete antibody
  E. Lack of precipitation with any type of antigen

5.3 The antigen binding sites of an antibody molecule are determined by the structure of the:
  A. Constant region of heavy chains
  B. Constant region of light chains
  C. Variable region of heavy chains
  D. Variable region of light chains
  E. Variable regions of both heavy and light chains

5.4 The immunoglobulin class that binds with very high affinity to membrane receptors on basophils and mast cells is:
  A. IgG_1
  B. IgA
  C. IgM
  D. IgD
  E. IgE

5.5 A preparation of pooled normal human IgG injected into rabbits will **NOT** induce the production of antibodies against:
   A. Gamma heavy chains
   B. Gamma 3 heavy chains
   C. J chain
   D. Kappa light chains
   E. Lambda light chains

5.6 Which feature of IgG3 molecules is believed to be related to their increased sensitivity to proteolytic enzymes?
   A. Extended, rigid hinge region
   B. Extra constant region domain
   C. High affinity for Fc receptors
   D. High carbohydrate content
   E. Tendency to form aggregates

5.7 The basic characteristic that defines a protein as belonging to the immunoglobulin superfamily is the:
   A. Ability to combine specifically with antigen substances
   B. Existence of β-pleated sheet regions in the polypeptide chains
   C. Homology of the NH-terminal regions
   D. Identification of variable and constant regions
   E. Sharing of common antigenic determinants

Questions 5.8–5.10

Match in Figure 5.12 the regions indicated by letters with the corresponding descriptions listed below.

5.8 The region involved in binding to antigen epitopes.
5.9 The region containing the binding sites for complement.
5.10 The region that determines the serological type of light-chain cells.

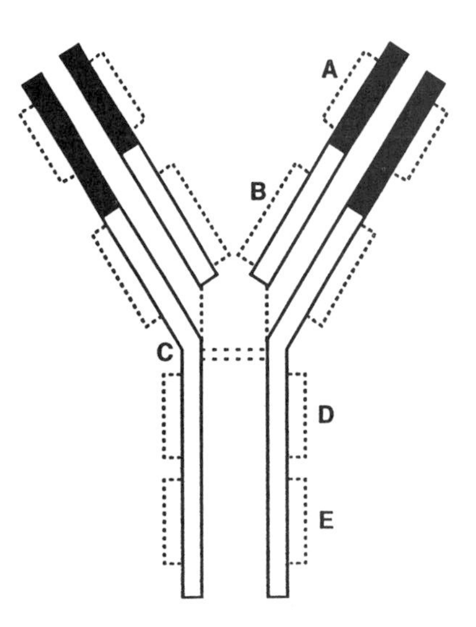

Figure 5.12

Answers

5.1 (E) The only specific characteristic of secretory IgA is the secretory component, which is not found in circulating dimeric IgA. Secretory IgA is predominantly of the IgA2 subclass and structurally polymeric, but IgA2 and dimerized IgA (with associated J chains) can also be found in circulation.

5.2 (A) The F(ab′)2 fragments obtained with pepsin are divalent and, therefore, can cross-link a multivalent antigen and lead to the formation of a precipitate.

5.3 (E) The variable regions of both light and heavy chains are believed to contribute to the formation of the "pouch" where the epitope of an immunogen will fit.

5.4 (E) The FcγRI binds IgG1 with high affinity, but is not expressed on basophils. Basophils express the FcϵRI, which binds IgE with very high affinity.

5.5 (C) J chain is only found in polymeric immunoglobulins (IgM and IgA); thus, immunization with polyclonal IgG is not likely to result in the formation of antibodies directed against this polypeptide chain.

5.6 (A) Proteolytic enzymes attack the hinge regions. IgG3 has an extended hinge region.

5.7 (B) The β-pleated sheet regions or "domains" are the structural hallmark common to all members of the Ig superfamily, which otherwise differ significantly among themselves.

5.8 (A) The variable regions of L and H chains form the antigen binding site

5.9 (D) The Cγ2 domain

5.10 (B) The CL domain

BIBLIOGRAPHY

Atassi, M.A., Van Oss, J., and Absolom, D.R. (eds.). *Molecular Immunology*. Marcel Dekker, Inc., New York, 1984.

Ban, N., Escobar, C., Garcia, R., Hasel, K., Day, J., Greenwood, A., and McPherson, A. Crystal structure of an idiotype-anti-idiotype Fab complex. *Proc. Natl. Acad. Sci. USA*, *1991*:1604, 1994.

Clark, W.R. *The Experimental Foundations of Modern Immunology*. Wiley & Sons, New York, 1980.

Day, E.D. *Advanced Immunochemistry, 2nd ed.* Wiley-Liss, New York, 1990.

Edmundson, A.B., Guddat, L.W., Rosauer, R.A., Anderson, K.N., Shan, L., and Rosauer, R.A. Three-dimensional aspects of IgG structure and function. In *The Antibodies—Vol. 1* (Zanetti, M., and Capra, D.J., eds.). Harwood Academic Publishers, New York, p. 41, 1995.

Nezlin, R. Internal movements in immunoglobulin molecules. *Adv. Immunol.*, *48*:1, 1990.

Nisonoff, A. *Introduction of Molecular Immunology*. Sinauer Ad. Inc., Sunderland, MA, 1982.

Underdown, B.J., and Schiff, J.M. Immunoglobulin A: Strategic defense initiative at the mucosal surface. *Annu. Rev. Immunol.*, *4*:389, 1986.

Van Oss, C.J., and van Regenmortel, M.H.V. *Immunochemistry*. Marcel Dekker, New York, 1994.

Williams, A.F., and Barclay, A.N. The immunoglobulin superfamily—domains for cell surface recognition. *Annu. Rev. Immunol.*, *6*:381, 1988.

6
Biosynthesis, Metabolism, and Biological Properties of Immunoglobulins

Gabriel Virella and An-Chuan Wang

I. IMMUNOGLOBULIN BIOSYNTHESIS

Immunoglobulin synthesis is the defining property of B lymphocytes and plasma cells.

A. **Resting B Lymphocytes** synthesize only small amounts of immunoglobulins that mainly become inserted into the cell membrane.

B. **Plasma Cells**, the most differentiated B cells, are considered as end-stage cells arrested at the late G1 phase. Plasma cells consequently show very limited mitotic activity but are specialized to produce and secrete large amounts of immunoglobulins. The synthetic capacity of the plasma cell is reflected by its abundant cytoplasm, extremely rich in endoplasmic reticulum (Fig. 6.1).

C. Normally, heavy and light chains are synthesized in separate polyribosomes of the plasma cell. The amounts of H and L chains synthesized on the polyribosomes are usually balanced so that both types of chains will be combined into complete IgG molecules, without surpluses of any given chain. The assembly of a complete IgG molecule can be achieved either by associating one H and one L chain to form an HL hemi-molecule, and joining in the next step two HL hemi-molecules to form the complete molecule (H2L2), or by forming H2 and L2 dimers that later associate to form the complete molecule.

D. When plasma cells undergo malignant transformation, this balanced synthesis of heavy and light chains persists in most cases, but in about one-third of the cases, synthesis may be grossly aberrant. The most common aberration is the synthesis of an excess of light chains. In human plasmacytomas, this is reflected by the elimination of the excessively produced light chains of a single isotype in the urine (**Bence Jones proteinuria**).

E. While free light chains can be effectively secreted from plasma cells, free heavy chains are generally not secreted. The heavy chains are synthesized and transported at the endoplasmic reticulum, where they are glycosylated, but secretion requires association to light chains to form a complete immunoglobulin molecule. If light chains are not synthesized or heavy chains are synthesized in excess, the free heavy chains associate via their CH1 domain with a heavy chain binding protein, which is believed to be responsible for their intracytoplasmic retention.

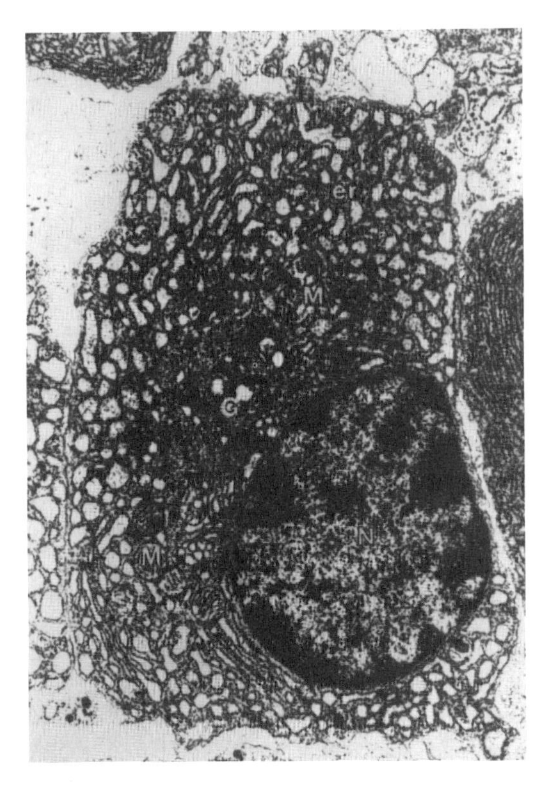

Figure 6.1 Ultrastructure of a mature plasma cell. Note the eccentric nucleus with clumped chromatin, the large cytoplasm containing a well-developed, perinuclear, Golgi apparatus (G), mitochondria (M), and abundant, distended, endoplasmic reticulum(er). The plasma cell, partially shown at the right, has abundant flattened endoplasmic reticulum. (Reproduced with permission, from Tanaka, Y., and Goodman, J.R. *Electron Microscopy of Human Blood Cells*. Harper & Row, New York, 1972.)

F. **Polymeric Immunoglobulins** (IgM, IgA) have one additional polypeptide chain, the **J chain**. This chain is synthesized by all plasma cells, including those that produce IgG. However, it is only incorporated to polymeric forms of IgM and IgA. It is thought that the J chain has some role in initiating polymerization, as shown in Figure 6.2. IgM proteins are assembled in two steps. First, the monomeric units are assembled. Then, five monomers and one J chain will be combined via covalent bonds to result in the final pentameric molecule. This assembly seems to coincide with secretion in some cells, in which only monomeric subunits are found intracellularly, while in other cells the pentameric forms can be found intracellularly.

G. **Secretory IgA** is also assembled in two stages, but each one takes place in a different cell.
   1. **Dimeric IgA**, containing two monomeric subunits and a J chain joined together by disulfide bridges, is predominantly synthesized by **submucosal plasma cells**, although a minor portion may also be synthesized in the bone marrow.

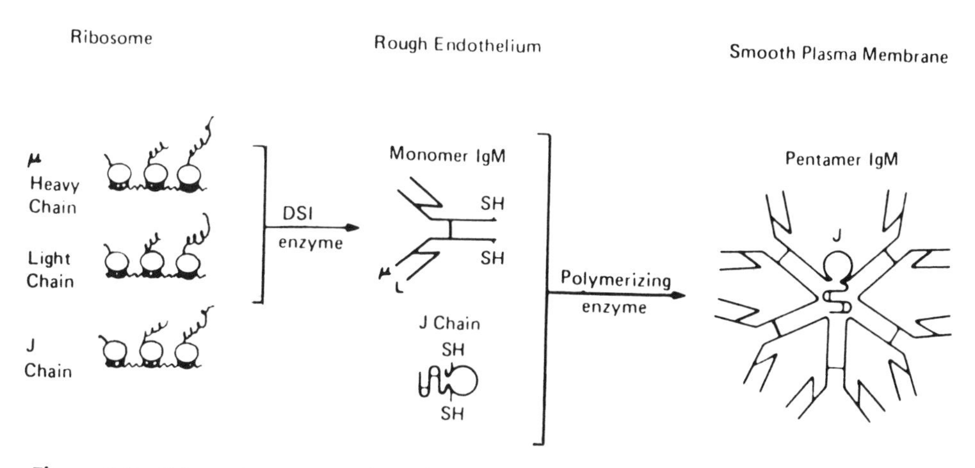

Figure 6.2 Schematic representation of IgM synthesis in a pentamer IgM-secreting cell. DSI, disulfide interchanging enzyme. (Reproduced with permission from Koshland, M.E. Molecular aspects of B cell differentiation. *J. Immunol. 131:*(6)i, 1983.)

2. **Secretory component (SC)**, on the other hand, is synthesized in the **epithelial cells**, where the final assembly of secretory IgA takes place. Two different **biological functions** have been postulated for the secretory component:
   a. First, SC is responsible for **secretion of IgA** by mucosal membranes. The process involves uptake of dimeric IgA, assembly of IgA-SC complexes, and secretion by the mucosal cells.
      i. The uptake of dimeric IgA by mucosal cells is mediated by a glycoprotein related to SC, called **polyimmunoglobulin receptor (Poly-IgR)**. **Poly-IgR** is constituted by a single polypeptide chain of approximately 95,000 daltons, composed of an extracellular portion with five immunoglobulin-like domains, a transmembrane domain, and an intracytoplasmic domain. It is expressed on the internal surface of mucosal cells and binds J-chain-containing polymeric immunoglobulins.
      ii. The binding of dimeric IgA to Poly-IgR seems to be the first step in the final assembly and transport process of secretory IgA. Surface-bound IgA is internalized and Poly-IgR is covalently bound to the molecule, probably by means of a **disulfide-interchanging enzyme** that will break intrachain disulfide bonds in both IgA and Poly-IgR and promote their rearrangement to form interchain disulfide bonds joining Poly-IgR to an α chain.
      iii. After this takes place, the transmembrane and intracytoplasmic domains of the receptor are removed by proteolytic cleavage, and the remaining five domains remain bound to IgA, as SC, and the complete secretory IgA molecule is secreted (Fig. 6.3).
      iv. Basically, the same transport mechanisms are believed to operate at the hepatocyte level. The hepatocytes produce Poly-IgR, bind and internalize dimeric IgA reaching the liver through the portal circulation, assemble complete secretory IgA, and secrete it to the bile.

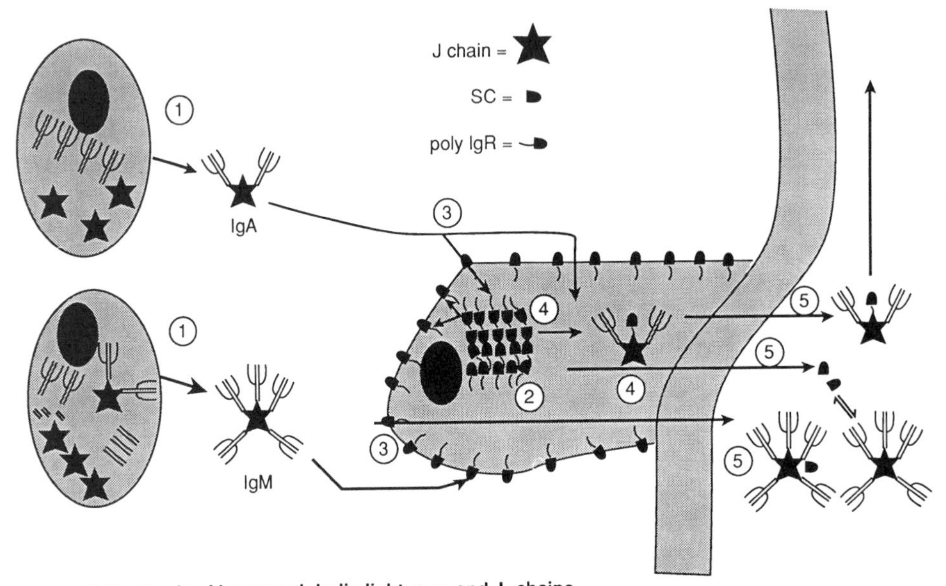

① Synthesis of immunoglobulin light, α, μ, and J- chains
② Synthesis and accumulation Poly IgR
③ Selective reception of IgA and IgM
④ Conjugation of Poly IgR with IgA and IgM; cleavage of Poly IgR
⑤ Secretion

Figure 6.3 Schematic representation of the mechanisms involved in the synthesis and external transfer of dimeric IgA. According to this model, a polyimmunoglobulin receptor is located at the membrane of mucosal cells and binds polymeric immunoglobulins in general, and dimeric IgA with the greatest specificity. The Poly-IgR-IgA complexes are internalized, and in the presence of a disulfide interchanging enzyme, covalent bonds are established between the receptor protein and the immunoglobulin. The transmembrane and intracytoplasmic bonds of the Poly-IgR are cleaved by proteolytic enzymes, and the extracellular portion remains bound to IgA, constituting the secretory component. The IgA-SC complex is then secreted to the gland lumen. If the individual is IgA-deficient, IgM may become involved in a similar process. (Modified from Brandzaeg, P., and Baklien, K. Intestinal secretion of IgA and IgM: A hypothetical model. In *Immunology of the Gut*, Ciba Foundation Symposium 46 (new series), Elsevier/Excerpta Medica/North-Holland, New York, 1977, p. 77.)

v. Secretory IgA must also flow back to the bloodstream, because small amounts are found in the blood of normal individuals. Higher levels of secretory IgA in blood are found in some forms of liver disease, when the uptake of dimeric IgA backflowing from the gut through the mesenteric lymph vessels takes place, but its secretion into the biliary system is compromised. Under those circumstances, secretory IgA assembled in the hepatocyte backflows into the systemic circulation.

vi. Among all J-chain containing immunoglobulins, the poly-IgR has higher binding affinity for dimeric IgA. In IgA-deficient individuals, IgM coupled with SC can be present in external secretions. It is believed that the same basic transport mechanisms are involved, starting by the binding of pentameric IgM to the Poly-IgR on a mucosal cell, and proceeding along the same lines outlined for the assembly and secretion of dimeric IgA. The fact that secretory IgM, with covalently bound SC, is detected exclusively in secretions of IgA-deficient individuals is believed to reflect the lower affinity of the interaction between poly-IgR and IgM-associated J chains (perhaps this as a consequence of steric hindrance of the binding sites of the J chain); consequently, the interaction between IgM and poly-IgR would only take place in the absence of competition from dimeric IgA molecules.

b. The second function proposed for SC is that of a **stabilizer of the IgA molecule**. This concept is based on experimental observations showing that secretory IgA or dimeric IgA to which SC has been noncovalently associated in vitro are more resistant to the effects of proteolytic enzymes than monomeric or dimeric IgA molecules devoid of SC. One way to explain these observations would be to suggest that the association of SC with dimeric IgA molecules renders the hinge region of the IgA monomeric subunits less accessible to proteolytic enzymes. From a biological point of view, it would be advantageous for antibodies secreted into fluids rich in proteolytic enzymes (both of bacterial and host origin) to be resistant to proteolysis.

II. IMMUNOGLOBULIN METABOLISM

A. **Immunoglobulin Half-Life**. One of the most commonly used parameters to assess the catabolic rate of immunoglobulins is the **half-life** (T 1/2), which corresponds to the time elapsed for a reduction to half of a circulating immunoglobulin concentration after equilibrium has been reached. This is usually determined by injecting an immunoglobulin labeled with a radioisotope (^{131}I is preferred for the labeling of proteins to be used for metabolic studies due to its fast decay rate), and following the plasma activity curve. Figure 6.4 reproduces an example of a metabolic turnover study. After an initial phase of equilibration, the decay of circulating radioactivity follows a straight line in a semilogarithmic scale. From this graph it is easy to derive the time elapsed between concentration n and n/2 (i.e., the half-life).

B. **Other Metabolic Parameters** that have been determined for immunoglobulins include the **synthetic rate** and the **fractional turnover rate** (which is the fraction of the plasma pool catabolized and cleared into urine in a day).

C. **Summary of the Metabolic Properties of Immunoglobulins**

1. **IgG** is the immunoglobulin class with the **longest half-life** (average of 21 days) and lowest fractional turnover rate (4–10%/12 h) with the exception of **IgG3**, which has a considerably **shorter half-life** (average of 7 days), close to that of **IgA** (5-6 days) and **IgM** (5 days). IgG catabolism is uniquely influenced by its circulating concentration of this immunoglobulin. At high protein concentrations, the catabolism will be faster, and at low IgG concentrations, catabolism will be slowed down. These differences are explained, according to **Brambell's theory**, by the protection of IgG bound to IgG-

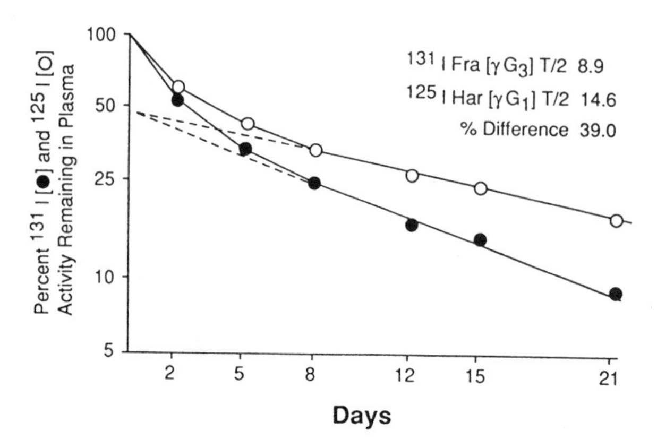

Figure 6.4 Plasma elimination curves of two IgG proteins, one typed as IgG1 (Har) and the other as IgG3 (Fra). The T 1/2 can be determined from the stable part of the curve and its extrapolation (dotted line) as the time necessary for a 50% reduction of the circulating concentration of labeled protein. (Reproduced with permission from Spiegelberg, H.L., Fishkin, B.G., and Grey, H.M. Catabolism of human G immunoglobulins of different heavy chain subclass. *J. Clin. Invest. 47*:2323, 1968.)

specific Fc receptors (FcγR) in the internal aspect of endopinocytotic vesicles from proteolytic enzymes. IgG is constantly pinocytosed by cells able to degrade it, but at low IgG concentrations most molecules are bound to the Fc receptors on the endopinocytotic vesicles and the fraction of total IgG degraded will be small. The undegraded molecules are eventually released back into the extracellular fluids. At high IgG concentrations, the majority of IgG molecules remain unbound in the endopinocytotic vesicle and are degraded, resulting in a high catabolic rate (Fig. 6.5).

2. While most immunoglobulin classes and subclasses are evenly distributed among the intra- and extravascular compartments, **IgM, IgD**, and to a lesser extent, **IgG3**, are predominantly concentrated in the **intravascular** space and **IgA2** is predominantly concentrated in secretions.
3. The synthetic rate of **IgA1** (24 mg/kg/day) is not very different from that of **IgG1** (25 mg/kg/day) but the serum concentration of IgA1 is about one-third of the IgG1 concentration. This is explained by a fractional turnover rate three times greater for IgA1 (24%/day).
4. The **highest fractional turnover rate** and **shorter half-life** are those of **IgE** (74%/day and 2.4 days, respectively).
5. The **lowest synthetic** rate is that of **IgE** (0.002 mg/kg/day compared to 20–60 mg/kg/day for IgG).

III. BIOLOGICAL PROPERTIES OF IMMUNOGLOBULINS

The antibody molecules have two major functions: binding to the antigen, a function that basically depends on the variable regions located on the Fab region of the molecule, and several other extremely important functions, listed in Table 6.1, which depend on the Fc

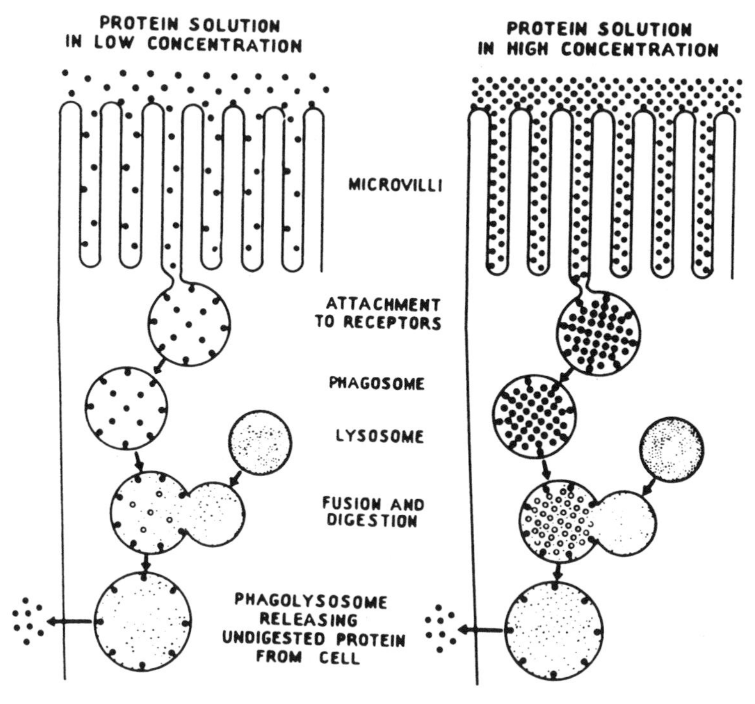

Figure 6.5 Schematic representation of Brambell's theory concerning the placental transfer of IgG and the relationship between concentration and catabolism of IgG. The diagram at the left shows that pinocytotic IgG will be partially bound to phagosome wall receptors and protected from proteolysis, being later released undigested. This mechanism would account for transplacental transfer. The diagram on the right side of the figure shows that if the concentration of IgG is very high, the number of IgG molecules bound to phagosome receptors will remain the same as when the concentration is low, while the number of unbound molecules will be much greater, and those will eventually be digested, resulting in a higher catabolic rate. (Reproduced with permission from Brambell, F.W.R. The transmission of immunity from mother to young and the catabolism of immunoglobulins. *Lancet ii*:1089, 1966.)

region. Of particular physiological interest are the placental transfer, complement fixation, and binding to Fc receptors.

A. **Placental Transfer**. In humans, the only major immunoglobulin transferred from mother to fetus across the placenta is IgG.

1. The placental transfer of IgG is an active process; the concentration of IgG in the fetal circulation is often higher than the concentration in matched maternal blood. It is also known that a normal fetus synthesizes only trace amounts of IgM, depending on placental transfer for acquisition of passive immunity against common pathogens.
2. The selectivity of IgG transport has been explained by **Brambell's receptor theory** for IgG catabolism. The trophoblastic cells on the maternal side of the placenta would endocytose plasma containing all types of proteins, but would have receptors in the endopinocytotic vesicles for the Fc region of IgG, and not for any other immunoglobulin. IgG bound to Fcγ receptors

Table 6.1 Biological Properties of Immunoglobulins

	IgG1	IgG2	IgG3	IgG4	IgA1	IgA2	IgM	IgD	IgE
Serum concentration (mg/dl)[a]	460–1140	80–390	28–194	2.5–16	50–200	0–20	50–200	0–40	0–0.2
Presence in normal secretions	–	–	–	–	+	+++	+	–	+
Placental transfer	+	+	+	+	–	–	–	–	–
Complement fixation									
classic pathway	+++	+	+++	–	–	–	+++	–	–
alternative pathway[b]	+	+	+	+	+	+	?	+	+
Reaction with Fc receptors on									
macrophages	++	–	++	–	+	+	–	–	–
neutrophils	++	–	++	–	–	–	–	–	–
basophils/mast cells	–	–	–	–	–	–	–	–	+++[c]
platelets	+	+	+	+	–	–	–	–	–
lymphocytes	++	?	++	?	–	–	+	–	–

[a]IgG subclass values (Shakib et al., *J. Immunol. Methods.*, *8*:17, 1975).
[b]After aggregation.
[c]High-affinity receptors.

would be protected from catabolism, and through active reverse pinocytosis would be released into the fetal circulation.

B. **Complement Activation**. The complement system can be activated by two pathways (see Chapter 9), and different structural areas of the immunoglobulin molecule are probably involved in complement fixation by either pathway. At the present time, all immunoglobulins have been found able to fix complement by one or the other pathway.

1. **Activation of the classical pathway**. **IgG1**, **IgG3**, and **IgM** molecules are the most efficient in fixing complement, all of them through the classical pathway.
2. **Consequences of complement activation**. Complement activation is an extremely important amplification mechanism, which mediates antibody-dependent neutralization and elimination of infectious agents. These effects depend on:
    a. **Generation of C3b** which, when deposited on the membrane of a microorganism, facilitates phagocytosis by cells with C3b receptors. For this reason, C3b is known as an **opsonin**.
    b. **Disruption of lipid bilayers**, which depends on the generation of the late complement components (C6-C9), which, when properly assembled on a cell membrane, induce the formation of transmembrane channels that result in cell lysis.

These mechanisms are discussed in greater detail in Chapter 9.

C. **Binding to Fc Receptors**. Virtually every type of cell involved in the immune response has been found to be able to bind one or more immunoglobulin isotypes through Fc receptors (Table 6.2). These receptors have been classified according to the isotype of immunoglobulin that they preferentially bind as Fcγ R (receptors for IgG), FcαR (receptors for IgA), FcεR (receptors for IgE), and FcμR (receptors for IgM). Subtypes of Fc receptors have been additionally defined, based on antigenic and structural differences:

- **FcγRI**, a high-affinity receptor able to bind monomeric IgG, and expressed exclusively by monocytes and macrophages.
- **FcγRII**, a low-affinity receptor for IgG expressed by phagocytic cells, platelets, and B lymphocytes.
- **FcγRIII**, a second low-affinity IgG receptor expressed by phagocytic and NK cells.

1. **Significance of FcγR binding**
    a. As discussed above, the **catabolic rate of IgG** and the **selective placental transfer of IgG** depend on the interaction with FcγR on pinocytotic vesicles.
    b. **IgG also mediates phagocytosis**, by all cells expressing FcγR on their membranes (granulocytes, monocytes, macrophages, and other cells of the same lineage). Thus, IgG is also considered as an **opsonin**. IgG and C3b have synergistic opsonizing effects, and their joint binding and deposition on the membrane of an infectious agent is a most effective way to promote its elimination.
    c. Granulocytes, monocytes/macrophages, and NK cells can destroy target cells coated with IgG antibody (**antibody-dependent cellular cyto-**

Table 6.2 Different Types of Fc Receptors Described in the Cells of the Immune System

Fc receptor	Characteristics	Cellular distribution	Function
FcγRI (CD64)[a]	Transmembrane and intracytoplasmic domains; high affinity; binds both monomeric and aggregated IgG	Monocytes, macrophages	ADCC (monocytes)
FcγRII (CD32)	Transmembrane and intracytoplasmic domains; low affinity	Monocytes/macrophages; Langerhans cells; granulocytes; platelets; B cells	IC binding; phagocytosis; degranulation; ADCC (monocytes)
FcγRIII (CD16)	Glycosyl-phosphatidyl inositol anchor in neutrophils; transmembrane segment in NK cells; low affinity	Macrophages; granulocytes; NK and K cells	IC binding and clearance; "priming signal" for phagocytosis and degranulation; ADCC (NK cells)
FcαR	Transmembrane and intracytoplasmic segments; low affinity	Granulocytes, monocytes/ macrophages, platelets, T and B lymphocytes	Phagocytosis, degranulation
FcεRI	High affinity	Basophils, mast cells	Basophil/mast cell degranulation
FcεRII (CD23)	Low affinity	T and B lymphocytes; monocytes/ macrophages; eosinophils; platelets	Mediate parasite killing by eosinophils
FcμR		T lymphocytes	

[a]The term CD designates the epitope(s) recognized by monoclonal antibodies raised against the different receptors (see Chapter 10).

toxicity; **ADCC**). In this case the destruction of the target cell does not depend on opsonization, but rather on the release of toxic mediators.

d. The specific elimination of target cells by opsonization and ADCC depends on the binding of IgG antibodies to those targets. The antibody molecule tags the target for destruction; phagocytic or NK cells mediate the destruction. Not all antibody molecules are able to react equally with the Fcγ receptors of these cells. The highest binding affinities for any of the three types of Fcγ receptor known to date are observed with IgG1 and IgG3 molecules.
e. Phagocytic cells also express the Fcα receptor, which in controlled experimental conditions also seems able to mediate attachment, degranulation, and/or phagocytosis of IgA-coated particles. However, the physiological role of this receptor has yet to be defined.
f. Two types of **Fcε receptors** specific for IgE have been defined:
   i. A low-affinity receptor (FcεRII), present in most types of granulo-

cytes, mediates ADCC reactions directed against helminths, which typically elicit IgE antibody synthesis.

ii. A high-affinity Fcε receptor (FcεRI) is expressed by basophils and mast cells. The basophil/mast cell-bound IgE functions as a true cell receptor. When an IgE molecule bound to a high-affinity FcεRI membrane receptor interacts with the specific antigen against which it is directed, the cell is activated and, as a consequence, histamine and other mediators are released from the cell. The release of histamine and a variety of other biologically active compounds is the basis of the immediate hypersensitivity reaction, which is discussed in detail in Chapter 19.

SELF-EVALUATION

Questions

Choose the ONE *best* answer.

6.1 A newborn's repertoire of antibodies is determined to include all of the following. Which one of the listed antibodies is most likely to have been synthesized by the newborn?
   A. IgG1 anti-Rh
   B. IgG2 anti-human immunodeficiency virus (HIV)
   C. IgG3 anti-tetanus toxoid
   D. IgG4 anti-*Haemophilus influenzae* type b
   E. IgM anti-toxoplasma

6.2 The Fab fragment of an IgG immunoglobulin includes the domain(s) responsible for:
   A. Antigen binding
   B. Binding to macrophage receptors
   C. Complement fixation
   D. Fixation to heterologous skin
   E. Placental transfer

6.3 Which of the following would be a major problem for the therapeutic administration of IgA antibodies to individuals congenitally unable to synthesize them?
   A. Affinity to mucosal cells
   B. Lack of complement-fixing capacity
   C. Low concentration in serum
   D. Molecular heterogeneity
   E. Short half-life

6.4 According to Brambell's theory, the key to the placental transfer of IgG is the:
   A. High diffusibility of monomeric antibodies
   B. Protection against proteolysis of IgG bound to receptors in the walls of pinocytotic vesicles that transport proteins across the trophoblast
   C. Local synthesis by subendothelial plasma cells on the maternal side of the placenta
   D. The existence of an uptake mechanism in trophoblast similar to that responsible for the transfer of IgA across mucosae
   E. Long half-life of maternal IgG

6.5 The relatively greater resistance of secretory IgA to proteolytic enzymes is assumed to be a consequence of the:
A. Anti-protease activity of secretory component
B. Increased resistance to proteolysis of IgA2 molecules
C. Predominantly dimeric nature of secretory IgA
D. Presence of J chain
E. "Wrapping" of secretory component around hinge region

6.6 Which immunoglobulin has a catabolic rate dependent on its circulating concentration?
A. IgG
B. IgA
C. IgM
D. IgD
E. IgE

Questions 6.7–6.10

Match the characteristic indicated below with the immunoglobulin to which it is most specifically associated.

A. Half-life of 7 days
B. Greatest fractional catabolic rate
C. Binding to FcγRI in its monomeric form
D. Predominantly B-cell bound
E. Contain J chain

6.7 IgG1
6.8 IgG3
6.9 Dimeric IgA
6.10 IgE

Answers

6.1 (E) IgM antibodies are not transmitted across the placenta, while all IgG subclasses are. Thus, the finding of specific IgM antibodies to any infectious agent in cord blood is considered as evidence of intrauterine infection.

6.2 (A) All biological properties of antibodies other than antigen binding are Fc-mediated.

6.3 (E) Although IgA is relatively low in the serum, it could be obtained in larger amounts from milk or colostrum. The main problem with replacing IgA in a deficient individual would be the short half-life (5–6 days), probably requiring weekly administration of a product made expensive by the need for special isolation protocols.

6.4 (B) Endocytosed and receptor-bound IgG will be protected from proteolysis.

6.5 (E) It is believed that SC sterically protects the hinge region by blocking access to proteolytic enzymes.

6.6 (A) Brambell's theory postulates that there is an inverse correlation between the circulating concentration of IgG and the fraction of IgG that is bound on the phagolysosome membranes and spared from degradation; hence, the fractional turnover rate will be directly proportional to circulating IgG concentrations.

6.7 (C) IgG1 binds with high affinity to the FcγRI in its native state.
6.8 (A) IgG3 is more susceptible to proteolytic enzymes than other IgG proteins of subclasses 1, 2, and 4. This could explain its faster catabolic rate and shorter half-life.
6.9 (E) Only polymeric immunoglobulins contain J chain in their molecules.
6.10 (B) Both IgE and IgD exist predominantly as cell-bound immunoglobulins; IgD is found on the membrane of B lymphocytes while IgE binds to FcϵR in mast cells and eosinophils.

BIBLIOGRAPHY

Brandtzaeg, P. Molecular and cellular aspects of the secretory immunoglobulin system. *APMIS, 103*:1, 1995.
Gergely, J., and Sarmay, G. The two binding-site models of human IgG binding Fcγ receptor. *FASEB J.*, *4*:3275, 1990.
Hendershot, L., and Kearney, J.F. A role for human heavy chain binding protein in the developmental regulation of protein transport. *Molecular Immunol.*, *25*:585, 1988.
Metzer, H. Fc receptors and membrane immunoglobulins. *Curr. Opin. Immunol.*, *3*:40, 1991.
Nezlin, R. Immunoglobulin structure and function. In *Immunochemistry* (van Oss, C.J., and van Regenmortel, M.H.V., eds.), Marcel Dekker, Inc., New York, 1994, p. 3.
Parkhouse, R.M.E. Biosynthesis of immunoglobulins. In *Immunochemistry: An Advanced Textbook* (Glynn, L.E., and Steward, M.W., eds.). Wiley & Sons, New York, 1977.
Sitia, R., and Cattaneo, A. Synthesis and assembly of antibodies in nature and artificial environments. In *The Antibodies—Vol. 1* (Zanetti, M., and Capra, J.D., eds.). Harwood Academic Publishers, New York, 1995, p. 127.
Turner, M.W. Structure and function of immunoglobulins. In *Immunochemistry: An Advanced Textbook* (Glynn, L.E., and Steward, M.W., eds.). Wiley & Sons, New York, 1977.
Underdown, B.J., and Schiff, J.M. Immunoglobulin A: Strategic defense initiative at the mucosal surface. *Annu. Rev. Immunol.*, *4*:389, 1986.
Waldmann, T.A., and Strober, W. Metabolism of immunoglobulins. *Prog. Allergy*, *13*:1, 1969.

7

Genetics of Immunoglobulins

Janardan P. Pandey

I. IMMUNOGLOBULIN GENES

Human immunoglobulin (Ig) molecules are coded by three unlinked gene families: two for light (L) chains located on **chromosomes 2 (κ chains)** and **22 (λ chains)**, and one for heavy (H) chains located on **chromosome 14**.

- **A. Variable Region Diversity**. As mentioned in the preceding chapters, each individual is able to produce several million antibody molecules with different antigenic specificities, and this diversity corresponds to the extreme heterogeneity of the variable (V) regions in those antibody molecules, implying that each individual must possess a large number of structural genes for Ig chains.
- **B. Constant Region Homogeneity**. The allotypic determinants on the constant (C) region (see following discussion) segregate as a single Mendelian trait, suggesting that there may be only one gene for each of the several Ig chain C regions.
- **C. Multiple Genes Code for Immunoglobulin Polypeptide Chains**. These seemingly contradictory observations were reconciled when it was proposed and later proved that the V and C regions are encoded by separate genes that are brought together by a translocation event during lymphocyte development.

II. IMMUNOGLOBULIN GENE REARRANGEMENT

Each immunoglobulin (Ig) polypeptide chain is coded by multiple genes scattered along a chromosome of the germ-line genome. These widely separated gene segments are brought together (recombined) during B-lymphocyte differentiation to form a complete Ig gene.

- **A. V-Region Gene Rearrangements**
  1. The **V regions of the immunoglobulin light chains** are coded by two gene segments, designated as V and J (J for joining, because it joins V and C region genes).
  2. **The V regions of the immunoglobulin heavy chains** are encoded by three gene segments: V, J, and D (D for diversity, corresponds to the most diverse region of the H chain).
  3. **To form a functional light or heavy chain gene, one or two gene rearrangements are needed**.

a. On chromosome 2 (where the κ-chain genes are located) or 22 (where the λ-chain genes are located), a V gene moves next to a J gene.
b. On chromosome 14 (where the heavy-chain genes are located), first the D and J regions are joined, and next one of the V genes is joined to the DJ complex.
c. The JV or VDJ segments, one of the C-gene complexes (C_κ or C_λ on chromosomes 2 or 22), and C_μ, C_δ, $C_\gamma 3$, $C_\gamma 1$, $C_\alpha 1$, $C_\gamma 2$, $C_\gamma 4$, C_ε, or $C_\alpha 2$ on chromosome 14 are then transcribed into nuclear RNA containing VJ or VDJ coding sequences, the interconnecting noncoding sequences, and the coding sequence of one of the C genes. The intervening noncoding sequences are then excised, making a contiguous VJC mRNA for an L chain and a contiguous VDJC mRNA for an H chain (Figs. 7.1 and 7.2).
d. **Gene rearrangements occur in a sequential order**: heavy-chain genes rearrange first, followed by κ-chain genes and, finally, by λ-chain genes.

4. It has been shown that the VDJ joining is regulated by two proteins encoded by two closely linked **recombination-activating genes**, RAG-1 and RAG-2, localized on the short arm of human chromosome 11. These genes have at least two unusual characteristics not shared by most eukaryotic genes:

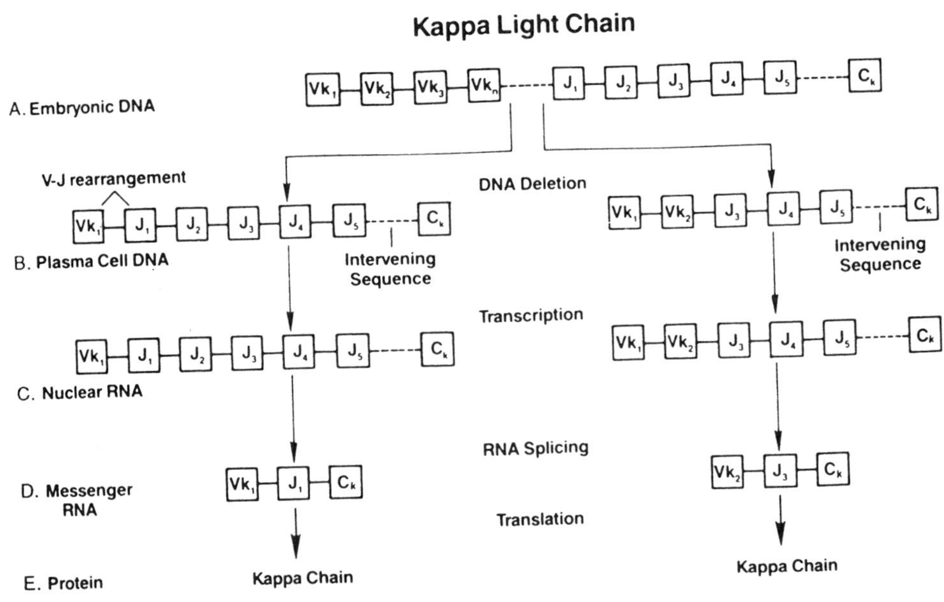

Figure 7.1 The embryonic DNA of chromosome 2 contains over 300 V genes, five J (joining) genes, and a C (constant) gene (A). The V and J gene code for the kappa chain's variable region, C for its constant region. In the left pathway, differentiation of the embryonic cell to a plasma cell results in deletion of the intervening V genes so that Vk_1 is joined with the J_1 gene (B). The linked Vk_1J_1 segment codes for one of over 1500 possible kappa light-chain variable regions. The plasma cell DNA is transcribed into nuclear RNA (C). Splicing of the nuclear RNA produces messenger RNAs with the Vk_1, J_1, and C genes linked together (D), ready for translation of a kappa light-chain protein (E). The alternate pathway at right (B-D) shows another of the many possible pathways leading to a different kappa light chain with a different variable region specificity. (Modified from David, J.R. Antibodies: Structure and Function. In *Scientific American Medicine*. Scientific American, Inc., New York, 1980.)

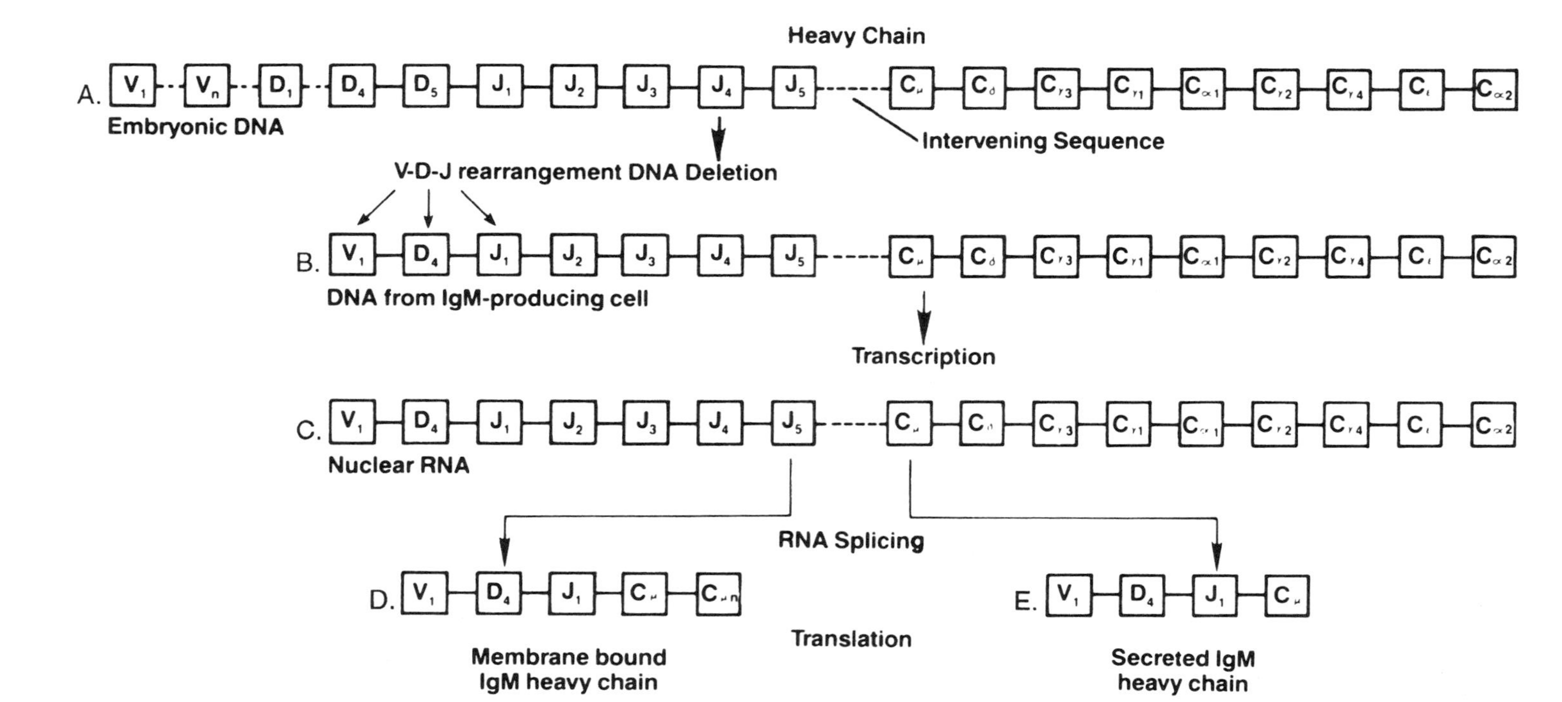

Figure 7.2 A stretch of embryonic DNA in chromosome 14 contains a section coding for the heavy-chain variable region; this DNA is made up of at least 100 V genes, 50 D genes, and four to six J genes. The section coding for the constant region is formed by nine C genes (A). In the pathway shown, when the embryonic cell differentiates into a plasma cell, some V and D genes are deleted so that V_1, D_4, and J_1 are joined to form one of many possible heavy-chain genes (B). The plasma-cell DNA is then transcribed into nuclear RNA (C). RNA splicing selects the C gene and joins it to the V_1, D_4, and J_1 genes (D). The resulting messenger RNAs will code for IgM heavy chains (E). If RNA splicing removes the Cμm piece from the Cμ gene, the IgM will be secreted. If the piece remains, the IgM will be membrane-bound. (Modified from David, J.R. Antibodies: Structure and Function. In *Scientific American Medicine*, Scientific American, Inc., New York, 1980.)

a. They are devoid of introns.
b. Although adjacent in location and synergistic in function, they have no sequence homology.

The lack of homology implies that, unlike the immunoglobulin and MHC genes, RAG-1 and RAG-2 did not arise by gene duplication. It has been speculated that these genes may have evolved from a more primitive prokaryotic recombination system. The exact mechanism by which the RAG gene products induce recombination is not well understood.

5. The transcription of Ig genes, like other eukaryotic genes, is regulated by promoters and enhancers. Promoters, located 5′ of the V segment, are necessary for transcription initiation. Enhancers, located in the introns between J and C segments, increase the rate of transcription. For this reason, **immunoglobulin synthesis** (H or L chains) **is only detected after the VDJ or VJ rearrangements, which bring the promoter in close proximity to the enhancer**.
6. During ontogeny and functional differentiation, the H-chain genes may undergo further gene rearrangements that result in **immunoglobulin class switching**.
   a. As the B lymphocytes differentiate into plasma cells, one heavy-chain C-gene segment can be substituted for another without alteration of the VDJ combination (Fig. 7.3). In other words, a given variable region gene can be expressed in association with more than one heavy-chain class or subclass so that at the cellular level the same antibody specificity can be associated with the synthesis of an IgM immunoglobulin (characteristic of the early stages of ontogeny and of the primary response) or with an IgG immunoglobulin (characteristic of the mature individual and of the secondary response).
   b. **Three mechanisms have been proposed to explain the heavy-chain class switch**
      i. Intrachromosomal DNA recombination (recombination between sister-chromatids that constitute a chromosome)
      ii. Interchromosomal recombination (recombination between nonsister-chromatids of the two homologous chromosomes)
      iii. Looping-out and deletion of intervening DNA sequences within a single chromosome

   The mechanism that contributes most substantially to isotype switching is probably a looping-out and deletion mechanism, in which the switched Ig sequences, C_{μ}, and other intervening CH genes loop out and are subsequently deleted. This mechanism would predict the formation of circular DNA molecules containing the sequences deleted by isotype switching. Such DNA circles have now been isolated.

III. GENETIC BASIS OF ANTIBODY DIVERSITY

It has been estimated that an individual is capable of producing up to 10^9 different antibody molecules. How this vast diversity is generated from a limited number of germline elements has long been one of the most intriguing problems in immunology. There are two possible

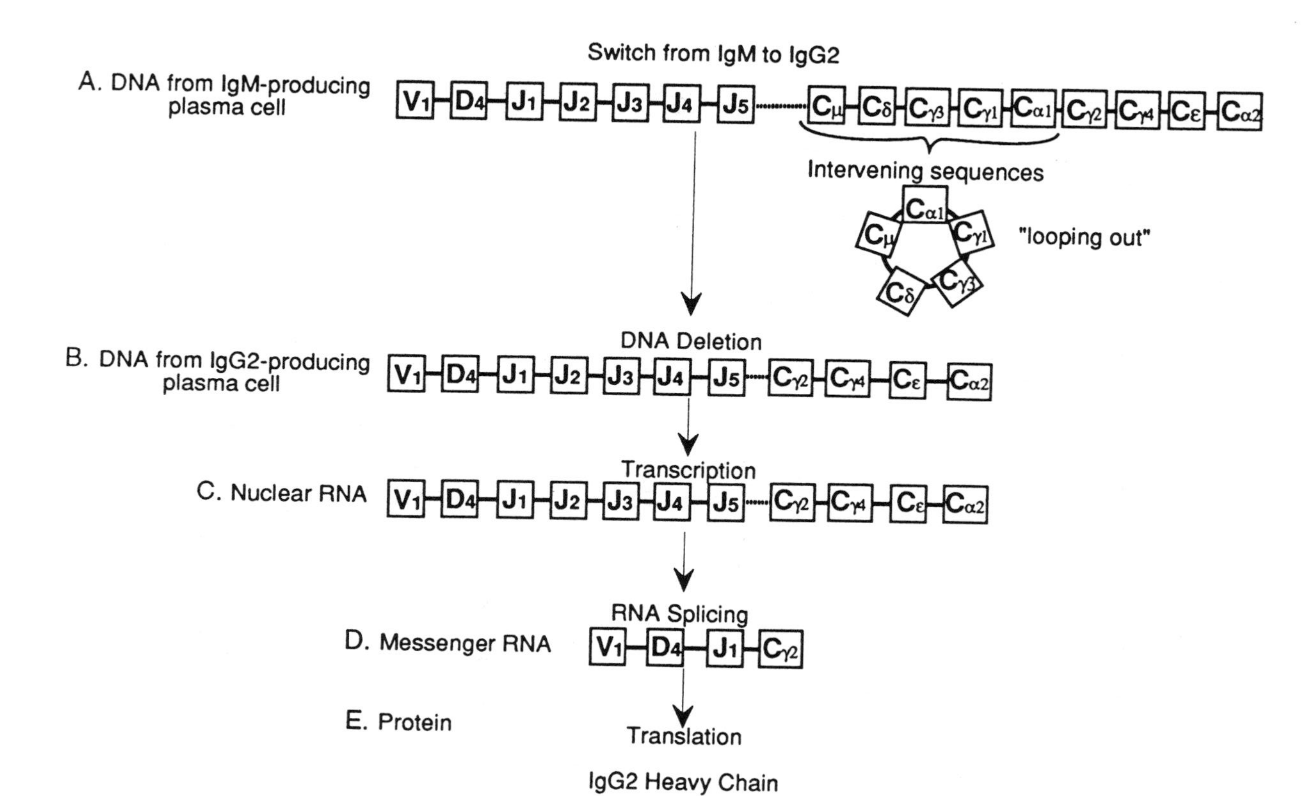

Figure 7.3 In the secondary response, a plasma cell switches from IgM production (A) to IgG2 production by deleting a DNA loop containing the constant-region genes Cμ, Cδ, Cγ3, Cγ1, and Cμ1 from the IgM heavy-chain gene (B). This DNA is now transcribed into nuclear RNA (C). RNA splicing links the Cγ2 gene with the J_1 gene (D), and then the mRNA is translated into an IgG2 heavy chain (E). (Modified from David, J.R. Antibodies: Structure and Function. In *Scientific American Medicine*, Scientific American, Inc., New York, 1980.)

mechanisms for this variability: Either the information is transmitted from generation to generation in the germ line, or it is generated somatically during B-lymphocyte differentiation. The following genetic mechanisms have been shown to contribute to the **generation of antibody diversity**.

A. **The existence of a large number of V genes and of a smaller set of D and J segments in the germline DNA**, which has probably been generated during evolution as a consequence of environmental pressure. In the mouse, sequence and cross-hybridization studies have shown that at least 100 VH genes exist in the germ line, which can be grouped into seven families, each of these associated with different antibody specificities. A similar situation is likely to exist in humans.

B. **Combinatorial Association**

1. As mentioned before, there are at least 100 V-region genes for the heavy chain, and this is probably a conservative estimate. The number of possible V regions resulting from these genes is increased by the recombinatorial events taking place prior to immunoglobulin synthesis.
   a. Any VH segment can combine, in principle, with any of a number (4 to 6) of J segments, and with one of 50 different D segments.
   b. In the case of the light chain, the number of V-region genes is estimated to be 300, and they can also recombine with different J-region genes.
   c. Imprecise joining of various V-gene segments, creating sequence variation at the points of recombination, augments diversity significantly.
   d. Random association of L and H chains adds diversity to the antibody binding site repertoire, because the configuration of the binding site is determined jointly by the variable regions of the light and heavy chains.
2. Accepting that all these recombinations are totally random, about 1000 H chains and 1000 L chains with different V regions would emerge from the recombinatorial process, and the random association of these L and H chains would produce 10^6 unique antibodies. However, there is increasing evidence suggesting that gene rearrangements are not totally random, a fact that could drastically reduce our estimates of diversity directly resulting from germ-line associations and which reinforces the argument for the role of somatic mutations in generating antibody diversity.

C. **Somatic Mutations**. Mutations occurring after embryonic differentiation were proposed as a source of antibody diversity in the 1950s. Experimental support for this hypothesis, however, was obtained three decades later.

1. Comparison of nucleotide sequences from murine embryonic DNA and DNA obtained from plasmocytomas revealed several base changes, suggesting occurrence of **point mutations** during lymphocyte differentiation. There appear to be some special mutational mechanisms involved in immunoglobulin genes since the mutation sites are clustered around the V genes and not around the C genes.
2. In addition to these point mutations, certain enzymes can randomly insert and/or delete DNA bases. Such changes can shift the reading frame for translation (**frameshift mutations**) so that all codons distal to the mutation are read out of phase and may result in different amino acids, thus adding to the antibody diversity.

D. **Somatic Mutations and Affinity Maturation**. Somatic mutations (sometimes termed hypermutations) play a very important role in **affinity maturation**—

production of antibodies with better antigen binding ability. During the initial exposure to an antigen, rearranged antibodies with appropriate specificity bind to the antigen. Late in the response, random somatic mutations in the rearranged V genes result in the production of antibodies of varying affinities. By a process analogous to natural selection, B cells expressing higher affinity antibodies are selected to proliferate and those with the lower affinity antibodies are eliminated.

E. **Gene conversion** describes a nonreciprocal exchange of genetic information between genes, which has also been shown to contribute to the antibody diversity. One of the two recombination activating genes described earlier, RAG-2, appears to be involved in gene conversion events.

IV. ANTIGENIC DETERMINANTS OF IMMUNOGLOBULIN MOLECULES

A. **Definitions**. Three main categories of antigenic determinants are found on immunoglobulin molecules:

1. **Isotypes** are antigenic determinants present on all molecules of each class and subclass of immunoglobulin heavy chains and on each type of light chain; they are defined serologically by antisera directed against the constant regions of H and L chains. The antisera are produced in animals which, upon injection of purified human immunoglobulins, recognize the structural differences between constant regions of H and L chains.
   a. Isotypic determinants are common to all members of a given species; hence they cannot be used as genetic markers. Their practical importance results from the fact that they allow the identification of classes and subclasses of immunoglobulins through the heavy-chain and light-chain (κ, λ) isotypes.
   b. The light-chain isotypes are shared by all classes and subclasses of normal immunoglobulins.
2. **Idiotypes** are antigenic determinants associated with hypervariable regions. The antigen combining site in the V region of the immunoglobulin molecule, in addition to determining specificity for antigen binding, can also act as an antigen and induce production of antibodies against it (anti-idiotypic antibodies). Antibodies of the same specificity share idiotypes.
3. **Allotypes** are hereditary antigenic determinants of Ig polypeptide chains that may differ between individuals of the same species. Because of their heterogeneity, they can be used as genetic markers.
   a. The loci controlling allotypic determinants are codominant (i.e., both are expressed phenotypically in a heterozygote) autosomal genes that follow Mendelian laws of heredity.
   b. All allotypic markers that have so far been identified on human immunoglobulin molecules, with one exception (see later), are present in the C regions of H chains of IgG, IgA, IgE, and on κ-type L chains.

B. **IgG Heavy-Chain Allotypes (GM Allotypes)**

1. **Distribution and nomenclature**. Allotypes have been found on γ1, γ2, and γ3 heavy chains, but not as yet on γ4 chains. They are denoted as G1M, G2M, and G3M, respectively (**G** for IgG, the numerals 1, 2, and 3 for the subclass, and the letter **M** for marker). At present 18 GM specificities can be

Table 7.1 Currently Testable GM Allotypes

Heavy-chain subclass		Numeric	Alphameric
γ1	GIM	1	a
		2	x
		3	f
		17	z
γ2	G2M	23	n
γ3	G3M	5	b1
		6	c3
		10	b5
		11	b0
		13	b3
		14	b4
		15	s
		16	t
		21	g1
		24	c5
		26	u
		27	v
		28	g5

defined (Table 7.1): four associated with IgG1 (G1M), one associated with IgG2 (G2M), and thirteen associated with IgG3 (G3M).

2. **Topography and structural basis**
   a. **G1M 3** and **G1M 17** are localized in the Fd portion of the IgG molecule, while the rest are in the Fc portion.
   b. In some cases, the antigenic differences recognized as allotypic are a consequence of a single amino acid substitution on the heavy chains. For instance, G1M 3 heavy chains have arginine at position 214 and G1M 17 heavy chains have lysine at this position. For this reason, these markers are designated as homoalleles or true alleles.
   c. A single heavy chain may possess more than one GM determinant; G1M 17 and G1M 1 are frequently present on the Fd and Fc portions of the same chain in Caucasians.
3. **Haplotypes**
   a. The four C-region genes on human chromosome 14, which encode the four IgG subclasses, are very closely linked. Because of this close linkage, GM allotypes of various subclasses are transmitted as a group called **haplotype**.
   b. Also, because of almost absolute linkage disequilibrium between the alleles of various IgG C-region genes, certain allotypes of one subclass are always associated with certain others of another subclass. For example, we should expect to find the IgG1 allotype G1M 3 associated with the IgG3 allotypes G3M 5 and G3M 21 with equal frequency; in fact, in Caucasians, a haplotype carrying G1M 3 is almost always associated with G3M 5 and not with G3M 21.

c. Every major ethnic group has a distinct array of GM haplotypes. GM*3 23 5, 10, 11, 13, 14, 26, and GM* 1, 17 5, 10, 11, 13, 14, 17, 26 are examples of common Caucasian and Negroid haplotypes, respectively. In accordance with the international system for human gene nomenclature, **haplotypes and phenotypes are written by grouping together the markers that belong to each subclass, by the numerical order of the marker and of the subclass; markers belonging to different subclasses are separated by a space, while allotypes within a subclass are separated by commas**. An asterisk is used to distinguish alleles and haplotypes from phenotypes.

B. IgA Heavy-Chain Allotypes (AM Allotypes)

1. Two allotypes have been defined on human IgA2 molecules: **A2M 1 and A2M 2**. They behave as alleles of one another.
2. No allotypes have been found on IgA1 molecules as yet.
3. Individuals lacking IgA (or a particular IgA allotype) have in some instances been found to possess anti-IgA antibodies directed either against one of the allotypic markers or against the isotypic determinant. In some patients, these antibodies can cause severe anaphylactic reactions, following blood transfusion containing incompatible IgA.

C. IgE Heavy-Chain Allotypes (EM Allotypes). Only one allotype, designated as EM 1, has been described for the IgE molecule. Because of a very low concentration of IgE in the serum, EM 1 cannot be measured by hemagglutination inhibition, the method most commonly used for typing all other allotypes. This marker is measured by radioimmunoassay using a monoclonal anti-EM 1 antibody.

D. Κ-Type Light-Chain Allotypes (KM Allotypes). Three KM allotypes have been described so far: **KM 1, KM 1,2, and KM 3**. They are inherited via three alleles, KM* 1, KM* 1,2, and KM* 3 on human chromosome 2. No allotypes have as yet been found on the λ-type light chains.

E. Heavy-Chain V-Region Allotype (HV 1). So far, HV 1 is the only allotypic determinant described in the V region of human immunoglobulins. It is located in the V region of H chains of IgG, IgM, IgA, and possibly also on IgD and IgE.

V. DNA POLYMORPHISMS: RFLPs

A. Several new genetic polymorphisms, detected directly at the DNA level, have been described in the Ig region. These are known as **restriction fragment length polymorphisms (RFLPs)** because they result from variation in DNA base sequences that modify cleavage sites for restriction enzymes. RFLPs have been described in both V and C regions. Their significance is under active investigation.

B. As mentioned before, the most widely used method for determining Ig allotypes is hemagglutination inhibition, and the antisera used for typing are obtained from fortuitously immunized human donors. Because of the scarcity of such antisera, investigations to examine the role of allotypes in immune responsiveness and disease susceptibility have been hampered. Recently, molecular methods that allow the detection of allotypes at the genomic level have been developed, thus circumventing the problems arising from the paucity of antisera.

VI. ALLELIC AND ISOTYPIC EXCLUSION

A. **Immunoglobulin Gene Rearrangements at a Single Cell Level.** One of the most fascinating observations in immunology is that immunoglobulin heavy-chain genes from only one of the two homologous chromosomes 14 (one paternal and one maternal) and immunoglobulin light chains from only one of chromosomes 2 or 22 are expressed in a given B lymphocyte.

1. Recombination of VDJC genes described earlier usually takes place on one of the homologous chromosomes. If this rearrangement is unproductive (i.e., it does not result in the secretion of an antibody molecule), the other homologue undergoes rearrangement. Consequently, of the two H-chain alleles in a B cell, one is productively rearranged and the other is either in the germline pattern or is aberrantly rearranged (in other words, excluded).
2. Involvement of the chromosomes in the rearrangement process is random; in one B cell, the paternal allele may be active, and in another, it may be a maternal allele. (Allelic exclusion is reminiscent of the X-chromosome inactivation in mammals, although it is genetically more complex.)

B. **Mechanism of Allelic Exclusion.** Two models have been proposed to explain allelic exclusion: **stochastic** and **regulated.**

1. The main argument favoring the stochastic model is the finding that a high proportion of VDJ or VJ rearrangements are nonproductive (i.e., they do not result in transcription of mRNA). Therefore, according to this model, allelic exclusion is achieved because of a very low likelihood of a productive rearrangement on both chromosomes.
2. According to the regulated model, a productive H- or L-chain gene arrangement signals the cessation of further gene rearrangements (feedback inhibition). Results from experiments with transgenic mice (mice in which foreign genes have been introduced in the germline) favor the regulated model. It appears that a correctly rearranged H-chain gene not only inhibits further H-chain gene rearrangements but also gives a positive signal for the κ-chain gene rearrangement.

C. **Light-Chain Isotypic and Allelic Exclusions**

1. As a rule, the rearrangement of the λ gene takes place only if both alleles of the κ gene are aberrantly rearranged. This mutually exclusive nature of a productive light-chain gene rearrangement results in **isotypic exclusion** (i.e., a given plasma cell contains either κ or λ chains, but not both).
2. In addition, only one of the two chromosomes 2 will rearrange, and this results in **allelic exclusion** (i.e., at a single-cell level, the kappa chains will express one of the two KM allotypes encoded in the genome of a heterozygous individual). However, a mixture of immunoglobulin molecules containing either one of the two κ-chain allotypes will be detected in serum, reflecting the random nature of chromosome 2 rearrangements.

D. **Heavy-Chain Allelic Exclusion.** Allelic exclusion is also evident at the level of the GM system. A given plasma cell from an individual heterozygous for G1M* 17/G1M* 3 will secrete IgG carrying either G1M 17 or G1M 3, but not both. Analogous to what is observed with κ-chain allotypes, serum samples from such an individual will have a mixture of G1M 17 IgG1 molecules and of G1M 3 IgG1 molecules, secreted by different immunoglobulin-producing cells.

E. **Antibody Specificity at the Single-Cell Level.** The constraints surrounding light-chain and heavy-chain chromosome rearrangements result in the synthesis of molecules with identical V regions in each plasma cell, because all expressed mRNA will have been derived from a single rearranged chromosome 14 and from a single rearranged chromosome 2 or 22. Therefore, the antibodies produced by each B lymphocyte will be of a single specificity.

VII. GM ALLOTYPES AND IgG SUBCLASS CONCENTRATIONS

Studies from several laboratories have found a correlation between certain GM allotypes or phenotypes and the concentration of the four subclasses of IgG. The results vary; however, virtually all studies report a significant association between the GM 3 5, 13 phenotype and a high IgG3 concentration and the G2M 23 allotype and an increased concentration of IgG2. These associations imply that a determination of whether or not a person's IgG subclass level is in the "normal" range should be made in the context of the individual's GM phenotype.

VIII. IMMUNOGLOBULIN ALLOTYPES, IMMUNE RESPONSE, AND DISEASES

A. Several investigators have shown that immune responsiveness (both humoral and cellular) to certain antigens is influenced by particular GM and KM allotypes or by genes in linkage disequilibrium with them. In addition, allotypes have been implicated in susceptibility to several diseases. Occasionally, the two unlinked genetic systems—human leukocyte antigen (HLA) on chromosome 6 and GM on chromosome 14—somehow interact to influence immune responsiveness and disease susceptibility.

B. The associations between allotypes and disease reported thus far are not strong enough to be of practical importance in medicine. This situation is likely to change when more RFLPs are used in association studies. For example, if the cleavage site for the restriction enzyme responsible for an RFLP is identical to the mutational site giving rise to the gene for disease susceptibility, we may have an absolute relationship between the presence of an RFLP allele and the disease. This information will be very valuable in diagnosis, prognosis, and prophylaxis of the disease in question.

C. The biological role and reasons for the extensive polymorphism of Ig allotypes remain unknown. The striking qualitative and quantitative differences in the distribution of these determinants among different races raise questions concerning the nature of the evolutionary selective mechanism that maintains this variation.

   1. One mechanism could be the possible association of these markers with immunity to certain lethal infectious pathogens implicated in major epidemics, and different races may have been subjected to different epidemics throughout our evolutionary history. After a major epidemic, only individuals with allotypic combinations conferring immunity to the pathogen would survive.

2. Supporting this theory are some interesting observations. For example, GM and HLA genes have been shown to influence the chance for survival in typhoid and yellow fever epidemics in Surinam, and a recent study has presented convincing evidence that certain HLA antigens are associated with protection from malaria in West Africa.
3. The role of immunoglobulin allotypes in susceptibility/resistance to malaria and other infectious diseases is currently being investigated.

SELF-EVALUATION

Questions

Choose the ONE *best* answer.

7.1 If we estimate that chromosome 14 has 100 V genes, 5 J segments, and 50 D segments, and that chromosome 2 has 300 V genes and 5 J segments, the number of possible different specificities that can be generated by random combinatorial associations of these genes and random associations of heavy and κ light chains with different variable regions is:
   A. 4×10^3
   B. 4.7×10^4
   C. 1.88×10^6
   D. 3.75×10^6
   E. 3.75×10^7

7.2 Which of the following is believed to be the major consequence of somatic mutations during a humoral immune response?
   A. Changes in the GM allotype expression at the single cell level
   B. Emergence of antibodies of higher affinity for the antigen
   C. Rearrangement of silent chromosomes
   D. Reduction in the repertoire of antibodies reacting with the antigen
   E. Switch from IgM to IgG synthesis

7.3 Which one of the following immunoglobulin polypeptide chains carries the structures recognized by the anti-isotypic antibodies which define the five classes of immunoglobulins?
   A. J chain
   B. T chain
   C. H chain
   D. L chain
   E. Secretory component

7.4 A human myeloma protein (IgM k) is used to immunize a rabbit. The resulting antiserum is then absorbed with a large pool of IgM purified from normal human serum. Following this absorption, the antiserum is found to react only with the particular IgM myeloma protein used for immunization; it is now defined as an anti-idiotypic antiserum. With what specific portion(s) of the IgM myeloma protein would this antiserum react?
   A. Constant region of the μ chain
   B. Constant region of the κ chain
   C. Variable regions of μ and κ chains
   D. J chain
   E. None of the above

In **Question 7.5**, the following data are presented at a paternity suit:

	G1M	KM
Mother	3	1,3
Mr. X	3,17	1
Child (3 years old)	3	1

7.5 Which of the following conclusions concerning paternity is correct:
A. Mr. X is the child's father
B. Mr. X is *not* the child's father
C. No conclusion about paternity is possible
D. The child should be retested later in life
E. The child is genetically unable to produce G1M 17 and KM 3 chains

7.6 Which of the following is characteristic of true alleles among immunoglobulin allotypes?
A. Association with a unique antibody specificity
B. Association with amino acid substitutions in the polypeptide chain
C. Association with the same immunoglobulin isotype
D. Co-expression in a single polypeptide chain
E. Different distribution among individuals

7.7 Which of the following is a definition of haplotype?
A. The sum of different allospecificities detected in a given individual
B. Half of the allospecificities particular to a given individual
C. A cluster of allospecificities transmitted with one of the parental chromosomes
D. The order of genes coding for allotypic specificities in chromosome 6
E. The sum of different genes involved in coding for an immunoglobulin polypeptide chain

7.8 The GM phenotype of a given individual is GM 1,3,17 5. A single IgG1 molecule of this individual may express the following specificities:
A. GM 1,3,17 5
B. GM 5
C. GM 1,3,17
D. GM 1,17
E. GM 3,17

7.9 The generation of antibody diversity is best explained by:
A. Somatic mutation
B. Germ line diversity
C. Combinatorial association
D. A combination of germ line diversity, recombination events, and somatic mutations
E. Adaptation to the environment

7.10 The circulating immunoglobulin G of a GM3, 17 individual will:
A. Carry KM3 light chains
B. Contain a mixture of GM3, GM17, and GM3,17 molecules
C. Express both GM3 and GM17 in the same molecules
D. Express either GM3 or GM17
E. Include a mixture of GM3 molecules and GM17 molecules

Answers

7.1 (E) Assuming that there are 300 V segments and 5 J segments coding for the VL region, and 100 V segments, 50 D segments, and 6 J segments coding for the VH region, and considering that both VH and VL play a role in determining antibody specificity and that the association of VH and VL regions is a random error, the total number of antibody specificities generated from this random association of different regions and segments can be calculated as 3.75×10^7. However, there is evidence suggesting that the association is not entirely a random process, so that the diversity generated by this mechanism may be considerably less than the theoretical prediction.

7.2 (B) Somatic mutations add to antibody diversity during the immune response. They do not affect the constant region genes, and are not involved in isotype switching. A major consequence is the emergence of B-cell clones with antibodies of higher affinity, which react more strongly with the antigen and tend to be strongly stimulated. As a consequence of somatic mutations, there is a gradual increase in the diversity of antibodies and in the average affinity of the antibody population during an immune response.

7.3 (C) The antibodies to immunoglobulin isotypes recognized structural epitopes in the immunoglobulin heavy chains.

7.4 (C) The idiotypic determinants are closely associated to the antigen binding site, which is defined by the VH and VL regions.

7.5 (C) The child is apparently homozygous for G1M 3 and KM 1. Since both the mother and Mr. X carried the G1M 3 and KM 1 markers, the child could belong to Mr. X, but could also belong to any other male carrying the same markers. In actuality, paternity can be excluded with certainty but only proved to a probability of about 95% by examining the traditional genetic markers: immunoglobulin allotypes, blood groups, and HLA antigens. By employing the modern DNA "fingerprinting" methods, however, paternity can be proven with 100% likelihood.

7.6 (C) True alleles are mutually exclusive (i.e., they cannot coexist in the same polypeptide chain). Among immunoglobulin allotypes, true alleles correspond to amino acid substitutions in one or two positions of the constant regions of heavy or light chains.

7.7 (C) Each child receives two haplotypes: one maternal and one paternal. His genotype will be the sum of the two haplotypes.

7.8 (D) GM 5 is a G3M allotype (present on IgG3 molecules rather than in IgG1 molecules); GM 3 and 17 are alleles of a single locus and are not present on the same molecule (allelic exclusion). GM 1 and 17 correspond to amino acid substitutions at two different sites for IgG1 and can be expressed simultaneously at different regions of a single $\gamma 1$ chain.

7.9 (D)

7.10 (E) As a consequence of allelic exclusion, each plasma cell only expresses one set of the alleles in the genome, but both sets of alleles will be expressed by different cells. Thus, both alleles will be represented in the circulating IgG population, where the product of all IgG1-synthesizing plasma cells is represented.

BIBLIOGRAPHY

de Vries, R.R.P., Meera Kahn, P., Bernini, L.F., van Loghem, E., and van Rood, J.J. Genetic control of survival in epidemics. *J. Immunogenet.*, *6*:271, 1979.

Dugoujon, J.M., and Cambon-Thomsen, A. Immunoglobulin allotypes (GM and KM) and their interactions with HLA antigens in autoimmune diseases: A review. *Autoimmunity*, *22*:245, 1995.

Hill, A.V.S., Allsopp, C.E.M., Kwiatkowski, D., Anstey, N.M., Twumasi, P., Rowe, P.A., Bennett, S., Brewster, D., McMichael, A.J., and Greenwood, B.M. Common West African HLA antigens are associated with protection from severe malaria. *Nature*, *352*:595, 1991.

Okada, A., and Alt, F.W. The variable region gene assembly mechanism. In *Immunoglobulin Genes*, 2nd ed. (Honjo, T., and Alt, F.W., eds.). Academic Press, San Diego, 1995, p. 205.

Pandey, J.P. and French, M.A.H. GM phenotypes influence the concentrations of the four subclasses of IgG in normal human serum. *Hum. Immunol.*, *51*:99, 1996.

Pandey, J.P., ed. Immunogenetic Risk Assessment in Human Disease. *Exp. Clin. Immunogenet.*, *12*:119, 1995.

Pandey, J.P., Elson, L.H., Sutherland, S.E., Guderian, R.H., Araujo, E., and Nutman, T.B. Immunoglobulin κ chain allotypes (KM) in onchocerciasis. *J. Clin. Invest.*, *96*:2732, 1995.

Seppälä, I.J.T., Sarvas, H., and Mäkelä, O. Low concentrations of Gm allotypic subsets G_3m^g and G_1m^f in homozygotes and heterozygotes. *J. Immunol.*, *151*:2529, 1993.

Wagner, S.D. and Neuberger, M.S. Somatic hypermutation of immunoglobulin genes. *Annu. Rev. Immunol.*, *14*:441, 1996.

Zhang, J., Alt, F.W., and Honjo, T. Regulation of class switch recombination of the immunoglobulin heavy chain genes. In *Immunoglobulin Genes*, 2nd ed. (Honjo, T., and Alt, F.W., eds.). Academic Press, San Diego, 1995, p. 235.

Zouali, M. Unraveling antibody genes. *Nature Genet.*, *7*:118, 1994.

8

Antigen–Antibody Reactions

Gabriel Virella

I. GENERAL CHARACTERISTICS OF THE ANTIGEN–ANTIBODY REACTION

A. **Physicochemical Nature**. Antigens and antibodies bind through noncovalent bonds in a similar manner to that in which proteins bind to their cellular receptors or enzymes bind to their substrates. The binding is reversible and can be prevented or dissociated by high ionic strength or extreme pH. The following intermolecular forces are involved in antigen–antibody binding:

1. **Electrostatic bonds**. Electrostatic bonds result from the attraction between oppositely charged ionic groups of two protein side chains, for example, an ionized amino group (NH^{4+}) on a lysine in the antibody, and an ionized carboxyl group (COO^-) on an aspartate residue in the antigen.
2. **Hydrogen bonding**. When the antigen and antibody are in very close proximity, relatively weak hydrogen bonds can be formed between hydrophilic groups (e.g., OH and C=O, NH and C=O, and NH and OH groups).
3. **Hydrophobic interactions**. Hydrophobic groups, such as the side chains of valine, leucine, and phenylalanine, tend to associate due to Van der Waals bonding and coalesce in an aqueous environment, excluding water molecules from their surroundings. As a consequence, the distance between them decreases, enhancing the energies of attraction involved. This type of interaction is estimated to contribute up to 50% of the total strength of the antigen–antibody bond.
4. **Van der Waals bonds**. These forces depend upon interactions between the "electron clouds" that surround the antigen and antibody molecules. The interaction has been compared to that which might exist between alternating dipoles in two molecules, alternating in such a way that at any given moment oppositely oriented dipoles will be present in closely apposed areas of the antigen and antibody molecules.

All these types of interactions depend on the close proximity of the Ag and Ab molecules. For that reason, the "good fit" between an antigenic determinant and an antibody combining site determines the stability of the antigen–antibody reaction.

B. **Specificity**. Most of the data concerning this topic was generated in studies of the immune response to closely related haptens.

1. If an animal is innoculated with a conjugate of *m*-aminobenzene sulfonate

(haptenic group) with an immunogenic carrier protein, it will produce antibody that recognizes this simple chemical group.

2. When the antibody to *m*-aminobenzene sulfonate is tested for its ability to bind to the *ortho*, *meta*, and *para* isomers of aminobenzene sulfonate and to related molecules in which the sulfonate group is substituted by arsonate or carboxylate, it can be seen, as shown in Table 8.1, that:
   a. Best reactivity occurs with *m*-aminobenzene sulfonate.
   b. *o*-Aminobenzene sulfonate reacts reasonably.
   c. *m*-Aminobenzene arsonate and *m*-aminobenzene carboxylate react very poorly or not at all.
3. Specificity is mainly determined by the overall degree of complementarity between antigenic determinant and antibody binding site (Fig. 8.1); differences in the degree of complementarity determine the affinity of the antigen–antibody reaction.

C. **Affinity**. Antibody affinity can be defined as the attractive force between the complementary conformations of the antigenic determinant and the antibody combining site.

1. Experimentally, the reaction is best studied with antibodies directed against monovalent haptens. The reaction, as we know, is reversible, and can be defined by the following formula

$$\text{Ab} + \text{Hp} \underset{k_2}{\overset{k_1}{\leftrightharpoons}} \text{Ab·Hp} \qquad (1)$$

   where k_1 is the association constant and k_2 is the dissociation constant.
2. In simple terms, The k_1/k_2 ratio is the **intrinsic association constant** or **equilibrium constant** (K). This equilibrium constant represents the intrinsic affinity of the antibody binding sites for the hapten. High values for K will reflect a predominance of k_1 over k_2, or, in other words, a tendency for the antigen–antibody complex to be stable and not to dissociate.
3. The equilibrium constant (K) can be defined by

$$k_1\,[\text{Ab}]\,[\text{Hp}] = k_2\,[\text{AbHp}] \qquad (2)$$

$$K = \frac{k_1}{k_2} = \frac{[\text{Ab·Hp}]}{[\text{Ab}][\text{Hp}]} \qquad (3)$$

   where [Ab] corresponds to the concentration of free antibody binding sites, [Hp] to the concentration of free hapten, and [Ab·Hp] to the concentration of saturated antibody binding sites.
4. K, also designated as the **affinity constant**, is usually determined by **equilibrium dialysis** experiments in which antibody is enclosed in a semipermeable membrane, and dialyzed against a solution containing known amounts of free hapten.
   a. Free hapten diffuses across the membrane into the dialysis bag where it will bind to antibody. Part of the hapten inside the bag will be free, part will be bound, and the ratio of free and bound haptens depends on the antibody affinity.
   b. When equilibrium is reached, the amounts of free hapten will be identical inside and outside the bag. The total amount of hapten inside the bag

Table 8.1 Antigens Tested with Immune Serum for *Meta*-Aminobenzene Sulfonic Acid (Metanilic Acid)

	Antigens		
	ortho-	*meta-*	*para-*
Aminobenzene sulfonic acid	++	+++	+
Aminobenzene arsenic acid	0	+	0
Aminobenzoic acid	0	±	0

Reproduced with permission from Landsteiner, K. *The Specificity of Serological Reactions*. Dover, New York, 1962.

minus the concentration of free hapten outside the bag will be equal to the amount of bound hapten. If the molar concentration of antibody in the system is known, it becomes possible to determine the values of r (number of hapten molecules bound per antibody molecule) and c (concentration of free hapten).

c. Taking Eq. (3) as a starting point, if [Ab·Hp] is divided by the total concentration of antibody, the quotient equals the number of hapten molecules bound per antibody molecule [r], and the quotient between the number of vacant antibody sites [Ab] divided by the total concentration of antibody equals the difference between the maximum number of antibody molecules that can be bound by antibody molecule [n or valency], and the number of hapten molecules bound per Ab molecule [r] at a given hapten concentration [c]. Equation (3) can be rewritten as

$$K = \frac{r}{(n - r)c} \tag{4}$$

d. Equation (4), in turn, can be rewritten as the Scatchard equation:

$$\frac{r}{c} = Kn - Kr \tag{5}$$

e. By determining r and c concentrations in a series of experiments with dialysis membranes carried out at different total hapten concentrations, it becomes possible, using Eq. (5), to construct what is known as a **Scatchard plot**, in which r/c is plotted vs. r (Fig. 8.2). It is also possible to determine the slope of the plot of r/c vs. r values, which corresponds to $-K$. The correlation between the slope and the affinity constant is also illustrated in Figure 8.2. With high-affinity antibodies, r will reach saturation ($r = n$) at relatively low concentrations of hapten, and the plot will have a steep slope, as shown in the left. With low-affinity antibodies, the stable occupancy of the antibody binding sites will require higher concentrations of free hapten, so the r/c quotients will be considerably lower and the slope considerably less steep, as shown on the right. Since the reactants (antibodies and haptens) are expressed as moles·liter^{-1}, the affinity constant is expressed as liters·mole^{-1} (l·mol^{-1}).

f. From this plot it is obvious that, at extremely high concentrations of unbound hapten (c), r/c becomes close to 0, and the plot of r/c vs. r will

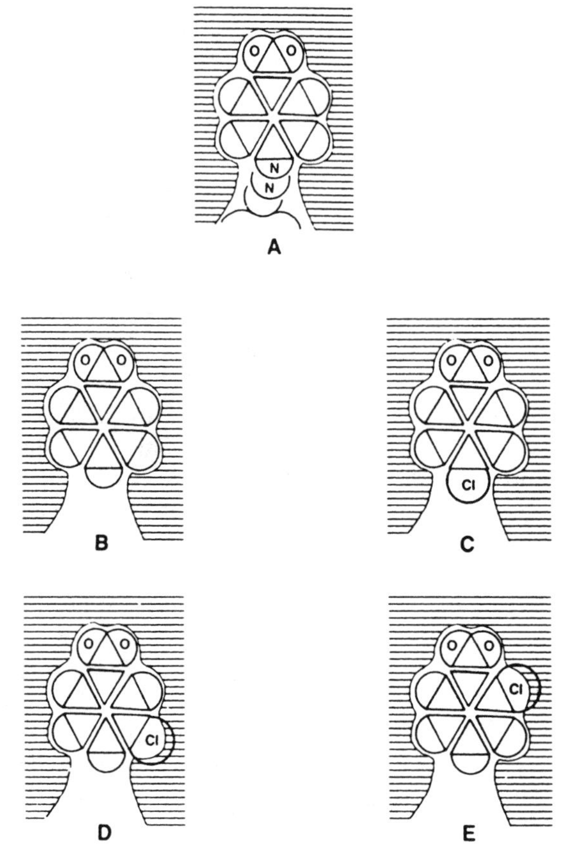

Figure 8.1 Diagrammatic representation of the "closeness of fit" between antigenic determinants and antibody binding sites. Antibodies were raised against the *p*-azobenzoate group of a protein-*p*-benzoate conjugate. The resulting anti-*p*-benzoate groups react well with the original protein-*p*-benzoate conjugate (A) and with *p*-benzoate itself (B). If a chlorine atom (Cl) is substituted for a hydrogen atom at the *p* position, the substituted hapten will react strongly with the original antibody (C). However, if chlorine atoms are substituted for hydrogen atoms at the *o* or *m* positions (D,E), the reaction with the antibody is disturbed since the chlorine atoms at those positions cause a significant change in the configuration of the benzoate group. Redrawn from Van Oss, C.J. In *Principles of Immunology*, 2nd ed. (Rose, N., Milgrom, F., and Van Oss, C., eds.). Macmillan, New York, 1979.

intercept r on the horizontal axis (the interception corresponds to n, the antibody *valency*). For an IgG antibody and all other monomeric antibodies, the value of n is 2; for IgM antibodies, the theoretical valency is 10, but the functional valency is usually 5, suggesting that steric hindrance effects prevent simultaneous occupation of the two binding sites of each subunit.

g. In most experimental conditions, an antiserum raised against one given hapten is composed of a restricted number of antibody populations with slightly different affinity constants. Under those conditions, it may be of practical value to calculate an **average intrinsic association constant** or

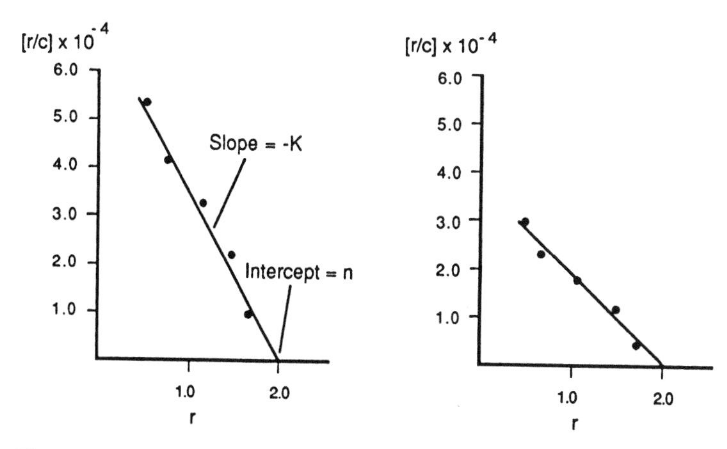

Figure 8.2 Schematic representation of the Scatchard plots correlating the quotient between moles of hapten bound per moles of antibody (r) and the concentration of free hapten (c) with the concentration of hapten bound per mole of antibody (r). The slopes of the plots correspond to the affinity constants, and the intercept with the horizontal axis correspond to the number of hapten molecules bound per mole of antibody at a theoretically infinite hapten concentration (n or valency of the antibody molecule). The plot on the left panel corresponds to a high-affinity antibody, and the slope is very steep; the plot on the right panel corresponds to a low-affinity antibody, and its slope is considerably less steep.

average affinity (K_0) that is defined as the free hapten concentration required to occupy half of the available antibody binding sites ($r = n/2$). Substitution of $r = n/2$ for r in Eq. (4) leads to the formula $K_0 = 1/c$. In other words, the average affinity constant equals the reciprocal of the free antigen concentration when half the antibody sites are occupied by antigens.

h. High-affinity antibodies have K_0 values as high as 10^{10} l·mol^{-1}. High-affinity binding is believed to result from a very close fit between the antigen binding sites and the corresponding antigenic determinants which favor the establishment of strong noncovalent interactions between antigen and antibody.

C. **Avidity**. Antibody avidity can be defined as the strength of the binding of the many different antibodies that are produced in response to an immunogen, which presents several different epitopes to the immune system. The strength of the Ag·Ab reaction is enhanced when several different antibodies bind simultaneously to different epitopes on the antigen molecule, cross-linking antigen molecules very tightly. Thus, a more stable bonding between antigen and antibody will be established, due to the "bonus effect" of multiple antigen–antibody bonds (Fig. 8.3); the increased stability of the overall antigen–antibody reaction corresponds to an increased avidity.

D. **Cross-Reactions**. When an animal is immunized with an immunogen, its serum will contain several different antibodies directed to the various epitopes presented by the immunizing molecule, reflecting the polyclonal nature of the response. Such serum from an immune animal is known as an *antiserum* directed against the immunogen.

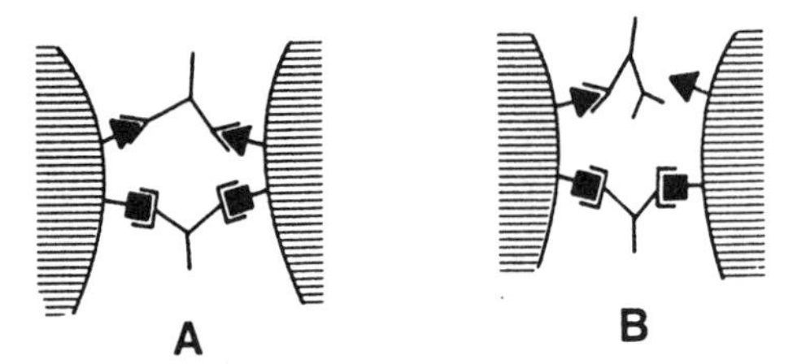

Figure 8.3 Diagrammatic representation of the avidity concept. The binding of antigen molecules by several antibodies of different specificities (A) stabilizes the immune complex, since it is highly unlikely that all Ag·Ab reactions dissociate simultaneously at any given point of time (B). Redrawn from Roitt, I. *Essential Immunology*, 4th ed. Blackwell Scientific Publications, Oxford, 1980.

1. Antisera containing polyclonal antibodies can often be found to *cross-react* with immunogens partially related to that used for immunization, due to the existence of common epitopes or of epitopes with similar configurations.
2. Less frequently, a cross-reaction may be totally unexpected, involving totally unrelated antigens that happen to present epitopes whose whole spatial configuration may be similar enough to allow the cross-reaction.
3. The avidity of a cross-reaction depends on the degree of structural similarity between the shared epitopes; when the avidity reaches a very low point, the cross-reaction will no longer be detectable (Fig. 8.4).
4. The differential avidity of given antiserum for the original immunogen and for other immunogens sharing epitopes of similar structure is responsible for the **specificity** of the antiserum (i.e., its ability to recognize only one single immunogen or a few, very closely related immunogens).

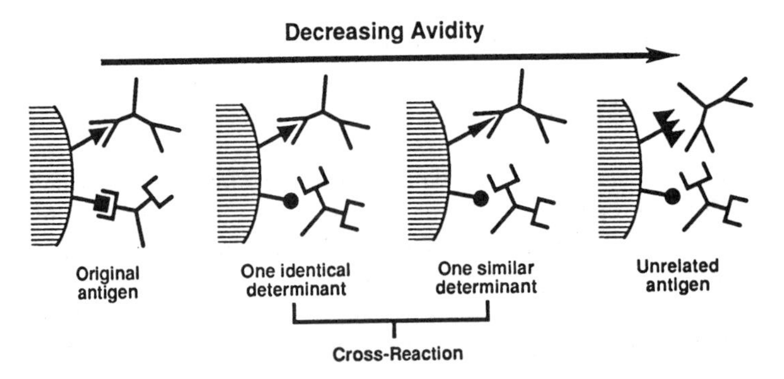

Figure 8.4 Diagrammatic representation of the concept of cross-reaction between complex antigens. An antiserum containing several antibody populations to the determinants of a given antigen will react with other antigens sharing common or closely related determinants. The avidity of the reaction will decrease with decreasing structural closeness, until it will no longer be detectable. The reactivity of the same antiserum with several related antigens is designated as a cross-reaction. Redrawn from Roitt, I. *Essential Immunology*, 4th ed., Blackwell Scientific Publications, Oxford, 1980.

II. SPECIFIC TYPES OF ANTIGEN–ANTIBODY REACTIONS

A. **Precipitation.** When antigen and antibody are mixed in a test tube, one of two things may happen. Both components will remain soluble, or variable amounts of Ag·Ab precipitate will be formed.

1. The **precipitin curve.** If progressively increasing amounts of antigen are mixed with a fixed amount of antibody, a **precipitin curve** can be constructed (Fig. 8.5). There are three areas to consider in a precipitin curve.
   a. **Antibody excess**—Free antibody remains in solution after centrifugation of Ag·Ab complexes.
   b. **Equivalence**—No free antigen or antibody remains in solution. The amount of precipitated Ag·Ab complexes reaches its peak at this point.
   c. **Antigen excess**—Free antigen is detected in the supernatant after centrifugation of Ag·Ab complexes.
2. The **lattice theory** explains why different amounts of precipitation are observed at different antigen–antibody ratios.

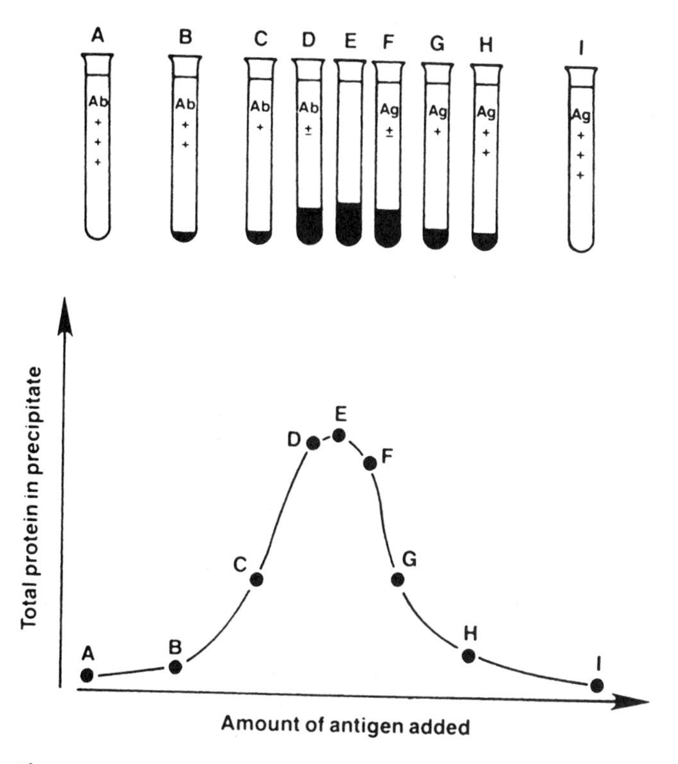

Figure 8.5 The precipitin curve. When increasing amounts of antigen are added to a fixed concentration of antibody, increasing amounts of precipitate appear as a consequence of the antigen–antibody interaction and, after a maximum precipitation is reached, the amounts of precipitate begin to decrease. Analysis of the supernatants reveals that at low antigen concentrations there is free antibody left in solution (**antibody excess**), at the point of maximal precipitation, neither antigen nor antibody are detected in the supernatant (**equivalence zone**); with greater antigen concentrations, antigen becomes detectable in the supernatant (**antigen excess**).

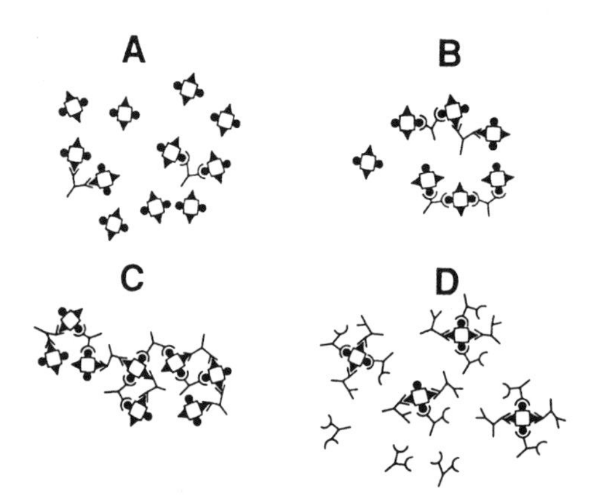

Figure 8.6 The lattice theory explaining precipitation reactions in fluid media: At great antigen excess (A), each antibody molecule has all its binding sites occupied. There is free antigen in solution, and the antigen–antibody complexes are very small ($Ag_2 \cdot Ab_1$, $Ag_1 \cdot Ab_1$). The number of epitopes bound per antibody molecule at great antigen excess corresponds to the antibody valency. With increasing amounts of antibody (B), larger Ag·Ab complexes are formed ($Ag_3 \cdot Ab_2$, etc.), but there is still incomplete precipitation and free antigen in solution. At equivalence, large Ag·Ab complexes are formed, in which virtually all Ab and Ag molecules in the system are cross-linked (C). Precipitation is maximal, and no free antigen or antibody are left in the supernatant. With increasing amounts of antibody (D), all antigen binding sites are saturated, but there is free antibody left without binding sites available for it to react. The Ag·Ab complexes are larger than at antigen excess [$Ag_1 \cdot Ab_{4,5,6(n)}$], but usually soluble. The number of antibody molecules bound per antigen molecule at great antibody excess allows an estimate of the antigen valency.

a. The equivalence point is characterized by maximum cross-linking between Ag and Ab (Fig. 8.6).
b. At great antibody excess, each antigen will tend to have its binding sites saturated, with antibody molecules bound to all its exposed determinants. *The number of antibody molecules bound to one single antigen molecule gives a rough indication of the valency of the antigen.*
c. At great antigen excess, the binding sites of the antibody molecule will be saturated by different antigen molecules and not much cross-linking will take place.

3. **Precipitation in agar**. Semisolid supports, such as agar gel, in which a carbohydrate matrix functions as a container for buffer that fills the interstitial spaces left by the matrix, have been widely used for the study of antigen–antibody reactions. Antigen and antibody are placed in wells carved in the semisolid agar, and allowed to passively diffuse. The diffusion of antigen and antibody is unrestricted, and in the area that separates antigen from antibody, the two reactants will mix in a gradient of concentrations. When the optimal proportions for Ag·Ab binding are reached, a precipitate will be formed, appearing as a sharp, linear opacity (Fig. 8.7).

B. **Agglutination**. When bacteria, cells, or large particles in suspension are mixed with antibodies directed to their surface determinants, one will observe the formation of large clumps; this is known as an **agglutination reaction**.

1. Agglutination reactions result from the cross-linking of cells and insoluble particles by specific antibodies. Due to the relatively short distance between

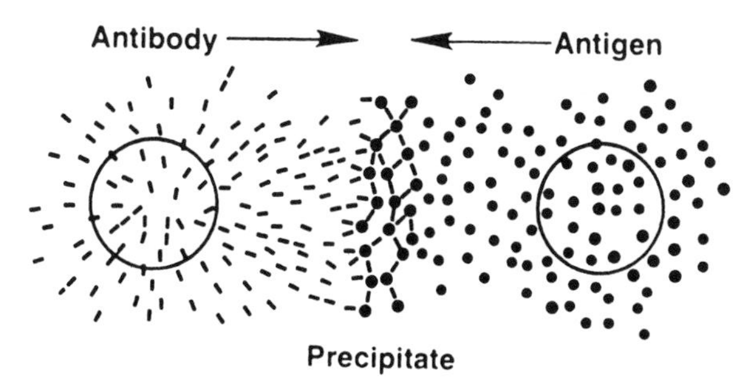

Figure 8.7 Diagrammatic representation of a reaction of double immunodiffusion. Antigen and antibody are placed in opposite wells carved in a semisolid medium (e.g., agarose gel). Both antigen and antibody diffuse in all directions and toward each other, reacting and eventually reaching equivalence, at which point a linear precipitate appears between the antigen and antibody wells.

the two Fab fragments, 7S antibodies (such as IgG) are usually unable to bridge the gap between two cells, each of them surrounded by an electronic "cloud" of identical charge that will tend to keep them apart. IgM antibodies, on the other hand, are considerably more efficient in inducing cellular agglutination (Fig. 8.8).

2. The visualization of agglutination reactions differs according to the technique used for their study. In slide tests, the nonagglutinated cell or particulate antigen appears as a homogeneous suspension, while the agglutinated antigen will appear irregularly clumped. If antibodies and cells are mixed in a test tube, the cross-linking of cells and antibodies will result in the diffuse deposition of cell clumps in the bottom and walls of the test tube, while the nonagglutinated red cells will sediment in a very regular fashion, forming a compact red button on the bottom of the tube.
3. Agglutination reactions follow the same basic rules of the precipitation reaction.
   a. When cells and antibody are mixed at very high antibody concentrations (low dilutions of antisera), antibody excess may result, no significant cross-linking of the cells is seen, and, therefore, the agglutination reaction may appear to be negative. Those dilutions at which antibody excess prevents agglutination constitutes the **prozone**.
   b. With increasing antibody dilutions, more favorable rations for cross-linking are reached, and very fine clumps cover the walls of the test tube or microtitration wells.
   c. When equivalence is approached, larger clumps of cells can be distinguished.
   d. At still higher dilutions, when the concentration of antibody is very low, the zone of antigen excess is reached, and agglutination is no longer seen (Fig. 8.9).

III. BIOLOGICAL CONSEQUENCES OF THE ANTIGEN–ANTIBODY REACTION

A. **Opsonization**. After binding to particulate antigens or after forming large molecular aggregates, antibodies unfold and may interact with Fc receptors on

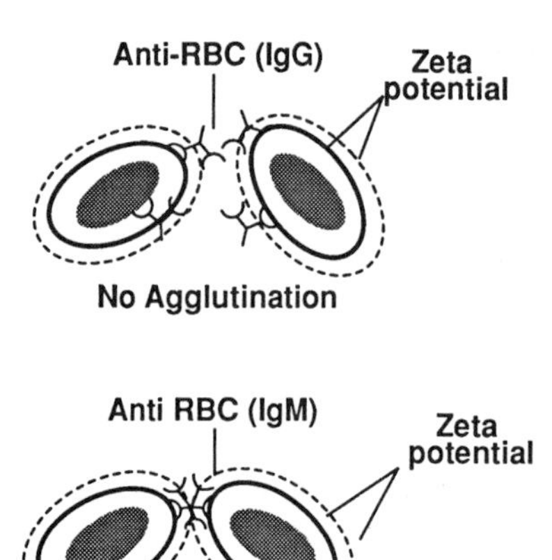

Figure 8.8 IgM antibodies are more efficient in inducing red cell agglutination. Red cells remain at the same distance from each other due to their identical electrical charge (zeta potential). IgG antibodies are not large enough to bridge the space between two red cells, but IgM antibodies, due to their polymeric nature and size, can induce red blood cell agglutination with considerable ease.

phagocytic cells (see Chap. 6). Such interaction is followed by ingestion by the phagocytic cell (phagocytosis). Substances that promote phagocytosis are known as opsonins.

B. **Fc-Receptor-Mediated Cell Activation**. The interaction of antigen–antibody complexes with phagocytic cells through their Fc receptors results in the delivery of activating signals to the ingesting cell.

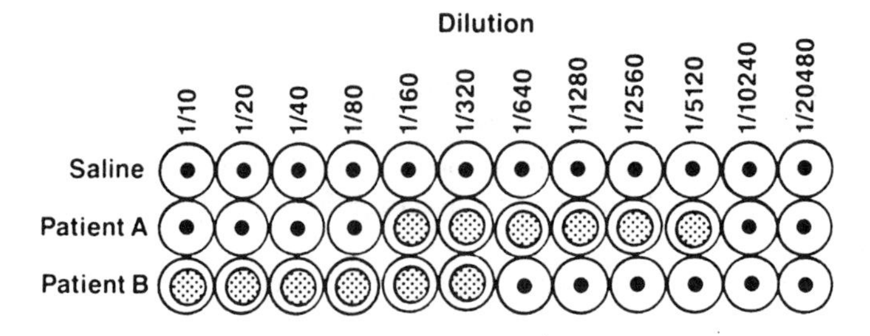

Figure 8.9 Diagrammatic representation of an hemagglutination reaction performed in a microtiter plate. The objective of the study is to determine the existence and titer of hemagglutinating antibodies in three different samples. In the first step, each sample is sequentially diluted from 1/10 to 1/20480 in a separate row of wells (A,B,C). In a second step, a fixed amount of red cells is added to each serum dilution with saline, a negative control. A, no agglutination can be seen. With patient A, the first three dilutions do not show agglutination (prozone), but the next dilutions, up to 1/5120 are positive; this sample is positive, and the tier is 5120. With patient B, the agglutination is positive until the 1/320 dilution; the titer of the sample is 320.

1. When the Fc-receptor-bearing cell is a phagocyte, the activation is usually associated with enhancement of its microbicidal activity.
2. Some of the toxic mediators generated in the phagocytic cell spill during phagocytosis. Spillage is maximal if the antigen–antibody complex is immobilized along a basement membrane or a cellular surface. The spillage of enzymes and oxygen active radicals can trigger an inflammatory reaction and cause tissue damage (see Chaps. 17 and 25).
3. The interaction of an antigen with IgE immobilized on a Fc receptor of a basophil or mast cell activates the release of potent mediators and triggers an allergic reaction (see Chap. 23).

C. **Complement Activation**. One of the most important consequences of antigen–antibody interactions is the activation (or "fixation) of the complement system (see Chap. 9).

1. The activation sequence induced by antigen–antibody reactions is known as the "classical" pathway. This pathway is initiated by the binding of Clq to the CH_2 domain of the Fc region of IgG and equivalent regions of IgM.
2. The complement binding sequences in IgG and IgM are usually not exposed in free antibody molecules. The antigen–antibody interaction causes configurational changes in the antibody molecule and the complement binding regions become exposed.
3. The activation of Clq requires simultaneous interaction with two complement binding immunoglobulin domains. This means that when IgG antibodies are involved, relatively large concentrations are required, so that antibody molecules coat the antigen in very close apposition allowing Clq to be fixed by IgG duplets. On the other hand, IgM molecules, by containing five closely spaced monomeric subunits, can fix complement at much lower concentrations. One IgM molecule bound by two subunits to a given antigen will constitute a complement binding duplet.
4. After binding of Clq, a cascade reaction takes place, resulting in the successive activation of eight additional complement components.
   a. Some of the components generated during complement activation are recognized by receptors on phagocytic cells and **promote phagocytosis**. C3b is the complement fragment with greater opsonizing capacity. An antigen coated with opsonizing antibodies and C3b is taken up with maximal efficiency by phagoctic cells (see Chaps. 9 and 13).
   b. The terminal complement components bind to cell membranes where they polymerize, forming transmembrane channels and eventually inducing **cell lysis**. These reactions have great biological significance and have been adapted to a variety of serological tests for diagnosis of infectious diseases, as will be discussed in Chapter 14.
   c. The activation of the complement system may have adverse effects, if it results in the destruction of host cells or if it promotes **inflammation**, which is beneficial with regard to the elimination of infectious organisms, but always has the potential for causing tissue damage.

D. **Neutralization**. The binding of antibodies to bacteria, toxins, and viruses has protective effects because it prevents the interaction of the microbial agents or their products with the receptors that mediate their ineffectiveness or toxic effects. As a consequence, the infectious agent or the toxin become harmless, or, in other words, are neutralized.

SELF-EVALUATION

Questions

Choose the ONE *best* answer.

8.1 The following precipitation curve is prepared by adding variable amounts of tetanus toxoid to a series of tubes containing 0.7 mg of antibody each.

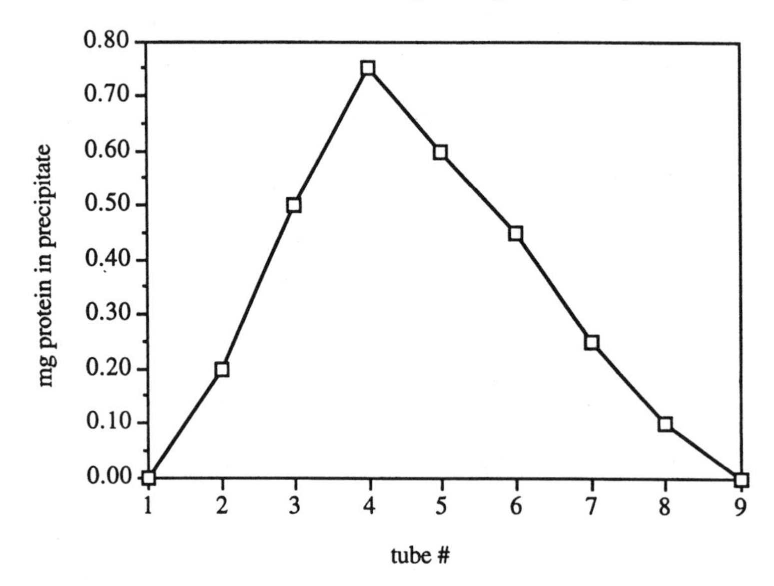

Precipitation observed in supernatant after addition of	Supernatant from tube # 1	2	3	4	5	6	7	8	9
Antibody	−	−	−	−	+	+	+	+	+
Antigen	+	+	±	−	−	−	−	−	−

(+ means visible precipitation; − means no visible precipitation)

What tube(s) correspond to the antigen excess zone:

A. 1 to 3
B. 1 to 4
C. 4
D. 4 to 9
E. 5 to 9

8.2 Using the data shown in Question 8.1, what was the concentration of antigen added to the tube corresponding to the equivalence point?

A. 0.005 mg
B. 0.01 mg
C. 0.05 mg
D. 0.1 mg
E. 0.5 mg

8.3 The affinity of an antigen–antibody reaction depends primarily on the:
A. Activation of the complement system
B. Antibody isotype
C. Closeness of fit between antibody-binding site and antigen epitope
D. Nature of the antigen
E. Valency of the antibody

8.4 Dr. I.M. Smart immunized a rabbit with sheep red cells and managed to separate anti-sheep red cell antibodies of several different isotypes. He then proceeded to mix 1 ml of saline containing 5×10^9 red cells and guinea pig complement with equimolecular concentrations of antibodies to sheep red cells of each different isotype. After incubating the mixture of red cells and anti-red cell antibodies for 5 minutes, he measured the amount of free hemoglobin in each tube. Which antibody was he likely to have added to the tube in which the concentration of free hemoglobin was highest?
A. IgA
B. IgD
C. IgG1
D. IgG4
E. IgM

8.5 Dr. Smart then proceeded to perform another experiment in which he incubated 1 ml of saline containing 5×10^9 red cells with equimolecular concentrations of antibodies to sheep red cells of each different isotype, in the complete absence of complement. After incubating the mixture of red cells and anti-red cell antibodies for 1 hour, he washed the red cells, added them to human monocytes, incubated for another hour, and then examined the monocytes microscopically to determine whether they had ingested the sheep red cells. Which antibody was most likely to have been added to the red cells that were more efficiently ingested?
A. IgA
B. IgD
C. IgE
D. IgG1
E. IgM

8.6 The protective effect of preformed antiviral antibodies is related to their ability to:
A. Agglutinate circulating viral particles
B. Form soluble immune complexes with viral antigens
C. Induce phagocytosis of the virus
D. Lyse the virus
E. Prevent the virus from infecting its target cell(s)

8.7 Which of the following hypotheses would sufficiently explain the nonprecipitation of an antigen–antibody system?
A. The antigen has only two determinants
B. The antigen has multiple, closely repeated determinants
C. The antibody has been cleaved with papain
D. The antibody has been cleaved with pepsin
E. Both C and D are correct

For **Questions 8.8–8.10** find the ONE lettered sentence most closely related to it. An animal is immunized with DNP-ovalbumin and 2 weeks later the animal is bled and the serum, containing antibodies to DNP-ovalbumin, is mixed with the hapten and two different hapten-carrier conjugates. What do you expect to observe with each mixture?

A. Moderate precipitation (+)
B. Heavy precipitation (+++)
C. No precipitation

8.8 Anti-DPN-ovalbumin + DNP
8.9 Anti-DNP-ovalbumin + DNP-ovalbumin
8.10 Anti DNP-ovalbumin + DNP-gammaglobulin

Answers

8.1 (E) If the tubes contain an excess of antigen, the supernatants will contain free antigen and visible precipitation will result from the addition of antibody to those supernatants.

8.2 (C) The amount of antibody in every tube was 0.7 mg and the amount of protein in the equivalence point (defined as the tube with maximal precipitation, whose supernatant contains neither free antigen nor free antibody) was 0.75 mg. Therefore, the difference between antibody concentration and total protein concentration was the antigen concentration (i.e., 0.05 mg).

8.3 (C)

8.4 (E) Molecule by molecule, IgM antibodies are more efficient complement fixators than IgG antibodies and would cause more red cells to lyse.

8.5 (D) IgG1 and IgG3 antibodies are the most efficient in what concerns inducing Fc-receptor-mediated phagocytosis.

8.6 (E) Preformed antiviral antibodies may have a neutralizing effect if they block the infection of the virus target cell(s). Phagocytosis of virus–antibody complexes often results in infection of the phagocytic cells.

8.7 (C) Papain digestion breaks the IgG molecule (which can cross-link multivalent antigen molecules to produce precipitation reactions) into one Fc and two Fab fragments. The Fab fragments have one single binding site and cannot cross-link antigens and cause precipitation. The $F(ab')_2$ fragment obtained with pepsin, on the contrary, is bivalent, able to cross-link multivalent antigens, forming large lattices and precipitants.

8.8 (C) The soluble DNP hapten, although able to bind to bivalent anti-DNP, will do so without forming a precipitate because DNP has one single binding site in its soluble form and cannot bind to more than one antibody molecule to cross-link and form a precipitate.

8.9 (B) The papain Fab fragments are monovalent, cannot cross-link, and precipitate antigen molecules; in contrast, the $F(ab')_2$ fragments obtained with pepsin are bivalent, can cross-link, and precipitate antigen molecules.

8.10 (A) It is expected that some precipitation will occur if anti-DNP is mixed with a different DNP-carrier combination than the one used for immunization, since each carrier molecule will express several DNP groups and cross-linking of the hapten-carrier conjugate can occur through anti-DNP antibodies. However, the amount of precipitate observed should be smaller than when anti-DNP-OVA is mixed with DNP-OVA, in which case both anti-DNP and anti-OVA antibodies will participate in precipitate formation.

BIBLIOGRAPHY

Atassi, M.-Z., van Oss, C.J., and Absolom, D.R. *Molecular Immunology*. Marcel Dekker, New York, 1984.

Day, E.D. *Advanced Immunochemistry*, 2nd ed. Wiley-Liss, New York, 1990.

Eisen, H.N. *Antibody-Antigen Reactions*. In *Microbiology* (Davis, B.D., Dulbecco, R., Eisen, H.N., and Ginsberg, H.S., eds.), Lippincott, Philadelphia, 1990.

Glynn, L.E., and Steward, M.W. (eds.). *Immunochemistry: An Advanced Textbook*. John Wiley & Sons, New York, 1977.

Landsteiner, K. *The Specificity of Serological Reactions*. Dover Publications, New York, 1962.

Steward, M.W. Introduction to methods used to study antibody-antigen reactions. In *Handbook of Experimental Immunology*, Vol. 1, 3rd ed. (Weir, D.M., ed.). Blackwell Scientific Publications, Oxford, 1978, p. 16.1.

Van Oss, C.J., and van Regenmortel, M.H.V. *Immunochemistry*. Marcel Dekker, New York, 1994.

9

The Complement System

Robert J. Boackle

I. INTRODUCTION

A. **Antigen–Antibody Reactions**. The interaction of antibody molecules with antigen is followed by conformational changes of the antibody, including changes in the spatial orientation and exposure of biologically active domains or segments located on the Fc region of the antibodies. The Fc regions of antigen-bound IgG or IgM are able to bind and activate the first component of a series of extremely powerful and rapidly acting plasma glycoproteins, known as the complement system.

B. **Complement System**. This system includes several proenzymes and components that exist in an inactive state in the plasma. When these complement components are activated, a sequential, rapid cascading pattern ensues.

1. **Synthesis and metabolism of complement components**. Complement glycoproteins are synthesized by liver cells, but macrophages and many other cell types are also a minor source of various complement components. All normal individuals always have complement components in their blood. The synthetic rates for the various complement glycoproteins increase when complement is activated and consumed.
2. **Activation of the complement system**. This system can be activated by:
   a. Antigen–antibody complexes containing IgG or IgM activate complement by the **classical pathway** that starts at C1.
   b. Membranes and cell walls of microbial organisms and many other substances activate complement by the **alternative pathway**.
   c. Proteolytic enzymes released either from microbes or from host cells can also activate the complement system by breaking down critical components.
3. **The complement cascade**. As a complement component is activated, it is either cleaved or becomes bound to a previously activated component or complex of complement components. Also, each component or complex of components, once activated, generally amplifies the cascading process by activating many molecules of the next component in the series.
4. **Function of the complement system**. The **primary function** of the complement system is to bind and neutralize foreign substances that activate it and to

promote the ingestion of those complement-coated substances by phagocytic cells.

II. THE CLASSICAL COMPLEMENT PATHWAY

Immunoglobulins and native complement components circulate in the blood and in the lymph, but these molecules do not interact with each other until antibody molecules interact with the corresponding antigen and undergo the necessary conformational changes. These conformational changes are the basis for specific activation of the **classical complement pathway**.

A. The Initiation of the Cascade: C1 Activation

1. Native, free IgG or IgM do not activate the complement system. A single, native IgG molecule will not bind and activate the **first component** in the complement pathway (**C1**). However, if antibodies of the IgG class are aggregated by antigen binding, this will result in C1 fixation and activation.
2. The **IgG subclasses** vary with regard to their efficiency in activating C1; IgG3 immunoglobulins are the most efficient, followed by IgG1 and IgG2. IgG4 immunoglobulins do not effectively activate the classical pathway.
3. For C1 to be activated, it must bind to at least two adjacent Fc regions (Fig. 9.1). This means that the concentration of antibody of the IgG class must be relatively high and that the specific antigenic determinants recognized by the IgG antibody must be in close proximity. For IgM, a pentamer, these logistic problems are of less importance.
4. Detailed chemical studies have revealed that C1 actually is a complex of three different types of molecules (C1q, C1r, and C1s) held loosely together through noncovalent bonds. In plasma, the components of C1 exist in conformations that partially limit their degree of association. Normal physiological levels of calcium and normal ionic strength are essential to stabilize the proper associations within the C1 macromolecular complex and prevent spontaneous activation.

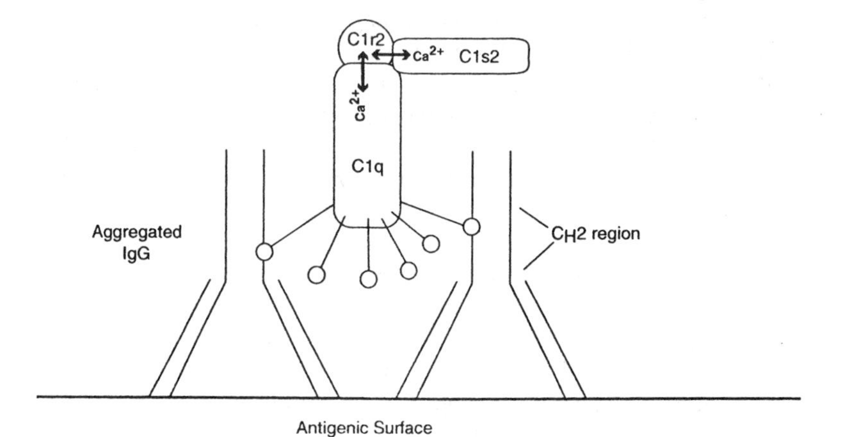

Figure 9.1 C1 activation by an antibody duplet formed by binding to the corresponding antigen.

5. One part of the C1 complex is **C1q**, which contains a collagen-like stem and an umbrella-shaped cluster of globular regions. In addition, C1 contains two **C1r** molecules and two **C1s** molecules.
6. The exposed aggregated Fc regions of the IgG molecule bind **C1q**. Upon contact with immune complexes, the C1q-C1r$_2$-C1s$_2$ components, which are loosely associated in plasma, become more tightly associated. As a consequence, C1q undergoes a conformational change that facilitates the cleavage and activation of the two C1r proenzymes by one another to form activated C1r$_2$ (Fig. 9.1).

B. The Early Stages: C1 to C3

1. Each activated C1r has protease activity and cleaves a peptide bond within the adjacent C1s molecule which in turn becomes activated. Activated C1s proteases are then able to cleave and activate the next component in the series, C4.
2. Native **C4** molecules are cleaved by activated C1s into a small fragment, which remains soluble (**C4a**), and a larger fragment, **C4b**.
3. Each C1s component is able to cleave and activate many C4 molecules. Upon activation, a short-lived and very reactive binding site appears on the C4b fragment. This active site allows C4b to bind **covalently** to the antigenic surface (e.g., a bacterial outer envelope) around the areas where the initial antigen-bound antibody-C1 complexes formed (Fig. 9.2).
4. Any activated C4b molecules that do not reach the antigenic surface and bind covalently within a few nanoseconds after activation lose their short-lived binding site. These fluid-phase C4b molecules bind to a circulating C4 binding protein (C4bp) and will become functionally inactivated as a consequence of digestion by serum proteases. The rapid loss of activity by the fluid phase C4b molecules is a very important control mechanism that protects the nearby membranes of the host's cells from "bystander" attack by C4b.

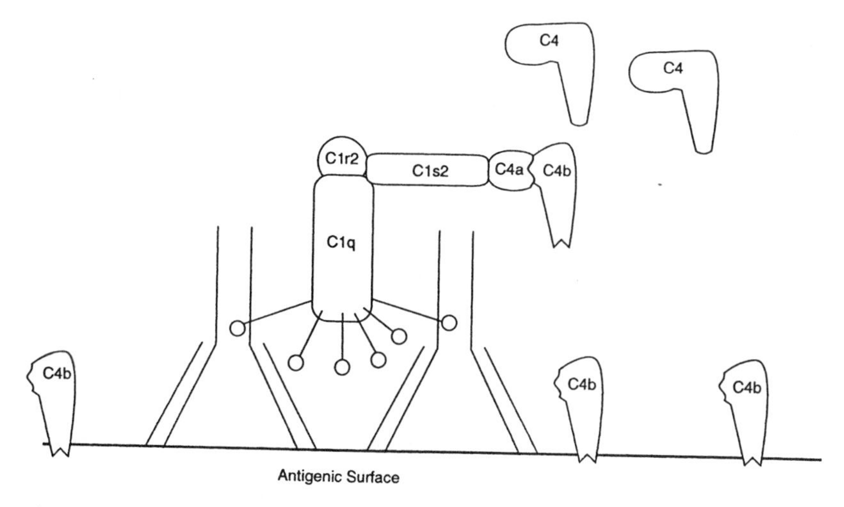

Figure 9.2 Binding of C4b to the cell membrane.

5. The activated Cls within the bound Cl macromolecular complex is also responsible for the activation of C2, the next component to be activated in the classical pathway. In the presence of magnesium (Mg^{2+}), C2 interacts with antigen-bound C4b and is, in turn, split by Cls into two fragments, termed C2b and C2a.
6. C2b fragments are released into the fluid phase, and **C2a** binds to C4b (Fig. 9.3). Thus, proper concentrations of $\mathbf{Ca^{2+}}$ ions are needed for optimal Clq-Clr-Cls interactions, and $\mathbf{Mg^{2+}}$ ions are required for proper C4b-C2a interactions. In the absence of Ca^{2+} and/or Mg^{2+} (due to the addition of metal chelators), the classical activation pathway is interrupted.
7. C2a contains the active enzyme site for the activation of **C3** and **C5** in the complement sequence. For C3 and C5 to be activated, C2a must remain as a stable complex with the activated membrane-bound C4b molecule. The active C4b2a complex is also known as **C3 convertase** since it causes the selective fragmentation and activation of the next component in the series, C3. Thus after rapid deposition of hundreds of active C4b2a complexes on the antigenic surface surrounding the immune complex, each C4b2a complex is capable of rapidly activating thousands of C3 molecules (Fig. 9.4).

C. **Regulatory Mechanisms of the Early Stages**. The activities of the different complement components activated at each stage of the sequence are regulated by several mechanisms. The spontaneous decay of C2a activity with time is such a mechanism. However, this particular regulatory event, while important, requires several minutes. Therefore, its role in restricting the range of complement effects *in vivo* may not be as critical as many of the other more rapid specific inhibitory reactions regulating the fast-acting complement system, such as the short-lived

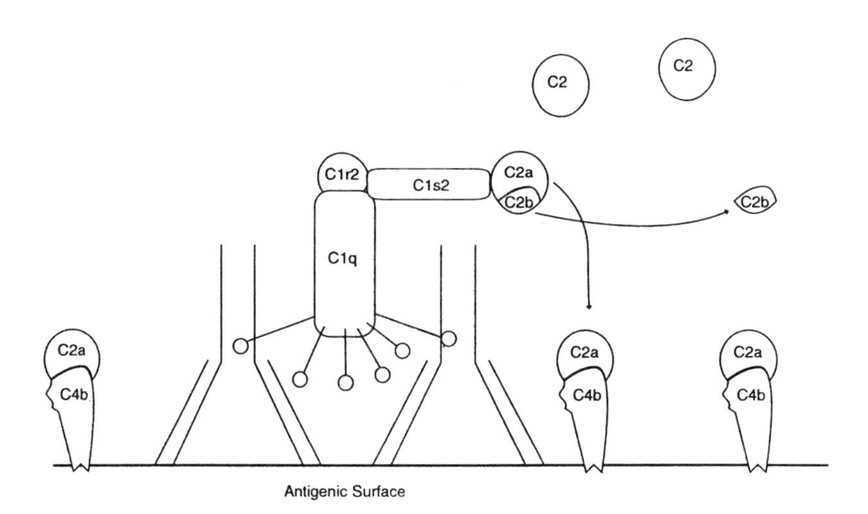

Figure 9.3 Formation of C4b2a complexes on cell membranes. Although activated Cls (within the bound Cl macromolecular complex) can cleave any C2 molecule that it encounters, the cleavage of a C2 molecule adjacent to membrane-bound C4b increases the probability of generating bound C4b2a complexes on the cell membrane.

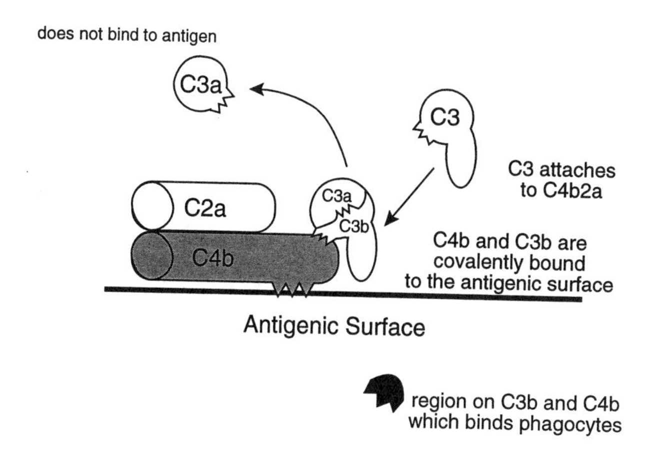

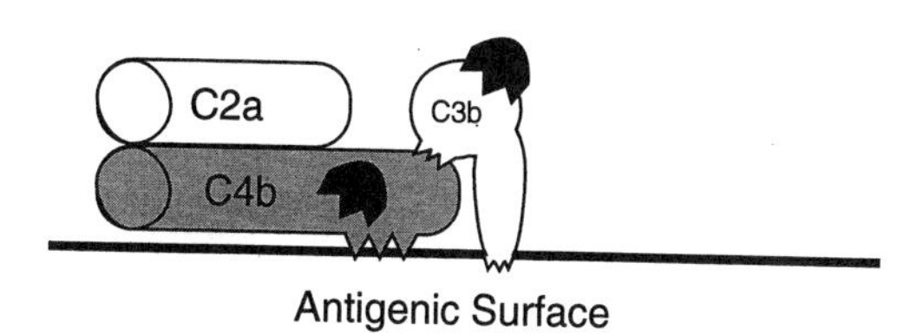

Figure 9.4 C3 activation by C4b2a.

active binding sites on activated complement fragments and the effects of the normally occurring serum complement inhibitors.

1. **Serum complement inhibitors** are especially effective in inhibiting activated complement components that fail to bind to the antigen and therefore enter the circulation in the free state. Normally, these serum inhibitors restrict the complement cascade to the surface of the foreign material, prevent bystander damage to the host, and limit unnecessary consumption of complement components.
2. The **C1 inhibitor** plays a critical role in limiting unnecessary complement activation.
   a. As complement proteins (i.e., C4b and C3b) deposit on the antigen, the antibody tends to dissociate from the antigen surface. This partial **dissolution of the immune complex** results in the partial separation or loosening of the C1 macromolecule from the immune complex. C1q-$C1r_2$-$C1s_2$ complexes separate from the immune complex, and the loosely associated $C1r_2$ and $C1s_2$ enzymes remain in their active form. The released C1q-$C1r_2$-$C1s_2$ complex reacts with C1 inhibitor and the $C1r_2$ and $C1s_2$ enzymes are irreversibly inhibited.
   b. C1 inhibitor may also aid in the removal of the entire C1 complex from the antigen–antibody complex.
   c. Thus, C1, once having performed its function while tightly bound to the

immune complex by cleaving substantial amounts of C4 and C2 (and forming many surface-bound C3 convertases, C4b2a), is quickly and irreversibly inhibited from activating any more C4 and C2. This C1 inhibitor function represents a very important regulatory mechanism that restricts the range of activated C1 action and prevents useless consumption of C4 and C2.

d. The ability of an immunoglobulin molecule to effectively activate C1 not only is a function of the binding of Fc to C1q globular heads, but also of the ability of that immunoglobulin molecule to protect the C1s from C1 inhibitor and allow the C1s to activate C4 and C2. Several host substances (such as serum mannan binding protein; MBP), upon binding to microbes, bind to C1q globular heads and activate C1. However, many of these substances are not effective C1 activators because they lack the ability to prevent the inactivation of C1s by C1 inhibitor.

e. The physiological importance of the C1 inhibitor is reflected by the existence of **C1 inhibitor deficiencies** clinically associated with a condition known as **angioneurotic edema**, characterized by unrestricted and unregulated systemic complement activation (see later in this chapter).

D. C3 Activation: Immune Adherence and Phagocytosis

1. The normal concentration of serum C3 is about 130 mg/dL, which is relatively high and indicates the importance and central role of this complement component.
2. When the classical pathway is activated, membrane-bound C4b2a complexes activate thousands of C3 molecules. C3 is activated through a process that involves its proteolytic cleavage and release of a small peptide, termed C3a (Fig. 9.4). The larger C3b fragment upon activation behaves very much like the C4b fragment in that it also has a short-lived, highly reactive binding site, which binds covalently to the nearest membrane surface, which is usually the complement-activating antigen.
3. The reactive binding site of C3b is formed by cleavage of an internal thiol ester bond during C3 activation to form an acylating group. This group is highly reactive with either an $-OH$ or $-NH_2$ group on the nearest molecule. C3b may bind to the antigen, to the bound antibody (Fab region), or remain in solution and react with H_2O. As C3b binds to the antigen, it forms a very stable bond with the activating surface, thereby changing the chemical nature of the antigen and reducing its toxicity.
4. As shown in Figure 9.5, bound C3b molecules associate with the activated C4b2a complexes and other C3b ($C3b_n$) independently bond to the antigenic membrane. The binding of C4b and C3b to the antigenic surface is followed by configurational changes that result in the exposure of biologically important regions on these two complement components, previously shielded in the folded molecules.
5. The biological importance of these exposed parts of the bound activated C4b and C3b lies in the fact that they are able to bind to C3b/C4b receptors (currently designated as **CR1 receptors**) located on almost all host cells, most notably phagocytes. The increased affinity of phagocytic cells for C3b (or iC3b)/C4b-coated particles is known as **immune adherence**, and its main consequence is a significant enhancement of **phagocytosis**, which is

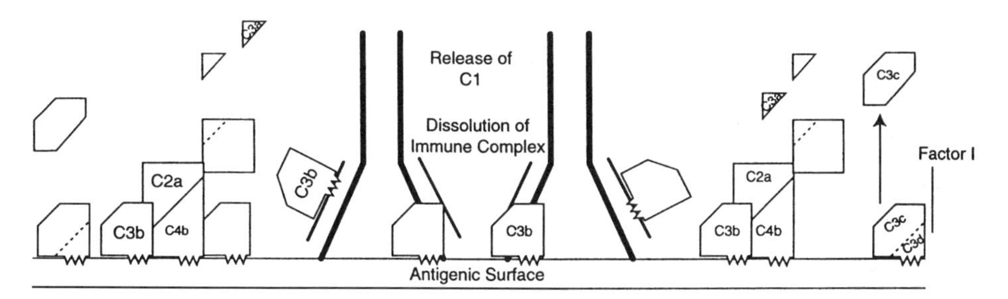

Figure 9.5 Binding of C3 fragments to cell membranes.

one of the most important fundamental defense mechanisms because it provides a direct way for the host to eliminate foreign substances (see Chaps. 13 and 17).

6. The sequence of events leading to **complement-induced phagocytosis** is summarized in Figure 9.6. In the figure, an antigenic cell or particle is represented as an oval structure, and the many areas on the foreign cell surface where antibody-mediated complement activation and deposition have occurred are represented as irregular blocks. The activated complement components form clusters around the antibodies bound to the foreign membrane surface. The large cells represent polymorphonuclear leukocytes, which, like other phagocytic host cells, have complement receptors on their membrane allowing them to bind, with increased affinity, to particles coated with C3b and its breakdown products (i.e., iC3b, see below).
7. **Antigen-bound C3b** is eventually catabolized and several fragments derived from C3b (**iC3b, C3dg, and C3d**) remain associated to the antigen and react with other receptors in phagocytic cells (Fig. 9.7).
   a. A serum protease known as **Factor I** slowly cleaves the alpha chain of bound C3b, transforming this component into **inactivated C3b** (**iC3b**). This fragment has lost the sites that allow C3b to bind to other complement components; consequently, iC3b has irreversibly lost the ability to participate in the complement activation sequence; yet, iC3b remains covalently bound to the foreign activator.
   b. Other very important reactive sites become exposed when Factor I cleaves C3b. The newly exposed regions on antigen-bound iC3b react with other cell receptors, CR2 and CR3. CR3, which avidly binds to iC3b, is not a ubiquitous complement receptor (like CR1); rather, it is expressed on phagocytes and is very important in enhancing phagocytosis.
   c. As iC3b continues to be degraded by Factor I, it breaks into two major fragments: C3dg and C3c. The **C3dg** fragment remains bound to the antigen and retains the site that interacts with CR2. With time, C3dg is further degraded into C3d by the continued action of plasma or tissue proteases; this fragment remains bound to the antigen and, like C3dg, continues to express the site for interaction with CR2, on B cells and on follicular dendritic cells (see Chap. 12).
   d. While some of the C3b molecules are being catabolized, other C3b

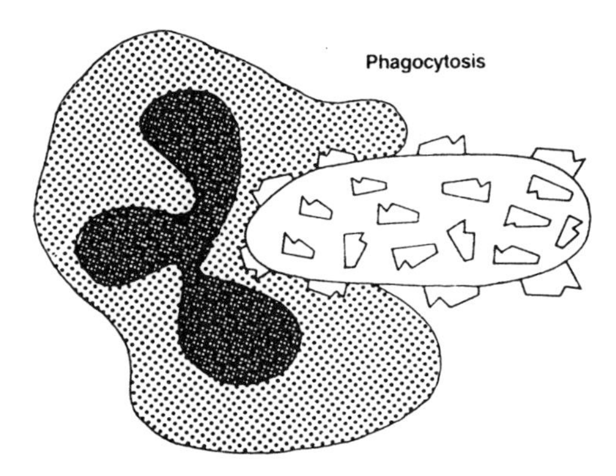

Figure 9.6 Opsonization and phagocytosis. The top panel shows immune adherence of phagocytic cells to antigenic cells coated with antibody and complement, and the bottom panel illustrates the ingestion (phagocytosis) of an opsonized cell.

molecules are just beginning to be deposited on the foreign surface at a rate that depends to some extent on the overall survival time of the previously deposited C3b and C3bBb.

E. The Late Stages of the Classical Pathway: C5 to C9

1. When the activated complement components are deposited on an antigenic cell membrane or on the surface of a microorganism, antigen-bound or membrane-associated $C4b2a3b_n$ will cleave C5 molecules.
2. Each **C5** molecule first binds to an activated $C4b2a3b_n$ complex and then is split into a small fragment (**C5a**), which is released into the fluid phase, and a large fragment (**C5b**). Unlike other complement fragments previously discussed, C5b does not bind immediately to the nearest cell membrane. A complex of C5b, C6, and C7 is first formed in the soluble phase, and it then attaches to the cell membrane through hydrophobic amino acid groups of C7 which become exposed as a consequence of the binding of C7 to the C5b-C6 complex.
3. The membrane-bound C5b-6-7 complex acts as a receptor for C8 and C9. C8, on binding to the complex, will stabilize the attachment of the complex

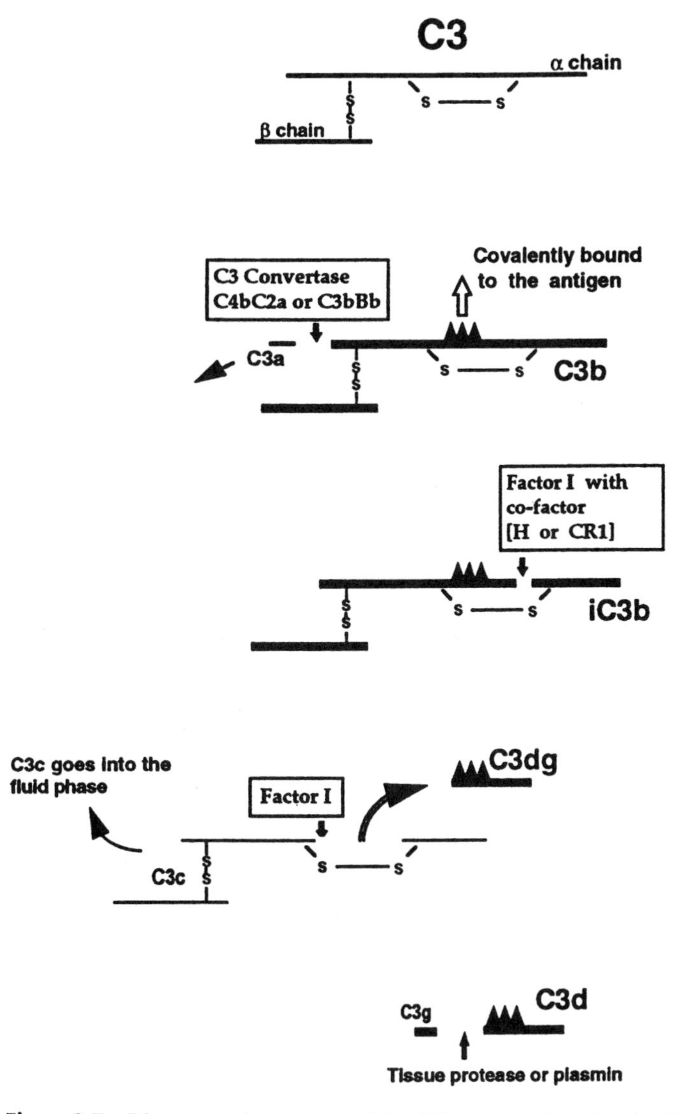

Figure 9.7 Diagrammatic summary of the different steps involved in C3 activation and inactivation.

to the foreign cell membrane through the transmembrane insertion of its alpha and beta chains. The C5b-8 complex acts then as a catalyst for C9, a single chain glycoprotein with a tendency to polymerize spontaneously (Fig. 9.8).

4. The C5b-9 complex is also known as the **membrane attack complex (MAC)**; this designation is due to the fact that on binding to C8, C9 molecules undergo polymerization, forming a transmembrane channel of 100 Å diameter, whose external wall is believed to be hydrophobic, while the interior wall is believed to be hydrophilic. This transmembrane channel will allow the free exchange of ions between the cell and the surrounding medium. Due to the rapid influx of ions into the cell and their association with cytoplasmic proteins, the osmotic pressure rapidly increases inside the

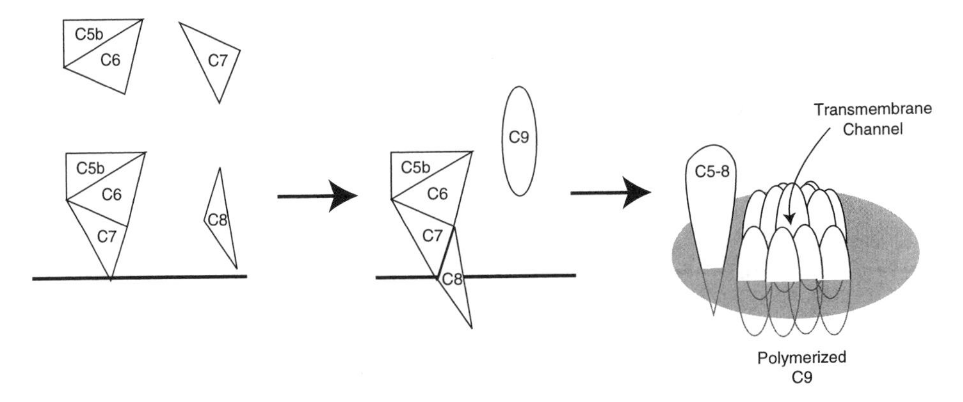

Figure 9.8 Formation of the "membrane attack complex."

cell. This results in an influx of water, swelling of the cell, and, for certain cell types, rupture of the cell membrane and lysis.

a. Sheep erythrocytes are exquisitely sensitive to complement-mediated lysis. Less than 20 seconds are required for lysis of 1 million sheep erythrocytes coated with excess IgG antibody when they are mixed with 1 ml of fresh undiluted human serum as a source of complement.
b. Many bacteria are not susceptible to damage by the MAC as long as their membrane is covered by an intact cell wall or by a polysaccharide capsule. Complement-mediated, enhanced phagocytosis is more important than complement-mediated lysis for elimination of bacteria and other infectious agents.
c. Normal human cells are generally resistant to lysis by human complement. Human cells express substances on their membranes which effectively inhibit the complement sequence.

F. **Inhibitory Substances for C3 and Later Complement Components**. Multiple inhibitory substances have inhibitory effects over different steps of the later activation sequence of the classical pathway. These are considered to be host cell protective mechanisms, which probably explain how a phagocyte approaching a complement-activating immune complex is itself resistant to bystander damage initiated by activated complement fragments (C3b and C4b) being formed on and near its surface.

1. **CR1 receptor**. This glycoprotein binds to activated C3b and acts as a cofactor for **Factor I**, which cleaves C3b into inactivated C3b (iC3b). The rate at which C3b is cleaved by Factor I is dependent upon availability of susceptible sites on C3b.
   a. If the C3b short-lived active binding site simply reacts with H_2O, it will never be able to bind covalently to any surface and will be very rapidly cleaved in its α-chain region by Factor I.
   b. If C3b is covalently bound to the surface of a foreign substance, the inactivation is slower.

2. **Decay-accelerating factor (DAF)** is another inhibitory substance located in a large variety of host cell membranes. The name of this factor derives from the fact that it can accelerate the dissociation of active C4b2a complexes, turning off their ability to continue activating native C3. In addition, DAF attaches to membrane-bound C4b and C3b and prevents the subsequent interaction with C2a and Factor B, respectively. As a consequence, the two types of C3 convertases, C4b2a and C3bBb, will not be formed, the rate of C3 breakdown will be limited, and the host cell will be spared from complement-mediated membrane damage.
3. Several additional complement inhibitory substances are also located on almost all host cell surfaces [e.g., homologous restriction factor (C8 binding protein), CD59 (membrane attack complex inhibitor), CD49 (membrane cofactor protein)].

G. Biologically Active Fragments of C3 and C5: C3a and C5a (Anaphylatoxins)

1. **Chemotactic properties**. The small complement fragments (particularly C5a) released into the fluid phase are chemotactic molecules specifically recognized by phagocytic cells (i.e., polymorphonuclear leukocytes) and cause them to migrate in the direction from which these small fragments originated, thereby attracting phagocytes to a tissue in which an antigen–antibody reaction is taking place. At that site, a phagocytic cell will recognize opsonized particles and ingest them.
2. **Neutrophil activation**. C5a not only has a chemotactic effect on neutrophils, but also activates these cells, causing their reversible aggregation and release of stored enzymes, including proteases.
3. **Anaphylatoxic properties**. The small C5a and C3a fragments bind to receptors on **basophils** and **mast cells** and induce the release of stored vasoactive amines (e.g., histamine) and heparin. The release of histamine into the tissues results in increased capillary permeability and smooth muscle contraction. Fluid is released into the tissue, causing edema and swelling. There is some evidence that the complement fragments **C3a** and **C5a** may also act directly on endothelial cells, causing increased vascular permeability. The end result is very similar to the classical anaphylactic reaction that takes place when IgE antibodies bound to the membranes of mast cells and basophils react with the corresponding antigens. For this reason, C3a and C5a are known as **anaphylatoxins**.

H. Summary. Some important concepts emerging from the study of the activation and regulation of the classical pathway need to be stressed.

1. **Complement exists in a stable and nonactivated form**, and its classical pathway is activated by antigen–antibody complexes.
2. **Complement is a biologically potent system**. Once it is activated, edema and contraction of smooth muscle may occur in the area of activation.
3. **Complement is a fast-acting, cascading (amplifying) system**, with most effects occurring within a few minutes.
4. **Each step in the complement sequence is tightly regulated and controlled** to maximize the damage to any foreign substance while sparing the nearby cells of the host and preventing unnecessary consumption of complement components.

III. PROTEOLYTIC ENZYMES AS NONSPECIFIC COMPLEMENT ACTIVATORS

A variety of proteolytic enzymes, some released by microbes at the site of an infection, and others released from host cells in areas of inflammation, can cleave individual complement components and cause at least partial activation of the system.

A. Polymorphonuclear leukocytes leak lysosomal proteolytic enzymes into the extracellular fluids during the phagocytosis process. Also, within inflamed or traumatized tissue, the damaged host cells release lysosomal proteases during their degeneration. Plasmin, a fibrinolytic enzyme activated during the clotting process, also activates certain complement components.

B. These bacterial and/or host proteases are able to directly cleave and thereby activate C1, C3, and C5. As a consequence of direct cleavage and activation of C3 and C5, biologically active peptides (C3a and C5a) are generated, contributing to a local inflammatory reaction by their direct action, and by the products released by polymorphonuclear leukocytes which C5a attracts to the area of tissue damage.

IV. THE ALTERNATIVE COMPLEMENT PATHWAY

A third group of activators of the complement system includes many types of aggregated (hydrophobic) proteins, microbial membranes, and cell walls (see Table 9.1). These activators affect the complement sequence via a mechanism termed the **alternative pathway**, which received this designation because its activation does not absolutely require antibody and can proceed in the absence of C1, C4, and C2, all essential for the classical pathway of complement activation. The complement sequence and its two activation pathways are schematically summarized in Figure 9.9.

A. Activators of the Alternative Pathway. A variety of substances have been found to be able to activate the alternative pathway (Table 9.1). Some of the most powerful activators of the alternative pathway are the bacterial membrane lipopolysaccharides characteristic of Gram-negative bacteria and the peptidoglycans

Table 9.1 Activators of the Alternative Pathway

Bacterial membranes (endotoxic lipopolysaccharides) and viral envelopes
Bacterial and yeast cell walls
Classical pathway (via C3b generation)
Proteases (i.e., via enhanced C3b generation), released by:
Polymorphonuclear leukocytes
Bacteria
Organ failure (pancreatitis)
Damaged tissue (burns, necrosis, trauma)
Fibrinolytic system (plasmin)
Aggregated immunoglobulins (including IgA, IgG4, and IgE)
Virus-transformed host cells (limited effects)

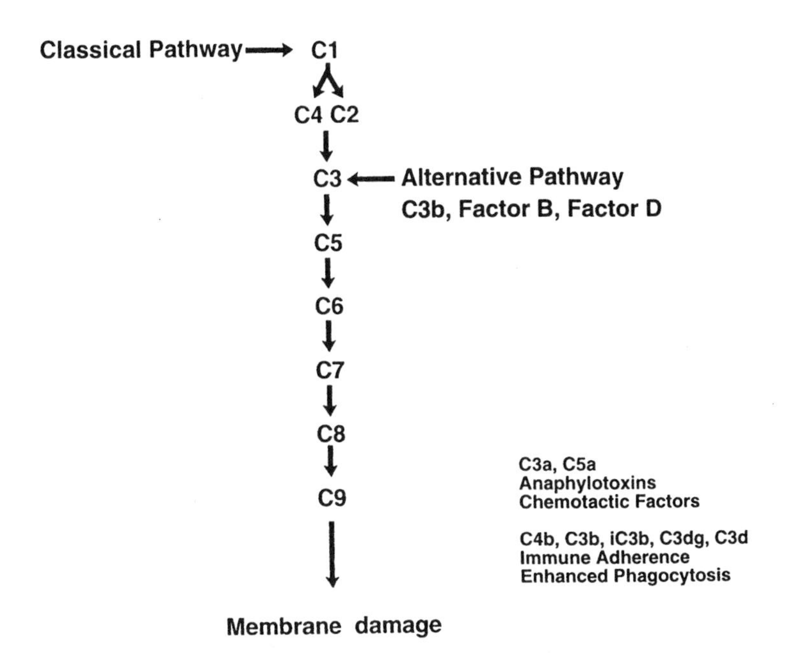

Figure 9.9 The sequence of complement activation.

and teichoic acids from the cell walls of certain Gram-positive bacteria. Artificially aggregated immunoglobulins of all classes and subclasses (including IgG4, IgA, and IgE) will weakly activate this alternative pathway, mediated in part by regions on their aggregated Fab fragments.

B. Sequence of Activation of the Alternative Pathway

1. The substances known to activate the alternative pathway are able to fix to their surfaces a group of several plasma glycoproteins, including **C3b**, **Factor B**, *Factor D*, and **properdin**, which constitute the initial portion of the alternative pathway sequence (see Table 9.2).
2. It is believed that generation of a "protected" (i.e., nondegraded), activated **bound form of C3b** must first occur in order for the alternative complement pathway to be initiated. So the alternative pathway begins with at least one protected C3b covalently bound to the activator, and ends with cleavage of hundreds of C3 molecules to form more bound C3b and soluble C3a.
3. Therefore, generation of C3b is essential for the activation of the alternative pathway. C3b is generated:
   a. During normal C3 turnover in blood
   b. In the presence of active host or bacterial proteases
   c. During classical pathway activation (for this reason activation of the classical pathway is always associated with activation of the alternative pathway which, in that case, functions as an **amplification loop** for the classical pathway, generating more activated C3).
4. Once formed, C3b has an opportunity to bind (by chance) via its short-lived labile binding site to the nearest surface. By definition, if the surface is an

Table 9.2 Alternative Pathway Sequence

1. C3 fragmentation (C3 cleavage via classical pathway or via natural turnover, tissue proteases, or bacterial proteases).
2. Deposition of C3b via its labile binding site on a surface that retards its rapid inactivation by Factor I and its cofactors (Factor H or CR1).
3. Binding of Factor B to C3b leading to C3bB.
4. Activation of the bound B by D leading to C3bBb.
5. C3bBb activation of more C3, leading to formation of $C3b_nBb$ (and liberation of more C3a).
6. Binding of properdin to C3b and stabilization of the association of Bb on C3b.
7. $C3b_nBb$ activation of C5 (with liberation of C5a); activation of the terminal sequence (i.e., membrane attack complex).

activator of the alternative pathway, then C3b will **not** be rapidly inactivated by natural plasma inhibitory systems and will survive on the activating surface long enough to bind to **Factor B** forming a **C3bB** complex.

5. The interaction between C3b and Factor B is stabilized by $\mathbf{Mg^{2+}}$, which is the only ion required for functional activation of the alternative pathway. Therefore, tests to discriminate between the two complement activation pathways are often based on the selective chelation of Ca^{2+} (to disrupt C1q, $C1r_2$, and $C1s_2$) and the addition of sufficient Mg^{2+} to allow activation of the alternative pathway.
6. Factor B within the C3bB complex is activated by a plasma enzyme, **Factor D**, to yield activated **C3bBb**. C3bBb is a **C3 convertase** that activates more C3, leading to the formation $C3b_nBb$, which in turn is capable of activating C5 and the **membrane attack complex**.
7. $C3b_nBb$ is stabilized by **properdin**, a plasma glycoprotein that binds to C3b. Since Factor D has never been isolated in its proenzyme form, it is generally believed to be activated immediately upon leaving the hepatocyte where it is synthesized.

C. **Control of the Alternative Pathway**

1. The nature of the surface to which the C3b binds regulates to a great extent C3b interaction with natural inactivating systems. If C3b binds to a surface or a molecule that is not an activator of the alternative pathway (i.e., host cells), then C3b is rapidly inactivated.
2. The inactivation of C3b occurs with the help of host cell surface substances (including DAF) and several cofactors working with the specific plasma inhibitor **Factor I**, a protease that cleaves C3b, rendering it incapable of properly binding to Factor B or C5, thereby stopping C3b participation in the complement cascade.
   a. Several types of **Factor I cofactors** have been identified: most notably, **CR1** (C3b/C4b receptor) on host cell membranes, **Factor H**, a plasma glycoprotein, and **membrane cofactor protein (MCP)**.
   b. When deposited C3b [or C4b] associates with these cofactors, **Factor I**-mediated cleavage of C3b [or C4b] renders C3b [and C4b] incapable of further involvement in either the classical or the alternative pathway.
   c. As CR1 is found on the surface of most host cells, it is believed that the

CR1 cofactor is most effective in binding and regulating C3b, which inadvertently binds to bystander host cells, while Factor H appears to bind and inactivate C3b, which escapes from the activating surface and enters the fluid phase.

d. Normal cell surface complement inhibitors such as DAF, etc., are upregulated on malignant tumor cells and are also present on the membrane of infected host cells, thus affording a degree of protection from the complement system and the membrane attack sequela, even in the presence of specific antibodies to cell surface tumor antigens or to antigens coded by the infectious agent.

D. **Biological Significance of the Alternative Complement Pathway.** The alternative pathway of complement activation is important especially during the early phase of an infection, when the concentrations of specific antibody are very low. Under those circumstances, some limited classical pathway activation occurs. However, in the presence of large numbers of bacteria, the relatively low levels of specific antibody may be effectively absorbed from the serum by antigens present on the proliferating bacteria, and many uncoated bacteria will escape destruction via antibody-mediated complement activation.

1. C3b molecules (produced via normal C3 turnover) interact with the outer surface of most infectious agents. Most bacteria, fungi, and viruses will activate the alternative complement pathway, but with varying efficiencies. That is, there is a large variability in the avidity and degree of the interaction with this pathway, depending on the species and strain of the microorganism.
2. Perhaps aiding the activation of the alternative pathway are the proteolytic enzymes being produced by the organisms, which directly activate components like C3. If C3 cleavage occurs near the membrane of the organism, C3b is more rapidly formed, and subsequently covalently deposited on the foreign surface via its highly reactive binding site, and the alternative complement sequence is more effectively initiated.
3. After antibodies are formed, the classical and alternative pathway work synergistically, the alternative pathway functioning as an **amplification loop** of the classical pathway.

V. COMPLEMENT RECEPTORS IN HUMAN CELL MEMBRANES

Several types of complement receptors have been identified on the membranes of mammalian cells.

A. **CR1 (Complement Receptor 1)** is a common membrane glycoprotein that can be detected on **almost all types of human cells including erythrocytes and cells from various tissues and organs**.

1. CR1 reacts with C3b until C3b becomes cleaved by Factor I. In phagocytic cells, the main biological role of CR1 is to enhance the phagocytosis of those antigenic substances to which C3b is covalently bound.
2. As discussed previously, cell surface CR1 molecules protect the cells on which they are expressed from complement-mediated bystander damage because they quickly bind to nearby C3b inadvertently deposited on host

cells and act as a cofactor for the serum Factor I enzyme, which cleaves C3b to form inactive C3b (iC3b) and thereby prevents further activation of the complement sequence.

3. CR1 is also expressed on erythrocytes, and binding of C3b-coated immune complexes by erythrocytes effectively removes those potentially inflammatory substances from the fluid phase of the blood. This removal helps to prevent accidental deposition of immune complexes into vital tissues and organs (see Chap. 25).

B. **CR2 (Complement Receptor 2, CD21)** is another important cell surface glycoprotein that has primary binding specificity for a molecular site on the α chain of C3, which is exposed on C3d, C3dg, and on iC3b.

1. **B lymphocytes** express **CR2** and **CR1** molecules on their surface. These receptors are involved in the reception of costimulatory signals to B cells. Thus, they act synergistically with other signals—including those delivered by helper T cells and selected interleukins—as enhancers of B-cell activation.
   a. In vitro, cross-linking of CR2 and CR1 by large-sized antigen–antibody complexes containing multiple C3b, iC3b, C3dg, and C3d molecules, which also interact (through the antigen moieties) with membrane immunoglobulin on B cells, delivers costimulating signals to B cells and antibody production is greatly enhanced.
   b. In animal models, when C3d alone was chemically linked to an antigen, and added to specific B cells in vivo, a thousandfold enhancement of antibody production occurred.
   c. The mechanism of B-cell signaling after CR2 occupancy seems to involve the activation of a transduction complex involving CR2, CD19, and two other molecules. Activation of CD19 is known to greatly stimulate antibody production (see Chap. 4).
2. Other cells important in induction of the immune response, most notably **follicular dendritic cells, also express CR2 receptors** (in addition to CR1 and CR3 receptors).

C. **CR3 (Complement Receptor 3)** is a cell surface glycoprotein that via Ca^{2+}-dependent interactions binds to site(s) exposed predominantly on iC3b; this CR3 receptor is expressed on **neutrophils, monocytes/macrophages**, follicular dendritic cells, certain natural killer (NK) cells, and on a low proportion of B and T lymphocytes.

D. **CR4 (Complement Receptor 4)** is expressed mainly on neutrophils, monocytes, and tissue macrophages. Like CR3, it binds to iC3b, which remains tightly bound to the antigenic surface.

E. **Other Complement Receptors**. As mentioned previously, certain host cells (neutrophils, mast cells, basophils, and certain lymphocyte populations) also have receptors for C3a and C5a.

1. The binding of C3a and/or C5a to these receptors stimulates several cellular functions, such as release of active mediators and upregulation of CR1 and CR3, leading to enhanced phagocytosis, etc.
2. Phagocytic cells, in addition to CR1, CR3, and CR4, have receptors for complement regulatory factors, such as C1 inhibitor. Their physiological role has yet to be defined.

Table 9.3 C3 Receptors on Human Peripheral Blood Cells

Complement receptor	Cellular distribution	Primary binding specificity
CR1	**Phagocytes (all types)**	**C3b and C4b**
(CD35)	**Erythrocytes**	
	Almost all human cell types	
CR2	**B lymphocytes**	**C3d, C3dg,**
(CD21)	**Follicular dendritic cells**	**and iC3b**
	Epithelial cells of the nasopharynx and cervix	
CR3	**Neutrophils**	**iC3b**
(CD11b/CD18)	**Macrophages**	
(Mac-1)	**Monocytes**	
	Follicular dendritic cells	
CR4	**Neutrophils**	**iC3b**
(CD11c/CD18)	**Monocytes**	
(p150,95)	**Macrophages**	
	B lymphocytes	
	Activated natural killer cells	

VI. PATHOLOGICAL SITUATIONS ASSOCIATED WITH EXAGGERATED COMPLEMENT ACTIVATION

A. **Introduction**. Once the complement cascade is activated, the complement components are under very tight regulation and control. An important aspect of this regulation is the constant presence of plasma **inhibitors** for the activated complement components. The tight regulation and rapid neutralization of the active fragments limit their range of action.

1. Serum inhibitors for C3a and C5a have been characterized. One of the inhibitors is believed to be a serum protease that cleaves off the carboxy-terminal arginine residue of the peptides and limits their ability to stimulate polymorphonuclear leukocytes, basophils, and mast cells.
2. The inhibitor for C3b, as described in detail in previous sections of this chapter, is **Factor I**, acting in conjunction with **Factor H** or **CR1**.
3. C4b is also inhibited by Factor I, and a cofactor termed C4 binding protein (C4bp). Factor I cleaves C4b and restricts its capacity to bind and activate C2.
4. The serum inhibitor for C1 is a serum protein, termed **C1 inhibitor (C1 INH)**, which tends to stabilize the nonactivated C1 macromolecular complex, preventing spontaneous activation. More importantly, C1 INH also binds covalently to activated C1r and C1s, at or near their active site and the inactivated complexes are released into the circulation.

A deficiency of any of these complement inhibitors or cofactors can lead to an imbalance in complement regulation, and disease may ensue.

B. **Hereditary Angioneurotic Edema**. This is a rare genetic disorder due to a **genetically inherited C1 INH deficiency**, of which two main variants are known. Also an acquired form of C1 INH deficiency can be detected in certain malignant diseases.

1. In the most common inherited form, the genetic inheritance of a silent gene results in a very low level of C1 inhibitor.
2. The second inherited variant is characterized by normal levels of C1 INH protein, but 75% of the molecules are dysfunctional (i.e., will not inhibit C1r or C1s).
3. While the lack of C1 INH can be easily detected by a quantitative assay, the synthesis of dysfunctional C1 INH can only be revealed by a combination of quantitative and functional tests.
4. Individuals with congenital C1 INH deficiency may present clinically with a disease known as **hereditary angioneurotic edema**, characterized by spontaneous swelling of the face, neck, genitalia, and extremities, often associated with abdominal cramps and vomiting. The disease can be life threatening if the airway is compromised by laryngeal edema, and tracheotomy may be a life-saving measure.
   a. This anaphylactoid reaction is due not to IgE-mediated reactions, but rather to spontaneous, uncontrolled activation of the complement system by C1. The reaction is usually self-limiting and will cease after all C4 and C2 have been consumed.
   b. Since most of the attacks in patients with C1 INH deficiency occur after surgical trauma, particularly after dental surgery, or after severe stress, it is tentatively postulated that C1 (C1r and C1s) may become activated not only by antigen–antibody complexes, but also indirectly by other serum enzymes with protease activity, such as the Hageman factor, kallikrein, or plasmin, which may be released and activated under circumstances of trauma or severe stress.
   c. It is notable that "activated" Hageman factor, kallikrein, and plasmin are also controlled by binding to C1 INH. Such binding further depletes the available C1 INH in deficient patients.
   d. In the absence of sufficient C1 INH, spontaneous activation of a limited number of C1 molecules will gradually accentuate the depletion of C1 inhibitor to a point that autocatalytic C1r-mediated activation of more C1 and other enzymes controlled by C1 INH will become unrestricted. C1r will activate C1s in the fluid phase. In turn, the continued presence of activated, uninhibited fluid phase C1s will cause spontaneous and continuous activation of the next two components in the sequence, C4 and C2, until their complete consumption. **Low C4 levels are considered diagnostic of C1 INH deficiency**, and they remain low even when the patients are not experiencing an attack, probably due to a continuously exaggerated C4 metabolism by activated C1.
   e. The responsible angioedema-producing peptide has been suggested to be the **C2b** fragment of C2, liberated by the action of C1 on C2 followed by the cleavage of C2 by plasmin. This theory has developed because serum C3 levels are not significantly altered during attacks of angioedema. However, in vitro evidence has shown that if appropriate levels of antibody to human C1 inhibitor are added to whole human serum, 100% C3 activation occurs. This complete C3 conversion is achieved because the function of C1 INH is blocked. Thus, the participation of low levels of C3a in angioneurotic edema cannot be ruled out when local C1 inhibitor levels approach zero.

C. **Paroxysmal Nocturnal Hemoglobinuria (PNH)** is a rare acquired disorder of the hemopoietic cells and erythrocytes.

1. Clinically, PNH is characterized by anemia associated with the intermittent passage of dark urine (due to elimination of hemoglobin), which usually is more accentuated at night.
2. The hemoglobinuria is due to an increased susceptibility of an abnormal population of erythrocytes to complement-mediated lysis. The erythrocytes are not responsible for the activation of the complement system; rather, they are lysed as innocent bystanders when complement is activated.
3. The **molecular basis of PNH** has been recently elucidated. Several membrane proteins are attached to cell membranes through **phosphatidylinositol "anchors."** The red cell membrane contains two such proteins: the **decay accelerating factor (DAF)**, whose controlling effect on the formation of C4b2a and C3bBb we have previously discussed, and a **C8-binding protein (C8bp)**, which prevents the proper assembly of the membrane attack complex. These two proteins, together with CR1, have an important protective role for the "bystander" erythrocytes by controlling the rate of complement activation on the erythrocyte membrane. The deficiency of the phosphatidylinositol anchoring system is reflected by deficiencies of DAF and C8bp. Red cells lacking both proteins are very sensitive to hemolysis; red cells lacking DAF have intermediate sensitivity to hemolysis.
4. Other proteins, including LFA-3 molecules (see Chap. 11), and the predominant type of Fc receptor in the neutrophil (FcγRIII), are also anchored to the membrane via phosphatidylinositol. In these patients, DAF and LFA-3 are deficient in a variety of cell types, and neutrophils are deficient both in DAF and in FcγRIII receptor expression. These deficiencies seem to be the basis of other abnormalities seen in PNH patients: thrombotic complications, attributed to increased complement-induced platelet aggregation secondary to DAF deficiency, and bacterial infections and persistence of immune complexes in the circulation, both attributed to a lack of Fc-mediated phagocytosis.

D. **Pulmonary Vascular Leukostasis as a Side Effect of Hemodialysis**

1. When fresh serum is passed over filters such as those used in hemodialysis, complement activation via both pathways may occur. The classical pathway may be activated by interfacially (solid–liquid or air–liquid) aggregated immunoglobulin or by direct binding of Clq. Alternatively, membrane filter-bound C3b may mediate activation of the alternative pathway.
2. In vivo, a blood anticoagulant such as heparin or citrate must be used whenever blood undergoes dialysis or extracorporeal circulation. However, anticoagulants not only alter blood clotting but also affect the complement system and interfere with its normal activation.
   a. Citrate, by chelating Ca^{2+}, partially restricts Cl activation because it disrupts Clq binding to Clr and Cls.
   b. At low concentrations, heparin has a limited direct inhibitory effect on Cl activity, but its major effect is to bind (and potentiate) Cl INH and Factor H.
   c. Passage of heparinized blood over a variety of filter materials (i.e., artificial hemodialysis membranes and nylon fiber substances used in the heart-lung machine) causes varying degrees of complement activa-

tion. In vitro, heparinized blood appears to generate more C3a and C5a fragments than citrated blood when passing across artificial filtration membranes.

3. Rapid generation of C5a causes a transient leukopenia secondary to sequestration of aggregated **polymorphonuclear leukocytes** in the blood capillaries of the lungs, where the aggregated PMN release oxygen-active radicals that damage the tissues surrounding the areas of PMN accumulation. As a result, repeated hemodialyses may lead to chronic fibrosis of the lung.

E. Pancreatitis, Severe Trauma, and Pulmonary Distress Syndrome

Any mechanism that causes a rapid release of high levels of C5a peptides into the blood may cause massive PMN aggregation and consequent **pulmonary distress syndrome**.

1. Pancreatitis or severe tissue trauma causes the release of large amounts of proteases into the blood. As a consequence of the peripheral sequestration of aggregated granulocytes, pulmonary distress syndrome and sometimes temporary blindness occur due to blockage of small blood vessels.
2. Similarly, in myocardial infarction, blockage of critical heart capillaries with PMN may extend cardiac damage. Steroids that prevent and reverse the PMN aggregation have been used to retard such damage in experimental animals.

VII. COMPLEMENT LEVELS IN DISEASE

The complement proteins have one of the highest turnover rates of any of the plasma components. At any one time, the level of a complement component is a direct function of its catabolic and synthetic rates.

A. Complement Metabolism

1. The **catabolic rates** of the complement system are a function of the extent of complement activation by the classical pathway, the alternative pathway, or direct proteolytic cleavage. Complement activation enhances catabolism because complement fragments are rapidly cleared from the circulation.
2. **Synthetic rates** of the complement proteins are controlled by ill-defined mechanisms but probably involve such variables as the levels of complement activators (i.e., immune complexes), the class and subclass of immunoglobulin within the immune complexes, the rate of complement activation, the steady-state level of complement fragments in the blood, and, in cases of certain chronic inflammatory diseases, the level of autoantibody to complement components. Because of this complex control, the synthetic rates of complement glycoproteins vary widely in disease states and during the course of a given disease.
3. The level of a complement component is a function of its metabolic rate (consumption versus synthesis) and of the stage and intensity of the inflammatory process associated with complement activation. Elevated levels of a given complement component in a disease state probably mean that there is both a rapid catabolic and synthetic rate. Low levels mean that consumption is greater than synthesis.

B. **Hypocomplementemia and Clearance of Immune Complexes.** As previously mentioned in our discussion of the classical pathway, activation of a normal complement system by immune complexes will eventually lead to partial dissolution of the immune complex. This phenomenon is due to the deposition of large complement fragments such as C4b and C3b on the antigen and on the Fab region of the antibody, which interferes with the antigen–antibody binding reaction (Fig. 9.6). Individuals with immune complex diseases often suffer with the inability to properly eliminate the immune complexes from the kidney and/or from the basement membrane of dermal tissues.
   1. In some cases, the lack of immune complex clearance is secondary to a deficiency in the early complement components. Such deficiency results in subnormal binding of C4b and C3b to the immune complex. As a result, the rate of formation of new immune complexes surpasses the inefficient rate of immune complex dissolution, and the generation of proinflammatory complement fragments will be possible for a longer period of time. The reasons for the lower serum levels of early complement components (i.e., Clq, C4, and/or C2) are multiple and include not only genetic factors but also a variety of metabolic control mechanisms mentioned above.
   2. In patients with systemic lupus erythematosus, a reduction in the levels of erythrocyte C3b receptors (CR1) has been reported. The implications of such deficiency are multifold.
      a. A partial CR1 deficiency may affect the generation of complement fragments that in turn may play a role in regulating complement metabolism.
      b. The binding of complement-coated immune complexes to erythrocytes (which express the CR1 receptor), which is believed to be an important physiological mechanism of immune complex removal from circulation, is likely to be adversely affected. Although human erythrocytes can adsorb immune complexes in the absence of complement, binding through CR1 may be important in stabilizing the interaction. Immune complexes that persist for longer periods in circulation may have a greater opportunity to be deposited in organs and tissues thereby causing inflammation.

C. **Complement Deficiencies**. Deficiencies of several of the components of the complement system have been reported by different groups. Basically, these deficiencies are associated with two types of clinical situations:
   1. **Chronic infections**, often by Neisseria species (usually associated with deficiencies of components of the terminal complement sequence)
   2. **Autoimmune disease**, mimicking systemic lupus erythematosus (usually associated with deficiencies in components of the earlier part of the complement sequence).

   Complement deficiencies and the pathology associated with them will be discussed in greater detail in Chapter 30.

VIII. MICROBIAL ANTICOMPLEMENTARY MECHANISMS

In general, complement-mediated phagocytosis is the most effective mechanism for elimination of infectious microorganisms. However, pathogenic organisms have evolved several mechanisms to circumvent either effective complement activation or effective complement deposition on their outer surface. These evasion strategies are most efficient during the early stages of an infection, when the levels of specific antibody are low.

A. In some cases, the microorganisms have on their surface a structural protein that mimics the protective effect of CR1, CR3, C4bp, or DAF. The subsequent deposition of complement components on the surface of the microorganism becomes less effective at promoting phagocytosis of the organism.

B. Certain microorganisms appear to adsorb normal complement regulators circulating in the peripheral blood. HIV-1 acquires DAF from host cell membranes to its surface as the particles leave the infected host cell. Then the HIV particles absorb Factor H from the plasma. The end result is, again, impaired opsonization.

C. Other less sophisticated anticomplementary mechanisms that different microorganisms have acquired include:

1. Shedding of MAC-coated pili
2. Restriction of the deposition of complement components by structural components of the bacterial cell wall (e.g., bacterial capsules)
3. Protection of the cytoplasmic membrane from the dissolving effect of the MAC by slime layers, peptidoglycan layers, and polysaccharide capsules
4. Proteolytic digestion of deposited complement components by bacterial proteases

D. In the presence of sufficient antibody levels, these protective mechanisms are usually overridden, and the microorganisms end up being phagocytosed, although some bacteria have acquired antiphagocytic capsules, which further complicate the job of the immune system. Some of these aspects will be discussed at greater detail in Chapter 13.

SELF-EVALUATION

Questions

Choose the ONE *best* answer.

9.1 Complement is responsible for:

A. Agglutination of incompatible red cells
B. Antibody-dependent cell-mediated cytotoxicity (ADCC)
C. Attraction of lymphocytes to inflammation sites
D. Cross-linking polysaccharide antigens
E. Enhanced phagocytosis of infectious agents

9.2 Which of the following is **only** observed when complement is activated by the classical complement pathway:

A. Breakdown of C3 into C3a and C3b
B. Breakdown of C4 into C4a and C4b
C. Breakdown of C5 into C5a and C5b
D. Activation of the membrane attack complex
E. Generation of anaphylatoxins

9.3 The mechanism leading to red-cell lysis in severe cases of paroxysmal nocturnal hemoglobinuria involves:

A. Complement activation induced by anti-red-cell antibodies

B. Deficiency of Factor I, limiting the ability to inactivate C3b bound to erythrocyte CR1 receptors
C. Opsonization and phagocytosis of red cells
D. Unchecked deposition of C3b on the red cell followed by activation of C5 to C9
E. Shedding of DAF into the circulation

9.4 Histamine is released from mast cells stimulated by:
A. C1q
B. C2a
C. C4b
D. C5a
E. C3b

9.5 The alternative pathway can be efficiently activated by:
A. Antigen–antibody complexes
B. C1
C. Chemically cross-linked Fc fragments of IgG
D. Mg^{2+}
E. Properdin

9.6 CR1 receptors on phagocytic cells have the greatest affinity for:
A. C3b
B. iC3b
C. C3dg
D. C3d
E. C3a

9.7 Which of the following is a major characteristic of hereditary angioneurotic edema?
A. Quantitative and/or functional deficiency of C1 INH
B. Induction by inhalation of complement-activating compounds
C. Normal levels of C2 and C4
D. Spontaneous breakdown of C3 and C5
E. Very high levels of IgE

9.8 Neutrophil aggregation as a consequence of hemodialysis is believed to result from:
A. Activation of the alternative pathways by heparin
B. Release of C3b
C. Excessive amounts of calcium
D. Generation of C5a
E. Retention of antigen–antibody complexes in the dialysis membrane and activation of the classical pathway

9.9 A deficiency of erythrocyte CR1 receptors is associated with:
A. Increased deposition of antigen–antibody complexes in tissues
B. Increased incidence of angioneurotic edema
C. Release of massive amounts of histamine
D. Accumulation of C3dg and C3d in circulation
E. Paroxysmal nocturnal hemoglobinuria

9.10 The "membrane attack complex" is formed by:
A. $C4b2a3b_n$
B. C56789
C. C4bC3b
D. C3bBb
E. C4b2a

Answers

9.1 (E) Complement and IgG antibodies are the major opsonins that enhance the uptake of infectious agents by phagocytic cells. The chemotactic effects of C5a and C3a are limited to phagocytic cells, particularly granulocytes.

9.2 (B) C4 is only activated through the classical pathway.

9.3 (D) In paroxysmal nocturnal hemoglobinuria, the red cells lack DAF and C8bp. Therefore, C3b deposited on the red cells as a consequence of complement activation by factors totally unrelated to the red cells can result in red cell lysis because the progression of the activation pathway from C3 to C9 can proceed unimpaired.

9.4 (D) C3a and C5a (and perhaps C2b) are the biologically active complement components able to trigger the release of histamine from mast cells (anaphylatoxins).

9.5 (A) Immune complexes, by activating the complement system by the classical pathway, will also activate the alternative pathway. Properdin is a stabilizer of the C3bBb complex, but by itself cannot activate the alternative pathway. Mg^{2+} is required for the activation of the alternative pathway (classical pathway activation, in contrast requires **both** Ca^{2+} and Mg^{2+}), but it is not an activator by itself. The Fc fragment of IgG is involved in activation of the classical pathway.

9.6 (A)

9.7 (A) The symptoms of hereditary angioneurotic edema are believed to be caused by C2 fragments. Limited breakdown of C3, leading to the generation of C3a, may take place, but only in response to the activation of C4 and C2. C4 and C2 levels are usually low since they are consumed as a consequence of the excessive activity of C1, while C3 levels tend to be less affected.

9.8 (D) Cellophane membranes used for hemodialysis are able to activate the alternative pathway, leading to the generation of C5a, which causes neutrophil aggregation.

9.9 (A) Soluble antigen–antibody complexes with attached C3b can be removed from the circulation through adsorption to the CR1 receptor on red cells. If those receptors are reduced in numbers, soluble antigen–antibody complexes will persist in circulation for longer periods, and will more likely be trapped eventually in tissues and cause inflammation.

9.10 (B)

BIBLIOGRAPHY

Asghar, S.S., Membrane regulators of complement activation and their aberrant expression in disease. *Lab. Invest.*, *72*:254–271, 1995.

Birmingham, D.J., Erythrocyte complement receptors. *Crit. Rev. Immunol.*, *15*:133–154, 1995.

Dempsey, P.W., Allison, M.E.D., Akkaraju, S., Goodnow, C.C., and Fearon, D.T. C3d of complement as a molecular adjuvant bridging innate and acquired immunity. *Science*, *271*:348–349, 1996.

Lokki, M.L., Colten, H.R., Genetic deficiencies of complement. *Ann. Med.*, *27*:451–459, 1995.

Sengelov, H., Complement receptors in neutrophils. *Crit. Rev. Immunol.*, *15*:107–131, 1995.

Seya, T., Human regulator of complement activation (RCA) gene family proteins and their relationship to microbial infection. *Microbiol. Immunol.*, *39*:295–305, 1995.
Volanakis, J,.E., Transcriptional regulation of complement genes. *Annu. Rev. Immunol.*, *13*:277–305, 1995.
Wetsel, R.A., Structure, function and cellular expression of complement anaphylatoxin receptors. *Curr. Opin. Immunol.*, *7*:48–53, 1995.

10
Lymphocyte Ontogeny and Membrane Markers

Jean-Michel Goust and Anne L. Jackson

I. INTRODUCTION

T and B lymphocytes start to develop during the genesis of the hematopoietic system from a common stem cell, but start to differentiate into separate lineages early in fetal life (Fig. 10.1).

A. In humans, the embryonic yolk sac of the developing embryo is the first structure that forms stem cells that develop into leukocytes, erythrocytes, and thrombocytes.

B. The fetal liver receives stem cells from the yolk sac at the sixth week of gestation and begins hematopoietic activity. In the twelfth week, a minor contribution is made to the production of blood cells by the spleen.

C. At 20 weeks of gestation, thymus, lymph nodes, and bone marrow begin hematopoietic activity and, at that time, two of the three major types of lymphocytes are already identifiable.

D. The bone marrow becomes the sole hematopoietic center after 38 weeks. At that time, differentiated T and B lymphocytes are present in the circulation.

E. The endpoint of this ontogenic process is to produce T and B lymphocytes able to recognize antigens using specific membrane receptors and to respond appropriately to the occupancy of the binding sites of those receptors. They also need special molecules to interact with each other and to migrate into areas of antigen challenge.

II. MAJOR STEPS IN THE DIFFERENTIATION OF T AND B LYMPHOCYTES

The differentiation of T and B lymphocytes proceeds in three basic steps.

A. **Rearrangement of the Genomic DNA Encoding Their Antigen Receptors.** As discussed in detail in Chapter 7, the generation of sufficient diversity of antigen binding sites on the immunoglobulin molecule requires random rearrangement of a large number of genes involved in determining the amino acid sequence of the

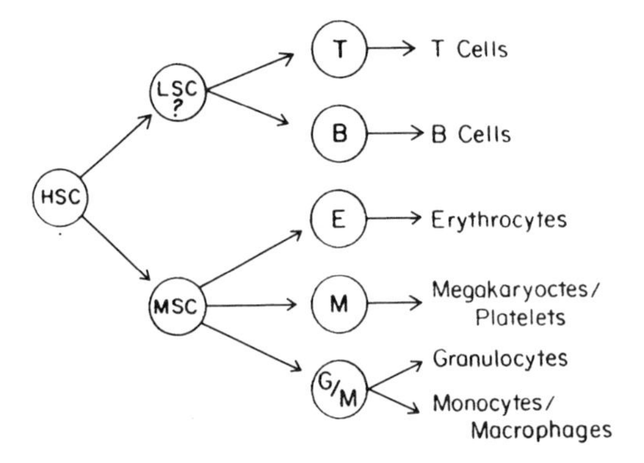

Figure 10.1 Hypothetical model of hemopoietic stem-cell differentiation. Multipotential hemopoietic stem cells (HSC) may give rise to more restricted progenitor cells with self-renewal capacity. Thus, lymphocyte-committed stem cells (LSC) give rise to T and B lymphocytes. Another group of hemopoietic cells, for which the immediate progenitor is a more differentiated myeloid stem cell (MSC), includes erythrocytes, megakaryocytes, and platelets as well as the granulocyte, monocyte/macrophage series. (Reproduced with permission from Cooper, M.D., Kearney, J., and Scher, I. B Lymphocytes. In *Fundamental Immunology* (W.E. Paul, ed.). Raven Press, New York, 1984.

antibody binding sites. A similar process takes place to generate the necessary diversity of binding sites on T-cell receptors (TcR).

B. **Association with Co-Receptors**. Both membrane-bound immunoglobulins (mIg) and the TcR have very short intracytoplasmic segments and are not able to transduce activation signals on their own. The products of the rearranged receptor genes (mIg, TcR) associate in the cytoplasm with other molecules, forming a **receptor complex** that is then transported to the cell membrane. The membrane-associated complexes of an antigen receptor and co-receptor molecules are able to transmit activating signals upon occupancy by an immunogenic peptide or epitope (see Chap. 11).

C. **Selection of the Right Receptors**. Because of its random nature, the rearranged receptor genes expressed on the surface of undifferentiated lymphocytes may include autoreactive receptors. A selection process has to exist to avoid the differentiation of competent autoreactive lymphocytes, otherwise they would quickly engage in autoimmune reactions against self cells and tissues.

III. APPROACHES TO THE STUDY OF LYMPHOCYTE DIVERSITY AND ONTOGENY

A. **The Production of Monoclonal Antibodies**. In 1975, **Kohler** and **Milstein** discovered that by fusing a malignant plasma cell with a nonantibody-producing B lymphocyte they could form a hybrid cell (hybridoma) that would constantly proliferate, like a malignant cell, but at the same time would conserve the antibody-producing ability of the normal B lymphocyte involved in the fusion. The process, summarized in Figure 10.2, involves two cell populations and three major steps:

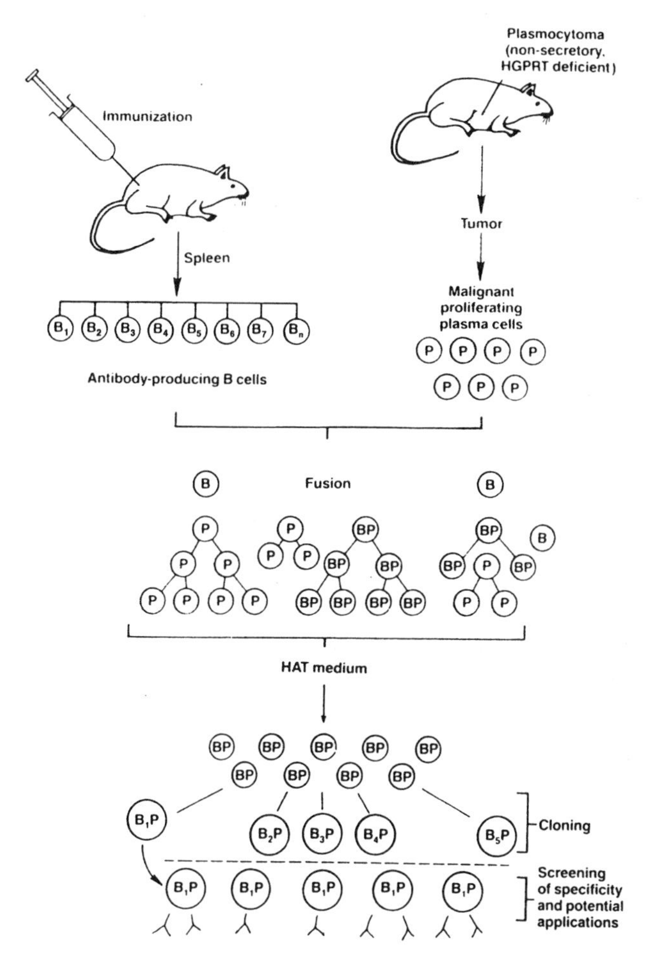

Figure 10.2 Schematic representation of the major steps involved in hybridoma production. First, antibody-producing lymphocytes are fused with nonsecretory malignant plasma cells, deficient in hypoxanthine guanine phosphoribosyl transferase (HGPRT). Nonfused lymphocytes will not proliferate, and nonfused plasma cells will die in a hypoxanthine-rich medium (HAT). The HGPRT-deficient plasma cells cannot detoxify hypoxanthine, while the hybrid cells have HGPRT provided by the antibody-producing B lymphocytes. The surviving hybrids are cloned by limiting dilution, and the resulting clones are tested for the specificity and potential value of the antibodies produced.

1. A malignant plasma cell line of mouse or rat origin, unable to secrete immunoglobulins and deficient for a critical enzyme that allows cells to synthesize purines from hypoxanthine [hypoxanthine guanine phosphoribosyl transferase (HGPRT)].
2. A suspension of spleen cells obtained from a normal mouse or rat harvested at the peak of an induced immune response. The suspension will contain a large number of B-lymphocyte clones, each clone producing a single antibody against one given epitope.
3. The two cell populations are fused, usually by incubation in the presence of **polyethylene glycol**. The fused B lymphocyte will provide the resulting

hybridoma with the capacity to produce a specific antibody and the capacity to produce HGPRT; the fused plasmocytoma cell will provide the hybridoma with the capacity to proliferate indefinitely.

4. Nonfused B cells will die after a few rounds of replication. To eliminate nonfused, HGPRT-deficient, malignant plasmocytoma cells, the mixture of fused and nonfused cells is grown in a medium (HAT) containing hypoxanthine and aminopterine (which blocks the remaining intracellular pathways for the synthesis of purines). Under those conditions, only fused cells, able to utilize hypoxanthine for the synthesis of purines, will survive.
5. A lengthy screening process follows, the aim of which is to select from the large number of hybrids produced by fusion those clones that produce antibody against antigens of interest. This **cloning** process involves the preparation of limiting dilutions or cell sorting of the hybrid cells, either process ensuring that single cells are seeded into individual receptacles containing tissue culture medium and allowed to grow into clones producing antibody of a single specificity (**monoclonal antibody**). Large quantities of the monoclonal antibodies generated in this fashion can be obtained, tested, and eventually used for diagnostic purposes and in some experimental therapeutic protocols.
6. The injection of mice with extracts of human lymphocyte membranes has resulted in the segregation of clones producing antibodies against a wide variety of membrane molecules. The same is true for many other cell types.

B. **Lymphocyte Membrane Markers**. Several hundred monoclonal antibodies identifying over 160 different membrane markers on human leukocytes have been produced. These membrane markers are complex antigens expressing many different epitopes, and monoclonal antibodies raised in different laboratories often recognize slightly different parts of the same molecule.

1. To uniformize nomenclature, the World Health Organization sponsors workshops determining which molecule is identified by a newly developed monoclonal antibody and assigns uniform designations to these molecules. A given antibody group is designated by the initials **CD**, for **clusters of differentiation** (a designation that recognizes the fact that each marker has multiple antigenic determinants), followed by a number.
2. The numerical designations are based partly on the order of discovery and partly on the ontogenic order of appearance. For instance, the monoclonal antibody recognizing what, at the time, was considered ontogenically as the most primitive T-lymphocyte membrane marker, was designated as CD1, and the T lymphocytes expressing it are known as CD1+.
3. As of October 1996, CD166 was last on the list, which includes interleukin receptors and molecules used for homing. In this chapter, we will limit our discussion to the lymphocyte markers most useful from the clinical point of view.

C. **Genetic Approaches**. Technical progress in molecular genetics has resulted in the cloning and sequencing of immunoglobulin genes, TcR genes, and a large number of other genes coding for molecules expressed on lymphocyte membranes. Often the genes are cloned and the sequence before the function of the molecule they code is defined. In addition, it has become possible to transfer human genes to experimental animals (i.e., transgenic mice) and it has also become possible to breed animals in which critical genes are deleted (i.e., knockout mice). These techniques have been widely used to investigate the development of T and B lymphocytes.

IV. B-LYMPHOCYTE ONTOGENY

B lymphocytes are relatively easy to individualize through the demonstration of their membrane-bound immunoglobulins by means of fluorescein-labeled anti-immunoglobulin antibodies. After adequate stimulation, B lymphocytes differentiate into plasma cells that secrete large quantities of immunoglobulins. Within a single cell or a clone of identical cells, the antibody binding sites of membrane and secreted immunoglobulins are identical. Therefore, studies of the genetic control of immunoglobulin synthesis and of the mechanisms responsible for the generation of binding-site diversity have been relatively easy (see Chap. 7). The following steps are involved in this process (Fig. 10.3).

A. **Rearrangements and Expression of the Immunoglobulin Genes**. The rearrangements bring together noncontiguous coding segments and delete intervening noncoding segments, resulting in the final coding sequence that is expressed.

1. The first rearrangement involves the DNA segment encoding the heavy chain of IgM. This rearrangement is followed by synthesis of μ chains that remain intracytoplasmic.

B Cell Ontogeny			Gestational Age	
Immunoglobulin Expression	Stage	DNA Rearrangement	Mice	Man
No	Stem Cell	None		
Cytoplasmic IgM No Light Chains	Large Pre-B Cell	VDJCμ	13 days Liver	9 weeks Liver
Cytoplasmic IgM No Light Chains	Small Pre-B Cell	VDJCμ	19 days Liver	11 weeks Liver
Surface IgM		VDJCμ + VJCk, λ	Bone marrow	
Surface IgM and IgD	Virgin B Cell	VDJCμ + VJCk, λ VDJCδ + VJCk, λ	21 days Bone marrow, spleen, lymph nodes	12 weeks Spleen
	Antigen Responding B Cell			

Figure 10.3 Diagrammatic representation of the early steps in B-lymphocyte ontogeny.

2. A few days later, the genes coding for either a kappa (κ) or a lambda (λ) light chain undergo rearrangement. At this point, association of the heavy and light chains yields complete IgM molecules. A similar sequence leads to the production of IgD.

B. **Addition of Co-Receptors and Assembly of a Competent B-Cell Receptor.** Fully assembled IgM and IgD need to associate with several other molecules, forming a complex that is then transported to the membrane surface and is competent to receive and transduce activation signals (Fig. 10.4). The following molecules are involved in the formation of this complex:

1. **CD79a and CD79b**, also known as Igα and Igβ. These two molecules are closely associated with the membrane immunoglobulin molecule and have numerous phosphorylation sites that serve as docking sites for tyrosine kinases used at the beginning of the activation cascade.
2. **CD19** is found on pre-B lymphocytes, all resting B lymphocytes, and almost all malignancies of B-lymphocyte origin (see Chap. 30). The CD19 molecule modulates Ca^{2+} influx into the cell and is a controlling element in B-lymphocyte activation.
3. **CD21** is expressed with high density on mature, resting B lymphocytes, which are also IgM^+, IgD^+. Activated B lymphocytes are, as a rule, $CD21^-$. CD21 is also known as CR2, because this molecule functions as a receptor for the C3d fragment of complement; in the right circumstances, the interaction between antigen-bound C3d and CD21 can deliver a co-stimulating signal to the B cell, which results in significant amplification of the humoral immune response.

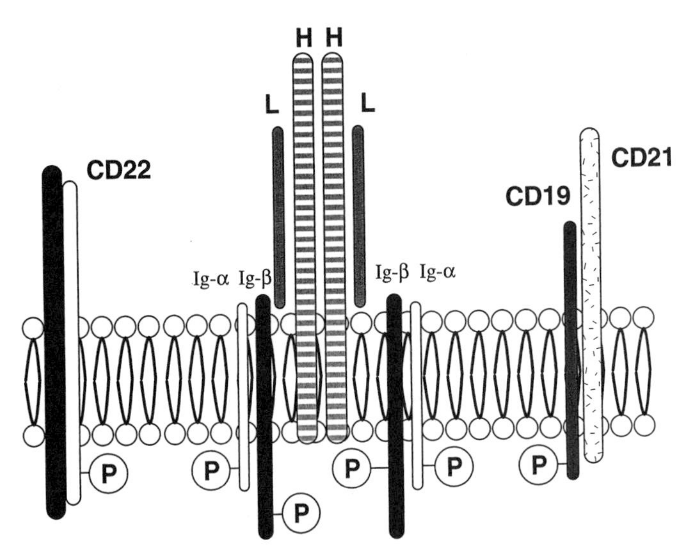

Figure 10.4. Diagrammatic representation of the B lymphocyte antigen-receptor complex (BcR), constituted by a membrane immunoglobulin molecule closely associated with Igα and Igβ molecules. Three additional molecules, CD19, CD20, and CD21 are associated with the BcR complex and play a role in B-cell signaling.

4. **CD22** is an integral part of the B-cell receptor complex, first detected in the cytoplasm of bone-marrow-derived B lymphocytes containing cytoplasmic μ chains. Later, it was found on the surface of 75% of sIgM$^+$ B lymphocytes, and, finally, on 90% of sIgM$^+$, sIgD$^+$ resting B lymphocytes. In the adult, CD22 is expressed at relatively high levels in tissue-based B lymphocytes (e.g., in the tonsils and lymph nodes), but not in circulating B lymphocytes and disappears from the cell membrane during B-lymphocyte activation. CD22 interacts with CD45RO, a marker expressed by memory helper T cells.

C. **Elimination or Down-Regulation of Autoreactive B Cells**. The membrane immunoglobulins of B-cell precursors include those able to combine with epitopes expressed on self antigens. As discussed in detail in Chapter 22, these autoreactive cells are either deleted or rendered unreactive by means of the inactivation of the signal transduction part of their B-cell receptor.

D. **Changes in the Expression of Membrane Immunoglobulins and CD Markers During B-Cell Maturation and Activation**

1. Around birth, differentiated but resting ("virgin") B lymphocytes will co-express IgM and IgD on their membranes. On individual B-lymphocyte clones, the co-expressed IgM and IgD have the same variable regions and the same antigenic specificity.
2. These mature B lymphocytes subsequently home to peripheral lymphoid organs where, upon antigenic challenge, they lose membrane IgD and, less constantly, IgM; the activated lymphocytes undergo further gene rearrangements, switching to the production of a different immunoglobulin isotype (IgG, IgE, or IgA). These immunoglobulins are then found on the membrane of nonoverlapping B-lymphocyte subsets, sometimes in association with membrane IgM.

E. **Ontogenic Development of Immunoglobulin Synthesis**. A normal newborn infant, though having differentiated B lymphocytes, produces very small amounts of immunoglobulins for the first 2 to 3 months of life. During that period of time, the newborn is protected by placentally transferred maternal IgG, which starts to cross the placenta at the twelfth week of gestation. By the third month of age, IgM antibodies produced by the newborn are usually detectable and the concentration of circulating IgM reaches adult levels by 1 year of age. It must be noted, however, that in cases of intrauterine infection, IgM antibodies are synthesized by the fetus and detected in cord blood. The onset of the synthesis of IgG and IgA occurs later and the concentration of these reaches adult levels at 6 to 7 years of age.

V. OTHER B-LYMPHOCYTE MARKERS

Many other membrane molecules have been identified in the membranes of B lymphocytes with monoclonal antibodies. Most are involved in cell-to-cell interactions, and several of them can also participate in the delivery of activating signals (Fig. 10.5). The most important are:

A. **CD20**, an antigenic cluster associated with the first membrane marker to be found on a developing B lymphocyte, originally designated as B1. It is detectable on

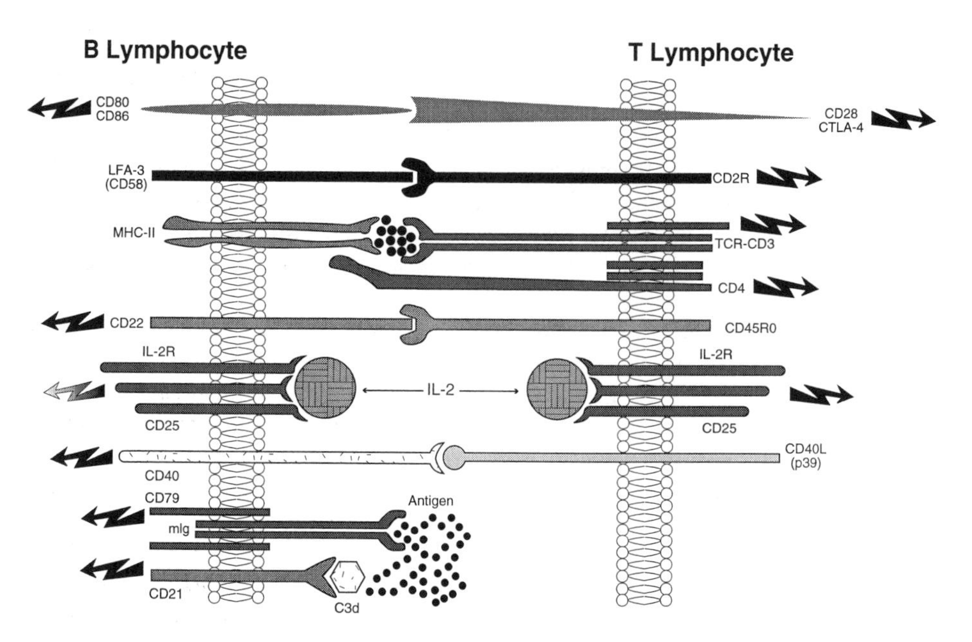

Figure 10.5 B- and T-lymphocyte membrane molecules involved in the delivery of activating signals. The role of IL-2 in B-cell signaling is not fully defined.

pre-B lymphocytes expressing cytoplasmic μ chains and remains expressed during maturation, on the mature B lymphocyte, but not on the plasma cell. It is an unusual molecule in that it crosses the membrane several times, has only 42 amino acid residues exposed on the outside, and both the amino and the carboxyl terminal ends are in the cytoplasm. The carboxyl terminal end has 15 serine and threonine residues, the hallmark of a protein susceptible to phosphorylation by protein kinases, which does occur after mitogenic stimulation. This suggests that CD20 may play an important role in the activation of mature B lymphocytes (see Chap. 11).

B. **CD45**, initially known as the leukocyte common antigen (LCA), is expressed by all leukocytes and their precursors. It is a major cell surface component of normal leukocytes where it occupies up to 10% of the surface, and exists in multiple isoforms generated by alternate splicing of nuclear RNA. B lymphocytes express only the highest molecular weight isoform of CD45. The most remarkable feature of CD45 is its cytoplasmic domain, which comprises 705 amino acids and is the largest intracytoplasmic domain of all the membrane proteins known to date. This intracytoplasmic domain has intrinsic tyrosine phosphatase activity and plays an essential role in lymphocyte activation.

C. **CD10**, also known as cALLA (**c**ommon **A**cute **L**ymphoblastic **L**eukemia **A**ntigen), was first detected on pre-B lymphocytes. It is subsequently turned off on mature B lymphocytes, but is expressed at high levels in some leukemic B lymphocytes and on neutrophils. This marker has endopeptidase activity, but its function is unknown.

D. **CD5** is expressed on most T lymphocytes as well as by a small subpopulation of B lymphocytes, and in most chronic lymphocytic leukemias. In nonleukemic individuals, B lymphocytes expressing CD5 appear to be activated and committed to the synthesis of IgM autoantibodies. However, there is no apparent correlation between disease activity and the numbers of circulating CD5+ cells; thus, the precise role of these cells remains speculative.

E. **Leukocyte Function-Associated Molecule 1 (LFA-1, CD11a)** and **Intercellular Adhesion Molecule-1 (ICAM-1, CD54)** are cell adhesion molecules (CAMs) expressed by both T and B lymphocytes, so that an LFA-1 in a T cell may interact with an ICAM-1 on the B cell and vice versa. Interactions involving complementary molecules equally expressed by the interacting cells are known as homotypic interactions.

F. **CD58 (Leukocyte Function-Associated 3; LFA3)** is expressed by a variety of cells, including B lymphocytes, and its complementary molecule, CD2, is found on T lymphocytes. Interactions such as these, involving two different molecules, each expressed in only one of the interacting cells, are known as **heterotypic interactions**.

G. **CD40, CD80 (B7-1)**, and **CD86 (B7-2)** are membrane proteins involved in signaling mediated by cell–cell contact. All are expressed at low levels on resting B cells and other APCs, but their expression increases rapidly after endocytosis of antigens.

1. **CD40** interacts with a molecule known as **gp39** or CD40 ligand (CD159) on helper T cells. This interaction is required for B-lymphocyte maturation and isotype switching. As discussed in Chapter 11, the lack of expression of the CD40 ligand is the molecular basis of most cases of a unique immunodeficiency disease known as hyper-IgM syndrome.
2. **CD80** and **CD86** interact, respectively, with **CD28** and **CTLA-4** (CD152) molecules, also expressed on T lymphocytes (see Chap. 11).

H. **Major Histocompatibility Complex (MHC) Antigens**. All B lymphocytes express high levels of both class I and class II MHC antigens. The presence of class II antigens is of particular importance because they enable B lymphocytes to serve as antigen-presenting cells. Thus, B lymphocytes are the only cells that combine antigen-specific effector properties (i.e., antibody synthesis) with the ability to present antigen to T lymphocytes.

VI. THE T-CELL RECEPTOR

A. Types and Basic Structure of the TcR

1. The TcR is a heterodimer with two unequal chains. Two different types of TcR have been identified.
   a. The first TcR to be detected during differentiation is known as **TcR1**. Its constituent chains are designated as γ and δ. The T cells expressing TcR1 are also known as γ/δ T cells. These T lymphocytes abound in the mucosal immune system and the skin, where they represent the dominant T-cell population. The precise biological role of $TCR1^+$ T lymphocytes is still a matter of considerable debate.
   b. The second type of TcR to differentiate is known as **TcR2** and its constituent chains are designated as α (acidic) (40 to 50 kD), and β (basic), slightly smaller (40 to 45 kD). The cells expressing TcR2 are

also known as α/β T lymphocytes and represent more than 95% of mature circulating T lymphocytes.

2. Both types of TcR are heterodimers, constituted by two unequal polypeptide chains consisting of a transmembrane domain and an intracytoplasmic tail, relatively constant and much shorter than the similar intracellular domains of immunoglobulins, and by an extracellular amino-terminal end.
3. The structure of TcR2 has been better characterized than the structure of TcR1. Both chains have constant (120-140 amino acids) and variable domains (100-120 amino acids), the variable domains located on the amino-terminal end of each chain. The length of the variable domains is similar to the length of the variable regions of the immunoglobulin heavy and light chains and hypervariable regions within the variable domain constitute the antigen binding structure, which is unique for each T-cell clone. Monoclonal antibodies directed against the antigen binding site of the TcR detect an idiotypic specificity unique to that particular clone.
4. Both the degree of genetic diversity and the surface area of the antigen binding site on the TcR2 are roughly identical to those of the antigen binding site of immunoglobulin molecules.
5. The antigen binding part of the TcR2 binds to the complex formed by an antigen-derived polypeptide and the groove of an MHC molecule. Since the TcR2 β chain is the most polymorphic, it seems likely to be predominantly responsible for the specificity of the interaction with the MHC-associated oligopeptides. Another area of the TcR2 interacts with neighboring but nonpolymorphic parts of the same MHC molecule. The less polymorphic α chain with its many J genes is believed to be involved in this interaction.

B. **TcR Genes**. Strategies similar to those used in the studies of the molecular genetics of immunoglobulin genes allowed the identification of the genes coding for the α, β, γ, and δ chains of the TcR. The β and γ chain genes are located in distant regions of chromosome 7, while the α and δ chain genes are located in close contiguity, on chromosome 14 (Fig. 10.6). Similarly to immunoglobulin heavy chains, each TcR chain is coded by four genes, designated as constant (C), diversity (D), joining (J), and variable (V); a Greek letter (α, β, γ, δ) is added to the capital letters designating each gene to identify the chain coded by each one. Like the immunoglobulin genes, the TcR genes undergo rearrangements through deletions of noncoding sequences and joining of segments of noncontiguous DNA. Comparison of the DNA sequences of the V-region genes showed that they could be grouped in families on the basis of sequence homologies.

C. **TcR Gene Binding-Site Diversity**. The random rearrangement of a large number of V, D, and J genes, and the random association of the two heterodimeric chains that constitute the TcR are the basis for binding-site diversity generation. In addition, during T-lymphocyte differentiation, random nucleotides are added at the D–J junction, further increasing the diversity. This process is catalyzed by **terminal deoxyribonucleotidyl transferase (Tdt)**, which adds deoxyribonucleotides to single-stranded DNA without a template. Calculations similar to those made to estimate the diversity of immunoglobulin genes were made to assess the potential diversity of the T-lymphocyte repertoire and reached similar conclusions. The number of possible recombinations between different V, D, and

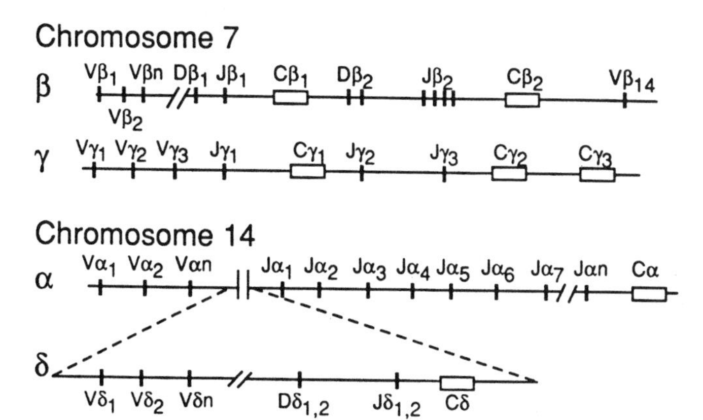

Figure 10.6 Genomic organization of the T-cell receptor genes. The genes for the β and γ chains rearrange independently at their own end of chromosome 7. The δ genes are in the middle of the α gene locus on chromosome 14. Rearrangement of the α genes leads to removal of the δ genes which are found as extrachromosomal DNA in cells that have productive rearrangements of the α and β genes yielding a $TcR2^+$ T cell. In $TcR1^+$ (γ/δ) T lymphocytes, the α genes are deleted.

J segments synthesized independently in the two heterodimer chains should yield at least 1×10^{10} different antigen binding sites. Therefore, a mature individual could possess at least that many different T-lymphocyte clones.

VII. T-LYMPHOCYTE ONTOGENY

A. **Role of the Thymus**. The thymus gland exists in all mammalian species, and its vital importance in the differentiation of T lymphocytes is reflected by the fact that when it is congenitally absent or surgically removed before immunological maturity is reached, a profound T-lymphocyte deficiency is observed. The T-lymphocyte deficiency in athymic and thymectomized mice is usually associated with severe immunocompromise and is incompatible with prolonged survival. T lymphocytes differentiate intrathymically from precursor cells originating in the bone marrow as early as the seventh week of gestation.

B. **Prethymocyte Markers**. The thymocyte precursors differentiated in the bone marrow express on their membrane the earliest T-lineage-specific marker, the **CD7** molecule. The next marker to be expressed is the **CD2** molecule, and both CD7 and CD2 will continue to be expressed by T lymphocytes at all subsequent stages of differentiation. The **CD1** marker, also expressed at high levels on prethymocytes, is barely detectable on mature T lymphocytes.

C. **Enzymatic Changes**. Once they reach the thymus, the T-lymphocyte-precursor stem cells proliferate very rapidly. Maturing thymocytes express several enzymes of the purine salvage pathway, such as **adenosine deaminase** (**ADA**), **purine nucleoside phosphorylase** (**PNP**), and **terminal deoxyribonucleotidyl transferase**, which undergo changes in the level of expression during ontogenic development (Fig. 10.7).

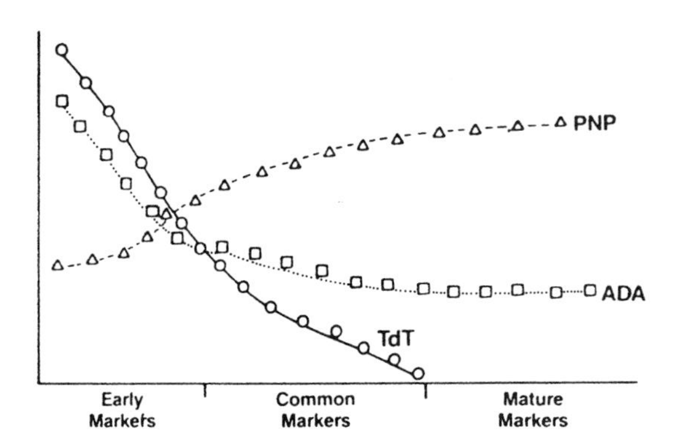

Figure 10.7 Longitudinal changes in the concentrations of purine nucleoside phosphorylase (PNP), adenosine deaminase (ADA), and terminal nucleotidyl transferase (TdT).

D. Rearrangement of the TcR Genes

1. **TcR1 formation**. Rearrangement of the γ and δ genes forming the TcR1 occurs first. The rearrangement of the δ gene in chromosome 14 cannot be done without eliminating the intervening α-chain genes. The differentiated γ/δ cells leave the thymus and home to the mucosal immune system and the skin.
2. **TcR2 formation**. The rearrangement of β-chain genes marks the onset of the development of a second set of T cells expressing a structurally different TcR (TcR2). The cells expressing this receptor are also known as α/β T lymphocytes.
   a. The process is initiated by an incomplete rearrangement that brings together the Dβ and any of the Jβ genes. This is rapidly followed by addition of a Vβ gene and a Cβ segment is added last. A complete β chain can be detected in the cytoplasm.
   b. Next, the genes coding for the α chain rearrange. The α chain has no D region but has at least 50 J genes, suggesting that its hinge region is the seat of great diversity.
   c. As a complete rearrangement of the TcR2 is generated, the recombinases involved in the processes become inactive, so that rearrangements on the second allele *does not* take place. Thus, the phenomenon of **allelic exclusion**, originally described for immunoglobulin synthesis, also exists for TcR synthesis.
   d. There are two significant differences between mIg and TcR, expressed on the membranes of B and T lymphocytes.
      i. Once the TcR has been completely rearranged, there will be no further change in their C gene expression (i.e., no isotype switching).
      ii. The TcR does not undergo somatic mutations or affinity maturation between primary and secondary immune responses. Table 10.1 summarizes the similarities and differences between mIg and TcR2.

Table 10.1 Comparison of the Similarities and Differences Between T- and B-Lymphocyte Receptors

During ontogeny	mIg (BcR)	TcR2
1st expressed chain	μ	β
2nd chain expressed later	κ or λ	α
Allelic exclusion	Yes	Yes
Short cytoplasmic domain	Yes	Yes
Association with co-receptors	Yes	Yes
Positive selection	No	Yes
After differentiation	**mIg (BcR)**	**TcR2**
Generation of new cells expressing the receptors	Continuous	Limited
Receptor rearrangements	Yes	No
Life span of memory cells	Weeks	Years

e. The differences in magnitude between primary and secondary cell-mediated immune responses results primarily from the increased number of specific memory cells that persists years after the primary immune response, and permits a faster and greater colonel expansion in the secondary response. In addition, a memory T cell also differs from a naive T cell in the set of co-receptors it expresses, as discussed later in this chapter.

E. **Addition of Co-Receptors and Formation of the TcR Complex**. After the rearrangement of the α- and β-chain genes of the TcR2 is complete, transport of the heterodimers to the membrane depends on their association to a second molecule, known as CD3 (Fig. 10.8).

1. **CD3** is a complex of five different subunits designated by the Greek letters γ, δ, ϵ, ζ, and η (note that the γ and δ chains of the CD3 unit are totally different from the γ/δ units of the TcR-1 heterodimer; the only common point is the use of the same Greek letters to designate them).
2. The CD3 γδϵ trimolecular complex is synthesized first and remains intracytoplasmic, where it becomes associated with TcR2 molecules. Soon thereafter, the CD3 ζ chains are synthesized and become associated to the CD3-TCR2 complex. Once the ζ chain has been added to the CD3 molecule, the whole CD3-TCR2 complex is transported from the Golgi apparatus to the cell membrane where it is inserted.
3. An important structural characteristic of the ζ chain is its long intracytoplasmic tail. This chain is involved in signal transduction. Occupancy of the TcR activates a protein kinase that rapidly binds to the ζ chain. It is known as the ζ associated protein or **ZAP70**.
4. Simultaneously with the differentiation of the TcR complex, two very important molecules (**CD4** and **CD8**) become simultaneously co-expressed on thymocytes. Thymocytes expressing both CD4 and CD8 are referred to as **double positives**.

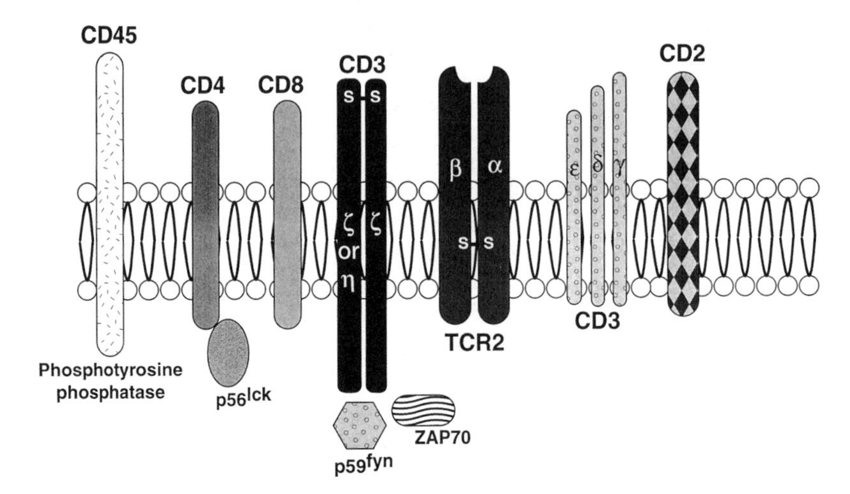

Figure 10.8 The TcR2/CD3 and associated membrane molecules in a pre-T lymphocyte. The α/β heterodimer is associated with the membrane to the CD3 complex constituted of the trimolecular complex CD3γδε and either a CD3ζζ dimer or a CD3ζη dimer. The CD45, CD2, CD4, and CD8 molecules, also expressed in the membrane of the pre-T lymphocyte, play important roles in T-lymphocyte differentiation, and later in T-lymphocyte activation.

F. **Thymocyte Selection**. The next steps in the differentiation of T lymphocytes will take place *in* the thymus, and basically ensure that useless or autoreactive pre-T lymphocytes will be eliminated before maturation. This process is based on the ability of TcR2 to interact with MHC-I or MHC-II molecules and the selection seems to be largely determined by the MHC molecules expressed by the thymic epithelial cells.

1. It is estimated that up to 99% of the $TcR2^{+}$ thymocytes will not interact with any MHC molecule. These lymphocytes will not be functionally useful, because signaling requires interaction with both MHC and its associated peptide. As a consequence of their lack of interaction with MHC molecules, these lymphocytes undergo **apoptosis** (programmed cell death, characterized by fragmentation of nuclear chromatin DNA and disintegration of the nucleus).
2. **Positive selection**. T lymphocytes with TcR2 which bind to MHC molecules expressed in the thymic environment with reasonable affinity (probably less than 1% of all the differentiating thymocytes) are rescued from death. The rescue requires adequate cell signaling, which is not delivered solely as a consequence of the MCH-TcR interaction. Other co-receptors must become involved to signal the cell toward positive selection.
   a. At this stage of development, the prethymocytes express both CD4 and CD8 molecules. As discussed in Chapter 3, CD4 interacts with class II MHC molecules, and CD8 with class I MHC molecules.
   b. When a TcR-2 interacts with MHC-I, two molecules on the T-lymphocyte membrane will be simultaneously engaged: TcR and CD8. The double activation signal will rescue the pre-T lymphocyte from its programmed death, turn off the CD4 gene, and promote the differentia-

tion of the pre-T lymphocyte into a CD8+ T lymphocyte that will rapidly leave the thymus.

c. Conversely, if the TcR-2 interacts with MHC-II, the CD4 molecule binds to a nonpolymorphic area and the TcR2. The resulting activating signal rescues the pre-T lymphocyte from its programmed death, turns off the expression of the CD8 molecule, and a CD4+ T lymphocyte will emerge and rapidly migrate away from the thymus.

d. Both CD4 and CD8 molecules play a role in the process of cellular signaling: their intracytoplasmic portions interact with $p56^{lck}$, a tyrosine kinase of the src family. It is believed that the ζ chain of the CD3 molecule may be the substrate for the action of the CD4 or CD8 linked tyrosine kinase $p56^{lck}$. Thus, the activation of the T lymphocyte would result from synergistic impulses derived from the TcR and the CD4 or CD8 molecules, whose converging point is the ζ chain of the CD3 molecule and its ZAP70 kinase (Fig. 10.9). The importance of T-cell signaling in positive T-lymphocyte selection and differentiation of the CD4+ and CD8+ subpopulations is underlined by the fact that in children with genetic abnormalities of the ZAP70 the lymphocytes fail to mature beyond the double positive stage.

3. **Negative selection**. While positive selection of MHC-interactive thymocytes is taking place, cells whose TcR would enable them to react with auto-antigens) are eliminated by negative selection (also known as **clonal deletion**).

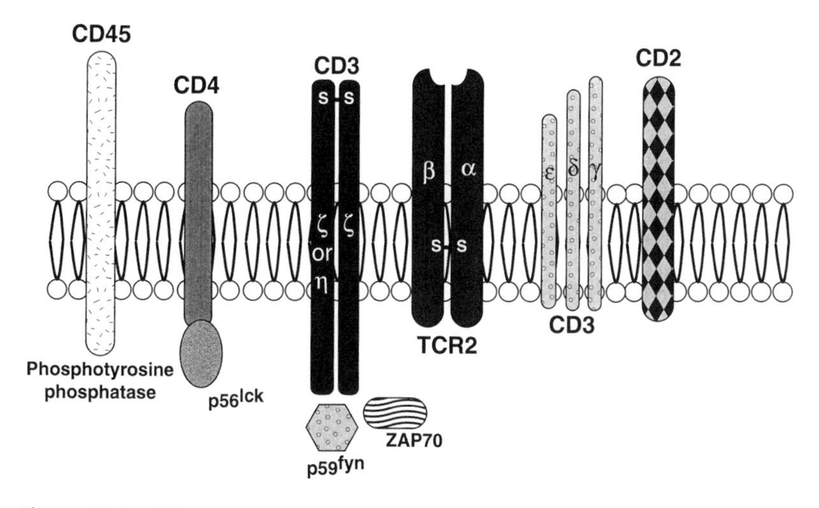

Figure 10.9 Diagrammatic representation of the TcR2 and its co-receptors on a mature CD4+ T lymphocyte. The α/β heterodimer is associated in the membrane to the CD3 complex constituted of the trimoleculer complex CD3γδε and either a CD3ζζ dimer or a CD3ζη dimer. It is believed that activating signals are transmitted by the ζ chain, but occupation of the TcR is not a very effective activator by itself. Other co-receptors play a role in the initial T-cell activation, such as CD45, which has intrinsic phosphatase activity, and CD4, which is associated to a $p56^{lck}$ tyrosine kinase. At least two more tyrosine kinases—$p59^{fyn}$ and zeta-associated p70 (ZAP 70) appear to play a role in the initial signaling cascade (see Chap. 11).

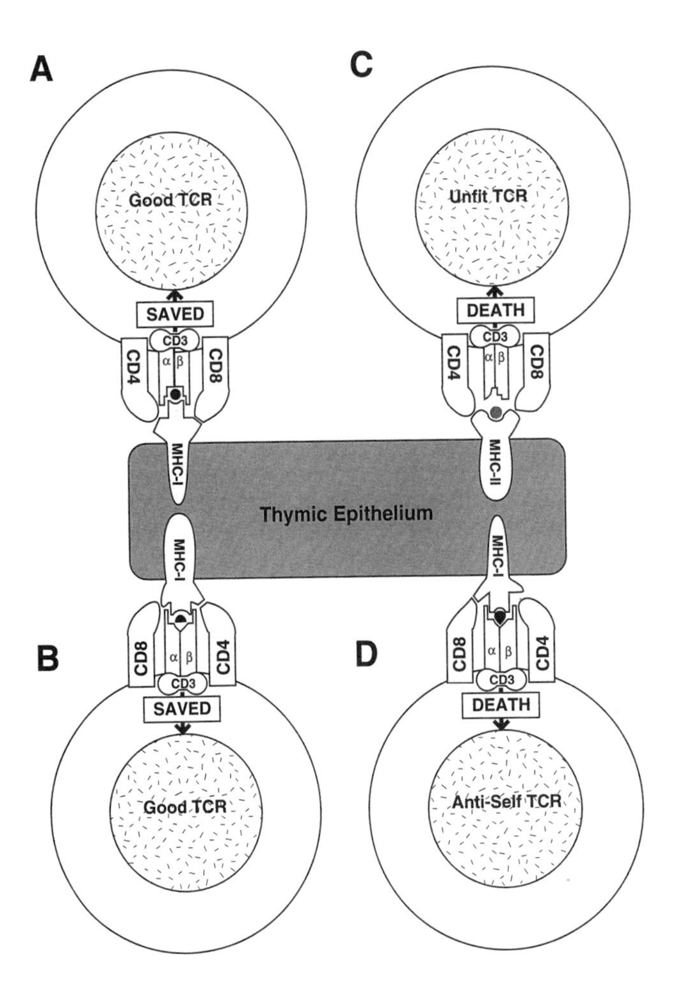

Figure 10.10 Diagrammatic representation of the role played by MHC in clonal selection during intrathymic lymphocyte differentiation. $TcR2^+$ pre-T lymphocytes, which co-express CD4 and CD8, interact with self-MHC class I and class II expressed on the thymic epithelium. (A) In this example, the $TcR2^+$ has a good fit with self-MHC class I, the CD8 molecule also interacts with the MHC-I, but the TcR does not interact with a self peptide associated with the MHC-I molecule. The dual interaction between CD8 and TcR with MHC-II will deliver a positive signal, transduced by the CD3, which will rescue the lymphocyte from programmed death. At the same time, the gene coding for the CD4 molecule, which did not interact with the MHC-I, will be repressed. Thus, a CD8+ T-lymphocyte clone will emerge from this positive selection process. (B) In this second example, the $TCR2^+$ has a good fit with self-MHC class II; the CD4 molecule also interacts with MHC-II, but the TcR does not interact with a self peptide associated with the MHC-II molecule. This lymphocyte will be rescued from programmed death by a positive signal transmitted by the CD3, and since the CD8 molecule did not interact with the MHC, its expression will be repressed. Thus, a CD4+ T-lymphocyte clone will emerge from the selection process. (C) If the TcR2 does not interact with either class I or class II MHC, that T cell does not receive any positive signal from CD3 and dies as programmed. (D) In this final example, a self-reactive pre-T cell is depicted. Its TcR interacts with MHC-I and with the self peptide associated with it. This results in a high-affinity interaction between the lymphocyte and the thymic epithelium, which accelerates the process of apoptosis, thus eliminating the autoreactive lymphocyte. This process of negative selection can eliminate both CD8+ and CD4+ autoreactive cells.

a. When random rearrangements have produced a TcR2 highly reactive with self-peptides expressed early in ontogeny, the MHC–TcR interaction delivers an extremely strong signal that triggers apoptosis in that T cell (Fig. 10.10).
b. If an autoreactive T cell escapes to the periphery, there are fail safe mechanisms that disable the TcR complex, turning the autoreactive T lymphocyte into an unresponsive cell.

G. **MHC Expression**. As the pre-T lymphocytes approach maturation, the gene products of the MHC become expressed on the membrane. The differentiated T lymphocytes that leave the thymus express only class I MHC antigens, but some subpopulations of T lymphocytes express class II MHC antigens after activation.

H. **The Differentiation of T Cells After Birth**. The thymus remains a very active site of T-cell production until early adulthood, becoming atrophic later on. The fact that memory T cells live much longer than memory B cells (up to decades) seems to ensure that an adequate supply of T lymphocytes will be available for the remaining adult life, but the possibility of extrathymic differentiation of T cells in adult life has not been ruled out.

VIII. MATURE T-LYMPHOCYTE MARKERS

A large number of different membrane molecules have been identified in T lymphocytes by means of monoclonal antibodies.

A. **Pan-T-Lymphocyte Antibodies** recognize structures common to most, if not all, T lymphocytes. Two major antibodies fall in this group:
1. **CD3** antibodies recognize the most specific T-lymphocyte marker, not expressed by any other type of leukocytes or by **NK cells**.
2. **CD2** antibodies recognize another pan-T marker, which was first identified by its interaction with sheep red cells, forming agglomerates of red cells surrounding single T lymphocytes known as **rosettes**. The red cell molecule responsible for the interaction with CD2 is homologous to the leukocyte function antigen (LFA)-3, expressed by human lymphocytes and accessory cells. It was later discovered that the LFA-3–CD2 interaction can stimulate T lymphocytes to proliferate and differentiate (see Chap. 11). NK cells also express CD2.

B. **T-Lymphocyte Subpopulation Markers** are defined by a variety of antibodies and combinations of antibodies. The two most important subpopulation markers are CD4 and CD8.
1. **CD4** antibodies identify a marker associated with the helper subpopulation of T lymphocytes.
2. **CD8** antibodies identify the cytotoxic subpopulation of T lymphocytes.
3. As shown in Figure 10.11, about three-quarters of peripheral blood mononuclear cells are T lymphocytes and, among T lymphocytes, CD4+ cells predominate over CD8+ cells.

C. **Markers Associated with T Lymphocyte Activation** include **HLA-DR** (**Class II MHC**), **CD25** (low-affinity **IL-2 receptor**), CD2R, CD71 (**transferrin receptor**), CD69 (**activation-inducer molecule**), CD54 (intercellular ad-

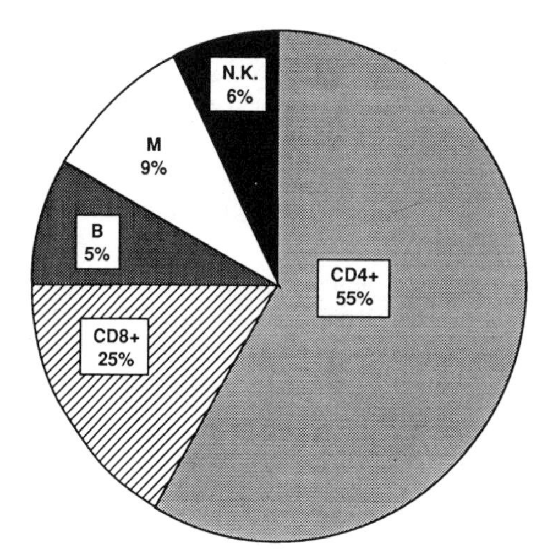

Figure 10.11 Graphic representation of the distribution of cell populations among peripheral blood mononuclear cells.

hesion molecule or **ICAM-I**), CD26 (**dipeptidylpeptidase IV**), CDw49b, CD30, and CD154 (CD40 ligand). Some of these markers are involved in different steps of the T-cell activation sequence.

1. The **low-affinity IL-2 receptor** (**CD25**) is expressed in low levels by about 30% of circulating (resting) lymphocytes, particularly on the CD4+/CD45RO$^+$ T-lymphocyte subset, which has been proposed to be associated with T-lymphocyte memory. Its level of expression is significantly increased after cell activation. Thus, monoclonal antibodies reacting with CD25 can be used to estimate the number of activated T lymphocytes, although it must be noted that activated B lymphocytes also express CD25.
2. **CD2R** is an epitope of CD2 expressed by activated T lymphocytes, and is probably the most specific of the activation markers. Its expression is likely to be associated with cell–cell interactions involving CD2 and LFA-3, which are believed to deliver additional activation signals to the T lymphocyte.
3. **CD45** is also known as the leukocyte common antigen (LCA), but its expression and function has been better characterized in T lymphocytes than in any other leukocyte. Three different isoforms (CD45RA, CD45RB, CD45RO) are detected on T lymphocytes. These isoforms share the intracellular domains but vary on their extracellular domains. CD45RA is the largest (220 kD), CD45RB is of intermediate size (205 kD), and CD45RO is the smallest (180 kD). The ontogenic development of the CD45 isoforms is not clear, but in mature lymphocytes their expression becomes restricted (as indicated by the letter **R** in the designation of these isoforms).
   a. The **CD45RA** marker (initially designated as 2H4 from the first monoclonal antibody that identified it) is expressed by a subset of

immunologically inexperienced (or naive) CD4+ T lymphocytes that have never encountered the antigen that they are programmed to recognize. After a first encounter with the antigen, the CD45RA marker ceases to be expressed and the CD45RO marker becomes detectable.

b. The **CD45RO$^+$** T lymphocyte is considered as either a **primed T lymphocyte** or a **memory T lymphocyte** (the expression of this marker seems to be maintained long after the primary response has waned), and cells with this phenotype seem to function as helper to B lymphocytes during humoral immune responses. The molecular basis for this role of CD45RO$^+$ lymphocytes has been suggested by the fact that the ligand for CD45RO is the CD22 molecule, expressed by activated B lymphocytes.

c. The CD45 molecule is believed to play an important role in T-lymphocyte activation due to the fact that its intracytoplasmic domain has tyrosine phosphatase activity.

4. **CD28** is expressed by mature thymocytes about to leave the thymus. In peripheral blood, CD28 is expressed on about 95-100% of CD4+ T lymphocytes, and on about 50% of CD8+ T lymphocytes. CD28 and a related molecule (CTLA-4 molecule) interact with the B-lymphocyte antigen BB7 (CD80, CD86), normally found at low levels on B cells and monocytes, but expressed at much higher levels within hours after endocytosis of antigens. Those interactions are believed to play important roles in the activation of cooperating T and B lymphocytes during the early stages of the immune response.
5. **CD40 ligand (CD154, CD40L, gp28)** is expressed transiently on activated CD40 helper T lymphocytes. The interaction between T cells expressing CD40L and CD4+ B lymphocytes plays a central role in the development and maturation of T-dependent B-lymphocyte responses.

IX. NATURAL KILLER CELLS

This small lymphocyte population has remained elusive in many respects. Morphologically, most NK cells fit into the population of large granular lymphocytes. Functionally, they are able to kill virus-infected or malignant cells with low or absent MHC molecules. NK cells are neither T nor B lymphocytes: TcR and immunoglobulin genes are in the unrearranged genomic configuration.

A. Ontogeny. Natural killer (NK) cells are most likely produced in the bone marrow from a common lymphocyte progenitor.

B. Membrane Markers

1. NK cells express some of the markers found in T cells, such as **CD2**, **CD7**, and **CD8**.
2. They also express markers found in phagocytic cells, such as the **CD16** molecule (low-affinity receptor for the Fc fragment of immunoglobulin G).
3. **CD56** antibodies recognize a cell-adhesion molecule unique to NK cells, among mononuclear cells, but also expressed is neural tissues. This last marker is phylogenetically very primitive, since it is expressed by the NK

cells of teleost fish. This suggests that NK cells are a primitive line of defense against infectious agents and appeared before the development of T and B lymphocytes.

Appendix Major Leukocyte Markers Recognized by Monoclonal Antibodies

CD designation	Major cell distribution	Synonyms, functional association(s)
CD2	All T lymphocytes; NK cells	LFA-3 receptor; rosette formation
CD3	All T lymphocytes	Transducing unit for TcR
CD4	Helper T lymphocytes, monocytes	Interaction with MHC-II
CD5	All T lymphocytes; activated autoreactive B lymphocytes	B-lymphocyte autoreactivity
CD7	Most T lymphocytes; NK cells; platelets	Unknown
CD8	T cytotoxic/suppressor cells; NK subset	Interaction with MHC-I
CD10 (cALLA)	Pre-B lymphocytes, graunlocytes	Endopeptidase
CD11a	Leukocytes	Cell-adhesion molecule (LFA-1 α chain); CD54 receptor
CD11b (Mac1)	Monocytes, granulocytes, NK cells	Cell-adhesion molecule; complement receptor 3
CD11c	Monocytes, granulocytes, NK cells	Cell-adhesion molecule
CD16	Neutrophils, monocytes, most NK cells	FCγRIII (low affinity)
CD18	Leukocytes	Cell-adhesion molecule (LFA-1 β chain)
CD19	All B lymphocytes	B-lymphocyte signaling
CD20	Mature B lymphocytes; dendritic cells	
CD21	Mature B lymphocytes; dendritic cells, T-lymphocyte subset	Complement receptor 2; Epstein-Barr virus receptor; B-lymphocyte activation
CD22	B lymphocytes	Interaction with CD45RO^{+} T lymphocytes
CD23	Activated B lymphocytes, eosinophils, platelets, and macrophages	FcεRII (low affinity)
CD25	Activated T and B lymphocytes	IL-2 receptor α chain
CC26	Activated T and B lymphocytes	Dipeptidylpeptidase IV
CD28	T lymphocytes, plasma cells	T-B interactions and activation
CDw29	Leukocytes	Integrin
CD30	Activated T and B lymphocytes	
CD32	B lymphocytes, monocytes, granulocytes	FcγRII (p40)
CD35	Granulocytes, monocytes, dendritic cells	Complement receptor 1
CD40	B lymphocytes	Binds to CD154 (CD40L, gp39) on activated helper T lymphocytes
CD44	All leukocytes	Adhesion molecule; mediates adhesion of lymphocytes to endothelial cells

Appendix Continued

CD designation	Major cell distribution	Synonyms, functional association(s)
CD45	All leukocytes	Leukocyte common antigen
CD45RA	T-lymphocyte subset, B and NK cells	"Naive" CD4+ cells
CD45RB	T-lymphocyte subset, B lymphocytes, granulocytes	"Memory" CD4+ lymphocytes
CD45RO	Memory CD4+ lymphocytes	Binds to CD22 on B lymphocytes
CD54	Leukocytes (weak), epithelial cells	Cell adhesion molecule (ICAM-1); ligand for CD11 a/18; Rhinovirus receptor
CD55	Hematopoietic cells	Decay-accelerating factor
CD56	NK cells, neural cells	Homotypic adhesion molecule (N-CAM)
CD57	NK subset, T-cell subset	
CD58	Most nucleated cells, erythrocytes	Ligand for CD2; intercellular adhesion molecule (LFA-3)
CD64	Monocytes	FcγRI (high affinity)
CD69	Activated lymphocytes	Activation-induced molecule
CD71	Activated lymphocytes and macrophages; proliferating cells	Transferrin receptor
CD79 a,b	B lymphocytes	Igα,β part of the BcR
CD80	Activated B lymphocytes; other APC	B7-1, interacts with CD28
CD86	Activated B lymphocytes; other APC	B7-2, interacts with CD152 (CTLA-4)
CD95	T and B lymphocytes	Fas antigen; interacts with the Fas ligand (FasL); apoptosis signaling
CD152	T lymphocytes	CTLA-4, interacts with CD86; down-regulates T lymphocytes
CD154	Activated helper T cells	CD40L, interacts with CD40 expressed by B lymphocytes; B-cell activation and differentiation

SELF-EVALUATION

Questions

Choose the ONE *best* answer.

10.1 Which of the following characteristics is first expressed by a pre-B lymphocyte?
- A. Cytoplasmic IgM
- B. Cytoplasmic light chains
- C. Membrane IgD
- D. Membrane IgM
- E. Rearranged V-D-J DNA regions

10.2 Which one of the following steps is used for the selection of hybridoma clones producing specific antibodies?
- A. Culture of the mixture of fused and nonfused cells in HAT medium

B. Fusion of splenocytes and plasmocytoma cells with polyethylene glycol
C. Harvesting of splenocytes from a mouse immunized with the relevant antigen
D. Limiting dilution of fused cells
E. Selection of a malignant plasma cell line producing antibodies of the desired specificity

10.3 Which of the following blocking monoclonal antibodies is likely to prevent the differentiation of IgG-producing cells in a mitogenically stimulated mononuclear cell culture?
A. Anti-CD19
B. Anti-CD2
C. Anti-CD3
D. Anti-CD40
E. Anti-CD8

10.4 An autoreactive CD4+/CD8+ pre-T lymphocyte is believed to be eliminated as a consequence of its interaction with a thymic epithelial cell
A. Expressing autologous MHC-I or MHC-II molecules
B. Presenting the self antigen recognized by the TcR in association with either the MHC-I or the MHC-II molecules
C. Presenting the self antigen recognized by the TcR in association with the MHC-I molecule
D. Presenting the self antigen recognized by the TcR in association with the MHC-II molecule
E. Presenting the self antigen recognized by the TcR on the cell membrane

10.5 Which of the following is a characteristic of the α/β T-lymphocyte receptor?
A. Association with the CD3 molecular complex on T-lymphocyte membranes
B. Early expression on pre-T lymphocytes lacking specific CD markers
C. High-affinity interaction with class I MHC molecules
D. Homodimeric structure
E. Up-regulation after T-lymphocyte activation

10.6 Which of the following lymphocyte markers is **NOT** likely to be detected on an activated B lymphocyte?
A. CD25 (IL-2 receptor)
B. CD2R
C. Class I MHC
D. Class II MHC
E. CR2

10.7 Which one of the following properties better defines CD4+ T lymphocytes?
A. Ability to deliver co-activating signals to other T and B lymphocytes
B. Expression of the γ/δ T-cell receptor
C. Function predominantly as antigen-specific cytotoxic T lymphocytes
D. High-affinity interaction with cells expressing class I MHC molecules
E. Susceptibility to infection by the Epstein-Barr virus

10.8 The first synthetic product of a B lymphocyte is(are):
A. IgM
B. IgD
C. μ chains
D. κ chains
E. J chains

10.9 After birth, human B-lymphocyte differentiation takes place in the
A. Bone marrow
B. Liver
C. Gut-associated lymphoid tissue
D. Spleen germinal center
E. Peyer's patches

10.10 Which of the following genes rearrange during the immune response?
A. Immunoglobulin heavy-chain C region genes
B. Immunoglobulin light- and heavy-chain C region genes
C. TcRα chain constant region genes
D. TcRβ chain constant region genes
E. Variable region genes of immunoglobulins and TcRs

Answers

10.1 (E) The first evidence for differentiation of a lymphocyte of the B-lymphocyte lineage is a DNA rearrangement that brings together the VDJ regions of chromosome 14. Soon thereafter, intracytoplasmic μ chains can be detected, preceding the rearrangement of chromosome 2 or 22 needed for light-chain synthesis and expression of membrane IgM.

10.2 (D) Although the antibody-producing cells are obtained from the spleens of mice immunized with the relevant antigen, the resulting hybrids contain a mixture of clones producing antibodies of different specificities. To obtain clones of cells producing antibodies of one single specificity, it is necessary to separate individual fused (hybrid) cells and allow them to proliferate into discrete clones. Such separation can be obtained by limiting dilutions or by cell sorting.

10.3 (E) A monoclonal antibody to CD40 will block the interaction between CD40 and its ligand. This interaction is essential for the switch from IgM to IgG synthesis.

10.4 (B) The differentiation of CD4+/CD8+ pro-T lymphocytes into CD4+ or CD8+ pre-T lymphocytes depends on the ability of T-lymphocyte precursors to interact with either MHC-I or MHC-II molecules expressed by the thymic epithelium. The cells not interacting with MHC are lost; those interacting with MHC and simultaneously recognizing an MHC-associated self peptide are deleted. This elimination of autoreactive T lymphocytes will be mediated by interactions with peptides associated either to MHC-I or to MHC-II molecules.

10.5 (A) The α/β TcR is a heterodimer constituted by one alpha and one beta chain which interact with both MHC-I and MHC-II molecules. Its expression on the differentiating T lymphocyte membrane is a relatively late event, since CD markers characteristic of the T-cell lineage are expressed at earlier stages of differentiation. The expression of TcR remains is not altered after T-cell stimulation. The transport of the α/β TcR receptor to the cell membrane of a differentiating T lymphocyte is believed to depend on its association with the CD3 complex, which also provides the TcR with a transducing unit required for cell signaling after antigen recognition.

10.6 (B) B lymphocytes express MHC-I, MHC-II, and CR2 regardless of their state of activation. The IL-2 receptor (CD25) is expressed by both activated T and B lymphocytes. In contrast, CD2 and its epitope associated with cell activation (CD2R) are expressed by T lymphocytes but not by B lymphocytes.

10.7 (A) The CD4 molecule identifies the helper T-lymphocyte subpopulation, which delivers co-activating signals to other T- and B-lymphocyte subpopulations.

10.8 (C) Before IgM is found on the cell membrane, intracytoplasmic m chains can be detected in B-lymphocyte precursors.

10.9 (A) Plasma cells apparently can differentiate both in the germinal centers of lymphoid organs and in the bone marrow, but resting B lymphocytes differentiate in the bone marrow.

10.10 (A) Gene rearrangements during the immune response are responsible for the "switch" between IgM and other isotypes during the immune response and are exclusive of the C-region genes of the immunoglobulin heavy chains.

BIBLIOGRAPHY

Banchereau, J., Bazan, F., Blanchard, D., et al. The CD40 antigen and its ligand. *Annu. Rev. Immunol.*, *12*:881, 1994.

Barclay, A.N., Birkeland, M.L., Brown, M.H., et al. *The Leukocyte Antigen Facts Book*. Academic Press, San Diego, CA, 1993.

Bierer, B.E., and Burakoff, S.J. T lymphocyte adhesion molecules. *FASEB J.*, *2*:2584, 1988.

Blackman, M., Kapler, J., and Marrack, P. The role of the T lymphocyte receptor in positive and negative selection of developing T lymphocytes. *Science*, *248*:1335, 1990.

Jameson, S.C., Hogquist, K.A., and Bevan, M.J. Positive selection of thymocytes. *Annu. Rev. Immunol.*, *13*:93, 1995.

Naor, D. A different outlook at the phenotype-function relationships of T cell subpopulations: Fundamental and clinical implications. *Clin. Immunol. Immunpathol.*, *62*:127, 1992.

Robey, E., and Fowles, B.J. Selective events in T cell development. *Annu. Rev. Immunol.*, *12*:675, 1994.

Scharff, M.D., Roberts, S., and Thammana, P. Hybridomas as a source of antibodies. In *The Biology of Immunologic Disease* (F.J. Dixon, and D.W. Fisher, eds.). Sinauer, Sunderland, MA, 1983.

von Boehmer, H., and Kisielow, P. How the immune system learns about self. *Sci. Am.*, *265*(4):74, 1991.

11
Cell-Mediated Immunity

Jean-Michel Goust and Barbara E. Bierer

I. INTRODUCTION

Immune responses have been traditionally subdivided into humoral (antibody-mediated) and cellular (cell-mediated). As a rule, humoral immune responses function predominantly in the elimination of soluble antigens and the destruction of extracellular microorganisms, while cell-mediated immunity is more important for the elimination of intracellular organisms (such as viruses). This compartmentalization of the immune response is an oversimplification, and it is now clear that there is significant interplay between the humoral and cellular arms of the immune response. The humoral response depends on lymphokine production ("help") by T lymphocytes, while some types of cell-mediated effector mechanisms depend on antibodies for target selection. Nevertheless, this subdivision of the immune response is useful because of its practical application.

II. T-CELL SIGNALING

A. **Antigen Recognition**. Early observations showed that T lymphocytes cannot respond to soluble, unmodified antigens to which B lymphocytes obviously respond with antibody synthesis. T cells recognize antigen-derived peptides associated to self-MHC molecules (see Chaps. 3 and 4). The tertiary structure of the oligopeptides generated as a consequence of processing by APCs has no resemblance to the tertiary structure of the native antigen. In other words, T and B cells respond to very different structures of one given antigen.

1. **TcR occupancy**. Antigen-derived oligopeptides fit into the binding sites of MHC-I or MHC-II molecules; MHC-II/peptide complexes expressed on the surface of APC are recognized by the T-cell receptor of CD4+ lymphocytes, while MHC-I/peptide complexes expressed by a variety of cells are recognized by the T-cell receptor of CD8+ lymphocytes. The recognition of MHC-associated oligopeptides is the first step in T-cell activation. While the following discussion applies mostly to CD4+ T cells, the general principles also apply to CD8+ T cells.
2. **Clonal Restriction of T-Cell Activation**. Each T cell carries a unique TcR which recognizes only one MHC antigen complex. This interaction delivers an antigen-specific signal that is essential, but not sufficient, for T-cell

activation. TcR occupancy, by itself, only induces limited T-cell activation and functional differentiation, probably because of the low affinity of the interaction between the MHC-II/peptide complex and the TcR.

B. **T Cell–APC Interaction**. The proper activation of CD4 T cells requires the establishment of strong bonds with APCs. A variety of complementary molecules on the membranes of APC and T lymphocytes stabilize the interaction between the two cells and strengthen the bond between the TcR and the MHC peptide complex (Fig. 11.1). Two sets of adhesion molecules (CD2/CD58 and LFA-1/ICAM-1) play a primary role in stabilizing the interaction between T lymphocytes and APC, and other sets may play ancillary roles:

1. **CD2** molecules, expressed by essentially all T lymphocytes, react with **CD58 (LFA-3)** molecules expressed by most nucleated cells as well as by erythrocytes. The initial interaction between the MHC/peptide complex and the TcR causes a conformational change on the CD2 molecule which then allows CD2 to interact with high avidity with the CD58 molecule expressed by the APC. In addition to its role in stabilizing cell–cell contact, the interaction between CD2R and CD58 delivers an activating signal to the T lymphocyte.
2. The **leukocyte function antigen-1 (LFA-1)** molecule, which interacts with the **intercellular adhesion molecule 1 (ICAM-1)**, also undergoes conformational changes in the early stages of T-cell activation. Both APC and T cells express LFA-1 and ICAM-1, and the homotypic interaction between pairs of these molecules results in the formation of strong intercellular bonds.
3. The **CD4** molecule on the lymphocyte membrane interacts with nonpolymorphic areas of the class II MHC molecules on the APC. This interaction

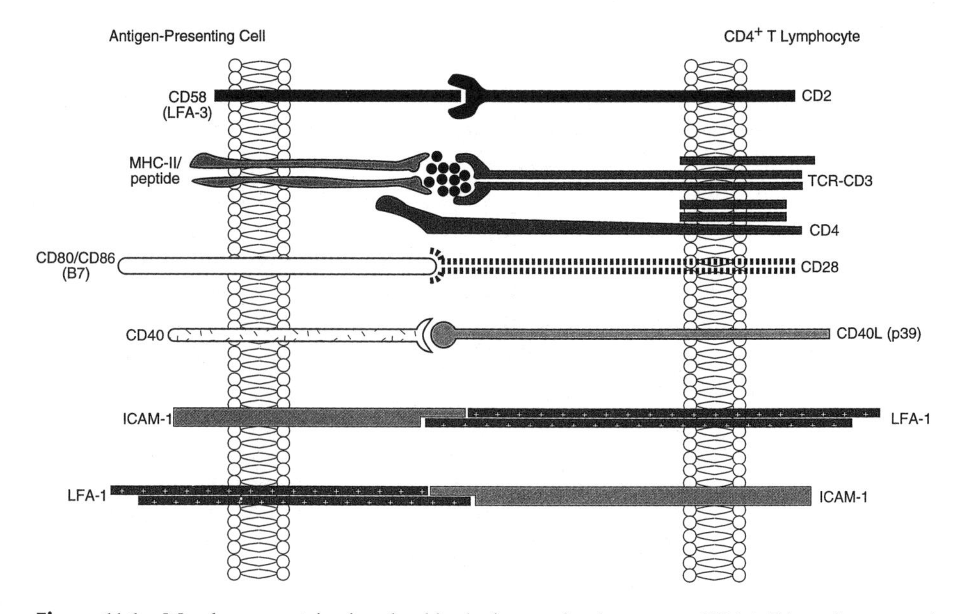

Figure 11.1 Membrane proteins involved in the interaction between a CD4+ T lymphocyte and an antigen-presenting cell (APC).

may not only help stabilize the contact between MHC-II and TcR, but it also seems involved in signal transduction in the early stages of T-lymphocyte activation, as discussed later in this chapter.

C. **Co-Stimulatory Signaling of T Lymphocytes**. The establishment of APC–T-lymphocyte interactions brings the membranes of the two cells into close proximity. Such close contact is critical for the delivery of additional activating signals to the T cell and allows for high local concentrations and maximal effects of the interleukins and cytokines released by APC and T lymphocytes.

1. Co-stimulatory signals dependent on cell–cell interactions. In general terms, receptor–ligand interactions appear to be associated with biochemical signals that modulate the signal delivered by the engagement of the antigen-specific TcR. The number, nature, and sequence of these co-stimulatory signals impact strongly on the T-cell response. The best characterized co-stimulatory interactions involve CD2 and CD58, MHC-II and CD4, CD28 and B7 family members, CD40L and CD40.
2. Co-stimulatory signals delivered by interleukins. A number of interleukins, including interleukin-1 (IL-1), interleukin-6 (IL-6), and interleukin-12 (IL-12), believed to be involved in the activation of helper T cells, are secreted by APC activated during ingestion and processing of immunogenic substances. These interleukins provide essential signals necessary for the proliferation of antigen-primed T lymphocytes, acting as *co-factors* in early T-cell activation.

D. **Active Response vs. Anergy**. Neither interleukins nor the signaling interactions involving different sets of membrane proteins can fully activate antigen-primed helper T cells by themselves; the two types of signals have a synergistic effect that is essential for T-cell activation, proliferation, and differentiation. It is believed that when a T cell is exposed to an MHC antigen complex and fails to receive an adequate complement of co-stimulatory signals, that T cell becomes unresponsive or anergic to the antigen in question (see Chap. 22).

III. INTRACELLULAR EVENTS ASSOCIATED WITH THE ACTIVATION OF CD4+ T LYMPHOCYTES

When the adequate complement of activating signals is delivered to a CD4+ T lymphocyte, a cascade of intracellular signals is triggered. The activation sequence can be divided into two distinct phases.

A. **First Phase: From TcR Occupancy to Interleukin-2 Gene Expression**

1. Signal transduction after TcR occupancy. As discussed in Chapter 10, the TcR heterodimer itself has no recognizable kinase activity, and the activation sequence must be triggered by kinases indirectly coupled to the TcR. Several tyrosine and serine/threonine kinases have been proposed to play a role in T-lymphocyte activation (Fig. 11.2).
    a. The src-related tyrosine kinase **p56^{lck}**, which is noncovalently associated with the cytoplasmic domains of CD4 and of CD8, is apparently activated by dephosphorylation mediated by CD45, a tyrosine phosphatase constitutively expressed on all hematopoietic cells. The nature of the physiological activating signal for CD45 phosphatase activity is not

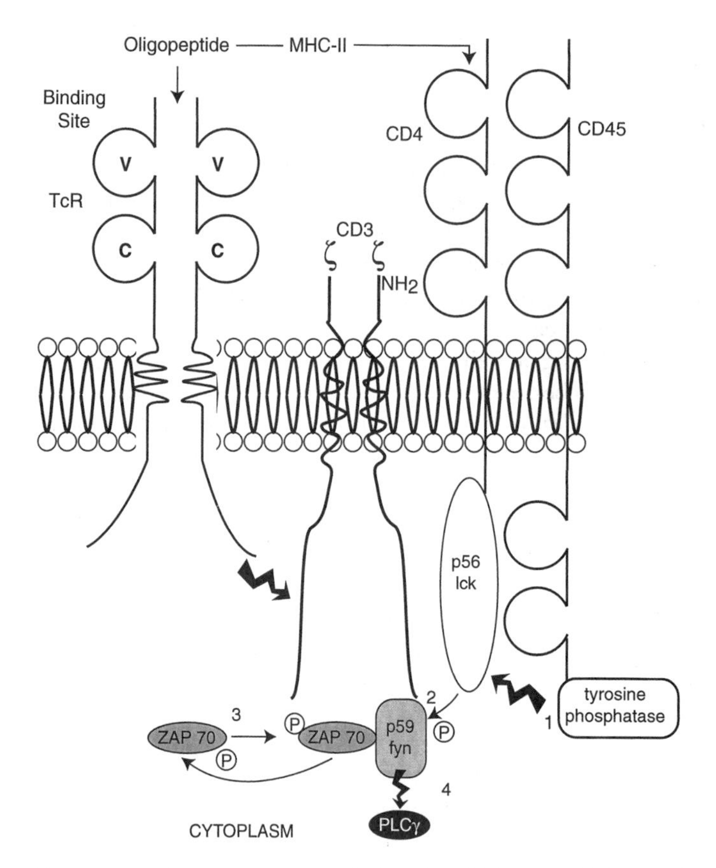

Figure 11.2 Sequence of events during the initial stages of T-cell activation. Antigen occupancy of the TcR induces modifications of the CD45 transmembrane protein. These modifications result in activation of its tyrosine phosphatase activity; one of the substrates of CD45 is a $p56^{lck}$ tyrosine kinase. Dephosphorylation of $p56^{lck}$ results in its enzymatic activation. Activated $p56^{lck}$ phosphorylates $p59^{fyn}$ and this is followed by recruitment and phosphorylation of ZAP70. The phosphorylation of ZAP70 is followed by activation and translocation of phospholipase C, which results in rapid membrane inositol turnover and generation of inositol 1,4,5, triphosphate (IP3) and diacylglycerol (DAG).

clear, but it may be related to either TcR occupancy or to the interaction between MHC and CD4 or CD8.

b. In addition to $p56^{lck}$ activation, a second src-related tyrosine kinase, **$p59^{fyn}$**, is activated; $p59^{fyn}$ is associated with the ζ chain of the CD3 molecule after TcR occupancy.
c. The activation of $p56^{lck}$ and of $p59^{fyn}$ is followed by recruitment and activation of a third tyrosine kinase, the **ZAP70 kinase**, which also appears to be associated with the ζ chain of the CD3 complex after activation.
d. The activation of ZAP70 leads to subsequent phosphorylation and activation of additional substrates, including **phospholipase Cγ (PLC-γ)**.

   e. Apparently, as a consequence of PLC-γ activation, membrane phosphoinositols are hydrolyzed yielding two potent intracellular second messengers. One of these messengers is **inositol 1,4,5, triphosphate (IP3)**, responsible for a sharp rise in intracellular ionized calcium concentration by mobilizing $\mathbf{Ca^{2+}}$ from intracytoplasmic stores. The rise in Ca^{2+} allows activation of the calcium and calmodulin-dependent serine/threonine phosphatase **calcineurin**. The second is diacylglycerol (DAG), responsible for the activation of **protein kinase C (PKC)**.
   f. Other pathways of phosphoinositol hydrolysis resulting from the engagement of co-stimulatory molecules are being intensively studied. One that has received considerable attention involves the activation of a **PI3 kinase**, triggered as a consequence of the engagement of the **CD28** molecule on the T-cell membrane by its ligand (**CD80**), expressed on the membrane of activated APCs.
2. **Synthesis and activation of transcription factors**. The exact sequence of events leading from the appearance of these cytoplasmic second messengers to the up-regulation of specific genes in the nucleus is largely unknown. Over 70 genes can be expressed in a characterizable sequence starting minutes after activation of a helper cell and continuing for the next several days. Several recently identified DNA-binding proteins which are part of those sequences are believed to play a significant role in gene activation and have been the object of considerable attention.
   a. **Nuclear factor κB (NFκB)** is a DNA-binding protein found complexed with a specific inhibitor (**IκB**) in the cytoplasm of resting T lymphocytes. Upon phosphorylation by an as yet unidentified tyrosine kinase activated at the early stage of T-cell activation, IκB dissociates from NFκB. The untethered NFκB is then able to move into the nucleus where it binds to the enhancers of many genes, including the IL-2 receptor gene and the *c-myc* gene, and plays a role in their transcriptional activation (Fig. 11.3).
   b. **Nuclear factor of activated T cells (NF-AT)** is almost exclusively found in hematopoietic cells. NF-AT binds to regions of specific DNA sequences, often contained within the 5′ promoter region of cytokine genes (e.g., the enhancer region of IL-2). Two components are required to form a complex capable of transcriptional activation.
      i. Prior to inactivation, an inactive form of NF-AT is located in the cytoplasm. This form is activated by dephosphorylation catalyzed by the calcium-dependent phosphatase known as calcineurin.
      ii. The activated cytoplasmic component of NF-AT translocates to the nucleus, where it becomes associated with AP-1. The synthesis of the nuclear component is activated by PKC.
      iii. Once the active NF-AT/AP-1 complex is assembled in the nucleus, it is able to bind to specific NF-AT sequences of a number of genes. Binding of the NF-AT/AP-1 complex to IL-2 promoter is essential for the expression of the Il-2 gene.
   c. **AP-1** is induced when T cells are stimulated via the TcR plus ancillary signals mediated by IL-1 or by CD28 cross-linking. It is formed by the association of two proto-oncogene products, **c-Fos** and **c-Jun**, whose

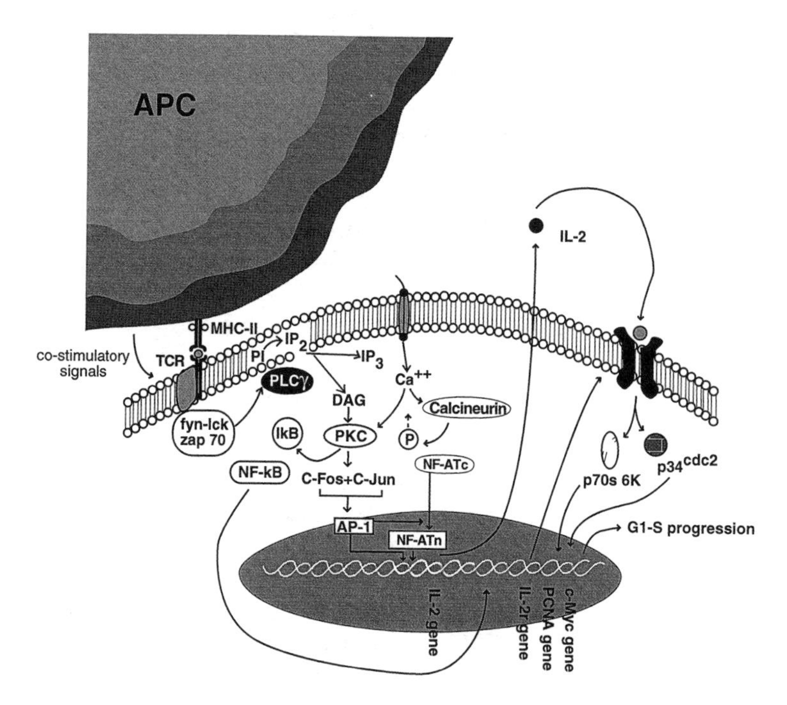

Figure 11.3 Diagrammatic representation of a simplified second messenger cascade involved in T-lymphocyte activation. After the sequence of events illustrated in Figure 4.8 takes place, two important enzymes are activated. Calcineurin, a serine/threonine phosphatase, dephosphorylates and activates the cytoplasmic component of the nuclear factor of activated T cells (NF-ATc). A complex of NF-ATc and calcineurin translocates to the nucleus where it forms a complex with AP-1, and the complex activates the expression of the IL-2 gene. Protein kinase C is also activated and the active form has at least two critical effects: (1) It promotes the association of c-Fos and c-Jun into AP-1, which translocates to the nucleus. In the nucleus AP-1 binds directly to the IL-2 gene enhancer region and combines with NF-ATc to form the active form of NF-AT (NF-ATn), which synergizes with AP-1, enhancing the expression of the IL-2 gene. (2) PKC activates a kinase that splits the complex formed between the nuclear factor kB (NFκB) and an inhibitory protein (IκB). The activated NFκB then translocates to the nucleus where it enhances the expression of several genes, including the one coding for the IL-2 receptor. The T cell will progress from G0 to G1, but additional signals are required for progression to S1. Synthesis and release of IL-2 associated with increased expression of the IL-2 receptor creates the conditions necessary for such progression. The occupancy of the IL-2 receptor is followed by activation of protein kinases and diacylglycerol kinase, followed by the activation of several transcription factors of the STAT (signal transducers and activators of transcription) family, and increased transcription of proteins controlling cell proliferation, such as the proliferating-cell nuclear antigen (PCNA), an obligate co-factor of DNA polymerase delta, which plays a significant role in DNA replication, and c-Myc, which controls cell proliferation.

synthesis is activated by PKC. The cytoplasmic form of c-Jun is activated by a specific kinase (**JNK kinase**), and forms a complex with activated c-Jun known as the AP-1 transcription factor. The AP-1 complex diffuses into the nucleus, where it binds to specific site DNA sequences leading to up-regulation of the expression of the IL-2 and IL-2R gene.

3. IL-2 synthesis and T-lymphocyte proliferation. Two early events in helper T-cell activation are the appearance of mRNAs for interleukin-2 (IL-2) and for the IL-2 receptor (IL-2R) in the cytoplasm. The expression of these two genes is critically important for the proliferation of activated T cells.
   a. **Regulation of IL-2 gene expression**. The IL-2 promoter region needs to be occupied by the NF-AT/AP-1 complex before the transcription of the IL-2 gene can start. If one of the nuclear factors is not generated, IL-2 will not be produced. Production of cytokines, including IL-2 regulated by the NF-AT/AP-1 complex is required for T-cell proliferation.
   b. **Consequences of incomplete signaling**. T lymphocytes receiving only a partially activating signal can either become anergic or undergo apoptosis. Apoptosis involves the expression of an alternative set of genes which ultimately causes cell death. The mechanisms controlling the evolution toward anergy vs. apoptosis are not fully understood.
   c. **Autocrine and paracrine effects of IL-2**. IL-2 stimulates lymphocyte proliferation both in autocrine and paracrine loops.
      i. Autocrine stimulation involves release of IL-2 from an activated T cell and binding to IL-2 receptors expressed by the same T cell.
      ii. Paracrine stimulation probably takes place as a consequence of the overproduction of IL-2, which takes place after persistent CD4 stimulation. The released IL-2 exceeds the binding capacity of the IL-2 receptors expressed by the producing cell and can stimulate other nearby cells expressing those receptors.
      iii. The targets of the paracrine effects of IL-2 are helper T lymphocytes, cytotoxic T lymphocytes, B lymphocytes, and NK cells, all of which express IL-2 receptors.
   d. **The IL-2 receptor**. The IL-2R expressed by T and B lymphocytes is constituted by three different polypeptide chains (Fig. 11.4).
      i. **CD25**, a 55-kD polypeptide chain (α), is expressed at very low levels in about 30% of the circulating (nonactivated) T cells but is sharply up-regulated a few hours after activation. CD25 binds IL-2 with low affinity and has a short intracytoplasmic domain unable to transduce growth signals upon binding to its ligand.
      ii. A 70–75-kD second chain (β) is expressed by activated T lymphocytes. The IL-2R α/β heterodimer has an increased affinity for IL-2. The β chain appears capable of signal transduction as its intracytoplasmic domain is 286 amino acids long. In contrast, the extracellular segments of the α and β chains have almost the same length.
      iii. A third polypeptide chain (γ) is also part of the trimeric IL-2 receptor expressed after T-cell activation. It, too, has a long intracytoplasmic segment. This chain is shared by a number of cytokine receptors, including those for IL-2, IL-4, IL-7, IL-9, and IL-15.
      iv. The binding of IL-2 to the trimeric high-affinity IL-2 receptor causes the association of the IL-2R to the src-related $p56^{lck}$ tyrosine kinase and to other specific tyrosine kinases (JAK family of

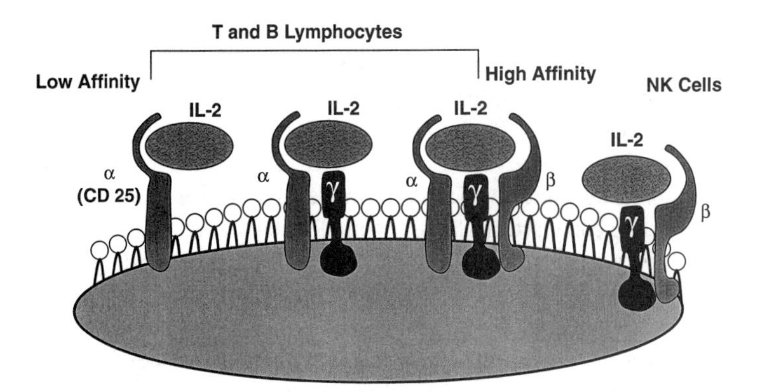

Figure 11.4 Diagrammatic representation of the different types of IL-2 receptors found in lymphoid cells. Resting T lymphocytes express the low-affinity IL-2 receptor, whose α chain is also known as CD25, while activated T and B lymphocytes express the trimeric high-affinity IL-2 receptor. NK cells express a CD25-negative IL-2 receptor constitutively, and the high-affinity trimeric receptor after activation.

kinases). As a consequence, the complex formed by IL-2 and the IL-2R heterodimer is internalized, and activating signals that lead to cell division are generated (Fig. 11.3).

B. **Second Phase: Progression to Mitosis**. IL-2 synthesis and IL-2R expression are sustained for about 48 hours and subsequently down-regulated. Activation of IL-2 gene expression is only a first phase. Signaling coming from other molecules is needed to maintain its expression and sustain the proliferation drive.
   1. CD86-CD28 interactions. Twelve hours after effective T-cell stimulation by APC, CD86 (B7-2)—a molecule related to CD80 (B7-1)—becomes expressed on the APCs, probably as a consequence of autocrine cytokine stimulation. CD86 activates T cells through interaction with a membrane protein known as CD28.
      a. The interaction between CD86 and CD28 results in the synthesis of a different set of DNA binding proteins that appear to block the rapid decay of the IL-2 mRNA.
      b. Preventing the CD80/CD86 CD28 interaction (for example, by blocking CD28) in the T-cell activation sequence has the same effect as the absence of full occupancy of the IL-2 promoter in the first phase, resulting in anergy.
      c. An alternative ligand for CD80/CD86, CD152 (CTLA-4), also expressed on activated T cells, delivers an inhibitory signal to the T cell. Thus, the differential expression of CD28 and CTLA-4 seems to play a critical regulatory role in the activation of helper T cells, but the factors controlling the expression of these two related molecules have not been defined.
   2. T-lymphocyte proliferation and differentiation. IL-2 synthesis and IL-2R expression are sustained for about 48 hours and are subsequently down-regulated. However, the T cell is committed to cell division within 8 hours of IL-2 interacting with IL-2R. Even if all triggering signals are withdrawn,

the T cell will divide. Clonal expansion of antigen-specific CD4+ T lymphocytes takes place as a consequence of the activation of this autocrine loop.

a. It is estimated that the initial helper T-lymphocyte population is capable of expanding 100-fold in 6 days, reaching more than 1000-fold its starting number by day 10, the time that it takes to elicit a detectable primary immune response to most infectious agents.
b. This expansion results in the differentiation not only of the short-lived effector cells needed for the ongoing immune response but also of long-lived memory cells. Following such rapid and robust T-cell expansion, many of the activated effector cells will undergo apoptosis. However, when the immune response subsides, an expanded population of memory T cells will remain and will be able to assist the onset of subsequent responses to the same antigen with greater efficiency and speed, which is characteristic of an anamnestic (memory) response.

IV. INTERLEUKINS AND CYTOKINES

One of the most remarkable consequences of the activation of CD4+ T cells is the synthesis of interleukins. Although activated CD8+ cells may produce some interleukins, and APCs themselves produce others, CD4+ T cells are their major source (Table 11.1). The effects of interleukins and cytokines are extremely diverse and they influence not only the immune response but inflammatory processes and hematopoiesis (Table 11.1).

A. **Proinflammatory Cytokines**. The group of soluble factors that influence inflammatory reactions include **interleukin-1 (IL-1)**, **tumor necrosis factor-α (TNF-α**, also known as **cachectin**), **interferon-γ**, **interleukin-8**, and **migration inhibition factor** (**MIF**). IL-1 and TNF-α seem to be directly or indirectly responsible for the systemic metabolic abnormalities and circulatory collapse characteristic of shock associated with severe infections. The main biological functions of these two cytokines are listed in Table 11.2.

1. **Interleukin-1 isoforms**. IL-1 exists in two molecular forms, IL-1α and IL-1β, coded by two separate genes and displaying only 20% homology to one another. In spite of this structural difference, both forms of IL-1 bind to the same receptor and share identical biological properties. IL-1α tends to remain associated to cell membranes, while IL-1β, synthesized as an inactive precursor, is released from the cell after being processed post-translationally by a cysteine protease (interleukin converging enzyme, ICE). This protease, or a very closely related enzyme, plays a crucial role in apoptosis signaling.
2. **Biological properties of IL-1 and TNF-α** (Table 11.2)
   a. **Metabolic effects**
      i. These interleukins affect the liver by inducing the synthesis of many proteins such as alpha-1 antitrypsin, fibrinogen, and C-reactive protein (known generically as acute phase reactants, because of their increase in situations associated with inflammatory reactions).
      ii. In cases of prolonged and severe infections, protracted TNF-α production results in negative protein balance, loss of muscle mass, and progressive wasting (cachexia).

Table 11.1 Major Interleukins and Cytokines

Interleukin/ cytokine	Predominant source	Main targets	Biological activity
IL-1α,β	Macrophages, monocytes, and other cell types	T and B lymphocytes	Stimulates T cells; activates several types of cells; proinflammatory mediator; pyrogen
IL-2	TH1 lymphocytes	T & B lymphocytes	Activates T cells and B cells
IL-3	T lymphocytes, mast cells	Hemopoietic stem cells; basophils	Hematopoietic growth factor; chemotactic for basophils
IL-4	TH2 lymphocytes, macrophages	B lymphocytes, TH1 lymphocytes, macrophages	Growth and differentiation factor for B lymphocytes; promotes IgE synthesis
IL-5	TH2 lymphocytes, macrophages	Eosinophils, lymphocytes	Chemotactic and activating factor for eosinophils; lymphocyte activation
IL-6	B & T lymphocytes, macrophages	B & T lymphocytes, others	B-cell differentiation factor; polyclonal B-cell activator; proinflammatory mediator; pyrogen
IL-7	Bone marrow, thymic epithelium	Pro-lymphocytes (B and T)	Lymphoid cell growth factor
IL-8	Macrophages	Neutrophils, T lymphocytes	Neutrophil and T-lymphocyte chemotactic factor
IL-9	Thymocytes, T lymphocytes	Hematopoietic stem cells	Hemopoietic growth factor
IL-10	TH2 lymphocytes	B lymphocytes, TH1 lymphocytes	Inhibits cytokine synthesis; activates B lymphocytes
IL-11	Mesenchymal cells	Hematopoietic stem cells	Induces megakaryocyte proliferation
IL-12	Macrophages, B lymphocytes	TH1 lymphocytes, NK cells	Natural killer cell stimulating factor; TH1 lymphocyte activation and proliferation; enhances the activity of cytotoxic cells
IL-13	Activated T lymphocytes	B lymphocytes, monocytes	Promotes immunoglobulin synthesis; suppresses monocyte/macrophage functions
IL-14	T lymphocytes	B lymphocytes	B-cell growth factor
IL-15	Monocytes/ macrophages	T lymphocytes, NK cells, B lymphocytes	Lymphocyte and NK cell growth factor; chemotactant for T lymphocytes
GM-CSF	TH2 lymphocytes	Hematopoietic stem cells	Promotes proliferation and maturation of granulocytes and monocytes
IFN-α/β	Leukocytes, fibroblasts	Lymphocytes	NK activator; proinflammatory

Table 11.1 Major Interleukins and Cytokines (Continued)

Interleukin/ cytokine	Predominant source	Main targets	Biological activity
IFN-γ	TH1 lymphocytes	Multiple	Macrophage and NK activator
TNF-α (cachectin)	APCs, CD4+ T lymphocytes	Multiple	Cytotoxic for some cells; cachexia; septic shock mediator; B-lymphocyte activator
TNF-β (lymphotoxin)	T & B lymphocytes	Restricted	Cytotoxic for some cells; PMN and NK activation

b. **Vascular effects**. IL-1β and TNF-α cause the up-regulation of cell adhesion molecules, particularly P-selectin and E-selectin, in vascular endothelial cells. This up-regulation has two types of consequences:
   i. Adherence of inflammatory cells, which eventually egress to the extravascular space, where they form tissue inflammatory infiltrates.
   ii. Endothelial cell damage by activated inflammatory cells, which is a major component of Gram-negative septic shock and toxic shock syndrome, dramatic examples of the adverse effects of massive stimulation of cells capable of releasing excessive amounts of interleukins (see Chap. 13).

c. **Central nervous system effects**. IL-1 does not cross the blood–brain barrier but acts on the periventricular organs where the blood–brain barrier is interrupted. It interacts with a group of nuclei in the anterior hypothalamus, causing fever and sleep, and also increases the production of ACTH.

Table 11.2 Cellular Sources and Biological Effects of IL-1 and TNF-α

	IL-1	**TNF-α**
Cellular source	Monocytes, macrophages, and related cells	Monocytes, macrophages, and related cells; CD4+ T lymphocytes
Biological property		
Pyrogen	+	+
Sleep inducer	+	+
Shock	+	+
Synthesis of reactive proteins	+	+
T-cell activation	+	+
B-cell activation	+	+
Stem-cell proliferation and differentiation	+	−

3. **Interferon-γ (IFN-γ)** is produced by CD4+ cells and NK cells and has a wide range of effects.
   a. In concert with IL-1β, it induces an increase in the expression of ICAM-1 on the cytoplasmic membrane of endothelial cells, enhancing **T-lymphocyte adherence to the vascular endothelium**, an essential first step for T lymphocyte egress from the vascular bed. Large numbers of T lymphocytes will thus exit the vascular bed in areas near the tissues where activated T cells are releasing interleukins and will form perivenular infiltrates, characteristic of delayed hypersensitivity reactions.
   b. The most important effect of IFN-γ seems to be the **activation of monocytes**. After exposure to IFN-γ, monocytes differentiate into phagocytic effector cells. Several major changes occur in monocytes activated with IFN-γ.
      i. The cell membrane becomes ruffled; the number of cytoplasmic microvilli increases by a factor of ten. This change reflects a considerable increase in phagocytic capacities.
      ii. The expression of MHC class II antigens and Fcγ receptors increases. The increase in the expression of MHC-II enhances the efficiency of monocytes as antigen-presenting cells, and the increased expression of Fc receptors further enhances their efficiency as phagocytic cells.
      iii. The production of TNF-α and of interleukin-12 are up-regulated.
      iv. There is an increased production of several different antibacterial products, including cathepsins (proteolytic enzymes), collagenases, superoxide radicals, and nitric oxide (NO). As a consequence, engulfed cells or proteins are rapidly killed or digested in the phagolysosomes.
   c. Excessive and protracted production of IFN-γ may have adverse effects. Hyperstimulated monocytes may become exceedingly cytotoxic, and may mediate tissue damage in inflammatory reactions and autoimmune diseases.
4. Chemokines. This designation is given to a group of cytokines with chemotactic properties. The following cytokines are included in this group:
   a. **Interleukin-8 (neutrophil activating factor)** is released by T lymphocytes and monocytes stimulated with TNF-α or IL-1. It functions as a chemotactic and activating factor for granulocytes, helping in their recruitment to areas of inflammation and increasing their phagocytic and proinflammatory abilities. It has also been demonstrated to be chemotactic for T lymphocytes.
   b. **RANTES** (regulated on activation, normal T cell expressed and secreted), released by T cells, attracts T cells with memory phenotype, NK cells, eosinophils, and mast cells.
   c. **Macrophage inflammatory proteins** (MIP), released by monocytes and macrophages, attract eosinophils, lymphocytes, NK, and LAK cells.
   d. **Macrophage chemotactic proteins** (MCP), produced by monocytes, macrophages, and related cells, attract monocytes, eosinophils, NK, and LAK cells.

e. **Migration-inhibition factor** (MIF), released by T cells, monocytes, and macrophages, keeps macrophages in the area where the reaction is taking place and contributes to their activation, promoting the release of TNF-α. Recently, it has been found that endotoxin stimulates the release of MIF by the pituitary gland, and its release seems associated with increased mortality in the postacute phase of septic shock.

B. **Interleukins and Hematopoiesis**. Several different interleukins and cytokines have significant hematopoietic effects, promoting the proliferation and differentiation of various cell types.

1. Interleukin-1 (IL-1), acting as a growth factor on the bone marrow, is a major factor leading to the general mobilization of leukocytes, objectively reflected by peripheral blood leukocytosis, characteristic of bacterial infections.
2. Granulocyte-monocyte colony-stimulating factor (GM-CSF), released by activated T lymphocytes, stimulates granulocyte and monocyte production and release from the bone marrow. It also stimulates monocyte cytotoxic activity and expression of MHC-II molecules, dendritic cell proliferation and differentiation, and B-cell proliferation and differentiation (at least in vitro).
3. Interleukin-3 (IL-3) and stem-cell factor, both released principally by activated T lymphocytes, are important growth factors for early bone marrow progenitors, inducing production of a wide variety of leukocytes.
4. Interleukin-6 (IL-6), synthesized primarily by monocytes, macrophages, and other antigen-presenting cells, promotes the differentiation of B-lymphocyte precursors, as well as the differentiation of T-lymphocyte precursors into cytotoxic T lymphocytes.
5. Interleukin-7 (IL-7) is produced in the bone marrow, thymus, and spleen. Its main role seems to be to induce the proliferation of B- and T-lymphocyte precursors.

C. **Cytokine Genes and Structure**. The precise role of individual cytokines in the immune and inflammatory responses is difficult to determine because so many cytokines are produced by the same cells, and each one of the cytokines can exert different and overlapping effects on multiple targets. Shared biological effects cannot be explained by structural similarities between molecules. However, many cytokine receptors activate different but overlapping sets of protein kinases and other signaling molecules. In addition, several interleukin genes appear to be under the influence of the same set of nuclear binding proteins, and several interleukin receptors are structurally similar. These may be keys to understanding the plurality of interleukin responses and apparent redundancy of effects among structurally different interleukins.

1. Interleukin genes. These genes differ from each other in their coding sequence but have ($> 70\%$) homology in their untranslated 5′ region suggesting the use of common transcription factors, such as NF-AT. In addition, the RNA transcripts have many AUUUA repeats near the 3′ end, which, in part, determine the rate of mRNA decay and probably explain why most interleukins are only transiently expressed. It needs to be noted that the stability of mRNA for several cytokines is under some type of regulation, since it is enhanced after CD28 engagement.

2. Tertiary structure. All monomeric interleukins have similar tertiary structure: at least four α helices arranged in pairs of parallel symmetry, each pair being antiparallel relative to the other. The helices are joined by connecting loops that contain β helical sheets of variable length. Of those, the one located near the aminoterminal end is the recognition helix that interacts with the cytokine receptor.

D. **Cytokine receptors**. Cytokine receptors can be grouped into several families, depending on structural characteristics (Fig. 11.5). The activation pathways triggered after receptor occupancy tend to be similar for receptors of the same family but different for receptors of different families. The fact that several cytokines may share a given receptor explains why some biological properties are common to several interleukins.

1. The most common type of receptor is the **hemopoietin-receptor family**, so called because it was initially characterized as the erythropoietin receptor.
   a. The receptors of this family are heterodimers or heterotrimers and always include an α and a β chain, the latter with a longer intracytoplasmic segment and signaling functions.
   b. The receptors for IL-2, IL-3, IL-4, IL-5, IL-6, IL-7, IL-9, IL-15, and GM-CSF are included in this family. Some of them share subunits:
      i. Receptors for IL-3, IL-5, and GM-CSF share a common β chain. A different β chain is shared by the receptors for IL-6 and IL-11.
      ii. Receptors for IL-2, IL-4, IL-7, IL-9, and IL-15 share a third chain (γ), which plays a significant role in signal transduction. One of the forms of severe combined immunodeficiency is secondary to abnormalities in the γ-chain gene (see Chap. 30).
2. Other receptor families include:
   a. The **tumor necrosis factor receptor family**, which, besides the receptor for TNF-α and TNF-β (lymphotoxin), includes CD40 and Fas.
   b. The **immunoglobulin superfamily receptors**, which include the IL-1 (α and β) receptors, a receptor for colony-stimulating factor 1, and, as a special subfamily, the type I interferon receptors, homodimers of two chains of the immunoglobulin superfamily, which bind interferon α and β.

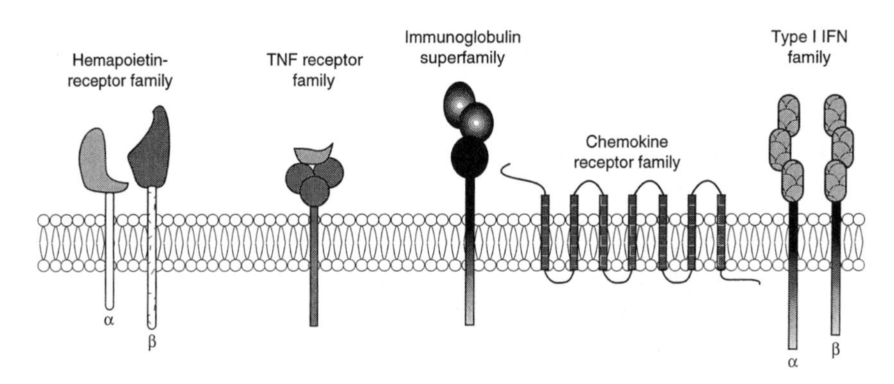

Figure 11.5 The main families of cytokine receptors: the chemokine receptors have seven transmembrane spans.

c. The **chemokine receptor family**, whose common features are seven transmembrane domains, including receptors for IL-8, platelet factor-4 (PF-4), RANTES, and macrophage chemotactic and activating proteins.

3. Up-regulation of a given subunit of these receptors is often a consequence of cellular activation and it usually results in the expression of a high-affinity receptor able to transduce activation signals.

V. T-HELPER-CELL SUBSETS AND THEIR REGULATORY ROLES

A. **TH1 and TH2 Populations**. Studies in experimental animals (particularly in mice) suggest that the release of cytokines may obey specific patterns that determine the type of effect observed on target cells. Murine antigen-specific T-helper lymphocyte clones can be divided into two groups according to the type of cytokines they release after repetitive stimulation. An analogous subclassification of T-helper lymphocytes has been more difficult to establish in humans, but recent work suggests that human antigen-specific CD4+ T-cell clones exhibit similar differences after long-term exposure to a stimulating antigen (Fig. 11.6).

1. TH1 clones produce predominantly IL-2, IL-3, IFN-γ, and TNFβ and assist the onset of delayed hypersensitivity, inflammatory, and cytotoxic reactions.
2. TH2 clones produce predominantly IL-4, IL-5, IL-6, and IL-10, and provide help to B lymphocytes.

B. **Factors Controlling TH1 and TH2 Differentiation and Activity**

1. Interleukin-12, produced by antigen-presenting cells, seems to play an important role in the differentiation and activation of TH1-helper lymphocytes. The activated TH1 cells release IFN-γ, which activates macrophages inducing additional release of IL-12 and a more efficient antimicrobial response, particularly against intracellular organisms.
2. Interleukin-4 is believed to play a critical role in the differentiation and activation of TH2 lymphocytes.
3. There is **reciprocal counter-regulation** of the differentiation and activation of TH1 and TH2 lymphocytes. Sustained production of high levels of IFN-γ suppress IL-4 production, down-regulating TH2 activity. In contrast, during a strong TH2 response, high levels of IL-4 and IL-10 have inhibitory effects on cytokine release, affecting both IFN-γ and IL-12. As a consequence, the expansion of TH1 cells is down-regulated.

VI. THE ROLE OF HELPER T CELLS IN INFLAMMATORY RESPONSES

A. **Delayed Hypersensitivity**. The understanding of the role of T lymphocytes in inflammation emerged with the study of a phenomenon known as delayed hypersensitivity (see Chap. 18). This type of abnormal immune reaction is easy to study when it is manifested as a cutaneous reaction, which is usually the case when the individual is sensitized and/or challenged by the introduction of a suitable antigen in the skin. For example, the tuberculin test is a skin test designed to determine if an individual is hypersensitive to antigenic products of *Mycobacterium tuberculosis*. A positive reaction to tuberculin is usually seen

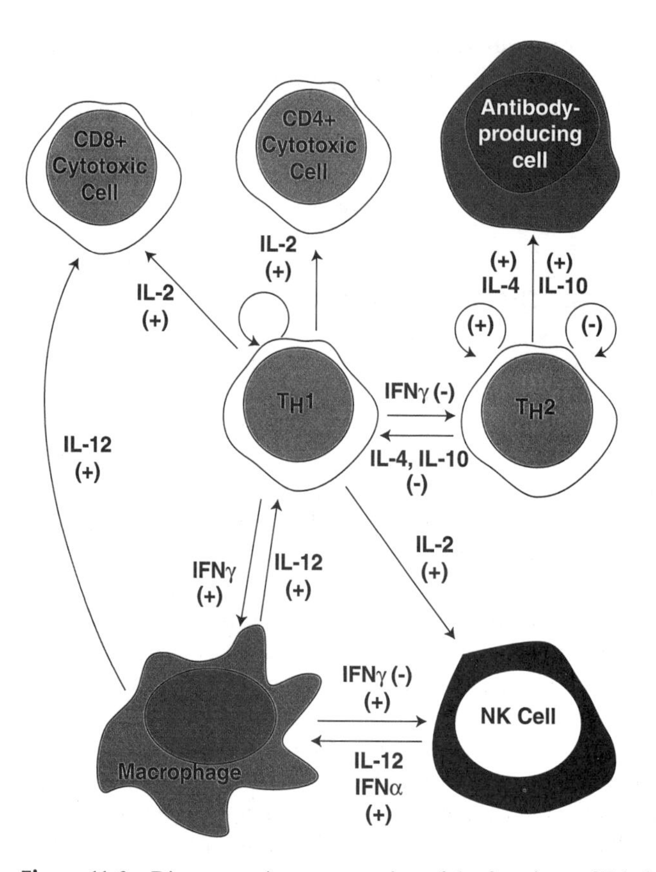

Figure 11.6 Diagrammatic representation of the functions of T-helper lymphocyte subpopulations.

24-48 hours after intradermal inoculation of the antigen, while patients allergic to pollens show a positive reaction to intradermally injected pollen extracts after a few minutes. For that reason, the tuberculin reaction is classified as a delayed hypersensitivity reaction.

B. **Role of T Lymphocytes in Delayed Hypersensitivity**. Studies with experimental animals showed that it was possible to transfer cutaneous delayed hypersensitivity reactions by transferring spleen cells from a sensitized animal to an MHC-identical nonimmune recipient. In this system, the cells primarily involved were CD4+ helper lymphocytes. Through the release of various lymphokines active on lymphoid and nonlymphoid cells, the activated CD4+ cells led to the formation of a cellular infiltrate around the area where they are reacting with the antigen. The cellular infiltrate was usually rich in lymphocytes and monocytes reflecting the proliferation of activated CD4+ lymphocytes and the release of chemotactic factors that attracted monocytes to the area.

C. **Other Pathological Consequences of T-Cell Hypersensitivity**

1. Granulomatous reactions are the expression of protracted T-cell activation in tissues, often caused by intracellular pathogens that have developed mechanisms of resistance to antimicrobial defenses. The granulomas con-

tain T lymphocytes, macrophages, histiocytes, and other cell types, often forming a barrier circumscribing a focus of infection. The formation of granulomas is due to the release of cytokines, which attract and immobilize other mononuclear cells. The activation of these cells in situ results in the release of enzymes that cause tissue destruction, and cytokines, which attract and activate additional inflammatory cells.

2. Graft rejection is also believed to be primarily mediated by activated CD4+ cells, through the attraction and activation of inflammatory cells to the grafted organ (see Chap. 27).

VII. T-CELL HELP AND THE HUMORAL IMMUNE RESPONSE

A. T-Dependent and T-Independent Antigens

1. As discussed in Chapter 4, antigens can be broadly subdivided into T-dependent and T-independent, according to the need for T-cell help in the induction of a humoral immune response. Most complex proteins are T-dependent antigens, while most polysaccharides can elicit antibody synthesis without T-cell help.
2. It is also clear that T and B lymphocytes cooperating in the inductive stages of an immune response do not recognize the same epitopes in a complex immunogen. The membrane immunoglobulin of the B cell reacts with epitopes expressed on the outside of the native antigen, whereas the cooperating T cell recognizes MHC-II-associated peptides derived from the processing of the antigen by accessory cells (a role that can be played both by monocytes/macrophages and by B lymphocytes).

B. Initial Activation of the Helper Cell (Fig. 11.7).

1. **Antigen presentation**. When an immunogen is introduced for the first time into an immunocompetent animal, the antigen is internalized, processed, and presented to helper T cells. Two types of cells can carry out this function:
   a. **Monocytes, macrophages**, and specialized cells such as the **interdigitating cells** found in the lymph node cortex (see Chap. 2) are most effective as antigen-processing cells. Their involvement in the primary immune response must involve the ingestion and processing of small quantities of antigen opsonized as a consequence of complement activation by the alternative pathway or to the binding of C-reactive protein (see Chap. 12).
   b. **Activated B lymphocytes** may also play the role of APC. Although B cells are uniquely suited for antigen recognition due to the existence of predifferentiated clones able to recognize epitopes in any newly introduced antigen in its native configuration, their role as APC is believed to require previous activation, which induces the expression of co-stimulatory molecules essential for the proper activation of CD4+ T lymphocytes.
2. **Antigen processing and presentation by B lymphocytes**. Once the membrane immunoglobulin of a B lymphocyte interacts with an epitope of a given immunogen, the complex of mIg-immunogen is internalized, and

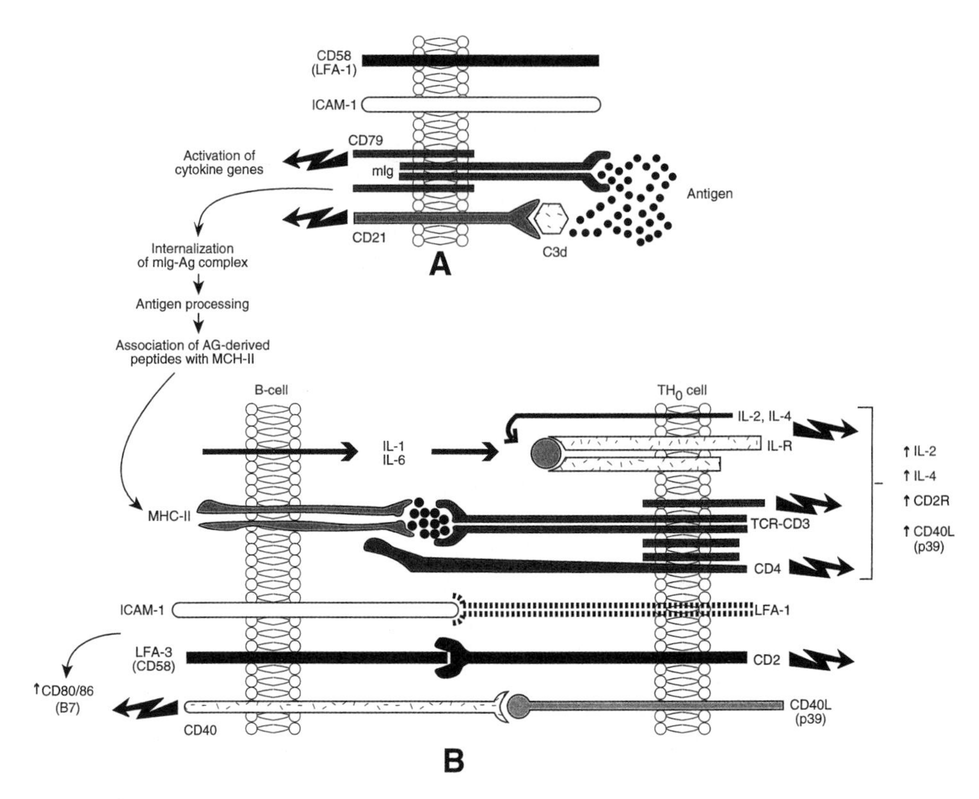
CD58
(LFA-1)
ICAM-1
CD79
Activation of
cytokine genes
mIg
Antigen
CD21
C3d
A
Internalization
of mIg-Ag complex
Antigen processing
Association of AG-derived
peptides with MCH-II
B-cell
TH0 cell
IL-2, IL-4
IL-1
IL-6
IL-R
↑IL-2
↑IL-4
↑CD2R
↑CD40L
(p39)
MHC-II
TCR-CD3
CD4
ICAM-1
LFA-1
LFA-3
(CD58)
CD2
↑CD80/86
(B7)
CD40
CD40L
(p39)
B

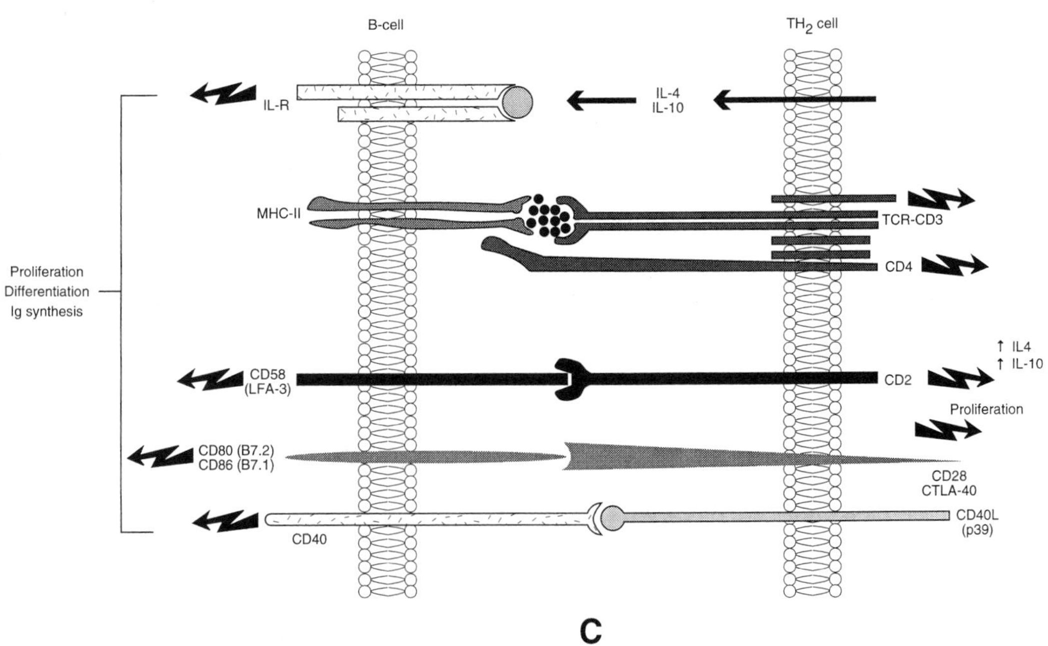
B-cell
TH2 cell
IL-R
IL-4
IL-10
MHC-II
TCR-CD3
CD4
Proliferation
Differentiation
Ig synthesis
CD58
(LFA-3)
CD2
↑ IL4
↑ IL-10
Proliferation
CD80 (B7.2)
CD86 (B7.1)
CD28
CTLA-40
CD40
CD40L
(p39)
C

mIg ceases to be expressed. Although B cells have considerably less cytoplasmic enzymes than professional phagocytes (monocytes, macrophages, and related cells), they are able (at least in vitro) to process the immunogen in an endosomal compartment, breaking it down into oligopeptides which become associated to MHC class II antigens. The MHC-II-oligopeptide complex is then transported to the membrane and presented to CD4+ T cells.

3. Delivery of co-stimulatory signals. In addition, these same activated B lymphocytes release IL-1 and IL-6 and can engage in co-stimulatory interactions with helper T cells mediated by up-regulated membrane molecules (see below). There is a need for high lymphocyte density and intense recirculation to fulfill the optimal conditions for T–B-cell cooperation, and those are most likely achieved in lymphoid tissues, such as the lymph nodes.

C. **T–B-Lymphocyte Interactions**. The cooperative interaction between T and B lymphocytes is believed to start outside the primary follicles, in the paracortical area. The subsequent interactions unfold in two consecutive phases.

1. Loose interaction. At this early phase, T–B-cell interactions are loose and involve molecules used by T cells in their interaction with APC, such as CD2, CD4, LFA-1, and ICAM-1. Other molecules involved in the early interactions between T and B cells include:
   a. CD40L (CD154) molecules on the T cell, which bind to CD40 on the B cell.
   b. CD5 molecules on the T cell, which may bind to CD72 molecules on the B cell.
   c. CD45RO molecules expressed exclusively by memory helper T lymphocytes may interact with CD22 molecules expressed by activated B lymphocytes.
2. Firm attachment. A few hours after this initial interactions, T and B cells become firmly attached due to the up-regulation of several sets of membrane molecules.
   a. Activated CD4 cells express increasing levels of CD40L (CD154).

Figure 11.7 Diagrammatic representation of the sequence of signals and interactions between TH2 cells and B cells taking place at the onset of an immune response. Panel A shows that the initial signal is provided by the recognition of an epitope of the native antigen by membrane immunoglobulin, the BcR. The occupancy of the BcR activates the synthesis of IL-1 and IL-6. After about 2 hours, the mIg-Ag complex is no longer detectable, having been internalized, and the B cells become mIg$^-$. The internalized Ag is processed and antigen-derived oligopeptides are loaded into nascent MHC-II. The MHC-II/peptide complexes are expressed in the B-cell membrane after about 4 hours. At that time, as shown in panel B, B lymphocytes will present the MHC-II-associated oligopeptide to the TcR. The interaction between T and B cells is stabilized by a variety of interactions, such as MHC-II/CD4, ICAM-1/LFA-1, and CD58/CD2. The T cell also receives activating signals in the form of IL-1 and IL-6, released by the B cell. The signals received by the T cell lead to the up-regulation of several interleukin genes, as well as to the expression of CD2R and CD154 (CD40L), which is detectable after about 12 hours. On the B-cell side, the expression of CD80/CD86 is also up-regulated. The up-regulation of these membrane molecules will allow a tighter interaction between the two cells, as well as the delivery of additional activating signals, as shown in panel C. The signals delivered to the B cell will result in proliferation and differentiation of antibody-producing plasma cells. At the same time, some of the interacting molecules (CD2, CD40L) deliver signals to the B cells that are essential for their continuing proliferation and activation.

b. Almost simultaneously, CD80 (B7-1), expressed at low or undetectable levels on B lymphocytes, is up-regulated. This protein and the constitutively expressed, closely related molecule, known as CD86 (B7-2) share CD28 and CTLA-4 (also inducibly expressed) as their counter-receptors.

c. As a consequence of these multiple interactions, the cells come to close apposition (the intercell distance is reduced to less than 12 nm), and activating signals can be transmitted both to T lymphocytes and B lymphocytes. There is also evidence suggesting that the membranes of contacting cells may fuse briefly, allowing direct trafficking of cytokines from the T cell to the B cell, a process that may have extraordinary significance in B-lymphocyte activation.

D. **B-Cell Proliferation and Differentiation**. After receiving all the above-mentioned activating signals, B cells enter mitosis and start the differentiation process toward antibody-producing plasma cells. This evolution is associated with migration through different areas of the lymph node.

1. First, the activated B cells separate from the helper T cell and migrate to the denser areas of the follicle, around the germinal centers, where they rapidly proliferate (the cycling time is about 7 hours). In 5 days, the antigen-stimulated B-lymphocyte population in a given germinal center increases by about 1000-fold. During this period, the Ig genes in the B cell undergo a series of rearrangements and differentiation steps.
2. Most resting B cells express IgM and IgD on the cell membrane, and in the initial stages of cell differentiation many cells will produce IgM antibody, characteristic of the early stages of the primary immune response.
3. As B cells continue to proliferate and differentiate, recombination genes will be activated (apparently as a consequence of cytokine-mediated signaling) and the process of class switch takes place. The constant region genes for μ and δ chains are looped out and one of the constant regions genes for IgG, IgA, or IgE moves into the proximity of the rearranged V-D-J genes (see Chap. 6). Subsequently, the synthesis of IgM antibody declines, replaced by antibody of the other classes, predominantly IgG.
4. The functional differentiation of B lymphocytes coincides with migration of the differentiating B cells and plasmablasts to different territories. First, the dividing B cells migrate into the clear areas of the germinal centers and into the mantle zone. Those areas are rich in CD4+, CD40L+ T cells which apparently deliver a critical signal to B cells, mediated by CD40–CD40L interaction.

   a. In murine models, activated B cells receiving a CD40-mediated signal together with co-stimulatory signals delivered by IL-2 and IL-10 become B memory cells. In contrast, if the CD40 ligand is blocked, inhibiting the delivery of the CD40-CD40L signal, the activated B cells differentiate into plasmablasts. In humans, the plasmablasts exit the lymph nodes through the medullary cords and migrate to the bone marrow, where they become fully differentiated, antibody-secreting plasma cells.

   b. Also in humans, CD40-mediated activating signals must be involved both in the differentiation of IgG-producing plasma cells and of memory B cells. Patients born with a defective CD40L gene suffer from an immunodeficiency known as the hyper-IgM syndrome in which B cells cannot switch from IgM to IgG production and immunological memory is not generated (see Chap. 30).

VIII. ACTIVATION AND DIFFERENTIATION OF CD8+ CYTOTOXIC T CELLS

A. **The Biological Significance of T-Cell-Mediated Cytotoxicity**. An essential function of cell-mediated immunity is the defense against intracellular infectious agents, particularly viruses. For example, circulating T lymphocytes isolated from individuals who are recovering from measles infection destroy MHC-identical fibroblasts infected with this virus in 2 to 3 hours. A number of experimental models has provided insights into the mechanisms of lymphocyte-mediated cytotoxicity against virus-infected cells.

B. **Recognition and Activation of Target Cells by Cytotoxic T Lymphocytes**
   1. **Initial stimulation of CD8+ cytotoxic T cells**. The first signal leading to the differentiation of cytotoxic T lymphocytes involves recognition of a viral peptide associated with an MHC-I molecule. As pointed out in Chapters 3 and 4, such recognition is only possible if the TcR and MHC-I molecules are able to interact weakly in the absence of the viral-derived oligopeptide, and this is only possible when the two interacting cells are MHC-identical, since only T lymphocytes able to interact with self-MHC molecules are selected during differentiation (see Chap. 10).
   2. **Requirement for helper T cells**. While strongly activated CD8+ cells have been shown to release some cytokines, including IL-2, simple contact with a virus-infected target cell does not trigger the proliferation of CD8+ T lymphocytes. Under these conditions, proliferation and differentiation of CD8+ cells appears to require IL-2 and other cytokines produced by helper CD4+ T lymphocytes. Thus, the responding T-lymphocyte population contains a mixture of CD4+ and CD8+ cells, and the stimulating population of viral-infected cells must present viral antigens in association with both MHC class I and MHC class II molecules: The CD4+ T lymphocytes will recognize viral antigens associated with class II MHC molecules while the CD8+ T lymphocytes recognize viral antigens complexed to class I MHC molecules.
   3. **Proliferation and differentiation of cytotoxic T cells**. The initial proliferation of helper and cytotoxic T cells takes place mostly during the prodromal stages, while symptoms are minimal. After a few days, the number of differentiated helper and cytotoxic T cells expands and clinical manifestations of the disease appear, resulting in part from the release of a variety of interleukins and cytokines from infected cells and helper T cells and in part from the destruction of viral-infected cells by the fully differentiated cytotoxic T cells.
   4. **Characteristics of cytotoxic T lymphocyte**
      a. Most cytotoxic T cells express the CD8 marker, but some CD4+ lymphocytes may also have cytotoxic properties.

b. In addition, cytotoxic T cells contain increased levels of perforins and esterases in cytoplasmic granules and have increased expression of the membrane-associated Fas ligand essential for their cytotoxic activity.

C. **Target Cell Killing**. The cytotoxic reaction takes place in a series of successive steps (see Fig. 11.8):

1. **Conjugation**. The first step in the cytotoxic reaction involves the conjugation of cytotoxic T cells with their respective targets. This conjugation involves, in the first place, the recognition by the TCR of a peptide-MHC-I complex. However, additional interactions are required to achieve the strong adhesion of the T cell to the target cell required for effective killing.
   a. The CD8+ molecule itself interacts with the nonpolymorphic part of the MHC class I molecule, specifically with its $\alpha 3$ domain. The avidity of this interaction increases after TcR occupancy.
   b. The CD2 molecule, present on all T cells, interacts with the CD58 (LFA-3) molecules expressed by the target cells. Any resting T cell can interact with CD58+ cells, but this interaction by itself is weak and does not lead to cell activation. The modification of the CD2 molecule after TcR engagement increases the affinity of the interaction between CD2 and CD58.
   c. In addition, the interaction between LFA-1 and ICAM-1 provides additional stabilizing bonds between the interacting cells. LFA-1 avidity for its ligand(s) is also up-regulated by T-cell activation.

   The conjugation of CD8+ T cell to its target cell is firm but transient, and, after about 30 minutes, the affinity of ICAM-1 for LFA-1 and of CD58 for

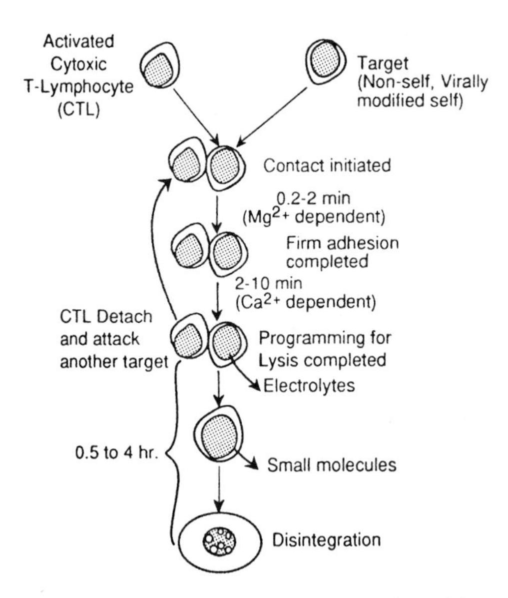

Figure 11.8 Diagrammatic representation of the sequence of events in a cytotoxic reaction. Notice that an activated cytotoxic (CD8+) T cell is able to kill several targets.

CD2 reverts to resting levels. The cytotoxic T cell can then move on to another antigen-bearing target with which it will develop the same interaction.

2. Target cell destruction. During the short period of intimate contact with its target, a series of reactions takes place, eventually resulting in the killing of the target cell (Fig. 11.9).
   a. First, the cytoskeleton of the cytotoxic cell reorganizes. The microtubule organizing center and the Golgi apparatus move to the area of contact with the target cell. This is associated with transport of cytoplasmic granules toward the target.
   b. When the cytoplasmic granules reach the membrane, their contents are emptied into the virtual space that separates the cytotoxic T cells and target cells. These granules contain a mixture of proteins:
      i. **Perforins** that polymerize as soon as they are released, forming "polyperforins." These polyperforins are inserted in the cell membrane where they form transmembrane channels.
      ii. **Granzymes A and B** are proteinases that penetrate the target cell (perhaps using the polyperforin channels) where they activate the IL-1 converting enzyme (ICE) that triggers apoptotic death of the target.
   c. **Apoptosis** of the target cell can also be triggered through a different pathway, which involves the Fas molecule (CD95), widely expressed

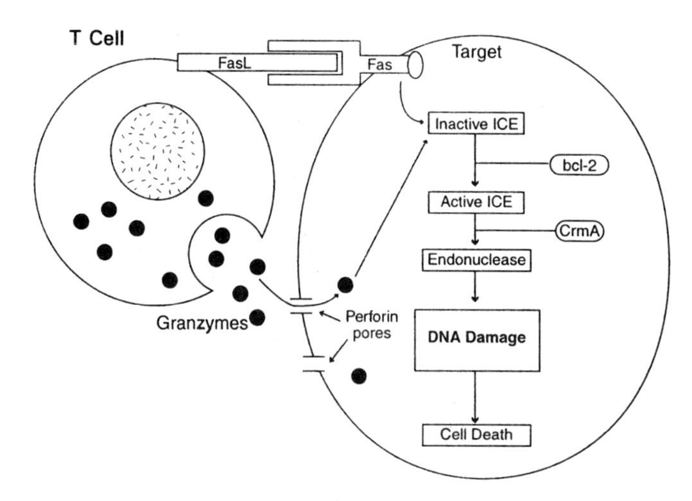

Figure 11.9 Diagrammatic representation of the sequence of events leading to apoptosis of a target cell after interaction with an activated cytotoxic T lymphocyte. Two pathways leading to activation of a critical enzyme (interleukin convertase enzyme, ICE) have been described: one involves the interaction between Fas and FasL. The other involves granzymes, which are released by the T lymphocytes and diffused into the cytoplasm through polyperforin clusters. The activation of ICE is followed by activation of an endonuclease and DNA breakdown, the hallmark of apoptosis. Two levels of control appear to exist: one, previous to ICE activation, is mediated by Bcl-2, a proto-oncogene product. The other, blocking the pathway between ICE and the activation of the endonuclease, involves CrmA, the product of a cytokine response modifier gene.

in nucleated cells, and its ligand (FasL), which is inducibly expressed in activated T cells. The apoptosis pathway triggered by the Fas–FasL interaction also appears to involve ICE activation. Since the expression of Fas and FasL is very widespread, even in resting cells, there must be negative control mechanisms to avoid unnecessary triggering of the pathway. A protein encoded by the proto-oncogene Bcl-2 and the product of a cytokine response modifier gene (CrmA) have been proposed to play this role, but there are still many gaps in our understanding of activation and control of apoptosis.

IX. ACTIVATION AND FUNCTION OF NATURAL KILLER CELLS

A. **The Biological Significance of NK Cells**. The mobilization of a T-lymphocyte-mediated cytotoxic response following an intracellular infection is a relatively ineffective process. In a resting immune system, the cytotoxic T-cell precursors exist in relatively low numbers ($< 1/10^5$). Proliferation and differentiation are required to generate a sufficient number of fully activated effector CD8+ T cells. Thus, an effective primary cytotoxic T-cell response is seldom deployed in less than 2 weeks. During this time, the host depends on defenses that can be deployed much more rapidly, such as the **production of type I interferons** (α and β), initiated as soon as the virus starts replicating, and the **activity of natural killer (NK) cells**. These two effector systems are closely related (type I interferons activate NK cell functions) and are particularly effective as defenses against viral infections.

B. **NK Cell Activation**. NK cells receive activation signals both from T lymphocytes and from monocytes/macrophages. Those signals are mediated by cytokines released from those cells.
   1. **Interleukin-2**. NK cells constitutively express a functional IL-2 receptor that is predominantly constituted by the p75 subunit (for this reason, NK cells are, for the most part, $CD25^-$). This variant of the IL-2 receptor enables NK cells to be effectively activated by IL-2 produced by activated helper T cells early in the immune response, without the need for any additional co-stimulating factors or signals. IL-2-activated NK cells are also known as **lymphokine activated killer cells (LAK cells)**.
   2. **Type I interferons**. Interferons α and β enhance the cytotoxic activity of NK cells. Thus, the release of these interferons from infected cells mobilizes nonspecific defenses before the differentiation of MHC-restricted cytotoxic T lymphocytes is completed.
   3. **Interleukin IL-12**, produced at the same time as interferons α and β, is believed to play a crucial role in the early activation of NK cells.

C. **Target Recognition by NK Cells**. The mechanism of recognition of target cells by NK cells has not been as clearly defined as the mechanism of antigen recognition for B and T lymphocytes. A major difference is that NK cells can recognize virus-infected cells and many different types of malignant cells without clonal restriction. In other words, their recognition mechanisms are relatively nonspecific and common to all NK cells. At this time, it is believed that two broadly reactive receptors are involved, one that delivers activating signals, and the other that delivers inhibitory signals (Fig. 11.10).

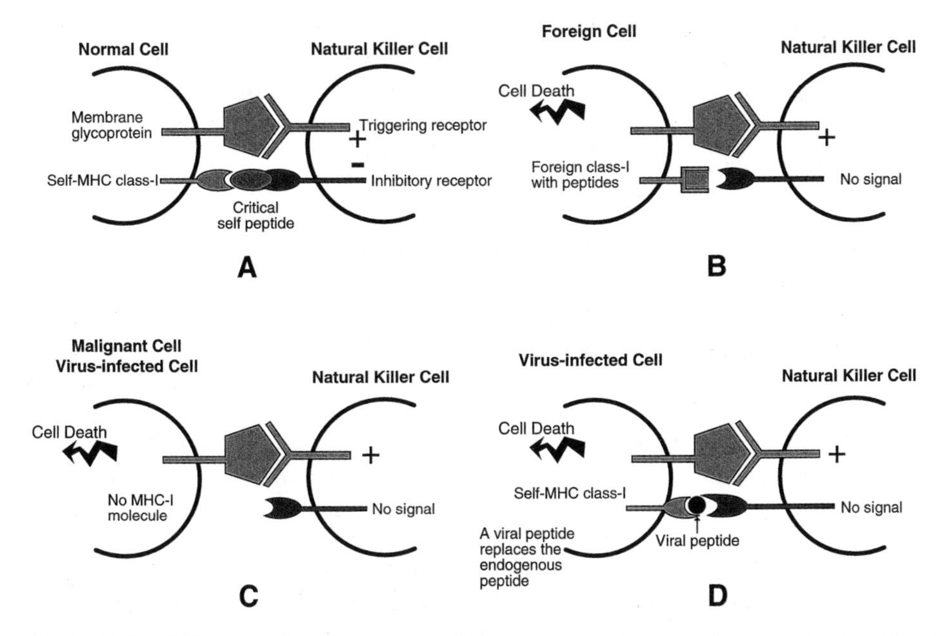

Figure 11.10 Diagrammatic representation of the mechanism of target cell recognition by NK cells. The activation or lack of activation of cytotoxic pathways depends on the balance between stimulatory and inhibitory receptors. If the inhibitory receptor is not triggered (due to either presentation of a non-self peptide, lack of interaction of the inhibitory receptor with a non-self MHC-I-peptide complex, or to down-regulation of expression of MHC-I molecules on the cell membrane), stimulatory activity prevails and the target cell is killed.

1. The **triggering or activating receptor** (NKAR, NKR-P1) recognizes membrane glycoproteins, probably with greater affinity when those glycoproteins are modified as a consequence of malignant transformation or viral infection.
2. The **inhibitory receptor** (NKIR, p58,70) prevents NK cells from killing normal cells. It has a molecular weight of 54 kD and interacts with the complex formed by HLA-C molecules (which have very little polymorphism) and their associated self-peptides.
3. **Activation of NK cells as a result of viral infection and malignant transformation**. Viral infection and malignant transformation not only can change the constitution of membrane glycoproteins, but can also interfere with the interaction of HLA-C with the inhibitory receptor. This interference can be due to:
   a. Replacement of a self-peptide on HLA-C by a non-self peptide.
   b. Down-regulation of the expression of MHC-I molecules on the infected cell membrane, which is frequently associated with viral infection.

   In either case, the end result will be that the triggering receptor will deliver a signal that will not be counteracted by a signal from the inhibitory receptor, and the NK cell will be able to proceed with the destruction of the target cell by similar mechanisms to those described for T-lymphocyte-mediated cytotoxicity.

D. NK Cells and ADCC. NK cells express the FcγRIII, a low-affinity Fc receptor that allows them to interact with IgG-coated cells. This interaction also results in the activation of the NK cell and lysis of the target, a process that is known as **antibody-dependent cellular cytotoxicity (ADCC)**. Activation of killing by this pathway is not inhibited by the p58,70 receptor.

X. SUPPRESSOR CELLS AND THE IMMUNE RESPONSE

A. Experimental Evidence for the Existence of Suppressor Cells. The knowledge that T lymphocytes can specifically down-regulate B-cell responses was obtained in studies on the regulation of the immune response of mice to keyhole limpet hemocyanin (KLH).

1. If mice are primed with KLH and later boosted with the same antigen, a population of T lymphocytes can be isolated 2 weeks after the boost that markedly suppresses the IgG anti-DNP response of genetically identical mice primed with DNP-KLH. However, this population will not suppress anti-DNP antibody production in animals immunized with DNP conjugated with a different carrier, such as bovine gamma globulin (BGG). The carrier specificity of the suppressive effect suggests that it is mediated by T lymphocytes.
2. Many other studies in mice have confirmed the existence of antigen-specific suppressor cells and suggested the existence of antigen-specific suppressor factors, factors that await definition at the molecular level.

B. The Nature of Suppressor Cells. In most experimental systems, suppressor activity is carried out by a subset of CD8+ lymphocytes, but rarely have CD4+ lymphocytes been shown to exert the suppressive effect. Suppression appears to be a negative feedback directly exerted by activated suppressor T lymphocytes on helper T lymphocytes, resulting indirectly in the inability of B lymphocytes to respond adequately to stimulation.

C. Induction of Suppressor Cells. Several lines of evidence suggest that the differentiation of suppressor T cells requires the assistance of a special suppressor-inducer CD4+ T-cell subpopulation, tentatively identified based on the expression of the CDw29 marker. This subpopulation must differentiate at later stages in the immune response, after the antigen has been eliminated and memory lymphocytes generated. Indeed, suppressor T cells appear at late stages of the immune response, when B-cell proliferation is no longer desirable.

D. Mechanism of Suppression. The mechanism of suppression by T lymphocytes remains uncertain. Several possibilities have been considered.

1. Experimental studies suggested the existence of antigen-specific suppressor factors produced by the suppressor cells, based on the fact that the suppressive activity could be transferred by cell-free supernatants from cultures of suppressor cells. However, these "suppressor factors" have resisted all attempts to be defined at the molecular level, and their existence and nature are still controversial.
2. Cross-regulation of T-helper subsets has emerged as an interesting possibility, at least to explain nonspecific suppression. If TH1 cells are strongly activated, they may release soluble factors (such as interferon-γ) that inhibit TH2 activity, and, consequently, interfere with B-cell activation. Conversely,

IL-4 and IL-10 may play a role in suppressing TH1 responses by down-regulating IL-12 synthesis. Other cytokines, such as transforming growth factor-β (TGF-β), which inhibits T-lymphocyte proliferation by blocking IL-2 transcription, and macrophage inflammatory proteins (e.g., MIP-1α) may play the role of soluble suppressor factor(s) in humans.

3. Direct cell-to-cell signaling. As noted in Chapter 3, activated T lymphocytes (CD4+ and CD8+) express MHC-II molecules. Because of this, activated CD8+ lymphocytes can interact with CD4+ helper cells. This cell–cell interaction mediated by class II MHC molecules expressed by the activated CD8+ cells and the CD4 molecule takes place independently of antigenic stimulation and has been proposed to be the structural basis for the down-regulation of helper T-cell functions by suppressor T cells.

SELF-EVALUATION

Questions

Choose the ONE *best* answer.

11.1 Which one of the following cytokines is able to enhance the activity of both TH1 lymphocytes and NK cells?
- A. Interferon-α
- B. Interferon-γ
- C. Interleukin-2
- D. Interleukin-4
- E. Migration inhibition factor (MIF)

11.2 Which of the following ligand-receptor interactions plays a major role in the effector stages of CD8-mediated cytotoxicity?
- A. CD2 : CD58 (LFA-3)
- B. CD8 : MHC-I
- C. CD95 (Fas) : Fas ligand
- D. IL-2 : IL-2R
- E. Interferon-γ (IFN-γ) : IFN-γR

11.3 The development of antigen-specific anergy is believed to result from:
- A. Apoptosis of the antigen-stimulated T lymphocytes
- B. Lack of delivery of co-stimulating signals by antigen-binding cells
- C. Lack of transport of antigen-derived peptides to the endoplasmic reticulum
- D. Predominant stimulation of TH2 helper cells
- E. Release of large concentrations of IL-10

11.4 Mice from strain A, expressing antigen H2-K on their lymphocytes, were immunized with influenza virus. Seven days later, T lymphocytes from these animals were mixed with ^{51}Cr-labeled influenza virus-infected T lymphocytes from H2-D mice, and [^{3}H] thymidine was simultaneously added to the system. Sixty minutes later you expect:
- A. Significant ^{51}Cr release into the supernatant
- B. Significant incorporation of [^{3}H] thymidine by the immune lymphocytes
- C. Significant incorporation of [^{3}H] thymidine by the target lymphocytes
- D. Transfer of ^{51}Cr from target cells into cytotoxic cells
- E. No significant release of ^{51}Cr

11.5 Which of the following mechanisms is responsible for the sparing of normal cells by NK cells?
A. Differential specificity of the NK cell Fas ligand for altered Fas molecules on malignant and virus-infected cells
B. Down-regulation of HLA-C molecules on normal cells
C. Lack of expression of Fas by normal cells
D. Lack of expression of the abnormal glycoproteins recognized by the NK triggering receptor
E. Recognition of self-peptides associated with HLA-C molecules by the NK cell

11.6 Which of the following is the *least* likely reason for the overlapping effects of some cytokines?
A. Reaction of related cytokines with common receptors
B. Sharing of transducing units by some cytokine receptor families
C. Similar receptor binding sites in related cytokines
D. Triggering of identical activation pathways by cytokine receptors of different families

11.7 Which of the following is a major characteristic of natural killer cells?
A. Constitutive expression of an IL-2 receptor
B. Inducible expression of Class II MHC antigens on their surface
C. Lack of expression of Fc receptors
D. Phenotypical identity to small T lymphocytes
E. Recognition of target cells expressing MHC-I molecules modified by viral oligopeptides

11.8 Which of the following factors is believed to determine the differentiation of B lymphocytes into plasma cells producing antibody of one isotype or another?
A. Cytokines released by TH2 helper T cells and macrophages
B. Feedback mechanisms involving soluble antigen–antibody complexes
C. Random rearrangements of germ line heavy- and light-chain genes
D. Somatic hypermutation of V region genes
E. The interaction between CD28 and CD80

11.9 Normal mononuclear cells were cultured in the presence of pokeweed mitogen (PWM) and monoclonal antibodies (MoAb) to IL-2 and the IL-2 receptor (IL-2R). The cultures were harvested after 6 days of culture and the concentrations of IgM measured by enzymoimmunoassay. The following results were obtained:

Culture conditions	**IgM (μg/mL)**
Mononuclear cells alone	0.01 ± 0.02
Mononuclear cells + PWM	5.5 + 1.5
Mononuclear cells + PWM + anti-IL-2 MoAb	6.5 ± 2.0
Mononuclear cells + PWM + anti-IL-2R MoAb	5.0 ± 2.2
Mononuclear cells + anti-IL-2 MoAb	0.02 ± 0.015
Mononuclear cells + anti-IL-2R MoAb	0.01 ± 0.01

Which of the following conclusions is supported by the data?
A. B lymphocytes do not express a transducing IL-2 receptor
B. IL-2 has a potentiating effect on B-cell stimulation by PWM
C. The differentiation of IgM-producing B cells is T-cell independent
D. The response of B cells to PWM stimulation is T-cell dependent
E. The stimulation of IgM synthesis by PWM is IL-2 independent

11.10 Which of the following is a well-known effect of interleukin-8?
 A. Attraction and activation of neutrophils
 B. Induction of B-cell proliferation
 C. Induction of the switch from IgM to IgG synthesis
 D. Induction of B-cell differentiation
 E. Suppression of interleukin release by helper T cells

Answers

11.1 (C) Among the listed interleukins, IL-2 is the only one that activates **both** TH1 lymphocytes and NK cells.

11.2 (C) The interaction between CD95 (Fas) and its ligand delivers a signal that induces the apoptotic death of target cells. The other listed interactions are either involved in the activation stages of cytotoxic T cells or mediate the cell–cell interactions that precede the delivery of signals leading to cell death.

11.3 (B) the lack of co-stimulation of a T lymphocyte that has received initial activating signals through the TcR may lead to anergy (lack of responsiveness) or to cell death (apoptosis). The release of large concentrations of IL-10 by activated TH2 cells could down-regulate an ongoing immune response, but the effects of IL-10 are not antigen-specific.

11.4 (E) The T lymphocytes from H2-K mice are programmed to kill cells expressing viral antigens in association with their own MHC molecules and will not kill infected cells from mice carrying different MHC antigens. On the other hand, lymphocytes from two different mouse strains are likely to interact and participate in a mixed lymphocyte reaction, but significant incorporation of [^{3}H] thymidine will only be observed after more than 3 days of co-culture.

11.5 (E) The cytotoxic reaction mediated by NK cells depends on the recognition of cellular membrane glycoproteins by an NK cell triggering receptor. However, if there is a simultaneous interaction between an inhibitory receptor and an HLA-C/endogenous peptide complex expressed by normal cells, cytotoxicity is inhibited. Malignant and virus-infected cells either have down-regulated expression of MHC molecules, or express a modified peptide in association with HLA-C. As a result, the inhibitory signal is not received by the NK cell and the target cell is killed.

11.6 (D) The least likely explanation for the overlapping biological properties of cytokines is the triggering of common activation pathways, since the current evidence suggests that there are multiple cellular activation pathways whose activation depends on the state of activation of the cell and the number and nature of signals received by the cell at any given point of time. In contrast, many cytokines have similar structures, and families of cytokines reacting with identical receptors or with receptors sharing transducing units have been identified.

11.7 (A) The constitutive expression of an IL-2 receptor able to transduce activating signals upon occupancy by IL-2 is responsible for the transformation of NK cells into LAK cells upon incubation with IL-2.

11.8 (A) The major factors controlling the differentiation of plasma cells synthesizing IgG or other immunoglobulins are believed to be cytokines released predominantly by TH2 cells (e.g., IL-4) and signals resulting from the interaction between B lymphocytes expressing CD40 and activated TH cells expressing the corresponding ligand.

11.9 (E) The results show that both immunoneutralization of IL-2 and blocking of the IL-2 receptor had no detectable effect on the stimulation of IgM synthesis on a mononuclear cell culture with PWM. Issues such as the T-cell dependence or independence of this response were not addressed by the experiment, since cytokines other than IL-2 may be involved in T-B cooperation and no effort was made to study the response of T-cell-depleted preparations. The experiment also was not designed to answer any questions about the IL-2 receptor of B cells, other than the fact that its block had no effect on the response to PWM.

11.10 (A) IL-8 is one of the proinflammatory cytokines, and has chemotactic effects of neutrophils and T lymphocytes.

BIBLIOGRAPHY

Berke, G. The CTL's kiss of death. *Cell, 81*:9, 1995.

Bierer, B.E., Sleckman, B.P., Ratnofsky, S.E., and Burakoff, S.J. The biologic roles of CD2, CD4, and CD8 in T cell activation. *Annu. Rev. Immunol., 7*:579, 1989.

Cantrell, D. T antigen receptor signal transduction pathways. *Annu. Rev. Immunol., 14*:259, 1996.

Davies, D.R., and Wlodawer, A. Cytokines and their receptor complexes. *FASEB J., 9*:50, 1995.

Dinarello, C.A. Interleukin-1 and interleukin-1 antagonism. *Blood, 77*: 1627, 1991.

Gold, M.R., and Matsuuchi, L. Signal transduction by the antigen receptor of B and T lymphocytes. *Int. Rev. Cytol., 157*:181, 1995.

Herman, A., Kapler, J.W., Marrack, P., and Pullen, A.M. Superantigens: Mechanism of T cell stimulation and role in immune responses. *Annu. Rev. Immunol., 9*:745, 1991.

Johnson, H.M., Russell, J.K., and Pontzer, C.H. Superantigens in human disease. *Sci. Am., 266*(4):92, 1992.

Kupfer, H., Monks, C.R., and Kupfer, A. Small splenic B cells that bind to antigen-specific T helper (Th) cells and face the site of cytokine production in the Th cells selectively proliferate: immunofluorescence microscopic studies of Th-B antigen-presenting cell interactions. *J. Exp. Med., 179*:1507, 1994.

Ling, C-C., Walsh, C.M., Young, J. D-E. Perforin: structure and function. *Immunol. Today, 16*:194, 1995.

Mosmann, T.R. Role of a new cytokine, interleukin-10, in the cross-regulation of helper T cells. *Ann. N.Y. Acad. Sci., 628*:337, 1990.

Smyth, M.J., and Trapani, J.A. Granzymes: exogenous proteinases that induce target cell apoptosis. *Immunol. Today, 16*:202, 1995.

Springer, T.A. The sensation and regulation of interactions with the extracellular environment: The cell biology of lymphocyte adhesion receptors. *Annu. Rev. Cell Biol., 6*:359, 1990.

Street, N.E., and Mosmann, T.R. Functional diversity of T interleukins due to secretion of different cytokine patterns. *FASEB J., 5*:171, 1991.

Szamel, M., and Resch, K. T cell antigen receptor-induced signal-transduction pathways. *Eur. J. Biochem., 228*:1, 1995.

Tracey, K.J., and Cerami, A. Metabolic responses to cachectin/tumor necrosis factor. A brief review. *Ann. N.Y. Acad. Sci., 587*:325, 1990.

Trowbridge, I.S., and Thomas, M.L. CD45: an emerging role as a protein tyrosine phosphatase required for lymphocyte activation and development. *Annu. Rev. Immunol., 12*:85, 1994.

Whiteside, T.L., and Herberman, R. The role of natural killer cells in human disease. *Clin. Immunol. Immunopathol., 53*:1, 1989.

12

The Humoral Immune Response and Its Induction by Active Immunization

Gabriel Virella

I. INTRODUCTION

The recognition of a foreign cell or substance triggers a complex set of events that results in the acquisition of specific immunity against the corresponding antigen(s). The elimination of "non-self" depends on effector mechanisms able to neutralize or eliminate the source of antigenic stimulation. While the inductive stages of most immune responses require T- and B-cell cooperation, the effector mechanisms can be clearly subdivided into cell-dependent and antibody-dependent (or humoral). The sequence of events that culminates in the production of antibodies specifically directed against exogenous antigen(s) constitutes the humoral immune response.

II. AN OVERVIEW OF THE INDUCTION OF A HUMORAL IMMUNE RESPONSE TO AN IMMUNOGEN

A. Exposure to Natural Immunogens

1. Infectious agents penetrate the organism via the skin, upper respiratory mucosa, and intestinal mucosa. In most cases, the immune system is stimulated in the absence of clinical symptoms suggestive of infection (subclinical infection).
2. The constant exposure to immunogenic materials penetrating the organism through those routes is responsible for continuous stimulation of the immune system and explains why relatively large concentrations of immunoglobulins can be measured in the serum of normal animals.
3. In contrast, animals reared in germ-free conditions synthesize very limited amounts of antibodies, and their sera have very low immunoglobulin concentrations.

B. Deliberate Immunization. Many infectious diseases can be prevented through active immunization.

1. When live, attenuated organisms are used for immunization they are usually delivered to the natural portal of entry of the organism. For example, a

vaccine against the common cold using attenuated rhinoviruses would be most effective if applied as a nasal aerosol.

2. When inert compounds are used as immunogens, they have to be introduced in the organism by injection, usually intramuscularly, subcutaneously, or intradermally.
3. In humans, immunization is usually carried out by injecting the antigen intradermally, subcutaneously, or intramuscularly, or by administering it by the oral route (attenuated viruses, such as poliovirus) or as an aerosol (some experimental vaccines). Usually injected immunogens are mixed or emulsified with **adjuvants**, compounds that enhance the immune response.
   a. In laboratory animals immunized to produce antisera, **complete Freund's adjuvant** (**CFA**), a water-in-oil emulsion containing killed mycobacteria, is one of the most widely used and most effective adjuvants.
   b. Adjuvants are also used in human immunizations, whenever the injected immunogens are nonreplicating (e.g., killed bacteria, toxoids). Inorganic gels, such as alum, and aluminum hydroxide are most frequently used.
   c. The mechanism of action of adjuvants usually involves two important factors:
      i. Slowing down the diffusion of the immunogen from the injected spot, so that antigenic stimulation will persist over a longer period of time.
      ii. Inducing a state of activation of antigen-presenting cells in the site of inoculation. This activation can be more or less specific: CFA causes a very intense local inflammation, while some bacterial compounds with adjuvant properties have a more targeted effect on macrophages and other APC and have strong adjuvant properties without causing a very intense inflammatory reaction.
      iii. Alum and aluminum hydroxide share both types of effects, but they do not induce an inflammatory reaction as intense as CFA and this is the reason for their use in human immunization.

C. **B-Cell Activation.** As discussed in greater detail in Chapters 4 and 11, B-cell activation requires multiple signals. The only specific signal is the one provided by the interaction between the antigen binding site of membrane immunoglobulins with a given epitope of an immunogen. The additional signals, provided by TH2 helper cells and APC, are nonspecific with regard to the antigen.

1. The recognition of an antigen by a resting B cell seems to be optimal when the immunogen is adsorbed to a follicular interdigitating cell or to a macrophage.
2. B lymphocytes recognize either unprocessed antigen or antigen fragments that conserve the configuration of the native antigen. All techniques used for measurement of specific antibodies use antigens in their original configuration as their basis, and succeed in detecting antibodies reacting with them. Whether or not some B cells may have membrane immunoglobulins reactive with immunogen-derived peptides associated with MHC-II molecules is not known. Such recognition would result in the synthesis of antibodies with no protective value, but their biological significance could be associated to immunoregulation, as discussed later in this chapter.
3. Helper TH2 lymphocytes provide other signals essential for B-cell proliferation and differentiation. The activation of this CD4 subpopulation is favored

by a low-affinity interaction between an immunogen-derived oligopeptide associated with an MHC-II molecule and the T-cell receptor, as well as additional signals, some derived from cell–cell interactions, such as the CD28/CD86 interaction, and others from cytokines, such as IL-4 and IL-1.

4. Activated TH2 cells, in turn, provide several co-stimulatory signals that promote B-cell proliferation and differentiation. Some of these signals derive from the interaction between cell membrane molecules that are up-regulated during the early stages of TH2 and B-cell activation (e.g., CD40L (gp39, on T cells) and CD40 (on B cells), while others are mediated by cytokines, such as IL-2, IL-3, IL-4, IL-6, IL-10, IL-13, IL-14, TNFα, and GM-CSF). (How many of these cytokines signal B cells in vivo and the physiological consequences of the signal they deliver are not known.)
5. When the proper sum of specific signals and co-stimulatory signals is received by the B cell, clonal proliferation and differentiation ensues. Since each immunogen presents a multitude of epitopes, a normal immune response is **polyclonal** (i.e., involves many different clones recognizing different epitopes of an immunogen).
6. Most activated B cells will become antibody-producing plasma cells; a few will become memory cells. What determines one or the other type of evolution is not known.
7. The induction of an immune response requires some time for activation of all the relevant cells and for proliferation and differentiation of B cells into plasma cells. Thus, there is always a **lag phase** between the time of immunization and the time when antibodies become detectable.
8. Experimental animals immunized with a given immunogen (e.g., tobacco mosaic virus) often show marked postimmunization hypergammaglobulinemia, but only a very small fraction of the circulating immunoglobulins react with the immunogen. In humans, the initial burst of IgE production, after first exposure to an allergen, seems to be mainly constituted by nonspecific antibodies. This apparent lack of specificity of the immune response is more obvious after the first exposure to an immunogen and can be explained by a variety of factors.
   a. The antibodies produced early in the immune response are of low affinity, and may not be detectable in assays that favor the detection of high-affinity antibodies.
   b. The co-stimulatory signals provided by T cells may enhance the immune response of any neighboring T cell. This may result in the enhanced synthesis of antibodies unrelated to the immunogen, reacting with other immunogens that the immune system is simultaneously recognizing.
   c. While the synthesis of unrelated antibodies is usually beneficial or inconsequential, it can also be the basis for at least some autoimmune reactions if strong help is provided to autoreactive B cells which otherwise would remain quiescent (see Chap. 22).

III. THE PRIMARY IMMUNE RESPONSE

The first contact with an antigen evokes a **primary response**, which has the following characteristics.

A. A relatively **long lag between the stimulus and the detection of antibodies** by current methods (varying between 3 to 4 days after the injection of heterologous erythrocytes and 10 to 14 days after the injection of killed bacterial cells). Part of this variation depends on the sensitivity of antibody detection methods, but it is also a reflection of the potency of the immunogen.

B. The first antibody class to be synthesized is usually **IgM.** Later in the response, **IgG** antibodies will predominate over IgM antibodies. This phenomenon is known as **IgM-IgG switch**, and it is believed to be controlled by different interleukins released by activated helper T lymphocytes.

C. After rising exponentially for some time, antibody levels reach a steady state, and then decline (Fig. 12.1). Adjuvant administration will keep the antibody levels high for months.

D. After the infectious agent (or any other type of immunogen) has been eliminated, several regulatory mechanisms will operate in order to turn off antibody production. Several negative signals are likely to contribute to the down-regulation of B-cell activity.

1. The elimination of the antigen will remove the most important positive signal. By far, this is believed to be the most important factor determining the decline of a humoral immune response.
2. The increase in IgG concentration during the later stages of the primary

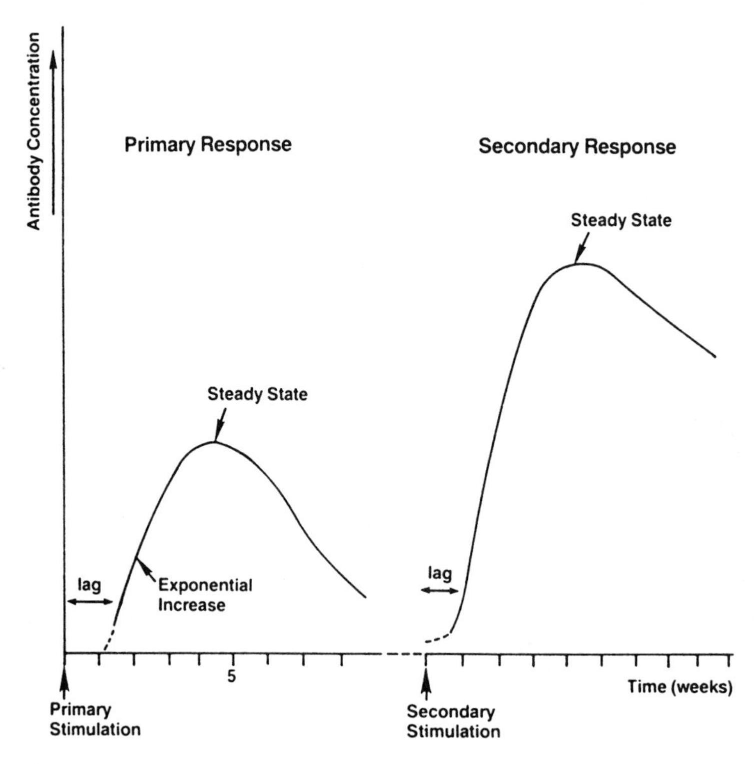

Figure 12.1 Diagrammatic representation of the sequence of events during a primary and a secondary immune response. (Modified from Eisen, H.N. *Immunology*, 2nd ed. Harper & Row, Cambridge, 1980.)

response may lead to a general depression of IgG synthesis as a consequence of negative feedback regulation (see Chap. 6).

3. Suppressor cell activity increases as the immune response evolves, and predominates after elimination of the antigen. The regulation of suppressor cells and the mechanisms of suppression are poorly understood aspects of the immune response. Several possibilities have been considered.
   a. As the TH2 activity predominates, several down-regulatory cytokines are produced. IL-4 down-regulates TH1 cells, and IL-10 down-regulates both TH1 and TH2 cells. T cells with suppressor activity persist after the antigen is eliminated, either as a consequence of their late activation or of a longer life span.
   b. As the immune response proceeds and IgG antibodies are synthesized, and IgG-containing antigen–antibody complexes have been proposed to have a direct down-regulating effect on B cells consequent to their binding through type II Fcγ receptors.
   c. Anti-idiotypic antibodies develop during normal immune responses, reacting with variable region epitopes or idiotypes presented by the antibodies produced against the antigen that elicited the immune re-

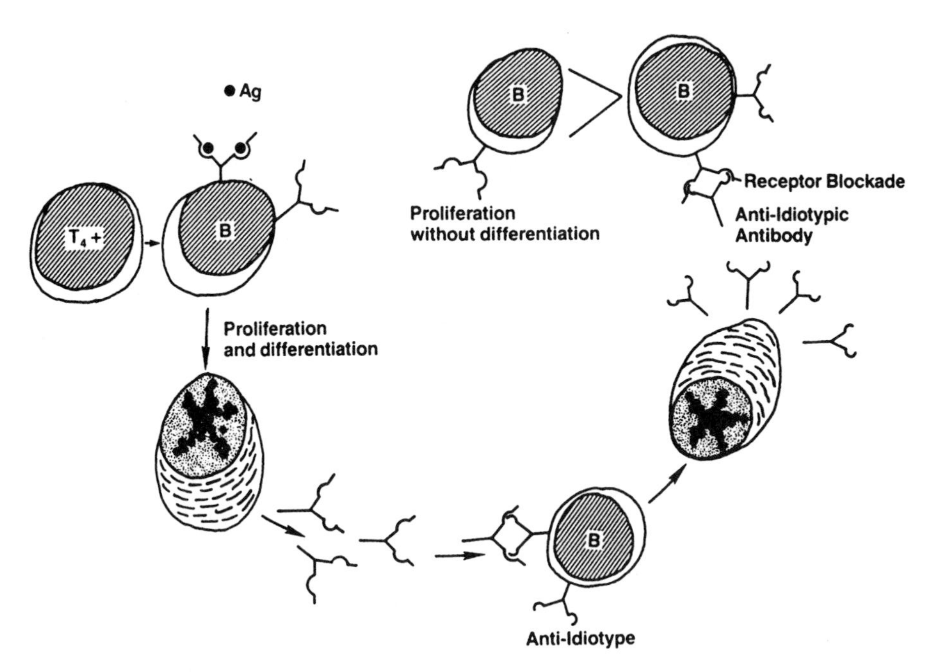

Figure 12.2 The role of anti-idiotypic antibodies in down-regulating the humoral immune response (I). From left to right, an antigen-stimulated B cell differentiates into a clone of plasma cells producing specific antibody. A second population of B cells, carrying a membrane immunoglobulin with specificity for the binding site of this first antibody (anti-idiotypic antibody) will be stimulated to proliferate and differentiate into a clone of plasma cells producing anti-idiotypic antibody. This second antibody will be able to bind to the antigen receptor of the cells involved in the initial responses, which will be stimulated to divide but not to differentiate into plasma cells; the antigen receptor will also be blocked from further reaction with the real antigen.

sponse. The easiest way to understand this response is to accept that a normal state of low zone tolerance to the millions of idiotypes that are presented by antigen receptors can be broken when one or a few specific antibodies are produced in large concentrations, suddenly exposing the immune system to large concentrations of molecules with unique idiotypes. These antibodies are believed to participate in negative regulation of the immune response.

i. One possible mechanism for this down-regulation would involve the binding of anti-idiotypic antibody to membrane immunoglobulins from antigen-specific B cells expressing variable regions of the same specificity as the antibody molecules that triggered the anti-idiotypic antibody. The binding of an antibody to a membrane immunoglobulin induces B-lymphocyte proliferation but the proliferating B cells fail to differentiate into antibody-producing cells. At the same time, the occupancy of the membrane immunoglobulin binding sites by the anti-idiotypic antibodies prevents the proper antigenic stimulation of the B lymphocyte (Fig. 12.2).

ii. A second mechanism proposes that the anti-idiotypic antibodies suppressor activity on T lymphocytes, which in turn would turn off activated TH2 helper lymphocytes (Fig. 12.3). This theory implies either that anti-idiotypic antibodies can be induced by the T-cell receptor, or that B lymphocytes can recognize the same processed antigens that stimulate T lymphocytes, and, as a result, antibodies sharing idiotypes with the T-cell receptor would be synthesized and induce the formation of anti-idiotypic antibodies cross-reactive with T-cell receptors. Neither hypothesis has been experimentally proved.

IV. THE SECONDARY OR ANAMNESTIC RESPONSE

The secondary or anamnestic response is elicited by reexposure of an immune animal or human being to the immunizing antigen. The capacity to mount a secondary immune response can persist for many years, providing long-lasting protection against reinfection. The secondary response has some important characteristics, some dependent on the existence of an expanded population of memory cells, ready to be stimulated, and others dependent on the prolonged retention of antigen in the lymph nodes with continuous stimulation of B cells over long periods of time.

A. **Differentiation of B Memory Cells.** During the peak of a primary response, there is a duality in the fate of activated B cells: while most will evolve into antibody-producing plasma cells, others will differentiate into memory B cells.

1. **Factors determining the evolution of antigen-stimulated T cells.** Although many questions remain to be answered, it is believed that signals delivered by helper T cells determine the outcome of the proliferation of antigen-stimulated B cells. Signaling through the CD40/CD40L interaction (see Chaps. 4 and 11) is critical for full differentiation of B cells into both antibody-producing plasma cells and into memory B-cell precursors.
2. **The germinal center reaction.** The differentiation of memory cells is be-

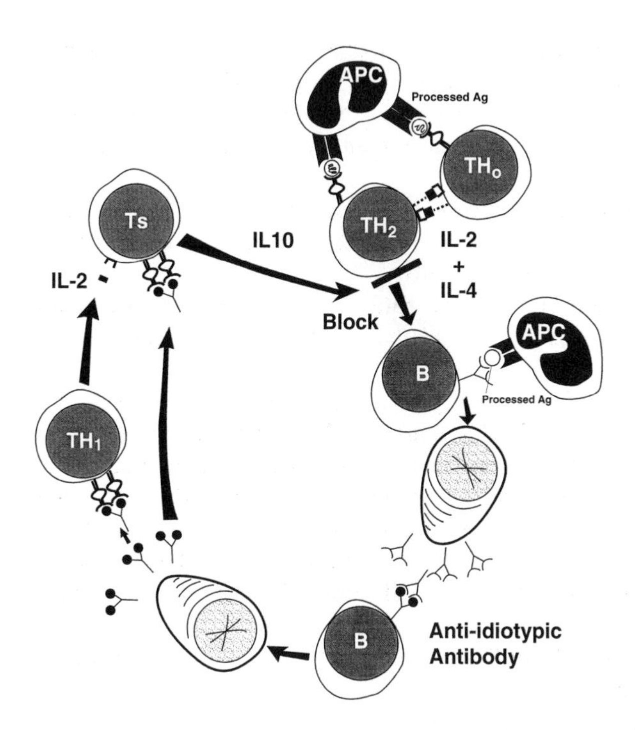

Figure 12.3 The role of anti-idiotypic antibodies in down-regulating the humoral immune response (II). In this case, a B-lymphocyte clone would be stimulated to produce antibodies recognizing the same processed antigen as the helper T lymphocyte. The corresponding anti-idiotypic antibody would cross-react with T-cell receptors of the same specificity in $CD8^+$ cells, and the occupancy of the TcR by the anti-idiotypic antibody would promote the functional differentiation of suppressor cells. The activated suppressor T cells would turn off the immune response by down-regulating the helper T lymphocytes involved in the co-activation of B lymphocytes.

lieved to take place in the germinal centers of secondary lymphoid tissues. As a prememory B cell enters a follicle, it migrates into the germinal center where it undergoes active proliferation. At this stage, the "switch" from IgM to IgG or other isotype synthesis is taking place and the V region genes undergo somatic hypermutation. After completing this round of proliferation, the precursors memory B cells need additional signals for full differentiation.

a. Clones with high-affinity mIg in the membrane will be able to interact with antigen molecules immobilized by follicular dendritic cells; clones with low-affinity mIg will not be able to compete with preformed antibody for binding to the immobilized antigen epitopes and will undergo apoptosis.
b. The evolution of this antigen-stimulated memory B-cell precursor into a memory B cell requires a second signal provided by a helper T cell, in the form of the CD40/CD40L interaction. Other signals, such as the one delivered by the CD21/CD23 interaction, may result in direct evolution of the prememory B cell into an antibody-producing plasma cell.

B. **Characteristics of the Secondary Immune Response Directly Resulting from the Existence of an Expanded Population of Memory Cells**

1. **Lower threshold dose of immunogen** (i.e., the dose of antigen necessary to induce a secondary response) is lower than the dose required to induce a primary response.
2. **Shorter lag phase** (i.e., the antibody concentration rises sooner). The capacity to respond faster and efficiently to antigenic stimulation depends on the existence of long-lived **memory cells**.
3. **Faster increase in antibody concentrations and higher titers of antibody** (Fig. 12.1), a direct consequence of the proliferation of a larger pool of antigen-specific cells.
4. **Predominance of IgG antibody**, probably a consequence of the fact that memory B cells have switched from IgM to other isotypes and most of them express IgG on their membranes and will produce IgG after stimulation.

Experimental data obtained mostly in studies carried out with some inbred strains of mice, congenitally or artificially deficient in either B or T cells, has indicated that no memory exists for T-independent antigens. Both memory T cells and B cells are required for the induction of a memory response to T-dependent antigens.

C. **Characteristics Resulting from Prolonged Retention of Antigen and Persistent B-Cell Stimulation**

1. **Longer persistence of Ab synthesis.**
2. **Increasing affinity, avidity, and cross-reactivity** of antibodies.
   a. **Affinity maturation.** It is known that the affinity of antibodies increases during the primary immune response and even more so in the secondary and subsequent responses. This maturation is a result of the selection of memory B cells with progressively higher affinity mIg antibodies during a persistent immune response.
      i. The persistent synthesis of antibodies for extended periods of time, which is characteristic of repeatedly immunized individuals, results in the formation of antigen-antibody complexes.
      ii. The follicular interdigitating dendritic cells express Fcγ receptors and bind IgG-containing immune complexes. Those antigen–antibody complexes are efficiently trapped and may persist in the lymph nodes for months to years, and the antigen moieties are effectively presented to the immune system for as long as the complexes remain associated to the dendritic cells.
      iii. As free antibodies and mIg compete for binding to the immobilized antigen, only B cells with mIg of higher affinity than the previously synthesized antibodies will be able to compete effectively and receive activation signals. Consequently, the affinity of the synthesized antibodies will show a steady increase.
   b. **Increased avidity.** During the secondary response to a complex immunogen, clones responding to minor determinants are stimulated and a wider range of antibodies is produced. This results in increased avidity. (As discussed in Chap. 8, avidity is the sum of binding forces mediated by different antibody molecules binding simultaneously to the same antigen.)

c. **Increased Cross-Reactivity.** As the repertoire of antibodies recognizing different epitopes of a given immunogen increases, so do the probabilities for the emergence of cross-reactive antibodies recognizing antigenic determinants common to other immunogens.

V. THE FATE OF ANTIGENS ON THE PRIMARY AND SECONDARY RESPONSES

Following intravenous injection of a soluble antigen, its concentration in serum tends to decrease in three phases (Fig. 12.4).

A. **Equilibration Phase.** This phase is characterized by a sharp decrease of brief duration corresponding to the equilibration of the antigen between intra and extravascular spaces.

B. **Metabolic Decay.** During this phase the antigen slowly decays, due to its catabolic processing by the host.

C. **Immune Elimination.** When antibodies start to be formed, there will be a phase of rapid immune elimination, in which soluble antigen–antibody complexes will be formed and taken up by macrophages. The onset of this phase of immune elimination is shorter in the secondary immune response, and virtually immediate if circulating antibody exists prior to the introduction of the antigen.

A similar sequence of events, with less distinct equilibration and metabolic decay phases, occurs in the case of particulate antigens. If the antigen is a live, multiplying organism, there might be an initial increase in the number of circulating or tissue-colonizing organisms, until the immune response promotes the elimination of the antigen by a variety of mechanisms (see Chap. 13).

VI. THE MUCOSAL HUMORAL IMMUNE RESPONSE

The gastrointestinal and respiratory mucosae are among the most common portals of entry used by infectious agents. Since this constant exposure only rarely results in clinical disease, it seems obvious that strongly protective mechanisms must exist at the mucosal level. Some of those protective mechanisms are nonspecific and of a physicochemical nature, including the integrity of mucosal surfaces, gastric pH, gastrointestinal traffic, proteases and bile present in the intestinal lumen, as well as the flow of bronchial secretions, glucosidases, and bactericidal enzymes (e.g., lysozyme) found in respiratory secretions. At the same time, cell-mediated and humoral immune mechanisms are also operative in mucosal membranes.

A. **Mucosal Cell-Mediated Immunity.** Progress in our understanding of cell-mediated immune mechanisms at the mucosal levels has been slow. Most evidence suggests that innate cell-mediated mechanisms predominate, including:

1. **Phagocytic cells** (particularly macrophages), which abound in the submucosa and represent an important mechanism for nonspecific elimination of particulate matter and microbial agents of limited virulence.
2. **γ/δ T lymphocytes**, which are also present in large numbers in the sub-

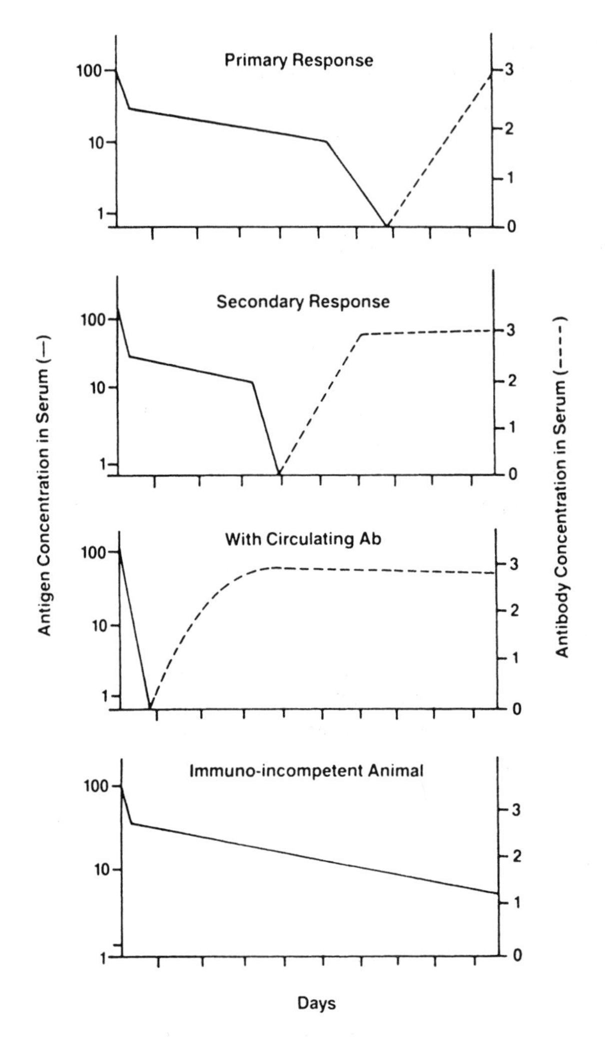

Figure 12.4 Diagrammatic representation of the fate of injected antigen in a nonimmune animal, which will undergo primary immune response; in an immune animal, which will show an accelerated, secondary response; in an animal with circulating antibodies, which will very rapidly eliminate the corresponding antigen from circulation; and in an immunoincompetent animal, which will slowly metabolize the antigen. (Modified from Talmage, D.F., Dixon, F.J., Bukantz, S.C., and Damin, G.J. *J. Immunol.* *67*:243, 1951.)

mucosal tissues, seem able to cause the lysis of infected cells by MHC-independent recognition of altered glycosylation patterns of cell membrane glycoproteins or of cell-associated microbial superantigens.

B. **Humoral Immunity at the Mucosal Level**

1. **Induction of mucosal immune responses.** Several lines of experimental work, both in animals and humans, have conclusively shown that the induction of secretory antibodies requires direct mucosal stimulation.
   a. Ogra and coworkers demonstrated that the systemic administration of an attenuated vaccine results in a systemic humoral response, while no

secretory antibodies are detected. In contrast, topical immunization with live, attenuated poliovirus results in both a secretory IgA response and a systemic IgM-IgG response (Fig. 12.5).

b. In addition, it has been demonstrated that the stimulation of a given sector of the mucosal system (GI tract) may result in detectable responses on nonstimulated areas (upper respiratory tract). This protection of distant areas is compatible with the unitarian concept of a **mucosal immunological network** with constant traffic of immune cells from one sector to another (Fig. 12.6).

c. Antigen-sensitized cells from the gut-associated lymphoid tissue (GALT), or from the peribronchial lymphoid tissues, enter the general circulation via the draining lymphatic vessels, and, subsequently, populate the remaining secretory-associated lymphoid tissues including the gastrointestinal tract, the airways, the urinary tract, and the mammary, salivary, and cervical glands of the uterus (Fig. 12.6).

2. **Passive transfer of mucosal immunity**

a. In some mammalian species, milk-secreted antibodies are actively absorbed in the newborn's gut and constitute the main source of adoptive

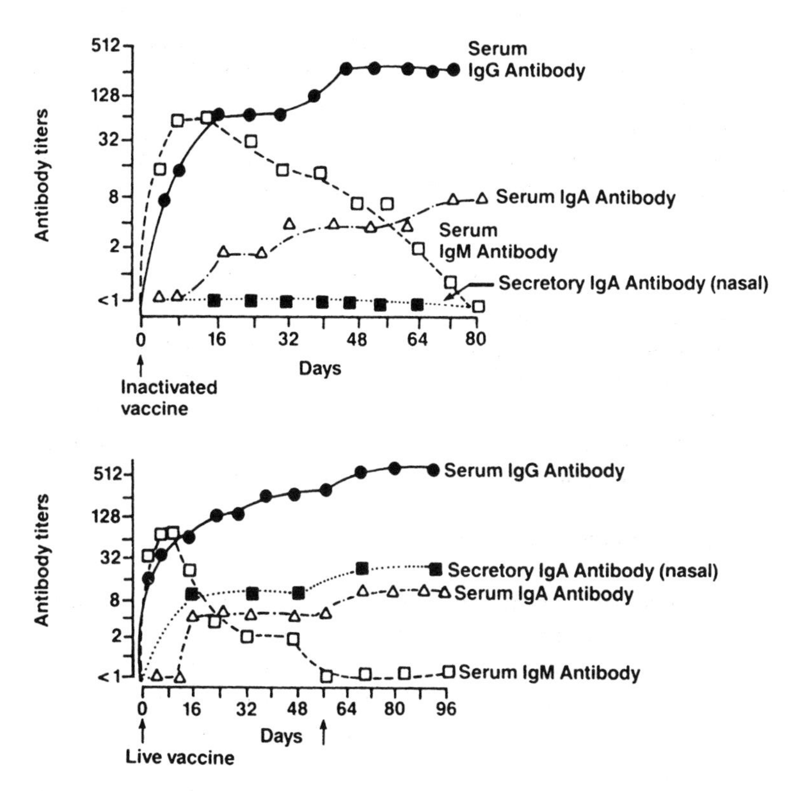

Figure 12.5 Comparison of the systemic and mucosal immune responses in human volunteers given killed polio vaccine (top) and live, attenuated polio vaccine (bottom). Note that secretory antibody was only detected in children immunized with live, attenuated vaccine. (Modified from Ogra, P.L., Karzon, D.T., Roghthand, F., and MacGillivray, M. *N. Engl. J. Med. 279*:893, 1968.)

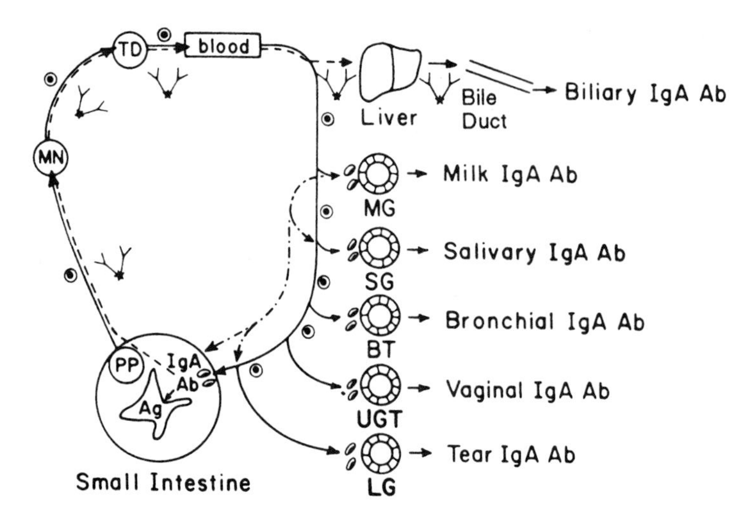

Figure 12.6 Diagrammatic representation of the pathways leading to the expression of IgA antibodies after antigenic stimulation of the GALT. IgA immunocytes (⊙) originating in Peyer's patches (PP) migrate to mesenteric lymph nodes (MN). Cells leave MN via the thoracic duct (TD) and enter circulation with subsequent homing to the mammary gland (MG), salivary gland (SG), lacrimal gland (LG), the lamina propria of the bronchial tree (BT), and intestinal or urogenital tract (UGT). The IgA antibodies are then expressed in milk, saliva, tears, and other secretions. IgA antibodies (Y) entering circulation (---) are selectively removed by the liver and subsequently expressed in bile. Cell traffic between peripheral mucosal sites (-·- MG to SG, LG and small intestine) is included in this scheme. (Reproduced with permission from Montgomery, P.C., Standera, C.A., and Majumdar, A.S. Evidence for migration of IgA bearing lymphocytes between peripheral mucosal sites. In *Protides of the Biological Fluids*, H. Peeters, ed. Pergamon, New York, 1985, p. 43.)

immunity in the neonate. This usually is observed in species in which there is limited or no placental transfer of antibodies.

b. In mammalians in which placental transfer of immunoglobulins is very effective (such as humans), the antibodies ingested with maternal milk are not absorbed. However, milk antibodies seem to provide passive immunity at the gastrointestinal level, which may be a very important factor in preventing infectious gastroenteritis in the newborn, whose mucosal immune system is not fully developed.

3. **Physiological significance of mucosal immunity**
   a. The main immunological function of secretory IgA is believed to be to prevent microbial adherence to the mucosal epithelia, which usually precedes colonization and systemic invasion. However, in several experimental models it has been demonstrated that disease can be prevented without interference with infection, so there are unresolved questions concerning the anti-infectious mechanism(s) of secretory antibodies.
   b. The absorption of immunogenic proteins is also believed to be inhibited by secretory antibodies.
   c. The relative importance of cellular vs. humoral mucosal defense mechanism has not been properly established. However, many IgA-deficient

individuals, with very low or absent circulating and secretory IgA, are totally asymptomatic, suggesting that cell-mediated mechanisms may play a significant protective role.

VII. IMMUNIZATION

A. **Historical Background** The concept of active immunization as a way to prevent infectious diseases is about two centuries old, if we consider the introduction of cowpox vaccination by **Jenner** in 1796 as the starting point. Jenner observed that milkmaids that had contracted cowpox were protected from smallpox and developed an immunization procedure based on the intradermal scarification of material from cowpox lesions. Empirically, he had discovered the principle of vaccine with live, attenuated microbes, which was later picked up by Louis Pasteur when he developed several of his vaccines. As infectious agents became better characterized, new vaccines were developed, some with inactivated organisms, others with microbial components, still others with attenuated infectious agents. Mass vaccination has had some remarkable successes, such as the eradication of smallpox and the significant declines in some of the most common or most serious infectious diseases of childhood, such as measles and polio. At the present time we are witnessing a new burst of progress in vaccine development, reflecting the application of molecular genetics techniques to the rational development of immunizing agents.

B. **Types of Vaccines.** A wide variety of immunizing agents have been developed. The following are some examples of the types of immunizing agents that are used for immunoprophylaxis in humans.

1. **Killed vaccines** are generally safe, but not as effective as attenuated vaccines.
   a. **Killed bacteria**, such as the traditional pertussis vaccine prepared with killed *Bordetella pertussis*, the etiological agent of whooping cough, the typhoid vaccine prepared with acetone-inactivated *Salmonella typhi*, and the cholera vaccine, prepared with killed *Vibrio cholerae*. The killed *B. pertussis* vaccine was reported to cause neurological reactions similar to autoallergic encephalitis, particularly in children with a history of neonatal or postnatal seizures. However, these reactions are extremely rare and were mostly observed in Great Britain.
   b. **Killed viruses**, such as Salk's polio vaccine, containing a mixture of the three known types of poliovirus, after inactivation with formalin. This vaccine has been as successful in the eradication of poliomyelitis as Sabin's attenuated oral vaccine. Its main advantage is safety, but is not as effective or amenable to mass immunizations as the oral vaccine (see below).
2. **Component vaccines**, which are even safer than killed vaccines, but their efficacy can be a problem.
   a. **Bacterial polysaccharides**, such as those used for *Streptococcus pneumoniae*, *Neisseria meningitidis*, and *Haemophilus influenzae* type b, and a vaccine for typhoid fever made of the Vi capsular polysaccharide.
   b. **Inactive toxins (toxoids)**, such as tetanus and diphtheria toxoids,

which are basically formalin-inactivated toxins that have lost their active site but maintained their immunogenic determinants and induce antibodies able to neutralize the toxins. More recently, *Clostridium perfringens* type C toxoid has been successfully used to prevent clostridial enteritis.

c. **Recombinant bacterial antigens**. Recently, a recombinant *Rickettsia rickettsii* antigen produced in *E. coli* has been proposed as a candidate vaccine for **Rocky Mountain spotted fever**.

d. **Mixed component vaccines**. The interest in developing safer vaccines for whooping cough has led to the introduction of **acellular vaccines**, constituted by a mixture of inactivated pertussis toxin or recombinant pertussis toxin (nontoxic due to the deletion of critical domains), a major determinant of the clinical disease, and one or several adhesion factors that mediate attachment to mucosal epithelial cells. These vaccines have been approved to replace the killed pertussis vaccine.

e. **Conjugate vaccines**. Most polysaccharide vaccines have shown poor immunogenicity, particularly in infants. This lack of effectiveness is a consequence of the fact that polysaccharides tend to induce T-independent responses with little immunological memory. This problem appears to be eliminated if the polysaccharide is conjugated to an immunogenic protein, very much like a hapten–carrier conjugate.

   i. The first conjugate vaccines to be developed involved the polyribositolribophosphate (PRP) of *Haemophilus influenzae* type b (HiB). Four conjugate vaccines have been successfully tested, the first three being currently approved by the F.D.A.:

      - PRP-OMPC, in which the carrier (OMPC) is an outer membrane protein complex of *Neisseria meningitidis*
      - HiB-OC, in which the carrier (OC) is a nontoxic mutant of diphtheria toxin
      - PRP-T, in which the carrier is diphtheria toxoid

      The introduction of these vaccines was followed by a 95% decrease in the incidence of *Haemophilus influenzae* type b infections affecting children of less than 5 years of age.

   ii. Conjugate vaccines prepared with *Streptococcus pneumoniae* and *Neisseria meningitidis* polysaccharides and suitable carrier proteins are currently being evaluated.

f. **Viral component vaccines** are based on the immunogenicity of isolated viral constituents.

   i. The best example is the hepatitis B vaccine, prepared originally with particles of the hepatitis virus outer coat protein (hepatitis B surface antigen or HBsAg) originally isolated from chronic carriers.

   ii. The current **hepatitis B vaccine** is produced by recombinant yeast cells. The gene coding for the hepatitis B surface antigen (HBsAg) was isolated from the hepatitis B virus and inserted into a vector, flanked by promoter and terminator sequences, which was used to transform yeast cells. HBsAg is obtained from disrupted yeast cells, and purified by chromatography.

iii. Some of the proposed HIV vaccines are component vaccines, constituted by envelope glycoproteins (**gp160** or its fragment, **gp120**) or peptides derived from these glycoproteins, produced in genetically engineered *E. coli*, insect cells, and mammalian cell lines. To date, these vaccines have not been proven to induce protective immunity. Two major problems have arisen when testing and evaluating these vaccines:
   - It is not certain that in vitro neutralization of an HIV isolate can be equated with protection from infection.
   - The HIV reverse transcriptase is error-prone and lacks copy-editing abilities. This results in a high frequency of mutations when the virus is transcribing RNA into DNA. Mutations affecting the epitopes of gp120 are not recognized by preexisting antibodies and eventually replace the antibody-sensitive strains. As a consequence, the virus "escapes" the effects of neutralizing antibodies by undergoing serial mutations. The sequential replacement of sensitive strains by their resistant counterparts plays a very important role in the progression of the disease and curtails the possible protective value of vaccination.

g. **Synthetic peptide vaccines**. The use of synthetic peptides for vaccination has the advantages of easy manufacture and safety. The goal is to synthesize the peptide sequences corresponding to known epitopes recognized by neutralizing antibodies and use them as vaccines. This theoretically appealing concept meets with two basic problems:
   i. It is questionable that a synthetic oligopeptide has the same tertiary configuration as the epitope expressed by the native antigen and that protective antibodies can be elicited in this manner
   ii. Small synthetic peptides are poorly immunogenic. This problem can be minimized by using peptide–protein (e.g., tetanus toxoid) conjugates.

   The most promising work with synthetic peptide vaccines has been carried out with *Plasmodium* peptides.
   i. In a murine malaria model, immunization with a tetanus toxoid–*Plasmodium berghei* peptide conjugate resulted in rates of protection ranging from 75 to 87%, identical to those observed with a killed vaccine made of the whole parasite.
   ii. In humans, efforts have been concentrated on the development of a vaccine against *Plasmodium falciparum*. A multirepeat region (approximately 40 repeats of the sequence Asn-Ala-Asn-Pro) of the circumsporozoite protein has been identified as the immunodominant B-cell epitope and used as a model for a peptide-based vaccine. However, the rate of protection obtained in the first trials with this vaccine was too low (two out of nine subjects immunized were protected).

h. **DNA vaccines.** It has recently been reported that intramuscular injection of nonreplicating plasmid DNA encoding the hemagglutinin (HA) or nucleoprotein (NP) of influenza virus elicited humoral and cellular protective reactions. It is not understood how DNA becomes expressed and its message translated into viral proteins, but the positive results obtained in this experiment have raised enormous interest in the scientific community.

3. **Attenuated vaccines.** Attenuated vaccines are generally very efficient, but in rare cases can cause the very disease they are designed to prevent, particularly in immunocompromised individuals.
   a. Most **antiviral vaccines** are made of viral strains attenuated in the laboratory, including the classic smallpox vaccine, the oral polio vaccine (a mixture of attenuated strains of the three known types of poliovirus), the mumps–rubella–measles vaccine, and the varicella-zoster vaccine, recently approved by the FDA.
   b. The application of molecular genetics techniques has allowed the development of **attenuated bacteria**, which are finding applications in immunoprophylaxis.
      i. A new bacterial vaccine against typhoid fever is under development, based on the use of an attenuated strain of *Salmonella typhi* that grows poorly and is virtually nonpathogenic but induces protective immunity in 90% of the individuals.
      ii. Mutant *Bordetella* strains that code for an immunogenic toxin lacking its binding site (thus devoid of pathogenic effects) are being field-tested.
4. **Recombinant organisms.** Recombinant technology has been used to delete the genes coding for virulence factors from bacteria, or to add genetic information to attenuated viruses or attenuated bacteria.
   a. **Recombinant vaccinia viruses**, in which the genetic information coding for relevant antigens of unrelated viruses has been added to the vaccinia virus genome, have been developed and used successfully.
      i. A recombinant vaccinia virus, carrying a retroviral *env* gene, protected mice against Friend leukemia virus.
      ii. Another type of recombinant vaccinia virus expressing an immunodominant region of streptococcal M protein has been shown to reduce streptococcal colonization in mice after intranasal immunization.
      iii. Experimental vaccines for AIDS were developed by incorporating parts of the *env* gene of HIV into the vaccinia virus genome. Initial data suggested that recombinant HIV coding for the gp120 vaccines of this type were more efficient in inducing cell-mediated immunity, and field trials were conducted in Africa. Unfortunately, several patients developed fatal forms of disseminated vaccinia infection and the trials were canceled. This accident illustrates the potential dangers of using an attenuated virus in large-scale vaccinations.
      iv. Since the genome of vaccinia virus is rather large, multiple recombinant constructs carrying simultaneously the genes for the

Table 12.1 Recommended Childhood Immunization Schedule for the United States, January 1997[a]

<table>
<tr><th>Age ®
Vaccine</th><th>Birth</th><th>1
mo.</th><th>2
mos.</th><th>4
mos.</th><th>6
mos.</th><th>12
mos.</th><th>15
mos.</th><th>18
mos.</th><th>4–6
yr.</th><th>11–12
yr.</th><th>14–16
yr.</th></tr>
<tr><td>Hepatitis B</td><td colspan="3">Dose 1</td><td></td><td></td><td></td><td></td><td></td><td></td><td></td><td></td></tr>
<tr><td>Hepatitis B (cont.)</td><td></td><td colspan="3">Dose 2</td><td colspan="3">Dose 3</td><td></td><td></td><td></td><td></td></tr>
<tr><td>DTaP[b] or DTP[c]</td><td></td><td></td><td>Dose 1</td><td>Dose 2</td><td>Dose 3</td><td></td><td colspan="2">Booster</td><td>DTaP</td><td colspan="2">Td booster</td></tr>
<tr><td>HiB[d]</td><td></td><td></td><td>Dose 1</td><td>Dose 2</td><td>Dose 3[f]</td><td colspan="2">Booster</td><td></td><td></td><td></td><td></td></tr>
<tr><td>Polio[e]</td><td></td><td></td><td>Dose 1</td><td>Dose 2</td><td></td><td colspan="2">Dose 3</td><td></td><td>Booster</td><td></td><td></td></tr>
<tr><td>MMR[g]</td><td></td><td></td><td></td><td></td><td></td><td colspan="2">Dose 1</td><td></td><td colspan="2">Booster</td><td></td></tr>
<tr><td>VZV[h]</td><td></td><td></td><td></td><td></td><td></td><td colspan="3">Dose 1</td><td></td><td></td><td></td></tr>
</table>

[a]Recommendations approved by the Advisory Committee on Immunization Practices, the American Academy of Pediatrics and the American Academy of Family Physicians (Modified from *Pediatrics*, 99; 135–138, 1997). Shaded boxes represent recommended schedules for vaccination of nonimmune individuals.

[b]DTaP = Diphtheria, tetanus, and acellular pertussis.

[c]DTP = Diphtheria, tetanus, and killed pertussis.

[d]*Haemophilus influenzae* type B PRP-conjugate vaccine.

[e]Both oral polio vaccine and killed polio vaccine are recommended for routine vaccination. Immunodeficient and immunosuppressed children and their close contacts should receive the killed polio vaccine.

[f]Third dose not needed if PRP-OMP is used.

[g]MMR = measles, mumps, rubella.

[h]Varicella-zoster virus vaccine.

HBsAg, glycoprotein D for herpes virus, and influenza virus hemagglutinin, have been successfully generated. The potential value of these constructs is obvious, since they could induce protection against multiple diseases after a single immunization.

b. **Recombinant BCG** organisms to whose genome the genes coding for IL-2, IL-4, IL-6, GM-CSF, and IFN-γ have been added have been shown to induce strong protective immunity in an animal model. If this approach could be safely translated to the human vaccine, it would constitute a major medical breakthrough.

C. **Recommended Immunizations.** At the present time, a wide variety of vaccines are available for protection of the general population or of individuals at risk for a specific disease due to their occupation or to other factors. Table 12.1 summarizes the recommended schedule for active immunization of normal infants and children. Additional information concerning recommended immunizations for adults, travelers, special professions, etc., can be obtained in a variety of specialized publications, including the *Report of the Committee on Infectious Diseases*, published annually by the American Academy of Pediatrics, the booklet *Health Information of International Travel*, also published annually by the U.S. Public Health Service, and the *Morbidity and Mortality Weekly Report*, published by the Centers for Disease Control in Atlanta.

D. **Vaccines as Immunotherapeutic Agents.** The use of vaccines to stimulate the immune system as therapy for chronic or latent infections is receiving considerable interest. Four areas of application have emerged.

1. Herpes virus infections, in which vaccination seems to reduce the rate of recurrence.
2. Leprosy, in which administration of BCG seems to potentiate the effects of chemotherapy.
3. Tuberculosis, in which a new vaccine made of killed *Mycobacterium vaccae* seems also to potentiate the effects of antituberculous drugs, even in patients resistant to therapy.
4. HIV infection, in which vaccination with killed HIV may alter the TH1/TH2 balance in favor of TH_1, more effective mediators of anti-HIV responses.

SELF-EVALUATION

Questions

Choose the ONE *best* answer.

12.1 Which one of the following most closely describes the biological basis of immunological memory?

A. Structural changes in the hypervariable region of IgG antibodies
B. Increased numbers of antigen-sensitive T and B cells
C. Increased synthesis of IL-2 by antigen-sensitive T cells
D. Predominant synthesis of IgG antibodies
E. Selection of B-cell clones producing high-affinity antibodies

12.2 In comparison with primary humoral immune responses, secondary humoral immune responses are characterized by:

A. A faster decline in antibody concentration after the steady state has been reached
B. Increased probability of detecting cross-reactive antibodies
C. Longer lag phases
D. Longer persistence of antigen in circulation
E. The need for greater doses of stimulating antigen

12.3 The half-life of an injected antigen is shortest when:
A. Preformed antibodies exist in circulation
B. T cells are not necessary to induce the immune response
C. The antigen is T-dependent
D. The synthesized antibody is of the IgM isotype
E. The injected animal has been previously immunized with the same antigen

12.4 The mucosal humoral immune response
A. Is characterized by predominant synthesis of IgG antibodies
B. Is essential for the maintenance of an infection-free state
C. Is localized to the mucosal segment that is directly challenged
D. Leads to complement-dependent killing of many infectious agents
E. Requires direct antigenic challenge of the mucosa

12.5 The main difference between killed polio vaccine (Salk) and attenuated polio vaccine (Sabin) is that only the latter induces:
A. Circulating complement-fixing antibodies
B. Circulating neutralizing antibiotic
C. Memory
D. Protection against viral dissemination through the bloodstream
E. Secretory IgA antibodies

12.6 Which of the following is the main advantage derived from the coupling of poorly immunogenic peptides or polysaccharides to bacterial toxoids?
A. Adjuvanticity of bacterial toxoids eliminating the need for the use of inorganic adjuvants
B. Conversion of poorly immunogenic compounds into integral components of an immunogenic "hapten"-carrier conjugate
C. Lesser toxicity of the toxoid-peptide or toxoid-polysaccharide conjugates
D. More efficient elicitation of cell-mediated immunity
E. Simultaneous immunization against more than one antigen

12.7 Which one of the following events is **less** likely to have a negative feedback effect on an ongoing humoral immune response?
A. Activation of TH1 lymphocytes
B. Elimination of the antigen
C. Formation of immune complexes
D. Production of cross-reactive antibodies directed against related antigens
E. Reaction of anti-idiotype antibodies with mIg on the B-cell membrane

12.8 The initial activation of a primary anti-infectious response is usually dependent on the:
A. Ability of B lymphocytes to act as antigen-presenting cells (APC)
B. Existence of cross-reactive antibodies
C. Nonimmunological uptake of the infectious agent by APC
D. Nonspecific activation of macrophages by microbial products
E. T-independent stimulation of B cells

12.9 Which one of the following is a unique consequence of the antigenic stimulation of the peri-intestinal immune system?
A. A systemic response of identical characteristics to that obtained when the antigen is injected by the intramuscular route
B. Diffusion of secretory IgA synthesized in the Peyer's patches into the systemic circulation
C. Increased differentiation of B cells with mIgA in the bone marrow
D. Migration of sensitized B cells from the peri-intestinal tissues to other peri-mucosal lymphoid areas
E. Production of secretory antibody limited to the intestine

12.10 Which one of the following is the most likely consequence of injecting human albumin intravenously into a previously immunized rabbit with antihuman albumin antibodies in circulation?
A. A new burst of IgM antibody synthesis
B. Formation of circulating immune complexes
C. Immediate increase in the levels of circulating antialbumin antibody
D. Immediate release of histamine from circulating neutrophils
E. Massive activation of T cells

Answers

12.1 (B) Memory cells include both T and B lymphocytes which will be able to respond faster and more energetically to a second antigenic challenge.

12.2 (B) In a secondary immune response, the antibody repertoire increases, and, as a result, the probability of a cross-reaction also increases.

12.3 (A) The immune elimination phase is immediately evident when the injected animal has preformed antibodies that will immediately combine with the antigen and promote its elimination from circulation.

12.4 (E) Systemic administration of antigen never stimulates mucosal responses. Many individuals with severely depressed levels of mucosal IgA live free of infections, which raises questions concerning whether IgA is the only, or the main, exponent of mucosal immunity.

12.5 (E) Only the attenuated vaccine, given orally, can induce mucosal immunity. However, both vaccines induce systemic immunity and elicit immunological memory.

12.6 (B) The toxoid-polysaccharide and toxoid-peptide vaccines are comparable to hapten-carrier conjugates. Toxoids, acting as carriers, are recognized by T cells and effectively recruit T-cell help, thus enhancing the immune response and ensuring the development of immunological memory.

12.7 (D) Cross-reactive antibodies against related antigens have not been implied in any theory concerning the down-regulation of the immune response. Activation of TH1 lymphocytes, on the other hand, would result in the synthesis of interferon-γ, which down-regulates TH2 cells, which are those primarily assisting the humoral immune response.

12.8 (C) The most potent APC are the follicular interdigitating cells, the interdigitating cells found in the paracortical area, and macrophages. All those cells have receptors that allow phagocytosis in the absence of antibody, and although the efficiency of nonimmunological phagocytosis is rather limited, it is sufficient to allow ingestion, processing, and presentation of microbial derived-peptides to the immune system.

12.9 (D) Mucosal stimulation is always associated with a local response. Systemic responses are only observed when a replicating organism is used as the immunizing agent. The local response propagates to areas not directly stimulated by traffic of sensitized B lymphocytes.

12.10 (B) Given the fact that circulating antibodies are already present, the injection of albumin will be associated with almost immediate formation of immune complexes, which will be quickly taken up by phagocytic cells. The concentration of albumin will fall rapidly, and the concentration of antialbumin antibody will show a transient decrease. Histamine could be released if the animal had produced IgE antibodies; however, the cells that release histamine are basophils, not neutrophils.

BIBLIOGRAPHY

Ahmed R., and Gray, D. Immunological memory and protective immunity: understanding their relation. *Science*, *272*:54, 1996.

Bektimirov, T., Lambert, P.H., and Torrigiani, G. Vaccine development: Perspectives of the World Health Organization. *J. Med. Virol.*, *31*:62, 1990.

Brown, F. From Jenner to genes—The new vaccines. *Lancet*, *i*:587, 1990.

Fujihashi, K., Yamamoto, M., McGhee, J.R., and Kiyono, H. Function of α/β TCR^+ and γ/δ TCR^+ IELs for the gastrointestinal immune response. *Int. Rev. Immunol.*, *11*:1, 1994.

Källberg, E., Jainandunsing, S., Gray, D., and Leanderson, T. Somatic mutation of immunoglobulin V genes in vitro. *Science*, *271*:1285, 1996.

Klaus, G.G.B., Humphrey, J.H., Kunkel, A., and Dongworth, D.W. The follicular dendritic cell: Its role in antigen presentation in the generation of immunological memory. *Immunol. Rev.*, *53*:3, 1980.

Kniskern, P.J., Marburg S., and Ellis, R.W. Haemophilus influenzae type b conjugate vaccines. *Pharm. Biotechnol.*, *6*:673, 1995.

Mestecky, J., Lue, C., and Russell, M.W. Selective transport of IgA. Cellular and molecular aspects. *Gastroenterol. Clin. North Am.*, *20*:441, 1991.

Sabin, A.B. Oral poliovirus vaccine: History of its development and use and current challenge to eliminate poliomyelitis from the world. *J. Infect. Dis.*, *151*:420, 1985.

Schaffner, W. Immunization in adults, I and II. *Infect. Dis. Clin. North Am.*, *4*(1,2), 1990.

13
Infections and Immunity

Gabriel Virella

I. INTRODUCTION

During evolution, an extremely complex system of anti-infectious defenses has emerged. But at the same time as vertebrates and mammals developed their defenses, microbes continued evolving as well, and many became adept at avoiding the consequences of the anti-infectious defense mechanisms. The interplay between host defenses, microbial virulence, and microbial evasion mechanisms determines the outcome of the constant encounters between humans and pathogenic organisms.

II. NONSPECIFIC ANTI-INFECTIOUS DEFENSE MECHANISMS

Nonspecific defense mechanisms play a most important role as a first line of defense, preventing penetration of microorganisms beyond the outer exposed surfaces of the body. The following is a brief description of the most important nonspecific defense mechanisms.

A. **Physical and Chemical Barriers**, including the integrity of the epithelial and mucosal surfaces, the flow of mucosal secretions in the respiratory tract, the acidity of the gastric contents, and the secretion of lysozyme in tears, saliva, and most secretions. The importance of these barriers is apparent from the prevalence of infections when their integrity is compromised.

B. **Inducible Nonspecific Responses**, including fever and release of interferons, activated when infectious agents manage to invade, particularly effective in preventing viral replication.

C. **Phagocytosis**. As a microbe penetrates beyond the skin or mucosal surface, it will encounter cells able to ingest it. Two types of cells are particularly adept at nonimmune phagocytosis: tissue macrophages and granulocytes (particularly neutrophils). This nonimmune phagocytosis involves a variety of recognition systems (Fig. 13.1).

   1. **CR1 and CR3 receptors** are able to interact with C3b and iC3b on the microbial membrane, generated as a consequence of complement activation by the alternative pathway, a property common to many bacteria (see below).
   2. **Mannose receptors** on phagocytic cells may mediate ingestion of organisms with mannose-rich polysaccharides, such as *Candida albicans*. Mannose-

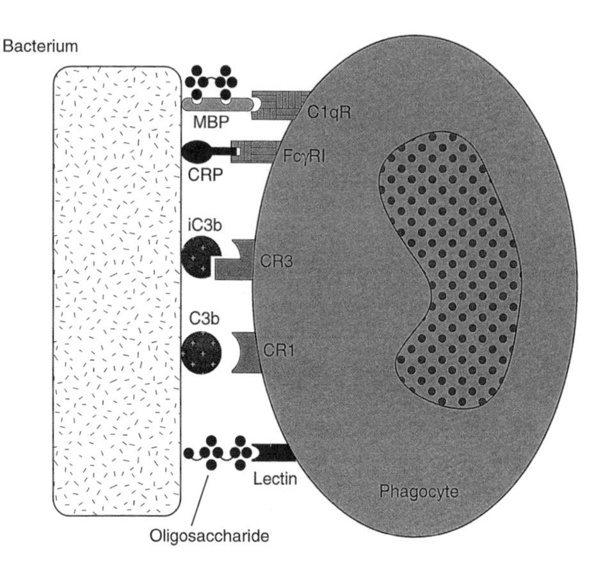

Figure 13.1 Diagrammatic representation of the different receptors that may mediate nonimmune phagocytosis (MBP = mannose-binding protein; CRP = C-reactive protein).

mediated phagocytosis is amplified by a **mannose-binding protein** that promotes phagocytosis through complement activation and direct interaction with the **C1q (collectin) receptor** on phagocytic cells.

3. **C-reactive protein** binds to certain bacterial polysaccharides and has very similar effects to the mannose-binding protein, activating complement and promoting phagocytosis, both through CR1 and CR3, as well as by other receptors, including the FcγRI and the C1q receptor, both of which bind this protein.

D. **Activation of the Complement System via the Alternative Pathway.** A variety of microorganisms (bacteria, fungi, viruses, and parasites) can activate complement by the alternative pathway (see Table 13.1). In most cases where adequate studies have been carried out, polysaccharidic structures have been proven to be responsible for complement activation of the alternative pathway. This activation will lead to phagocytosis, through the generation of C3b, and to chemotaxis, through the release of C3a.

E. **Acute Phase Reaction and Leukocyte Chemotaxis.** The initial recognition by phagocytes and the activation of the complement system by the alternative pathway, by themselves, may not be sufficient to eradicate the invading microorganism, but the response is quickly amplified by a multitude of cytokines released by macrophages activated as a consequence of phagocytosis, including IL-1, IL-8, and TNF-α.

1. IL-1 and TNF-α cause an increase in body temperature, mobilize neutrophils from the bone marrow, and up-regulate the synthesis of a variety of proteins known as **acute phase reactants**, including C-reactive protein and the mannose-binding protein mentioned above.
2. TNFα and IL-1 up-regulate the expression of cell adhesion molecules in the endothelial cells of neighboring endothelial cells, thus promoting adherence

Table 13.1 Examples of Infectious Agents Able to Activate the Alternative Pathway of Complement Without Apparent Participation of Specific Antibody

a. Bacteria	c. Parasites
H. influenzae Type b	*Trypanosoma cyclops*
Streptococcus pneumoniae	*Schistosoma mansoni*
Staphylococcus aureus	*Babesia rodhaini*
S. epidermidis	
b. Fungi	d. Viruses
Candida albicans	Vesicular stomatitis virus

of leukocytes, and increase vascular permeability. Both factors facilitate the migration of leukocytes off the vessels, toward the focus of infection.

3. IL-8 has chemotactic properties. Together with other chemotactants, such as C5a and bacterial peptides, it attracts neutrophils toward the focus of infection.

F. **Natural Killer** (**NK**) **Cells** are able to destroy viral infected cells as a consequence of the delivery of an activating signal in the absence of an inhibitory signal.
   1. The activating signal is delivered as a consequence of the interaction between a recognition molecule (NKR-P1) on the NK cells and glycoproteins expressed on the membrane of the infected cells.
   2. The inhibiting signal is usually delivered as a consequence of the interaction between a membrane protein in the NK cell membrane and MHC-I molecules on the membrane of normal cells. Viral infected cells often have a reduced expression of MHC-I molecules due either to a down-regulation of cellular protein synthesis or a specific inhibition of MHC-I transport to the cell membrane, discussed later in this chapter.

G. **γ/δ T Lymphocytes** are predominantly localized to the mucosal epithelia, where they appear to recognize infected epithelial cells by a nonimmunological mechanisms (i.e., not involving the T-cell receptors), which are subsequently destroyed.

III. NATURAL ANTIBODIES

Preexisting antibodies may play a very important anti-infectious protecting role. Natural antibodies may arise as a consequence of cross-reactions, as exemplified in the classic studies concerning the isoagglutinins of the ABO blood group system (i.e., circulating antibodies that exist in a given individual and are able to agglutinate erythrocytes carrying alloantigens of the ABO system different from those of the individual himself).

A. **The Origin of the AB Isoagglutinins**
   1. Chickens are able to produce agglutinins recognizing the AB alloantigens, but only when fed conventional diets; chicks fed sterile diets do not develop agglutinins. Furthermore, anti-A and anti-B agglutinins develop as soon as chicks fed sterile diets after birth are placed on conventional diets later in life.
   2. These observations pointed to some dietary component as a source of

immunization. It was eventually demonstrated that the cell wall polysaccharides of several strains of enterobacteriaceae and the AB oligosaccharides of human erythrocytes are structurally similar. Thus, cross-reactive antibodies to enterobacteriaceae are responsible for the "spontaneous" development of antibodies to human red cell antigens in chickens.

3. Newborn babies of blood groups A, B, or O do not have either anti-A or anti-B isohemagglutinins, but will develop them during the first months of life, as they get exposed to common bacteria with polysaccharide capsules. However, newborns are tolerant to their own blood group substance, so they will only make antibodies against the blood group substance that they do not express. Blood group AB individuals never produce AB isoagglutinins.

B. **Other Mechanisms for the Generation of Natural Antibodies.** Cross-reaction is probably the most common explanation for the emergence of "natural" antibodies, but other mechanisms, such as the mitogenic effects of T-independent antigens and the nonspecific stimulatory effects of lymphokines released by antigen-stimulated T lymphocytes, which could activate B cells responding to other antigens, could explain the rise of "nonspecific" immunoglobulins that is observed in the early stages of the humoral response to many different antigens. It is only a matter of random probability that some of those "nonspecific" immunoglobulins may play the role of "natural antibody" relative to an unrelated antigen.

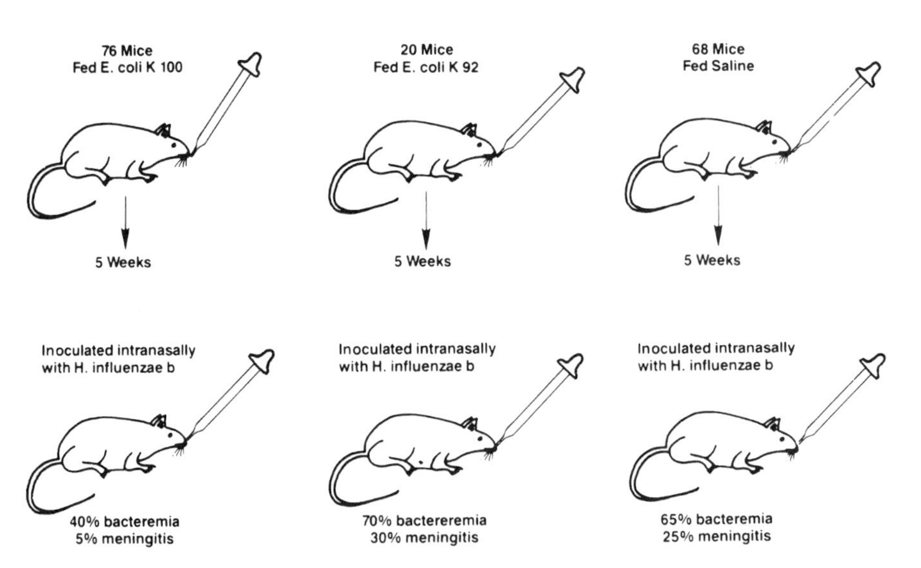

Figure 13.2 Diagrammatic representation of an experiment proving the anti-infectious protective role of cross-reactive "natural" antibodies. Three groups of mice were orally immunized with *E. coli* K100, *E. coli* K92, and saline, as a control. Five weeks later all animals were challenged with *H. influenzae* type b by the intranasal route. *E. coli* K100 has cross-reactivity with *H. influenzae* type b, but the same is not true for *E. coli* K92. The animals immunized with *E. coli* K100 showed significantly lower rates of bacteremia and meningeal infection than the animals immunized with *E. coli* K92 or controls fed with saline. (Based on Moxon, E.R., and Anderson, P. *J. Inf. Dis. 140*:471, 1979.)

C. **The Significance of "Natural" Antibodies.** "Natural" antibodies may play an important protective role, as shown by the experiments summarized in Figure 13.2. Antibodies elicited to *E. coli* K100 cross-react with the polyribophosphate of *Haemophilus influenzae* and can protect experimental animals against infection with the latter organism. It is logical to assume that such cross-immunizations may be rather common and play an important protective role against a variety of infectious agents.

IV. THE PROTECTIVE ROLE OF SPECIFIC ANTIBODIES

A. **The Humoral Immune Response.** If a pathogen is not eliminated by nonimmunological means and continues to replicate, it will eventually spread through the blood and lymph, and will usually be trapped by macrophages and follicular interdigitating cells in the lymph nodes and spleen. Those cells are able to internalize antigens interacting with receptors such as the mannose receptor (see above), process the antigen in the endosomic compartment, and express MHC-II-associated antigen-derived peptides. This creates the ideal conditions for the onset of an immune response: B lymphocytes can interact with membrane-bound antigen while helper T cells recognize MHC-II with antigen-derived peptides presented by the same APC. The antigen-recognizing cells will interact and costimulate each other and, after a time lag necessary for proliferation and differentiation of B cells into antibody-producing cells, circulating antibody will become detectable.

1. **A primary immune response** will take 5 to 7 days (some times as long as 2 to 3 weeks) to be detected. The predominating isotype of the antibodies made early in a primary immune response is IgM.
2. **A secondary immune response** has a shorter lag phase (as short as 3–4 days) and the predominant isotype of the antibodies is IgG (see Chap. 12).
3. The differentiation between IgM and IgG antibodies is a useful index for discriminating between recent and past infections, particularly in the case of viral infections.

B. **Antibody-Dependent Anti-Infectious Effector Mechanisms.** As soon as specific antibodies become available, they can protect the organism against infection by several different mechanisms.

1. **Complement-mediated lysis** of a microorganism or virus-infected cell can result from the activation of the complete sequence of complement. However, as discussed below, both mammalian cells and most pathogenic microorganisms have developed mechanisms that allow them to resist complement-mediated lysis.
2. **Opsonization and phagocytosis** are, in contrast, of extreme physiological significance. Several proteins can opsonize and promote phagocytosis, as discussed above, but IgG antibodies are the most efficient opsonins, particularly when complement is activated and C3b joins IgG on the microbial cell membrane.
   a. **Opsonins.** Because of their role in promoting opsonization, IgG antibodies and C3b are considered "**opsonins.**" Both C3b and IgG alone are efficient opsonins, but their effects are synergistic.

b. **Fc and CR1 receptors**, expressed on the membranes of all phagocytic cells, mediate phagocytosis and also deliver activating signals to the cells, inducing release of cytokines, as mentioned above.
c. Killing through opsonization has been demonstrated for bacteria, fungi, and viruses, while phagocytosis of antibody/complement-coated unicellular parasites has not been clearly demonstrated.
d. The biological significance of phagocytic cells as ultimate mediators of the effects of opsonizing antibodies is obvious; the protective effects of antibodies are lost in patients with severe neutropenia or with severe functional defects of their phagocytic cells, and those patients have increased incidence of infections with a variety of opportunistic organisms.

3. **Antibody-dependent cell-mediated cytotoxicity (ADCC).** Cells with Fc receptors may be able to participate in killing reactions that target antibody-coated cells. IgG1, IgG3, and IgE antibodies and cells with FcγR or FcϵR are usually involved. Large granular lymphocytes or monocytes are the most common effector cells in ADCC, but in the case of parasitic infections, **eosinophils** play the principal role in cytotoxic reactions (Fig. 13.3). Different effector mechanisms are responsible for killing by different types of cells: large granular lymphocytes kill through the release of granzymes and signaling for apoptosis; monocytes kill by releasing oxygen active radicals and nitric oxide; eosinophil killing is mostly mediated by the release of a "**major basic protein**," which is toxic for parasites.
4. **Toxin neutralization.** Many bacteria release toxins, which are often the major virulence factors responsible for severe clinical symptoms. Antibodies to these toxins prevent their binding to cellular receptors and promote their elimination by phagocytosis.
5. **Virus neutralization.** Most viruses spread from an initial focus of infection to a target tissue via the blood stream. Antibodies binding to the circulating virus change its external configuration and prevent either its binding to cell receptors or its ability to release nucleic acid into the cell.
6. **Mucosal protection.** Secretory antibodies seem to play their protective role by preventing the attachment and penetration of microbial agents through mucosal surfaces.

C. **Factors Influencing the Effectiveness of an Anti-Infectious Humoral Response.** The effectiveness of the humoral immune response in eliminating or preventing the proliferation of an infectious agent or in neutralizing a toxin before it causes severe disease depends on whether or not antibodies can be available in time to prevent the infection to develop or the toxin to reach its cellular target.

1. If the relevant antibodies are present in circulation, as a result of vaccination, previous infection, or cross-reaction between different microorganisms, protection is most effective, the microorganism or its toxin(s) will be almost immediately neutralized, and the infection will remain subclinical.
2. If preformed antibodies are not available, protection will depend on whether antibody synthesis can take place before the "incubation period" (period of time during which the infectious agent is multiplying but has not yet reached sufficient mass to cause clinical disease) is over.

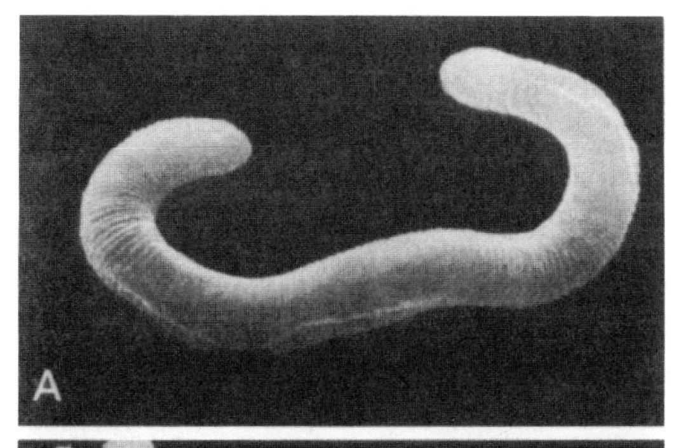

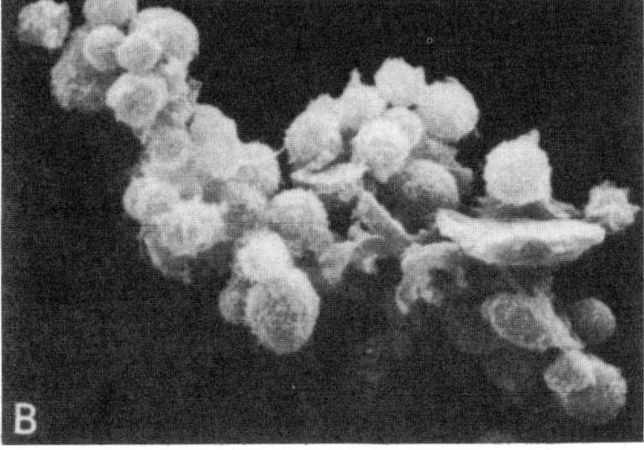

Figure 13.3 (A) Scanning electron microphotograph of a *Trichinella spiralis* larva incubated with eosinophils and complement-depleted normal (nonimmune) mouse serum for 4 hours; (B) scanning electron microphotograph of a *Trichinella spiralis* larva incubated with eosinophils and complement-depleted immune serum for 4 hours. Notice that the attachment of eosinophils happened only when eosinophils were added to *T. spiralis* larvae in the presence of immune sera containing antibodies directed against the parasite. (Reproduced with permission from Kazura, J.W., and Aikawa, M. *J. Immunol. 124*:355,1980.)

a. Some infectious agents, such as the influenza virus, have very short incubation periods (about 2 to 3 days) and, in such cases, not even a secondary immune response can be protective.
b. In most infections, the duration of the incubation period is sufficiently long as to allow a secondary immune response to provide protective antibodies. Thus, for many infections, particularly the common viral diseases of childhood, previous exposure and acquisition of memory ensure that antibody will be produced in time to maintain subsequent exposures, which play the role of natural "booster" doses, probably explaining the "immunity for life" associated with them.
c. The goal of prophylactic immunization may vary from case to case. In diseases with very short incubation periods, it is essential to maintain the levels of neutralizing antibody in circulation necessary to immediately abort infection. In most other diseases, it may be sufficient to induce immunological memory, since once memory has been induced, the immune system will be able to respond in time to prevent the development of clinical infection.

d. Protection by humoral immunity is only possible if the infectious agent is easily available to the antibodies produced against it. Thus, organisms that grow intracellularly are not easy to eliminate by antibodies. In addition, as discussed in detail later in this chapter, those organisms able to change their antigenic make-up during the course of an infection can persist in spite of a vigorous humoral response.

V. THE PROTECTIVE ROLE OF CELL-MEDIATED IMMUNITY

Many organisms have the ability to grow and replicate intracellularly, some as an absolute requirement, others as an option that allows them to survive after phagocytosis. Antibodies are largely ineffective against those organisms, and T lymphocytes play a major role in their elimination, as reflected by the finding of lymphocytic infiltrates in tissues infected by intracellular infectious agents, such as viruses. The immune system has two basic options to eliminate those organisms: to kill the infected cell, or to enhance the infected cell's ability to kill intracellular organisms. Either option requires the persistent activation of the TH1 subpopulation of helper cells.

A. **Lymphocyte-Mediated Cytotoxicity.** It can be easily demonstrated that viral-infected cells are lysed as a consequence of their incubation with "immune" lymphocytes, obtained from an animal previously exposed to the same virus.
   1. **Recognition of infected cells**
      a. Virus-infected cells that express MHC-I molecules with associated viral peptides can be easily recognized by CD8+ cytotoxic T cells. While CD8+ lymphocytes, when strongly stimulated, are able to release IL-2 and differentiate into effector cytotoxic T cells, usually that differentiation requires the collaboration of TH1 helper T cells. The activation of CD4+ TH0 cells and their differentiation into TH1 helper T cells is likely to take place in lymphoid tissues, since macrophages are often infected by viruses and virus-derived peptides are likely to be presented by those cells both in association with MHC-I and MHC-II molecules.
      b. Macrophages are often infected by intracellular organisms of all kinds, including bacteria, fungi, and protozoa. Thus, recognition by CD4+ and CD8+ T lymphocytes is likely to take place, resulting in the differentiation of effector CD8+ cytotoxic T cells.
   2. **Killing** involves recognition of the target cell, release of perforins and esterases, and delivery of apoptosis signals through the *fas-fas* ligand interaction, as described in Chapter 11.

B. **Lymphocyte-Mediated Activation of Macrophages and Other Inflammatory Cells.** In many cases, the cells infected by intracellular organisms are the tissue macrophages, and the persistence of the infection depends on a delicate balance between a state of relative inactivity by the macrophage and mechanisms that allow the infectious agent to escape proteolytic digestion once inside the cytoplasm (see below).
   1. **Macrophage activation**
      a. **CD4+ TH1 lymphocytes** are activated as a consequence of their interaction with infected macrophages expressing MHC-II-associated peptides and releasing IL-12. Once activated, TH1 lymphocytes release a variety

of lymphokines, particularly **interferon-γ**, which activates macrophages, enhancing their ability to kill intracellular organisms (see Chap. 17) and **GM-CSF**, which promotes differentation and release of granulocytes and monocytes from the bone marrow and delivers costimulatory signals to B cells.

b. Once the initial activation signals are delivered, macrophages and lymphocytes enter in a complex cycle of self and mutual activation involving a variety of cytokines. In addition, several of these cytokines may activate other types of cells and have chemotactic properties, for which reason they are known as **chemotactic cytokines** or **chemokines** (see Chap. 11). This group of cytokines includes **interleukin-8** (IL-8), **RANTES, macrophage inflammatory proteins** (MIP), **macrophage chemotactic proteins** (MCP), and **migration-inhibition factor** (MIF). Collectively the chemokines attract, activate, and retain leukocytes to the area where a cell-mediated immune reaction is taking place.

2. **Inflammatory reaction.** In concert with the release of chemotactic cytokines, the expression of CAMs in neighboring microvessels is up-regulated, favoring adherence and migration of monocytes and granulocytes to the extravascular space. Inflammatory cells accumulate in the area of infections and, as a consequence of cross-activation circuits involving phagocytes and TH1 lymphocytes, the localized macrophages and granulocytes become activated. Consequently,

   a. the enzymatic contents and respiratory burst of the cells become more intense, making them better suited for killing intracellular organisms.
   b. Phagocytosis is enhanced.
   c. Activated macrophages release higher levels of cytokines, including IL-12, which continues to promote the differentiation of TH1 cells, as well as IL-1, IL-6, and TNF-α, which in association with IL-8, play the major role in inducing the synthesis of reactive proteins, increasing body temperature, and several other metabolic effects characteristic of the inflammatory reaction.

VI. THE IMMUNE DEFICIENCY SYNDROMES AS MODELS FOR THE STUDY OF THE IMPORTANCE OF IMMUNE DEFENSES AGAINST INFECTIONS IN HUMANS

Most of our information about the immune system in humans has been learned from the study of patients with immunodeficiency diseases (see Chap. 30). The most characteristic clinical features of immunocompromised patients are the repeated or chronic infections, often caused by opportunistic agents. There are some characteristic associations between specific types of infections and generic types of immune deficiency, which provide the best data about the role of the different components of the immune system (Table 13.2).

A. Patients with antibody deficiencies and conserved cell-mediated immunity suffer from repeated and chronic infections with pyogenic bacteria.

B. Patients with primary deficiencies of cell-mediated immunity usually suffer from chronic or recurrent fungal, parasitic, and viral infections.

C. Neutrophil deficiencies are usually associated with bacterial and fungal infec-

Table 13.2 Common Patterns of Infection in Patients with Immunodeficiency

Type of immunodeficiency	Common infections
Combined (cellular and humoral) deficiency	All types of agents, including bacteria, viruses, fungi, and parasites
Humoral (antibody deficiency)	Mostly pyogenic bacteria
Cellular (T-cell deficiency)	Intracellular bacteria, viruses, and protozoa; fungi
Complement deficiencies	Pyogenic bacteria (particularly *Neisseria sp.*)

tions caused by common organisms of low virulence, usually kept in check through nonimmune phagocytosis and other inate resistance mechanisms.

D. Isolated complement component deficiencies are also associated with bacterial infections, most frequently involving *Neisseria gonorrhoeae* and *N. meningitidis*, whose elimination appears to require the activation of the membrane attack complex.

VII. ESCAPE FROM THE IMMUNE RESPONSE

Many infectious agents have developed the capacity to avoid the immune response. Several mechanisms are involved.

A. **Anti-Complementary Activity** has been characterized for bacterial capsules and outer proteins of some bacteria. For example, the M protein of group A *Streptococcus* inactivates complement convertases, preventing activation of the alternative pathway. The anti-complementary activity of bacterial outer components has as a net result a decreased level of opsonization by C3b and other complement fragments.

B. **Resistance to Phagocytosis** can be evident at different levels of the sequence of events associated with ingestion and digestion of microorganisms:

1. **Resistance to ingestion** is usually associated with **polysaccharide capsules**, which repeal and inhibit the function of phagocytic cells. *Haemophilus*, *Pasteurella*, *Klebsiella*, *Pseudomonas*, and some yeasts (e.g., *Cryptococcus neoformans*) have such capsules.
2. **Ability to survive after ingestion** is characteristic of the group of bacteria known as facultative intracellular (*Mycobacteria*, *Brucella*, *Listeria*, and *Salmonella*), as well as of some fungi and protozoa (*Toxoplasma*, *Trypanosoma cruzi*, and *Leishmania*). These microorganisms survive inside phagocytosis by several different strategies.
   a. Secretion of molecules that prevent the formation of phagolysosomes; these bacteria survive inside phagosomes that are relatively devoid of toxic compounds (*Mycobacterium tuberculosis*, *Legionella pneumophila*, and *Toxoplasma gondii*).
   b. Synthesis of outer coats that protect the bacteria against proteolytic enzymes and free toxic radicals (such as the superoxide radical).
   c. Depression of the response of the infected phagocytic cells to cytokines which usually activate their killing functions, such as interferon-γ.

d. Exit from the phagosome into the cytoplasm, where the bacteria can live and multiply unharmed (*Trypanosoma cruzi*).
e. The resistance to phagocytosis of several organisms seems to involve a combination of these strategies.
   i. *Mycobacterium leprae* is coated with a phenolic glycolipid layer that scavenges free radicals and releases a compound that inhibits the effects of interferon-γ. In addition, the release of IL-4 and IL-10 is enhanced, contributing to the down-regulation of TH1 lymphocytes.
   ii. Two glycolipids present in the membrane of *Leishmania* parasites inhibit the response of monocytes and neutrophils to cytokines and a variety of stimuli; in addition, *Leishmania* seems to resist the effects of proteolytic enzymes in the phagolysosome.
f. Viruses taken up by phagocytes as antigen–antibody complexes often infect the phagocytic cells. The infection of phagocytic cells is frequent when viruses disseminate through the blood stream.
g. Phagocytic cells are also infected by the HIV virus, either as a consequence of ingestion of virus-IgG antibody complexes or as a consequence of interactions with a CD4+ related molecule which is expressed in their membrane. In either case, the infection of the monocyte/macrophage is not cytotoxic, but the virus continues to replicate in small amounts and these cells become permanent sources of reinfection.

C. **Ineffective Immune Responses**

1. **Polysaccharide capsules** are not only antiphagocytic, but poorly immunogenic and elicit predominantly IgM and IgG2 antibodies, which are inefficient as opsonins (the FcγR of phagocytic cells recognize preferentially IgG1 and IgG3 antibodies).
2. **Neisseria meningitidis** often induces the synthesis of IgA antibodies. In vitro data suggest that IgA can act as a weak opsonin or induce ADCC (monocytes/macrophages and other leukocytes express Fcα receptors on their membranes), but the physiological protective role of IgA antibodies is questionable. Patient sera with high titers of IgA antibodies to *N. meningitidis* failed to show bactericidal activity until IgA-specific anti-*Neisseria meningitidis* antibodies were removed. This observation suggests that the IgA antibodies acted as a "**blocking factor**," preventing opsonizing IgG antibodies from binding to the same epitopes.
3. **Antiviral antibodies** are not always neutralizing and protective.
   a. **Retroviruses** induce both neutralizing and non-neutralizing (blocking) antibodies. Both seem to recognize the same epitopes; when non-neutralizing antibodies predominate, they block the binding of neutralizing antibodies.
   b. **Hepatitis B virus** induces the synthesis of large quantities of soluble antigenic proteins by the infected host cells. These circulating antigens will block the antiviral antibodies before they reach the infected cells, in a sense, acting as a "deflector shield" that protects the infected tissues from "antibody aggression."

D. **Loss and Masking of Antigens** have been demonstrated with schistosomula (the larval forms of schistosoma).

1. In the few hours after hatching, there is a rapid decrease in the surface antigens, as reflected by a progressive decrease in the binding of purified and fluorescein-labeled heterologous antischistosomular antibodies to the parasite membrane. This is observed even in parasites cultured in serum-free medium.
2. Schistosomula recovered from infected animals have host proteins bound to the outer layers, masking the remaining parasite antigens.

E. **Antigenic Variation** has been characterized in bacteria (*Borrelia recurrentis*), parasites (trypanosomes, the agents of African sleeping sickness; *Giardia lamblia*), and viruses (human immunodeficiency virus, HIV).
1. African **trypanosomes** have a surface coat constituted mainly of a single glycoprotein (variant-specific surface glycoprotein or VSG), for which there are about 10^3 genes in the chromosome. At any given time, only one of those genes is expressed, the others remain silent. For every 10^6 or 10^7 trypanosome divisions, a mutation occurs that replaces the active VSG gene on the expression site by a previously silent VSG gene. The previously expressed gene is destroyed, and a new VSG protein is coded, which is antigenically different. The emergence of a new antigenic coat allows the parasite to multiply unchecked. As antibodies emerge to the newly expressed VSG protein, parasitemia will decline, only to increase as soon as a new mutation occurs and a different VSG protein is synthesized.
2. ***Giardia lamblia*** has a similar mechanism of variation but the rate of surface antigen replacement is even faster (once every 10^3 divisions).
3. ***Borrelia recurrentis***, the agent of **relapsing fever** carries genes for at least 26 different variable major proteins (VMP) which are sequentially activated by duplicative transposition to an expression site. The successive waves of bacteremia and fever correspond to the emergence of new mutants which, for a while, can proliferate unchecked until antibodies are formed.
4. **HIV** exhibits a high degree of antigenic variation, which seems to be the result of errors introduced by the reverse transcriptase when synthesizing viral DNA from the RNA template. The mutation rate is relatively high (one in every 10^3 progeny particles), and the immune response selects the mutant strains that present new configurations in the outer envelope proteins, allowing the mutant to proliferate unchecked by preexisting neutralizing antibodies.

F. **Cell-to-Cell Spread** allows infectious agents to propagate without being exposed to specific antibodies or phagocytic cells.
1. ***Listeria monocytogenes***, after becoming intracellular, can travel along the cytoskeleton and promote the fusion of the membrane of an infected cell with the membrane of a neighboring noninfected cell, which is subsequently invaded.
2. **Herpes viruses, retroviruses, and paramyxoviruses** cause the fusion of infected cells with noninfected cells allowing viral particles to pass from cell to cell.

G. **Integration of Microbial Genomes** is a tactic exclusive of viruses, particularly DNA viruses and retroviruses. The integration of viral genomes in the host genome allows the virus to cause persistent or latent infections, with minimal replication of the integrated virus and minimal expression of viral proteins in the membranes of infected cells.

H. **Immunosuppressive Effects of Infection.** Although immunosuppressive effects have been described in association with bacteria and parasitic infections, the best documented examples of infection-associated immunosuppression are those described in viral infections.

1. **Measles.** Patients in the acute phase of measles are more susceptible to bacterial infections, such as pneumonia. Both delayed hypersensitivity responses and the in vitro lymphocyte proliferation in response to mitogens and antigens are significantly depressed during the acute phase of measles and the immediate convalescence period, usually returning to normal after 4 weeks. Recent investigations suggest that infection of monocytes/macrophages with the measles virus is associated with a down-regulation of interleukin-12 synthesis, which can explain the depression of cell-mediated immunity associated with measles.
2. **Cytomegalovirus.** Mothers and infants infected with cytomegalovirus show depressed responses to CMV virus, but normal responses to T-cell mitogens, suggesting that, in some cases, the immunosuppression may be antigen-specific, while in measles it is obviously nonspecific.
3. **Influenza virus** has been found to depress CMI in mice, apparently due to an increase in the suppressor activity of T lymphocytes.
4. **Epstein-Barr virus** releases a specific protein that has extensive sequence homology with interleukin-10. The biological properties of this viral protein are also analogous to those of interleukin-10; both are able to inhibit lymphokine synthesis by T-cell clones.
5. **Human immunodeficiency virus.** HIV infection is associated with depletion of CD4+ cells, which are the primary target of the virus. The depletion of CD4 cells results mostly from viral replication itself and by the priming of infected cells for apoptosis, which is triggered by stimuli that normally would cause T-cell proliferation. In addition, FcR-dependent phagocytosis is depressed in HIV-infected monocyte-derived macrophages, further contributing to the immunological compromise of the infected patients.

X. ABNORMAL CONSEQUENCES OF THE IMMUNE RESPONSE

A. The Activation of T lymphocytes by Bacterial Superantigens

1. A variety of bacterial exotoxins, such as **staphylococcal enterotoxins**-A and -B (SE-A and SE-B), **staphylococcal toxic shock syndrome toxin-1** (TSST-1), exfoliating toxin, and streptococcal exotoxin A, as well as other unrelated bacterial proteins (such as streptococcal M proteins) have been characterized as "superantigens."
2. Superantigens are defined by their ability to stimulate T cells without being processed. The stimulation of T cells is polyclonal; thus the designation of "superantigen" is a misnomer, but it has gained popularity and is widely used in the literature.
3. The best studied superantigens are the staphylococcal enterotoxins, which

are potent polyclonal activators of murine and human T lymphocytes, inducing T-cell proliferation and cytokine release. TSST-1 also appears to activate monocytes and is a potent B-cell mitogen, inducing B-cell proliferation and differentiation.

4. **Mechanism of action**. Superantigens bind directly and simultaneously to the nonpolymorphic area of class II MHC on professional accessory cells (macrophages and related cells) and to the Vβ chain of the α/β TcR (Fig. 13.4).
   a. For example, staphylococcal enterotoxins bind exclusively to specific subfamilies of Vβ chains that are expressed only by certain individuals. When expressed, these Vβ chain regions can be found on 2–20% of a positive individual's T cells, and the cross-linking of the TcR2 and of the APC by the enterotoxin activates all T cells (both CD4 and CD8+) expressing the specific Vβ region recognized by the enterotoxin.
   b. The massive T-cell activation induced by superantigens results in the release of large amounts of IL-2, interferon-γ, lymphotoxin (TNF-β), and TNF-α.
   c. Patients infected by bacteria able to release large amounts of superantigens (e.g., *S. aureus*-releasing enterotoxins or TSST-1 and Group A *Streptococcus*-releasing) may develop septic shock as a consequence of the systemic effects of these cytokines, which include fever, endothelial damage, profound hypotension, disseminated intravascular coagulation, multiorgan failure, and death.
   d. After the initial burst of cytokine release, the stimulated T cells either undergo apoptosis or become anergic. This effect could severely disturb

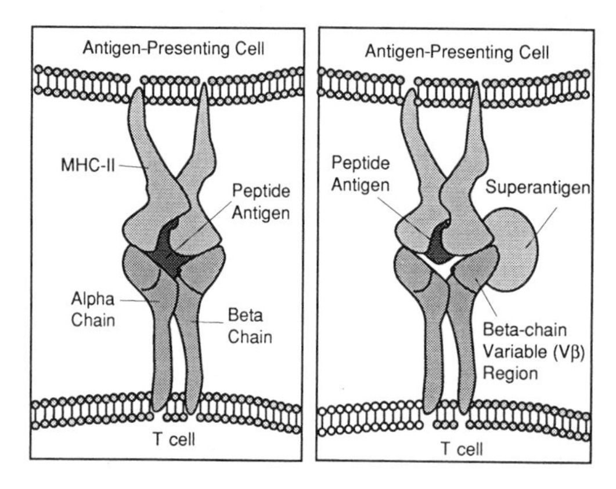

Figure 13.4 Diagrammatic comparison of the mechanisms of T-cell stimulation by conventional antigens and staphylococcal enterotoxins. While conventional antigens are processed into oligopeptides, which bind to MHC-II molecules, and then bind specifically to a TcR binding site (left panel), bacterial superantigens interact with nonpolymorphic areas of the Vβ chain of the TcR and of the MHC-II molecule on an APC (right panel). Notice that superantigen binding overrides the need for TcR recognition of the MHC-II-associated oligopeptide, and thus T cells of many different specificities can be activated. It is also important to note that both APC and T cells can be stimulated as a consequence of the extensive cross-linking of membrane proteins. (Modified from Johnson, H.M., Russell, J.K., and Pontzer, C.H. *Sci. Am. 266* (4):92, 1992.)

the ability of the immune system to adequately respond to superantigen-releasing baceria.

B. **Infection as a Consequence of the Uptake of Antigen–Antibody Complexes.** The immune response, in some cases, facilitates the access of infectious agents to cells in which they will be able to proliferate.
   1. Macrophages are often infected by intracellular organisms that are ingested as a consequence of opsonization.
   2. *Babesia rodhaini*, a bovine intraerythrocytic parasite, penetrates the host's cells after it has bound complement, particularly C3. Absorption of C3b-containing circulating antigen–antibody complexes to the CR1 expressed by red cells allows the parasite to gain access to the red cell, which becomes its permanent location.

C. **Post-Infectious Tissue Damage.** Several examples of deleterious consequences of the immune response have to do with the predominant role of the immune response in the pathogenesis of the disease.
   1. **Immune-complex-induced inflammation.** Antigen–antibody complexes, if formed in large amounts, can cause disease by being trapped in different capillary networks and leading to inflammation. The clinical expression of immune complex-related inflammation depends on the localization of the trapped complexes: vasculitis and purpura, when the skin is predominantly affected, glomerulonephritis if trapping takes place on the glomerular capillaries, arthritis when the joints are affected (see Chap. 21). Viruses are often involved in the formation of circulating antigen–antibody complexes.
   2. **Immune destruction of infected cells and tissues.** An immune response directed against an infectious agent may be the main cause of damage to the infected tissue.
      a. In **subacute sclerosing panencephalitis**, a degenerative disease of the nervous system associated with persistent infection with the measles virus, the response against viral epitopes expressed in infected neurons is believed to be the primary mechanism of disease.
      b. In some forms of **chronic active hepatitis** (see Chap. 23), the immune response directed against viral epitopes expressed by infected hepatocytes seems to cause more tissue damage than the infection itself.
      c. **Cross-reactions with tissue antigens** have been proposed as the basis for the association of streptococcal infections with rheumatic carditis and glomerulonephritis. Antibodies to type 1 streptococcal M protein cross-react with epitopes of myocardium and kidney mesangial cells and cause inflammatory changes in the heart and glomeruli, respectively.

X. EPILOGUE

The outcome of an infectious process depends on a very complex set of interactions with the immune system. A successful pathogen has developed mechanisms that avoid fast elimination by an immunocompetent host. These mechanisms allow the infectious agent to replicate, cause disease, and spread to other individuals before the immune response is induced. The immune response, on the other hand, is a powerful weapon that, once set in motion, may destroy friendly targets. Thus, the therapeutic strategies in infectious diseases have to consider all these questions, such as the particular survival strategy of the infectious agents, the effects of the infection on the immune system, and the possibility that the immune response may be more of a problem than the infection itself.

SELF-EVALUATION

Questions

Choose the ONE *best* answer.

13.1 Which one of the following variables is most important as a determinant of the possible use of active immunization to protect an individual from an infectious disease after a known exposure?
A. Existence of memory cells stimulated in a previous immunization
B. Immunogenicity of the antigen used for vaccination
C. Length of the incubation time of the disease
D. Preexisting levels of protective antibody
E. Resistance of the infectious agent to nonimmune phagocytosis

13.2 The elimination of an intracellular organism from an infected human macrophage is most likely to depend on the:
A. Formation of syntitia by fusion of infected and noninfected cells
B. Intracellular diffusion of complement-fixing antibodies
C. Phagocytosis of the infected cells by noninfected macrophages
D. Release of enzymes and superoxide radicals directly into the cytoplasm
E. Sensitization of T cells against microbial-derived peptides presented in association with MHC-II molecules on the membrane of the infected cell

13.3 Which of the following mechanisms is the basis for the massive activation of the immune system caused by staphylococcal enterotoxins?
A. Ability to interact nonspecifically with MHC-II molecules and α/β TcR
B. Activation of APC, followed by release of massive amounts of lymphocyte-activating cytokines
C. Binding to APC, followed by activation of all T cells expressing TcR with specific Vβ regions
D. Mitogenic activation of the TH2 helper subpopulation
E. Promiscuous binding to the peptide-binding grooves of MHC-II molecules

13.4 The finding of a generalized depression of cellular immunity during the acute phase of measles should be considered as:
A. A poor prognosis indicator
B. Evidence of a primary immune deficiency syndrome
C. Reflecting a transient state of immunosuppression
D. The result of a concurrent bacterial infection
E. The result of a laboratory error

13.5 Which of the following approaches would be useful to treat an intracellular infection such as the one caused by *Mycobacterium tuberculosis*?
A. Administration of a mixture of interleukins 4, 5, and 6 to activate a B-cell response
B. Administration of in vitro–activated NK cells
C. Administration of interferon-γ to activate macrophages

D. Administration of interleukin-2 to induce a predominant response of TH1 cells
E. Transfusion of T lymphocytes from patients who survived *M. tuberculosis* infections

13.6 An antibody to tetanus toxoid will prevent the clinical manifestations of tetanus by:
A. Binding to the antigenic portion of the toxin molecule and inhibiting the interaction between the toxin and its receptor
B. Causing the destruction of *Clostridium tetani* before it releases significant amounts of toxin
C. Facilitating the onset of a recall response to the toxin
D. Inhibiting the binding of the toxin to its receptor by blocking the toxin's active site
E. Promoting ADCC reactions against *C. tetani*

13.7 Herpes simplex virus escapes immune defenses by:
A. Being a very weak immunogen
B. Causing immunosuppression
C. Infecting immunocompetent cells
D. Producing an excess of soluble antigen that "blocks" the corresponding antibody
E. Spreading from cell to cell with minimal exposure to the extracellular environment

13.8 The ABO isohemagglutinins are synthesized as a result of:
A. Blood transfusions with incompatible blood
B. Cross-immunization with polysaccharide-rich enterobacteriaceae
C. Genetic predisposition
D. Mixing of placental and fetal blood at the time of birth
E. Repeated pregnancies

13.9 In a newborn baby with blood typed as A, Rh positive, the lack of anti-B isoagglutinins is:
A. A reflection of lack of exposure to intestinal flora during intrauterine life
B. A very exceptional finding, identifying the baby as a nonresponder to blood group B substance
C. Evidence of maternal immunoincompetence
D. Evidence suggestive of fetal immunoincompetence
E. Unlikely; the test should be repeated

13.10 The parasite-killing properties of eosinophils are linked to the production and secretion of:
A. Histamine
B. Leukotrienes
C. Major basic protein
D. Perforin
E. Prostaglandins

Answers

13.1 (C) If the incubation period exceeds that of the time necessary for eliciting an immune response (primary or secondary), then active immunization can be used to prevent the development of the disease after a known exposure.

13.2 (E) Sensitized T cells will become activated and release interferon-γ, which can enhance the killing properties of macrophages and related cells.

13.3 (C) After binding to MHC-II molecules, staphylococcal enterotoxins are able to interact and activate T cells expressing TcR with specific Vβ regions.

13.4 (C) Measles is characteristically associated with a transient depression of cellular immunity during the acute phase; a patient with such immune depression should not be considered as having a primary immune deficiency or as necessarily having a poor prognosis.

13.5 (C) Administration of interferon-γ, the natural macrophage activator, would be the most promising of the listed approaches. For activation of TH1 cells, IL-12 would be the indicated cytokine. Activated NK cells have not been proven to be important in the elimination of cells infected by intracellular bacteria. Transfusion of T cells from recovering patients has the disadvantage over interferon-γ administration of potential reactions due to donor–recipient histoincompatibility.

13.6 (A) An antitetanus toxoid antibody is neither cytotoxic to *Clostridium tetani*, since it reacts with the toxin, which is an exotoxin, nor able to bind to the active site of the toxin, since the toxoid used for immunization has lost (as a consequence of detoxification) the active site. Memory responses depend on an increased number of memory cells, and not on the presence of circulating antibody.

13.7 (E) Herpes simplex viruses are immunogenic, do not infect immunocompetent cells, do not cause immunosuppression, and do not induce the release of large amounts of soluble antigens from infected cells. However, they can cause fusion of infected and noninfected cells, and this allows them to spread from cell to cell without being exposed to humoral defense mechanisms.

13.8 (B) Mixing of maternal and fetal blood at birth may cause maternal sensitization to Rh blood groups, but does not appear to cause sensitization of the newborn, probably due to a combination of factors such as the immaturity of the immune system and the fact that the maternal red cells directly enter the fetal circulation, which is not a very immunogenic route of presentation for any given antigen.

13.9 (A) Since the newborn's intestine has not been colonized by enterobacteriaceae, the antigenic stimulation for production of isoagglutinins has not taken place before birth and negative titers of isohemagglutinins are normal. Maternal isoagglutinins, being in a large majority of cases of the IgM class, do not cross the placenta.

13.10 (C) Major basic protein is the main parasiticidal compound released by eosinophils recognizing IgE-coated parasites.

BIBLIOGRAPHY

Borst, P., and Graves, D.R. Programmed gene rearrangements altering gene expression. *Science*, *235*:658, 1987.

Cook, D.N., Beck, M., Coffman, T.M., et al., Requirement of MIP-1a for an inflammatory response to viral infection. *Science*, *269*:1583, 1995.

Eze, M.O. Avoidance and inactivation of reactive oxygen species: Novel microbial immune evasion strategies. *Med. Hypotheses*, *34*:252, 1991.

Karp, C.L., Wysocka, M., Wahl, L.M., Ahearn, J.M., Cuomo, P.J., Sherry, B., Tirnchieri, G., and Griffin, D. Mechanism of suppression of cell-mediated immunity by measles virus. *Science*, *273*:228, 1996.

Kim, J., Urban, R.G., Strominger, J.L., and Wiley, D.C. Toxic shock syndrome toxin-1 complexed with a class II major histocompatibility molecule HLA-DR1. *Science*, *266*:1870, 1994.

Kraus, W., Dale, J.B., and Beachey, E.H. Identification of an epitope of type 1 streptococcal M protein that is shared with a 43-kDa protein of human myocardium and renal glomeruli. *J. Immunol.*, *145*:4089, 1990.

Louzir, H., Ternynck, T., Gorgi, Y., Tahar, S., Ayed, K., and Avrameas, S. Autoantibodies and circulating immune complexes in sera from patients with hepatitis B virus-related chronic liver disease. *Clin. Immunol. Immunopathol.*, *62*:160, 1992.

Mauel, J. Macrophage-parasite interactions in *Leishmania* infections. *J. Leukoc. Biol.*, *47*:187, 1990.

Miethke, T., Wahl, C., Heeg, K., and Wagner, H. Superantigens: The paradox of T-cell activation vs. inactivation. *Int. Arch. Allergy Immunol.*, *106*:3, 1995.

Research Brief. Disarming the immune system: HIV uses multiple strategies. *J. NIH Res.*, *8*:33, 1996.

14
Immunoserology

Gabriel Virella

I. INTRODUCTION

The exquisite sensitivity and specificity of antigen–antibody reactions has been utilized as the basis for many diagnostic procedures. Depending on the test design, one can detect or assay either specific antigens or specific antibodies. Because serum is most often used in these assays, they are also known by the generic designation of immunoserological tests.

A. **General Principles**

1. For detection (or quantitation) of antibody in body fluids by immunological methods, a purified preparation of antigen must be available. For instance, when human serum is tested for antibody to diphtheria toxoid, a purified preparation of the toxoid must be available. Second, a method for detecting the specific antigen–antibody reaction must be developed.
2. To detect antigens in biological fluids, specific antibody must be available.
3. In both types of assays, positive and negative controls are required for proper interpretation of the results.

B. **Applications**

1. Antigen-detection tests have been applied to the diagnosis of infectious diseases, the monitoring of neoplasms, the quantitation of hormones and drugs, and pregnancy diagnosis.
2. Quantitation of specific antibodies has found wide application in the diagnosis of infectious, allergic, autoimmune, and immunodeficiency diseases.

II. PRECIPITATION ASSAYS

Several of the commonly used serological assays are based on the detection of antigen–antibody aggregates, either through visualization of precipitates, or by measuring the light dispersed by antigen–antibody complexes in suspension.

A. **Double Diffusion Method (Ouchterlony's Technique)**

1. **Principle.** Wells are punched in an agar gel; antigen and antibody placed in separate wells diffuse toward each other and precipitate at the point of antigen–antibody equivalence.

2. **Advantages and Limitations**
   a. The main advantages are simplicity, minimal equipment requirements, and specificity. If properly carried out, double immunodiffusion assays are 100% specific as far as detecting an antigen–antibody reaction.
   b. The main limitations are lack of sensitivity, and the time required for full development of visible precipitation (up to 72–96 hours of diffusion in systems when either reactant is present in small quantities).
3. **Applications.** In general, double diffusion assays are useful for determining the presence or absence of a given antigen or antibody in any kind of biological fluid.

B. **Counterimmunoelectrophoresis**
1. **Principle.** This technique is a variation of double immunodiffusion, in which antigen and antibody are forced to move toward each other with an electric current (antibodies move to the cathode while most antigens are strongly negatively charged and move toward the anode). Precipitin lines can be visualized between the antigen and antibody wells.
2. **Advantages and Limitations.** The method is faster and more sensitive than double immunodiffusion. Visible precipitation can be usually observed after 1–2 hours, although for maximal sensitivity it is necessary to wash, dry, and stain the agar gel in which the reaction took place, a process that extends the time for final reading of the results to several days.
3. **Applications.** Counterimmunoelectrophoresis has been used for detection of fungal or bacterial antigens in CSF (in patients with suspected meningitis), and for the detection of antibodies to *Candida albicans* (Fig. 14.1), DNA, and other antigens.

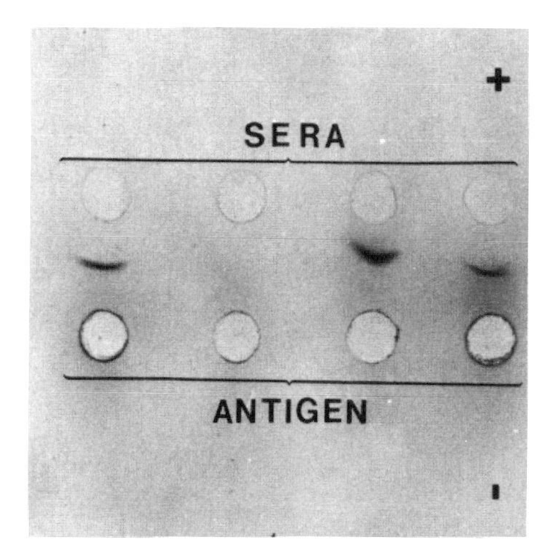

Figure 14.1 An example of the use of counterimmunoelectophoresis for the detection of antibodies to microbial antigens. In this case, the test was designed to detect anti-*C. albicans* antibodies. The wells on the cathodal side were filled with an antigenic extract of *C. albicans*, and the wells on the anodal side were filled with patient's sera. The appearance of a precipitation line between a given serum well and the antigen well directly opposed to it identifies the patient as positive.

C. Immunoelectrophoresis

1. **Principle.** Immunoelectrophoresis (**IEP**) is a two-step technique; first, a small sample of serum is applied in a well and electrophoresed through a support medium, usually agarose. After electrophoresis is completed, specific antisera are deposited in a trough cut into the agarose, parallel to the axis of the electrophoretic separation. The electrophoretically separated proteins and the antisera diffuse toward one another and, at the zone of antigen–antibody equivalence, a precipitin pattern in the form of an arc will appear (Fig. 14.2).
2. **Advantages and Limitations.** The problems of poor sensitivity and long waiting time for the development and visualization of precipitin lines discussed above also apply to immunoelectrophoresis. In addition, the interpretation of the precipitin patterns requires an experienced interpreter.
3. **Applications.** Immunoelectrophoresis is particularly useful for analytical studies in patients suspected of plasma cell malignancies. Quantitative assays of immunoglobulins in these patients show a marked increase in one immunoglobulin class. IEP analysis can establish or disprove the monoclonal nature of the immunoglobulin increase by determining whether or not the quantitative increase is due to a homogeneous population, with a single light chain type. Monoclonal free light chains (Bence–Jones proteins) may also be present in the urine of these patients, and IEP of urine is the method classically used for their detection.

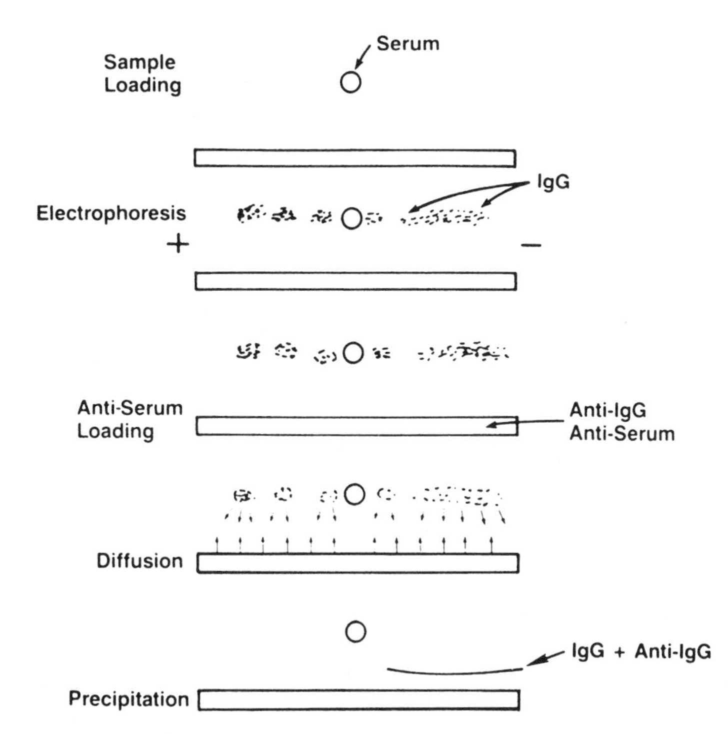

Figure 14.2 Diagrammatic representation of the basic steps in immunoelectrophoresis.

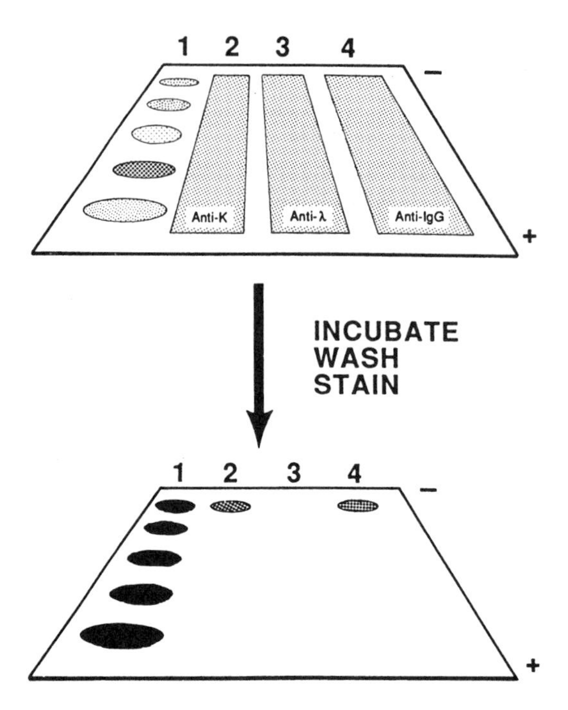

Figure 14.3 Diagrammatic representation of the typing of a monoclonal IgG protein by immunofixation. The top panel illustrates the electrophoretic separation of serum proteins, revealed by staining in lane 1, and the overlay with three different antisera (anti-κ, anti-λ, and anti-IgG) of lanes 2, 3, and 4, where the same proteins were separated. After incubation to allow antigens and antibodies to react, the unreacted proteins and antibodies are washed off, and the precipitates formed with the different antibodies are stained. In the example illustrated, precipitates were formed in the cathodal region (where immunoglobulins are separated by electrophoresis) with anti-IgG and anti-κ, confirming that the serum contained a monoclonal protein typed as IgGκ.

D. Immunofixation (Immunoblotting)

1. **Principle.** The principle of the technique is diagrammatically represented in Fig. 14.3.
   a. In the first step, several aliquots of the patient's serum are simultaneously separated by electrophoresis. One of the separation lanes is stained as reference for the position of the different serum proteins, while paper strips imbedded with different antibodies are laid over the remaining separation lanes.
   b. The antibodies diffuse into the agar and react with the corresponding immunoglobulins. After washing off unbound immunoglobulins and antibodies, the lanes where immunofixation takes place are stained, revealing whether the antisera did or did not recognize the proteins they are directed against.
2. **Advantages and Limitations.** This technique has largely replaced immunoelectrophoresis, mainly because the interpretation of results is considerably easier. The time requirements remain identical, and the technique requires considerable expertise for its performance.

3. **Applications.** The main application of immunoblotting is the detection of monoclonal proteins. In the case illustrated in Figure 14.3, a patient's serum was being tested for the presence of an IgG monoclonal protein. Such protein, by definition, has to be homogeneous in mobility, has to react with anti-IgG antibodies, and either with anti-kappa or with anti-lambda antibodies. The diagram shows stained precipitates corresponding to the lanes that were overlaid with anti-IgG and with anti-κ light chains. This result would prove that an IgGκ monoclonal protein existed in this patient's serum.

E. **The Western Blot** is a variation of immunoblotting that has been popularized by its use in the diagnosis of HIV infection.

1. **Principle** (Fig. 14.4)
   a. The first step in the preparation of an immunoblot (also known as a **Western blot**) is to separate the different viral antigens (gp160 to p16) according to their molecular size [the numbers to the right of "gp" or "p" refer to the protein mass in kilodaltons (kD)]. This is achieved by performing electrophoresis in the presence of a negatively charged detergent (such as sodium dodecyl sulfate), which becomes associated with the proteins and obliterates their charge differences, using as support for the separation a medium with sieving properties. The result is the separation of a protein mixture into components of different sizes.
   b. After the separation of the viral proteins is completed, it is necessary to transfer the separated proteins to another support, in order to proceed with the remaining steps. This transfer or "blotting" is easily achieved by forcing the proteins to migrate into a nitrocellulose membrane by a second electrophoresis step (electroblotting).
   c. The nitrocellulose membrane, to which the viral antigens have been transferred, is then impregnated with the patient's serum. If antibodies to any or several of these antigens are contained in the serum, they will combine with the antigen at the point where it is blotted.
   d. The following steps are designed to detect the antigen–antibody complexes formed in the cellulose membrane. First, all the unprecipitated antigens and antibodies are washed off; then the protein binding sites still available on the membrane are blocked to prevent false positive reactions; finally, the membrane is overlaid with a labeled antibody to human immunoglobulins. This antibody will react with the complexes formed by viral antigens and human antibodies, and can be later revealed either by adding a color-developing substrate (if the antihuman immunoglobulin is labeled with an enzyme) or by autoradiography (if the antihuman immunoglobulin is labeled with ^{125}I).
   e. The retention of the labeled antibody in site(s) known to correspond to the separation of viral proteins indicates that the patient's serum contained antibodies reacting with those antigens.
2. **Advantages and Limitations.** This technique has the advantage over other screening assays of not only detecting antibodies, but also identifying the antigens against whom the antibodies are directed. This results in increased specificity. The downside is the time consumed running the assay, which requires very careful quality control, and needs to be carried out by specialized personnel in certified laboratories.

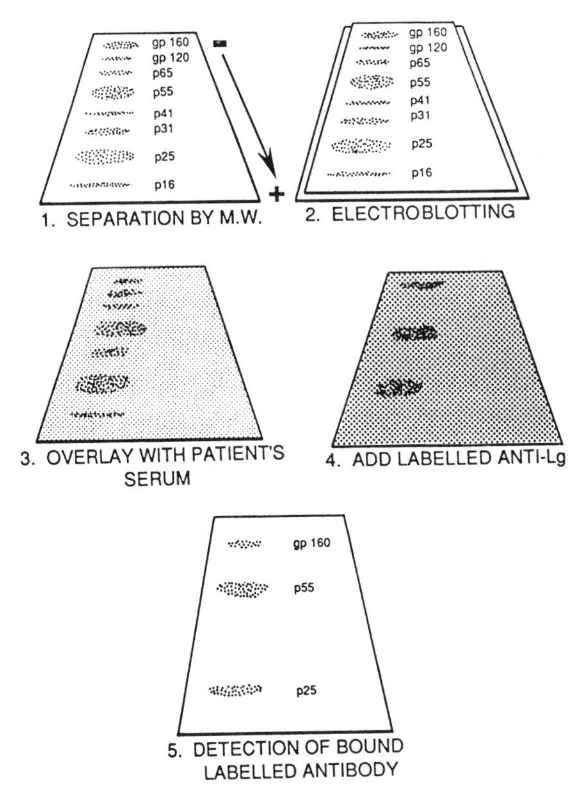

Figure 14.4 Diagrammatic representation of a Western blot study to confirm the existence of anti-HIV antibodies. In the first step, a mixture of HIV antigens is separated by size (large antigens remain close to the origin where the sample is applied, smaller antigens move deep into the acrylamide gel used for the separation). In the second step, the separated antigens are electrophoresed into a permeable nitrocellulose membrane (electroblotting). Next, the patient's serum is spread over the cellulose membrane to which the antigens have been transferred. If antibodies to any of these antigens are present in the serum, a precipitate will be formed at the site where the antigen has been transferred. After washing excess of unreacted antigens and serum proteins, a labeled second antibody is overlaid on the membrane; if human antibodies precipitate by reacting with blotted antigens, the second antibody (labeled antihuman immunoglobulin) will react with the immunoglobulins contained in the precipitate. After washing off the excess of unreacted second antibody, its binding to an antigen–antibody precipitate can be detected either by adding a color-developing substrate (if the second antibody is labeled with an enzyme) or by autoradiography (if the second antibody is labeled with an adequate isotope, such as ^{125}I).

F. **Radial Immunodiffusion** represents a hallmark in the evolution of immunoserology because it represents the first successful attempt to develop a precise quantitative assay suitable for routine use in the diagnostic laboratories.

1. **Principle.** Radial immunodiffusion received its designation from the fact that a given antigen is forced to diffuse concentrically on a support medium to which antiserum has been incorporated (Fig. 14.5).
   a. A polyclonal antiserum, known to precipitate the antigen, is added to molten agar and an agar plate containing the antibody is then prepared.

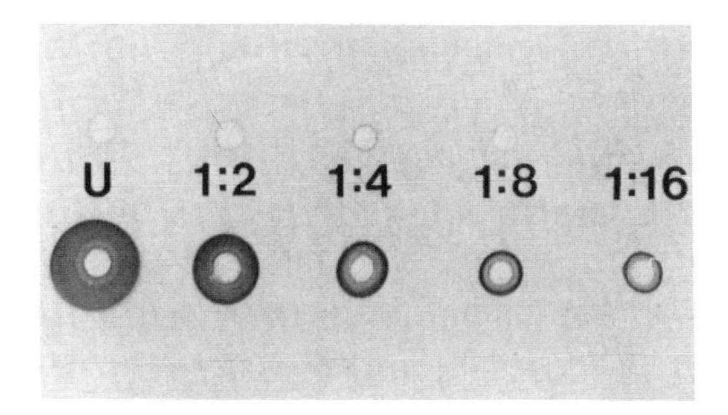

Figure 14.5 The principle of radial immunodiffusion: if five wells carved into antibody-containing agar are filled with serial dilutions of the corresponding antigen, the antigen will diffuse and eventually be precipitated in a circular pattern; the diameters or areas of these circular precipitates are directly proportional to the concentration of antigen in each well.

b. Identical wells are cut in the antisera-containing agar and those wells are filled with identical volumes of samples containing known amounts of the antigen (calibrators) and of unknown samples where the antigen needs to be assayed.
c. After 24–48 hours it is possible to measure the diameters of circular precipitates formed around the wells where antigens were placed. Those diameters are directly proportional to antigen concentration.
d. A plot of precipitation ring diameters vs. concentrations is made for the samples with known antigen concentrations. This plot, known as a calibration curve, is used to extrapolate the concentrations of antigen in the unknown samples, based on the diameter of the corresponding precipitin rings.

2. **Advantages and Limitations.** Radial immunodiffusion, while not as sensitive as some newer techniques and relatively slow, is nonetheless reliable for routine quantitation of many serum proteins.
3. **Applications.** Quantitation of immunoglobulins, complement components, and, in general, any antigenic protein that exists in concentrations greater than 5–10 mg/dL in any biological fluid.

G. **Quantitative Immunoelectrophoresis (Rocket Electrophoresis)** is an adaptation of radial immunodiffusion in which the antigen is actively driven through an antibody-containing matrix by an electric potential. As the equivalence zone of antigen–antibody reaction is reached, the reaction ceases and elongated precipitin arcs ("rockets") become visible (Fig. 14.6). The lengths of those "rockets" are proportional to the antigen concentrations in the wells. Rocket electrophoresis can be used for the quantitative assay of many proteins, including immunoglobulins. The reverse modality with antigen incorporated in the agar can be used for the quantitation of specific antibodies. This technique is faster and more sensitive than radial immunodiffusion.

H. **Immunonephelometry.** This quantitative technique has replaced radial immunodiffusion, particularly when large numbers of samples are to be assayed.

1. **Principle**

a. When dilute solutions of antigen and antibody are mixed under antibody

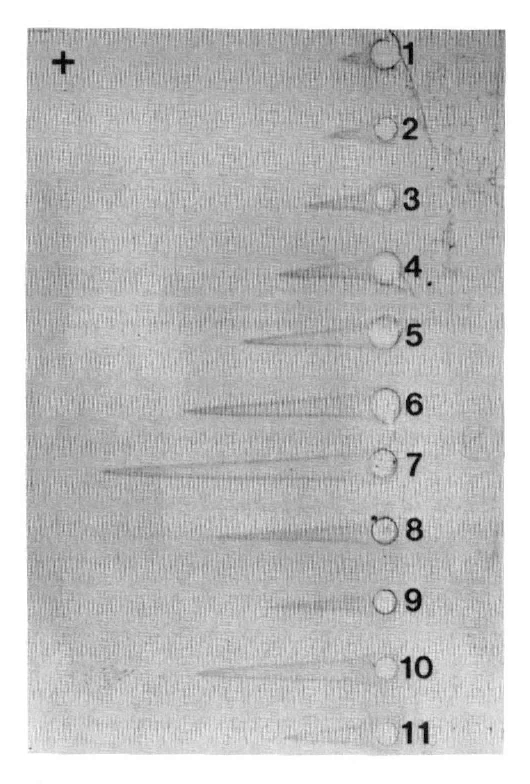

Figure 14.6 Quantitative electrophoresis. In this assay, antibody to apolipoprotein B was incorporated into the agar. Reference standards with known concentrations of apolipoprotein B were used to fill wells 1–7; the remaining wells were filled with patient's sera. (Courtesy of Dr. Maria F. Lopes-Virella, VAMC, Charleston, SC.)

excess conditions, soluble antigen–antibody complexes are formed that remain in suspension.

b. The relative amount of soluble aggregates can be assayed by measuring nephelometrically the amount of light dispersed by the mixture. This is most frequently done by passing a beam of light through tubes containing mixtures of fixed amounts of antibody and variable concentrations of antigen. The scattered light is measured at angles varying from near 0° to 45° relative to the incident light.

c. If the antibody concentration is kept constant, the amount of dispersed light will be directly proportional to the antigen concentration. The principles of calibration and extrapolation of unknown values are identical to those of radial immunodiffusion.

2. **Advantages and Limitations.** Immunonephelometry is superior to radial immunodiffusion in terms of sensitivity, speed, and automation. However, the preparation of samples and cost of equipment and reagents make it suitable mostly for large volume assays.
3. **Applications.** Immunonephelometry is the technique of choice for the measurement of immunogenic proteins (such as immunoglobulins or comple-

ment components) in concentrations greater than 1 mg/dL when the volume of samples to be analyzed is large.

III. METHODS BASED ON AGGLUTINATION

When bacteria, antigen-coated particles, or cells in suspension are mixed with antibody directed to their surface determinants, the antigen–antibody reaction leads to the clumping (agglutination) of the cells or particles carrying the antigen. Quantitative analysis to determine the agglutinating antibody content of an antiserum involves dilution of the serum and determination of an end point, which is the last dilution at which agglutination can be observed. The reciprocal of this last agglutinating dilution is designated as antibody titer.

A. Agglutination of Whole Microorganisms

1. **Applications.** Agglutination of whole microorganisms is commonly used for serotyping isolated organisms and, less commonly, for the diagnosis of some infectious diseases (e.g., the Weil–Felix test for the diagnosis of typhus) based on the fact that certain strains of *Proteus* share antigens with several *Rickettsia* species.
2. **Advantages and Limitations.** Agglutination is rapid (takes a matter of minutes) and, by being visible with the naked eye, does not require any special instrumentation other than a light box. Its disadvantages are the need for isolated organisms and poor sensitivity, requiring relatively large concentrations of antibody.

B. Agglutination of Inert Particles Coated with Antigen or Antibody

1. **Principle.** Latex particles and other inert particles can be coated with purified antigen and will agglutinate in the presence of specific antibody. Conversely, specific antibodies can be easily adsorbed by latex particles and will agglutinate in the presence of the corresponding antigen.
2. **Applications**
   a. Immunoglobulin-coated latex particles are used for the detection of anti-Ig factors (rheumatoid factors) in the **rheumatoid arthritis (RA) test** (Fig. 14.7).

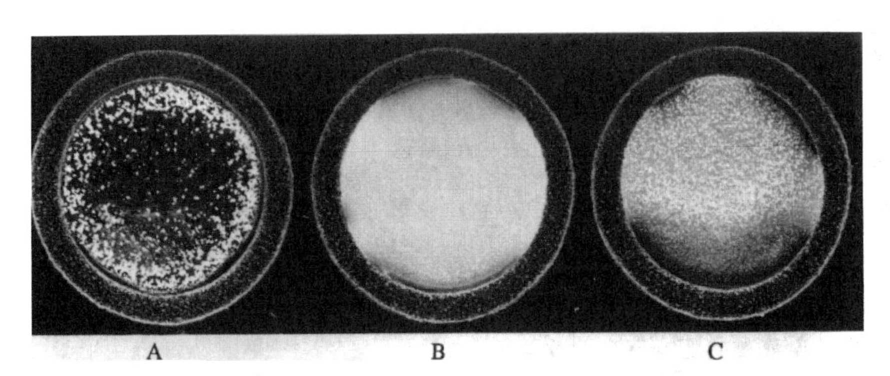

Figure 14.7 Detection of the rheumatoid factor by the latex agglutination technique. A suspension of IgG-coated latex particles is mixed with a 1:20 dilution of three sera. Obvious clumping is seen in A, corresponding to a strongly positive serum; no clumping is seen in B, corresponding to a negative serum; very fine clumping is seen in C, corresponding to a weakly positive serum.

b. Latex particles coated with thyroglobulin are used in a **thyroglobulin antibody test**.

c. A wide variety of diagnostic tests for infectious diseases has been developed based on latex agglutination.

   i. In some cases, the antigen is bound to latex, and the test detects specific antibodies (e.g., tests for histoplasmosis, cryptococcosis, and trichinosis).

   ii. **Rapid diagnosis** tests for bacterial and fungal meningitis have been developed by adsorbing the relevant specific antibodies to latex particles. The antibody-coated particles will agglutinate if mixed with CSF containing the relevant antigen. This procedure allows a rapid etiological diagnosis of meningitis which is essential for proper therapy to be initiated.

3. **Advantages and Limitations.** The main advantages of these tests are their simplicity and quick turn-around of results. The main disadvantages are the need for large amounts of reagents, cost, and relatively low sensitivity, particularly in the case of the tests for diagnosis of infectious diseases.

C. **Hemagglutination.** Hemagglutination tests can be subclassified depending on whether they detect antibodies against red cell determinants (direct and indirect hemagglutination) or against compounds artificially coupled to red cells (passive hemagglutination).

1. **Direct hemagglutination**

   a. **Principle.** Red cells are agglutinated when mixed with IgM antibodies recognizing membrane epitopes.

   b. Applications

      i. Determination of the **ABO blood group** and titration of isohemagglutinins (anti-A and anti-B antibodies).

      ii. Titration of **cold hemagglutinins** (IgM antibodies which agglutinate RBC's at temperatures below that of the body), as illustrated in Figure 14.8.

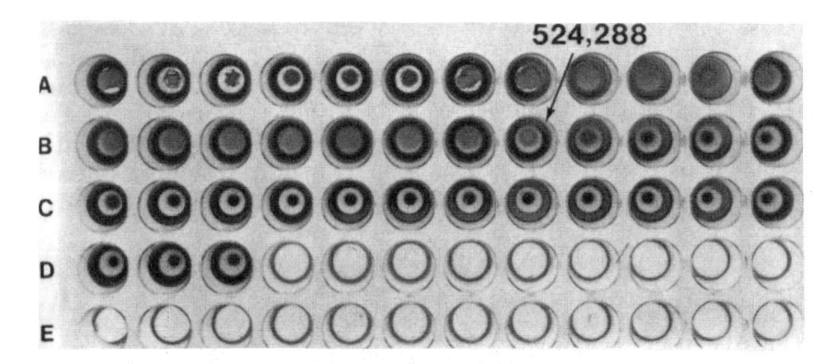

Figure 14.8 Detection of cold agglutinins by direct hemagglutination. The wells in the microtiter plate were first filled with serial dilutions of a patient's serum (rows A and B), serial dilutions of a control serum (row C) and saline (row D), and then with O Rh positive red cells, incubated at 4°C and examined for agglutination. The normal control and saline control do not show agglutination. The patient's serum shows a prozone followed by agglutination up to a dilution of 1:512,000.

iii. The **Paul-Bunnell test**, useful for the diagnosis of infectious mononucleosis, detects circulating heterophile antibodies (cross-reactive antibodies that combine with antigens of an animal of a different species) that induce the agglutination of sheep or horse erythrocytes.

c. **Advantages and limitations.** Hemagglutination is simple to execute and requires very simple materials. However, the tests for cold agglutinins associated with infectious diseases (see Chap. 24) and the Paul Bunnell test lack specificity.

2. **Indirect hemagglutination**
   a. **Principle.** Indirect hemagglutination detects antibodies that react with antigens present in the erythrocytes but which by themselves cannot induce agglutination. Usually, these are IgG antibodies that are not as efficient agglutinators of red cells as polymeric IgM antibodies. A second antibody directed to human immunoglobulins is used to induce agglutination by reacting with the red-cell bound IgG molecules, and consequently, cross-linking the red cells.
   b. **Applications.** The best known example of indirect agglutination is the **antiglobulin** or **Coombs' test**, which is used in the diagnosis of **autoimmune hemolytic anemia**.
   c. **Advantages and limitations**. The technique is simple to perform. Poor sensitivity is the main limitation.
3. **Passive hemagglutination**
   a. **Principle.** Passive hemagglutination techniques use red blood cells as a substrate much as latex is used in tests involving inert particles. Antigen can be coated onto the red cells by a variety of methods, and the coated cells will agglutinate when exposed to specific antibody (Fig. 14.9).
   b. **Applications**
      i. This system has been used as a basis for a variety of diagnostic procedures such as a test to detect antithyroid antibodies, the **Rose Waaler** test for anti-Ig factors present in the serum of patients with

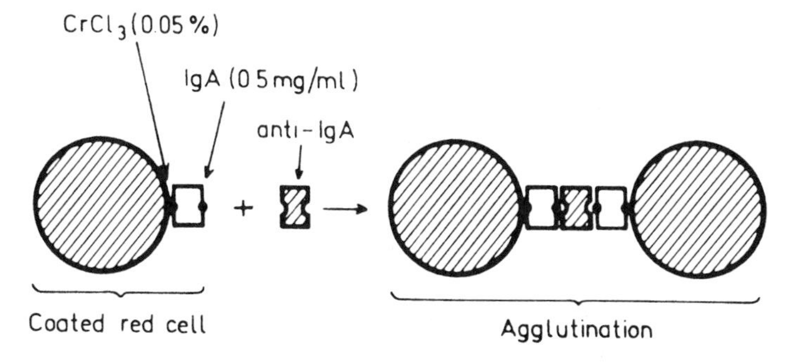

Figure 14.9 Diagrammatic representation of a passive hemagglutination test to detect anti-IgA antibodies. Purified IgA was coated to chromium chloride-treated RBC; the IgA-coated red cells will be agglutinated by anti-IgA antibodies. (Original diagram from Dr. Jukka Koistinen, Finnish Red Cross, Helsinki, Finland.)

rheumatoid arthritis, and many tests to detect anti-infectious antibodies.

ii. Antigens can also be detected by this technique by determining whether a biological fluid suspected of containing them can reduce the agglutinating capacity of an antiserum (**hemagglutination inhibition**). In a first step, the antiserum and the biological fluid suspected of containing the antigen are mixed. In a second step, red cells coated with antigen are added to dilutions of the mixture. The agglutinating titer of the antiserum is known, and if a four-fold or greater decrease in titer is observed, it can be concluded that antigen was present in the tested biological fluid.

c. **Advantages and limitations.** This technique shares the relative simplicity of all hemagglutination procedures, and it is considerably more sensitive than regular hemagglutination. However, passive hemagglutination tests are difficult to standardize and reproduce, specificity is often problematic, and not too accurate (as all titration tests).

IV. COMPLEMENT FIXATION

A. **Principle.** When antigen and antibodies of the IgM or the IgG classes are mixed, complement is "fixed" to the antigen–antibody aggregate. If this occurs on the surface of a red blood cell, the complement cascade will be activated and hemolysis will occur.

1. The method actually involves two antigen–antibody complement systems: a test system and an indicator system (Fig. 14.10).

   a. The **indicator system** consists of red blood cells that have been preincubated with a specific anti-red cell antibody, in concentrations that do not cause agglutination, and in the absence of complement to avoid hemolysis; these are designated as **"sensitized" red cells**.

   b. In the **test system**, patient's serum is first heated to 56°C to inactivate the native complement and adsorbed with washed sheep RBC to eliminate broadly cross reactive anti-red-cell antibodies (also known as Forssman-type antibodies) which could interfere with the assay. Then the serum is mixed with purified antigen and with a dilution of fresh guinea pig serum, used as a controlled source of complement. The mixture is incubated for 30 minutes at 37°C to allow any antibody in the patient's serum to form complexes with the antigen and fix complement. "Sensitized" red cells are then added to the mixture.

      i. If the red cells are lysed, it indicates that there were no antigen-specific antibodies in the serum of the patient, so complement was not consumed in the test system and was available to be used by the anti-RBC antibodies, resulting in hemolysis. This reaction is considered **negative**.

      ii. If the red cells are not lysed, it indicates that antibodies specific to the antigen were present in the test system, "fixed" complement, but none were available to be activated by the indicator system. This reaction is considered **positive**.

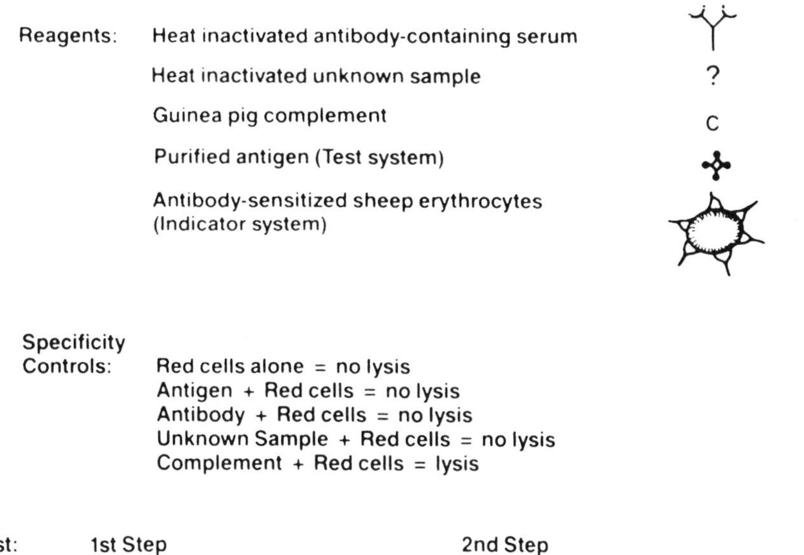

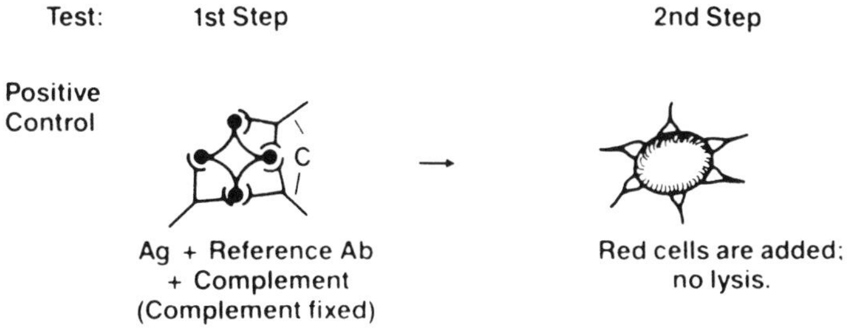

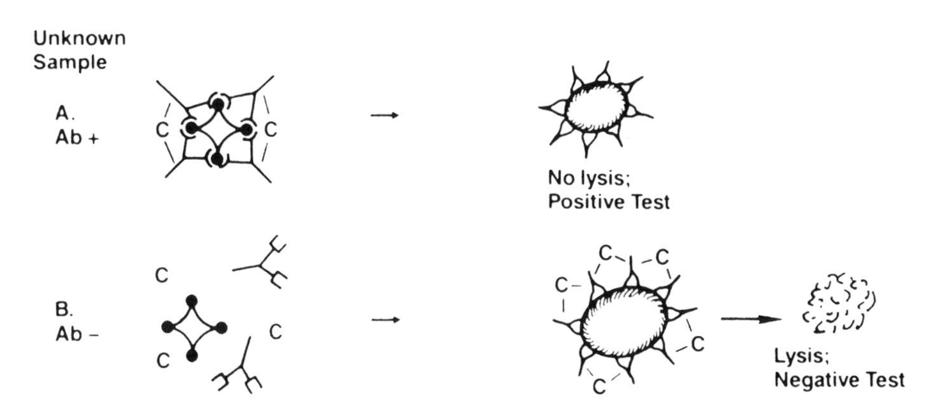

Figure 14.10 Diagrammatic representation of the general principles of a complement fixation test.

2. **Applications.** Complement fixation has the advantage of being widely applicable to the detection of antibodies to almost any antigen. Thus, complement fixation reactions have been widely used in a large number of tests designed to assist in the diagnosis of specific infections, such as the **Wassermann test** for syphilis and tests for antibodies to *Mycoplasma pneumoniae*,

Bordetella pertussis, many different viruses, and to fungi such as *Cryptococcus*, *Histoplasma*, and *Coccidioides immitis*.

3. **Limitations.** Complement fixation tests are riddled with technical difficulties and have been progressively replaced by other methods.

V. IMMUNOFLUORESCENCE

A. **Principle.** The primary reaction between antibodies chemically combined with fluorescent dyes and cells or tissue-fixed antigens can be visualized in a suitable microscope. There are several variations of immunofluorescence that can be used to detect the presence of unknown antigen in cells or tissues and the presence of unknown antibodies in patient's serum (Fig. 14.11).

1. **In direct immunofluorescence**, antigen in a cell or tissue is visualized by direct labeling with fluorescent antibody. Similarly, tissue-deposited

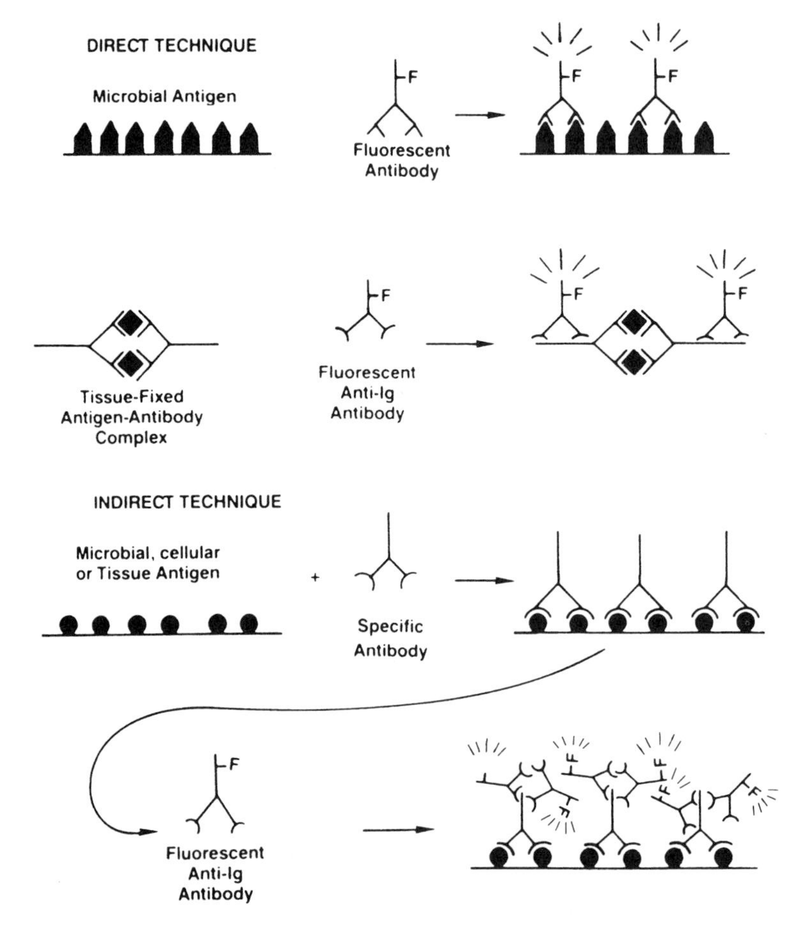

Figure 14.11 Diagrammatic representation of the general principles of direct and indirect immunofluorescence.

antigen–antibody complexes can be revealed by reaction with a fluorescent anti-immunoglobulin antibody.

2. **Indirect immunofluorescence** (as all indirect tests) involves two steps.
   a. Incubation of a substrate containing a fixed antigen (e.g., in a cell or tissue) with unlabeled antibody, which becomes associated with the antigen.
   b. After careful washing, a fluorescent antibody (e.g., fluorescent labeled anti-IgG) is added to the substrate. This second antibody will become associated to the first and the complex antigen–antibody 1–antibody 2 can be visualized on the fluorescence microscope. The indirect method has the advantage of using a single labeled antibody to detect many different specific antigen–antibody reactions.

B. **Applications.** Immunofluorescence has been widely applied in diagnostic tests.
   1. In microbiology, it can be used to identify isolated organisms, and also to diagnose an infection through the demonstration of the corresponding antibodies.
      a. An example is the use of the indirect fluorescence test for the diagnosis of **syphilis**. In a first step, the patient's serum is incubated with killed *Treponema pallidum*; in the second step, a fluorescent-labeled antihuman antibody is used to determine if antibodies from the patient's serum become bound to *Treponema*.
      b. Similar techniques have been used for the diagnosis of some viral diseases using virus-infected cells as substrate.
      c. Using fluorescent antibodies specific for different immunoglobulin isotypes, it is possible to identify the class of a given antibody. While **IgG antibodies** can be present in circulation for extended periods of time, **IgM antibodies** are characteristic of the early stages of the primary immune response. Thus, tests specifically designed to detect IgM antibodies are particularly useful for the diagnosis of acute infections.
      d. Through the use of fluorescent monoclonal antibodies and cytofluorographs, immunofluorescence is used to identify and enumerate **lymphocyte subpopulations** (Chap. 16).
      e. Immunofluorescence is also the technique of choice for the detection of autoantibodies such as *antinuclear antibodies*. Classically, the suspect serum is incubated with an adequate tissue (rat kidney, HeLa cells), and indirect immunofluorescence is performed in a second step to detect antibodies fixed to the substrate. In a positive test, the nuclei of the cells used as substrate will be fluorescent.
      f. Anti-double-stranded (ds) **DNA antibodies** can also be detected by immunofluorescence using a parasite (*Chritidia lucilliae*), which has a kinetoplast composed of pure ds DNA. Fixed parasites are incubated with patients seen in the first step. Anti-ds DNA antibodies, if present, will bind to the kinetoplast and will be revealed with fluorescein-labeled anti-IgG antibody.
      g. With the introduction of fluorometers, a wide new array of **quantitative immunofluorescence** assays has been developed. The principles are similar to those just mentioned: antigen is bound to a solid phase, exposed to a serum sample containing specific antibody, unbound immu-

noglobulins are rinsed off, and a fluorescein-labeled antibody is added to reveal the antibody that reacted specifically with the immobilized antigen. A fluorometer is used to assay the amount of fluorescence emitted by the second antibody. Since the amount of fluorescent antibody added to the system is fixed, the amount that remains bound is directly proportional to the concentration of antibody present in the sample. Thus, a quantitative correlation can be drawn between the intensity of fluorescence and the concentration of antibody added in the first step.

C. **Advantages and Limitations.** One of the main advantages of immunofluorescence techniques is the fact that they allow the identification of antigenic structures. The whole field of flow cytometry has also hinged in the availability of fluorescent antibodies. One of the major limitations of immunofluorescence is that all the techniques require expensive equipment and reagents. Also, the assays based on microscopy require trained personnel and have a factor of subjectivity that may result in erroneous results.

VI. RADIOIMMUNOASSAY (RIA)

A. Principle

1. The classical radioimmunoassay methods were based on the principle of **competitive binding**: unlabeled antigen competes with radiolabeled antigen for binding to antibody with the appropriate specificity. Thus, when mixtures of radiolabeled and unlabeled antigen are incubated with the corresponding antibody, the amount of free (not bound to antibody) radiolabeled antigen is directly proportional to the quantity of unlabeled antigen in the mixture.
   a. Mixtures of known variable amounts of cold antigen and fixed amounts of labeled antigen and mixtures of samples with unknown concentrations of antigen with identical amounts of labeled antigen are prepared in the first step.
   b. Identical amounts of antibody are added to the mixtures.
   c. Antigen–antibody complexes are precipitated either by cross-linking with a second antibody or by means of the addition of reagents that promote the precipitation of antigen–antibody complexes.
   d. Counting radioactivity in the precipitates allow the determination of the amount of radiolabeled antigen co-precipitated with the antibody.
   e. A standard curve is constructed by plotting the percentage of antibody-bound radiolabeled antigen against known concentrations of a standardized unlabeled antigen (Fig. 14.12), and the concentrations of antigen in patient samples are extrapolated from that curve.
2. The separation of free and antibody-bound radiolabeled antigen becomes considerably more simple if the antibody is immobilized. This is the hallmark of **solid-phase RIA** (Fig. 14.13).
   a. Antibody is adsorbed or coupled to a solid surface (test tube wall, polystyrene beads, etc.).
   b. Mixtures of labeled and unlabeled antigen and samples with added labeled antigen, prepared as described above, are added to the solid phase containing immobilized antibody.

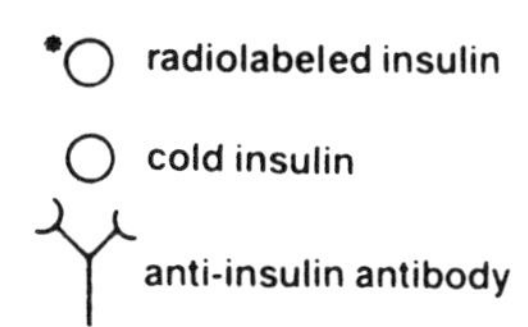

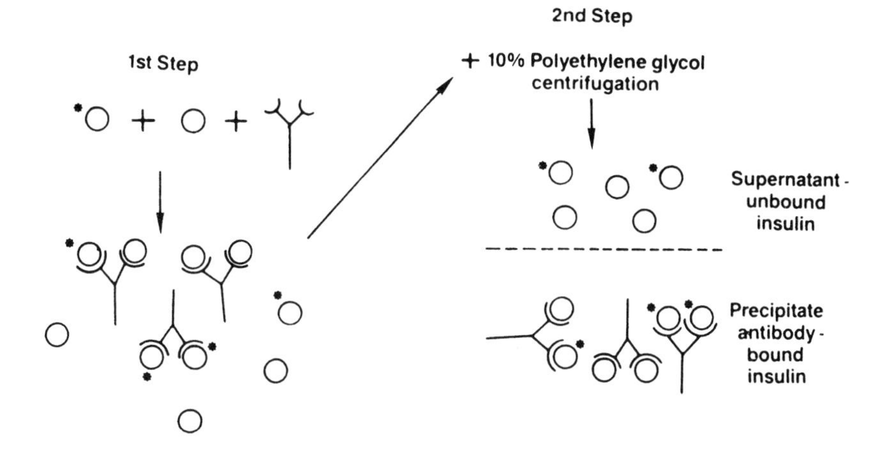

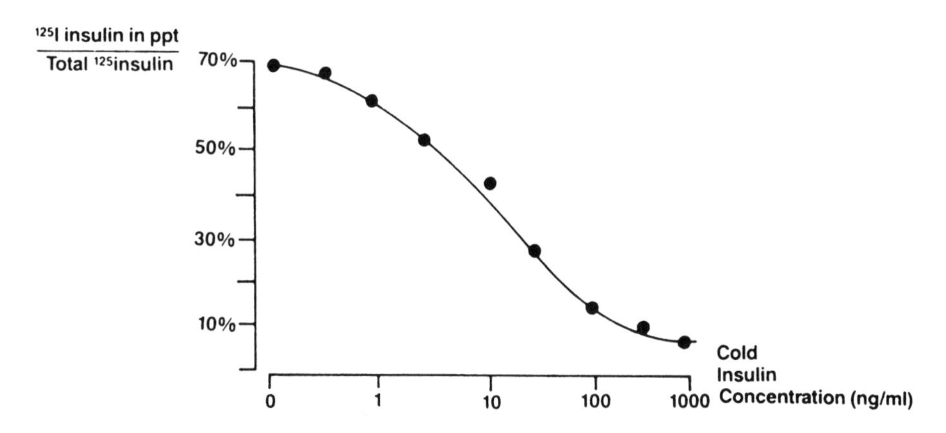

Figure 14.12 Diagrammatic representation of the general principles of competitive radioimmunoassay in fluid phase. In this type of assay (of which the quantitation of insulin levels is a good example), the free and antibody-bound antigens are separated either by physicochemical techniques, as shown in the diagram, or by using a second antibody (anti-immunoglobulin) to precipitate the antibody molecules that will carry with them any antigen with which they may have reacted.

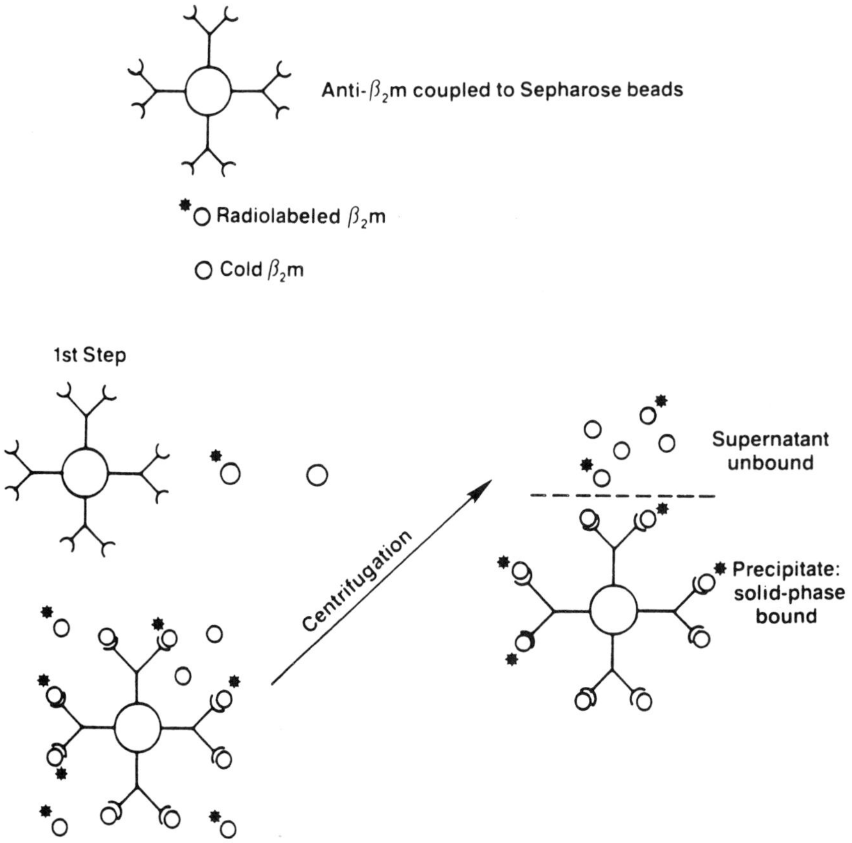

Figure 14.13 Diagrammatic representation of the general principles of competitive radioimmunoassays in solid phase. In this example (an assay for β2 microglobulin), since the antibody is bound to an insoluble matrix, the separation of free antigen from antibody-bound antigen is achieved by simple centrifugation. The calibration of the assay follows the simple principles shown in Figure 14.12.

c. Unbound reagents are washed off from the solid phase and radioactivity is counted to determine the amount of labeled antibody retained in each one of the different mixtures tested.

d. A calibration curve is constructed and unknown concentrations determined.

3. **Noncompetitive Solid-Phase RIA** for the detection of specific antibodies have also been described. The antigen is bound to solid phase, the antigen-coated solid phase is exposed to a sample containing antibody, and a radiolabeled antihuman immunoglobulin is used to assay the antibody that becomes bound to the immobilized antigen.

B. **Applications.** RIA have been used with extremely good results in the assay of many different hormones (insulin, aldosterone, human FSH, progesterone, testosterone, thyroxin, vasopressin, etc.), proteins (alpha-fetoprotein, carcinoembryonic antigen, IgE, hepatitis B surface antigen), nucleic acids (DNA), vitamins (vitamin B12), drugs (digoxin, LSD, barbiturate derivatives), enzymes (pepsin, trypsin), etc.

C. **Advantages and Limitations.** The extremely high sensitivity of RIA is its major advantage, allowing the assay on the nanogram/milliliter range or even lower. Their main drawbacks lie in the cost of equipment and reagents, short shelf-life of radiolabeled compounds, and in the problems associated with the disposal of radioactive waste. In recent years, RIA have been virtually replaced by quantitative fluorescence assays or by enzymoimmunoassays.

VII. ENZYMOIMMUNOASSAY

A. **Principle.** Conceptually, EIA is very close to solid-phase RIA; one of the components of the reaction (antigen if we want to assay antibodies; antibody if we want to assay antigens) is adsorbed onto a solid phase (e.g., polystyrene tubes, plastic microtiter plates, plastic or metal beads, plastic disks, absorbent pads, etc.).

1. The second step will differ depending on the type of assay.
    a. In a **competitive EIA** for antigen, a mixture of enzyme-labeled and unlabeled antigen is added to immobilized antibody.
    b. In a **"sandwich" assay**, the second step consists of adding the antigen-containing sample.
    c. In a **direct antibody assay**, the second step is adding the antibody-containing sample (Fig. 14.14).
2. These last two types of assays have a common third step, which is adding a labeled antibody.
    a. In the sandwich assay, the labeled antibody is of identical specificity to the antibody bound to the solid phase.
    b. In the direct antibody assay, the labeled antibody is an anti-immunoglobulin antibody.
3. Finally, an enzyme-labeled component (antigen or antibody) is added; its binding to the previous components is measured by adding an adequate substrate which, upon reaction with the enzyme (usually peroxidase or alkaline phosphatase), develops a color whose intensity can be measured by spectrophotometry.

B. **Applications**

1. **Enzymoimmunoassay (EIA)** has become perhaps the most widely used immunological assay method. Its sensitivity allows the assay of nanogram amounts without great difficulty and can be further increased by modifications of the technique. Enzymoimmunoassays for antimicrobial antibodies, antigen detection, and for hormone and drug quantitation have been successfully developed and commercialized.
2. **Rapid diagnosis EIA.** In recent years, a variety of very simple home and office tests based on EIA have been developed and successfully introduced.
    a. Some of the most popular rapid diagnosis EIA tests have been developed for pregnancy diagnosis. Most of these tests are sandwich assays that use two monoclonal antibodies recognizing two different, non-competing epitopes in human chorionic gonadotrophin. One antibody

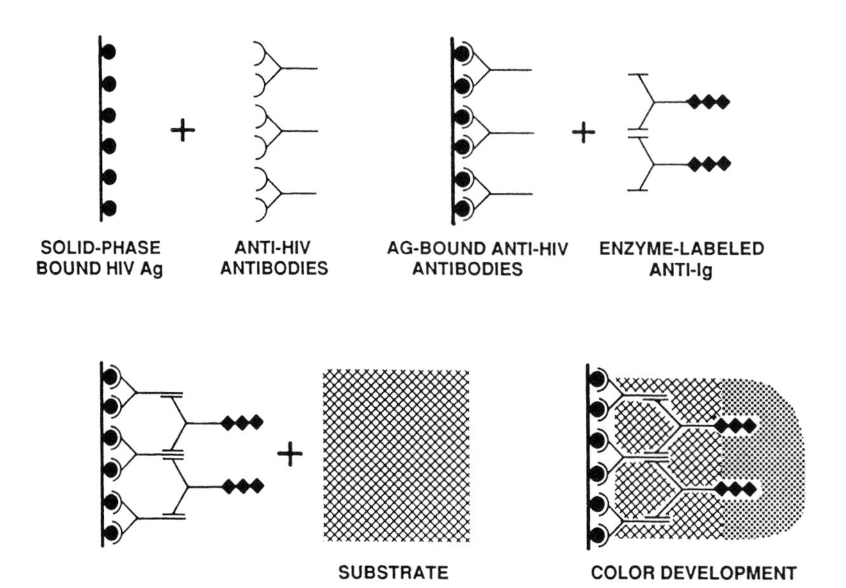

Figure 14.14 Diagrammatic representation of an enzymoimmunoassay test for the diagnosis of HIV infection. HIV antigens are adsorbed to the solid phase, then incubated with patient's sera and calibrated positive sera; After washing unbound proteins, a second enzyme-labeled, antihuman immunoglobulin antibody is added to the reactants. This second antibody will bind to the patient's immunoglobulins that had reacted with the immobilized antigen in the first step. After washing the unbound second antibody, a substrate is added; the substrate is usually colorless, but in the presence of the enzyme-labeled second antibody, it will develop a color. The development of color indicates positivity, and it allows quantitative measurements, since its intensity is proportional to the amount of bound enzyme-conjugated antibody which, in turn, is proportional to the concentration of anti-HIV antibodies bound to the immobilized antigen. The reactivity of calibrator samples with known antibody concentration is used to draw a calibration curve of color intensity versus antibody concentration. From this calibration the antibody concentrations in unknown samples are calculated.

is immobilized onto a solid phase and its function is to capture the antigen (hCG). The second is labeled with an enzyme and will be retained on the solid surface only if antigen has been captured by the first antibody. The retention of labeled antibody is detected by a color reaction secondary to the breakdown of an adequate substrate.

b. Rapid diagnosis EIA have also been developed for the diagnosis of infectious diseases, including streptococcal sore throat, respiratory syncytial virus infections, viral influenza, HIV infection (not approved for use in the U.S.), etc. The convenience and simplicity of these tests have ensured their rapid diffusion, and it is expected that many other tests based on similar principles will continue to be introduced in the immediate future.

C. **Advantages and Limitations.** Enzymoimmunoassays combine the high sensitivity common to all labeled immunoassays (quantitative immunofluorescence, RIA) with some specific advantages, such as the lack of risk involved with the use of enzyme-labeled compounds, with technical flexibility that results in a wide range of techniques, ranging from those rapid diagnostic assays which require minimal to no equipment to large-volume, automated procedures. The most important limitations of EIA are related to cross-reactions and nonspecific reactions, both of which lead to false positive results, which is a greater problem in assays of enhanced sensitivity.

SELF-EVALUATION

Questions

Choose the ONE *best* answer.

14.1 A possible cause for false positivity in a complement fixation test could be the:
  A. Excess of complement in the guinea pig serum
  B. Existence of antigen–antibody complexes in the unknown sample
  C. Inefficient complement inactivation in the unknown sample
  D. Involuntary omission of antigen in the test
  E. Use of old red cells

14.2 The main drawback of quantitative assays based on precipitation is:
  A. Difficulty in obtaining reagents
  B. Lack of specificity
  C. Poor reproducibility
  D. Poor sensitivity
  E. Short shelf-life of reagents

14.3 The antibodies detected in passive hemagglutination are:
  A. Directed against antigens chemically coupled to red cells
  B. Directed against antigens of ABO system
  C. Directed against antigens of the Rh complex
  D. Incomplete red-cell agglutinins
  E. The result of passive immunization

14.4 In an enzymoimmunoassay for antitetanus antibodies, the intensity of color measured after adding the substrate in the final step is:
  A. Directly proportional to the concentration of antibody in the patient's serum
  B. Directly proportional to the concentration of antigen in the solid phase
  C. Directly proportional to the concentration of enzyme-labeled antibody
  D. Inversely proportional to the concentration of antibody in the patient's serum
  E. Inversely proportional to the concentration of substrate

14.5 Which of the following immunoassays is most sensitive?
  A. Agglutination
  B. Immunodiffusion
  C. Immunoelectrophoresis
  D. Quantitative immunofluorescence
  E. Radial immunodiffusion

14.6 In a competitive radioimmunoassay:
  A. Radiolabeled antibody is immobilized in a solid phase
  B. Free antigen is preferentially precipitated with polyethylene glycol
  C. The radiolabeled antigen is first incubated with immobilized antibody
  D. The calibration curve is established by measuring the amount of labeled antigen bound in the presence of increasing concentrations of cold antigen
  E. The amount of radiolabeled antigen complexed with the antibody is directly proportional to the concentration of antigen in the unknown sample.

14.7. The serum of a patient is sent to the laboratory for a complement fixation test. The serum is heated at 56° C for 30 minutes, adsorbed with washed sheep red cells, mixed with the corresponding antigen, and later with a dilution of fresh guinea-pig serum. The final mixture is incubated for 30 minutes at 37°C, and then "sensitized" sheep red cells are added to the mixture. After 15 minutes you can see obvious hemolysis in the test tube. All controls reacted adequately. This constitutes a:

A. Negative reaction; there was no complement fixation
B. Negative reaction; the antibodies did not react with the antigen
C. Positive reaction; complement was fixed by antigen–antibody complexes
D. Positive reaction; the red cells fixed complement
E. Positive reaction; there was antigen in the patient's serum

14.8. Which of the following steps is an essential part of a solid-phase competitive enzymoimmunoassay for insulin?

A. Calibration based on the binding of different concentrations of cold insulin by immobilized antibody
B. Coupling of cold insulin to the solid phase
C. Coupling of enzyme-labeled insulin to the solid phase
D. Preparation of mixtures containing identical concentrations of enzyme-labeled insulin and variable concentrations of cold insulin
E. Separation of insulin–antibody complexes from free insulin by precipitation with polyethylene glycol

14.9. To develop a test to screen circulating antitreponemal antibodies by immunofluorescence, you need all of the following EXCEPT:

A. A fluorescence microscope
B. A positive serum control, known to contain antitreponemal antibodies
C. A suspension of formalin-killed *Treponema pallidum*
D. Fluorescein-labeled anti-human IgG
E. Isolated patient's IgG labeled with a fluorescent compound

14.10 A positive EIA for HIV is routinely repeated and confirmed by Western blot before any given individual is considered HIV infected. Which one of the following problems is most likely the cause of the need for such a careful approach to the diagnosis of HIV infection?

A. Daily variations in antibody titer
B. Frequent technical errors
C. Possible false positive reactions
D. Subjectivity in the interpretation of EIA results
E. Variability of the results obtained by different techniques

Answers

14.1 (B) A positive CF test is the one in which complement is consumed, and the red cells are NOT lysed. Excess of complement, endogenous or exogenous, would cause lysis (false negative). Omission of antigen would also result in nonconsumption of complement. Old red cells often lyse spontaneously or with minimal concentration of complement, also causing false negative results. In contrast, antigen–antibody complexes in the test sample could cause complement activation and a false positive result.

14.2 (D)

14.3 (A) In passive hemagglutination, the red cells are used as indicators of the reaction between an antigen artificially coupled to the red cells and the corresponding antibodies.

14.4 (A) In ELISA assays, the concentrations of immobilized antigen, enzyme-labeled second antibody, and substrate are kept constant; therefore, the intensity of color developed when the reaction is completed is directly proportional to the concentration of antibody in the unknown sample (patient's sera) and in the samples with known antibody concentrations used to calibrate the assay.

14.5 (D) All immunoassays using labeled antibodies are more sensitive than immunoassays based on direct observation of precipitation or agglutination.

14.6 (D)

14.7 (A) The mixture of serum with antigen obviously did not consume complement, since complement was available to cause the lysis of "sensitized" red cells.

14.8 (D) Such mixtures are used for calibration of the system; when incubated with immobilized antibody, the amount of labeled insulin that is bound is inversely proportional to the concentration of free insulin in the mixture. Linear regression analysis of the correlation between bound labeled insulin and unlabeled insulin present in the mixtures prepared with known amounts of unlabeled insulin allows the extrapolation of concentrations of insulin in unknown samples to which the same concentration of labeled insulin was added.

14.9 (E) Although it could be possible to design an assay using patient's IgG, the steps involved in isolation of IgG from patient's serum and labeling of the isolated IgG would render the test too costly and cumbersome. In addition, it would be very difficult to ensure that the yield of antibody at the end was an accurate reflection of the serum concentration of antibody.

14.10 (C) False positive reactions are one of the major limitations of some EIA procedures, and in the case of a diagnosis with such serious consequences as HIV infection it is essential to carefully confirm a result before notifying a patient.

BIBLIOGRAPHY

Bogulaski, R.C., Maggio, E.T., and Nakamura, R.M., eds. *Clinical Immunochemistry: Principles of Methods and Applications*. Little, Brown & Co., Boston, 1984.

Bryant, N.J. *Laboratory Immunology and Serology*, 3rd ed. W.B. Saunders, Philadelphia, 1992.

Collins, W.P., ed. *Alternative Immunoassays*. John Wiley & Sons, New York, 1985.

Collins, W.P., ed. *Complementary Immunoassays*. John Wiley & Sons, New York, 1988.

Cronenberg, J.H., and Jennette, J.C., eds. *Immunology. Basic Concepts, Diseases, and Laboratory Methods*. Appleton & Lange, Norwalk, CT, 1988.

Grieco, M.H., and Meriney, D.K., eds. *Immunodiagnosis for Clinicians*. Year Book Medical Publishers, Chicago, 1983.

Kaplan, L.A., and Pesce, A.J., eds. *Non-Isotopic Alternatives to Radioimmunoassay*. Marcel Dekker, New York, 1981.

Phillips, T.M. *Analytical Techniques in Immunochemistry*. Marcel Dekker, New York, 1992.

Turgeon, M.L. *Immunology and Serology in Laboratory Medicine*, 2nd ed. Mosby, St. Louis, 1996.

Wreghitt, T.J., and Morgan-Capner, P. *ELISA in the Clinical Microbiology Laboratory*. Public Health Laboratory Service, Whaddon, England, 1990.

15

Diagnostic Evaluation of Humoral Immunity

Gabriel Virella

I. INTRODUCTION

One of the most frequent problems faced by clinical immunologists is the investigation of a possible immunodeficiency. The attention of the physician is often aroused by a patient who presents with a history of repeated infections that may subside after adequate antibiotic therapy, but recur a short time after antibiotics are withdrawn. As discussed in Chapters 13 and 30, the nature of the infection may give some preliminary indication about whether humoral or cell-mediated immunity are depressed. If a humoral immunodeficiency is suspected, it is usual to start by the simplest procedures and to proceed to more sensitive and complex tests as needed. The same sequence will be followed in this chapter.

II. QUANTITATION OF IMMUNOGLOBULINS AND ANTIBODIES

A. **Serum Immunoglobulin Levels.** This is the most frequently performed screening test for humoral immunity. Usually it is sufficient to assay the three major immunoglobulin classes (IgG, IgA, and IgM) since there is no proof that deficiencies of IgD or IgE might have any pathological consequences. However, in interpreting immunoglobulin levels in children, it is very important to remember that normal values vary with age, as shown in Table 15.1.

B. **Classification of Humoral Immunodeficiencies Based on Immunoglobulin Levels.** Immunoglobulin assay is a fundamental element in the classification of immunodeficiencies (Table 15.2).

1. A quantitative depression of one or more of the three major immunoglobulin isotypes is considered as compatible with a diagnosis of humoral immunodeficiency.
2. If all immunoglobulin classes are depressed, the condition is designated as **hypogammaglobulinemia**.
   a. If the depression is very severe, and the combined levels of all three immunoglobulins are below 200 mg/dL, the patient is considered as having **severe hypogammaglobulinemia** or **agammaglobulinemia**.

Table 15.1 Normal Values for Human Immunoglobulins[a]

	IgG	IgA	IgM
Newborn	636–1606	0	6–25
1–2 mo.	250–900	1–53	20–87
4–6 mo.	196–558	4–73	27–100
10–12 mo.	294–1069	16–84	41–150
1–2 yr.	345–1210	14–106	43–173
3–4 yr.	440–1135	21–159	47–200
5–18 yr.	630–1280	33–200	48–207
8–10 yr.	608–1572	45–236	52–242
Greater than 10 yrs.	639–1349	70–312	57–352

[a]In mg/dL, as determined by immunonephelometry in the Department of Laboratory Medicine, Medical University of South Carolina.

b. When only one or two immunoglobulin classes are depressed, we designate the condition as **dysgammaglobulinemia.**
c. **Hypogammaglobulinemia** is often found in association with secondary immunodeficiencies, such as those of patients with plasma cell dyscrasias, the nephrotic syndrome, intestinal malabsorption, or long-term immunosuppression. In contrast, **agammaglobulinemia** is usually seen in cases of primary immunodeficiency. Only in very rare cases will an agammaglobulinemic individual be asymptomatic.
d. Among **dysgammaglobulinemias**, the most frequent by far is **IgA deficiency**, which is also the most frequent form of human immune deficiency.
   i. The frequency estimates of IgA deficiency vary according to the criterion used to define it and with the sensitivity of the methodology used:
      - If IgA deficiency is defined as total absence of IgA, and methods of very high sensitivity are used for IgA assay, less individuals are shown to be IgA deficient.
      - For practical purposes, one should consider as IgA deficient those cases where IgA levels are very low (e.g., <99th percentile of normals of same age as the patient), as measured by

Table 15.2 Ig Levels in Immune Deficiency[a]

Patient	IgG	IgA	IgM	Interpretation
A	850	2.8	128	IgA deficiency
B	1990	39.4	145	IgA deficiency
C	131	28.2	Traces	Severe hypogammaglobulinemia
D	690	16.0	264	IgA deficiency
E	154	60.0	840	Hyper IgM syndrome

[a]In mg/dL.

routine methods for immunoglobulin assay (such as radial immunodiffusion or immunonephelometry), with the understanding that only a small proportion of such cases will have an absolute lack of IgA.

Based on this criteria, the frequency of IgA deficiency varies between 1:500 and 1:800 individuals.

ii. At least 50% of IgA-deficient individuals are clinically asymptomatic. Those with recurrent infections often have associated IgG subclass deficiencies (see below).

iii. Patients with IgA deficiency often have anti-IgA antibodies of variable specificities (anti-isotypic or anti-allotypic). The origin and role of such antibodies are controversial. Their presence in serum will result in rapid binding and elimination of exogenous or endogenous IgA, in such a way that antibodies to IgA could perpetuate the IgA deficiency. Another consequence of the existence of such antibodies may be a severe reaction after blood transfusion or administration of gammaglobulin preparations. Usually, severe reactions occur only when the antibody titers are high (greater than 80 when measured by passive hemagglutination).

e. One other relatively frequent form of dysgammaglobulinemia is characterized by a combined deficiency of IgG and IgA, with elevated IgM (**hyper-IgM syndrome**). This situation reflects the inability of the patient to switch from IgM to IgG/IgA synthesis, due to the lack of signaling from helper T cells (see Chap. 30).

f. **IgG subclass deficiencies** have also been reported. Total IgG concentration might be normal to slightly depressed and one or two of the minor subclasses may be deficient.

   i. Particular attention has been give to **IgG2 subclass deficiency**, which may be associated to bacterial infections with polysaccharide capsules.

   ii. Combined deficiencies of IgG2 and IgA are relatively frequent, and individuals with this combined deficiency are often symptomatic, suffering mostly from bacterial infections.

C. **Determination of Common Antibodies.** Not infrequently, the assay of serum immunoglobulins may fail to support a clinical diagnosis of humoral immunodeficiency. The next step may be the assay of antibodies that are found in most normal individuals, either as a result of normal exposure to the antigen or as a result of vaccination:

1. Anti-A and anti-B isohemagglutinin titers
2. Anti-streptolysin 0 titer
3. Anti-tetanus toxoid antibody concentration
4. Anti-*Haemophilus influenzae* polysaccharide antibody concentration
5. Antibodies to common viruses (mumps, measles, polio)
   a. Abnormally low levels of one or more of these antibodies will support a diagnosis of humoral immunodeficiency, providing that exposure to the corresponding antigens can be unquestionably documented.
   b. The determination of preformed antibodies to common organisms and agents used in routine immunizations is mainly indicative of **past immunoreactivity** and may not provide useful information in cases where

the deficiency is more subtle or when the immunodeficiency is of recent onset, in which the first function to be lost is the ability to mount a primary immune response.

D. **Determination of Antibodies Against Infectious Agents Known to Infect the Patient.** The need to identify the etiological agents in patients with recurrent and unusual infections cannot be overstressed. Besides providing very useful treatment for the selection of the most adequate antimicrobial(s), it allows the most informative test for the diagnosis of a humoral immunodeficiency to be carried out (i.e., the assay of antibodies against the infectious agent). If the patient fails to produce antibodies, the diagnosis is obvious.
    1. In some patients, all investigations might be inconclusive except for the measurement of antibodies to the infecting microorganism(s), which might reveal an "**antigen-selective**" **immunodeficiency**.
    2. Unfortunately, proper microbiological studies are seldom done in patients with suspected humoral immunodeficiency, and adequate assays for antibodies to some common organisms (e.g., *Staphylococci*) are not commonly available.

III. QUANTITATION OF ANTIBODIES AFTER ANTIGENIC CHALLENGE

This is the ideal approach for the investigation of the humoral immune response, since it determines very specifically the ability of the patient to sustain a functional antibody response after adequate challenge. This type of investigation can be carried out from two different perspectives: (a) to determine whether a patient is able to develop an immune response; and (b) to determine whether a patient can synthesize antibodies against a specific, infecting microorganism. The antigens chosen for this investigation should follow the following criteria: lack of risk for the patient; availability of techniques for the measurement of the corresponding antibodies; and adequacy of the antigen for the purpose in mind.

A. For evaluation of **primary immune responsiveness**, one needs to use an antigen to which the individual has never been exposed.
    1. In immunodeficient children, for whom good records of previous immunizations and infections are available, any component or killed vaccine (**never live attenuated**) to which there has been no previous exposure can be used.
    2. Proteins extracted from lower animals, such as keyhole limpet hemocyanin, can also be used. However, it is very difficult to exclude the possibility of having been exposed to a cross-reactive immunogen in the past, leading to detectable titers of "natural" antibodies prior to immunization.
    3. The best approach devised so far is the immunization with bacteriophage ØX174. Both the evolution of antibody levels (Fig. 15.1) and the clearance of the injected phage (significantly delayed in immunodeficiency patients) can be followed with very sensitive techniques to determine whether antibodies are synthesized and whether or not there is an effective immune elimination of the bacteriophage. Phage immunization has been carried out by several groups in different countries and has been proven to be a harmless procedure. At this point, the main problem preventing the widespread use of this

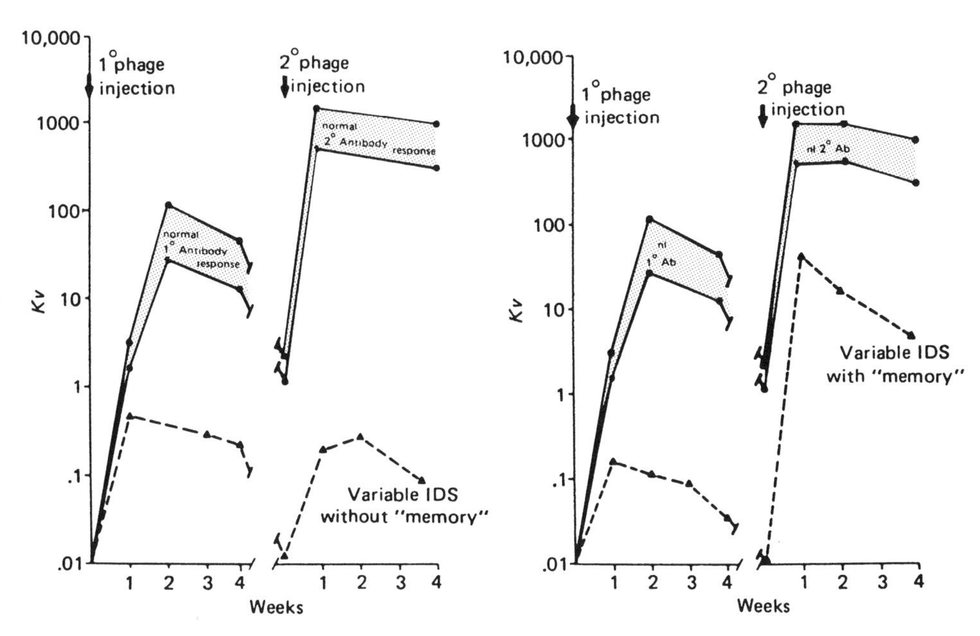

Figure 15.1 Primary and secondary response to bacteriophage ØX 174 in a patient with a variable immunodeficiency syndrome. The shaded area between the solid lines indicates the range for normal responses. The patient's response is the interrupted line. The patient studied on the left panel showed a definite, but diminished, antibody response; the secondary response was not greater than the primary—no memory/amplification occurred. The immunoglobulin class of antibody in both primary and secondary responses was entirely IgM. The patient studied on the right panel showed greater response to secondary immunization than to primary (memory/amplification) although both were diminished in comparison to the normal range. The immunoglobulin class of the secondary response was entirely IgM. (Reproduced with permission from Wedgewood et al. *Birth Defects: Original Article Series*, *11*:331, 1975.)

approach is the need for development of phage inactivation assays for the measurement of the antiphage antibody response, which is beyond the scope of possibilities for many clinical diagnostic laboratories.

B. The evaluation of the **secondary immune response** does not raise so many problems, but is less informative, since the capacity to initiate a primary immune response seems to be the first (and sometimes the only) function affected by immunosuppressive agents or in diseases associated with immunodepression.

1. **Diphtheria** and **tetanus toxoids** are frequently used and, because these are strongly immunogenic proteins, their use has only minimal risk for the patient and specific antibodies can be assayed by a variety of techniques, such as enzymoimmunoassay.

   a. Given the lack of information concerning normal values for these antibodies, and the fact that the abnormality searched for is the lack of an active response rather than a low level of antibody, the best approach is to collect blood for baseline study prior to a booster with the corresponding antigen and repeat the collection 2 to 3 weeks later. Following this protocol, we detected active responses in all but two of a group of

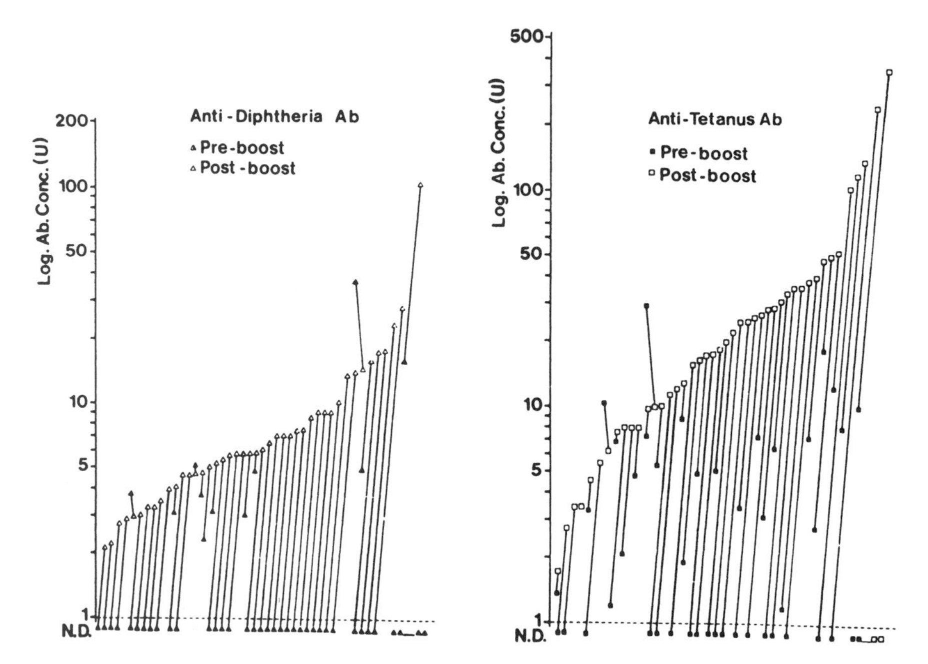

Figure 15.2 Pre- and post-boost antibody titers to diphtheria and tetanus toxoid determined in 46 healthy children between 16 and 33 months of age. (Reproduced with permission from Virella, G., Fudenberg, H. H., Kyong, C U., Pandey, J.P., and Galbraith, R.M., *Z. Immunitatsforsch.*, *155*:80, 1978.)

children randomly selected from the population of a rural county of South Carolina (Fig. 15.2). The existence of normal nonresponders needs to be considered when evaluating a patient suspected of having an immunodeficiency.

b. An interesting example of the application of tetanus toxoid immunization is the follow-up of the immune response after a bone marrow graft. Patients receiving bone marrow grafts are given large doses of immunosuppressive drugs before the graft, in the hope of avoiding rejection, and continue to be immunosuppressed after the graft to avoid a graft-vs.-host reaction (see Chap. 27). When the evolution of the patient is uneventful, the immunosuppressive drugs are stopped, and the patient is then immunized for tetanus toxoid to determine whether the immune system regains its ability to mount an active immune response. As illustrated in Figure 15.3, this recovery may only be observed several months after the suspension of immunosuppressive therapy. It needs to be noted that the immunoglobulin levels may be normal while the patient shows a complete lack of response to immunization.

2. **Challenge with polysaccharide vaccines** (such as the *Haemophilus influenzae* or the *Streptococcus pneumoniae* vaccines) is indicated not only when this type of organism is predominantly involved but in cases of IgG2 subclass deficiency, in which patients often show subnormal response to polysaccharides.

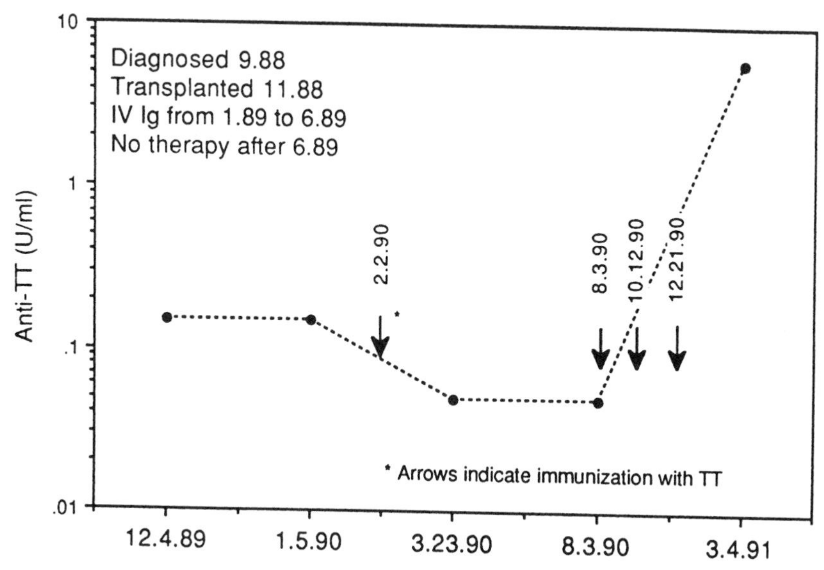

Figure 15.3 Graphic representation of a longitudinal study of the serum levels of antitetanus toxoid (anti-TT) antibodies in a patient who received a bone marrow graft in December, 1988. The patient received intravenous gamma globulin from January to June, 1989, as anti-infectious prophylaxis. The clinical evolution was excellent, and all therapy (immunosuppressive and immunoprophylactic) was discontinued in June, 1989. The first assay of anti-TT antibody in December, 1989 showed that the patient had low levels of antibody (0.15 U/ml). A first immunization with TT in February, 1990 was followed by a paradoxical decrease of anti-TT antibody concentration. Only after three additional boosters given between August and February of 1991 was there a significant increase in anti-TT antibody concentration. At that time, the patient could be judged as immunologically recovered.

IV. ADDITIONAL INVESTIGATIONS

The finding of an immunoglobulin deficiency or of the inability to mount a humoral immune response does not give many clues as to the pathogenesis of the defect. Besides the possibility of dealing either with a primary or secondary immunodeficiency, which implies the need to investigate known causes of secondary immunodeficiency, primary humoral immunodeficiencies may result from a variety of defects, such as absence or lack of differentiation of B cells, defects in intracellular synthesis, assembly, or secretion of immuoglobulins, hyperactivity of suppressor cells, deficient helper T-cell function, etc. (discussed in greater detail in Chap. 30). Several investigations can be performed in order to clarify the nature of the defect, which, although they may be mostly of academic significance, provide very valuable data for our understanding of the human immune system.

A. **Analysis of Lymphocyte Membrane Markers.** The development of monoclonal antibodies and automated flow cytometry techniques has resulted in an explosion of studies attempting to define lymphocyte subpopulations and their functional properties.

1. **Quantitation of B lymphocytes in the peripheral blood.** This is usually investigated by determining the percentage of peripheral blood lymphocytes with specific B-lymphocyte markers (CD19, CD20). The proportion of lymphocytes identifiable as B cells in the peripheral blood by such techniques is

between 4 and 10%, corresponding to a range of 96–421 CD19+ cells/μL. If the number of B cells in the peripheral blood is significantly depressed, the immunodeficiency is most likely to result from lack of B-cell differentiation. Such lack of B-cell differentiation is the rule in infantile agammaglobulinemia (see Chap. 30).

2. **Quantitation of helper T lymphocytes in the peripheral blood.** Although the correspondence between membrane markers and function is not perfect, the quantitation of cells expressing CD4 is believed to be an adequate assessment of the number of helper T cells. Very low numbers of $CD4^+$ lymphocytes are a major hallmark of the acquired immunodeficiency syndrome, in which humoral immunity (particularly the ability to mount a primary immune response) is severely compromised (see Chap. 30).
3. **Determination of markers involved in the co-stimulation of B cells.** Of the variety of membrane molecules involved in T–B-cell interactions and signaling, the CD40 ligand (CD40L, CD154, gp39) is the most clinically relevant. The deficient expression of this molecule in helper T cells is the most common molecular abnormality in patients with the hyper-IgM syndrome (see Chap. 30).

B. **Study of the Differentiation of B Cells In Vivo.** The best approach is to look for germinal centers and immunoglobulin-producing cells in a lymph node biopsy from an area draining the site where an antigenic challenge has been carried out a week earlier (for example, with diphtheria or tetanus toxoids). The main drawback is the need for surgical excision of a lymph node, which often may be quite difficult to localize. An alterative recommended by some groups is to perform a rectal biopsy and look for the presence of germinal centers in the submucosa, as evidence for the normal differentiation of B cells in the peri-intestinal tissues.

C. **Investigation of B-Cell Function In Vitro.** The investigation of B-cell function in vitro requires separation of peripheral blood lymphocytes (PBL) and their stimulation with substances known to induce the proliferation and/or differentiation of B lymphocytes.

1. **Pokeweed mitogen** is a plant lectin that induces proliferation and differentiation of T and B lymphocytes. The differentiation of B lymphocytes is easy to measure, using immunoglobulin synthesis as the endpoint. The effects of this mitogen on B lymphocytes are T-cell-dependent.
2. **Killed *Staphylococcus aureus*** can stimulate both proliferation and differentiation of B lymphocytes in the presence of T cells, but can also stimulate proliferation (but not the differentiation) of B lymphocytes in the absence of T cells.
3. **Killed *Salmonella paratyphi B*** can induce the differentiation of B lymphocytes without T-cell help.
4. Using combinations of these mitogens, it is possible to analyze in some detail the type of B-cell defect responsible for the humoral immunodeficiency state (see Fig. 15.4).

D. **In Vitro Testing of Helper and Suppressor Functions**

1. To determine whether the immunodeficiency is secondary to a defect in the immunoregulatory circuits, peripheral lymphocytes from the patient are cocultured with lymphocytes from a normal individual, and the response to

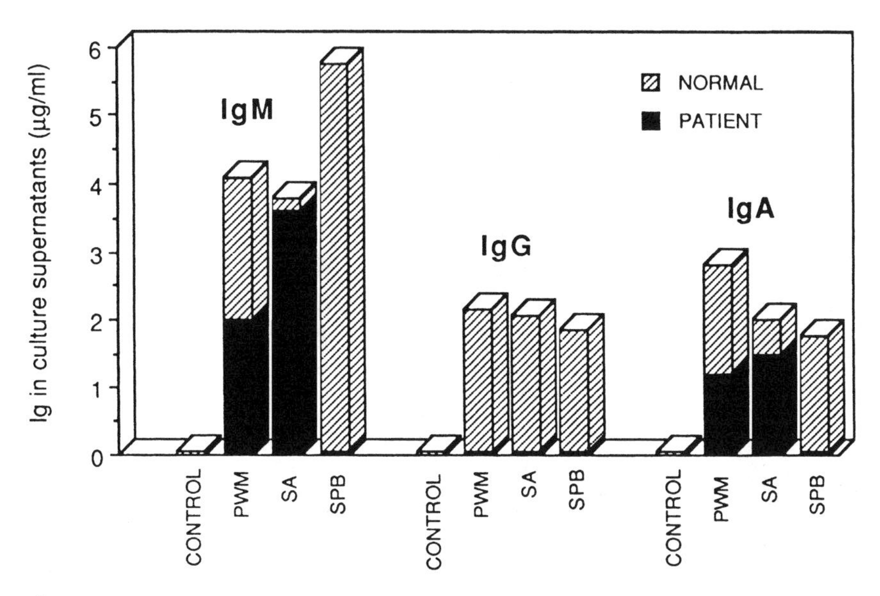

Figure 15.4 Study of the differential stimulation of PBL isolated from a patient, and a normal control with common variable immunodeficiency with pokeweed mitogen (PWM), *S. aureus* (SA), and *Salmonella paratyphi B* (SPB). The levels of immunoglobulin M, G, and A quantitated on 7-day culture supernatants are shown by open bars for the control and closed bars for the patient. The patient's PBL responded with IgM and IgA production to stimulation with pokeweed mitogen and *S. aureus*, but failed to respond to *Salmonella paratyphi B*. The induction of immunoglobulin secretion in vitro by PWM and *S. aureus* is believed to be T-dependent, while SPB is believed to induce immunoglobulin synthesis without T-cell help. Thus, the results show two significant abnormalities in the response of this patient's B cells: first, it was not possible to stimulate B cells without T-help; second, even when stimulated, the patient's B cells failed to produce IgG, pointing to a defect in the switch mechanism.

pokeweed mitogen compared with that obtained with cultures of patient and control lymphocytes alone.

a. A reduction in the stimulation indices in such mixed cultures below that observed in control cultures of lymphocytes from normal donors indicates suppressor activity by the patient's peripheral lymphocytes.
b. An increased response in comparison to the patient's control culture would indicate a deficiency in the patient's helper cells.

2. A better characterization of the cellular defects involved will require purification of cell populations (monocytes, B lymphocytes, and T lymphocytes) from the patient and a normal control, and their co-culture.
   a. In the example shown in Figure 15.5, co-culture of a patient's B cells and T cells isolated from a normal donor failed to reconstitute the ability of the patient's B cells to produce and secrete immunoglobulins, pointing to an intrinsic B-cell defect.
   b. If the patient's T cells have excessive suppressive activity or defective helper activity, the response of a mixture of normal B cells and the patient's T cells will be deficient.

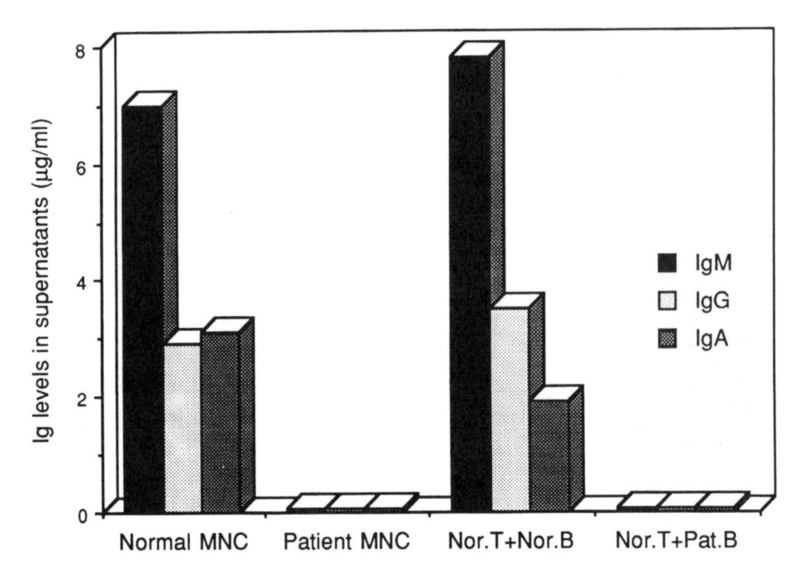

Figure 15.5 Co-culture experiment designed to evaluate a possible immunoregulatory defect in a patient with common variable immunodeficiency. Unfractionated patient's PBL failed to respond to pokeweed mitogen (PWM), while both unfractionated PBL and 1:1 mixture of T and B cells separated from the normal control responded to PWM stimulation by producing detectable amounts of IgM, IgG, and IgA (measured on 7-day culture supernatants). A co-culture of patient's B cells and normal T cells (1:1) failed to produce immunoglobulins, suggesting that the lack of response to PWM stimulation lies in a defect of the patient's B cells, unable to synthesize or to secrete immunoglobulins in a normal fashion even in the presence of normal helper T cells.

i. A lack of helper T cells can be assumed if the patient's B cells recover their ability to produce immunoglobulins when mixed with normal T cells.
ii. Excessive suppressor activity is suggested if a mixture of patient's T cells plus normal T and B cells results in depressed B-cell response.

E. **Limitations of Functional Studies.** Functional studies are not routinely done in most diagnostic laboratories. They are carried out either by specialized laboratories or by research laboratories. Consequently, the cost of the tests is high and it cannot be covered as a diagnostic procedure. In addition, the tests require relatively large volumes of blood which may not be ethical to draw from small infants. However, these types of investigations are desirable and ethically justified, since they will have a positive impact on the treatment of future patients. Thus, when ethically permissible, investigational tests should be carried out providing the patient is not placed in physical jeopardy or economically encumbered with their cost. It always needs to be kept in mind that most of the data concerning the organization and regulation of the immune system in humans have been derived from the study of patients with immune deficiencies and that progress in our knowledge will always result in better patient care.

SELF-EVALUATION

Questions

Choose the ONE *best* answer.

15.1 An IgA-deficient patient is scheduled for surgery and in the preoperatory workup found to be A, Rh-positive and to have a high titer of anti-IgA antibodies. You should alert the blood bank for the possible need for:
- A. A, Rh-positive blood from an IgA-deficient donor
- B. AB, Rh-positive blood
- C. O, Rh-positive blood
- D. Frozen plasma
- E. Packed A, Rh-positive red cells

15.2 Which one of the following procedures is preferred for the evaluation of the ability of a patient to mount a secondary immune response
- A. Determination of antibody titers to measles pre- and postadministration of a booster injection
- B. Determination of antibody titers to tetanus toxoid pre- and postadministration of a booster injection
- C. Phage immunization
- D. Quantitation of serum immunoglobulin levels (G, A, and G)
- E. Titration of isohemagglutinins

15.3 A 21-year-old female is seen as an outpatient due to a history of repeated chronic pyogenic infections. Preliminary investigations show that the patient's blood group is O, with no detectable isoagglutinins; her immunoglobulin levels are strongly depressed, but her peripheral blood contains 89 CD19+ cells/μl. Which *one* of the following tests would be the *least* useful for the characterization of this patient's immunodeficiency?
- A. Co-culture of T and B lymphocytes isolated from the patient's peripheral blood and from the peripheral blood of a normal control
- B. Enumeration of T-lymphocyte subpopulations
- C. Quantitation of secretory IgA
- D. Study of the ability of the patient's lymphocytes to secrete immunoglobulins in vitro after stimulation with PWM, *S. aureus* protein A, and SPB
- E. Determination of anti-*S. pneumoniae* antibody levels before and after a booster with Pneumovax

15.4 A 4-year-old boy has had recurrent pneumonia since 3 months of age. Bacterial examinations have been repeatedly positive for *Haemophilus influenzae*. Immunological levels are: IgG: 400 mg/dL; IgA: 40 mg/dL; IgM: 50 mg/dL. Which one of the following do you expect to give the most useful information?
- A. Assay of antibodies to tetanus and diphtheria toxoids prior to and 3 weeks after a DTP booster
- B. Assay of secretory IgA
- C. Determination of antibody titers to *Haemophilus influenzae* capsular polysaccharide
- D. Study of the primary immune response to a bacteriophage
- E. Titration of isohemagglutinins

15.5 Consider the following data from a co-culture experiment in which B-cell activity was evaluated by assaying immunoglobulins in culture supernatants harvested 7 days after pokeweed mitogen (PWM) stimulation using PBL from a patient with common variable immune deficiency and from a normal control.

	IgG[a]	IgA[a]	IgM[a]
Patient's unfractionated PBL	0	0	0
Control's unfractionated PBL	4	2	9
Patient's T + patient's B cells (1:1)	0	0	0
Control's T + control's B cells (1:1)	3	1.5	9
Patient's T + control's B cells (1:1)	0	0	1
Patient's T + control's unfractionated PBL (1:1)	0	0	0
Control's T + patient's B cells (1:1)	2	3	5

[a]Values in μg/mL of culture containing 1×10^6 PBL/mL.

These results indicate that the patient's humoral immune deficiency is due to:
A. Abnormal biosynthesis of immunoglobulins in the patient's B cells
B. Excess of suppressor monocytes
C. Excess of suppressor T cells
D. Lack of B-cell differentiation
E. Lack of helper T cells

15.6 In a 4-year-old child with a history of repeated pyogenic infections caused by bacteria with polysaccharide-rich capsules, which of the following deficiency(ies) should be investigated?
A. IgA deficiency
B. IgG1 deficiency
C. IgG2 deficiency
D. IgA and IgG2 deficiency
E. IgG and IgA2 deficiency

15.7 In a normal 6-month-old child, you expect to see:
A. A concentration of IgG close to that of an adult because of the persistence in circulation of maternally transferred IgG
B. A total immunoglobulin concentration (IgG + IgA + IgM) below 200 mg/dL
C. IgM as the quantitatively preponderant serum immunoglobulin
D. Lower levels of all immunoglobulin classes relative to normal adult levels
E. Undetectable IgA

15.8 All of the following are adequate antigens for use in evaluating the humoral immune response of a suspected immune deficiency *except*:
A. Bacteriophage ØX174
B. Diphtheria and/or tetanus toxoids
C. Keyhole limpet hemocyanin (KLH)
D. Oral polio vaccine
E. *S. typhi* vaccine

15.9 Which one of the following investigations would you rank with the *lowest* priority in a patient found to have a level of serum IgA below 10 mg/dL?
A. Assay of IgG subclasses
B. Determination of anti-IgA antibodies
C. Enumeration of lymphocyte subsets
D. Measurement of antibody levels to common antigens
E. Study of the response to active immunization with toxoids and polysaccharides

15.10 The earliest functional evidence of secondary (or acquired) immunodeficiency in most individuals is:
A. A depression in the peripheral lymphocyte count
B. A depression of serum immunoglobulin levels
C. An antigen-specific immune deficiency
D. Lack of capacity to initiate a primary immune response
E. Loss of immunological memory

Answers

15.1 (A) In the presence of high titers of anti-IgA antibody, the possibility that IgA-containing blood products may cause a serious transfusion reaction needs to be considered. Ideally, IgA-deficient A, Rh-positive blood should be used. If this is not possible, extensively washed and packed A, Rh^+ cells should be used (packed, not washed red cells contain at least 20% plasma).

15.2 (B) Toxoids are fully adequate for this purpose; attenuated vaccines, on the contrary, should always be avoided. Phage neutralization, the technique needed to evaluate the effects of phage immunization, is not available in most medical centers.

15.3 (C) Co-cultures of T and B lymphocytes isolated from the patient and a normal control are useful to define abnormalities in the immunoregulatory cell populations. The enumeration of different T-lymphocyte subpopulations could reveal a CD4 T-cell deficiency. The study of in vitro immunoglobulin synthesis after stimulation with B-cell mitogens would help determine if the patient's B cells are totally unresponsive to stimulation. The study of the immune response to Pneumovax (the vaccine for *Streptococcus pneumoniae*) could help to confirm a diagnosis of humoral immunodeficiency. In contrast, the assay of secretory IgA is likely to give a low value, which will not contribute in any significant way to our understanding of the immunodeficiency state affecting this patient.

15.4 (C) When a patient suffers from repeated infectious bouts caused by one given microorganism, the most informative studies of humoral immunity are those measuring the response to the infecting agent.

15.5 (C) The patient's T cells suppressed the response of normal B cells, even when normal T cells were present in the co-culture (patient's T + control's unfractionated mononuclear cells). In contrast, normal T cells helped the patient's B cells to produce normal amounts of immunoglobulins. These two observations, taken together, point to an excess of T lymphocytes with suppressor activity in the patient's peripheral blood.

15.6 (D) Combined IgA and IgG2 deficiency is frequently associated with increased frequency of infections with encapsulated pyogenic bacteria.

15.7 (D)

15.8 (D) Live, attenuated vaccines should be avoided in immunodeficient patients unless the risks are outweighed by potential benefits, which is not the case when an investigation of immune responsiveness is being carried out.

15.9 (C) There is no evidence that IgA deficiency is associated with an imbalance of lymphocyte subpopulations. All other investigations would be useful to define the extent of the patient's immunodeficiency.

15.10 (D)

BIBLIOGRAPHY

Gergen, P., McQuillan, Geraldine M., Kiely, Michele, Ezzati-Rice, Trena M., Sutter, Roland W., and Virella, G. Serologic immunity to tetanus in the U. S. population: Implications for national vaccine programs. *N. Engl. J. Med.*, *332*(12):761–766, 1995.

Hanson, L.Å., Söderström, R., Nilssen, D.E., Theman, K., Björkander, J., Söderström, T., Karlsson, G., and Brandtzaeg, P. IgG subclass deficiency with or without IgA deficiency. *Clin. Immunol. Immunopathol.*, *61*(2):S70, 1991.

Rose, N.R., de Marcario, E.C., Fahey, J.L., Friedman, H., and Penn, G.M. (eds.) *Manual of Clinical Immunology*, 4th ed. American Society of Microbiologists, Washington, DC, 1992.

Virella, G., and Hyman, B. Quantitation of anti-tetanus and anti-diphtheria antibodies by enzymoimmunoassay: Methodology and applications. *J. Clin. Lab. Analysis*, *5*:43, 1991.

WHO Scientific Group on Immunodeficiency. Primary immunodeficiency diseases. *Clin. Exp. Immunol.*, *99*(Suppl. 1):1, 1995.

16

Diagnostic Evaluation of Lymphocyte Functions and Cell-Mediated Immunity

Gabriel Virella and Jean-Michel Goust

I. INTRODUCTION

It is traditional to designate as cell-mediated immunity (CMI) the complex network of interrelated cellular reactions often resulting in the production and release of soluble factors that appear to mediate the cooperation between different mononuclear cell populations and the expression of a variety of lymphocyte effector functions. A variety of tests, some performed in vivo and others in vitro, have been shown to correlate with different parameters or functions believed to depend primarily on lymphocyte stimulation and activation and, in some cases, to reflect the adequacy of immune responses primarily mediated by T lymphocytes.

II. IN VIVO TESTING OF DELAYED-TYPE HYPERSENSITIVITY

Delayed hypersensitivity responses, which will be discussed in greater detail in Chapter 18, are primarily mediated by T lymphocytes, and thus can be considered as manifestations of a hyperstimulation of cell-mediated immunity. Using controlled conditions, it is possible to challenge individuals with antigens known to cause this type of reaction as a way to explore their cell-mediated immunity. The two classical approaches to measure delayed-type hypersensitivity responses (DTH) in vivo are **skin testing** and **induction of contact sensitivity**.

A. **Skin Testing**. First described by Koch in 1891, skin testing is based on eliciting a secondary response to an antigen to which the patient was previously sensitized.
   1. A small amount of soluble antigen is injected intradermally on the extensor surface of the forearm. The antigens used are usually microbial in origin [e.g., purified protein derivative (PPD) of tuberculin, tetanus toxoid, mumps antigens, and a variety of fungal extracts, including Candidin (from *Candida albicans*), coccidioidin (from *Coccidioides immitis*), and histoplasmin (from *Histoplasma capsulatum*)].
   2. The area of the skin receiving the injection is observed for the appearance of **erythema** and **induration**, which are measured at 24 and 48 hours. The designation of **delayed hypersensitivity** is based on the contrast with a

totally different type of skin reaction which develops in a matter of minutes after antigen inoculation, lasts only for a few hours, and is characterized by erythema and localized swelling, but without induration (**immediate hypersensitivity reactions** are discussed in Chaps. 22 and 23).

3. A positive skin test is usually considered to be associated with an area of induration greater than 10 mm in diameter. If no reaction is observed, the test may be repeated with a higher concentration of antigen.
4. If a patient has no reaction after being tested with a battery of antigens, it is assumed that a state of **anergy** exists. Anergy can be caused by immunological deficiencies and infections (such as measles or chronic disseminated tuberculosis), but it can also be the result of errors in the technique of skin testing.
5. Since the capacity to demonstrate a delayed hypersensitivity reaction may persist for long periods of time, a positive skin test may indicate either a past or present infection.
6. Although these tests have the theoretical advantage of testing the function of the T-cell system in vivo, they meet with a variety of problems:
   a. Poor reproducibility due to the difficulty in obtaining consistency among different sources and batches of antigens and variations in the technique of inoculation among different investigators.
   b. The interpretation of negative tests has to be carefully weighed. Negative results after challenge with antigens to which there is no record of previous exposure can always be questioned, while a negative result with an antigen extracted from a microbial agent that has been documented as causing disease in the patient has a much stronger diagnostic significance, implying a functional defect in cell-mediated immunity.

B. Contact Sensitivity

1. Some compounds (nickel, dyes, etc.) induce hypersensitivity reactions after repeated skin contact. In the investigation of a suspected case of **contact hypersensitivity** (see Chap. 22), it is often necessary to determine the identity of the sensitizing substance. This can be accomplished by the **patch test**.
   a. Several potential antigens are applied in a dilute, nonirritating form to the patient's back, and covered with an occlusive dressing.
   b. Forty-eight hours later, the dressing is removed and, if a reaction consisting of erythema, edema, and vesiculation is seen at the point of application of one of the antigens, that antigen is identified as the cause of the hypersensitivity reaction.

   This test, therefore, is not used to investigate the status of cell-mediated immunity, but rather to identify a compound to which a patient has become hypersensitive.
2. *De novo* induction of contact sensitivity by application of low-molecular-weight chemical compounds (e.g., dinitrochlorobenzene or DNCB) to intact skin was used as a way of evaluating the ability to induce a primary cell-mediated immune response in vivo.
   a. The chemical applied to the skin diffuses to the dermis, where it reacts with skin proteins and elicits an immune response. After the initial application, a period of 7 to 10 days is needed for contact sensitivity to develop, whereupon a challenge dose may be given.

b. As opposed to skin testing, induration does not usually occur, but instead a flare response is noted. Lack of a reaction means inability to initiate a cell-mediated immune response.

This type of testing has been avoided in recent years due to the carcinogenic properties of DNCB and related substances.

III. ENUMERATION OF T AND B LYMPHOCYTES AND THEIR SUBPOPULATIONS

A. Serological Definition of T-Lymphocyte Subpopulations

1. The first successful attempts to define T-lymphocyte subpopulations based on antigenic differences were made using sera obtained from patients with systemic lupus erythematosus or juvenile rheumatoid arthritis containing autoantibodies recognizing a variety of leukocyte antigens. However, those antisera presented problems of multispecificity and were not easily available.
2. The development of **murine hybridomas** producing **monoclonal antibodies** to different leukocyte markers resulted in the identification of about 160 different cell-membrane-associated antigens. Many of those are expressed by lymphocytes, and a select few are known to identify subpopulations with different functions or on different stages of activation (see Chap. 10).

B. Flow Cytometry. A variety of T-cell subsets can be identified and enumerated by immunofluorescence procedures in which fluorescent-labeled monoclonal antibodies are used to detect the different T-lymphocyte antigens (Table 16.1). The enumeration of lymphocytes carrying a given specific marker that is recognized by a corresponding monoclonal antibody can be done manually by immunofluorescence microscopy, or, in a more automated and reliable fashion, using **flow cytometry**.

Table 16.1 Major T Lymphocyte Markers Recognized by Monoclonal Antibodies

CD marker	Cell distribution	Functional association(s)
CD2	All T lymphocytes	T–B lymphocyte interaction
CD3	All T lymphocytes	Transducing unit for TcR
CD4	Helper T cells, monocytes	Interaction with MHC-II
CD5	All T cells; activated autoreactive B cells	B-cell autoreactivity
CD8	Cytotoxic T lymphocytes; NK cells	Interaction with MHC-I
CD25	Activated T and B lymphocytes; monocytes and macrophages[a]	Low-affinity IL-2 receptor (β-chain)
CD26	Activated T lymphocytes	Dipeptidylpeptidase IV
CD45RA	T and B lymphocytes; monocytes	"Naive" CD4+ lymphocytes
CD45RO	T lymphocyte subset; granulocytes; monocytes	Activated helper/memory CD4+ lymphocytes
CD71	Activated lymphocytes; macrophages	Transferrin receptor

[a]The expression is maximal on mitogenically stimulated T lymphocytes.

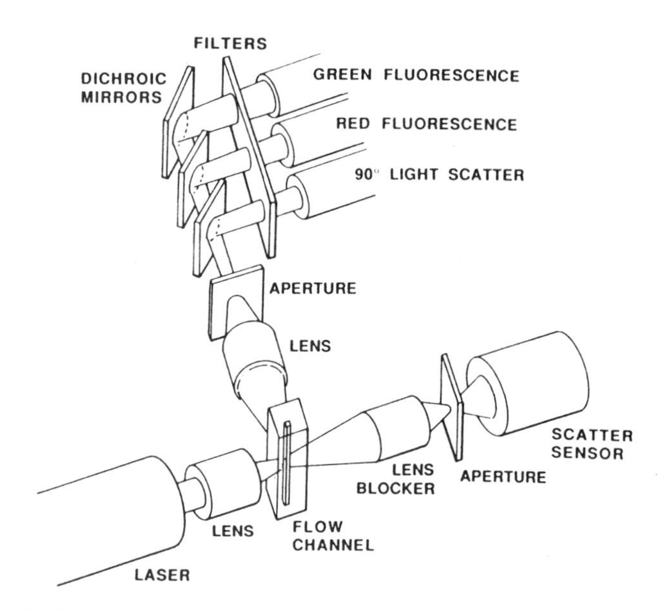

Figure 16.1 Diagrammatic representation of the principle of flow cytometry. A cell suspension is premixed with one or two different monoclonal antibodies to cell surface markers, and as it flows on an optic channel, several parameters are analyzed: light scattered (forward and at a 90° angle) and emission of fluorescence light at two different wavelengths. The light scattering data are processed by a microcomputer and used to discriminate different cell populations according to size. The ability to analyze fluorescence at two or more different wavelengths enables the simultaneous analysis with two or more different antibodies, providing each one is labeled with a different fluorochrome emitting fluorescence at different wavelengths. The simultaneous consideration of fluorescence data and size data enables the discrimination of cell populations by size and presence or absence of markers (as shown in Fig. 16.2).

1. The principle of flow cytometry, as illustrated in Figure 16.1, is relatively simple.
    a. A blood sample is incubated with one or several fluorescent-labeled monoclonal antibodies (if more than one antibody is used, each one is labeled with a different fluorochrome).
    b. Lymphocytes at high dilution flow into a cell as a unicellular stream, and are analyzed through light scattering and fluorescence.
        i. Light-scattering measurements allow sizing of the cells and determination of their granularity.
        ii. Fluorescence measurements allow the determination of the number of cells expressing markers recognized by specific, fluorescent-labeled, monoclonal antibodies.
        iii. By means of computer-assisted analysis, cell populations of homogeneous size can be segregated from cross-reactive cell types that may express identical markers but be of a totally different nature, and analyzed for the expression of one or several markers.
        iv. Using monoclonal antibodies labeled with different fluorochromes and flow cytometers able to measure fluorescence at different

wavelengths, it is possible to analyze cells for the simultaneous expression of two to five different markers (Fig. 16.2).

2. The study of T-lymphocyte subsets has found its main clinical application in the characterization of T-lymphocyte development abnormalities (such as in primary immune deficiencies), acquired T-lymphocyte subpopulation deficiencies (such as in AIDS), abnormalities in the quantitative distribution of T-lymphocyte subpopulations (as in certain autoimmune diseases), and expression of markers not usually detected in normal peripheral blood lymphocytes (as in many types of leukemia).
3. Flow cytometry can also be used to detect cytoplasmic proteins. A common example is the determination of cells containing **terminal deoxynucleotidyl transferase (Tdt)**, a nuclear enzyme present in immature blood cells. It catalyzes the polymerization of deoxynucleotide triphosphates in the

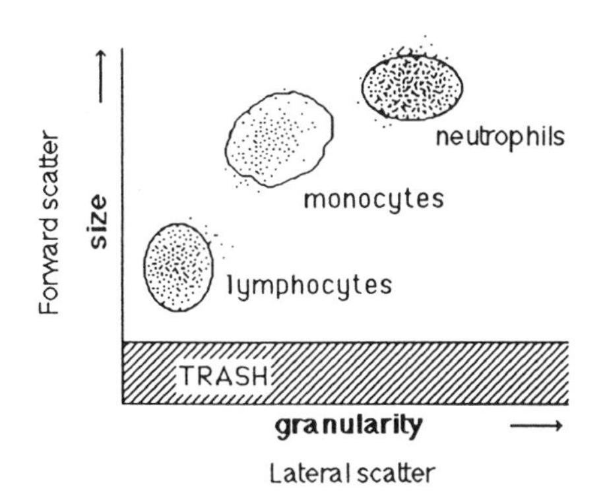

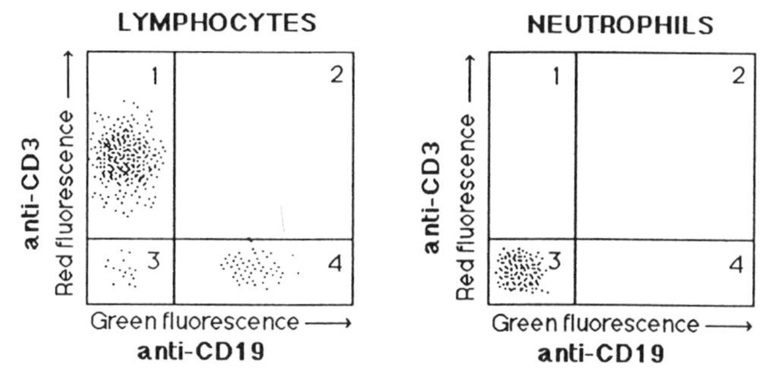

Figure 16.2 Diagrammatic representation of the data processing sequence in flow cytometry. The top drawing illustrates a plot of forward versus lateral light scattering of peripheral blood mononuclear cells. Three leukocyte populations can be segregated by their size and granularity: neutrophils (the largest and most granular), monocytes, and lymphocytes (the smallest and less granular). The lines surrounding these populations correspond to "gating" (i.e., instructing the computer to process separated data from these three populations). The two lower graphs represent analysis of fluorescence for two of the leukocyte populations. The lymphocytes can be separated into two different subpopulations: Those reacting with rhodamine-labeled anti-CD3 antibodies, which emit red fluorescence and correspond to T lymphocytes, and those that treat with fluorescein-labeled anti-CD19, which emits green fluorescence and correspond to B lymphocytes. The analysis of neutrophils shows lack of reactivity with both antisera.

absence of a template. Its presence, usually detected by immunofluorescence, is characteristic of immature T and B lymphocytes, and of the differentiated lymphocytes in leukemias and lymphomas. Its interest, therefore, is mostly related to the investigation of lymphocyte ontogeny and to the classification of lymphocytic malignancies.

C. **Enumeration of B Cells**

1. **Assay for surface immunoglobulins.** The first technique to identify and enumerate B lymphocytes was based on the fact that these cells have immunoglobulin (Ig) molecules attached to their membranes. The assay is essentially based on the incubation of isolated mononuclear cells with fluorescein-labeled anti-immunoglobulin antibodies [aggregate-free or, preferably, $F(ab')_2$ fragments]. The actual counting of B lymphocytes can be done manually, using a fluorescence microscope, or by flow cytometry, as described earlier in this chapter.
2. **B-lymphocyte membrane markers.** Once the production of monoclonal antibodies became easily accessible, antibodies identifying B-cell markers were soon introduced.
   a. A variety of antibodies react with either all B lymphocytes or with B-cell subpopulations. Some of the most commonly used monoclonal antibodies reacting with B lymphocytes are listed in Table 16.2. Of those, **CD19** and **CD20** are used to enumerate B cells by cytofluorography (see Table 16.2). The fact that CD20 is also expressed by dendritic cells is inconsequential when phenotyping peripheral blood B lymphocytes.
   b. Circulating, mature B cells express MHC class II antigens. Monoclonal antibodies to the constant regions of these antigens are available, but cannot be used to enumerate B lymphocytes because MHC-II molecules are also expressed by monocytes and activated T lymphocytes in the peripheral blood.
   c. It should be noticed that CD19, CD20, and MHC-II are not expressed by plasma cells; these cells express specific antigens, such as PCA-1.

D. **Enumeration of NK Cells.** Phenotyping of NK cells with monoclonal antibodies (Table 16.3) is the most popular parameter to assess this cell subpopula-

Table 16.2 Major B-Lymphocyte Markers Recognized by Monoclonal Antibodies

CD marker	Cell distribution	Functional association(s)
CD5	All T cells; activated autoreactive B cells	B-cell autoreactivity
CD10 (CALLA)	Pre-B cells, granulocytes	B-cell leukemia marker
CD19	All B cells	B-cell activation
CD20	All B cells; dendritic cells	
CD21	All B cells; dendritic cells	CR2; Epstein-Barr virus receptor
CD22	All B cells	Interaction with CD45RO on T cells
CD23	Activated B cells, eosinophils, macrophages	FcεRII (low affinity)
CD25	Activated T and B cells	Low-affinity IL-2 receptor (β chain)

Table 16.3 Major NK Cell Markers Recognized by Monoclonal Antibodies

CD marker	Cell distribution	Functional association(s)
CD7	All T cells; NK cells	Unknown
CD8	T cytotoxic/suppressor cells; NK cells	Interaction with MHC-I
CD16	Neutrophils, monocytes, NK cells	FcγRIII (low affinity)
CD56 (N-CAM)	NK cells, neural cells	Intercellular adhesion molecule
CD57	NK cells	Unknown

tion in a clinical context. However, the precise phenotype associated with a fully activated NK cell has not been established and the enumeration of the resting NK cell population may not be very informative.

E. **Normal Distribution of Lymphocyte Subpopulations.** The normal distribution of lymphocyte subpopulations in a normal adult is shown in Table 16.4.

F. **Isolation Techniques Based on the Use of Monoclonal Antibodies**

1. **Cell sorting.** Special cell sorters have been developed which take advantage of changes in lymphocyte properties induced by the binding of monoclonal antibodies to separate those lymphocytes which express the membrane marker recognized by the monoclonal antibody from the remaining lymphocytes. This approach has limitations, including time consumption, cost, sharing of markers by different cell populations, and relatively low yields.
2. **Magnetic sorting.** An alternative technique for separation uses magnetic beads to which a monoclonal antibody is chemically attached. After batch incubation of mononuclear cells with the beads, the tube containing the mixture is placed in a magnetic field. Those cells with the marker recognized by the monoclonal antibody attached to the beads stick to the wall of the tube. Cells not recognized by the monoclonal antibody remain in suspension and can be aspirated and further fractionated or discarded. To separate the cells bound to the magnetic beads, the tubes are carefully rinsed while under the influence of the magnetic field. When the tubes are moved away from the magnetic field, the beads and attached cells sediment. If the sedimented cells are then incubated at 37°C, the cells dissociate from the beads and can be recovered from the supernatant.

Table 16.4 Distribution of the Major Human Lymphocyte Subpopulations in Peripheral Blood, as Determined by Flow Cytometry[a]

CD marker	Lymphocyte subpopulation	Normal range (%)	Normal range (absolute count)
CD19, CD20	B lymphocyes	4–20	96–421 cells/μl
CD3	T lymphocytes	62–85	700–2500 cells/μl
CD2	T lymphocytes; NK cells	70–88	840–2800 cells/μl
CD4	Helper T cells	34–59	430–1600 cells/μl
CD8	Suppressor/cytotoxic T cells	16–38	280–1100 cells/μl

[a]Values obtained at the Flow Cytometry Laboratory, Department of Pathology and Laboratory Medicine, Medical University of South Carolina, Charleston, South Carolina.

III. FUNCTIONAL ASSAYS

There are limitations to the interpretation of numerical data, given the very loose correlation between membrane markers and biological function. However, numerical data are simpler and cheaper to obtain than functional data, which usually require cell isolation and which are obtained in conditions that are anything but physiological.

A. **Isolation of Mononuclear Cells.** Highly enriched preparations of mononuclear cells (T and B lymphocytes plus monocytes and NK cells) are used in most lymphocyte functional assays.

1. Mononuclear cells are separated by density gradient centrifugation, usually in Ficoll–Hypaque. This separation medium has a specific gravity of 1.077, which lies between the density of human erythrocytes (1.092) and the density of human lymphocytes (1.070). By carefully centrifuging blood in Ficoll–Hypaque, a gradient is formed with erythrocytes and PMN leukocytes sedimented at the bottom of the tube, a thin layer containing lymphocytes and monocytes appears in the interface between Ficoll–Hypaque and plasma, and platelet-rich plasma fills the rest of the tube from the mononuclear cell layer to the top.
2. Approximately 80% of the cells recovered in the mononuclear cell layer are lymphocytes and 20% are monocytes. Of the lymphocytes, approximately 80% are T cells, 4–10% are B cells, and the remaining are NK cells and other non-T, non-B lymphocytes.
3. The concentration of lymphocytes is usually adjusted to 1×10^6 lymphocytes/mL, and a stimulating compound (usually a mitogen or an antigen) is then added to the culture, usually in two or three different concentrations. Upon stimulation, T lymphocytes are activated and eventually undergo differentiation. Several endpoints are used to detect lymphocyte activation.
   a. **IL-2 secretion and expression of IL-2 receptors** can be detected 24 hours after mitogenic stimulation. The expression of IL-2 receptors can be measured by flow cytometry, using fluorescent-labeled anti-CD25 (TAC) antibodies, and the release of IL-2, can be measured by enzymoimmunoassay (see later in this chapter).
   b. **Blastogenic transformation.** After 3 days, large numbers of lymphocytes are dividing and the microscopic examination of the culture shows large numbers of lymphoblasts. Counting the proportion of lymphoblasts in the culture gives an indication of the magnitude of the lymphocyte response, but this method is inaccurate and gives inconsistent results.
   c. **Incorporation of tritiated thymidine** into dividing cells is the most common endpoint used to measure lymphocyte proliferation. In the stage of blastogenic transformation there is intense lymphocyte proliferation with active DNA synthesis. Tritiated thymidine [^{3}H-Tdr] is added to the culture after 72 hours of incubation with the mitogen, and the dividing cells remain exposed to ^{3}H-Tdr for 6 to 8 hours. The lymphocytes are then harvested, washed, and the amount of radioactivity incorporated into DNA by the dividing cells is determined with a scintillation counter. A stimulation index (S.I.) can be calculated as follows:

$$\text{S.I.} = \frac{\text{cpm in mitogen-stimulated lymphocytes}}{\text{cpm in unstimulated control lymphocytes}}$$

B. Mitogenic Stimulation Assays. Human lymphocytes can be stimulated in vitro by specific antigens or by mitogenic substances. Although testing the response to specific antigens should be the preferred approach to the study of lymphocyte function, the likelihood of success in such studies is limited by the fact that very few T cells (and even fewer B cells) in the peripheral blood will carry specific receptors for any antigen, even if the individual has already developed a memory response to that particular antigen. In contrast, mitogenic responses are easier to elicit, because mitogenic substances are able to stimulate nonspecifically large numbers of peripheral blood lymphocytes, and therefore lymphocyte proliferation becomes much easier to detect. Some of the most commonly used lymphocyte mitogens are listed in Table 16.5.

1. The first widely used mitogens were mainly **plant glycoproteins (lectins)**, such as **phytohemagglutinin (PHA)**, **concanavalin A (ConA)**, and **pokeweed mitogen (PWM)**. PWM stimulates both B cells and T cells, while PHA and ConA stimulate T cells only.
2. **Mitogens of bacterial origin** include two proteins from *Staphylococcus aureus*, protein A and enterotoxin A, and an enteric pathogen, *Salmonella paratyphi B*.
   a. **Protein A** is only active when presented either on the membrane of killed *Staphylococci*, or as polymers of the isolated protein. Since this protein has specific binding activity for the Fc region of human IgG, it is possible that its mitogenic effect may be mediated by the cross-linking of membrane immunoglobulins. Protein A is able to stimulate B cells in T-cell-depleted lymphocyte preparations but B-cell differentiation is only observed when protein A is used to stimulate unfractionated mononuclear cells so that T-cell help can be elicited.
   b. **Salmonella paratyphi B (SPB)** is not a true mitogen, since this bacteria binds to an unknown receptor on B cells and promotes their differentiation into antibody-producing cells with minimal, if any, cell proliferation.
   c. **Staphylococcal enterotoxin A (SEA)** is a unique T lymphocyte mitogen, included in the group of substances known as "superantigens" (see Chap. 11). These substances bind directly to the nonpolymorphic

Table 16.5 Lymphocyte Mitogens

Mitogen	T lymphocyte	B lymphocyte
Phytohemagglutinin (PHA)	+	–
Concanavalin A (ConA)	+	–
S. aureus enterotoxin A	+	–
Anti-CD3 monoclonal antibody	+	–
Pokeweed mitogen (PWM)	+	–[a]
S. aureus protein A	±[b]	+
Salmonella paratyphi B (SPB)	±	+

[a]PWM is mainly a T-cell mitogen, inducing also B-cell proliferation and differentiation through the release of soluble factors by T cells.
[b]T-cell cooperation is not required to induce B-lymphocyte division, but it is required for the functional activation of B lymphocytes into Ig-secreting cells.

area of class II MHC on APC and to specific alleles of the Vβ chain of the TcR. The cross-linking of T-cell receptors caused by the interaction with SEA induces T-cell proliferation. As many as 20% of T lymphocytes in peripheral blood respond to this type of stimulation with superantigens.

   d. **Toxic shock syndrome toxin** (**TSST**) is another staphylococcal superantigen; an energetic lymphocyte proliferative response induced with this toxin indicates that the individual who donated the lymphocytes expresses one or several of the specific Vβ regions to which TSST binds. Therefore, this test could indicate which patients are susceptible to the development of the toxic shock syndrome.

3. Immobilized **anti-CD3 monoclonal antibodies** cross-link multiple TcR complexes on the T-lymphocyte membrane and deliver a mitogenic signal to CD3+ T lymphocytes. The immobilization of the monoclonal antibody can be easily achieved by spontaneous adsorption to the walls of the microculture plates used in proliferation assays, or by binding to monocytes present in the mononuclear cell culture through Fc receptors. The use of anti-CD3 as a mitogen has the advantage of probing the function of the transducing component of the T-cell receptor.
4. **Advantages and limitations of mitogenic assays.** Mitogenic responses are relatively simple to study, but the assays have a variety of problems, such as poor reproducibility and individual variations among normals. In addition, these assays only measure the proliferative capacity of lymphocytes, and are not very informative about their functional activity. Nevertheless, the finding of a very low mitogenic index after stimulation with a T-lymphocyte mitogen suggests a deficiency of T-cell function.

C. **Response to Antigenic Stimulation.** The study of the response of lymphocytes to antigenic stimulation in vitro is functionally more relevant than the study of mitogenic responses. However, even in the best possible circumstances (i.e., when the antigen can be recognized by T lymphocytes, which predominate in peripheral blood, and the lymphocyte donor has developed memory to the antigen in question), the proportion of cells responding to stimulation is not likely to exceed 0.1%, and the proportion of responding B lymphocytes is even lower.

1. To maximize the probability of obtaining a measurable response, the lymphocytes are stimulated with antigens to which the lymphocyte donor has been previously exposed, and the cultures are incubated with the antigen for 5 to 7 days prior to addition of tritiated thymidine.
2. The elicitation of B-lymphocyte responses in vitro is considerably more difficult. Most studies in which positive results have been reported have used heterologous red cells or tetanus toxoids as antigens. The in vitro response to tetanus toxoid is easier to elicit using peripheral blood mononuclear cells separated from donors who had received a booster 1 to 3 weeks earlier. Usually, the incubation periods in studies of B-cell activation have to be increased even further, up to 9 to 11 days.
3. **Advantages and limitation of antigenic stimulation assays**
   a. As noted, the main advantage is biological relevance. The assays test the ability to respond to a specific antigen, and the end point may be as

specific as the synthesis of antibody directed against the stimulating antigen.

b. Several factors limit the clinical usefulness of antigenic stimulation studies, besides the already noted problem of a low number of responding cells.

   i. The number of antigens available for these studies is very limited, usually including Purified Protein Derivative (PPD), *Candida albicans* antigens, and tetanus toxoid (which stimulates both T and B cells), and none of them has been properly standardized for this type of in vitro study.

   ii. Second, the end points for T-cell activation (thymidine incorporation, IL-2 release, expression of IL-2 receptors) are not antigen specific and there is always the possibility that what one measures is a nonspecific mitogenic response induced by some of these antigens or by contaminants that may be present in the antigen preparations.

D. **Cytokine Assays.** Since one of the most biologically significant consequences of T-lymphocyte activation is the release of cytokines, it is not surprising that immunologists have used cytokine measurements as indices of T-cell activation.

1. **Migration inhibitory factor (MIF)** is a proinflammatory cytokine that is released by T cells, monocytes, and macrophages, and one of its main effects is to retain monocytes and macrophages around the area where it is being released. Its measurement usually implies comparing the area of diffusion of stimulated versus nonstimulated mononuclear cells from a capillary tube. The release of MIF keeps the cells from diffusing away from the tip of the capillary tube. The MIF assays are difficult to standardize and reproduce.

2. **Cytokine assay in the supernatants of stimulated lymphocytes.** The identification of cytokines was quickly followed by the development of monoclonal antibodies, which became the basis for enzymoimmunoassays which have rapidly replaced most functional assays. The availability of these assays has provided a physiological end point for studies of T-lymphocyte activation.

   a. **IL-2 assay.** The assay of IL-2 by EIA is probably the method of choice for the evaluation of the functional response of helper T cells. The most common approach to this assay consists of incubating mononuclear cells with several concentrations of mitogenic substances for 24 hours, and measuring IL-2 concentrations in the supernatants. Low or absent release of IL-2 has been observed in a variety of immunodeficiency states, particularly in patients with AIDS. It is also a good parameter to follow longitudinally when the effects of a drug or substance over the immune system are being studied, as illustrated in Figure 16.3.

   b. **Enzymoimmunoassays for IL-4, IL-5, IL-6, IL-10, IL-12, GM-CSF, TNFα and β, and interferon-γ** are also available, and their judicious use allows a more complete picture of the functional response of T lymphocytes to be obtained. For example, predominant release of IL-4, IL-5, and IL-10 is characteristic of TH2 responses, while predominant release of IL-2, GM-CSF, and interferon-γ is characteristic of TH1 responses.

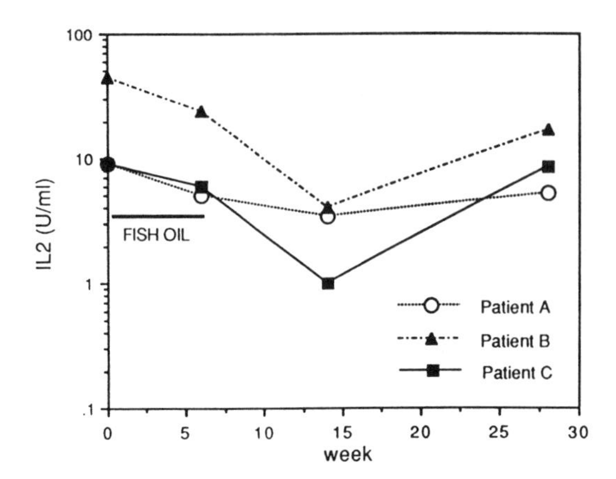

Figure 16.3 Longitudinal study of the release of IL-2 by peripheral blood lymphocytes stimulated with pokeweed mitogen in three volunteers who ingested 8 g/day of a fish oil extract for 6 weeks (as indicated by the horizontal bar in the figure). The release of IL-2 was reduced in the three volunteers at the end of the 6 weeks of dietary supplementation with fish oil, but the reduction became more accentuated at week 15. By week 30, IL-2 release was back to normal in patient C, but still depressed in patients A and B. (Based on results published by Virella, G., Fourspring, K., Hyman, B., Haskill-Stroud, R., Long, L., Virella, I., La Via, M., Gross, A.J., and Lopes-Virella, M. *Clin. Immunol. Immunopathol. 61*:161, 1991.)

c. **Cytokine mRNA assays** are gaining popularity particularly in the analysis of TH1 versus TH2 responses. There are several techniques available, all requiring suitable cDNA probes, which can be isotopically or nonisotopically labeled. Some techniques are based on hybridization of PCR-amplified mRNA obtained from nonstimulated T-lymphocyte clones, others are based on in situ hybridization performed on slides of stimulated cells. The current techniques are semiquantitative at best, but progress in this area has been very rapid.

3. **Assay of cytokines and cytokine receptors in serum and urine**
   a. **Serum IL-2 levels** have been measured as a way to evaluate the state of T-lymphocyte activation in vivo. Increased levels of circulating IL-2 have been reported in multiple sclerosis, rheumatoid arthritis, and patients undergoing graft rejection, situations in which T-cell hyperactivity would fit with the clinical picture. However, a significant problem with these assays is the existence of a serum factor that interferes with the assay of IL-2 by EIA, and the results are often very imprecise.
   b. **Urinary levels of IL-2** can also be measured by enzymoimmunoassay and are not affected by inhibitory substances. Increased urinary levels of IL-2 have been reported in association with kidney allograft rejection and proposed as a parameter that may help differentiate acute rejection from cyclosporine A toxicity (see Chaps. 26 and 27).
   c. **Serum levels of IL-2 receptor.** Activated T cells shed many of their membrane receptors, including the IL-2 receptor. Elevated levels of circulating soluble receptors (shed by activated T lymphocytes) exist in patients with hairy cell leukemia, AIDS, rheumatoid arthritis, graft

rejection, etc. In general, the results of assays for the IL-2 receptor show parallelism with the results of assays for IL-2.

d. **Serum levels of IL-6.** The measurement of serum IL-6 correlates with B-lymphocyte activity. High levels of IL-6 have been detected in patients with AIDS and systemic lupus erythematosus. Both types of patients have increased levels of circulating immunoglobulins, reflecting hyperactivity of the B-cell system.

4. **Limitations of serum and urine cytokine assays.** These assays have not proven to be as useful as expected when first introduced. A major limiting factor is their lack of specificity, which makes their correlation with specific clinical conditions rather difficult. In addition, the high cost of reagents is a significant limiting factor that has prevented their widespread use and proper evaluation of their usefulness.

E. **Assays for Cytotoxic Effector Cells.** The functional evaluation of cytotoxic effector cells, which include cytotoxic T lymphocytes, natural killer cells (NK-cells), and cells mediating antibody-dependent cell cytotoxicity (ADCC) reactions is based on cytotoxicity assays.

1. The cellular targets for cytotoxic cells vary according to cytotoxic cell population to be evaluated.
   a. The evaluation of T-cell-mediated cytotoxicity requires mixing sensitized cytotoxic T cells with targets expressing the sensitizing antigen.
   b. NK cell activity is usually measured with transformed cell lines known to be susceptible to NK cell killing.
   c. ADCC is measured using antibody-coated target cells.
2. The assessment of cell death is usually done by one of two techniques:
   a. **Counting of dead cells.** Dead target cells are differentiated from live target cells by the uptake of vital dyes, such as trypan blue. This technique, however, is mostly used for the study of antibody-mediated cytotoxicity.
   b. **Release of radiolabeled chromium (^{51}Cr)** from its previously labeled target cells is preferred when evaluating T-lymphocyte or NK cell cytotoxity.
3. **Advantages and limitations of cytotoxicity assays.** The functional interest of cytotoxicity assays is evident.
   a. In the case of T cells, the cytotoxicity assays (see below) measure the functional adequacy of one of the major effector T-cell subpopulations. It needs to be stressed, however, that an abnormal result in a cytotoxicity assay does not necessarily mean that the cytotoxic population is abnormal; it could also reflect, for example, a defect in the helper T-cell population, which would prevent the proper differentiation of effector cells.
   b. As for NK and ADCC effector cells, their definition is basically dependent on cytotoxicity assays. Questions remain, however, about the physiological significance of ADCC and NK cells, since there are no well-defined cases of immunodeficiency disease that can be explained on the basis of a deficiency of these cell subpopulations.

F. **Mixed-Lymphocyte Reaction (MLR).** One of the best ways to study the function of the T-lymphocyte system in vitro is the mixed-lymphocyte reaction

(MLR). However, this is a difficult and expensive procedure, which is not often performed for the diagnostic evaluation of patients with suspected CMI deficiencies.

1. The basis of the MLR is the recognition of antigenic differences mostly related to the expression of class II MHC antigens on the membrane of mononuclear cells.
2. For many years, **one-way MLR reactions** were used to type HLA-D locus specificities, which were the first class II HLA specificities described. This was usually accomplished by mixing mitomycin-treated mononuclear cells from a donor of known HLA-D specificity with untreated mononuclear cells of an untyped individual. The mitomycin-treated cells (particularly the B-cell and monocyte populations that express MHC-II antigens) acted as stimulator cells, while the untreated cells were the responder cells (Fig. 16.4). A response in this system, as measured by ^{3}H-Tdr incorporation, was considered as indicating lack of identity of the D locus. This cumbersome approach has been made obsolete by the development of antibodies and DNA probes to type MHC-II specificities (see Chap. 3).

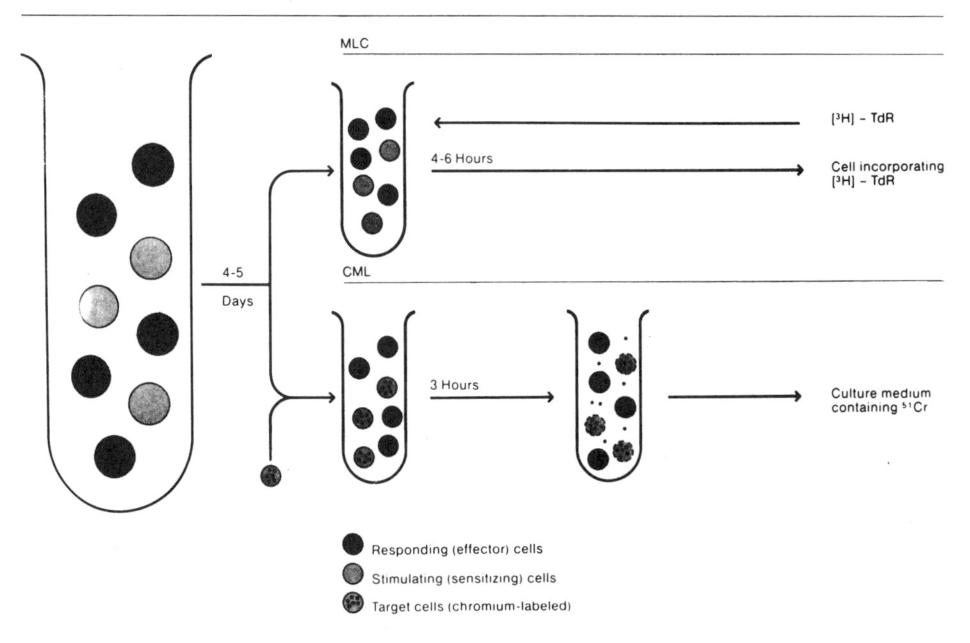

Figure 16.4 (top) Diagrammatic representation of a one-way mixed-lymphocyte culture reaction. A mixture of viable (responder) lymphocytes and heterlogous mitomycin-treated lymphocytes (stimulator cells) is incubated, and, as a result, the viable T lymphocytes will proliferate. At the end of 4–5 days the proliferating lymphocytes will incorporate measurable amounts of tritiated thymidine (^{3}H-Tdr). (bottom) Diagrammatic representation of a cytotoxicity assay. The initial steps are identical to those of a mixed lymphocyte culture, but, after 4–5 days of incubation, viable target cells labeled with chromium (^{51}Cr) are added to the culture. The ^{51}Cr-labeled target cells are genetically identical to the stimulator cells. If the culture contains differentiated cytotoxic T lymphocytes, those will lyse the target cells and cause the release of ^{51}Cr. (Reproduced with permission from R.D. Guthmann, ed., *Immunology—A Scope Publication*, Upjohn Co., Kalamazoo, MI, 1981.)

3. The one-way MLR can be used to evaluate T-cell function by using mitomycin-treated mononuclear cells from a genetically unrelated donor as stimulators of a patient's lymphocytes. As an endpoint, one can use ^{3}H-Tdr incorporation, or the generation of cytotoxic T cells. For this last assay, a mixed culture of inactivated stimulator cells and responder T cells is set. After 5 days, ^{51}Cr-labeled viable "target" lymphocytes, obtained from the same individual that provided the cells used as stimulators, are added to the culture. If cytotoxic T cells were generated during the previous 5 days of culture, the viable cells added in the second step will become their targets and will be killed in a few hours, releasing significant amounts of ^{51}Cr (Fig. 16.4)

IV. GUIDELINES FOR THE EVALUATION OF A PATIENT WITH A SUSPECTED CELL-MEDIATED IMMUNITY DEFECT

A. **Clinical Clues.** A defect in cell-mediated immunity is suggested by an abnormal frequency of infections with agents that are known to be more efficiently controlled by CMI mechanisms than by circulating antibodies. These include viruses, particularly those that propagate from cell to cell, such as the viruses of the herpes family, intracellular bacteria, such as *Mycobacterium tuberculosis*, parasites such as *Toxoplasma*, and fungi (particularly *Candida albicans*, *Pneumocystis carinii*, *Coccidioides immitis*, and *Histoplasma capsulatum*).

B. **Lymphocyte Counts.** Since T cells represent 75–80% of the peripheral blood lymphocyte population, **a significant decrease of T cells will always be reflected by lymphopenia in the peripheral blood**; hence, a simple leukocyte count with differential helps in detecting a T-lymphocyte deficiency. If the deficiency is only functional, it may not be apparent or may be suggested by abnormal lymphocyte morphology.

C. **Skin tests.** The performance of **skin tests** using preferentially an antigen to which the patient was exposed, either as a result of infection or as a result of active immunization, can be the initial step in the evaluation of a patient with suspected CMI deficiency. An alternative approach is to skin test with a battery of common antigens, and it is expected that over 95% of normal individuals will react to at least 3 out of 5 such antigens.

D. **The Enumeration of T Cells and Their Subpopulations** can either be the starting point or one of the earlier steps in the investigation of a suspected CMI deficiency. The methods are standard and the results quantitative, but there are several problems with this approach:

   1. The counts of critical subpopulations, such as the CD4+ helper T cells, are known to be very variable, and several determinations may be needed before a conclusion is reached.
   2. There is always some uncertainty about the perfect correlation between the cell markers that allow the enumeration of T cells and their subpopulations and the functional capabilities of the cells carrying those markers. It is clear that a marked decrease of CD4+ (helper) T cells is generally associated with clinical immunodeficiency, but examples of dissociation between cell markers and function have been reported. Therefore, the significance of

moderate deviations from normality in the distribution of T-cell subpopulations may not be paralleled by the expected functional abnormalities.

E. **The Evaluation of T-Cell Function** by in vitro assays should be the hallmark of the investigation of T-cell deficiencies, since many believe the in vitro tests to be more reproducible than skin tests. However, these tests also have problems.
   1. Their performance requires specialized laboratories, often use costly reagents, require relatively large volumes of blood, and the results take several days to be obtained.
   2. The easier in vitro ass ays to perform are the studies of the response to T-cell mitogens (PHA, ConA, anti-CD3 monoclonal antibodies), but their physiological relevance is open to question, particularly when plant mitogens are exclusively used.
   3. Assays measuring the generation of cytotoxic T cells or the production of cytokines tend to be considered the best available choices for the functional evaluation of T lymphocytes, including the assessment of THl versus TH2 functional predominance. Unfortunately, those assays are expensive and complex and have met with limited clinical application.

F. **In Vitro Assessment of Helper and Suppressor Cells.** The functional evaluation of defective helper function or excessive suppressor function, as discussed in Chapter 15, is one of the most complex tests to perform. These assays require the separation of T and B lymphocytes from a patient and a normal control and their co-culture in different combinations to study the effect of normal cells on patient cells and vice-versa. For these reasons, the tests are not routinely used to evaluate patients with suspected CMI deficiencies.

SELF-EVALUATION

Questions

Choose the ONE *best* answer.

16.1 Which one of the following tests would you consider as the LEAST adequate for the characterization of T-lymphocyte abnormalities?
   A. Enumeration of CD3+ cells
   B. Enumeration of CD4+ cells
   C. Mitogenic response to *S. aureus* protein A
   D. One-way mixed-lymphocyte reaction
   E. Assay of cytokines released after mitogenic stimulation

16.2 A patient with chronic pulmonary tuberculosis is found to have negative skin tests to coccidioidin, candidin, and tetanus toxoid. On the basis of this observation, you can state that the patient:
   A. Has a generalized cellular immune deficiency
   B. Is in a state of anergy associated with chronic active tuberculosis
   C. Is unable to respond to polysaccharides
   D. Needs to be vaccinated against tetanus
   E. Needs further evaluation of his(her) cell-mediated immunity

16.3 Which of the following tests gives more significant information concerning T-lymphocyte function?
   A. Number of CD4/CD25 (IL-2 receptor)$^+$ cells

B. Number of $CD3^+$ cells
C. Release of IL-2 after mitogenic stimulation
D. Tritiated thymidine incorporation by anti-CD3-stimulated mononuclear cells.
E. Tritiated thymidine incorporation by PHA-stimulated mononuclear cells.

16.4 The need to use mitomycin-treated, unfractionated mononuclear cells as stimulators in a one-way mixed-lymphocyte reaction results from the:
A. Inactivation of suppressor cells by mitomycin
B. Key stimulating role played by $MHC\text{-}II^+$ cells able to interact with viable CD4+ T lymphocytes
C. Mitogenic properties of mitomycin
D. Need to expand the population of stimulator lymphocytes
E. Requirement for phagocytic cells to process and present non-self MHC molecules

16.5 Which of the following monoclonal antibodies should be coupled to magnetic beads to remove activated T cells from a PHA-stimulated mononuclear cell preparation?
A. CD3
B. CD4
C. CD5
D. CD19
E. CD25

16.6 Which of the following manipulations is likely to inhibit a mixed-lymphocyte reaction between lymphocytes of two genetically unrelated individuals?
A. Adding anti-CD8 antibodies to the culture
B. Adding anti-MHC-I antibodies to the culture
C. Eliminating all MHC-II positive cells
D. Eliminating CD25+ cells prior to the culture
E. Treating one set of lymphocytes with mitomycin

16.7 NK cells are functionally defined by their:
A. Ability to kill target cells incubated with specific antibody
B. CD phenotype
C. Cytotoxic effect on specific transformed cell lines
D. Proliferation in one-way mixed-lymphocyte reactions
E. Relatively large size and granular cytoplasm

16.8 In an infant who suffers from repeated bacterial infections and fails to form specific antibodies after initial immunization with DTP, it is important to:
A. Determine the serum levels of IL-6
B. Enumerate the proportion of CD25+ B lymphocytes
C. Measure the concentrations of circulating immunoglobulins
D. Enumerate CD19+ lymphocytes in the peripheral blood
E. Give another DTP booster and repeat antibody determinations

16.9 The following cell markers were longitudinally determined in a normal lymphocyte culture stimulated with pokeweed mitogen (PWM). Which one of these markers is likely to be undetectable at the onset of the experiment and easily detectable 6 days after?
A. CD2
B. CD4
C. CD19

D. CD20
E. PCA-1

16.10 When the MLC is followed by a cytotoxicity assay, the target cells must be:
A. Labeled with ^{125}I
B. Stained with trypan blue
C. Syngeneic with the responding cells
D. Syngeneic with the stimulating cells
E. Virus-infected

ANSWERS

16.1 (C) The mitogenic response to *S. aureus* protein A involves predominantly B lymphocytes.

16.2 (E) A negative reaction to candidin, coccidioidin, and tetanus toxoid could reflect generalized immunodeficiency or a state of anergy, but could also reflect lack of sensitization of a perfectly immunocompetent individual. Also, anergy is more common in patients with disseminated (miliary) tuberculosis. The most important skin test to be performed in this patient would have been the tuberculin test, which was omitted. It is not possible to claim that this patient does not respond to polysaccharides, since no such antigens were used (tetanus toxoid is a protein). Protection against tetanus toxoid is antibody mediated, and a negative skin test has no meaning as far as the degree of protection. In conclusion, if the purpose of the skin test was to evaluate this patient's cell-mediated immunity status, additional tests need to be considered.

16.3 (C) A CD3 count only gives an indication of the total number of T lymphocytes, but is not informative about their function. The co-expression of CD25 (IL-2 receptor) by CD4+ cells is typical of activated helper T lymphocytes, but does not prove whether the labeled cells are functionally competent. The mitogenic responses to PHA and anti-CD3 only give an indication about the general ability of T cells to proliferate in response of different types of stimulation. IL-2 release is probably the major determinant of the initial expansion of T cells (particularly of the TH1 subpopulation) during an immune response, and therefore, is a better index of the functional status of T lymphocytes than any other of the listed alternatives.

16.4 (B) The induction of the one-way MLR requires MHC-II$^+$ cells (monocytes, B lymphocytes) to stimulate helper T cells, which recognize MHC-II-associated non-self endogenous peptides. The purpose of treating the stimulator cells with mitomycin (a DNA synthesis inhibitor) is to inhibit their ability to proliferate.

16.5 (E) Although CD25 (IL-2 receptor α chain) is expressed by both activated T and B lymphocytes, only T lymphocytes respond to PHA stimulation.

16.6 (C) Cells expressing MHC-II are absolutely essential for the induction of a MLR. Adding MHC-I or anti-CD8 antibodies to the culture could only interfere with the cytotoxicity reaction, but not with the proliferative stage that characterizes the MLR. Few CD25+ are likely to be present prior to stimulation, and their elimination is unlikely to have a measurable effect. Treatment with mitomycin of one set of cells would not prevent the proliferation of the other set.

16.7 (C)

16.8 (D) Although repeating the immunization could be considered, in a child with repeated bacterial infections and apparently impaired humoral immunity, the assay of circulating B cells could rapidly help establish a diagnosis of B lymphocyte deficiency, which could explain the infections and lack of response to immunization.

16.9 (E) PWM stimulation activates the proliferation of both T and B cells. However, the B and T lymphocyte markers listed are expressed both by resting and by activated cells. However, the differentiation of B cells into antibody-producing cells caused by PWM is associated with the loss of B-cell markers (CD19, CD20) and with the expression of plasma cell markers (PCA-1) not detectable in resting B lymphocytes.

16.10 (D) The responding cells in MLR proliferate and differentiate after specifically recognizing heterologous MHC antigens. The effector cells generated in the MLR can only destroy target cells genetically identical (syngeneic) to those used to stimulate them. Trypan blue staining is not used for assessment of target cell death in MLR because of the subjectivity inherent in techniques based on microscopic observation, which are also extremely time consuming and require experienced personnel.

BIBLIOGRAPHY

Barclay, A.N., et al. *The Leukocyte Antigen Facts Book.* Academic Press, Oxford (U.K.), 1993.

Denny, T.N., and Oleske, J.M. Flow cytometry in pediatric immunologic diseases. *Clin. Immunol. Newslett.*, *11*:65, 1991.

Honda, M., Kitamura, K., Mizutani, Y., Oishi, M., Arai, M., Okura, T., Igarahi, K., Yasukawa, K., Hirano, T., Kishimoto, T., Mitsuyasu, R., Cherman, J.-C., and Tokunaga, T. Quantitative analysis of serum IL-6 and its correlation with increased levels of serum IL-2R in HIV-induced diseases. *J. Immunol.*, *145*:4059, 1990.

Linker-Israeli, M., Deans, R.J., Wallace, D.J., Prehn, J., Ozeri-Chen, T., and Klinenberg, J.R. Elevated levels of endogenous IL-6 in systemic lupus erythematosus. A putative role in pathogenesis. *J. Immunol.*, *147*:117, 1991.

Mishell, B.B., and Shiigi, S.M., eds. *Selected Methods in Cellular Immunology.* W.H. Freeman & Co., San Francisco, CA, 1980.

Rose, N.R., de Macario, E.C., Fahey, J.L., Friedman, H., and Penn, G.M., eds., *Manual of Clinical Immunology*, 4th ed. American Society of Microbiology, Washington, DC, 1992.

WHO Scientific Group on Immunodeficiency. Primary immunodeficiency diseases. *Clin. Immunol. Immunopathol.*, *28*:450, 1983.

Zielinski, C.C., Pesau, B., and Müller, C. Soluble interleukin-2 receptor and soluble CD8 antigen in active rheumatoid arthritis. *Clin. Immunol. Immunopathol.*, *57*:74, 1990.

17
Diagnostic Evaluation of Phagocytic Function

Gabriel Virella

I. INTRODUCTION

The failure or success of an antibody response directed against an infectious agent depends entirely on its ability to trigger the complement system and/or to induce phagocytosis. Most mammals, including humans, have developed two well-defined systems of phagocytic cells: the polymorphonuclear leukocyte system (particularly the neutrophil population) and the monocyte/macrophage system. Both types of cells can engulf microorganisms and cause their intracellular death through a variety of enzymatic systems, but the two cell systems differ considerably in their biological characteristics.

II. PHYSIOLOGY OF THE POLYMORPHONUCLEAR LEUKOCYTE

Neutrophils and other polymorphonuclear (PMN) leukocytes are "wandering" cells, constantly circulating around the vascular network, able to recognize foreign matter by a wide variety of immunological and nonimmunological mechanisms. Their main biological characteristics are summarized in Table 17.1. Their effective participation in an anti-infectious response depends on the ability to respond to chemotactic signals, ingest the pathogenic agent, and kill the ingested microbes.

A. Chemotaxis and Migration to the Extravascular Compartment
   1. In normal conditions, the interaction between leukocytes and endothelial cells is rather loose and involves a family of molecules, known as selectins, which are constitutively expressed on endothelial cells and glycoproteins, which are expressed on the leukocyte cell membrane. There interactions cause the slowing down ("rolling") of leukocytes along the vessel wall, but do not lead to firm adhesion of leukocytes to endothelial cells.
   2. Several chemotactic stimuli, which in most cases will be of bacterial origin, but can also be released as a consequence of tissue necrosis, as a result of monocyte and lymphocyte activation, or as a by-product of complement activation, are involved in the recruitment of leukocytes in the extravascular space.

Table 17.1 Comparison of the Characteristics of PMN Leukocytes and Monocytes/Macrophages

Characteristic	PMN leukocytes	Monocyte/macrophage
Numbers in peripheral blood	3 − 6×10³/μL	285–500/μL
Resident forms in tissues	−	+ (macrophage)
Nonimmunological phagocytosis	++	+
Fc receptors	FcγRII,III	FcγRI,II,III
C3b receptors	++	++
Enzymatic granules	++	++
Bactericidal enzymes	++	++
Ability to generate superoxide and H_2O_2	+++	++
Synthesis and release of leukotrienes	+ (B4)	++ (B4, C4, D4)
Synthesis and release of prostaglandins	−	++
Release of PAF	++	+
Release of interleukins	+	++
Response to nonimmunological chemotactic factors	+	−
Response to C5a/C3a	+	−
Response to lymphokines	+ (IL-8)	++ (IFN-γ)
Antigen processing	−	++
Expression of HLA class II antigens	−	++
Phagocytosis-independent enzyme release	++	−

a. Among bacterial products, formyl-methionyl peptides, such as f-methionine-leucine-phenylalanine (f-met-leu-phe), are extremely potent chemotactic agents.
b. Tissue damage may result in the activation of the plasmin system that may, in turn, initiate complement activation with generation of C5a, another extremely potent chemotactic agent.
c. Many microorganisms can probably generate C5a by activation of the complement system through the alternative pathway.
d. After an inflammatory process has been established, proteases released by activated neutrophils and macrophages can also split C5, and the same cells may release leukotriene B4, another potent chemotactic factor, attracting more neutrophils to the site.
e. Chemokines such as **IL-8, monocyte chemotactic protein-1, and RANTES** are also chemotactic for neutrophils and monocytes.

3. After receiving a chemotactic stimulus, leukocytes undergo changes in the cell membrane, which is smooth in the resting cell, and becomes "ruffled" after the cell receives the chemotactic signal. The activated PMN has a marked increase in cell adhesiveness, associated with increased expression of adherence molecules, namely, integrins of the **CD11/CD18 complex**, which includes:
   a. **CD11a** [the α chain of **LFA (leukocyte function antigen)-1**]
   b. **CD11b** (the **C3bi receptor** or **CR3**, also known as **Mac-1**) molecule
   c. **CD11c** (also known as protein **p150,95**)
   d. **CD18** [the β chain of **LFA (leukocyte function antigen)-1**]
4. These cell adhesion molecules (CAM) are common to the majority of

leukocytes, but their individual density and frequency may vary in the two main groups of phagocytic cells.
   a. CD11a and CD18 are expressed virtually by all monocytes and granulocytes
   b. CD11b is more prevalent among granulocytes
   c. CD11c is more frequent among monocytes
5. The expression of these CAM mediates a variety of cell–cell interactions such as those that lead to neutrophil aggregation, and, most importantly, those that mediate firm adhesion of leukocytes to endothelial cells. For example, CD11a (LFA-1) and CD11b interact with molecules of the immunoglobulin gene family, such as ICAM-1, ICAM-2 and VCAM-1, expressed on the endothelial cell membrane.
6. Cytokines released by activated monocytes and lymphocytes, such as IL-1 and TNF-α can up-regulate the expression of VCAM-1 and, to a lesser degree, of ICAM-1 and -2. This further enhances the adhesion of leukocytes to endothelial cells.
7. After adhering to endothelial cells, leukocytes migrate to the extravascular compartment. The transmigration involves interaction with a fourth member of the immunoglobulin gene family—platelet endothelial cell adhesion molecule 1 (PECAM-1)—which is expressed at the intercellular junctions between endothelial cells. The interaction of leukocytes with PECAM-1 mediates the process of diapedesis, by which leukocytes squeeze through the endothelial cell junctions into the extravascular compartment.
8. The diapedesis process involves the locomotor apparatus of the leukocyte, a contractile actin–myosin system stabilized by polymerized microtubules. Its activation is essential for the leukocyte to move to the extravascular space and an intact CD11b protein seems essential for the proper modulation of microtubule assembly, which will not take place in CD11b-deficient patients.

B. **Phagocytosis and Intracellular Killing.** At the area of infection, PMN leukocytes recognize the infectious agents, which are ingested and killed intracellularly. The sequence of events leading to opsonization and intracellular killing is summarized in Figure 17.1.
1. Several recognition systems appear to be involved in the phagocytosis step:
   a. The best defined of which is the reaction with the Fc fragment of opsonizing antibodies. Neutrophils express two types of Fcγ receptors, FcγRII and FcγRIII, both of which are involved in phagocytosis.
   b. The CR1 (C3b) receptor is also expressed by neutrophils, and the binding and ingestion of microorganisms through this receptor has been well established.
   c. Opsonization with both IgG antibodies and C3b seems associated with maximal efficiency in ingestion.
   d. As pointed out in Chapter 13, opsonization is not an absolute requirement for ingestion by neutrophils. Nonimmune recognition systems leading to phagocytosis are believed to be responsible for the ingestion of microorganisms with polysaccharide-rich outer layers. The neutrophil is also able to ingest a variety of particulate matter, such as latex beads, silicone, asbestos fibers, etc., in the absence of opsonizing antibodies or complement.

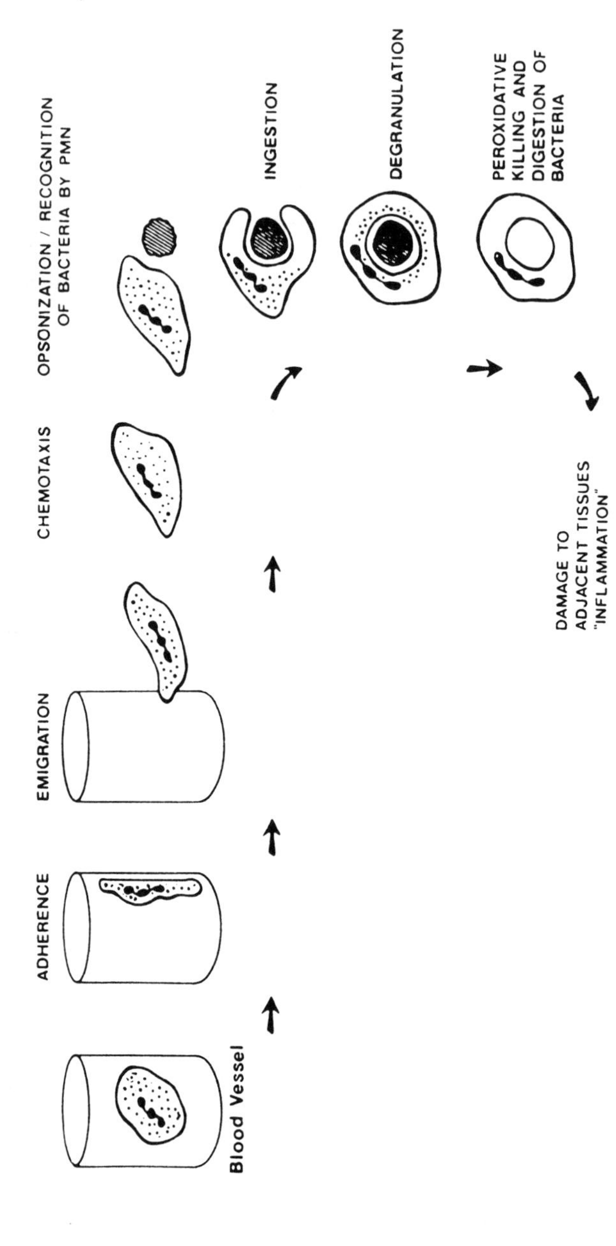

Figure 17.1 Diagrammatic representation of the sequence of events that takes place during PMN leukocyte phagocytosis. (Reproduced with permission from Wolach, B., Baehner, R.L., and Boxer, L.A., *Isr. J. Med. Sci.*, *18*:897, 1982.)

2. Ingestion is achieved through formation of pseudopodia that surround the particle or bacteria, with pseudopodia eventually fusing at the distal pole to form a **phagosome**. The cytoplasmic granules of the neutrophil (lysosomes) then fuse with the phagosomes, and their contents empty inside the phagosomes (degranulation). This degranulation process is very rapid and delivers a variety of antimicrobial substances to the phagosome.
   a. The azurophilic or **primary granules** contain, among other substances, **myeloperoxidase**, **lysozyme**, acid hydrolases (such as **β-glucuronidase**), **cationic proteins**, and neutral proteases (including **collagenase**, **elastase**, and **cathepsin C2**).
   b. The **secondary granules** or lysosomes contain **lysozyme** and **lactoferrin**.
3. Killing of ingested organisms involves the effects of cationic proteins from the primary granules and lysosomal enzymes, such as lysozyme and lactoferrin, as well as of the by-products of the **respiratory burst**, activated as a consequence of phagocytosis.
   a. **Cationic proteins** can bind to negatively charged cell surfaces (such as the bacterial outer membrane) and interfere with growth.
   b. **Lactoferrin** has antimicrobial activity by chelating iron and preventing its use by bacteria that need it as an essential nutrient.
   c. **Lysozyme** splits the β-1,4 linkage between the N-acetylmuramic acid peptide and N-acetylglucosamine on the bacterial peptidoglycan. However, the importance of this enzyme as a primary killing mechanism has been questioned due to the relative inaccessibility of the peptidoglycan layer in many microorganisms, which may be surrounded by capsules or by the lipopolysaccharide-rich outer membrane (Gram-negative bacteria).
   d. From the bactericidal point of view, however, the activation of the superoxide-generating system (**respiratory burst**) appears considerably more significant. This system is activated primarily by opsonization but also by a variety of PMN-activating stimuli, ranging from bacterial peptides, such as f-met-leu-phe, to C5a.
      i. A key enzymatic activity (**NADPH oxidase**) is activated and results in the transfer of a single electron from NADPH to oxygen, generating superoxide (O_2^-).
      ii. NADPH oxidase is a molecular complex located on the cell membrane, constituted by:
         - cytochrome B, which is a heterodimer formed by two polypeptide chains (91 kD and 22 kD, respectively), believed to play the key role in the reduction of oxygen to **superoxide**, possibly by being the terminal electron donor.
         - two cytosolic proteins—p47 and p67—one of which (p47) is a substrate for protein kinase C.

         In a resting cell the complex is inactive and its components are not associated.
      iii. After the cell is activated, p47 is phosphorylated, and it becomes associated with p67 (and possibly with a third protein, $p21^{rac}$. The phosphorylated complex binds to cytochrome B in the phagocytic cell membrane, which is considered to be the active oxidase.
      iv. Cytochrome b transfers one electron from NADPH to oxygen through at least three steps:

- reduction of a flavin adenine dinucleotide (FAD) bound to the high-molecular-weight subunit of cytochrome b
- transfer of an electron from FADH2 to ferric iron in a heme molecule associated with the low-molecular-weight subunit of cytochrome B
- transfer of an electron from reduced iron to oxygen, generating superoxide

v. Since at the time this oxidase is fully activated the cell membrane is invaginating around the particle that stimulated the phagocytic process, the brunt of the active oxygen radicals generated by this system is delivered to the phagosome (Fig. 17.2).

vi. The respiratory burst generates two toxic compounds essential for intracellular killing of bacteria: **superoxide** and $\mathbf{H_2O_2}$. Through myeloperoxidase, H_2O_2 can be peroxidated and led to form hypochlorite and other halide ion derivatives, which are also potent bactericidal agents.

vii. These compounds are also toxic to the cell, particularly superoxide, which can diffuse into the cytoplasm. The cell has several detoxifying systems, including superoxide dismutase, which converts superoxide into H_2O_2 and, in turn, H_2O_2 is detoxified by **catalase** and by the oxidation of reduced **glutathione**, which requires activation of the hexose monophosphate shunt.

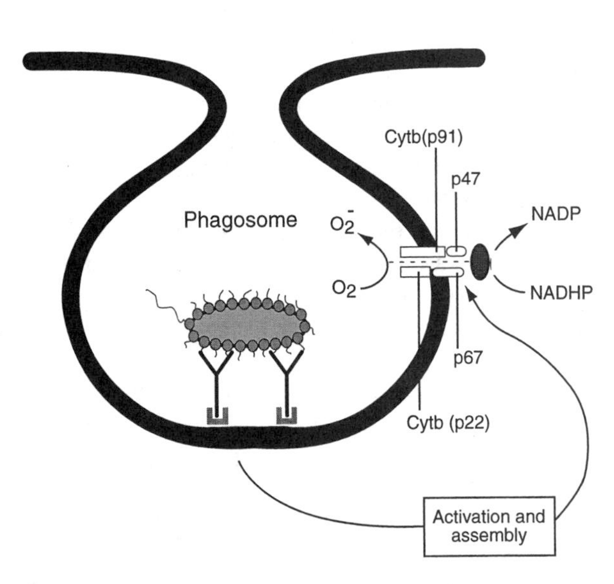

Figure 17.2 Diagrammatic representation of the major events involved in the respiratory burst of phagocytic cells. The occupancy of Fc and/or CR1 receptors triggers the activation sequence, which involves protein kinase activation, enzyme activation, and phosphorylation of at least one cytosolic protein (p47). As a result, a molecular complex, constituted by cytochrome B (Cytb), p47, and p67, is assembled on the cell membrane, which is folding to constitute a phagosome. This complex has NADPH oxidase activity, oxidizes NADPH, and transfers the resulting electron to an oxygen molecule, resulting in the formation of superoxide (O_2^-).

III. PHYSIOLOGY OF THE MONOCYTE/MACROPHAGE

A. **A Comparison of PMN Leukocytes and Monocyte/Macrophages.** The two populations of phagocytic cells share many common characteristics, such as
1. Presence of Fc and C3b receptors on their membranes
2. Ability to engulf bacteria and particles
3. Metabolic and enzymatic killing mechanisms and pathways

In contrast, other functions and metabolic pathways differ considerably between these two types of cells (see Table 17.1).
1. One important distinguishing feature is the involvement of the monocyte/macrophage series of cells in the inductive stages of the immune response, due to their ability to process antigens and present antigen-derived peptides to the immune system.
2. The monocyte/macrophage is also involved in immunoregulatory signals, providing both activating signals (in the form of IL-1, IL-6, and IL-12) and down-regulating signals (in the form of PGE_2) to T lymphocytes.
3. These two types of phagocytic cells have different preferences as far as phagocytosis. For example, PMN leukocytes are able to ingest inert particles such as latex, but have very little ability to engulf antibody-coded homologous erythrocytes, while the reverse is true for the monocyte/macrophage.
4. While neutrophils seem to be constitutively ready to ingest particulate matter, the circulating monocytes and the tissue-fixed (resident) macrophages are usually resting cells that need to be activated by several types of stimuli, including microorganisms or their products and cytokines, before they can fully express their phagocytic and killing properties.

B. **The Activated Macrophage** has unique morphological and functional characteristics.
1. Morphologically, the activated macrophage is larger, and its cytoplasm tends to spread and attach to surfaces.
2. The composition of the plasma membrane is changed, and the rates of pinocytosis and engulfment are increased (phagocytosis through C3b receptors is only seen after activation).
3. Intracellularly, there is a marked increase in enzymatic contents, particularly of plasminogen activator, collagenase, and elastase, and the oxidative metabolism (leading to generation of superoxide and H_2O_2) is greatly enhanced.

III. LABORATORY EVALUATION OF PHAGOCYTIC FUNCTION

The evaluation of phagocytic function is usually centered on the study of neutrophils, which are considerably easier to isolate than monocytes or macrophages. Phagocytosis by neutrophils can be depressed as a result of reduction in cell numbers or as a result of a functional defect. Functional defects affecting every single stage of the phagocytic response have been reported and have to be evaluated by different tests. The following is a summary of the most important tests used to evaluate phagocytic function.

A. **Neutrophil Count.** This is the simplest and one of the most important tests to perform since phagocytic defects due to neutropenia are, by far, more common than the primary, congenital, defects of phagocytic function. As a rule, it is believed that a neutrophil count below 1000/μL represents an increased risk of infection, and when neutrophil counts are lower than 200/μL, the patient will invariably be infected.

B. **Adherence.** The increased adherence of activated phagocytic cells to endothelial surfaces is critical for the migration of these cells to infectious foci. Although specialized tests to measure aggregation and adherence of neutrophils in response to stimuli such as $C5a_{desarg}$ (a nonchemotactic derivative of C5a), presently this property is evaluated indirectly, by determining the expression of the different components of **CD11/CD18** complex which mediate adhesion by flow cytometry.

C. **Chemotaxis and Migration.** The migration of phagocytes in response to chemotactic stimuli can be studied in vitro, using the Boyden chamber, or in vivo, by means of the Rebuck's skin window technique.

1. **Chemotaxis assays using the Boyden chamber**
   a. The basic principle of all versions of the Boyden chamber is to have two compartments separated by a membrane whose pores are too tight to allow PMN leukocytes to passively diffuse from one chamber to the other, but large enough to allow the active movement of these cells from the upper chamber, where they are placed, to the lower chamber.
   b. The movement of the cells is stimulated by adding to the lower chamber a chemotactic factor such as C5a or the tetrapeptide f-met-leu-phe.
   c. The results are usually based either on counting the number of cells that reached the bottom side of the membrane, or on the indirect determination of the number of cells reaching the bottom chamber using ^{51}Cr-labeled PMN (as illustrated in Fig. 17.3).
   d. All versions of this technique are difficult to reproduce and standardize and are not used clinically.
2. The **Rebuck's skin window** technique is used to evaluate the capacity to recruit PMN into an area of inflammation in vivo.

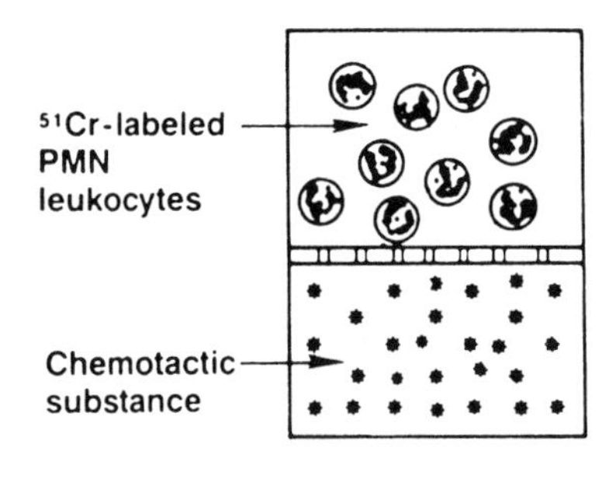

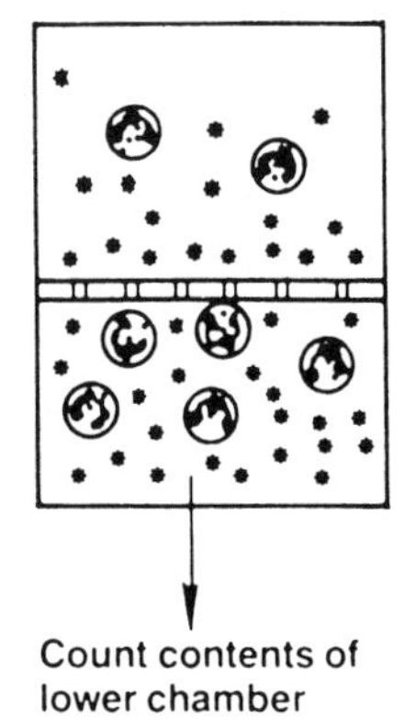

Figure 17.3 Schematic representation of the principle of chemotaxis assays using the Boyden chamber and ^{51}Cr-labeled PMN leukocytes.

a. A superficial abrasion of the skin is covered with a glass cover slip forming a small diffusion chamber ("skin window").
b. Inflammatory cells reaching it will adhere to the glass and can be stained and counted.
c. This technique is also not used routinely.

D. Ingestion. Ingestion tests are relatively simple to perform and reproduce.

1. They are usually based on incubating PMN with opsonized particles, and, after an adequate incubation, determining either the number of ingested particles or a phagocytic index:

$$\text{Phagocytic index} = \frac{\text{No. of cells with ingested particles}}{\text{Total no. of cells}} \times 100.$$

2. Several types of particles have been used, including latex, zymosan (fragments of fungal capsular polysaccharidic material), killed *C. albicans*, and IgG-coated beads (immunobeads). All these particles will activate complement by either one of the pathways and become coated with C3, although opsonization with complement is not the major determinant of phagocytosis.
3. The easiest particles to visualize once ingested are fluorescent latex beads; their use considerably simplifies the assay (Fig. 17.4), particularly if performed in a flow cytometer.
4. This test is not routinely used because others are available (e.g., the nitroblue tetrazolium reduction test, see below) which test both for ingestion and for the ability to mount a respiratory burst

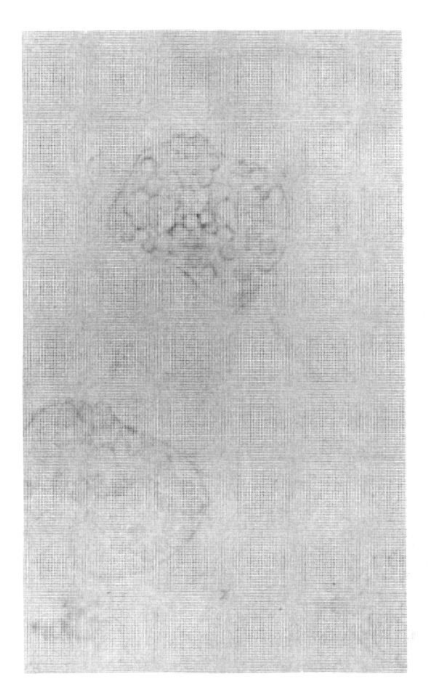
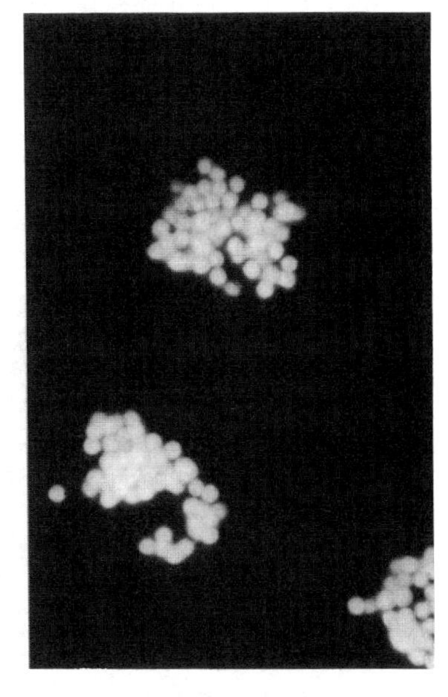

Figure 17.4 Use of fluorescent latex beads for evaluation of phagocytosis. The panel on the left reproduces a photograph of microscopic field showing the phagocytic cells that have ingested latex beads under visible light. The panel on the right shows the same field under UV light.

E. **Degranulation.** When the contents of cytoplasmic granules are released into a phagosome, there is always some leakage of their contents into the extracellular fluid. The tests to study degranulation involve ingestion of particulate matter, as mentioned above, but in this case the supernatants are analyzed for their contents of substances released by the PMN granules such as myeloperoxidase, lysozyme, β-glucuronidase, and lactoferrin.

F. **Measurement of the Oxidative Burst.** Several different techniques have been proposed to measure the oxidative burst.

1. **Chemiluminescence.** The chemiluminescence assay is based on the fact that the superoxide ion is unstable, and that its dissociation can be measured either directly or indirectly after addition of luminol that is activated during superoxide dissociation. This is perhaps the most sensitive and directly quantitative assay for the oxidative burst, but it has a major drawback in that it requires special and complex instrumentation.
2. **Reduction of cytochrome C.** The reduction of the cytochrome C can be used to measure superoxide release because this pigment, when reduced by superoxide, will change its light absorbance properties. The change in color of cytochrome C can be measured with a conventional spectrophotometer. The main drawbacks of the assay are its relatively low sensitivity and difficulties in reproducibility.
3. **Fluorescence assays.** Several techniques for the measurement of the superoxide burst are based on the oxidation of 2′,7′-dichlorofluorescein diacetate (nonfluorescent), which results in the formation of 2′,7′-dichlorofluorescein (highly fluorescent).
   a. The respiratory burst is induced with phorbol myristate acetate (or any other soluble PMN activator)
   b. The numbers of fluorescent cells and fluorescence intensity of activated and nonactivated PMN suspensions from patients and suitable controls can be determined by flow cytometry.
   c. In patients with primary defects of the enzymes responsible for the respiratory burst, both the mean fluorescence intensity and the numbers of fluorescent cells after stimulation are considerably lower than those determined in normal, healthy volunteers.
4. **Nitroblue tetrazolium (NBT) reduction tests.** Tests based on NBT reduction are the most commonly used for the evaluation of neutrophil function.
   a. **Principle.** Oxidized NBT, colorless to pale yellow in solution, is transformed by reduction into blue formazan. The test usually involves incubation of purified neutrophils, NBT, and a stimulus known to activate the respiratory burst. Two types of stimuli can be used.
      i. Opsonized particles, which need to be ingested to stimulate the burst. In this way the test examines both the ability to ingest and the ability to produce a respiratory burst.
      ii. Diffusible activators, such as phorbol esters. Those compounds diffuse into the cell and activate protein kinase C, which in turn activates the NADPH-cytochrome B system and induce the respiratory burst directly, bypassing the ingestion step.
   b. **Microscopic technique.** The simplest NBT reduction assays rely on conventional microscopy to count the number of PMN with blue-

stained cytoplasm after incubation with opsonized particles. This microscopic assay is difficult to standardize, and its interpretation can be affected by subjectivity.

c. **Quantitative techniques**

   i. The classic quantitative technique involves the extraction of intracellular NBT with pyrimidine and measures its absorbance at 515 nm (which corresponds to the absorbance peak of reduced NBT). This modality of the NBT test is extremely sensitive and accurate but is difficult to perform because the reagents used to extract the dye from the cells are highly toxic.

   ii. An alternative are tube tests in which the PMN are simultaneously exposed to opsonized particles and NBT, and the change of color of the supernatant from pale yellow to gray or purple (as a result of the spillage of oxidizing products during phagocytosis) is measured. This assay, however, is not very sensitive, because it relies on the spillage of active oxygen radicals rather than on intracellular reduction.

   iii. With the introduction of kinetic colorimeters, it has been possible to develop an assay in which the color change of NBT can be measured without need to extract the dye from the cells or to separate the cells from the supernatant (Fig. 17.5).

d. **Result interpretation**

   i. A patient whose neutrophils respond to stimulation with phorbol ester but not to stimulation with opsonized beads is likely to have an ingestion defect.

   ii. Neutrophils from a patient with a primary defect in the ability to generate the respiratory burst will not respond to any kind of stimulus.

G. Killing Assays

1. The main protective function of the neutrophil is the ingestion and killing of microorganisms. This ability can be tested using a variety of bacteria and fungi that are mixed with PMN in the presence of normal human plasma (a source of opsonins) and, after a given time, the cells are harvested, lysed, and the number of intracytoplasmic viable bacteria is determined.
2. The assays are difficult and cumbersome, and require close support from a microbiology laboratory, and for this reason have been used less than the indirect killing assays based on detection of the oxidative burst of the PMN mentioned in the previous section.
3. Alternative and simpler approaches to the evaluation of intracellular killing are based on the differential uptake of dyes (such as acridine orange) between live and dead bacteria.

IV. DISEASES OF PHAGOCYTIC FUNCTION

Phagocytic function can be negatively affected by a variety of factors, some of a quantitative nature, some of a qualitative nature. Figure 17.6 diagrammatically illustrates the aspects of PMN function that can be affected in different pathological situations.

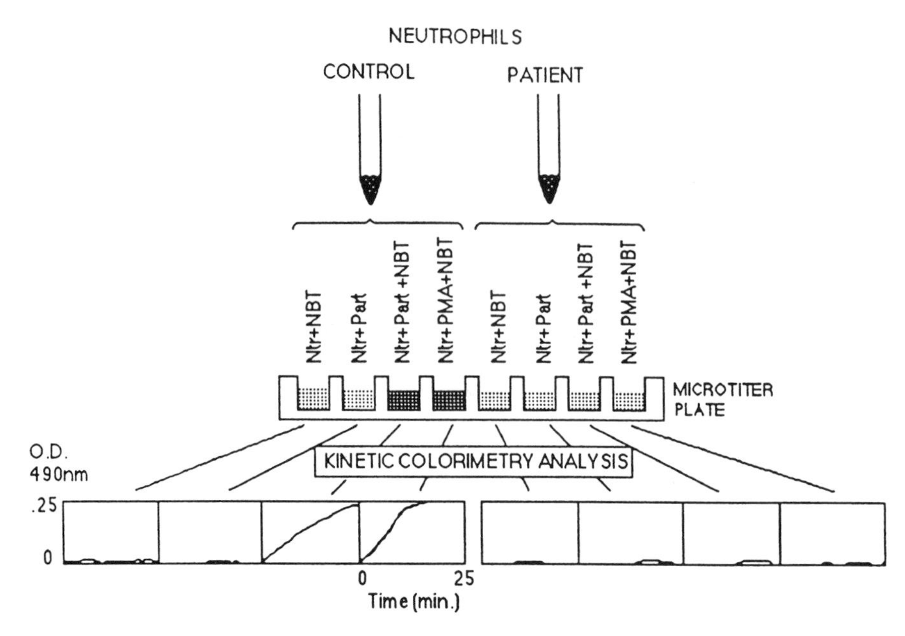

Figure 17.5 Diagrammatic representation of a quantitative NBT assay carried out by kinetic colorimetry. Neutrophils are isolated from a patient and a normal control and incubated separately in a microtiter plate with NBT (to check for spontaneous activation of neutrophils), with opsonized particles (to check for interference of cells and particles with the colorimetric assay), and with opsonized particles and phorbol myristate acetate (PMA) in the presence of NBT (to check for the induction of the respiratory burst). A kinetic colorimeter is used to monitor changes in O.D. due to the reduction of NBT over a 25-minute period, and the results are expressed diagrammatically and as an average of the variation of the O.D./unit of time. The graphic depiction of the results obtained with neutrophils from a normal control and from a patient with chronic granulomatous disease is reproduced in the lower part of the diagram.

A. **Neutropenia.** The reduction of the total number of neutrophils is the most frequent cause of infection due to defective phagocytosis. Although there are congenital forms of neutropenia of variable severity, neutropenia is most frequently secondary to a variety of causes (see Table 17.2).

B. **Disorders of Adherence.** A rare congenital disease, characterized by the lack of expression of the CD11/CD18 complex has recently been described. This disease is usually inherited as an autonomal recessive trait, and the first clinical manifestation, in many instances, is a delayed separation of the umbilical cord. During childhood, these individuals suffer from repeated pyogenic infections and, with less frequency, fungal infections.

C. **Job's Syndrome (Hyper-IgE Syndrome).** This syndrome is characterized by dermatitis, very high levels of serum IgE, and recurrent staphylococcal infections of the lungs and cutaneous abscesses. Other types of pyogenic infections, particularly of the upper airways, and chronic candidiasis can also be present.
   1. The mechanism of this disease has not been well defined. A defect of monocyte chemotaxis has been reported in most patients, but its severity is quite variable, and many doubt that it is, indeed, the primary defect. An

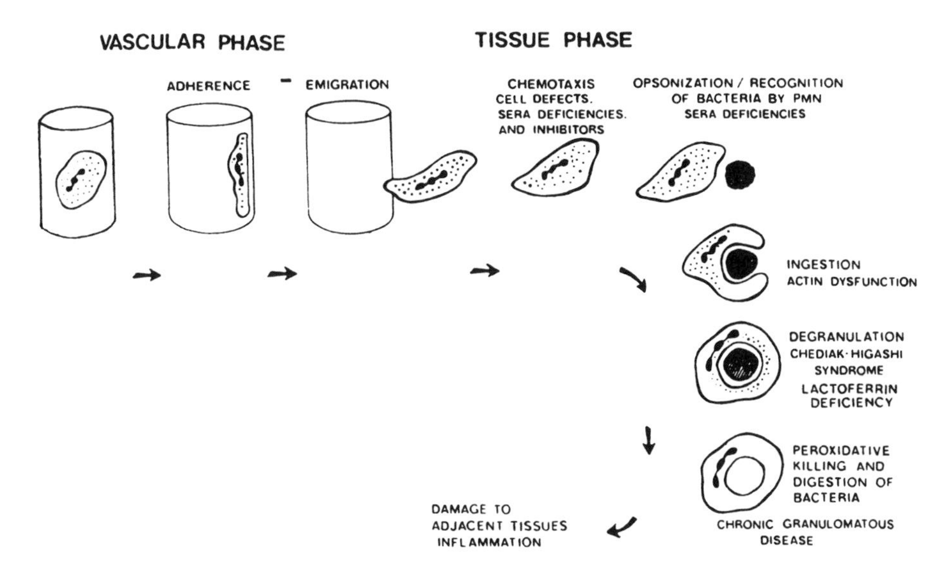

Figure 17.6 Diagrammatic representation of the major primary functional derangements of neutrophils that have been characterized in humans. (Reproduced with permission from Wolach, B., Baehner, R.L., and Boxer, L.A., *Isr. J. Med. Sci.*, *18*:897, 1982.)

alternative hypothesis to explain the syndrome would be the generation of inhibitor(s) of chemotaxis by mononuclear cells.

2. The high levels of IgE correspond, at least in part, to the production of IgE anti-*S. aureus* antibodies; in contrast, IgA antibodies to *S. aureus* are abnormally low, and other indices of humoral immune function (responses to toxoid boosters and to in vitro stimulation with PWM) are also depressed.

D. Disorders of Phagocyte Killing

1. **Chronic granulomatous disease** (**CGD**) is a rare, inherited disease, affecting about one in one million persons, characterized by recurrent life-threatening pyogenic infections. In the majority of cases, the disease seems

Table 17.2 Causes of Neutropenia

I. Congenital
II. Secondary (acquired)
    A. Depressed bone marrow granulocytosis
        1. Drug-induced
        2. Tumor invasion
        3. Nutritional deficiency
        4. Unknown cause (idiopathic)
    B. Peripheral destruction of neutrophils
        1. Autoimmune (Felty's syndrome)[a]
        2. Drug-induced

[a]An association of rheumatoid arthritis, splenomegaly, and neutropenia.

to be X-linked, but autosomal recessive inheritance seems to be involved in 25–35% of the cases.

a. **Pathogenesis.** The molecular basis of CGD is heterogeneous.
   i. About 60% of the cases are X-linked, and in the vast majority of those the basic defect affects the heavy chain (91 kD) of cytochrome B. Recent molecular genetic studies have shown that while about half of the cases of X-linked CGD fail to express message for the heavy chain of cytochrome B, the other half transcribes mRNA for this chain, which may even be synthesized, but fails to be transported or properly inserted in the cytoplasmic membrane.
   ii. The remaining cases of CGD are autosomal recessive, and the vast majority of these lack p47.
   iii. Other cases lack either p67 or the light chain (22 kD) of cytochrome B.
   iv. These molecular defects prevent the assembly of the functional oxidase at the cell membrane level, the generation of superoxide and H_2O_2 are grossly impaired, and, consequently, intracellular killing is defective. Both types of phagocytic cells (PMN leukocytes and monocytes) are affected by the defect.
   v. The killing defect affects mostly the elimination of catalase positive organisms such as *Staphylococcus spp.*, *Serratia marscecens*, *Klebsiella spp.*, *Aerobacter spp.*, *Salmonella spp.*, *Chromobacterium violaceum*, *Pseudomonas cepacia*, *Nocardia spp.* and *Aspergillus sp.* Catalase negative, peroxide-generating microorganisms such as *Streptococcus pneumoniae*, are not usually involved in these patients' infections, due to the fact that this group of catalase-negative organisms, when ingested, continue to generate H_2O_2 which they cannot break down. The H_2O_2 generated by the bacteria progressively accumulates in the phagosome, eventually reaching bactericidal levels.

b. **Clinical picture.** Recurrent bacterial and fungal infections are the prominent clinical feature in this disease.
   i. The most frequent infection sites are the lungs, followed by the lymph nodes, liver, skin, and soft tissues. It should be noted that in over 50% of febrile episodes suffered by these patients, no microorganism is recovered from any suspected site of infection.
   ii. The infections are characterized by microabscess and granuloma formation, and the most typically described include suppurative lymphadenitis, pyoderma, pneumonia with suppurative complications, liver abscesses, osteomyelitis, and severe periodontal disease.
   iii. Generalized lymphadenopathy, hepatosplenomegaly, and hypergammaglobulinemia are frequent.

c. **Diagnosis** is confirmed by abnormal results in one of the variations of the NBT reduction test.

d. **Treatment**
   i. Infections are treated with antimicrobials chosen on the basis of studies if bacteria have been recovered from infection sites.

ii. Prophylactic administration of **trimethoprim-sulfamethoxazole** is generally recommended.
iii. **Interferon-γ** administration has been found to result in a decrease of the frequency of infectious episodes in patients with CGD. Early trials suggested that interferon-γ enhanced the neutrophil respiratory burst and that the effect was more pronounced in patients with the autosomal variety of CGD. However, recent trials suggest that the beneficial effect is unrelated to the respiratory burst.
iv. The identification of the genes coding for the different molecular components of the oxidase system has raised the possibility of trying **gene therapy**, by inserting the relevant genes into the patient's stem cells, which would be returned to the patient in the hope that the defect would be corrected in at least part of the mature phagocytes.

2. **Chediak-Higashi syndrome**
   a. **Pathogenesis.** This rare disease is due to abnormalities in the cytoplasmic granules, so that the killing of certain microorganisms is impaired. It is believed that the primary defect may be in the regulation of membrane activation. The PMN leukocytes are able to ingest microorganisms, but the cytoplasmic granules tend to coalesce into giant secondary lysosomes, with reduced enzymatic contents, that are inconsistently delivered to the phagosome. As a consequence, intracellular killing is slow and inefficient. NK cell function has also been reported to be impaired in these patients.
   b. **Clinical picture.** Clinically, the syndrome is characterized by mucocutaneous albinism, recurrent neutropenia, and unexplained fever and peripheral neuropathy. Later, patients may develop hepatosplenomegaly and lymphadenopathy, and this is associated with recurrent bacterial and viral infection, fever, and prostration. At that stage, the prognosis is very poor.
   c. **Diagnosis.** The diagnosis is usually confirmed by the morphological features (giant lysosomes) and abnormal results in microbial killing tests.
   d. **Therapy.** Infections are treated symptomatically with antibiotics. Ascorbic acid administration is associated with increased bactericidal activity at least in some patients. This improvement may be related to an effect of ascorbic acid on membrane fluidity, which is abnormally high in the patient's PMN leukocytes and is normalized by ascorbic acid.

SELF-EVALUATION

Questions

Choose the ONE *best* answer.

17.1 The most frequent cause of depressed phagocytic function is:
A. CD11/CD18 deficiency
B. Chediak-Higashi syndrome

C. Chronic granulomatous disease
D. Drug-induced neutropenia
E. Job's disease

17.2 The most common molecular defect in the X-linked variety of chronic granulomatous disease is a deficiency of:
A. Myeloperoxidase
B. Phagolysosome formation
C. Superoxide dismutase
D. The cytosolic protein p47
E. The heavy chain of cytochrome B

17.3 Which of the following enzymes is most likely to be absent in organisms that are very seldom isolated from patients with chronic granulomatous disease?
A. Catalase
B. Coagulase
C. Hyaluronidase
D. NADPH oxidase
E. Superoxide dismutase

17.4 Which one of the following compounds has been found to increase intracellular killing in patients with chronic granulomatous disease?
A. Ascorbic acid
B. Interferon-γ
C. Interleukin-2
D. Penicillin
E. Trimethoprim-sulfamethoxazole

17.5 The significance of the expression of CD11/CD18 markers on phagocytic cells is related to the involvement of these molecules on:
A. Adhesion to endothelial cells
B. Antigen presentation to helper T lymphocytes
C. Recognition of chemotactic substances
D. Recognition of opsonized particles and microbes
E. Signal transduction after occupancy of Fc and CR1 receptors

17.6 Which of the following is the reason why the NBT test is considered as an indirect measurement of killing capacity?
A. NBT is oxidized in the presence of lactoperoxidase, and this enzyme is the major killing mechanism in PMN leukocytes
B. The reduction of NBT is a major step in the bactericidal pathways
C. The reduction of NBT reflects the adequacy of the oxidative metabolic pathways
D. The test determines the viability of intracellular bacteria previously incubated with PMN leukocytes and NBT
E. This test reflects the ability to form phagolysosomes

Questions 17.7–17.10

Directions: This group of questions consists of a set of lettered headings followed by a list of numbered words or phrases. For each numbered word or phrase, select the ONE lettered heading that is most closely related to it. The same heading may be used once, more than once, or not at all.

A. Boyden chambers
B. Chemiluminescence
C. Flow cytometry
D. Latex particles
E. Rebuck skin window

17.7 Used for in vitro measurement of chemotaxis
17.8 Used to measure the generation of superoxide
17.9 Used to assess ingestion
17.10 Used to detect cell adhesion molecules (CD11/CD18 complex) on peripheral blood monocytes and neutrophils

Answers

17.1 (D) While many drugs cause granulocytopenia and agranulocytosis, primary phagocytic disorders are extremely rare.
17.2 (E) A deficiency in cytosolic protein p47 is the most common abnormality in the autosomal recessive forms of CGD.
17.3 (A) Catalase negative organisms such as *S. pneumoniae* are not usually involved as pathogens in chronic granulomatous disease.
17.4 (B) Interferon-γ has been used in the treatment of chronic granulomatous disease and it is known to increase intracellular killing independently of the superoxide burst. Ascorbic acid appears to stabilize the fluidity of cell membranes and this appears to correct (at least in some cases) the phagocytic defect associated with Chediak-Higashi.
17.5 (A) The CD11/CD18 family of proteins are considered as CAMs (cell adhesion molecules); in the case of phagocytic cells, these CAMs are particularly significant because they mediate their attachment to endothelial cells, which is the first step that precedes migration of circulating phagocytes to infected tissues.
17.6 (C) NBT is used as a visible substrate for reduction; hence, the reduction of NBT is considered as an indirect verification of the adequacy of the oxidative metabolism in a phagocytic cell.
17.7 (A)
17.8 (B) Chemiluminescence, cytochrome C reduction, and NBT reduction assays are the most frequently used assays for the respiratory burst of phagocytic cells.
17.9 (D)
17.10 (C) Flow cytometry is the method of choice for detection of any membrane associated molecules on peripheral blood leukocytes.

BIBLIOGRAPHY

Albelda, S.M., and Buck, C.A. Integrins and other cell adhesion molecules. *FASEB J.*, *4*:2868, 1990.
Anderson, D.C., Schmalsteig, F.C., Finegold, M.J., Miller, J., Kohl, S. et al. The severe and moderate phenotypes of heritable Mac-1, LFA-1 deficiency; their quantitative definition and relation to leukocyte dysfunction and clinical features. *J. Inf. Dis.*, *152*:668, 1985
Collins, T. Adhesion molecules in leukocyte emigration. *Sci. Am. Med.*, *2(6)*:28, 1995.
Gallin, J.I., and Malech, H.L. Update on chronic granulomatous diseases of childhood. *JAMA*, *263*:1533, 1990.

Gallin, J.I. Interferon-gamma in the treatment of the chronic granulomatous diseases of childhood. *Clin. Immunol. Immunopathol., 61(Pt.2)*:S100, 1991.

Malech, H.L. and Gallin, J.I. Neutrophils in human diseases. *N. Engl. J. Med., 317*:687, 1987.

Pallister, C.J. and Hancock, J.T. Phagocytic NADPH oxidase and its role in chronic granulomatous disease. *Br. J. Biomed. Sci., 52*:149, 1995.

Roos, D. The genetic basis of chronic granulomatous disease. *Ann. Hematol., 68*:267, 1994.

Thrasher, A.J., Keep, N.H., Wientjes, F. and Segal, A.W. Chronic granulomatous disease. *Biochim. Biophys. Acta, 1227*:1, 1994.

Umeki, S. Mechanisms for the activation/electron transfer of neutrophil NADPH-oxidase complex and molecular pathology of chronic granulomatous disease. *Ann. Hematol., 68*:267, 1994.

18
Tolerance and Autoimmunity

Jean-Michel Goust, George C. Tsokos, and Gabriel Virella

I. HISTORICAL INTRODUCTION

A. **Ehrlich and the "horror autotoxicus."** In 1901 Ehrlich postulated that "organisms possess certain contrivances by means of which the immune reaction.[...] is prevented from acting against (its) own elements." Such "contrivances" constitute what in modern terms is designated as "tolerance," and, still in Ehrlich's words "... are of the highest importance for the individual." Several decades later, when autoimmune diseases were described, they were interpreted as the result of a breakdown or failure of the normal tolerance to self, resulting in the development of an autoimmune response. Ehrlich's hypothesis was apparently supported by the definition of pathogenic mechanisms for different diseases considered as autoimmune in which the abnormal anti-self immune reaction played the main role.

B. **Chimerism as a Model of Tolerance.** In the 1940's Owen, a British biologist, was involved in ontogeny studies using bovine dizygotic twins, which share the same placenta. Under these circumstances, each animal is exposed to cells expressing the genetic markers of the nonidentical twin during ontogenic development. When the animals are born, they often carry two sets of antigenically distinct red cells in circulation—one of the best examples of natural **chimerism**. With time, the red cell set acquired from the twin calf will disappear, but the "chimeric" calves will remain tolerant to each other's tissues for the rest of their lives. Thus, these experiments seem to prove that there is a critical period during development during which the immune system becomes tolerant to any antigen it encounters.

C. **Mouse Models.** Brent, Medawar, and co-workers were the first to use mice as experimental models in the study of tolerance. Mice are born with an incompletely developed immune system, and these investigators discovered that mice can be rendered tolerant to neonatally injected antigens, corroborating and expanding Owen's observation with chimeric animals.

D. **The Clonal Deletion Theory of Tolerance.** The first theory concerning tolerance, subscribed to by Burnet, Fenner, and Medawar, stated that self tolerance is achieved by the **elimination of autoreactive clones during the differentiation of the immune system**. However, the development of autoimmune diseases proved that deletion of these clones was not absolute but the remaining clones must be **silenced** or **anergized** to self antigens. None of these mechanisms of tolerance is foolproof for all individuals.

II. DEFINITION AND GENERAL CHARACTERISTICS OF TOLERANCE

Tolerance is best defined as a state of **antigen-specific immunological unresponsiveness**. This definition has several important implications.

A. When tolerance is experimentally induced it does not affect the immune response to antigens other than the one used to induce tolerance. This is a very important characteristic which differentiates tolerance from generalized immunosuppression, in which there is a depression of the immune response to a wide array of different antigens. Tolerance may be transient or permanent, while immunosuppression is usually transient.

B. Tolerance Must Be Established at the Clonal Level. In other words, if tolerance is antigen-specific, it must involve the T- and/or a B-lymphocyte clone(s) specific for the antigen in question and not affect any other clones.

C. Tolerance Can Result from Clonal Deletion or Clonal Anergy

1. **Clonal deletion** involves different processes for T and B lymphocytes.
   a. **Clonal deletion of T lymphocytes.** T lymphocytes are massively produced in the thymus and, once generated, will not rearrange their receptors. Memory T cells are very long lived, and there is no clear evidence that new ones are generated after the thymus ceases to function in early adulthood. Therefore, elimination of autoreactive T cells must occur at the production site (thymus), at the time the cells are differentiating their TcR repertoire. Once a T-cell clone has been eliminated, there is no risk of reemergence of that particular clone.
   b. **B-cell clonal deletion** involves different mechanisms than T-cell clonal deletion. B cells are continuously produced by the bone marrow and initially express low-affinity IgM on their membranes. In most instances, interaction of these resting B cells with circulating self molecules neither activates them nor causes their elimination. Selection and deletion of autoreactive clones seem to take place in the peripheral lymphoid organs during the onset of the immune response. At that time, activated B cells can modify the structure of their membrane immunoglobulin as a consequence of somatic mutations in their germ-line Ig genes. B cells expressing self-reactive immunoglobulins of high affinity can emerge from this process and their elimination takes place in the germinal centers of the peripheral lymphoid tissues.
2. **Clonal anergy.** Both T-cell and B-cell clonal deletion fail to eliminate all autoreactive cells. In the case of T cells, those that recognize self-antigens not expressed in the thymus will eventually be released and will reach the peripheral lymphoid tissues. The causes of B-cell escape from clonal deletion are not as well defined, but they exist nonetheless. Thus, **peripheral tolerance** mechanisms must exist to ensure that autoreactive clones of T and B cells are neutralized after their migration to the peripheral lymphoid tissues. Clonal anergy is one such mechanism.
   a. **Clonal anergy** can be defined as the process that incapacitates or disables autoreactive clones that escape selection by clonal deletion. Anergic clones lack the ability to respond to stimulation with the corresponding antigen.
   b. The most obvious manifestation of clonal anergy is the inability to

respond to proper stimulation. Anergic B cells carry IgM autoreactive antibody in their membrane but are not activated as a result of an antigenic encounter. Anergic T cells express TcR for the tolerizing antigen, but fail to properly express the IL-2 and IL-2 receptor genes and to proliferate in response to it.

c. Anergy results from either an internal block of the intracellular signaling pathways, or from down-regulating effects exerted by suppressor cells, and it can be experimentally induced after the ontogenic differentiation of immunocompetent cells has reached a stage in which clonal deletion is no longer possible.

3. There is now ample evidence suggesting that **tolerance results from a combination of clonal deletion and clonal anergy.** Both processes must coexist and complement each other under normal conditions so that autoreactive clones that escape deletion during embryonic development may be down-regulated and become anergic. The failure of either one of these mechanisms may result in the development of an autoimmune disease.

III. ACQUIRED TOLERANCE—TOLEROGENIC CONDITIONS

A. **Acquired Tolerance** can be induced in experimental animals, under the right conditions, known as **tolerogenic conditions** (Table 18.1).

1. **Immune competence of the host.** Newborn inbred mice of strain A injected with lymphoid cells from mice of a different genetic strain (strain B), on reaching adult life and immunological maturity, can tolerate a skin graft from mice of the donor strain (strain B). Therefore, exposure to a given antigen very early in life results in acquisition of long-lasting tolerance.
2. **Pharmacological immunosuppression.** An extension of the concept that an immunoincompetent host is predisposed to develop tolerance led to experiments that demonstrated that tolerance can be achieved in animals whose degree of immune competence is artificially lowered (e.g., by drug-induced immunosuppression).
3. **General structure and configuration of the antigen.** An antigen that induces tolerance is termed a **tolerogen**. Size and molecular complexity are among the most important factors determining whether a substance is antigenic or tolerogenic.

a. The response of the immune system to the injection of aggregates

Table 18.1 Factors Influencing the Development of Tolerance

Immune competence of the host
Genetic predisposition
Soluble, small-sized antigen
Antigen structurally similar to self protein
Intravenous administration of antigen
High or low dose of antigen

versus soluble monomers of a given protein is drastically different. When the aggregated protein is injected, an active immune response is elicited. If, instead, all protein aggregates are removed from the suspension by high-speed centrifugation and only soluble protein monomers are injected, it is easier to achieve a state of tolerance.

i. Large or complex antigens are usually not tolerogenic because they are phagocytosed and processed by macrophages, creating optimal conditions for stimulation of an immune response.
ii. Small, soluble antigens may not be taken up by the macrophages and thus fail to be adequately presented to helper T cells. The resulting lack of co-stimulation signals will favor the development of tolerance.

b. Exceptions to these rules have been noted. Some autoantigen-derived peptides have been used to induce tolerance in laboratory animals. It is possible that such peptides are able to bind directly to MHC-II molecules and may deliver tolerogenic signals to T lymphocytes with the corresponding autoreactive TcR.

4. **Degree of structural homology.** Antigens with a high degree of structural homology with endogenous proteins of the animal into whom they are injected are more likely to induce tolerance. For example, tolerance to human immunoglobulins should be easier to induce in primates than in rodents, since the primary structure of human and primate immunoglobulins is considerably more similar than the primary structure of human and rodent immunoglobulins.
5. **Degree of immunogenicity.** The immunogenicity of a given antigen is the result of several factors, some antigen-related (such as the degree of structural homology with host proteins and the chemical complexity of the antigen, discussed above) and some related to the genetic constitution of the animal (some antigens are strongly immunogenic in a given species or strain and not in another). It is extremely difficult or impossible to induce tolerance against a strong immunogen.
6. **Route of antigen administration.** Tolerance is achieved more easily when antigens are injected intravenously rather than intramuscularly or subcutaneously, probably as a result of dilution in the systemic circulation.
7. **Antigen dosage.** Experiments designed to determine the relationship between dosage and the induction of tolerance showed that tolerance can be

Table 18.2 A Comparison of the Characteristics of High-Zone and Low-Zone Tolerance

	High-zone tolerance	Low-zone tolerance
Antigen dose	High	Low
Cells involved	T and B cells	T cells only
Onset	Slow	Quick
Duration	Short	Long-lasting
Physiological significance	Questionable	Important

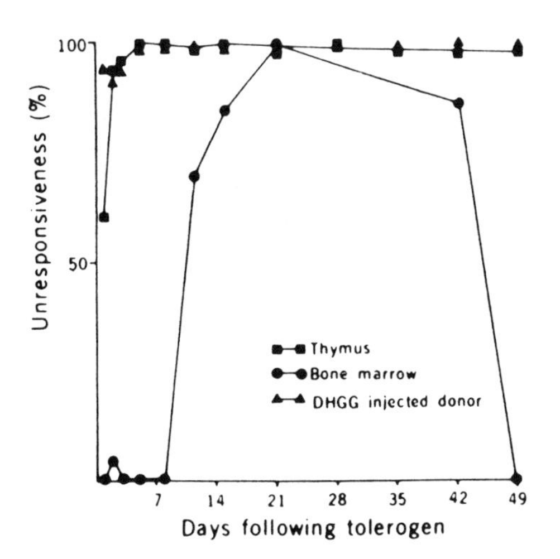

Figure 18.1 Induction and persistence of tolerance in B- and T-cell populations. Thymus (T) and bone marrow (B) lymphocytes were removed at various times from mice rendered tolerant with 2.5 mg of aggregate-free human gammaglobulin (HGG), mixed with complementary lymphocytes from normal donors, and transferred to irradiated syngeneic mice that were subsequently challenged with 0.4 mg of aggregated HGG to test the ability of the transferred cells to cooperate in Ab formation. Results are given as percent of antibody levels obtained in controls using B and T cells obtained from untreated donors. Tolerance appeared sooner and lasted longer in T lymphocytes than in B lymphocytes. (Reproduced with permission from Chiller, J.M., Habicht, A.S., and Weigle, W.O. *Science*, *171*:813, 1971.)

induced with antigen doses well below or well above those that are optimal for the induction of a response. Thus, tolerance can be classified into two major types: high zone and low zone (Table 18.2). Physiological tolerance is, in most cases, low-zone tolerance, primarily affecting T lymphocytes.

8. **Long duration of T-cell tolerance.** Weigle and co-workers induced tolerance with a single large dose of aggregate-free human gamma globulin (HAGG) and followed longitudinally the ability of various combinations of T and B lymphocytes of the tolerized animals and normal controls to reconstitute the immune response in sublethally irradiated mice (Fig. 18.1). T-cell unresponsiveness developed after 1 to 2 days and lasted over 49 days, while B-cell unresponsiveness was obvious only after more than a week from the time of injection of HAGG and was of shorter duration, since by day 49 the B cells of the injected animals had recovered their ability to respond to an HAGG challenge. These studies are the basis for the concept that the physiological state of tolerance of nondeleted self-reactive clones results from the establishment of low-zone, long-lasting T-cell tolerance.

IV. EXPERIMENTAL APPROACHES TO THE DEFINITION OF THE MECHANISMS OF LYMPHOCYTE TOLERANCE

A. **Transgenic mice.** The understanding of the mechanisms involved in tolerance has received a significant boost through the use of transgenic mice. These mice are obtained by introducing a gene in the genome of a fertilized egg, which is subsequently implanted in a pseudo-pregnant female in which it develops. The new gene introduced in the germ line is passed on, allowing the study of the acquisition of tolerance to a defined antigen under physiological conditions. Double transgenic mice, expressing a given antigen and a predetermined antibody, have been constructed by breeding transgenic mice. The tissue expression of the transgene can be manipulated by coupling a tissue-specific promoter to the gene in question.

B. **B-Lymphocyte Tolerance Models.** The main characteristics of B-cell tolerance are summarized in Table 18.3. Experimental evidence supporting both anergy and clonal deletion as mechanisms leading to B-cell tolerance has been obtained in transgenic mouse models.

1. **Models for B-cell anergy.** A most informative model for the understanding of B-cell tolerance was obtained by breeding double transgenic mice from animals transgenic for hen egg lysozyme, which develop tolerance to this protein during development, and animals of the same strain carrying the gene coding for IgM egg lysozyme antibody.
   a. The double transgenic Fl hybrids express the gene coding for egg lysozyme in nonlymphoid cells.
   b. B lymphocytes of these mice also express IgM anti-egg lysozyme antibody, and these lymphocytes are present in large numbers in the spleen. The predominance of B cells with membrane IgM specific for lysozyme is a consequence of allelic exclusion: the insertion of a completely rearranged Ig transgene blocks rearrangement of the normal immunoglobulin genes.
   c. The double transgenic Fl hybrids failed to produce anti-egg lysozyme antibodies after repeated immunization with egg lysozyme.
   d. Thus, these animals have B lymphocytes carrying and expressing a gene that codes for a self-reactive antibody, but cannot respond to the antigen. Experiments on these cells suggest that one or several of the kinases

Table 18.3 B-Cell Tolerance

B-cell anergy:	Antigen is soluble Reactivation may occur Direct proof (a) Double transgenic animals (soluble egg lysozyme and anti-egg lysozyme Ab genes): B cells synthesize egg lysozyme but do not secrete antilysozyme Ab (b) Transgenic animals (anti-DNA Ab gene on B cells): B cells do not secrete anti-DNA Ab
B-cell deletion:	Antigen is surface bound Direct proof (a) Double transgenic mice (genes coding for surface bound lysozyme and antilysozyme Ab): B cells do not produce lysozyme or antilysozyme Ab (b) Transgenic mice with B cells transfected with genes coding for anti-H2-K^k antibody mated with H2-K^k mice produce offspring which lack H2-K^k antibody-positive B cells

activated during the response of a normal B cell to antigenic stimulation remain in an inactive state, interrupting the activation cascade.

2. **Reversibility of B-cell anergy.** By definition, a state of anergy should be reversible. Reversibility was experimentally proven as follows:
   a. Lymphocytes from double transgenic F1 hybrids expressing the gene coding for anti-egg lysozyme antibody were transferred to irradiated nontransgenic recipients of the same strain. In this new environment, from which egg lysozyme was absent, the transferred B lymphocytes produced anti-egg lysozyme antibodies upon immunization. These experiments suggest that continuous exposure to the circulating self-antigen is necessary to maintain B-cell anergy.
   b. Another approach to activate anergic cells is to separate peripheral blood B lymphocytes from an anergic animal and stimulate them in vitro with lipopolysaccharide, which is a polyclonal B-cell mitogen for murine cells. As a consequence of this stimulation, the signaling block that characterizes anergy is overridden, and autoreactive B cells secreting antilysozyme antibody can be detected.
3. **Models for B-lymphocyte clonal deletion.** Evidence supporting clonal deletion in B-cell tolerance has also recently been obtained in transgenic animal models.
   a. F1 double transgenic mice were raised by mating animals that expressed egg lysozyme not as a soluble protein but as an integral membrane protein, with transgenic mice of the same strain carrying the gene for IgM anti-egg lysozyme antibody. In the resulting double transgenic F1 hybrids, B lymphocytes carrying IgM anti-egg lysozyme antibody could not be detected.
   b. Additional experiments have proven that stimulation of an immature IgM/IgD autoreactive B-cell clone by a self-antigen abundantly expressed on a cell membrane leads to clonal deletion by apoptosis. The elimination of autoreactive clones seems to take place in the lymph node germinal centers.
4. **Conclusions**
   a. B-cell tolerance can result both from clonal anergy and clonal deletion, and the choice of mechanism depends on whether the antigen is soluble or membrane-bound.
   b. Clonal deletion involves apoptosis of the self-reactive cells, but we do not know why membrane-bound antigens trigger apoptosis, and we also do not understand how the autoreactive clones are prevented from re-emerging in adult life, since they should be regenerated from multipotent stem cells.
   c. B-cell anergy is associated with a block in the transduction of the activating signal resulting from the binding of antigen to the membrane immunoglobulin, probably consequent to the lack of co-stimulatory signals which are usually delivered by activated TH2 cells (see Chaps. 4 and 11).

C. T-Lymphocyte Tolerance Models. The main characteristics of T-cell tolerance are summarized in Table 18.4. As in the case of B-cell anergy, experimental

Table 18.4 T-Cell Tolerance

Clonal deletion:	Ag presented in the thymus
	T cells die by apoptosis
	TcR repertoire bias
	Never absolute (residual autoreactive cells seem to persist)
Clonal anergy:	Occurs in periphery
	Stimulation of T cells in the absence of proper co-stimulation leads to anergy

evidence supporting both anergy and clonal deletion as mechanisms leading to T-cell tolerance has been obtained in transgenic mouse models.

1. **Models for T-lymphocyte clonal deletion.** There is solid evidence supporting clonal deletion as a mechanism involved in T-cell tolerance:
   a. Transgenic mice were transfected with the gene coding for the T-cell receptor (TcR) cloned from a MHC-I restricted CD8+ cytotoxic T-cell clone specific for the male HY antigen (Fig. 18.2). This TcR is able

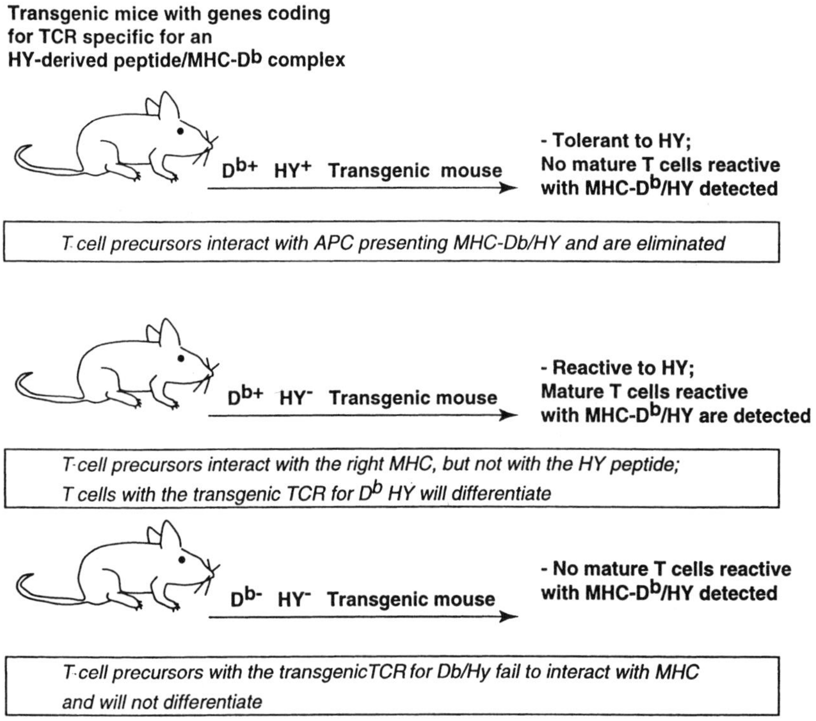

Transgenic mice with genes coding for TCR specific for an HY-derived peptide/MHC-D^b complex

D^{b+} HY^+ Transgenic mouse

- Tolerant to HY; No mature T cells reactive with MHC-D^b/HY detected

T cell precursors interact with APC presenting MHC-Db/HY and are eliminated

D^{b+} HY^- Transgenic mouse

- Reactive to HY; Mature T cells reactive with MHC-D^b/HY are detected

T cell precursors interact with the right MHC, but not with the HY peptide; T cells with the transgenic TCR for D^b HY will differentiate

D^{b-} HY^- Transgenic mouse

- No mature T cells reactive with MHC-D^b/HY detected

T cell precursors with the transgenicTCR for Db/Hy fail to interact with MHC and will not differentiate

Figure 18.2 Diagrammatic representation of an experiment in which transgenic animals expressing a TcR specific for an HY-derived peptide-MHC-D^b complex were shown to become tolerant to the HY peptide, but only in D^{b+} animals. The tolerance in this model was apparently due to clonal deletion, since no mature cytotoxic T cells reactive with MHC-D^b/HY were detected in the tolerant animals.

to mediate a cytotoxic reaction against any cell expressing the HY antigen.

i. Female transgenic mice (HY^-) were found to have mature CD8+ cells expressing the TcR specific for HY.
ii. None of the transgenic male animals (HY^+) had detectable mature CD8+ cells expressing the anti-HY TcR. However, functionally harmless CD4+ cells with the autoreactive TcR could be detected in male animals.

b. These observations are interpreted as meaning that those lymphocytes expressing the autoreactive TcR and the CD8+ antigen interacted effectively with a cell presenting an immunogenic HY-derived peptide in association with an MHC-I molecule, and those cells were deleted. CD4+ lymphocytes, even if carrying the same TcR, cannot interact effectively with MHC-I-associated HY-derived peptides and are spared (see Chapter 10).
c. Similar experiments using mice transfected with genes coding for MHC class II restricted TcRs showed that the CD4+ lymphocytes were selectively deleted, as expected from the fact that the reaction between a TcR and an MHC-II-associated peptide is stabilized by CD4 molecules. In other words, the role of CD4 and CD8 as stabilizers of the reaction between T lymphocytes and APC is not only important for antigenic stimulation, but is also critical for clonal deletion.
d. It must be noted that **absolute** clonal deletion remains difficult to prove. A small residual population of T lymphocytes with autoreactive TcRs (for example, representing 1 of 500,000 T lymphocytes) could remain undetected in the assays currently available, but have not done any damage in the transgenic mice where they could have persisted.

2. **Mechanisms of T-cell clonal deletion.** A common point to all types of clonal deletion is that cell death is due to **apoptosis** (see Chapter 10) involving interaction of the Fas molecule with its ligand. However, may details concerning T-cell apoptosis remain unexplained.
3. **Models for T-lymphocyte anergy.** Experimental models addressing the question of how T cells become tolerant to tissue-specific determinants that are not expressed in the thymus have provided evidence for the role of clonal anergy. Transgenic mice were constructed in which the transgene was coupled to a tissue-specific promoter which directed their expression to an extrathymic tissue. For example, heterologous MHC class II (I-E) genes were coupled to the insulin promoter prior to their injection to fertilized eggs. Consequently, the MHC class II antigens coded by the transfected genes were expressed only in the pancreatic islet β cells. Class II (I-E) specific helper T cells were detectable in the transgenic animals, but they could not be stimulated by exposure to lymphoid cells expressing the transfected MHC-II genes. Thus, tolerance to a peripherally expressed MHC-II self antigen can be due to clonal anergy.
4. **Mechanisms of T-cell anergy.** Proper stimulation of mature CD4+ T lymphocytes requires at least two signals; one delivered by the interaction of the TcR with the MHC-II-Ag complex, while the other signal(s) is (are) delivered by the accessory cell. Both signals require cell-cell contact, in-

volving a variety of surface molecules, and the release of soluble cytokines (see Chap. 11). When all these signals are properly transmitted to the T lymphocyte, a state of activation ensues. Several experiments suggest that the state of anergy develops when TcR-mediated signaling is not followed by co-stimulatory signals.

a. If T lymphocytes are stimulated with chemically fixed accessory cells (which cannot release cytokines or upregulate membrane molecules involved in the delivery of co-stimulatory signals) or with purified MHC II-antigen complexes (which also cannot provide co-stimulatory signals), anergy results.
b. From the multitude of co-stimulatory pairs of molecules that have been described, the CD28/CTLA4-B7 family is the most significant in the physiology of T-cell anergy.
    i. CD28-mediated signals are necessary for the production of IL-2, which seems to be critical for the initial proliferation of TH0 cells and eventual differentiation of TH1 cells. If the interaction between CD28 and its ligand is prevented at the onset of the immune response, anergy and tolerance ensue.
    ii. If CD28 interacts with CTLA-4 (CD152) rather than with CD80 or CD86, a down-regulating signal is delivered to the T lymphocyte (see Chap. 11). Obviously, a better understanding of the regulatory mechanisms controlling the expression of alternative CD28 ligands is needed for our understanding of the how a state of anergy is induced and perpetuated.
c. It is possible that a parallel could be defined for B-cell anergy. The CD40 (B cells)-CD40 ligand (T cells) interaction is critically important for B-cell differentiation. In the absence of CD40 signaling, B cells are easy to tolerize.
d. It is possible to interpret the differences between high-zone vs. low-zone tolerance as a result of differences in the degree of co-stimulation received by T cells.
    i. In high-zone tolerance, the co-stimulatory signals are excessively strong and both T cells and B cells are down-regulated.
    ii. Very low antigen doses fail to induce the delivery of co-stimulatory signals to T cells, and low-zone T-cell tolerance ensues.

5. **Conclusions.** Clonal deletion seems extremely efficient during embryonic differentiation, but a large number of potentially autoreactive clones seem to escape deletion.
    a. Whether those autoreactive clones are activated may just depend on whether the autoantigens against which they are directed are ever presented in a context able to induce an active immune response (i.e., by activated APC able to deliver co-stimulatory signals to the autoreactive T and/or B cells).
    b. On the other hand, the recognition of autoantigens in the absence of co-stimulatory signals by helper T cells is likely to contribute to the perpetuation of a state of **T-cell anergy**, which seems to have an important contribution to perpetuate tolerance in adult life.

V. DOWN-REGULATION OF THE IMMUNE RESPONSE

Many different mechanisms are involved in the down-regulation of the immune response. Extreme situations of down-regulation may very well be indistinguishable from tolerance. This similarity between activation of suppressor circuits and clonal anergy is made more obvious by the fact that the experimental protocols used to induce one or the other are virtually identical. Several mechanisms have been proposed to explain the down-regulation of the immune response.

A. **Suppressor Cells.** Two types of mononuclear cells appear to be able to have suppressor activity: lymphocytes and monocytes.
   1. Monocyte-mediated suppression is usually due to the release of PGE_2 and is nonspecific.
   2. Lymphocytes with suppressor activity are usually CD8+ and mediate antigen-specific suppression, at least in the mouse.
      a. The antigen-specific suppressor effect is exerted via **suppressor factors** that may simply represent released TcR molecules. However, attempts to isolate and sequence those suppressor factors or to clone their genes have met with failure.
      b. It is possible that suppressor CD8+ lymphocytes, after antigen-specific stimulation, may exert their effect by releasing nonspecific suppressor factors such as TGF-β and interleukin-10. Since the trigger for the release of these factors would be the specific recognition of an antigen, and the effects of soluble factors must be limited to cells in the immediate vicinity of the stimulated cell, the suppression would predominantly affect helper T cells recognizing the same antigen in association with an MHC-II molecule.

B. **Oral Tolerance.** Recent reports of therapeutic benefit of oral administration of collagen to patients with rheumatoid arthritis has revived considerable interest in the concept of oral tolerance. This phenomenon had been experimentally observed at the turn of the century, when it was demonstrated that guinea pigs sensitized to hen albumin developed systemic anaphylaxis after reexposure to the antigen, unless they were previously given the antigen with their food.
   1. The administration of large doses of oral antigen is believed to cause TH1 anergy-driven tolerance. However, this seems to be a rather exceptional mechanism, with little clinical application.
   2. The administration of low doses of antigen is believed to stimulate TH2 responses and cause bystander suppression of TH1 cells. A proposed framework for this type of suppression is as follows:
      a. The ingested antigen is transported to submucosal accessory cells, which process it and present it to CD8+ T cells, which after proliferation and differentiation become functionally suppressor. The suppressor effect is mediated by secretion TGF-β and IL-10 after reexposure to the tolerizing antigen.
      b. Some activated CD8+ suppressor cells enter the circulation and are attracted to areas of ongoing reactivity. On those areas, the release of IL-10 and TGF-β suppresses the activity of helper cells assisting the autoimmune local process, resulting in a state of relative tolerance.

c. The antigen used to induce oral tolerance does not need to be identical to that recognized by the autoreactive T cells in vivo, since the suppressor effects of IL-10 and TGF-β are nonspecific and can affect cells reacting with other antigens (**bystander suppression**). However, the best results with oral tolerization protocols are obtained when antigens structurally related to the autoantigens are given orally. Thus, cross-reactivity between the two antigens may be important in localizing the activated suppressor CD8+ cells to the right tissue.

C. **Anti-Idiotypic Immune Responses.** The role of anti-idiotypic circuits in the downregulation of T and B cells has been the object of considerable interest (see Chap. 12).

VI. TERMINATION OF TOLERANCE

If tolerance depends on the maintenance of a state of anergy, there are several possible scenarios that could explain the termination of tolerance.

A. **Clonal regeneration.** Because new B lymphocytes are constantly produced from stem cells, if a tolerogenic dose of antigen is not maintained, the immune system will eventually replace aging tolerant cells by young, nontolerant cells, and recover the ability to mount an immune response.

B. **Cross-Immunization.** Exposure to an antigen that cross-reacts with a tolerogen may induce the activation of T helper lymphocytes specific for the cross-reacting antigen are activated and provide autoreactive B lymphocytes with the necessary co-stimulatory signals necessary to initiate a response against the tolerogen.

C. **Co-Stimulation of anergic clones.** It was discussed above that anergy may be the result of incomplete stimulation of T lymphocytes. Thus, it can be postulated that proper stimulation of T lymphocytes by reestablishing the co-stimulatory pathway can terminate anergy and initiate the autoreactive process.

1. Experimental evidence supporting this concept was obtained in studies of transgenic mice expressing a lymphocytic choriomeningitis viral glycoprotein on the pancreatic cells.
    a. These transgenic mice have T lymphocytes that recognize the glycoprotein but remain anergic.
    b. The state of anergy in these transgenic mice can be terminated by an infection with the lymphocytic choriomeningitis virus. The infection stimulates the immune system, terminates the state of peripheral T-cell unresponsiveness, and the previously tolerant animals develop inflammatory changes in the Langerhans islets (insulitis) caused by lymphocytes reacting with the viral antigen expressed on the pancreatic cells. Those changes precede the development of diabetes.
    c. This model supports the concept of tolerance resulting from the lack of co-stimulatory signals and that infections, by generating a microenvironment favorable to the induction of an active immune response, may be indirectly responsible for the activation of autoreactive clones.
2. Infections caused by superantigen-producing bacteria are effective anergy breakers. Those superantigens cross-link a large numbers of T lymphocytes

(carrying TcR of specific variable region families) with antigen-presenting cells and deliver strong activating signals to T lymphocytes, to the point that previously anergic self-reactive T cells will be activated (as evidenced by the active expression of the IL-2 receptor gene).

VII. AUTOIMMUNITY

A. **Introduction.** Failure of the immune system to "tolerate" self tissues may result in pathological processes known as **autoimmune diseases.** At the clinical level, autoimmunity is apparently involved in a variety of apparently unrelated diseases such as systemic lupus erythematosus (SLE), insulin-dependent diabetes mellitus, myasthenia gravis, rheumatoid arthritis, multiple sclerosis, and hemolytic anemias. There are at least 40 diseases known or considered to be autoimmune in nature, affecting about 5% of the general population. Their distribution by sex and age is not uniform. As a rule, autoimmune diseases predominate in females and have a bimodal age distribution. A first peak of incidence is around puberty, whereas the second peak is in the forties and fifties.

B. **Classification of the Autoimmune Diseases.** There are several different ways to classify autoimmune diseases. Because several autoimmune diseases are strongly linked with MHC antigens, one of the most recently proposed classifications, shown in Table 18.5, groups autoimmune diseases according to their association with class I or with class II MHC markers. It is interesting to notice that although both sexes may be afflicted by autoimmune diseases, there is female preponderance for the class II-associated diseases and a definite increase in the prevalence of class I-associated diseases among males.

C. **Pathophysiology of Autoimmune Diseases.** The autoimmune pathological process may be initiated and/or perpetuated by autoantibodies, immune complexes (IC) containing autoantigens, and autoreactive T lymphocytes. Each of these immune processes plays a preponderant role in certain diseases or may be synergistically associated, particularly in multiorgan, systemic autoimmune diseases.

1. **The role of autoantibodies in autoimmune diseases.** B lymphocytes with autoreactive specificities remain nondeleted in the adult individuals of many species. In mice, polyclonal activation with lipopolysaccharide leads

Table 18.5 Classification of Autoimmune Diseases

I. **MHC class II-associated**
   A. **Organ-specific** (autoantibody directed against a single organ or closely related organs)
   B. **Systemic** (systemic lupus erythematosus—variety of autoantibodies to DNA, cytoplasmic antigens, etc.)

II. **MHC class I-associated**
   A. **HLA-B27-related spondyloarthropathies** (ankylosing spondylitis, Reiter's syndrome, etc.)
   B. **Psoriasis vulgaris** (which is associated with HLA-B13, B16, and B17)

to production of autoantibodies. In humans, bacterial and viral infections (particularly chronic) may lead to the production of anti-immunoglobulin and antinuclear antibodies. In general, it is accepted that polyclonal B-cell activation may be associated with the activation of autoreactive B lymphocytes.

a. Autoantibody-associated diseases are characterized by the presence of autoantibodies in the individual's serum and by the deposition of autoantibodies in tissues. The pathogenic role of autoantibodies is not always obvious and depends on several factors, such as the availability and valence of the autoantigen and the affinity and charge of the antibody.
   i. Antibodies with high affinity for the antigen are considered to be more pathogenic because they form stable IC, which can activate complement more effectively.
   ii. Anti-DNA antibodies of high isoelectric point, very prevalent in SLE, have a weak positive charge at physiological pH, bind to the negatively charged glomerular basement membrane, which also binds DNA. Such affinity of antigens and antibodies for the glomerular basement membrane creates the ideal conditions for in situ IC formation and deposition, which is usually followed by glomerular inflammation.
b. Autoantibodies may be directly involved in the pathogenesis of the disease, while in others, they may serve simply as disease markers without a known pathogenic role. For example, the anti-Sm antibodies that are found exclusively in patients with systemic lupus erythematosus (SLE) are not known to play a pathogenic role. However, in may other situations, autoantibodies can trigger various pathogenic mechanisms leading to cell or tissue destruction (Table 18.6).
   i. Complement-fixing antibodies to red cells (IgG and IgM) may cause intravascular red cell lysis, if the complement activation sequence proceeds all the way to the formation of the membrane attack complex, or may induce phagocytosis and extravascular lysis if the sequence is stopped at the C3 level, due to the accumulation of C3b fragments on the red cell membrane.
   ii. If the antigen–antibody reaction takes place in tissues, proinflammatory complement fragments (C3a, C5a) are generated and attract granulocytes and mononuclear cells that can release proteolytic enzymes and toxic radicals in the area of IC deposition, causing tissue damage.
   iii. Other autoantibodies may have a pathogenic role dependent not on causing cell or tissue damage, but on the interference with cell functions resulting from their binding to physiologically important cell receptors.
c. Representative human autoimmune diseases in which autoantibodies are believed to play a major pathogenic role are listed in Table 18.7. It must be noted in some of these diseases there is also a cell-mediated immunity component. For example, in **myasthenia gravis**, autoreac-

Table 18.6 Pathogenic Mechanisms Triggered by Autoantibodies

Mechanism	Disease	Comments
C′ mediated cell lysis	Autoimmune cytopenias	C′ activating Ig binds to cell membrane antigen; C′ is activated; membrane attack complex is formed; cell is lysed.
Tissue destruction by inflammatory cells	SLE	Antinuclear antibodies bind to tissue-fixed antigens; C′ is activated; C3a and C5a are produced; PMNs are attracted; inflammation develops.
Blockage of receptor	Insulin-resistant diabetes mellitus (acanthosis nigricans)	Anti-insulin receptor antibodies bind to insulin receptor and compete with insulin.
Charge-facilitated	Lupus nephritis	Cationic anti-DNA tissue deposition; antibodies bind to glomerular basement membrane.
Activation of C′	Membranoproliferative glomerulonephritis	Anti-C3bBb antibodies (nephritic factors) bind to and stabilize the C3 convertase (C3bBb) which cleaves C3.
Phagocytosis and intracellular lysis	Autoimmune cytopenias	Antibody binds to cell; may or may not activate C′; cell–antibody (C3b, C3d) complexes are phagocytosed by Fc receptor and/or complement receptor-bearing cells.

tive T lymphocytes have been described, and both autoreactive cell lines and clones have been successfully established from patients' lymphocytes. These T lymphocytes may provide help to autoreactive B lymphocytes producing antiacetylcholine receptor antibodies. In such cases, autoreactive T lymphocytes could be more central in the pathogenesis of the disease than autoantibody-producing B lymphocytes. However, the pathogenic role of autoantibodies is evident from the fact that newborns to mothers with myasthenia gravis develop myasthenia-like symptoms for as long as they have maternal autoantibodies in circulation.

Table 18.7 Antibody-Mediated Autoimmune Diseases

Disease	Antigen
Autoimmune cytopenias (anemia, thrombocytopenia, neutropenia)	Erythrocyte, platelet, or neutrophil cell surface determinant
Goodpasture's syndrome	Type IV collagen
Pemphigus vulgaris	Cadherin on epidermal keratinocytes
Myasthenia gravis	Acetylcholine receptor
Hyperthyroidism	Thyroid-stimulating hormone receptor (Graves disease)
Insulin-resistant diabetes (acanthosis nigricans)	Insulin receptor
Pernicious anemia	Intrinsic factor, parietal cells

2. **The pathogenic role of immune complexes (IC) in autoimmune diseases.** In autoimmune diseases, there is ample opportunity for the formation of IC involving autoantibodies and self antigens. However, not all IC are pathogenic.
   a. Several factors determine the pathogenicity of IC.
      i. Size (intermediate size IC are the most pathogenic).
      ii. The ability of the host to clear IC (individuals with low complement levels or deficient Fc receptor and/or complement receptor function have delayed IC clearance rates and are prone to develop autoimmune diseases).
      iii. Physicochemical properties of IC (i.e., charge, affinity, and isotype of the antibody moiety) which determine the ability to activate complement and/or the deposition in specific tissues as discussed earlier in this chapter.
   b. On many occasions, IC are formed in situ, activate the complement system, complement split products are formed, and neutrophils are attracted to the area of IC deposition where they will mediate the IC-mediated tissue destruction.
   c. **SLE** and **polyarteritis nodosa** are two classic examples of autoimmune diseases in which IC play a major pathogenic role. In SLE, DNA and other nuclear antigens are predominantly involved in the formation of IC, while in polyarteritis nodosa, the most frequently identified antigen is hepatitis B surface antigen.
3. **The role of activated T lymphocytes in the pathogenesis of autoimmune diseases.** Typical T-cell-mediated autoimmune diseases are summarized in Table 18.8. T lymphocytes that are involved in the pathogenesis of such autoimmune diseases may be autoreactive and recognize self antigens; recognize foreign antigen associated with self determinants (modified self); or respond to foreign antigens but still induce self tissue destruction.
   a. **Cytotoxic CD8+ lymphocytes** play a pathogenic role in some autoimmune diseases, usually involving the recognition of non-self peptides

Table 18.8 Examples of T-Cell-Mediated Autoimmune Diseases

Disease	Specificity of T-cell clone/line[a]	T-cell involved
Experimental allergic encephalomyelitis	Myelin basic protein	CD4+
Autoimmune thyroiditis	Thyroid follicular epithelial cells	?
Insulin-dependent diabetes mellitus	Pancreatic islet beta cells	CD8+ (CD4+)
Viral myocarditis	Coxsackie B virus	CD8+

[a]Derived from cells isolated from tissue lesions or peripheral blood of patients and animals affected by the experimental disease. Some of these T-cell lines have been used for adoptive transfer of the disease. In experimental animals, treatment with anti-T-cell antibodies may improve the clinical manifestations of the disease.

expressed in the context of self MHC and destroying the cell expressing such "modified" self. For example, coxsackie B virus-determined antigens expressed on the surface of myocardial cells may induce CD8+-mediated tissue destruction, causing a viral-induced autoimmune myocarditis.

b. **Activated CD4+ helper cells** appear to be frequently involved in cell-mediated autoimmune reactions. Their pathogenic effects are mediated by the release of cytokines (IFN-γ, IL-1, L-2, IL-4, etc.) that can either trigger inflammatory reactions (if TH1 cells are predominantly involved) or activate autoantibody-producing B lymphocytes (if TH2 cells are predominantly involved). As a rule, TH1 cells appear to play a dominant role in many organ-specific autoimmune diseases, and TH2 cells are predominantly involved in systemic autoimmune diseases, such as SLE or rheumatoid arthritis.

D. **Pathogenic Factors Involved in the Onset of Autoimmune Diseases.** Multiple factors have been proposed as participating in the pathogenesis of autoimmune diseases. These factors can be classified as immunological, genetic, environmental, and hormonal. Each group of factors is believed to contribute in different ways to the pathogenesis of different diseases.

1. **Abnormal immunoregulation.** Multiple lymphocyte abnormalities have been described in patients with autoimmune diseases. Prominent among them are B-lymphocyte hyperactivity, presence of spontaneously activated T and B lymphocytes, and decreased suppressor T-cell function. These abnormalities are typified in SLE and will be discussed in Chapter 24.

2. **Anti-idiotypic antibodies.** It has been postulated that anti-idiotypic immune responses may play an important role in autoimmunity.

   a. For example, during a normal immune response to a virus, an immune response directed against the viral structures mediating attachment to its target cell is likely to be triggered. As the immune response evolves, anti-idiotypic antibodies reacting with the antigen binding site of the antiviral antibodies may develop. These anti-idiotypes, by recognizing the "internal image of the antigen" (which is the configuration of the binding site of the first antibody), may be able to combine with the virus receptor protein in the cell. If the membrane protein used as binding site by the virus happens to be a receptor with important physiological functions, the synthesis of anti-idiotypic antibodies may have adverse effects by either activating or blocking the physiological activation of these functions.

   b. Such a mechanism could explain the origin of antibodies against the acetylcholine and the TSH receptors detected in myasthenia gravis and Graves disease, respectively. A hypothetical viral infection would trigger the synthesis of anti-idiotypic antibodies cross-reacting with the acetylcholine receptor at the neuromuscular junction or with the TSH receptor on thyroid cells; these antibodies could interfere with normal functions, cause cell death, or induce cell stimulation depending upon their isotype and the receptor epitopes to which they would bind.

c. An alternative theory linking anti-idiotypic responses to autoimmunity postulates that autoreactive cells are stimulated by anti-idiotypic antibodies emerging in the wake of an anti-infectious response, due to the cross-reactivity of these anti-idiotypic antibodies with the membrane immunoglobulins of autoreactive B lymphocytes. In other words, the anti-infectious antibody that stimulated the anti-idiotypic response and the antigen receptors of B lymphocytes would share **cross-reactive idiotypes**.

3. **Genetic factors.** Clinical observations have documented increased frequency of autoimmune diseases in families, and increased rates of clinical concordance in monozygotic twins. Several studies have also documented associations between HLA antigens and various diseases (see Chap. 3). As stated earlier in this chapter, autoimmune diseases can be classified into two groups, one apparently associated with MHC-I genetic markers, and the other associated with MHC-II genetic markers.
   a. **Linkage with MHC-I markers.** The classic example is the association between HLA-B27 and inflammatory spondyloarthropathies (ankylosing spondylitis, Reiter's syndrome, etc.), which has been discussed in Chapter 3.
      i. The link between HLA-B27 and these diseases has been strengthened by experiments in which transgenic mice carrying the gene for HLA-B27 were observed to spontaneously develop inflammatory disease involving the gastrointestinal tract, peripheral and vertebral joints, skin, nails, and heart. The disease induced in transgenic mice resembled strikingly the B27-associated disorders that afflict humans with that gene.
      ii. It has been postulated that the autoimmune reaction is triggered by an infectious peptide presented by HLA-B27 and followed by cross-reactive lymphocyte activation by an endogenous collagen-derived peptide, equally associated with HLA-B27.
   b. **Linkage with MHC-II markers.** In recent years the definition of MHC-II alleles has undergone a rapid expansion, due to the development of antisera and DNA probes, as well as to the successful sequencing of the genes coding for the constitutive polypeptide chains of MHC-II molecules. For example, insulin-dependent diabetes mellitus (IDDM) is strongly associated with serologically defined MHC-II markers (HLA-DR3 and HLA-DR4), but is even more strongly correlated with the presence of uncharged amino acids at position 57 of the β chain of DQ (DQβ) (see Chap. 19).
   c. **Mechanisms explaining the association between HLA alleles and disease susceptibility.** Although the exact mechanisms are unknown, two have been hypothesized, both based on the persistence and later activation of autoreactive clones.
      i. **Molecular mimicry**, i.e., cross-reactivity between peptides derived from infectious agents and peptides derived from autologous proteins that are expressed by most normal resting cells in the organism. Anergic autoreactive T-cell clones would be activated

by an immune response against an infectious agent due to this type of cross-reactivity.

ii. **Lack of expression of MHC alleles able to bind critical endogenous peptides.** Under those circumstances, potentially autoreactive T-lymphocyte clones would not be eliminated and would remain available for later activation due to an unrelated immune response or by the presentation of a cross-reactive peptide.

d. **Associations with specific TcR variable region types.** The specific recognition of different oligopeptides by different T lymphocytes depends on the diversity of the TcR (see Chaps. 10 and 11). The TcR repertoire and, particularly, the TcR Vβ gene polymorphism may determine autoimmune disease susceptibility. In other words, if the genome of an individual includes a particular V-region gene that encodes a TcR that can combine with an autologous peptide, and the clones expressing such receptors are not eliminated during embryonic differentiation, the individual could be susceptible to a given autoimmune disease.

i. Immunogenetic studies in different animals and humans with different manifestations of autoimmunity suggest that linkages between specific TcR V-region genes and specific autoimmune diseases may exist (for example, insulin-dependent diabetes mellitus, multiple sclerosis, and SLE). But, even in identical twins, concordance for a particular autoimmune disease never exceeds 40%, suggesting that the presence of these TcR V-region genes is not sufficient to cause disease by itself.

ii. Experimental corollaries of the postulated positive association between specific TcR V-region genes and disease are found in experimental allergic encephalomyelitis and murine collagen-induced arthritis. The lymphocytes obtained from arthritic joints of mice susceptible to collagen-induced arthritis have a very limited repertoire of Vβ genes. In mouse strains that do not develop collagen-induced arthritis, there are extensive deletion of Vβ genes, including those preferentially expressed by susceptible mice.

4. **Environmental factors.** Autoimmunity can result from exposure to foreign antigens sharing structural similarity with self determinants. The term **molecular mimicry** is used to describe identity or similarity of either amino acid sequences or structural epitopes between foreign and self antigens.

a. The cardiomyopathy that complicates many cases of **acute rheumatic fever** is one of the best-known examples of autoimmunity resulting from this mechanism. Group A β-hemolytic streptococci have several epitopes cross-reactive with tissue antigens. One of them cross-reacts with an antigen found in cardiac myosin. The normal immune response to such a cross-reactive strain of *Streptococcus* will generate lymphocyte clones that will react with myosin and induce myocardial damage long after the infection has been eliminated.

b. Several other examples of molecular mimicry have been described, as

Table 18.9 Human Proteins with Structural Homology to Human Pathogens

Disease	Human protein	Pathogen
Ankylosing spondylitis, Reiter's syndrome	HLA-B27	*Klebsiella pneumoniae*
Rheumatoid arthritis	HLA-DR4	Epstein-Barr virus
IDDM	Insulin receptor	Papilloma virus
	HLA-DR	Cytomegalovirus
Myasthenia gravis	Acetylcholine receptor	Poliovirus
Ro-associated clinical syndromes	Ro/SSA antigen	Vesicular stomatitis virus
Rheumatic heart disease	Cardiac myosin	Group A *Streptococci*
Celiac disease	A-gliadin or wheat gluten	Adenovirus type 12
Acute proliferative glomerulonephritis	Vimentin	*Streptococcus pyogenes* type 1

summarized in Table 18.9, and additional ones await better definition. For example, molecular mimicry between the envelope glycolipids of Gram-negative bacteria and the myelin of the peripheral nerves may explain the association of the **Guillain-Barré syndrome** with *Campylobacter jejuni* infections.

c. Viruses can precipitate autoimmunity by inducing the release of **sequestered antigens**. In **autoimmune myocarditis** associated with coxsackie B3 virus, the role of the virus is to cause the release of normally sequestered intracellular antigens as a consequence of virus-induced myocardial cell necrosis. Autoantibodies and T lymphocytes reactive with sarcolemma and myofibril antigens or peptides derived from these antigens emerge and the autoreactive T lymphocytes are believed to be responsible for the development of persistent myocarditis.

d. Physical trauma can also lead to immune responses to sequestered antigens. The classic example is **sympathetic ophthalmia**, an inflammatory process of apparent autoimmune etiology affecting the normal eye after a penetrating injury to the other.

e. Latent viral infections are believed to be responsible for the development of many autoimmune disorders. Latent infection is commonly associated with integration of the viral genome into the host chromosomes, and while integrated viruses very seldom enter a full replicative cycle and do not cause cytotoxicity, they can interfere, directly or indirectly, with several functions of the infected cell. For example, T-cell activation by an unknown nonlytic virus has been proposed to explain the onset of **autoimmune thyroiditis**. The infection would lead to T-lymphocyte activation and, as a consequence, to the release of interferon-γ and TNF-α, both known to be potent inducers of MHC-II antigen expression. The increased expression of class II MHC antigens in the thyroid gland would create optimal conditions for the onset of an autoimmune response directed against MHC-II self peptide complexes.

E. **Animal Models of Autoimmunity.** Our understanding of autoimmune disease has been facilitated by studies in animal models. Several animal models have been developed, each sharing some characteristics of a human disease of autoimmune etiology. These animal models often provide the only experimental approaches to the study of the pathogenesis of autoimmune diseases.

1. In some animal models autoimmune diseases are induced by injecting normal animals with antigens extracted from the human target tissues, and the resulting diseases are characterized by rapid onset and an acute course.
2. Most useful for the study of autoimmunity are animals who develop autoimmune disease spontaneously, whose course is protracted and parallels closely the disease as seen in humans. Representative animal models of different autoimmune diseases are listed in Table 18.10.
3. **Experimental allergic encephalomyelitis (EAE)** in mice and rats is the best characterized experimental model of organ-specific autoimmune disease.
    a. The disease is induced by immunizing animals with myelin basic protein and adjuvant. One to two weeks later, the animals develop encephalomyelitis characterized by perivascular mononuclear cell infiltrates and demyelination.
    b. The mononuclear cell infiltrates show a predominance of CD4+ helper T lymphocytes, which upon activation release cytokines that attract phagocytic cells to the area of immunological reaction; those cells are, in turn, activated and release enzymes that are responsible for

Table 18.10 Representative Autoimmune Disease Models and Their Human Analogs

	Animal model	Human disease analog
A. Antigen-induced		
Myelin basic protein	Experimental allergic encephalomyelitis	Multiple sclerosis
Collagen type II	Collagen-induced arthritis	Rheumatoid arthritis
B. Induced by injecting mycobacterial extract	Adjuvant arthritis	Rheumatoid arthritis
C. Chemically-induced		
$HgCl_2$	Nephritis in rats	Nephritis
D. Spontaneous models		
NZB, (NZB × NZW)F_1 MRL lpr/lpr BXSB murine strains	Murine lupus	SLE
Nonobese diabetic mice and rats Inbred BB rats	Diabetes	IDDM
E. Transgenic animals		
HLA-B27 transgenic mice	Spondyloarthropathy	Inflammatory spondyloarthropathies

the demyelination. CD4+ T-lymphocyte clones from animals with EAE disease can transfer the disease to normal animals of the same strain.

c. Genetic manipulations leading to deletion of the genes coding for two specific variable regions of the TcR β chain (Vβ8 and Vβ13) prevent the expression of disease. These two Vβ regions must obviously be involved in the recognition of a dominant epitope of human myelin.

4. **Diabetes** develops spontaneously in inbred BB rats, as well as in nonobese diabetic mice. In both species, the onset of the disease is characterized by T-cell-mediated insulitis, which evolves into diabetes. This disease demonstrates H-2 linkage remarkably similar to that observed between human insulin-dependent diabetes mellitus (IDDM) and HLA-DR3, DR4, and other MHC-II alleles. In non-obese diabetic mice, a decreased expression of MHC-I genes, secondary to a TAP-1 gene deficiency, has recently been characterized. Such deficiency would prevent these animals from deleting autoreactive clones during the differentiation. Transgenic BB mice expressing TAP-1 and MHC-I do not develop the disease.
5. **Systemic lupus erythematosus.** A number of murine strains spontaneously develop autoimmune disease that resembles human SLE.

a. (NZB × NZW)F_1 female mice develop glomerulonephritis, hemolytic anemia, and anti-DNA antibodies. Numerous alterations in the release of T-cell lymphokines and macrophage functions have been described in these animals.

b. MRL-lpr/lpr mice produce autoantibodies and develop arthritis and kidney disease but they also develop massive lymphadenopathy, which is not seen in human disease.

c. BXSB mice develop anti-DNA antibodies, nephritis, and vasculitis. In this strain, in contrast to the others, disease susceptibility is linked to the Y chromosome.

VIII. TREATMENT OF AUTOIMMUNE DISEASES

Standard therapeutic approaches to autoimmune disease usually involve symptomatic palliation with anti-inflammatory drugs and attempts to down-regulate the immune response. Corticosteroids, which have both anti-inflammatory and immunosuppressive effects, have been widely used, as well as immunosuppressive and cytotoxic drugs. However, the use of these drugs is often associated with severe side effects and are not always efficient. Other therapeutic approaches that have been tried include:

A. **Plasmapheresis**, which consists of pumping the patient's blood through a special centrifuge in which plasma and red cells are separated. The red cells and plasma substitutes are pumped back into the patient, while the plasma is discarded. The rationale of plasmapheresis in autoimmune diseases is to remove pathogenic autoantibodies and immune complexes from the circulation. This is an expensive therapeutic modality, but when it is coupled with immunosuppression to prevent or reduce production of new autoantibodies, it is an efficient therapeutic measure in a number of diseases.

Table 18.11 Summary of Interventions Aimed at Disrupting Co-Stimulation of T Cells in Animal Models of Autoimmune Diseases

Model	Anti-CD80 Ab	Anti-CD86	CTLA4-Ig
(NZB × NZW)F1 lupus nephritis			Benefit
Nonobese diabetic mice	Worsening	Prevention	
Experimental allergic encephalomyelitis	Benefit (↑ TH2)	Worsening	

B. Injection of a normal pool of immunoglobulin (IVIG) has been tried in a number of human autoimmune diseases and proved to be of definite help in a form of pediatric vasculitis (Kawasaki's syndrome) as well as in many cases of **idiopathic thrombocytopenic purpura**. The mechanism of action is not clear, but it is believed that IVIG administration has immunomodulating effects that result in the down-regulation of the synthesis of autoantibodies.

C. Elimination of T cells by injection of monoclonal anti-T-cell antibodies has been shown to be therapeutic in a number of animal models. However, this approach is riddled with difficulties when it comes to its application to humans, including the immunogenicity of murine monoclonals, the adverse effects of generalized T-cell elimination (see Chaps. 26 and 27), and unexpected results, such as generalized T-cell activation.

D. Immunotoxins have been prepared by combining either monoclonal antibodies or IL-2 with cytotoxins, hoping to increase the destruction rate of the cells responsible for the autoimmune process. These approaches have not met with great success. It is hoped that better definition of cell markers for lymphocytic subsets that are involved in the pathogenesis of autoimmune diseases may lead to the introduction of more specific and more effective antibodies or immunotoxins.

E. Blocking of Co-Stimulatory Signals. The knowledge that co-stimulatory signals are essential for T-cell activation has led to attempts to induce anergy by disrupting co-stimulatory interactions, with variable but promising results in animal models (see Table 18.11).

F. Induction of Tolerance to the responsible antigen is a logical approach that is hampered by the fact that the identity of the antigen is not known with certainty in many diseases. However, this may not be an insurmountable obstacle, due to the phenomenon of **bystander tolerance**, discussed on an earlier section of this chapter. There is great interest in protocols of oral tolerization because of the encouraging results observed in experimental allergic encephalomyelitis with oral administration of basic myelin protein and in rheumatoid arthritis with oral administration of collagen type II. In addition, oral tolerization is devoid of side effects.

IX. CONCLUSION

The study of the mechanisms leading to tolerance and autoimmunity has attracted immunologists during the last two decades, but our knowledge is still very fragmentary, as reflected in this chapter. The rewards are obvious, ranging from the design of approaches to minimize graft rejection to the introduction of new forms of therapy (or even prevention) of autoimmune disorders. Two areas have exploded with new and challenging information. The experiments with transgenic animals are contributing to the clarification of many points that were previously the object of controversy or speculation. New approaches to immunosuppression aimed at blocking the sensitization or promoting the elimination of autoreactive lymphocytes are being actively tried. It is not entirely unlikely that these two lines of work may soon converge and result in total new approaches in the management of transplantation and autoimmune disorders.

SELF-EVALUATION

Questions

Choose the ONE *best* answer.

18.1 In low zone tolerance:
- A. Both T and B lymphocytes are unresponsive
- B. Only B lymphocytes are unresponsive
- C. T lymphocytes are predominantly affected
- D. The autoreactive clones are permanently deleted
- E. The duration of the tolerant state is relatively short

18.2 To induce tolerance, you would prefer to immunize with:
- A. Aggregated antigens, intradermally
- B. Aggregated antigens, intramuscularly
- C. Aggregated antigens, intravenously
- D. Aggregate-free antigens, intramuscularly
- E. Aggregate-free antigens, intravenously

18.3 Tolerance due to clonal anergy is likely to be:
- A. Irrelevant in human autoimmunity
- B. Irreversible, due to antigen-specific helper T-cell deficiency
- C. Irreversible, due to lack of antigen processing
- D. Reversible, due to the lack of peripheral mechanisms ensuring the persistence of the tolerant state
- E. Reversible, if the autoreactive clones are effectively stimulated

18.4 If an animal is made tolerant to low doses of dinitrophenyl (DNP)-BGG, the injection of an immunogenic dose of nitrophenyl acetyl (NP)-BGG will be followed by:
- A. Antibody response to BGG only
- B. Antibody response to NP and BGG
- C. Antibody response to NP only
- D. Antibody response to NP, DNP, and BGG
- E. No apparent response to either NP or BGG

18.5 Tolerance to self antigens is favored by all of the following factors ***except***:
- A. Binding of fragments derived from endogenous antigens to MHC-II molecules
- B. Continuous exposure to low doses of circulating antigen
- C. Cross-reactivity with microbial antigens

D. Exposure to endogenous antigens during the differentiation of the immune system
E. Release to the circulation of cellular antigens or their fragments

18.6 The carditis associated with rheumatic fever is believed to be caused by:
A. Activation of T lymphocytes carrying a TcR which cross-reacts with membrane structures of infected myocardial cells
B. An autoimmune reaction triggered by the release of myosin from cardiac cells directly damaged by infection with group A *Streptococcus*
C. Antibodies to group A *Streptococcus* which cross-react with cardiac myosin
D. Deposition of immune complexes in the myocardium
E. Increased expression of MHC-II molecules in myocardial tissue infected with group A *Streptococcus*

18.7 An animal injected with 2 mg of DNP-BSA I.V. while immunosuppressed with cyclophosphamide fails to show antibody responses to either DNP or BSA. Cyclophosphamide is stopped, and 2 weeks later the animal is challenged with 2 mg of DNP-HGG mixed with complete Freund's adjuvant I.M. Ten days later you would expect to detect:
A. Anti-DNP antibodies only
B. Anti-HGG and anti-DNP antibodies
C. Anti-HGG antibodies only
D. Anti-HGG, anti-BSA, and anti-DNP antibodies
E. No antibodies at all

In Questions 18.8–18.10, match the numbered phrase with the letter heading that is most closely related to it. The same heading may be used once, more than once, or not at all. Match the autoimmune disease with the mechanism most likely involved in its pathogenesis:
A. Cross-reactive anti-idiotypic antibodies
B. Cross-reactive idiotypes
C. Increased antigen presentation
D. Molecular mimicry
E. Release of sequestered antigens

18.8 Myocarditis associated to coxsackie-B 3 virus infection
18.9 Sympathetic ophthalmia
18.10 Guillain-Barré syndrome

Answers

18.1 (C) Low zone tolerance (induced with small concentrations of antigen) affects predominantly T cells and is of long duration, but the autoreactive clones are not deleted.

18.2 (E)

18.3 (E) Clonal anergy is believed to be significant in human tolerance, particularly because anergy (in contrast to clonal deletion) is potentially reversible, and, therefore, this type of anergy is compatible with autoimmunity, while clonal deletion is not. The reversibility, however, is not believed to be spontaneous, as if it resulted from the lack of peripheral mechanisms to ensure the persistence of the tolerant state, but rather to result from vigorous stimulation of anergic cells, most likely by cross-reactive antigens properly presented in association with fully functional accessory cells.

18.4 (E) The animal has been made tolerant to BGG by injection of low doses of antigen (low zone tolerance). Thus, T lymphocytes are tolerant and will not assist the response of B cells to either hapten or carrier.

18.5 (C) The cross-reactivity with microbial antigens is actually believed to be an important factor leading to loss of tolerance. All the other listed factors may be relevant in the inductive stages of tolerance.

18.6 (C) Rheumatic fever is believed to be an example of cross-reactive autoimmunity. Some strains of group A *Streptococcus* elicit antibodies that cross-react with myosin, leading to the development of the cardiac lesions characteristic of the disease. The arthritis component is more likely to be due to the deposition of immune complexes.

18.7 (B) The conditions described in the stem are likely to result in T-cell tolerance. This type of tolerance is long lasting and, in the case of a hapten-carrier reaction, the tolerance is specifically directed against the carrier. When the tolerant animal is stimulated with a different hapten-carrier combination, a response to both is expected, since the B lymphocytes are not tolerant, and the T lymphocytes are only tolerant to BSA and not to HGG.

18.8 (E) Coxsackie-B 3 virus is believed to damage the myocardium, induce the release of myosin, which will induce an autoimmune reaction that will perpetuate the myocarditis.

18.9 (E)

18.10 (D) The Guillain-Barré syndrome is believed to result from an autoimmune response triggered by cross-reactivity between glycolipids in the membrane of *Campylobacter jejuni* and neuronal cells.

BIBLIOGRAPHY

Blackman, M., Kapler, J., Marrack, P. The role of the T lymphocyte receptor in positive and negative selection of developing T lymphocytes. *Science*, *248*:1335, 1990.

Braden, B. C., and Poljak R. J. Structural features of the reaction between antibodies and protein antigens. *FASEB J.*, *9*:9, 1995.

Burkly, L. C., Lo, D., and Flavell, R. A. Tolerance in transgenic mice expressing major histocompatibility molecules extrathymically on pancreatic cells. *Science*, *248*:1364, 1990.

Cyster, J. G., Hartley, S. B., and Goodnow, C. C. Competition for follicular niches excedes self-reactive cells from the recirculating B-cell repertoire. *Nature*, *371*:389, 1994.

Fields, P. E., Gajewski, T. F., and Fitch, F. W. Blocked Ras activation in anergic CD4+ T cells. *Science*, *271*:1276, 1996.

Goodnow, G. C. Transgenic mice and analysis of B-cell tolerance. *Annu. Rev. Immunol.*, *489*:10, 1992.

Harlan, D. M., Abe, R., Lee, K. P., and June, C. H. Potential roles of the B7 and CD28 receptor families in autoimmunity and immune evasion. *Clin. Immunol. Immunopathol.*, *75*:99, 1995:

Mountz, J. D., Zhou, T., Su, X., et al. The role of programmed cell death as an emerging new concept for the pathogenesis of autoimmune diseases. *Clin. Immunol. Immunopathol.*, *80(part 2)*:S2, 1996.

Quill, H. Anergy as a mechanism of peripheral T cell tolerance. *J. Immunol.*, *156*:1325, 1996.

Rose, N. R. Defining criteria for autoimmune diseases (Witebsky's postulates revisited). *Immunol. Today*, *14*:1993.
Thomson, C. B. Distinct roles for the costimulatory ligands B7-1 and B7-2 in helper cell differentiation. *Cell*, *81*:979, 1995.
Weiner, H. L., Friedman, A., Miller, A., et al. Oral tolerance: immunologic mechanisms and treatment of animal and human organ-specific autoimmune diseases by oral administration of auto antigens. *Annu. Rev. Immunol.*, *12*:809, 1994.

19
Organ-Specific Autoimmune Diseases

Christian C. Patrick, Jean-Michel Goust, and Gabriel Virella

I. INTRODUCTION

Autoimmune diseases can be roughly divided into organ-specific and systemic, based both on the extent of their involvement and the type of autoantibodies present in the patients. The systemic forms of autoimmune diseases best exemplified by **systemic lupus erythematosus (SLE)** and **rheumatoid arthritis** will be discussed in Chapters 20 and 21. Less generalized autoimmune processes may affect virtually every organ system (Table 19.1); in many instances, only certain cell types within an organ system will be affected in a particular disease (i.e., gastric parietal cells in pernicious anemia). In this chapter we will restrict our discussion to the major autoimmune diseases that affect specific organs and the associated autoantibodies, with the understanding that it is not clear whether or not these antibodies are the cause of the disease or just a secondary manifestation.

II. AUTOIMMUNE DISEASES OF THE THYROID GLAND

Autoimmune responses have been implicated in two major thyroid diseases, Graves' disease and Hashimoto's disease.

A. **Graves' Disease** also known as thyrotoxicosis, diffuse toxic goiter, and exophthalmic goiter, is the result of the production of antibodies against the thyrotrophic hormone (TSH) receptor (**TSH receptor antibodies**).
   1. **Pathogenesis.** The TSH receptor antibodies detected in patients with Graves' disease stimulate the activity of the thyroid gland. For that reason they have been known by a variety of descriptive terms, including: (a) **long-acting thyroid stimulator (LATS)**; (b) **thyroid-stimulating immunoglobulin**; and (c) **thyroid-stimulating antibodies (TSI)**.
      a. Thyroid-stimulating antibodies are detected in 80–90% of patients with Graves' disease, are usually of the IgG isotype, and have the capacity to stimulate the production of thyroid hormones by activating the adenylate cyclase system after binding to the TSH receptor.
      b. A major sign is protrusion of the eyeball (exophthalmos), classically attributed to an increased volume of extraocular tissues due to edema

Table 19.1 Representative Examples of Organ-Specific and Systemic Autoimmune Diseases

Disease	Target tissue	Antibodies mainly against
Organ-specific diseases		
Graves' disease	Thyroid	TSH Receptor
Hashimoto's thyroiditis	Thyroid	Thyroglobulin
Myasthenia gravis	Muscle	Acetylcholine receptors
Pernicious anemia	Gastric parietal cells	Gastric parietal cells Intrinsic factor (IF) B_{12}-IF complex
Addison's disease	Adrenals	Adrenal cells Microsomal antigen
Insulin-dependent diabetes mellitus	Pancreas	Pancreatic islet cells; insulin
Primary biliary cirrhosis	Liver	Mitochondrial antigens
Autoimmune chronic active hepatitis	Liver	Nuclear antigens, smooth muscle, liver–kidney microsomal antigen, soluble liver antigen, etc.
Autoimmune hemolytic anemia	RBC	RBCs
Idiopathic thrombocytopenic purpura	Platelets	Platelets
Systemic diseases		
Systemic lupus erythematosus	Kidney, skin, lung, brain	Nuclear antigens, microsomes, IgG, etc.
Rheumatoid arthritis	Joints	IgG, nuclear antigens
Sjögren's syndrome	Salivary and lacrimal glands	Nucleolar mitochondria
Goodpasture's syndrome	Lungs, kidneys	Basement membranes

and/or to deposition of mucopolysaccharides. Recently, it has been proposed that an autoimmune response to a tissue antigen expressed on both thyroid and eye muscle membrane could induce orbital inflammation and exophthalmos.

2. **Clinical presentation**
   a. Graves' disease has its peak incidence in the third to fourth decade and has a female-to-male ratio of 4–8 : 1.
   b. Patients usually present with diffuse goiter and 60–70% of patients have ocular disturbances.
   c. Symptoms of hyperthyroidism include increased metabolic rate with weight loss, nervousness, weakness, sweating, heat intolerance, and loose stools.
   d. Abnormalities on physical examination include diffuse and nontender enlargement of the thyroid, tachycardia, warm and moist skin, tremor, exophthalmos, and pretibial edema. The ophthalmopathy can be unilateral or bilateral and may be associated with proptosis, conjunctivitis, and/or periorbital edema.

3. **Diagnosis**
   a. **Biopsy** of the thyroid gland shows diffuse lymphoplasmocytic interstitial infiltration.
   b. **Laboratory** data shows increased levels of thyroid hormones (triiodothyronine or T_3 and thyroxine or T_4); increased uptake of T_3; and antithyroid receptor antibodies.
   Two types of assays can be used to demonstrate **thyroid receptor antibodies**:
      i. Those based on the inhibition of TSH binding by TSI antibodies (TSH-binding inhibition assay). This assay is relatively simple and precise, but the results obtained with it do not always correlate with disease activity, since nonstimulating antibodies can also block TSH binding.
      ii. Those based on the functional consequences of thyroid receptor antibody binding to TSH. This last group of assays measures true thyroid-stimulating antibodies and includes:
         - tests in which the end point for thyroid activation is the accumulation of intracellular colloid droplets or the penetration of thrombogenic substrates into lysosomal membranes
         - tests in which the activation of adenylate cyclase is measured
         - tests in which cAMP accumulation is measured.
   The functional assays correlate better with disease activity, but are difficult to calibrate and reproduce, and when heterologous thyroid is used as substrate, there is always the possibility that some human TSI antibodies might not react across species.
4. **Therapy** is directed at reducing the thyroid's ability to respond to stimulation by antibodies. This can be achieved:
   a. Surgically, by subtotal thyroidectomy
   b. Pharmacologically, either by administration of radioactive iodine (^{131}I) (which is difficult to dose), or by the use of antithyroid drugs such as propylthiouracil and methimazole, which are useful but slow in their effects.

B. **Hashimoto's Thyroiditis (Autoimmune Thyroiditis)** is primarily a subclinical disease in which no thyroid dysfunction is evident and no therapy is needed until the late stages of disease.
   1. **Pathogenesis.** A cell-mediated autoimmune reaction triggered by unknown factors is believed to be primarily responsible for the development of the disease. Several lines of evidence support this conclusion.
      a. The inflammatory infiltrate of the thyroid gland shows predominance of activated, lymphokine-secreting T lymphocytes. Numerous plasma cells can also be seen.
      b. Experimental thyroiditis can be easily transferred by infusing lymphocytes from sick to healthy animals.
      c. Infants of mothers with active disease-carrying IgG autoantibodies reacting with thyroglobulin (which cross the placenta) are unaffected.
   Whether or not autoantibodies against thyroglobulin and microsomal antigens frequently detected in those patients play any pathogenic role is unclear. The main argument supporting their involvement is a good correlation

between their titers and disease activity. However, this relationship is also expected if those antibodies arise as a consequence of the activation of helper T cells and presentation of high levels of MHC-II/endogenous peptide complexes to previously tolerant B lymphocytes. Furthermore, these autoantibodies are detected in low titers in up to 15% of the normal adult female population.

2. **Clinical presentation.** It is the most common form of thyroiditis, and it usually has a chronic evolution. Its incidence peaks during the third to fifth decades, with a female–male ratio of 10:1.
   a. Hashimoto's thyroiditis is functionally characterized by a slow progression to hypothyroidism, and symptoms develop insidiously. Patients often present with dysphagia or a complaint that their clothes are too tight around the neck. Most patients become hypothyroid with symptoms of malaise, fatigue, cold intolerance, and constipation.
   b. Signs include dry, coarse hair and a diffuse enlarged goiter, usually not tender.
3. **Diagnosis.** The diagnosis is usually confirmed by the detection of antithyroglobulin antibodies
   a. 60–75% of patients show a positive reaction by passive hemagglutination using thyroglobulin-coated erythrocytes (titers higher than 25, while normals usually have titers up to 5).
   b. Although these antibodies are also found in other autoimmune disorders such as pernicious anemia, Sjogrën's syndrome, and in 3–18% of normal individuals, the titer of autoantibodies is lower in all other groups with the exception of patients with Sjogrën's syndrome.
   c. In patients with hypothyroidism, T_3 and T_4 levels and T_3 uptake are low and TSH is increased.
5. **Therapy**
   a. Corticosteroids may be used as mild immunosuppressants, with the aim of reducing the autoimmune response.
   b. In patients who develop hypothyroidism, thyroid hormone replacement is indicated.
   c. In patients with large goiters, the administration of thyroxine and triiodothyronine may reduce the size of the thyroid gland.

III. ADDISON'S DISEASE (CHRONIC PRIMARY HYPOADRENALISM)

This disease can either be caused by exogenous agents (e.g., infection of the adrenals by *Mycobacterium tuberculosis*) or be idiopathic. The idiopathic form is believed to have an immune basis, since 50% of patients have been found to have antibodies to the microsomes of adrenal cells (as compared to 5% in the general population) by immunofluorescence.

A. Pathogenesis

   a. The autoantibodies directed against the adrenals react mainly in the zona glomerulosa, zona fasciculata, and zona reticularis and are believed to play the main pathogenic role in this disease.
   b. The end result of the autoimmune reaction against the adrenal cortex is atrophy and loss of function.

c. This autoimmune form of Addison's disease is found frequently in association with other autoimmune diseases, such as thyroiditis, pernicious anemia, and diabetes mellitus; autoantibodies to adrenal cortex are not found in Addison's disease caused by tuberculosis of the adrenal glands.

B. **Clinical Presentation.** Symptoms of Addison's disease or adrenal insufficiency include weakness, fatigability, anorexia, nausea, vomiting, weight loss, and diarrhea. Signs include increased skin pigmentation, vascular collapse, and hypotension. Addison's disease is most commonly found in the fourth and fifth decades of life and is two to three times more frequent in females.

C. **Diagnosis**

1. **Biopsy.** The adrenal glands show marked cortical atrophy with an unaltered medulla. Abundant lymphocytes are seen between the residual islands of epithelial cells.
2. **Laboratory** data show: metabolic acidosis; hyperkalemia, hyponatremia, and low levels of chloride and bicarbonate; hypoglycemia; lymphocytosis with eosinophilia; and low plasma cortisol levels and low levels of urine 17-ketosteroids and 17-hydroxycorticoids.
3. The diagnosis is confirmed by demonstration of antiadrenal antibodies by indirect immunofluorescence.

D. **Therapy.** Replacement of glucocorticoids and mineralocorticoids.

IV. AUTOIMMUNE DISEASES OF THE PANCREAS

A. **Diabetes Mellitus (DM)** is a multiorgan disease, perhaps with multiple etiologies, and certainly with more than one basic abnormality. In insulin-dependent diabetes (IDDM, type I diabetes), the basic defect is a decreased-to-absent production of insulin, while in type II, or non-insulin-dependent diabetes, there is a decrease of the effect of insulin at the target cell level. In this chapter we will limit our discussion to IDDM, which is believed to be an autoimmune disorder.

1. **Pathogenesis**

a. **Autoimmunity in diabetes.** Many different types of autoantibodies have been detected in patients with IDDM, but a considerable degree of uncertainty remains as to their precise pathogenic role.

i. **Anti-islet cell antibodies (ICA)** reacting against membrane and cytoplasmic antigens of the islet cells are detected in as many as 90% of type I diabetic patients at the time of diagnosis, but they diminish in frequency to 5–10% in patients with long-standing diabetes. ICA are usually of the IgG2 and/or IgG4 subclasses, can be detected months or years before the appearance of clinical symptoms, and have been suggested as being involved in the pathogenesis of islet cell damage.

ii. **Anti-insulin autoantibodies** are detected in as many as one-third of noninsulin-treated patients with IDDM at the time of diagnosis. The pathogenic significance of insulin autoantibodies is not clear, but the coexistence of anti-insulin antibodies and ICA has a strong predictive value for the future development of diabetes. Two theories have been proposed to explain the emergence of anti-insulin antibodies:

- During destruction of the islet cells, insulin may be exposed to the immune system in a form that may be recognized as foreign. This could explain the development of anti-insulin antibodies in type I patients with insulitis (islet cell inflammation).
- A recent finding of antigenic mimicry between insulin and a retroviral antigen, apparently leading to the spontaneous emergence of anti-insulin antibodies in nonobese diabetic-prone mice, supports the alternative possibility that anti-insulin antibodies may be triggered as a result of infection with an agent expressing cross-reactive antigen(s).

iii. **Anti-insulin antibodies** are found in a large majority of patients (both of Types I and II) treated with heterologous insulin and are basically induced by the repeated injection of bovine or porcine insulin. However, antihuman insulin antibodies can also be detected (less frequently) in patients treated with recombinant human insulin, whose tertiary configuration differs from that of the insulin released by the human pancreas. The antibodies directed against therapeutically administered insulin appear to be predominantly of the IgG2 and IgG4 isotypes, which do not activate complement very efficiently.

iv. **Cell-mediated immunity** is believed to have a more important pathogenic role in causing islet cell damage than the autoantibodies to islet cells and insulin.

- The pathological hallmark of recent-onset diabetes is the mononuclear cell infiltration of the islet cells, known as **insulitis**. Similar observations can be made in animals with experimentally induced forms of diabetes.
- The predominant cells in the islet cell infiltrates are **T lymphocytes**, with predominance of activated CD8+ T lymphocytes. It is thought, based on data generated by in vitro experiments and observations made in animal models of the disease, that a direct cytotoxic effect of T cells is highly unlikely, but the T cells infiltrating the islets are involved in active secretion of cytokines, including IL-2 and IFN-γ.
- IL-2 causes the up-regulation of MHC-II in islet β cells, which creates favorable conditions for the induction of autoreactive cells. In experimental animal models, this change precedes the development of insulitis.
- IFN-γ activates macrophages, causing the release of cytokines, such as IL-1 and toxic oxygen radicals (such as superoxide and nitric oxide) known to be toxic to islet cells in vitro, and IL-12, which could activate additional TH1 cells as well as NK cells. NK cells are also known to cause islet cell death in vitro, probably by inducing apoptosis.

v. **Initiating factors.** Of crucial importance to our understanding of the pathogenesis of DM is the definition of the insult(s) that may activate autoreactive T lymphocytes and trigger the disease. Viral infections have been suspected for a number of years. There is evidence suggesting this in some patient populations: for example, 12–15% of patients with congenital rubella develop type I diabetes, particularly when they are DR3 or DR4 positive. But for the majority of diabetic patients, a link with any given viral infection remains elusive.

2. **Genetic factors**. IDDM is a polygenic disease and three major susceptibility loci have been identified:
   a. The **IDDM1 locus**, which includes the MHC genes determining resistance/susceptibility to diabetes. Several MHC genes and gene alleles have been linked to diabetes.
   b. The **IDDM2 locus**, which includes a region near the insulin and insulin-growth factor II on chromosome 11.
   c. **MHC alleles.** Several DP and DQ alleles are associated with predisposition or resistance to diabetes.
      i. 95% of diabetics express DR3 and/or DR4, compared to 42–54% of nondiabetics. This corresponds to a disease risk of 2 to 5.
      ii. Several DQ and DR alleles associated with susceptibility or resistance to diabetes have been identified. In Caucasians, resistance is associated with the presence of Asp (a negatively charged amino acid) in DQβ 57. The presence of a neutral, amino acid in that same position as well as the presence of arginine in position 52 of DQα are associated with predisposition for diabetes.
   d. The **Tap genes**, genes controlling the synthesis of transport-associated proteins responsible for the transport of endogenous peptides to the endoplasmic reticulum, where those peptides become associated with MHC-I peptides.
      i. The Tap genes are located on chromosome 6, near the MHC-region,

Table 19.2 Insulin-Dependent Diabetes-Related MHC Markers

Markers associated with protection	Markers associated with predisposition
DR2	DR3
DR5	DR4[a]
DQβw3.1[b]	DQβw3.2[b]
Asp 57^{+} DQb	Asp 57^{-} DQβ[c]
	Arg52^{+} DQα

[a]Maximal risk in DR3/DR4 heterozygous individuals.
[b]DR4 subtypes; DQβw3.2 is associated with predisposition mainly in DR3/DR4 heterozygotes.
[c]Maximal predisposition is associated either with the expression of two Asp-57 alleles of DR3 or DR4 or with the expression of an Asp 57-, Arg52+ DQαβ heterodimer. Maximal protection is associated with expression of two Asp 57+ DQβ alleles.

and may be transmitted in linkage disequilibrium with MHC-II genes.

ii. One Tap-2 gene allele (Tap-2*0101) is associated with susceptibility to diabetes (relative risk of 3.4) while another allele has been defined as protective in relation to diabetes.

3. **How genetic factors influence the development of diabetes.** Several hypotheses have been advanced to explain how MHC markers and TAP proteins influence the development of diabetes, all of them hinging on their ability to bind β-cell-derived peptides that would trigger the autoimmune response resulting in diabetes (diabetogenic peptides).
   a. Protective markers would bind and present diabetogenic peptides (or peptides closely resembling those diabetogenic peptides) to the immune system during embryonic differentiation, causing the deletion or down-regulation of self-reactive clones.
   b. The reverse could also be true (i.e., protective alleles would be those that would not accommodate the self-reactive peptide or any structurally related peptides, thus eliminating the possibility of activating an autoreactive clone).
   c. Similarly, it has been proposed that Tap genes may determine susceptibility or resistance depending on whether or not they can transport the relevant endogenous peptide(s) which trigger(s) the autoimmune reaction to the endoplasmic reticulum.
4. **Sequence of pathogenic events leading to the development of IDDM.** Based on our current knowledge of the control of immunological responses and tolerance and on data accumulated from studies of IDDM patients and experimental animal models, the following hypothetical sequence of events leading to the development of IDDM can be proposed.
   a. Autoreactive clones potentially able to be engaged in autoimmunity against pancreatic β cells persist in adult life. However, in normal conditions, the interaction of those self-reactive TcR with MHC-diabetogenic peptide complexes results in weak, tolerogenic, or apoptotic signaling of the autoreactive T cells.
   b. The level of expression of diabetogenic peptides is genetically determined by the structure of MHC-II molecules and the function of Tap proteins. Those individuals able to express high levels of those peptides in association with MHC-II are at greater risk to develop IDDM.
   c. A viral infection affecting the β cells or neighboring tissues leads to the activation of TH cells involved in the antiviral response. Those cells will deliver co-stimulatory signals to the autoreactive TH cells, pushing them into a state of activation rather than anergy. IL-2 and other cytokines released by activated T cells induce the expression of MHC-II and CAMs in B cells.
   d. The activated autoreactive T cells accumulate in the pancreatic islets and release chemotactic cytokines and interferon-γ, which will attract and activate monocytes/macrophages to the area, where interactions with islet cells overexpressing CAMs will contribute to their fixation in the islets.
   e. The activated monocytes/macrophages release cytokines such as IL-1, IL-12, TNF-α, and toxic compounds such as oxygen-active radicals and

nitric oxide. The cytokines contribute to the damage by increasing the level of activation of TH1 cells (IL-12), monocytes, and macrophages (IL-1, TNF-α). β-cell death is a consequence of the release of toxic radicals, leading to oxidative changes affecting membrane lipids.

f. In addition, activated T cells expressing Fas may also contribute to β-cell death by delivering apoptotic signals.

5. **Immunotherapy.** The understanding of the pathogenesis of IDDM has led several groups to use immunosuppressive drugs, particularly cyclosporine A, to prevent the full development of the disease.
   a. To be effective, immunosuppressive therapy needs to be instituted to recently diagnosed patients with residual β-cell function.
   b. The treatment is effective while cyclosporine A is administered, but in the vast majority of cases progression to diabetes is seen soon after immunosuppressive therapy is discontinued.

B. Acanthosis Nigricans is a rare syndrome that received its name because of thickening and hyperpigmentation of the skin in the flexural and intertriginous areas, in which patients develop a particularly labile form of diabetes associated with antibodies directed against the insulin receptor.

1. **Pathogenesis.** The anti-insulin receptor antibodies block the binding of insulin to the receptor. If the antibodies themselves are devoid of activating properties, they induce insulin-resistant diabetes. On the other side, the antibodies may stimulate the insulin receptor and cause hypoglycemia.
2. **Clinical symptoms**
   a. Blocking antibodies to the insulin receptor causes hyperglycemia, which does not respond to the administration of insulin (insulin-resistant diabetes)
   b. Antireceptor antibodies with stimulating properties may induce the cellular metabolic effects usually triggered by insulin, albeit in an abnormal and unregulated fashion. The clinical picture is one of hyperinsulinism.
   c. The same patient may undergo cycles of predominance of hypo- and hyperinsulinism-like symptoms, mimicking an extremely brittle and difficult to control form of diabetes.

V. AUTOIMMUNE DISEASES OF THE GASTROINTESTINAL TRACT AND LIVER

A. Pernicious Anemia. Pernicious or megaloblastic anemia is a severe form of anemia secondary to a special type of chronic atrophic gastritis associated with lack of absorption of vitamin B_{12}.

1. **Pathogenesis.** Two major abnormalities underlie pernicious anemia: chronic atrophic gastritis and failure of production of intrinsic factor, which is required for the absorption of vitamin B_{12}.
   a. Three major types of autoantibodies related to these abnormalities are present in most patients:
      i. **Type I (blocking) antibody**, which is present in 75% of patients, binds to intrinsic factor (IF) and prevents its binding to vitamin B_{12}.
      ii. **Type II (binding) antibody**, which reacts with the IF–vitamin B_{12}

complex and inhibits IF action. The type II antibody is found in 50% of patients, and it does not occur in the absence of antibody I.

iii. **Type III (parietal canalicular) antibody** is present in the microvilli of the canalicular system of the gastric mucosa. It is found in 85–90% of patients and reacts with the parietal cell, inhibiting the secretion of IF.

In 10–15% of patients with pernicious anemia, no antibody can be detected with currently available techniques.

b. Other autoimmune diseases such as thyroiditis and Addison's disease are diagnosed with abnormally high frequency in patients with pernicious anemia.

c. Patients with pernicious anemia develop severe **neuropathy**. This is a consequence of the fact that vitamin B_{12} is an essential coenzyme for the metabolism of homocysteine, the metabolic precursor of methionine and choline. Choline is required for the synthesis of choline-containing phospholipids, and methionine is also needed for the methylation of basic myelin. The synthesis of fatty acids is also abnormal, and those abnormal fatty acids are incorporated into neural tissues. Therefore, the metabolism of myelin is abnormal in patients with vitamin B_{12} deficiency resulting in demyelination and nerve tissue damage.

d. **Hemopoiesis** is also abnormal, because vitamin B_{12} is required for the normal cellular metabolism of tetrahydrofolate; if tetrahydrofolate is not properly synthesized, folate will not be properly conjugated, and a tissue folate deficiency will ensue. In turn, purine metabolism is impaired, DNA metabolism is abnormal, and hemopoiesis cannot proceed normally.

2. **Clinical symptoms** are insidious in onset and multisystemic.

a. **Neurological symptoms**, particularly those secondary to loss of myelin on the dorsal and lateral spinal tracts, can be most striking. The patient may experience weakness and numbing of the extremities. Signs consist of loss of vibratory sense, ataxia, incoordination, and impaired mentation.

b. **Gastrointestinal abnormalities** include atrophic glossitis and gastritis.

c. **Hematological signs** are predominantly those associated with anemia.

3. **Laboratory data**

a. **Hematological findings**

i. **Megaloblastic anemia** with large, abnormal red cells in the peripheral blood and hypercellularity with numerous megaloblasts on the bone marrow (from which the term megaloblastic anemia derives).

ii. The hemopoietic abnormalities also affect other blood cells; hypersegmented neutrophils and mild-to-moderate thrombocytopenia are often present in these patients.

iii. Parenteral administration of vitamin B_{12} is followed by a marked increase in reticulocyte count, which is considered diagnostic.

b. **Other findings**

i. Plasma levels of vitamin B_{12} are low, due to decreased absorption.

ii. A histamine stimulation test of the gastric cells will show achlorhydria.

4. **Treatment.** Intramuscular injection of vitamin B_{12} will correct both hematological and early neurological manifestations.

B. **Primary Biliary Cirrhosis (Autoimmune Cholangitis)** is a chronic granulomatous inflammatory liver disease that results in destruction of the intrahepatic biliary tree, specifically affecting the epithelium of the small intrahepatic bile ducts. This disease is often associated with other autoimmune diseases such as Sjögren's syndrome and scleroderma.
   1. **Pathogenesis.** Although the true pathogenic process is not known, several immunological abnormalities can be demonstrated:
      a. **Antimitochondrial antibodies** are detected in over 99% of patients.
      b. Circulating serum immune complexes.
      c. Increased levels of serum immunoglobulins.
      d. Abnormal counts and function of CD4+ and CD8+ T lymphocytes.
   2. **Clinical presentation.** The disease is mainly a disease of middle-aged women. The onset is insidious and is heralded by pruritus and symptoms of cholestasis. Jaundice is a late sign. Patients have a large, nontender liver.
   3. **Laboratory findings.** The most significant findings from the diagnostic point of view include:
      a. Positive test for antimitochondrial antibodies.
      b. Increase in serum alkaline phosphatase with normal transaminases and bilirubin.
   4. **Treatment.** There is no satisfactory treatment for this disease. Penicillamine, a heavy metal chelating agent, has been used with some success. By unknown mechanisms, this drug is known to reduce the activity of helper T lymphocytes, which is reflected into a depression of humoral immune responses both in experimental animals and humans. However, penicillamine is nephrotoxic, and its use may be associated with severe side effects.

C. **Chronic Active Hepatitis (CAH)** is a disease characterized by persistent hepatic inflammation, necrosis, and fibrosis, which often leads to hepatic insufficiency and cirrhosis. It can be subclassified by its etiology as viral-induced, drug- or chemical-induced, autoimmune, and cryptogenic (cases that do not fit into any of the other groups).
   1. **Pathogenesis**
      a. **Viral chronic active hepatitis** can be caused by a variety of hepatotropic viruses, namely, hepatitis B, C, and D viruses.
         i. Serological evidence of chronic persistent infection with the hepatitis B virus is present in 20–30% of the patients (these patients are positive in a variety of tests that detect different hepatitis B antigens).
         ii. Superinfection of a chronic hepatitis B carrier with the hepatitis D virus is also believed to be associated with the development of CAH.
         iii. The liver disease is often accompanied by extrahepatic manifestations suggestive of immune complex disease, such as arthralgias, arthritis, skin rash, vasculitis, and glomerulonephritis. These manifestations are believed to result from chronic viral antigen release, eliciting an antibody response and consequent immune complex formation and deposition in different tissues and organs.

b. **Autoimmune chronic active hepatitis** is characterized by the presence of autoantibodies and by lack of evidence of viral infection. Based on the pattern of autoantibodies detected in different patients, CAH can be subclassified into four types.
   i. Classic autoimmune chronic hepatitis (also known as "lupoid hepatitis") is defined by the detection of antinuclear antibodies. The term "lupoid" is used to stress the common feature (i.e., antinuclear antibodies) between this type of CAH and systemic lupus erythematosus. The antinuclear antibodies in autoimmune CAH are heterogeneous and are not directed against any specific nuclear antigen. In addition, autoantibodies to liver membrane antigens and to smooth muscle are also detected in patients with this type of CAH.
   ii. The other types of autoimmune chronic active hepatitis are characterized by different patterns of detection of autoantibodies to smooth muscle, liver–kidney microsomal antigens, and soluble liver antigens.
   iii. The autoimmune form of CAH affects predominantly young or postmenopausal women. A genetic predisposition is suggested by the strong association with certain MHC antigens, particularly HLA-B1, B8, DR3, and DR4, which are also found in association with other autoimmune disorders. In addition, relatives may suffer from a variety of autoimmune diseases, such as thyroiditis, diabetes mellitus, autoimmune hemolytic anemia, and Sjögren's syndrome.
   iv. Evidence suggesting a dysregulation of the immune system in these patients includes marked hypergammaglobulinemia and detection of multiple autoantibodies.

c. **Liver damage** in all forms of CAH is believed to be the result of a cell-mediated immune response against altered hepatocyte membrane antigens.
   i. Both circulating and liver-derived lymphocytes from these patients have been shown to be cytotoxic for liver cells in vitro. Antibody-dependent cell-mediated cytotoxicity has also been suggested as playing a pathogenic role.
   ii. In the case of viral infections, the expression of viral proteins in the cell membrane of infected cells could be the initiating stimulus for the response.
   iii. The trigger of most autoimmune forms of CAH remains unknown. In some cases, drugs, particularly α-methyldopa, may play the initiating role. α-Methyldopa is believed to modify membrane proteins of a variety of cells and induce immune responses that cross-react with native membrane proteins and perpetuate the damage, even after the drug has been removed.
   iv. The pathogenesis of cryptogenic CAH, in which there is no evidence of viral infection, exposure to drugs known to be associated with CAH, or autoimmune responses, remains unknown. It is possible that most of these cases may have been caused by an unde-

tected viral infection or by exposure to an unsuspected drug or chemical agent.

2. **Diagnosis** of CAH is usually established by liver biopsy.
   a. Typically, the biopsy will reveal a picture of "piecemeal necrosis," characterized by marked mononuclear cell infiltration of the periportal spaces and/or paraseptal mesenchymal–parenchymal junctions, often expanding into the lobules. Plasma cells are often prominent in the infiltrate.
   b. There is also evidence of hepatocyte necrosis at the periphery of the lobules, with evidence of regeneration and fibrosis. It is believed that this picture reflects an immune attack of the infiltrating lymphocytes directed against the periportal and paraseptal lymphocytes.
   c. In one-quarter to one-half of the patients (depending on the study), evidence of postnecrotic cirrhosis is detected; and in some patients, the evolution toward cirrhosis is progressive.
3. **Treatment**
   a. Glucocorticoids are indicated in the autoimmune forms, but not in cases associated with viral infection.
   b. α-Interferon seems beneficial for patients with viral CAH who can complete several months of therapy without severe side effects. In some cases, interferon administration is associated with the emergence of antinuclear antibodies, which usually disappear after therapy is discontinued, but rarely may evolve toward a complete picture of autoimmune CAH requiring glucocorticoid therapy.

VI. AUTOIMMUNE DISEASES OF THE NEUROMUSCULAR SYSTEMS

A. **Myasthenia Gravis (MG)** is a chronic autoimmune disease caused by a disorder of neuromuscular transmission.

1. **Pathogenesis and pathology**
   a. Two main pathological findings are characteristic of myasthenia gravis.
      i. The production of antinicotinic-acetylcholine receptor antibodies, detected in 85–90% of the patients.
      ii. A 70–90% reduction in the number of acetylcholine receptors in the neuromuscular junctions.
   b. The reduction in the number of acetylcholine receptors is believed to be due to their destruction by the immune system. This could be a consequence of direct cytotoxicity by complement, opsonization, ADCC, activation of phagocytic cells, or T-cell-mediated cytotoxicity.
   c. Cell-mediated immunity has been suggested as playing the major pathogenic role due to the lymphocytic infiltration that is often seen at the neuromuscular junction level, and because blast transformation can be achieved by stimulating T lymphocytes isolated from myasthenia gravis patients with acetylcholine receptor protein. However, the lymphocytic infiltrates are not detected in a significant number of patients clinically indistinguishable from those with infiltrates.
   d. Thymic abnormalities are frequent in myasthenia gravis. 70% of the

patients have increased numbers of B-cell germinal centers within the thymus, which some authors have suggested to be the source of autoantibodies. About 10% of the patients develop malignant tumors of the thymus (thymomas).

2. **Clinical presentation**
   a. **Symptoms** of myasthenia gravis are increased muscular fatigue and weakness especially becoming evident with exercise. Weakness is usually first detected in extraocular muscles resulting in diplopia or ptosis. The face, tongue, and upper extremities are also frequently involved. Skeletal muscle involvement is usually proximal.
   b. The disease is usually marked by spontaneous remission periods.
3. **Laboratory diagnosis.** The diagnosis is confirmed by the finding of anti-acetylcholine receptor antibodies.
4. **Treatment**
   a. Clinical symptoms can be at least partially relieved by the administration of acetylcholinesterase inhibitors, such as neostigmine and pyridostigmine (Mestinon), in combination with atropine.
   b. **Thymectomy** is undertaken with improvement in 75% of patients and with remission in the other 25%, although it may be several months after surgery that clinical improvement starts to be obvious.
   c. **Glucocorticoids** are used in patients who do not respond to anticholinesterase drugs or thymectomy and can induce clinical improvement in 60–100% of the patients, depending on the series.
   d. **Plasmapheresis** and **thoracic duct drainage** can also be effective by removing circulating antibodies. However, the benefits of this type of therapy are very short-lived, unless the synthesis of autoantibodies is curtailed with glucocorticoids or immunosuppressive drugs.

B. Multiple Sclerosis (MS) is an autoimmune disease that results from the destruction of the myelin sheath in the central nervous system.

1. **Epidemiology**
   a. Multiple sclerosis is more frequent in northern latitudes. In the USA it is more prevalent north of the Mason Dixon line. In Europe, Scandinavian countries and Scotland have the highest incidence.
   b. MS occurs mostly in young adults between the ages of 16 and 40 with a 3-to-1 female predominance.
2. **Pathology and Pathogenesis**
   a. MS lesions observed at autopsy are characterized by areas of myelin loss surrounding small veins in the deep white matter. A perivenous cuff of inflammatory cells is associated with acute lesions but is absent from old lesions where gliosis replaces myelin and the oligodendrocytes that produce and support it.
   b. The inflammatory cells found in MS lesions are a mixture of T and B lymphocytes and macrophages (which are known as microglial cells in the central nervous system). The T lymphocytes are mostly CD4+, express IL-2R and secrete IL-2 and IFN-γ. A smaller proportion of CD4+ lymphocytes produces IL-4 and IL-10, suggesting that TH1 activity predominates over TH2 activity. A few CD8+ lymphocytes are also present in the lesions.

c. The importance of T lymphocytes in the pathogenesis of MS is supported by several lines of evidence.
   i. Experimental allergic encephalomyelitis (EAE), the best animal model for MS, is transferred by CD4+ T lymphocytes but not by serum. Injection of T-cell clones specific for the immunodominant epitope of myelin basic protein (MBP) derived from sick animals is the most efficient protocol to transmit the disease to healthy animals.
   ii. MBP-specific CD4 clones can be established from lymphocytes isolated from the spinal fluid of MS patients. These clones generally recognize an epitope located at amino acids 87 to 99, but clones specific for other groups of 12 amino acids in the MBP molecule and to other myelin components, such as proteolipid protein and myelin-associated glycoprotein, are also expanded. Therefore, many different T-cell clones with different TcRs appear to be involved in the autoimmune response.
d. As in many other autoimmune diseases, the role of genetic factors was suggested by the finding that some HLA alleles are over-represented among MS patients, particularly HLA-DR2 and HLA- DQ1, which are found in up to 70% of the patients. These class-II MHC molecules are likely to be involved in peptide presentation to CD4+ lymphocytes.
e. It has been demonstrated that normal individuals have myelin-specific T cells in their blood, but do not develop MS, even when they are HLA-DR2. MBP-specific T lymphocytes are not deleted during differentiation, probably because myelin antigens are not expressed in the thymus. However, in normal individuals these clones remain in a state of anergy or tolerance.
f. Very little is known about what activates previously unreactive MBP-specific clones and other autoreactive clones involved in MS. Viral infections have been proposed as the trigger for MS, perhaps as a consequence of molecular mimicry.
   i. Many viral antigens from corona viruses, Epstein-Barr virus, hepatitis B virus, herpes simplex virus, and others, have sequences identical to MBP epitopes. Consequently, the immune response to the virus would activate a set of T cells whose TcR would cross-react with myelin basic protein peptides.
   ii. Another possibility is that a viral superantigen could inadvertently activate an MBP-specific T lymphocyte, and cause its expansion.
g. Autoreactive T lymphocytes, by themselves, are incapable of damaging the myelin sheath. However, autoreactive T lymphocytes secrete interferon-γ, which activates the macrophages found in the lesions. Some of these activated macrophages are seen attached to the sheath, which they actively strip of myelin, becoming lipid laden. In addition, once they have engulfed myelin, they present myelin-derived antigens to T cells, contributing to the perpetuation of the immune reaction.

3. **Clinical presentation and evolution**
   a. The **clinical manifestations** of the disease are very protean and often include visual abnormalities, abnormal reflexes, sensory and motor

abnormalities. This variety of manifestations reflects the fact that lesions can occur **anywhere** in the white matter of the brain, cerebellum, pons, or spinal cord, at **any time**. The multiplicity and progression (both in number and extent) of MS lesions is the major clinical diagnostic criterion for this disease.

b. The course of MS is characterized by relapses and remissions in about 60% of the patients, but each new attack may bring additional deficits when the myelin sheaths are incompletely or imperfectly replaced. Frequently, after 5 to 15 years of evolution, these patients enter a phase of relentless chronic progression and become wheelchair bound, bedridden, and totally dependent for all activities of daily living. In the remaining 40% of the cases, MS is steadily progressive from the onset.

3. **Diagnosis**
   a. **Magnetic resonance imaging (MRI)** is most informative. It demonstrates the breakdown of the blood–brain barrier, which is always present at the beginning of a new attack, and can also document the spatial dissemination of MS lesions. However, MRI abnormalities alone are not diagnostic, because several other diseases can produce similar abnormalities.
   b. **Spinal fluid electrophoresis** is another valuable diagnostic test, based on the detection of oligoclonal bands (multiple electrophoretically homogeneous bands) of IgG in the spinal fluid. It is not specific for multiple sclerosis, because this can be observed in other neurological illnesses associated with an intrathecal immune response.
4. **Treatment**
   a. **Glucocorticoids** have been used extensively during the past 20 years. Usually high doses are required (to seal the blood–brain barrier), not suitable for long-term administration. In addition, glucocorticoid administration does not affect disease progression.
   b. Recombinant **interferon-β** and the closely related **interferon-β1** are recommended for the treatment of relapsing-remitting MS. These interferons act by down-regulating IFN-γ production and class-II expression on antigen-presenting cells. Interferon-β administration has been shown to slow down the progression of MS.
   c. **Copolymer-1** (COP-1, copaxone), a synthetic molecule designed to resemble MBP epitopes, without the ability to induce T-cell proliferation, will be available very soon. It is probably the first attempt to apply the principle of molecular mimicry to reinduce a lost tolerance.

VII. AUTOIMMUNE DISEASES OF THE BLOOD CELLS

Virtually all types of blood cells can be affected by autoantibodies. Autoimmune hemolytic anemia (AHA) is discussed in detail in Chapter 24 and autoimmune neutropenia is discussed in Chapter 17. In this chapter the discussion is limited to idiopathic thrombocytopenic purpura (ITP).

A. **Idiopathic Thrombocytopenic Purpura (ITP)** is an autoimmune disease associated with low platelet counts (thrombocytopenia).

1. **Pathogenesis.** The low platelet counts result from a shortened platelet half-life caused by antiplatelet antibodies that cannot be compensated by increased release of platelets from the bone marrow.
   a. Antiplatelet autoantibodies have been detected in 60–70% of patients with the "immunoinjury" technique that relies on the release of platelet factors, such as serotonin, following exposure to sera containing such antibodies.
   b. Competitive binding assays and antiglobulin assays can also be used to demonstrate antiplatelet antibodies.
   c. Idiopathic thrombocytopenic purpura can present as an acute or as a chronic form.
      i. **Acute ITP** is due to the formation of immune complexes containing viral antigens that become adsorbed to the platelets or to the production of antiviral antibodies that cross-react with platelets. Platelet destruction can occur due to irreversible aggregation caused by immune complexes, when antiplatelet antibodies are involved, or when there is complement-induced lysis or phagocytosis.
      ii. **Chronic ITP** is caused by autoantibodies that react with platelets and lead to their destruction by phagocytosis.
2. **Clinical presentation.** ITP is characterized by easy and exaggerated bleeding secondary to thrombocytopenia. Patients with ITP usually present with subcutaneous and mucosal bleeding. The subcutaneous bleeding is clinically described as petechiae or ecchymosis, appearing without obvious cause. When the bleeding is profuse, it will lead to the appearance of areas of purple discoloration, from which the designation of "purpura" derives. Epistaxis and gingival, genitourinary and gastrointestinal tract bleeding are the common manifestations of mucosal bleeding.
   a. **Acute ITP** is seen mainly in children, often in the phase of recovery after a viral exanthem or an upper respiratory infection and is usually self-limited.
   b. **Chronic ITP** is an adult disease, often associated with other autoimmune diseases.
3. **Laboratory diagnosis.** The platelet count in the acute form is usually less than 20,000/mm^3, whereas the chronic form has platelet counts between 30,000–100,000/mm^3. The white cell count is usually normal. The bone marrow is also usually normal, but in some cases an increase in megakaryocytes may be seen, representing an attempt to compensate for the excessive destruction in the peripheral blood. The spleen may be enlarged due to platelet sequestration by phagocytic cells.
4. **Treatment**
   a. **Glucocorticoids** have been used but their efficiency in severe cases is questionable.
   b. Administration of **intravenous gammaglobulin** is the therapy of choice. It is associated with a prolongation of platelet survival and improvement in platelet counts. Two mechanisms have been postulated to explain this effect of intravenous gammaglobulin in ITP.
      i. Competition with immune complexes for the binding to platelets

(immune complexes would cause irreversible aggregation or complement-mediated cytolysis and intravenous gammaglobulin would not).

ii. Blocking of Fc receptors in phagocytic cells, which would inhibit the ingestion and destruction of antibody-coated platelets (for which there is experimental documentation).

The long-term, beneficial effects of intravenous gammaglobulin administration seem to result from a not-so-well understood immunomodulatory effect that results in a decrease of the titers of antiplatelet antibodies.

B. **Splenectomy** is indicated in severe cases that do not respond to other forms of treatment. The removal of the spleen, usually the major site of platelet sequestration and destruction, is often associated with prolonged platelet survival.

SELF-EVALUATION

Questions

Choose the ONE *best* answer.

19.1 Which one of the following characteristics best defines thyroid-stimulating antibodies?
A. Activation of thyroid function after binding to the thyroglobulin receptor
B. Association with Graves' disease
C. Blocking the binding of TSH to its receptor
D. IgG isotype
E. Mimicking the effects of TSH upon binding to the TSH receptor

19.2 Which one of the following genetic markers is associated with protection against the development of diabetes in Caucasians?
A. Arginine at DQα52
B. Aspartic acid at DQβ57
C. B27 positivity
D. D3/DR4 positivity
E. Valine at DQβ57

19.3 What is believed to be the mechanism responsible for the rapid increase in platelet counts observed in patients with idiopathic thrombocytopenic purpura after starting therapy with intravenous gammaglobulin?
A. Blockade of Fc receptors on phagocytic cells
B. Down-regulation of autoantibody-producing plasma cells
C. Inactivation of helper T cells
D. Induction of suppressor cell activity
E. Stimulation of megakaryocyte release from the bone marrow

19.4 Which of the following is believed to be the pathogenic mechanism of the neurological abnormalities associated with pernicious anemia?
A. Abnormal synthesis of phospholipids and fatty acids
B. Autoimmune demyelination affecting both motor and sensory tracts
C. Lack of oxygen supply to the dorsal and lateral spinal tracts
D. Loss of acetylcholine receptors at the neuromuscular junctions
E. Progressive and multifocal degeneration of neural tissue

19.5 Which of the following is believed to be the mechanism underlying systemic symptoms (such as arthritis and glomerulonephritis) in patients with hepatitis B virus-associated chronic active hepatitis?
A. Deposition of circulating immune complexes
B. Deposition of unconjugated bilirubin in tissues
C. Exacerbation of the liver disease by co-infection with hepatitis D virus
D. Production of antiviral antibodies cross-reactive with tissue antigens
E. Reaction of autoantibodies with tissue antigens

19.6 The use of immunosuppressive drugs in recently diagnosed type I diabetic patients is aimed at:
A. Avoiding further destruction of islet cells by islet cell antibodies
B. Preventing further loss of islet cells due to cell-mediated cytotoxicity
C. Reducing the expression of "predisposing" MHC-II haplotypes
D. Reducing the synthesis of anti-insulin antibodies
E. Reducing the synthesis of anti-islet cell antibodies

In **Questions 19.7–19.10** match EACH numbered word or phrase with the ONE lettered heading that is most closely related to it. The same heading may be selected once, more than once, or not at all.
A. Anti-islet cell antibodies
B. Antiadrenal microsomal antibodies
C. Antinuclear antibodies
D. Antimitochondrial antibodies
E. Anti-insulin receptor antibodies

19.7 Addison's disease
19.8 Primary biliary cirrhosis
19.9 Autoimmune chronic active hepatitis
19.10 *Acanthosis nigricans*

Answers

19.1 (E) All antibodies to the TSH receptor, including TSI and those antibodies devoid of stimulating effects, bind to the TSH receptor and block the binding of the hormone, which is the natural substrate for the receptor. In addition, TSI are so designed because they induce the same type of stimulating effects over thyroid cells as TSH.

19.2 (B) The presence of aspartic acid at DQβ57 is associated with protection against the development of diabetes, probably by interfering with the binding of peptides associated with the autoimmune reaction against β cells.

19.3 (A) The fast increase in platelet count seen in patients with ITP after therapy with intravenous gammaglobulin instituted is believed to be due to the blockade of the Fc receptors of reticuloendothelial phagocytic cells by the infused IgG.

19.4 (A) The neuropathy associated with pernicious anemia is secondary to demyelination caused by an abnormal synthesis of phospholipids and fatty acids affecting predominantly the dorsal and lateral spinal tracts.

19.5 (A) In hepatitis B-associated CAH, there is continuing synthesis and release of viral antigens that become associated in circulation with the corresponding antibodies, forming soluble immune complexes which eventually can be trapped at the level of small vessels in the skin, kidneys, and joints, initiating inflammatory processes that lead to tissue damage.

19.6 (B) At the time of diagnosis of type I diabetes, there is active insulitis, but not all islet cells are destroyed; often a brief spontaneous remission (honeymoon period) can be observed shortly after the initial diagnosis. The rationale for the use of immunosuppressive drugs at that time is to reduce the attack of cytotoxic cells to the B cells of the Langerhans islets, hoping to preserve their function and to avoid progression toward diabetes.

19.7 (B)

19.8 (D)

19.9 (C)

19.10 (E) Anti-insulin receptor antibodies are found in rare cases of noninsulin-dependent diabetes, often associated with *acanthosis nigricans* which characteristically shows a marked clinical instability.

BIBLIOGRAPHY

Acha-Orbea, H., and McDevitt, H.O. The role of class II molecules in development of insulin-dependent diabetes mellitus in mice, rats and humans. *Curr. Topics Microbiol. Immunol.*, *156*:103, 1990.

Caillat-Zucmann, S., Bertin, E., Timsit, J., et al. Protection from insulin-dependent diabetes mellitus is linked to a peptide transporter gene. *Eur. J. Immunol.*, *23*:1784, 1993.

Ehrlich, H.A. HLA class II polymorphism and genetic susceptibility to insulin-dependent diabetes mellitus. *Curr. Topics Microbiol. Immunol.*, *164*:41, 1990.

Griffin, A.C., Zhao, W., Wegmann, K.W. and Hickley, W.F. Experimental autoimmune insulitis. Induction by T lymphocytes specific for a peptide of proinsulin. *Am. J. Pathol.*, *147*:845, 1995.

Imbach, OP. Immune thrombocytopenic purpura and intravenous immunoglobulin. *Cancer*, *68*:1422, 1991

Lindstrom, J., Shelton, D., and Fujii, Y. Myasthenia gravis. *Adv. Immunol.*, *42*:233, 1988.

Mariotti, S., Chiovato, L., Vitti, P., Marcocci, C., Fenzi, G.F., Del Prete, G.F., Tiri, A., Romagnani, S., Ricci, M., and Pinchera, A. Recent advances in the understanding of humoral and cellular mechanisms implicated in thyroid autoimmune disorders. *Clin. Immunol. Immunopathol.*, *50*:S73, 1989.

McDougall, I.R. Graves' disease. Current concepts. *Med. Clin. North Am.*, *75*:79, 1991.

Mondelli, M.U., Manns, M., and Ferrari, C. Does the immune response play a role in the pathogenesis of chronic liver disease? *Arch. Pathol. Lab. Med.*, *112*:489, 1988.

Pozzilli, P., Visalli, N., Boccuni, M.L. et al. Randomized trial comparing nicotinamide and nicotinamide plus cyclosporine in recent-onset insulin-dependent diabetes. *Diabetic Med.*, *11*:98, 1994.

Rapoport, B. Pathophysiology of Hashimoto's thyroiditis and hypothyroidism. *Annu. Rev Med.*, *42*:91, 1991.

Wall, J.R., Bernard, N., Boucher, A., et al. Pathogenesis of thyroid associated ophthalmopathy. *Clin. Immunol. Immunopathol.*, *68*:1, 1993.

20
Systemic Lupus Erythematosus

Jean-Michel Goust and George C. Tsokos

I. INTRODUCTION

Systemic lupus erythematosus (SLE) is a generalized autoimmune disorder associated with multiple cellular and humoral immune abnormalities and protean clinical manifestations. It is most common in females of child-bearing age.

Case 20.1

A 25-year-old female was taken to the emergency room by her husband after having a generalized seizure. When she regained consciousness, she gave a history of feeling tired over the previous 3 weeks, to the point where she was breathless and exhausted after climbing to her second-floor apartment. She also complained of the progressive development of pain and stiffness in her wrists and fingers. She had noticed what looked like a bad sunburn on her cheeks and had been running a slight fever (99.9 F). Her past history was significant for two previous episodes of moderate wrist and finger pain which was relieved by aspirin. She had experienced two miscarriages in the past 3 years. Social history revealed that she did not smoke, drank only in moderation, and did not use recreational drugs. On physical examination, this well-nourished Caucasian female had a temperature of 101.5 F, pulse 90/min., BP 135/60, and weight 142 lb. representing a loss of 7 lb. in 3 weeks. She had what appeared to be a second-degree burn on her face. Hands and wrists were warm, painful to pressure, with a limited range of motion that slowly increased with mobilization. Neurological examination revealed no evidence of a focal deficit. Optic fundi were normal. Remaining PE was unremarkable.

This case history raises the following questions:

- What is the most likely explanation for this patient's fatigue and seizures?
- Is there any likely relation between her disease and the three miscarriages?
- What is the pathogenesis of this patient's skin lesions?
- What complications would one be most concerned about?
- What test(s) would help confirm the diagnosis and evaluate the cause of this patient's symptoms?
- What triggered this disease in this patient?

II. CLINICAL MANIFESTATIONS

The clinical expression of SLE varies among different patients. The kind of organ (vital versus nonvital) that becomes involved determines the seriousness and the overall prognosis of the disease. The average frequency of some main clinical manifestations of SLE that may be observed during the entire course of SLE is shown in Table 20.1

A. **Diagnosis.** The diagnosis is based on the verification that any four of the manifestations that are listed in Table 20.2 are present simultaneously or serially during a period of observation. Those manifestations are both clinical, indicating multisystemic involvement, and laboratory demonstration of autoantibodies.

B. **Course.** Exacerbations and remissions, heralded by the appearance of new manifestations and worsening of preexisting symptoms, give the disease its fluctuating natural history.

C. **Overlap Syndrome.** Occasionally, physicians observe clinical situations in which the differentiation between SLE and another connective tissue disease is difficult. In some patients, the distinction may be impossible, and they are classified as having an overlap syndrome. This syndrome represents the association of SLE with another disorder such as scleroderma or rheumatoid arthritis.

III. IMMUNOLOGICAL ABNORMALITIES IN SLE

A. Autoantibodies

1. **The LE cell** is a peculiar-looking polymorphonuclear leukocyte that has ingested nuclear material. It was possible to reproduce this phenomenon in vitro by incubating normal neutrophils with damaged leukocytes preincubated with sera obtained from SLE patients. Investigations concerning the

Table 20.1 Main Clinical Manifestations of SLE

Manifestation	Patients (%)
Musculoarticular	95
Renal disease	60
Pulmonary disease (pleurisy, pneumonitis)	60
Cutaneous disease (photosensitivity, alopecia, etc.)	80
Cardiac disease (pericarditis, endocarditis)	20
Fever of unknown origin	80
Gastrointestinal disease (hepatomegaly, ascites, etc.)	45
Hematological/reticuloendothelial (anemia, leukopenia, splenomegaly)	85
Neuropsychiatric (organic brain syndrome, seizures, peripheral neuropathy, etc.)	20

Table 20.2 Diagnostic Features of Systemic Lupus Erythematosus[a]

Facial erythema (butterfly rash)
Discoid lupus
Photosensitivity
Oral or nasopharyngeal ulcers
Arthritis without deformity
Pleuritis or pericarditis
Psychosis or convulsions
Hemolytic anemia, leukopenia, lymphopenia, or thrombocytopenia
Heavy proteinuria, or cellular casts in the urinary sediment
Positive anti-dsDNA or anti-Sm antibodies
False positive syphilis serologies
Antinuclear antibodies

[a]Established by the American College of Rheumatologists.

nature of this phenomenon led to the discovery that antibodies directed to nuclei could promote the formation of LE cells, and subsequently to the definition of a heterogeneous group of antinuclear antibodies.

2. **Antinuclear antibodies (ANA).** ANAs are detected by an indirect immunofluorescence assay using a variety of tissues and cell lines as substrates. A positive result is indicated by the observation of nuclear fluorescence after incubating the cells with the patient's serum, and, after thorough washing to remove unbound immunoglobulins, with an antihuman immunoglobulin serum. Four patterns of fluorescence can be seen, indicating different types of antinuclear antibodies (see Table 20.3). The test for antinuclear antibodies is not very specific, but is very sensitive. A negative result virtually excludes the diagnosis of SLE (95% of patients with SLE are ANA positive), while high titers are strongly suggestive of SLE but not confirmatory, since ANAs can be detected in other conditions including other systemic autoimmune/collagen diseases and chronic infections.

Table 20.3 Immunofluorescence Patterns of Antinuclear Antibodies

Pattern	Antigen	Disease association(s)
Peripheral	Double-stranded DNA	SLE
Homogeneous	DNA-histone complexes	SLE and other connective tissue diseases
Speckled	Non-DNA nuclear antigens	
	Sm	SLE
	ribonucleoprotein	Mixed connective tissue disease, SLE, scleroderma, etc.
	SS-A, SS-B	Sjögren's disease
Nucleolar	Nucleolus-specific RNA	Scleroderma

3. **Anti-DNA antibodies** are the most important in SLE. They can react with single-stranded DNA (ssDNA) or with double-stranded DNA (dsDNA). Two-thirds of patients with SLE have circulating anti-DNA antibodies.
   a. Anti-ssDNA may be found in many diseases besides SLE.
   b. Anti-dsDNA antibodies are found almost exclusively in SLE (60–70% of the patients). They are most commonly detected by immunofluorescence, using as a substrate a noninfectious flagellate, *Crithidia lucilliae*, which has a kinetoplast packed with double-stranded DNA. This test is very specific, and the antibodies can be semiquantitated by titration of the serum (to determine the highest serum dilution associated with visible fluorescence of the kinetoplast after addition of a fluorescent-labeled anti-IgG antibody).
   c. Levels of serum anti-DNA antibodies may fluctuate with disease activity, but they are poor predictors of disease activity.
4. **Antibodies to the DNA-histone complex** are present in over 65% of patients with SLE. The use of enzyme-linked immunosorbent assays has permitted the identification of antibodies to all histone proteins including H1, H2A, H2B, H3, and H4. Antihistone antibodies are also present in patients with drug-induced SLE, most frequently associated with hydralazine and procainamide therapies.
5. **Antibodies to nonhistone proteins** have been studied intensely lately. The nonhistones against which antibodies have been described include:
   a. **Anti-Sm.** Antibodies to the Sm antigen are present in one-third of patients with SLE. Anti-Sm antibodies have not been found in other conditions. The antigenic determinant is on a protein that is conjugated to one of six different small nuclear RNAs (snRNA).
   b. **Anti-U1-RNP.** The antigenic epitope is on a protein conjugated to U1-RNA. Antibodies to this antigen are present in the majority of patients with SLE and in mixed connective tissue disease, which is included in the overlap syndromes.
   c. **Anti-SS-A/Ro.** These antibodies are present in one-third of patients with SLE and two-thirds of patients with Sjögren's syndrome (SS). Antibodies to the Ro antigen are frequently found in patients with SLE who are ANA-negative. Babies born to mothers with anti-Ro antibodies may have heart block, leukopenia, and/or skin rash.
   d. **Anti-SS-B/La.** The antigenic epitope recognized by this antibody is on a 43-kD protein conjugated to RNA. Antibodies to La antigen are present in about one-third of patients with SLE and in approximately one-half of the patients with Sjögren's syndrome.
6. **Cross-reactive autoantibodies.** Patients with SLE frequently have anti-phospholipid antibodies and antibodies cross-reactive with cardiolipin. The anticardiolipin antibodies recognize a cryptic epitope on β_2-glycoprotein I that is exposed after it binds to anionic phospholipids.

B. **The Pathogenic Role of Autoantibodies in SLE.** Classically, it has been accepted that autoantibodies do not play the initiating role in the pathogenesis of SLE, but that they are likely to play either an important role as co-factors in the pathogenesis of the disease, or play a direct role in the pathogenesis of some of the manifestations of the disease. This dogma, based on the belief that autoantibodies cannot enter living cells, has recently been challenged. Not only can anti-RNP, anti-DNA, and anti-Ro antibodies enter live cells, but also they can induce apoptosis. Thus, the concepts about the pathogenic role of autoantibodies are likely to change in the near future.

1. **Anti-T-cell antibodies** are believed to bind and eliminate certain regulatory subsets of T cells; as a consequence, the normal negative feedback circuits controlling B-cell activity may not be operational, explaining the uncontrolled production of autoantibodies by the B cells.
2. **Antibodies against CR1** (complement receptor 1) and against the **C3 convertase** are occasionally detected. CR1 antibodies may block the receptor and interfere with the clearance of immune complexes. Antibodies to the C3 convertase, by stimulating its function, may contribute to increased C3 consumption.
3. Anti-red-cell antibodies and antiplatelet antibodies are the cause, respectively, of **hemolytic anemia and thrombocytopenia**.
4. Autoantibodies directed against central nervous system antigens may be detected in the serum and the cerebrospinal fluid of patients with SLE who have CNS involvement and have also been considered pathogenic.
5. **Anti-DNA antibodies** form immune complexes by reacting with DNA and are implicated in the pathogenesis of glomerulonephritis (see below).
6. **Anticardiolipin antibodies** cause false positives in serological tests for syphilis, seen frequently in SLE patients. They are also associated with miscarriages, thrombophlebitis and thrombocytopenia, and various central nervous system manifestations because of vascular thrombosis. The constellation of these symptoms is known as **antiphospholipid syndrome** and, although it was first recognized in lupus patients, the majority of the cases do not fulfill the diagnostic criteria for SLE. Anticardiolipin antibodies are detected in up to 8% of the normal population.

C. **The Diagnostic Value of Autoantibodies.** Some autoantibodies may **not be linked with any specific** clinical manifestations but are very useful as disease markers. For example, **Anti-dsDNA** and **anti-Sm antibodies** are diagnostic of SLE. Most other autoantibodies are present in more than one clinical disease or syndrome. However, even if many of the patients share some common immunological abnormalities, particularly the presence of antinuclear antibodies or of rheumatoid factor, specific disorders can usually be individualized by the presence of a specific set of autoantibodies, as illustrated diagrammatically in Figure 20.1.

D. **Immune Complexes in SLE**

1. Marked elevations in the levels of circulating immune complexes can be detected in patients with SLE sera during acute episodes of the disease by a variety of techniques (see Chap. 25). Since patients with active SLE have high levels of free circulating DNA and most have also anti-DNA antibodies, DNA-anti-DNA IC are likely to be formed either in circulation or in collagen-rich tissues and structures such as the glomerular basement membrane, which have affinity for DNA.
2. Besides the fact that immune complexes are formed at increased rates in

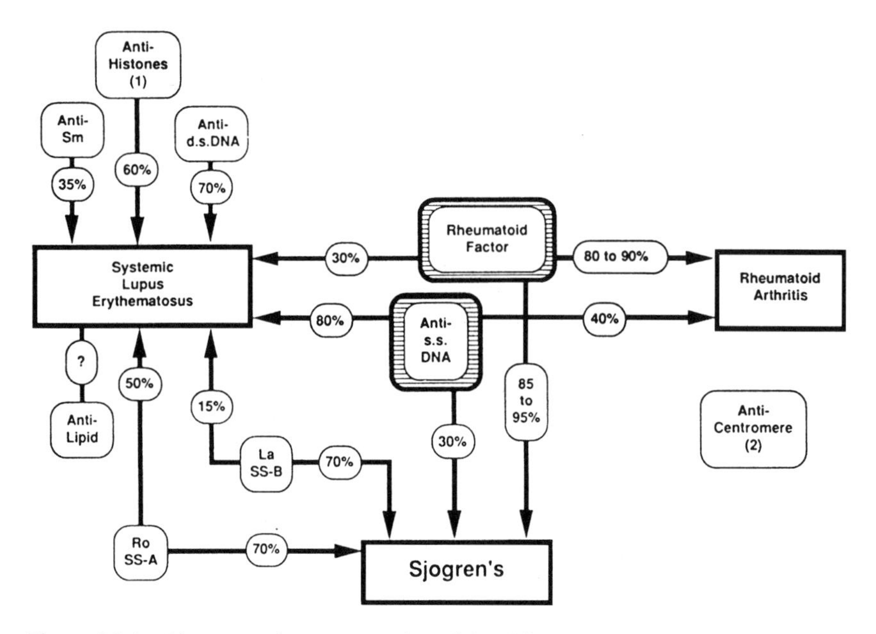

Figure 20.1 Diagrammatic representation of the different types of autoantibodies detected in SLE and related diseases. The percentages refer to the positivity rate in the disease to which the arrows point. Note that (1) antihistone antibodies are frequently found as the only serological abnormality in cases of SLE-like syndromes, and (2) anticentromere antibodies are characteristically associated to the "CREST" syndrome (associated of calcinosis, Raynaud's syndrome, scleroderma involving the skin and the esophagus, and telangiectasia).

patients with SLE, the clearance rate of circulating immune complexes is decreased. Several factors seem to contribute to the impaired clearance of immune complexes:

a. Immune complexes are cleared by the Fc-receptor-bearing cells of the reticuloendothelial system. This function has been found to be decreased in patients with SLE.
b. Immune complexes are transported to the reticuloendothelial system by red blood cells that bind them through their CR1 (as noted earlier, IC often have adsorbed complement components and split products, including C3b, which reacts with CR1), and patients with SLE have decreased numbers of CR1, a fact that may compromise the clearance of immune complexes.
c. Immune complexes are partially solubilized as a consequence of complement activation, a process that contributes to their inactivation and clearance. Individuals with C4 deficiency develop a disease with clinical features resembling those of SLE. This observation may be explained by the fact that immune complexes are cleared at slower rates in C4-deficient individuals, perhaps due to the role of C4 fragments in the solubilization and clearance of circulating immune complexes.

2. The pathogenic role of immune complexes is suggested by a variety of observations:

a. Rising levels of anti-DNA antibodies in conjunction with falling serum C3 levels (reflecting consumption by antigen-antibody complexes) are good predictors of an oncoming flare of lupus nephritis.
b. Patients with IgG1 and IgG3 (complement fixing) anti-DNA antibodies develop lupus nephritis more frequently than those patients in whom anti-DNA antibodies are of other isotypes.
c. There is ample evidence for complement activation via the classic and the alternative pathway in patients with active nephritis. Circulating levels of C3 and C4 are usually decreased, whereas plasma levels of complement breakdown products such as C3a, C3d, and Bb are increased. C1q, C3b, and complement split products such as C3d, C3b, and C3c can be detected on circulating immune complexes.

3. It is believed that tissue-fixed immune complexes cause inflammatory changes secondary to complement activation. Glomerulonephritis, cutaneous vasculitis, arthritis, and some of the neurological manifestations of SLE are fully explainable by the local consequences of immune complex formation or deposition. It is unclear whether tissue-fixed IC results mostly from the deposition of circulating IC or from formation of antigen–antibody complexes in situ.
   a. **Glomerulonephritis.** Immunofluorescence studies indicate that the capillary tufts of renal glomeruli in patients with **lupus nephritis** contain deposits of immunoglobulins and complement. Elution studies have shown that DNA and anti-DNA antibodies are present in these deposits, confirming that they correspond to antigen–antibody complexes. The deposition of IC in the glomerular basement membrane can be explained in three different ways:
      i. Deposition of soluble, circulating IC (as discussed in Chap. 25)
      ii. Formation of immune complexes in situ. DNA has affinity to glomerular basement membranes and, once immobilized, may react with circulating anti-DNA antibodies to form antigen–antibody complexes.
      iii. Cross-reaction of anti-DNA antibodies with collagen and cytoskeleton proteins.

      Currently, in situ formation of DNA–anti-DNA antibodies appears as the most likely initiating event, but regardless of how they are formed, IC in the basement membrane are considered nephritogenic because they can activate complement and cause inflammation.
   b. Deposition in the dermo-epidermal junction. In SLE patients, immune complex deposits have also been noted on the dermo-epidermal junction of both inflamed skin and normal skin, appearing as a fluorescent "band" when a skin biopsy is studied by immunofluorescence with antisera to immunoglobulins and complement components (band test).

IV. PATHOGENESIS OF SLE

A. General Considerations. At the present time, we believe that multiple factors contribute to the pathogenesis of SLE. Environmental, hormonal, genetic, and immunoregulatory factors are involved in the expression of the disease. The various etiological factors contribute differentially to the expression of the disease in individual patients.

B. **Insights from Animal Models.** The understanding of the pathogenic mechanisms underlying the progression of SLE has been helped by the discovery of spontaneously occurring disease in mice which resembles SLE in many respects. During the inbreeding of mice strains, it was observed that the F1 (first generation) hybrids obtained by mating white and black mice from New Zealand [(NZB × NZW)F1] spontaneously developed a systemic autoimmune disease involving a variety of organs and systems. Throughout the course of their disease, the mice develop hypergammaglobulinemia, reflecting a state of hyperactivation of the humoral immune system. The animals have a variety of autoantibodies and manifestations of autoimmune disease and immune complex disease similar to those seen in humans with SLE. As the disease progresses, they develop nephritis and lymphoproliferative disorders and die.

C. **Genetic Factors in Murine and Human SLE**

1. **Genetic determinants of disease in NZB mice.** The importance of genetic factors is underlined by the observation that the parental NZB mice have a mild form of the disease manifested by autoimmune hemolytic anemia, but that the introduction of the NZW genetic background made the disease accelerate and worsen. Genetic linkage studies and microsatellite gene marker analysis indicate that many of the immunological abnormalities are under multigenic control, one gene(s) controlling the animal's ability to produce anti-DNA antibodies, another the presence of anti-erythrocyte antibodies, and still other genes controlling high levels of IgM production and lymphocytic proliferation.
2. **Genetic determinants of disease in MRL mice.** Two other mouse strains that develop an SLE-like disease spontaneously have been identified: MRL lpr/lpr and MRL gld. The first strain has a defect in the FAS gene, whereas the second has a defect in the FAS ligand gene. The products of these two genes are responsible for the programmed cell death of cells also known as apoptosis, which is critical for the control of undesirable immune responses. Similar abnormalities have not been clearly established in humans.
3. **Genetic factors in human SLE.** Several pieces of evidence indicate that genetic factors also play a role in the pathogenesis of human SLE.
   a. Presence of serum anti-DNA and anti-T-cell antibodies as well as cellular abnormalities in healthy relatives of SLE patients.
   b. Moderate degree of concordance among monozygotic twins. The fact that the clinical concordance between twins is only moderate strongly indicates that genetic factors alone may not lead to the expression of the disease and that other factors are needed. The genes that could play a role, probably in synergy with environmental factors, have not been identified.
   c. Current evidence indicates that in humans, as in mice, these genes are probably linked to the MHC. For example, the HLA-DR2 haplotype is overrepresented in patients with SLE. Also, as mentioned before, an SLE-like disease develops frequently in individuals with **C4A deficiency** (C4A genes are located in chromosome 6, in close proximity to

the MHC genes). At present, lack of one of the C4A genes represents the highest single genetic predisposing factor for the development of SLE in humans.

D. Immune Response Abnormalities in SLE

1. **B-cell abnormalities.** Increased numbers of B cells and plasma cells are detected in the bone marrow and peripheral lymphoid tissues secreting immunoglobulins spontaneously. The number of these cells correlates with disease activity.
   a. In murine lupus, the B cells responsible for the production of anti-DNA antibodies express the CD5+ marker, which has been suggested to identify activated B cells. However, in humans, anti-DNA antibodies are secreted by both CD5+ and CD5− B cells.
   b. In addition, only a limited number of light- and heavy-chain genes are used, demonstrating that the autoantibody response involves only a few of all B-cell clones available. Furthermore, the changes appearing in their sequence over time strongly suggest that they undergo affinity maturation, a process that requires T-cell help. It also suggests that the response is driven by a few antigens.
   c. Immunosuppressive drug treatment of both murine and human lupus causes clinical improvement associated with decreased B-cell activity.
   d. Any infection that induces B-cell activation is likely to cause a clinical relapse in patients with inactive SLE.
2. **T-cell abnormalities.** From our knowledge of the biology of the immune response, it can be assumed that the production of high titers of IgG anti-dsDNA antibodies in patients with SLE must depend upon excessive T-cell help and/or insufficient control by suppressor T cells.
   a. In both human and murine lupus, a new subset of CD3+ cells that express neither CD4 nor CD8 has been found to provide help to autologous B cells synthesizing anti-DNA antibodies.
   b. The finding of anti-T-cell antibodies in the serum of (NZB × NZW)F1 mice and in the sera of humans with SLE raised the possibility that another factor contributing to the inordinate B-cell activity associated with the development of autoimmunity could be the deletion of a specific subset of regulatory cells. Obviously, defective suppressor T-cell function may further enhance the helper T-cell-mediated B-cell overactivity.
   c. In humans, the anti-T-cell antibodies are also responsible for the lymphopenia that is frequently seen in patients with SLE. This lymphopenia is often associated with findings suggestive of a generalized depression of cell-mediated immunity, such as decreased lymphokine production (IL-1 and IL-2) and lack of reactivity (anergy) both in vivo and in vitro to common recall antigens, particularly during active phases of human SLE. The impairment of cell-mediated immunity may explain the increased risk of severe opportunistic infections in patients with SLE.
   d. Extensive **deletions in the T-cell repertoire** have been found in NZW mice, in which the $C_{\beta2}$ and $D_{\beta2}$ genes of the T-cell antigen receptor are missing. These deletions could be associated with a faulty establishment of tolerance to self-MHC during intrathymic ontogeny.

e. In humans, restriction fragment length polymorphism studies of the constant region of the TcR demonstrated an association between TcRα chain polymorphism and SLE and TcRβ chain polymorphism and production of anti-Ro antibodies. More recently, sequence information of the TCR chains of pathogenic human T-cell clones demonstrated bias in the T-cell repertoire selection process, whose meaning is still to be defined. The immune response is thus "oligoclonal" in both T and B compartments.

D. Nonimmune Factors Influencing the Course of SLE

1. **Hormonal effects.** The expression of the genetic and immunological abnormalities characteristic of murine lupus like disease is influenced by female sex hormones.
   a. In (NZB × NZW)F1 mice, the disease is more severe females. Administration of estrogens aggravates the evolution of the disease, which is only seen in castrated male mice and not in complete males.
   b. The extent of the hormonal involvement in human SLE cannot be proven so directly, but the large (9:1 female:male ratio) female predominance as well as the influence of puberty and pregnancies at the onset of the disease, or the severity of the disease's manifestations, indicates that sex hormones play a role in the modulation of the disease.
2. **Environmental factors**
   a. **Sunlight exposure** was the first environmental factor influencing the clinical evolution of human SLE to be identified. Exposure to sunlight may precede the clinical expression of the disease or disease relapse. This could be related to the fact that the Langerhans cells of the skin and keratinocytes release significant amounts of interleukin-1 upon exposure to UV light, and could thus represent the initial stimulus tipping off a precarious balance of the immune system.
   b. **Infections** also seem to play a role. The normal immune response to bacterial and viral infections may spin off into a state of B-cell hyperactivity, triggering a relapse.
   c. **Drugs**, particularly those with DNA binding ability, such as *hydantoin*, *isoniazide*, and *hydralazine*, can cause a **drug-induced lupus-like syndrome**.
      i. These drugs are known to cause DNA hypomethylation. Because hypomethylated genes are transcribed at higher rates, it is theoretically possible that they cause SLE by increasing the transcription rate of genes that are involved in the expression of the disease.
      ii. Antinuclear antibodies appear in 15 to 70% of patients treated with any of these drugs for several weeks. These antinuclear antibodies belong, in most cases, to the IgM class and react with histones. Only when the antibodies switch from IgM to IgG does the patient become symptomatic. These antinuclear antibodies usually disappear after termination of the treatment.
      iii. Patients with drug-induced SLE usually have a milder disease, without significant vital organ involvement.

Case Revisited

- The patient's fatigue could just be a reflection of a systemic inflammatory disease, but could also be due to anemia. Hemolytic anemia is not infrequent in SLE. Seizures and other neurological symptoms may be due either to the deposition of immune complexes in CNS tissues or to the binding of antineuronal autoantibodies, with or without complement activation, the effect of infiltrating autoreactive T cells, and to the effects of neurotoxins (such as quinolinic acid) released by activated immune cells.
- Patients with SLE often develop antiphospholipid antibodies. Although the exact pathogenic sequence is not known, these antibodies interfere with clotting factors causing vascular thrombosis usually without vasculitis.
- The pathogenesis of skin lesions is likely to involve several factors. Deposition of IC at the dermoepidermal junction is likely to play a role, but not the only one, since this deposition can be observed in normal skin. Exposure to sunlight is likely to play a significant role as well, perhaps because the Langerhans cells and keratinocytes of the skin release significant amounts of interleukin-1 upon exposure to UV light, and could thus add an additional trigger to a local inflammatory reaction involving both the effects of UV exposure and the effects of IC deposition.
- In this patient, several complications would cause concern: autoimmune hemolytic anemia may be extremely difficult to treat, and the progression of her CNS involvement would also raise considerable therapeutic problems. In addition, deterioration of the kidney function, due to the development of lupus nephritis, is always a major concern in any patient with SLE.
- This patient presented with several clinical features typical of SLE: facial erythema, arthritis, seizures, and possible anemia. The detection of autoantibodies such as those directed against dsDNA or the Sm antigen would be confirmatory, but even if only nonspecific antinuclear antibodies were detected, a diagnosis of SLE should still be entertained. To evaluate the cause of some of the most striking symptoms of this patient, a complete blood count should be performed and, if anemia is present, Coombs tests should be ordered. Other important tests to be ordered included X-rays of the hands, rheumatoid factor, anticardiolipin and antiphospholipid antibodies, serum creatinine, and urine protein (to evaluate kidney function).
- One of the most difficult questions to answer is what triggers the onset of an autoimmune disease. This patient fits in the age and sex group in which SLE is more prevalent. Other than hormonal influences, it is likely that this patient carries a genetic predisposition to develop SLE (although the precise marker and nature of such predisposition are not yet clearly identified). Most unclear of all is what is the initial stimulus. An infection leading to cross-reactive autoimmunity is a very appealing hypothesis, but we have no clue about the nature of such infection.

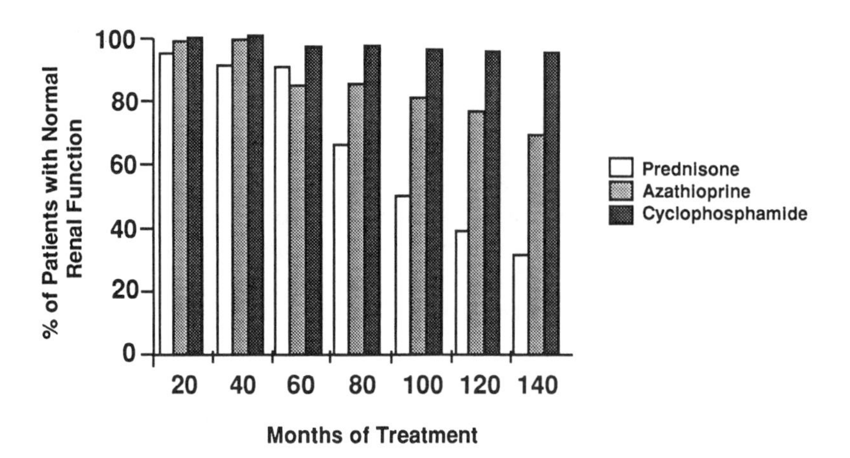

Figure 20.2 Comparison of three different types of immunosuppressive therapy in the evolution of renal disease in patients with SLE. Three different groups of patients with three different types of immunosuppressive therapy: prednisone, oral azathioprine, and cyclophosphamide administered intravenously. Cyclophosphamide appears to be more successful in delaying the onset of renal failure.

V. TREATMENT

Improvement of our understanding of the pathogenesis of SLE has led to reasonable therapeutic strategies that have dramatically improved the well being and life expectancy of patients with SLE, whose survival rate at 10 years is now 80%. The therapeutic approach to each patient is determined by the extent of the disease and, most importantly, by the nature and extent of organ involvement.

A. **Corticosteroids** combine anti-inflammatory effects with a weak immunosuppressive capacity. The anti-inflammatory effect is probably beneficial in disease manifestations secondary to immune complex deposition, while the immunosuppressive effect may help to curtail the activity of the B-cell system.
B. **Nonsteroidal Anti-Inflammatory Drugs** are frequently used in order to control arthritis and serositis.
C. In patients with vital organ involvement (i.e., glomerulonephritis, CNS involvement), immunosuppressive drugs may be indicated. **Cyclophosphamide**, given intravenously, has been successfully used to prolong adequate renal function with only few side effects (Fig. 20.2). Maximal benefit is achieved when long-term treatment is started early, with relatively good renal function.
D. Careful avoidance of factors implicated in the induction of relapses—such as high-risk medications, exposure to sunlight, infections, etc.—is also indicated.
E. A number of **experimental therapeutic** approaches have been used, or are under study, in patients with SLE. These include *plasmapheresis* (removal of autoantibodies and immune complexes from the circulation), *total lymphoid irradiation*, and administration of *antilymphocyte antibodies* (to eliminate cells that produce autoantibodies or provide help to autoantibody-producing B lymphocytes).

SELF-EVALUATION

Questions

Choose the ONE *best* answer.

20.1 High titers of circulating anti-dsDNA in the serum of patients who have systemic lupus erythematosus are:
- A. Associated with the formation of mixed cryoglobulins
- B. Frequently associated with hemolytic anemia
- C. Predictors of the development of inflammatory complications
- D. Rarely associated with high titers of antinuclear factor
- E. Suggestive of a drug-induced etiology

Questions 20.2–20.3 refer to the following case history: A 28-year-old woman has a history of weight loss, intermittent fever, and pains in several joints, mainly hands, wrists, and knees. Physically she shows an erythematous rash on the malar regions and enlarged, nontender, lymph nodes in the axillary and inguinal regions. Laboratory tests showed anemia, a positive indirect Coombs test, positive antinuclear antibodies, and proteinuria (3 g/24 h).

20.2 The proteinuria in this patient is likely to result from:
- A. Bacterial infections
- B. Idiopathic glomerulonephritis
- C. Inflammation triggered by autoantibodies reactive with the glomerular basement membrane
- D. Tubular damage secondary to hemolysis
- E. Type III hypersensitivity reaction at the glomerular level

20.3 Which of the following antibodies would be most valuable from the diagnostic point of view?
- A. Anti-dsDNA
- B. Antinuclear antibodies (diffuse pattern)
- C. Rheumatoid factor
- D. Anti-SS-A
- E. Anticentromere antibodies

20.4 The exaggerated synthesis of autoantibodies in SLE is believed to result from:
- A. Excessive activity of CD4+ helper lymphocytes
- B. Genetic abnormalities resulting in the overexpression of TCR able to react with autologous DNA
- C. Massive stimulation of the immune system by bacterial DNA
- D. Exaggerated release of interleukins 1 and 2 during the active states of the disease
- E. Uncontrolled activation of B lymphocytes

20.5 A direct immunofluorescence study of a normal skin biopsy obtained from a patient with systemic lupus erythematosus (SLE) using fluorescein-labeled antihuman IgG antiserum shows deposition of irregularly stained material at the dermoepidermal junction. This result is considered as indicative of the presence of:
- A. Antinuclear antibodies
- B. Anticollagen antibodies
- C. Antielastin antibodies
- D. Antiepithelial antibodies
- E. Antigen–antibody complexes

20.6 The rationale for the use of plasmapheresis in the treatment of systemic lupus erythematosus is the removal of:
A. Autoreactive B lymphocytes
B. Autoantibodies and immune complexes from the circulation
C. Damaged red cells and leukocytes
D. Excessive amounts of IgG and other immunoglobulins
E. Free nucleic acids

20.7. Soluble immune complexes with highest potential pathogenicity are likely to have the following characteristic(s):
A. Contain IgG1 antibodies
B. Be formed in antibody excess
C. Contain ssDNA
D. Contain IgM antibodies
E. Be of extremely large size

20.8 Anticardiolipin antibodies in SLE are associated with:
A. Autoimmune hemolytic anemia
B. Central nervous system involvement
C. Congenital heart block
D. False-positive serologies for Lyme's disease
E. Thrombophlebitis

20.9 The pathogenic role of anti-DNA antibodies has best been proven in association to:
A. Pleurisy
B. Hemolytic anemia
C. Cutaneous vasculitis
D. Glomerulonephritis
E. Pericarditis

20.10 The first autoantibodies synthesized by (NZB/NZW)F1 mice react with:
A. T-lymphocyte subsets
B. Neutrophils
C. Nucleic acids
D. Red cells
E. All of the above

Answers

20.1 (C) Soluble IC formed at slight antigen excess between dsDNA and its corresponding antibody appear to play the main pathogenic role in SLE-associated glomerulonephritis and other inflammatory complications. Mixed cryoglobulins usually correspond to cold-precipitable immune complexes involving a first antibody (IgG or IgA) and a second anti-immunoglobulin antibody (IgG in most cases) and are more frequently observed in patients with chronic infections or rheumatoid arthritis than in SLE patients.

20.2 (E) The sum of clinical and laboratory data in this patient is strongly suggestive of SLE. In those patients, proteinuria is usually caused by glomerulonephritis secondary to the deposition of immune complexes in (or around) the basement membrane. Antibasement membrane antibodies can also cause glomerulonephritis, but these antibodies are found in Goodpasture's syndrome and not in SLE. The hemolytic anemia in SLE is usually due to extravascular hemolysis (the anti-RBC antibodies are of the IgG isotype, not very efficient in causing intravascular hemolysis).

20.3 (A) Anti-dsDNA antibodies are virtually diagnostic of SLE.

20.4 (E) The production of autoantibodies in SLE (both human and experimental) is secondary to a state of hyperactivation of B cells, which is partly due to autocrine or paracrine B-cell stimulation, increased help from a poorly characterized double-negative T-cell subpopulation, and decreased T-cell suppressor activity.

20.5 (E) The study described in this question is known as the "band test" and a positive result is considered as indicating the deposition of antigen–antibody complexes at the dermoepidermal junction.

20.6 (B) Plasmapheresis consists on the ex vivo separation of the patient's blood into cells and plasma, the cells to be reinfused, and the plasma to be discarded. The main reason to discard the plasma is to reduce the concentrations of autoantibodies and soluble immune complexes. This removal, associated with immunosuppressive therapy, may have dramatic beneficial effects, although in most cases the benefits may not be long-lasting.

20.7 (A) IgG1 is complement-fixing and able to interact with the Fc receptors of phagocytic cells; both properties appear to be important in the induction of inflammatory changes secondary to IC deposition.

20.8 (C) The presence of anticardiolipin autoantibodies is associated with miscarriages, thrombophlebitis, thrombocytopenia, and false positivity in serological tests for syphilis (not Lyme's disease). Anti-Ro antibodies are passively transmitted transplacentally from mothers with SLE, which seems to be associated with the development of heart block in newborns.

20.9 (D)

20.10 (C) The antibodies to red cells, causing immune hemolytic anemia, appear later in the evolution of the disease.

BIBLIOGRAPHY

Alarcon-Segovia D., Ruiz-Arguelles A., Llorerente L. Broken dogma, penetration of autoantibodies into living cells *Immunol. Today*, *17*:163–164, 1996.

Boumpas, D.T., Fessler, B.J., Austin, H.A. III, Balow, J.E., Klippel, J.H., Lockshin, M.D. Systemic lupus erythematosus: emerging concepts. Part 1: Renal, neuropsychiatric, cardiovascular, pulmonary, and hematologic disease. *Ann. Intern. Med.*, *121*:940, 1995; Part 2: Dermatologic and joint disease, the antiphospholipid antibody syndrome, pregnancy and hormonal therapy, morbidity and mortality, and pathogenesis. *Ann. Intern. Med.*, *123*:42, 1995.

Hochberg, M.C. Systemic lupus erythematosus. *Rheum. Dis. Clin. North Am.*, *16*:617, 1990.

Radic M.Z., and Weigert M. Genetic and structural evidence for antigen selection of anti-DNA antibodies. *Annu. Rev. Immunol.*, *12*:487–520, 1994.

Steinberg, A.D., Gourley, M.F., Klinman, D.M., Tsokos, G.C., Scott, D.E., and Krieg, A.M. NIH Conference. Systemic lupus erythematosus. *Ann. Intern. Med.*, *115*:548, 1991.

Tan, F.J., Chan, E.K.L., Sullivan, K.F., and Rubin, R.L. Antinuclear antibodies (ANAs): Diagnostically specific immune markers and clues toward the understanding of systemic autoimmunity. *Clin. Immunol. Immunopathol.*, *47*:121, 1988.

Theofilopoulos, A.N. The basis of autoimmunity: Part I. Mechanisms of aberrant self-recognition. *Immunol. Today*, *16*:90–98, 1995; Part II. Genetic predisposition. *Immunol. Today*, *16*:150–159, 1995.

Tsokos, G.C. (moderator). Pathogenesis of Systemic Lupus Erythematosus. A Symposium-in-Writing. *Clin. Immunol. Immunopath.*, *63*:3, 1992.

Wallace, D. (ed.) Systemic Lupus Erythematosus and Sjögren's Syndrome. *Curr. Opin. Rheumatol.*, *8*:393–445, 1996.

21
Rheumatoid Arthritis

Jean-Michel Goust

I. INTRODUCTION

Rheumatoid arthritis (RA) is a chronic autoimmune disease characterized by inflammatory and degenerative lesions of the distal joints, frequently associated with multiorgan involvement. This disease affects just under 1% of the population and its etiology is complex; immunological, genetic, and hormonal factors are thought to determine its development. Rheumatoid arthritis waxes and wanes for many years, but the attacks progressively run into one another, setting the stage for the chronic form of the disease which is associated with deformity and functional impairment.

Case 21.1

A 28-year-old married African–American seeks medical attention because of exacerbation of joint pains. She had been diagnosed as having rheumatoid arthritis 4 years earlier. She had been doing very well during her second pregnancy and delivered her son 6 months ago but during the past 2 months started to suffer from progressively severe pains affecting her hands, feet, and right knee. The pain is particularly severe in the morning, but decreases around noon. She has been taking ibuprofen since the beginning of her last relapse, but the medication seems to have lost its effectiveness. On examination, the mucosae appear slightly pale and the distal joints of both hands are swollen, erythematous, and warm. There is a moderate amount of synovial fluid in the right knee. The spleen is palpable 3 cm below the rib cage. A subcutaneous nodule is visible just below the left elbow. The rest of the physical examination is within normal limits. A hemogram shows a red cell count of 3.9×10^6/mm^3, white blood cell count of 5200/mm^3 with 20% neutrophils and 70% lymphocytes. Hematocrit was 38%, hemoglobin 11 g/dL. A rheumatoid factor titer was of 2560 and antinuclear antibodies were positive (homogeneous pattern, titer of 320). X-rays of the hand showed cartilage erosion and fluid in the distal phalanges of both hands.

This case history raises the following questions:

- Should the diagnosis of rheumatoid arthritis be revised because of the positive antinuclear antibody assay?
- What is the significance of the low neutrophil count in this patient's blood?
- What is the pathogenesis of the joint lesions?

- Is the subcutaneous nodule seen below the left elbow related to rheumatoid arthritis?
- Why did the patient do better during pregnancy?
- Should the therapy be modified?

II. CLINICAL AND PATHOLOGICAL ASPECTS OF RHEUMATOID ARTHRITIS

A. Joint Lesions

1. **Clinical presentation.** The most common clinical presentation of rheumatoid arthritis is the association of **pain**, **swelling**, and **stiffness of the metacarpo-phalangeal and wrist joints**, often associated with pain in the sole of the foot, indicating metatarso-phalangeal involvement.
   a. The disease is initially limited to small distal joints.
   b. With time, rheumatoid arthritis progresses from the distal to the proximal joints so that in the late stages joints such as the ankles, knees, and elbows may become affected.
   c. Rheumatoid arthritis is characterized by a chronic inflammatory process of the joints (see below), which progresses through different stages of increasing severity (Table 21.1). The damage is reversible until cartilage and bones become involved (stages 4 and 5). At that time the changes become irreversible and result in severe functional impairment.
2. **Pathological manifestations**
   a. In the **early stages** the inflammatory lesion is limited to the lining of the normal diarthrodial joint. The normal synovial lining is constituted by a thin membrane composed of two types of synoviocytes:
      i. The **type A synoviocyte**, which is a phagocytic cell of the monocyte-macrophage series with a rapid turn-over.
      ii. The **type B synoviocyte**, which is believed to be a specialized fibroblast.

      This cellular lining sits on top of a loose acellular stroma that contains many capillaries.

Table 21.1 Stages of Rheumatoid Arthritis

Stage	Pathological process	Symptoms	Physical signs
1	Antigen presentation to T lymphocytes	None	None
2	Proliferation of T and B lymphocytes	Malaise, mild joint stiffness	Swelling of small joints
3	Neutrophils in synovial fluid; synovial cell proliferation	Joint pain and morning stiffness, malaise	Swelling of small joints
4	Invasive pannus; degradation of cartilage	Joint pain and morning stiffness, malaise	Swelling of small joints
5	Invasive pannus; degradation of cartilage; bone erosion	Joint pain and morning stiffness, malaise	Swelling of small joints; deformities

b. The earliest pathological changes, seen at the time of the first symptoms, affect the endothelium of the microvasculature, whose permeability is increased, as judged by the development of edema and of a sparse inflammatory infiltrate of the edematous subsynovial space, in which polymorphonuclear leukocytes predominate. Several weeks later, hyperplasia of the synovial lining cells and perivascular lymphocytic infiltrates can be detected.

c. In the **chronic stage**, the size and number of the synovial lining cells increases and the synovial membrane takes a villous appearance. There is also subintimal hypertrophy with massive infiltration by lymphocytes, plasmablasts, and granulation tissue (forming what is known as pannus).

d. This thick pannus behaves like a tumor and in the ensuing months and years continues to grow, protruding into the joint. The synovial space becomes filled by exudative fluid, and this progressive inflammation causes pain and limits motion.

e. With time, the cartilage is eroded and there is progressive destruction of bones and tendons, leading to severe limitation of movement, flexion contractures, and severe mechanical deformities.

B. **Systemic Involvement**

1. **Clinical presentation.** Frequently some signs and symptoms more indicative of a systemic disease are observed, particularly those that are indicative of **vasculitis**. The most frequent sign is the formation of the **rheumatoid nodules** over pressure areas, such as the elbows. These nodules are an important clinical feature because with rare exceptions they are pathognomonic of RA in patients with chronic synovitis, and generally indicate a poor prognosis.

2. **Pathological manifestations.** In contrast with the necrotizing vasculitis associated with systemic lupus erythematosus (SLE), due almost exclusively to immune complex deposition, the vasculitis seen in rheumatoid arthritis is associated with granuloma formation.

   a. Histopathological studies of **rheumatoid nodules** show fibrinoid necrosis at the center of the nodule surrounded by histiocytes arranged in a radial palisade. The central necrotic areas are believed to be the seat of immune complex formation or deposition.

   b. When the disease has been present for some time, small brown spots may be noticed around the nail bed or associated with nodules. These indicate small areas of **endarteritis**.

   c. Rheumatoid patients with **vasculitis** usually have persistently elevated levels of circulating immune complexes, and generally, a worse prognosis.

C. The **Overlap Syndrome** describes a clinical condition in which patients show variable degrees of association of rheumatoid arthritis and systemic lupus erythematosus. The existence of this syndrome suggests that the demarcation between SLE and RA is not absolute, resulting in a clinical continuum between both disorders.

1. Clinically these patients present features of both diseases.
2. Histopathological studies show lesions with associations of the two basic

pathological components of RA and SLE (necrotizing vasculitis and granulomatous reactions).

3. Serological studies in patients with the overlap syndrome demonstrate both antibodies characteristically found in SLE (e.g., anti-dsDNA) and antibodies typical of RA (rheumatoid factor, see below).

D. Other Related Diseases

1. **Sjögren's syndrome** can present as an isolated entity or in association with rheumatoid arthritis, SLE, and other collagen diseases. It is characterized by dryness of the oral and ocular membranes (**Sicca syndrome**), and the detection of rheumatoid factor is considered almost essential for the diagnosis, even for those cases without clinical manifestations suggestive of rheumatoid arthritis.
2. **Felty's syndrome** is an association of rheumatoid arthritis with neutropenia caused by antineutrophil antibodies. The spleen is often enlarged, possibly reflecting its involvement in the elimination of antibody-coated neutrophils.

III. AUTOANTIBODIES IN RHEUMATOID ARTHRITIS

A. Rheumatoid Factor and Anti-Immunoglobulin Antibodies. The serological hallmark of rheumatoid arthritis is the detection of **rheumatoid factor (RF)** and other anti-immunoglobulin antibodies. By definition, **classic RF is an IgM antibody to autologous IgG**. The more encompassing designation of **anti-immunoglobulin (Ig) antibodies** is applicable to anti-IgG antibodies of IgG or IgA isotypes.

1. **Immunochemical characteristics of rheumatoid factor**
   a. As a rule, the affinity of IgM rheumatoid factor for the IgG molecule is relatively low and does not reach the mean affinity of other IgM antibodies generated during an induced primary immune response.
   b. Rheumatoid factors from different individuals show different antibody specificity, reacting with different determinants of the IgG molecule.
      i. In most cases, the antigenic determinants recognized by the antigen-binding sites of these IgM antibodies are located in the Cγ2 and Cγ3 domains of IgG; some of these determinants are allotype-related.
      ii. Circulating RF reacts mostly with IgG1, IgG2, and IgG4; in contrast, RF detected in synovial fluid reacts more frequently with IgG3 than with any other IgG subclasses. The significance of this difference is unknown, but suggests that circulating RF and synovial RF may be produced by different B-cell clones.
      iii. Other RF react with determinants which are shared between species, a fact that explains the reactivity of the human RF with rabbit IgG as well as with IgG from other mammalians.
   c. The frequent finding of RF reactive with several IgG subclasses in a single patient suggests that the autoimmune response leading to the production of the RF is polyclonal. This is supported by the fact the idiotypes of RF are heterogeneous, being obviously the product of several different V-region genes.

2. **Methods used for the detection of rheumatoid factor.** Rheumatoid factor and anti-Ig antibodies can be detected in the serum of affected patients by a variety of techniques.
   a. The **Rose–Waaler test** is a passive hemagglutination test that uses sheep or human erythrocytes coated with anti-erythrocyte antibodies as indicators. The agglutination of the IgG-coated red cells to titers greater than 16 or 20 is considered as indicative of the presence of RF. These tests detect mostly the classic IgM rheumatoid factor specific for IgG.
   b. The **latex agglutination test**, in which IgG-coated polystyrene particles are mixed with serum suspected of containing RF or anti-Ig antibodies. The agglutination of latex particles by serum dilutions greater than 1:20 is considered as a positive result. This test detects anti-immunoglobulin antibodies of all isotypes.
3. **Diagnostic specificity of anti-immunoglobulin antibodies**
   a. As with many other autoantibodies, the titers of RF are a continuous variable within the population studied. Thus, any level intending to separate the seropositive from the seronegative is arbitrarily chosen to include as many patients with clinically defined RA in the seropositive group, while excluding from it as many nonrheumatoid subjects as possible.
   b. RF is neither specific nor diagnostic of RA. First, it is found in only 70 to 85% of RA cases, while it can be detected in many other conditions, particularly in patients suffering from **Sjögren's syndrome**. Also, RF screening tests can be positive in as many as 5% of apparently normal individuals, sharing the same V-region idiotypes (and by implication, the same V-region genes) as the antibodies detected in RA patients.
4. **Physiological role of anti-immunoglobulin antibodies.** The finding of RF in normal individuals raises the concept that RF may have a normal, physiological role, such as to ensure the rapid removal of infectious antigen–antibody complexes from circulation. This is a direct challenge to the postulate that autoantibodies emerge only as a result of loss of tolerance to self-antigens. The synthesis of anti-Ig antibodies in normal individuals follows some interesting rules:
   a. Anti-Ig antibodies are detected transiently during anamnestic responses to common vaccines, and in these cases, are usually reactive with the dominant immunoglobulin isotype of the antibodies produced in response to antigenic stimulation.
   b. Anti-Ig antibodies are also found in relatively high titers in diseases associated with persistent formation of antigen–antibody complexes such as subacute bacterial endocarditis, tuberculosis, leprosy, and many parasitic diseases.
   c. The titers of vaccination-associated RF follow very closely the variations in titer of the specific antibodies induced by the vaccine; similarly, the levels of RF detected in patients with infections associated with persistently elevated levels of circulating immune complexes decline once the infection has been successfully treated. In contrast, the anti-immunoglobulin antibodies detected in patients with rheumatoid ar-

thritis persist indefinitely, reflecting their origin as part of an autoimmune response.

d. Infection-associated RF bind to IgG molecules whose configuration has been altered as a consequence of binding to exogenous antigens. The resulting RF-IgG-Ag complexes are large and quickly cleared from circulation. The adsorption of IgG to latex particles seems to induce a similar conformational alteration of the IgG molecule as antigen binding, and, as a result, IgG-coated latex particles can also be used to detect this type of RF.
e. The transient detection of anti-Ig antibodies in normal individuals suggests that:
   i. The autoreactive clones responsible for the production of autoantibodies to human immunoglobulins are not deleted during embryonic differentiation. The persistence in adult life of such autoreactive clones is supported by the observation that the bone marrow contains precursors of RF-producing B cells. Their frequency is surprisingly high in mice where it is relatively easy to induce the production of RF in high titers after polyclonal B-cell stimulation. Human bone marrow B lymphocytes can also be stimulated to differentiate into RF-producing plasmablasts by mitogenic stimulation with PWM or by infection with Epstein-Barr virus.
   ii. Tolerance to self-IgG must be ensured by a strong negative feedback mechanism, since tolerance is broken only temporarily.
f. Both in mice and humans, the B lymphocytes capable of differentiating into RF-producing plasmablasts express CD5 in addition to the classic B-cell markers, such as membrane IgM and IgD, CR2, CD19, and CD20. CD5 is expressed by less than 2% of the B lymphocytes of a normal individual and was first detected in patients suffering from very active rheumatoid arthritis. It is considered as characteristic of autoimmune situations.

6. **Pathogenic role of rheumatoid factor and anti-immunoglobulin antibodies.**
   a. Although fluctuations in RF titers often seen in patients with RA do not seem to correlate with the activity of the disease, high titers of RF tend to be associated with a more rapid progression of the articular component and with systemic manifestations, such as subcutaneous nodules, vasculitis, intractable skin ulcers, neuropathy, and Felty's syndrome. Thus, the detection of RF in high titers in a patient with symptomatic RA is associated with a poor prognosis.
   b. IgM RF activates complement via the classic pathway. The ability of RF to fix complement is of pathogenic significance, because it may be responsible, at least in part, for the development of rheumatoid synovitis.
   c. Locally produced anti-Ig antibodies are likely to play an important role in causing the arthritic lesions. The joints are the principal site of RF production in RA patients, and it should also be noted that in some individuals the locally produced anti-Ig antibodies are of the IgG isotype. When this is the case, the joint disease is usually more severe,

because anti-IgG antibodies of the IgG isotype have a higher affinity for IgG than their IgM counterparts; consequently, they form stable immune complexes that activate complement very efficiently. In some seronegative patients (see below), RF and immune complexes may be only detectable in synovial fluid.

7. **Seronegative rheumatoid arthritis.** Some patients with RA may have negative results on the screening tests.
   a. In many instances, such results may be false negatives. Three different mechanisms may account for false-negative results in the RA test:
      i. The presence of anti-Ig antibodies of isotypes other than IgM, less efficient than IgM RF in causing agglutination (particularly in tests using red cells) and therefore more likely to be overlooked.
      ii. The reaction between IgM RF and endogenous IgG results in the formation of soluble immune complexes that, if the affinity of the reaction is relatively high, will remain associated when the RF test is performed. Under these conditions, the RF binding sites are blocked, unable to react with the IgG coating indicator red cells or latex particles.
      iii. RF may be present in synovial fluid but not in peripheral blood.

      In clinical practice, it is very seldom necessary to investigate these possibilities, since a positive test is not necessary for the diagnosis.
   b. True seronegative RA cases exist, particularly among agammaglobulinemic patients. In spite of their inability to synthesize antibodies, these patients develop a disease clinically indistinguishable from RF-positive rheumatoid arthritis. This is a highly significant observation since it argues strongly against the role of the RF or other serological abnormalities as a major pathogenic insult in rheumatoid arthritis and suggests that the inflammatory response in the rheumatoid joint could be largely cell-mediated.

B. **Anticollagen Antibodies.** Antibodies reacting with different types of collagen have been detected with considerable frequency in connective tissue diseases such as **scleroderma**. In rheumatoid arthritis, considerable interest has been aroused by the finding that antibodies elicited by injection of type II collagen with complete Freund's adjuvant into rats is associated with the development of a rheumatoid-type disease. However, the frequency of these antibodies in RA patients has recently been estimated to be in the 15–20% range, which is not compatible with a primary pathogenic role. It is probable that the anticollagen antibodies in RA arise as a response to the degradation of articular collagen, which could yield immunogenic peptides.

C. **Antinuclear Antibodies**
   1. Antibodies against native, double-stranded DNA are conspicuously lacking in patients with classic RA, but antibodies against single-stranded DNA can be detected in about one-third of the patients. The epitopes recognized by anti-ssDNA antibodies correspond to DNA-associated proteins.
   2. The detection of anti-ssDNA antibodies does not have diagnostic or prognostic significance because these antibodies are neither disease-specific nor involved in immune complex formation.
   3. The reasons for the common occurrence of anti-ssDNA in RA and in many

other connective tissue diseases is unknown, but these antibodies may represent an indicator of immune abnormalities due to the persistence of abnormal B-lymphocyte clones that have escaped the repression exerted by normal tolerogenic mechanisms and are able to produce autoantibodies of various types.

IV. GENETIC FACTORS IN RHEUMATOID ARTHRITIS

A. **HLA-Associations.** The incidence of familial rheumatoid arthritis is low, and only 15% of the identical twins are concordant for the disease. However, 70–90% of Caucasians with rheumatoid arthritis express the **HLA DR4** antigen, which is found in no more than 15–25% of the normal population. Individuals expressing this antigen are 6 to 12 times more at risk for having RA, but HLA-DR 1 was also found to increase susceptibility to RA and wide fluctuations in the frequency of these markers are seen between different patient populations.

B. **HLA-DR4 Subtypes.** DNA sequencing of the β chain of the DR4 molecules defined 5 HLA-DR4 subtypes: Dw4, Dw10, Dw13, Dw14, and Dw15. The same technique has allowed the identification of HLA-DR1 subtypes. While Dw4, Dw10, Dw13, and Dw15 differ from each other in amino acid sequence at positions 67, 70, and 74 of the third hypervariable region of the β1 domain of the β chain, Dw4 and Dw14 have identical amino acid sequences at these positions and are associated with RA. The same amino acids are present in the Dw1 subtype of HLA-DR1.

1. The prevalence of Dw4, Dw 14, or Dw1 in the general population is 42%. Of these individuals, 2.2% develop RA. In contrast, the frequency of RA in individuals negative for these markers is only 0.17%, a 12.9-fold difference.
2. Since most humans are heterozygous, a given individual may inherit more than one susceptibility allele. Individuals having both Dw4 and Dw14 have a much higher risk (7:1) of developing severe RA. In contrast, individuals with the Dw10 and Dw13 markers, whose sequence differs in the critical residues (Table 21.2), seem protected against RA.
3. The interpretation of these findings hinges on the fact that amino acids 67, 70, and 74 are located on the **third hypervariable region of the DR4 and**

Table 21.2 HLA-DR Subtypes and Rheumatoid Arthritis

Subtype	Critical residues on the third diversity region of β1				Predisposition to rheumatoid arthritis
	67	70	71	74	
DRB1*0101 (Dw1)	L	Q	R	A[a]	+
DRB1*0401 (Dw4)	L	Q	K	A	+
DRB1*0404 (Dw14)	L	Q	R	A	+
DRB1*0403 (Dw13)	L	Q	R	E	−
DRB1*0402 (Dw10)	I	D	E	A	−

[a]LQRA = Leu, Gln, Arg, Ala.

DR1 β chains. This region is part of a helical region of the peptide-binding pouch (see Chap. 3), which interacts both with the side chains of antigenic peptides and with the TcR. Its configuration, rather than the configuration of any other of the hypervariable regions of the DR4 and DR1 β chains, seems to determine susceptibility or resistance to RA, depending on the charge of amino acids located on critical positions.

a. It has been postulated that the structure of those DR4 and DR1 molecules which are associated with increased risk for the development of RA is such that they bind very strongly an "arthritogenic epitope" derived from an as yet unidentified infectious agent. The consequence would be a strong and prolonged immune response cross-reactive with tissue antigens expressed predominantly in the joints. The reverse would be the case for those DR4 molecules associated with protection against the development of RA.

b. Supporting this interpretation are several observations concerning the severity of the disease in patients bearing those HLA antigens and subtypes. For example, DR4 positivity reaches 96% in patients suffering from Felty's syndrome, the most severe form of the disease. More recent studies showed that RA patients who are DR4-Dw14 positive have a faster progression to the stages of pannus formation and bone erosion.

4. The most significant discrepancy in this apparent consensus sequence between DR sequence and RA susceptibility was found in African Americans with RA; in this group, only 20% are DR4+. In this ethnic group, predisposition and severity appear independent of the presence and dose of the "arthritogenic" DR alleles identified in Caucasians.

V. CELL-MEDIATED IMMUNITY ABNORMALITIES IN RHEUMATOID ARTHRITIS

All the essential cellular elements of the immune response are present in the hypertrophic synovium of the rheumatoid joints, where they are easily accessible to study by needle biopsy or aspiration of the synovial fluid. An important pathogenic role for T lymphocytes is suggested by the association with DR alleles, given the role that these molecules have in presenting antigen to the helper subpopulation, and by the finding of increased concentrations of many cytokines in the synovial fluid, probably reflecting the activation of infiltrating lymphocytes.

A. T-Lymphocyte Abnormalities. Immunohistological studies of the inflammatory infiltrates of the synovial membrane show marked lymphocytic predominance (lymphocytes may represent up to 60% of the total tissue net weight).

1. Among the lymphocytes infiltrating the synovium, **CD4+ helper T lymphocytes** outnumber CD8+ lymphocytes in a ratio of 5:1. Most of the infiltrating CD4+ lymphocytes have the phenotype of a memory helper T cell (CD4+, CD45RO+), which represent 20–30% of the mononuclear cells in the synovium. They also express class-II MHC, consistent with chronic T-cell activation, but only 10% express CD25, suggesting that they do not proliferate actively in the synovial tissues.

2. In situ hybridization studies performed in biopsies of synovial tissue obtained at late stages of the disease disclosed that these chronically activated T cells express mRNAs for IL-2, IFN-γ, IL-7, IL-13, IL-15, and GM-CSF. One conspicuously missing interleukin is IL-4, suggesting that the CD4+, CD45RO+ T lymphocytes in the synovial tissues are predominantly of the THl type, which produce predominantly IL-2 and IFN-γ.
3. A variety of chemokines (Rantes, MIPl-α, MIPl-β, and IL-8), most of them produced by lymphocytes, can also be detected in the synovial fluid. These chemokines are probably responsible for the attraction of additional T lymphocytes, monocytes, and neutrophils to the rheumatoid joint.
4. A most significant question that remains unanswered is what is the nature of the stimulus responsible for the activation of the T lymphocytes found in the synovial infiltrates.
   a. Studies of the TcR Vβ genes expressed by the infiltrating T lymphocytes has shown that the repertoire is limited (i.e., only some T-cell clones appear to be activated and the same clones are found in several joints of the same patient). An even more restricted profile was observed when the analysis was confined to the antigen-binding area of the Vβ chain (the so-called CDR3 region). These findings suggest that antigenic stimulation through the TcR plays a critical role.
   b. However, these studies have not yet found a defined correlation between the Vβ chains expressed by the infiltrating T lymphocytes and the patient's MHC-II alleles, a correlation that would be expected because of the role played by the MHC molecules in selecting the T-cell repertoire of any given individual (see Chap. 11). Thus, HLA-linked RA susceptibility alleles could introduce a first bias in TCR selection, but there is no evidence supporting this hypothesis.
   c. The long-term goal of these approaches, to identify the actual targets of the immune attack, remains elusive at this time. A major obstacle is our limited knowledge about antigens recognized by different TcR Vβ region families. Until some breakthrough happens in that area, our understanding will remain fragmentary and highly speculative.

B. **Monocyte/Macrophage Abnormalities.** The synovial infiltrates are rich in activated monocytes and macrophages, which are believed to play several critical pathogenic roles.
   1. One of the significant roles played by this cell is antigen presentation to CD4 lymphocytes. It is not unusual to see macrophage–lymphocyte clusters in the inflamed synovial tissue and in those clusters, CD4+ lymphocytes are in very close contact with large macrophages expressing high levels of class II MHC antigens. In addition, IL-12 mRNA and secreted IL-12 are found at biologically active concentrations, and could play an important role in THl expansion.
   2. The other critical role of synovial monocytes and macrophages is to induce and perpetuate local inflammatory changes. Several lines of evidence support this role.
      a. The synovial fluid of patients with RA contains relatively large concentrations of **phospholipase A2** (PLA2), an enzyme that has a strong chemotactic effect on lymphocytes and monocytes. Moreover, this en-

zyme is actively involved in the metabolism of cell membrane phospholipids, particularly in the early stages of the cyclooxygenase pathway, which leads to the synthesis of eicosanoids such as **PGE2**, one of a series of proinflammatory mediators generated from the breakdown of arachidonic acid. Thus, it is possible that these high levels of PLA2 reflect a hyperactive state of infiltrating macrophages, engaged in the synthesis of PGE2 and other eicosanoids.

b. Other factors locally released by activated macrophages include:
   i. **Transforming growth factor-β (TGF-β)**, which further contributes to the suppression of TH2 activity at the inflammatory sites.
   ii. **GM-CSF**, which induces the proliferation of several cell types in the monocyte–macrophage family, including dendritic cells. It has been suggested that this overproduction of GM-CSF by all CD4 cells (macrophages and T lymphocytes) is responsible for the relatively large number of dendritic cells found in the inflammatory lesion. This is a significant finding, because activated dendritic cells release a variety of proinflammatory lymphokines, such as IL-1.
   iii. Another prominent role of GM-CSF is to be a very strong inducer of the expression of MHC-II molecules (stronger than interferon-γ). Increased MHC-II expression is believed to be an important factor leading to the development of autoimmune responses, and could help perpetuate a vicious cycle of anti-self immune response by facilitating the persistent activation of TH1 cells and, consequently, the stimulation of synovial cells, monocytes, and dendritic cells.

c. Activated macrophages secrete a variety of **proteolytic enzymes** (such as collagenase, matrix metalloproteinase-1, and stromelysin), particularly when stimulated with IL-1 and TNF-α. Studies of biopsies of the rheumatoid synovium discussed above found these two cytokines at levels high enough to deliver such stimulatory signals to monocytes and fibroblasts.

VI. A SUMMARY OVERVIEW OF THE PATHOGENESIS OF RHEUMATOID ARTHRITIS

A. **Predisposing Factors.** Two important types of factors seem to have a strong impact in the development of rheumatoid arthritis.

1. **Genetic factors.** The link to HLA-DR4, and particularly with subtypes Dw4 and Dw14, as well as with the structurally related Dw1 subtype of HLA-DR1, as has been previously discussed in this chapter. It is currently accepted that such DR subtypes may be structurally fitted to present a cross-reactive peptide to helper T lymphocytes, thus precipitating the onset of the disease.
2. **Hormonal factors.** The role of hormonal factors is suggested by two observations:
   a. RA is three times more frequent in females than in males, predominantly affecting women from 30 to 60 years of age.

b. Pregnancy produces a remission during the third trimester, sometimes followed by exacerbations after childbirth.

The mechanism by which hormonal factors would determine the increased risk for development of rheumatoid arthritis (as well as of other autoimmune diseases) is not known. A possible mechanism has been recently suggested by the observation that estrogens potentiate B-lymphocyte responses in vitro.

B. **Precipitating Factors.** Three main mechanisms responsible for the escape from tolerance that must be associated with the onset of RA have been proposed.
   1. Decreased suppressor cell activity.
   2. Nonspecific B-cell stimulation by microbial products (e.g., bacterial lipopolysaccharides) or infectious agents (e.g., viruses).
   3. Activation of self-reactive T lymphocytes as a consequence of the presentation of a cross-reactive peptide (possibly of infectious origin) by an activated antigen-presenting cell.

   The last theory is supported by the genetic linkages discussed earlier in this chapter. Also, the key role of T lymphocytes is supported by histological data (discussed earlier), when it was observed that HIV infection and immunosuppression for bone marrow transplantation, two conditions that affect T helper lymphocytes function very profoundly, are associated with remissions of RA.

C. **Self-Perpetuating Mechanisms.** Once helper T cells are activated, a predominantly TH1 response develops. Activated TH1 cells release interferon-γ and GM-CSF, which activate macrophages and related cells, inducing the expression of MHC-II molecules, creating conditions for continuing and stronger stimulation of helper T lymphocytes. As this cross-stimulation of TH1 lymphocytes and macrophages continues, chemotactic factors are released and additional lymphocytes, monocytes, and granulocytes are recruited into the area. As inflammatory cells become activated, they release collagenases, proteases, and proinflammatory mediators, such as PGE2. The release of collagenases and proteases will cause damage on the synovial and perisinovial tissues, while the activation of osteoblasts and osteoclasts by mediators released by activated lymphocytes and macrophages is the cause of bone damage and abnormal repair.

VII. THERAPY

It is not surprising that our very incomplete knowledge of the pathogenesis of rheumatoid arthritis is reflected at the therapeutic level. Most of our current therapeutic approaches are symptomatic, aiming to reduce joint inflammation and tissue damage. Under these conditions, RA therapy is often a frustrating experience for patients and physicians.

A. **Nonsteroidal Anti-Inflammatory Drugs.** This group of anti-inflammatory drugs, which includes, among others, **aspirin, ibuprofen, naproxen, and indomethacin,** have as a common mechanism of action the inhibition of the cyclooxygenase pathway of arachidonic acid metabolism, which results in a reduction of the local release of prostaglandins. Their administration is beneficial in many patients with rheumatoid arthritis.

B. **Glucocorticoids.** In more severe cases, in which the cyclooxygenase inhibitors are not effective, glucocorticoids are indicated.

1. The use of glucocorticoids in RA raises considerable problems, because in most instances, their administration masks the inflammatory component only as long as it is given. Thus, glucocorticoid therapy needs to be maintained for long periods of time.
2. The side effects observed with high doses of glucocorticoids include muscle and bone loss and may become more devastating than the original arthritis. To avoid this problem, very low doses on alternate days are now used.

C. **Immunosuppression** by drugs such as methotrexate or azathioprine or by total lymphoid irradiation is usually reserved for the most severe cases and may be considerably more efficient than the administration of anti-inflammatory drugs but carries the long-term risk of malignancy. However, **methotrexate** administered in low weekly doses is not associated with long-term side effects, while controlling the inflammatory component of the disease and delaying the appearance of the chronic phase.

D. **Reinduction of Tolerance.** Attempts to reinduce tolerance to cartilage antigens postulated to be involved in the autoimmune response by feeding animal cartilage extracts to RA patients have yielded promising results. However, the clinical benefits reported so far have been observed in short-term studies, and research is necessary to determine if the benefits persist in the long run. Also, additional studies are needed to better define the mechanism(s) involved in oral tolerization.

Case 21.1 Revisited

- Antibodies against single-stranded DNA can be detected in about one-third of the patients. The epitopes recognized by anti-ssDNA antibodies are DNA-associated proteins and correspond to the homogeneous pattern seen by immunofluorescence (see Chap. 20). The detection of anti-ssDNA antibodies does not have diagnostic nor prognostic significance because these antibodies are neither disease specific nor involved in immune complex formation.
- The low neutrophil count and splenomegaly seen in this patient are suggestive of Felty's syndrome, the association of rheumatoid arthritis with autoimmune neutropenia.
- The pathogenesis of rheumatoid arthritis is surrounded by questions. It is believed that activation of self-reactive T lymphocytes is a consequence of the presentation of a cross-reactive peptide (possibly of infectious origin) by an activated antigen-presenting cell. Such peptide would be mostly expressed in the synovial tissues, and the activation of helper T cells (predominantly TH1) would be followed by the release of interferon-γ and GM-CSF, which activate macrophages and antigen-presenting cells. Activated macrophages overexpress MHC-II molecules, creating conditions for continuing and stronger stimulation of helper T lymphocytes. As this cross-stimulation of TH1 lymphocytes and macrophages continues, chemotactic factors are released and additional lymphocytes, monocytes, and granulocytes are recruited into the area. As inflammatory cells become activated, they release

collagenases, proteases, and proinflammatory mediators, such as PGE2. The release of collagenases and proteases will cause synovial and cartilage damage. When the process evolves to this level of joint damage, the prognosis is poor.

- The subcutaneous nodule seen below the left elbow is a rheumatoid nodule, the clinical expression of vasculitis associated with RA.
- It is frequent to observe some signs and symptoms more indicative of a systemic disease, particularly those that are indicative of vasculitis, secondary to immune complex deposition in the vessels of the dermis. Rheumatoid nodules are virtually pathognomonic of RA and indicate a poor prognosis.
- RA predominantly affects women from 30 to 60 years of age. It has been suggested that this association is a consequence of the fact that estrogens potentiate B-lymphocyte responses. Pregnancy is often associated with remissions, probably reflecting hormonal changes that have down-regulating effects on the immune system.
- In a case of severe RA, such as the one described, administration of immunosuppressive drugs is indicated. Methotrexate administered in low weekly doses controls the inflammatory component of the disease, delays the onset of the chronic phase, and is not associated with long-term side effects.

SELF-EVALUATION

Questions

Choose the ONE *best* answer.

21.1 The classic rheumatoid factors predominantly detected by techniques based on red-cell agglutination are anti-IgG antibodies of the:
A. IgA isotype
B. IgD isotype
C. IgE isotype
D. IgG isotype
E. IgM isotype

21.2 Felty's syndrome is a severe form of rheumatoid arthritis associated with:
A. Anti-IgA antibodies
B. Autoimmune neutropenia
C. HLA DR4/Dw14
D. HLA-B27
E. Mixed cryoglobulinemia

21.3 Which one of the following observations provided the strongest argument against a primary pathogenic role for rheumatoid factor in rheumatoid arthritis?
A. Detection of other autoantibodies in patients with RA
B. Detection of RF in patients with chronic infections
C. Development of a clinically indistinguishable form of RA in seronegative agammaglobulinemic patients
D. Negativity of standard RF assays in about 25-30% of the patients with classic RA
E. Presence of RF in low titers in asymptomatic individuals

21.4 Which one of the following HLA-haplotypes is associated with the highest susceptibility to rheumatoid arthritis?
A. Dw4, Dw10
B. Dw1, Dw13
C. Dw1, Dw10
D. Dw4, Dw14
E. Dw10, Dw13

21.5 Which of the following is a common cause of seronegativity in patients with rheumatoid arthritis?
A. Agammaglobulinemia
B. Anti-immunoglobulin antibodies reacting exclusively with IgA
C. Circulating IgG–anti-IgG immune complexes
D. IgG anti-immunoglobulin antibodies
E. IgM anti-immunoglobulin antibodies

21.6 What is the significance of detecting a high titer of rheumatoid factor in a patient with suspected rheumatoid arthritis?
A. A diagnosis of Felty's syndrome should be considered
B. A diagnosis of Sjögren's syndrome should be considered
C. High probability for the development of systemic complications
D. Large concentrations of circulating immune complexes are likely to exist in circulation
E. No special meaning

21.7 Which of the following cell populations is most likely to play the key role in the release of mediators responsible for the inflammatory process in the joints of patients with RA?
A. CD4+, CD45RO+ T lymphocytes
B. CD5+, CD22+ B lymphocytes
C. CD22+ B lymphocytes
D. MHC-II+ macrophages
E. RF-producing plasmablasts

21.8 Which one of the following cytokines is usually undetectable in rheumatoid synovial fluid?
A. GM-CSF
B. IL-2
C. IL-4
D. IL-8
E. Interferon-γ

21.9 Which of the following membrane markers is shared by most T lymphocytes and a B-lymphocyte subpopulation supposedly involved in autoimmune reactions?
A. CD5
B. CD11/18
C. CD25
D. CD45RO
E. MHC-II

21.10 Which of the following is the mechanism of action responsible for the beneficial effects of aspirin and indomethacin in rheumatoid arthritis?
A. Suppression of the immune response
B. Depression of phagocytic cell functions

C. Down-regulation of the release of proinflammatory cytokines by macrophages
D. Inhibition of platelet aggregation
E. Reduction of the synthesis and release of prostaglandins

Answers

21.1 (E) The classic rheumatoid factors are IgM antibodies reacting with IgG immunoglobulins and are preferentially detected by passive hemagglutination tests using IgG-coated red cells as an indicator system.

21.2 (B)

21.3 (C) In most patients with seronegative rheumatoid arthritis, the test is false negative for a variety of reasons. In contrast, agammaglobulinemic patients who develop rheumatoid arthritis are truly seronegative, and provide the best argument against a primary pathogenic role of rheumatoid factor.

21.4 (D) Heterozygous twins who have inherited a double dose of susceptibility alleles have the highest susceptibility; in contrast, individuals whose haplotypes include protective markers, such as Dw10 or Dw13, are somewhat protected.

21.5 (C) The serum of most patients with seronegative rheumatoid arthritis contains circulating immune complexes involving anti-IgG antibodies of the IgG isotype and IgG. The anti-Ig antibodies in these complexes have their binding sites blocked, and the IC do not dissociate as readily as those involving IgM. As a result, false-negative results are obtained in the screening tests.

21.6 (C) Positive RF tests with high titers can be observed in RA patients, and are often associated with systemic complications. Similar high titers of rheumatoid factor can be detected in patients with Sjögren's or Felty's syndrome, but the high titers of RF by themselves do not indicate a higher probability for any of these last two diagnoses over RA. The titer of RF cannot be considered as a direct indication of the levels of circulating IC, because the titer of the RF tests really depends on the amount of free antibody binding sites that can become involved in the cross-linking of indicator particles or red cells. Thus, antigen–antibody ratios, RF isotype, and RF affinity or IgG will have a stronger impact in the results of the RF test than the concentration of circulating IC.

21.7 (D) These cells appear to be activated and to release large amounts of cytokines and inflammatory mediators. The contribution of T lymphocytes can only be indirect, by inducing macrophage activation. B lymphocytes appear to be stimulated secondarily, and the synthesis of RF is only likely to play a limited proinflammatory role in synovial tissues.

21.8 (C) Of the listed cytokines, IL-4 is the only one not found in rheumatoid synovium indicating a predominance of helper T lymphocytes with TH1 functions. GM-CSF, released by chronically activated macrophages and T lymphocytes, induces the proliferation of several cell types in the monocyte-macrophage family, the dendritic cells among them, which release a variety of lymphokines contributing to the migration of inflammatory cells into the synovial tissue. In addition, GM-CSF is a very strong inducer of the expression of MHC-II molecules, facilitating the persistent activation of autoreactive TH1 lymphocytes.

21.9 (A)

21.10 (E) Aspirin and indomethacin inhibit the cyclooxygenase pathway of arachidonic acid metabolism that leads to the synthesis of prostaglandins E_2 and F_2.

BIBLIOGRAPHY

Carsons, A., Chen, P.P., and Kipps, T.J. New roles for rheumatoid factor. *J. Clin. Invest.*, *87*:379, 1991.

Feldman, M., Brennan, F.M., and Maini R. N. Role of cytokines in rheumatoid arthritis. *Annu. Rev. Immunol.*, *14*:397, 1996.

Firestein, G.S., and Zvaifler, N. How important are T cells in chronic rheumatoid synovitis? *Arthr. Rheum.*, *33*:768, 1990.

Harris, E.A. Rheumatoid arthritis. Pathophysiology and implications for therapy *N. Engl. J. Med.*, *322*:1277, 1990.

Kingsley, G., Lanchbury, J., and Panayi, G. Immunotherapy in rheumatic disease: an idea whose time has come—or gone. *Immunol. Today*, *17*:9 1996.

Kirwan, J.R. and the ARC low dose glucocorticosteroid study group. The effects of glucocorticoids on joint destruction in rheumatoid arthritis. *N. Eng. J. Med.*, *333*:142, 1995.

McDaniel, D.O., Alarcon, G.S., Pratt, P.W., and Reveille, J.D. Most African-American patients with rheumatoid arthritis do not have the rheumatoid antigenic determinant. *Ann. Intern. Med.*, *123*:1812, 1995.

Schur, P.H. Serologic tests in the evaluation of rheumatic diseases. *Immunol. Allergy Pract.*, *13*:138, 1991.

Struik, L., Hawes, J.E., Chatila, M.K., et al. T cell receptors in rheumatoid arthritis. *Arthr. Rheum.*, *38*:577, 1995.

Winchester, R. The molecular basis of susceptibility to rheumatoid arthritis. *Adv. Immunol.*, *56*:389, 1994.

Wordsworth, P., and Bell, J. Polygenic susceptibility in rheumatoid arthritis. *Ann. Rheum. Dis.*, *50*:343, 1991.

Ziff, M. The rheumatoid nodule. *Arthr. Rheum.*, *33*:768, 1990.

22
Hypersensitivity Reactions

Gabriel Virella

I. INTRODUCTION

The immune response is basically a mechanism used by vertebrates to eradicate infectious agents that succeed in penetrating the natural barriers. However, in some instances the immune response can be the cause of disease, both as an undesirable effect of an immune response directed against an exogenous antigen, or as a consequence of an autoimmune reaction.

A. **Hypersensitivity and Allergy.** Hypersensitivity can be defined as an abnormal state of immune reactivity that has deleterious effects for the host. A patient with hypersensitivity to a given compound suffers pathological reactions as a consequence of exposure to the antigen to which he or she is hypersensitive. The term "allergy" is often used to designate a pathological condition resulting from hypersensitivity, particularly when the symptoms occur shortly after exposure.

B. **Classification of Hypersensitivity Reactions.** Hypersensitivity reactions can be classified as **immediate** or as **delayed**, depending on the time elapsed between the exposure to the antigen and the appearance of clinical symptoms. They can also be classified as **humoral** or **cell-mediated**, depending on the arm of the immune system predominantly involved. A classification combining these two elements was proposed in the 1960s by Gell and Coombs, and although many hypersensitivity disorders may not fit well into their classification, it remains popular because of its simplicity and obvious relevance to the most common hypersensitivity disorders.

C. **Gell and Coombs' Classification of Hypersensitivity Reactions.** This classification considers four types of hypersensitivity reactions. Type I, II, and III reactions are basically mediated by antibodies with or without participation of the complement system; type IV reactions are cell-mediated (see Table 22.1). While in many pathological processes mechanisms classified in more than one of these types of hypersensitivity reactions may be operative, the subdivision of hypersensitivity states into four broad types aids considerably in the understanding of their pathogenesis.

Table 22.1 General Characteristics of the Four Types of Hypersensitivity Reactions as Defined by Gell and Coombs

Type	Clinical manifestations	Lag between exposure and symptoms	Mechanism
I (immediate)	Anaphylaxis, asthma, hives, hay fever	Minutes	Homocytotropic Ab (IgE)
II (cytotoxic)	Hemolytic anemia, cytopenias, Goodpasture's syndrome	Variable	Complement-fixing/ opsonizing Ab (IgG, IgM)
III	Serum sickness, Arthus reaction, vasculitis	6 hours[a]	Immune complexes containing complement-fixing Ab (mostly IgG)
IV (delayed)	Cutaneous hypersensitivity reactious; graft rejection	12–48 hours	Sensitized lymphocytes

[a]For the Arthus reaction.

II. TYPE I HYPERSENSITIVITY REACTIONS (IgE-MEDIATED HYPERSENSITIVITY, IMMEDIATE HYPERSENSITIVITY)

A. **Experimental Models and Historical Background.** Much of our early knowledge about immediate hypersensitivity reactions was derived from studies in guinea pigs. Guinea pigs immunized with egg albumin, frequently suffer from an acute allergic reaction upon challenge with this same antigen. This reaction is very rapid (observed within a few minutes after the challenge) and is known as an **anaphylactic reaction**. It often results in the death of the animal in **anaphylactic shock**.

1. **Passive transfer of anaphylactic reactions.** If serum from a guinea pig sensitized 7 to 10 days earlier with a single injection of egg albumin and adjuvant is transferred to a nonimmunized animal which is challenged 48 hours later with egg albumin, this animal develops an anaphylactic reaction and may die in anaphylactic shock. Because hypersensitivity was transferred with serum, this observation suggested that antibodies play a critical pathogenic role in this type of hypersensitivity.
2. **Passive Cutaneous Anaphylaxis.** The passive transfer of hypersensitivity can take less dramatic aspects if the reaction is limited to the skin.
   a. In these experiments, nonsensitized animals are injected intradermally with the serum from a sensitized donor.
   b. The serum from the sensitized donor contained **homocytotropic antibodies** which became bound to the mast cells in and around the area where serum was injected.
   c. After 24 to 72 hours the antigen in question is injected intravenously, mixed with Evans blue dye. When the antigen reaches the area of the skin where antibodies were injected and became bound to mast cells, a localized type I reaction takes place, characterized by a small area of vascular hyperpermeability that results in edema and redness. When Evans blue is injected with the antigen, the area of vascular hyper-

permeability will have a blue discoloration due to the transudation of the dye.

3. The **Prausnitz-Kustner reaction** is a reaction with a similar principle that was practiced in humans and helped our understanding of the immediate hypersensitivity reaction. Serum from an allergic patient was injected intradermally into a nonallergic recipient. After 24 to 48 hours, the area of skin where the serum was injected was challenged with the antigen that was suspected to cause the symptoms in the patient. A positive reaction consisted of a wheal and flare appearing a few minutes after injection of the antigen. The reaction can also be performed in primates, which are injected intravenously with serum of an allergic individual and challenged later with intradermal injections of a battery of antigens that could be implicated as the cause of the allergic reaction. Both of these reactions are no longer used for any clinical purpose.

B. **Clinical Expression.** A wide variety of hypersensitivity states can be classified as immediate hypersensitivity reactions. Some have a predominantly cutaneous expression (**hives** or **urticaria**), others affect the airways (**hay fever**, **asthma**), while others still are of a systemic nature. The latter are often designated as **anaphylactic reactions**, of which **anaphylactic shock** is the most severe form.

1. The expression of anaphylaxis is species specific. The guinea pig usually has bronchoconstriction and bronchial edema as predominant expression, leading to death in acute asphyxiation. In the rabbit, on the contrary, the most affected organ is the heart, and the animals die of right heart failure. In humans, bronchial asthma in its most severe forms closely resembles the reaction in the guinea pig.
2. Most frequently, human type I hypersensitivity has a localized expression, such as the bronchoconstriction and bronchial edema that characterize **bronchial asthma**, the mucosal edema in **hay fever**, and the skin rash and subcutaneous edema that defines **urticaria** (**hives**). The factor(s) involved in determining the target organs that will be affected in different types of immediate hypersensitivity reactions are not well defined, but the route of exposure to the challenging antigen seems to be an important factor.
    i. Systemic anaphylaxis is usually associated with antigens that are directly introduced into the systemic circulation, such as in the case of hypersensitivity to insect venoms or to systemically administered drugs.
    ii. Allergic (extrinsic) asthma and hay fever are usually associated with inhaled antigens
    iii. Urticaria is seen as a frequent manifestation of food allergy.
3. **Systemic anaphylactic reactions** in humans usually present with itching, erythema, vomiting, abdominal cramps, diarrhea, respiratory distress, and in severe cases, laryngeal edema and vascular collapse leading to shock that may be irreversible.

C. **Pathogenesis.** Immediate hypersensitivity reactions are a consequence of the predominant synthesis of specific **IgE** antibodies by the allergic individual; these IgE antibodies bind with high affinity to the membranes of **basophils** and **mast cells**. When exposed to the sensitizing antigen, the reaction with cell-bound IgE triggers the release of **histamine** through degranulation, and the synthesis of

leukotrienes C4, D4, and E4 (this mixture constitutes what was formerly known as **slow reacting substance of anaphylaxis or SRS-A**). These substances are potent constrictors of smooth muscle and vasodilators and are responsible for the clinical symptoms associated with anaphylactic reactions (see Chap. 23).

D. **Atopy.** In medicine, the term atopy is used to designate the tendency of some individuals to become sensitized to a variety of **allergens** (antigens involved in allergic reactions) including pollens, spores, animal danders, house dust, and foods. These individuals, when skin tested, are positive to several allergens and successful therapy must take this multiple reactivity into account. A genetic background for atopy is suggested by the fact that this condition shows familial prevalence.

III. CYTOTOXIC REACTIONS (TYPE II HYPERSENSITIVITY)

In its most common forms, this second type of hypersensitivity involves complement-fixing antibodies (IgM or IgG) directed against cellular or tissue antigens.

A. **Autoimmune Hemolytic Anemia and Other Autoimmune Cytopenias.** Autoimmune hemolytic anemia, autoimmune thrombocytopenia, and autoimmune neutropenia (discussed in greater detail in Chaps. 19, 20, and 24) are clear examples of type II (cytotoxic) hypersensitivity reactions. Autoimmune hemolytic anemia is the best understood of these conditions.
   1. Patients with autoimmune hemolytic anemia synthesize antibodies directed to their own red cells. Those antibodies may cause hemolysis by two main mechanisms:
      a. If the antibodies are of the IgM isotype, complement is activated up to C9, and the red cells can be directly hemolysed (**intravascular hemolysis**).
      b. If, for a variety of reasons, the antibodies (usually IgG) fail to activate the full complement cascade, the red cells will be opsonized with antibody (and possibly C3b) and are taken up and destroyed by phagocytic cells expressing Fcγ and C3b receptors (**extravascular hemolysis**).
      c. When rapid and massive, intravascular hemolysis is associated with release of free hemoglobin into the circulation (hemoglobinemia), which eventually is excreted in the urine (hemoglobinuria). Hemoglobinuria can induce acute tubular damage and kidney failure, usually reversible.
      d. Extravascular hemolysis is usually associated with increased levels of bilirubin, derived from cellular catabolism of hemoglobin.
      e. All hemolytic reactions usually lead to the mobilization of erythrocyte precursors from the bone marrow to compensate for the acute loss. This is reflected by reticulocytosis and, in severe cases, by erythroblastosis (see Chap. 24).

B. **Goodpasture's Syndrome.** The reactions of circulating antibodies with tissue antigens have been traditionally classified as cytotoxic or type II reactions. The classic example is Goodpasture's syndrome.

1. The **pathogenesis** of Goodpasture's syndrome involves the spontaneous emergence of **basement membrane autoantibodies** that bind to antigens of the glomerular and alveolar basement membranes. Those antibodies are predominantly of the IgG isotype.
2. Using fluorescein-conjugated antisera, the deposition of IgG and complement in patients with Goodpasture's syndrome usually follows a linear, very regular pattern, corresponding to the outline of the glomerular or alveolar basement membranes.
3. The pathogenic role of antibasement membrane antibodies is supported by two types of observations:
   a. Elution studies yield immunoglobulin-rich preparations that, when injected into primates, can induce a disease similar to human Goodpasture's syndrome.
   b. Goodpasture's syndrome recurs in patients who receive a kidney transplant and the transplanted kidney shows identical patterns of IgG and complement deposition along the glomerular basement membrane.
4. Once antigen-antibody complexes are formed in the kidney glomeruli or in the lungs, **complement** will be activated and, as a result, C5a and C3a will be generated. These complement components are chemotactic for **PMN leukocytes**; C5a also increases vascular permeability directly or indirectly (by inducing the degranulation of basophils and mast cells) (see Chap. 9).
5. Furthermore, C5a can up-regulate the expression of **cell adhesion molecules** of the CD11b/CD18 family (see Chap. 17) in PMN leukocytes and monocytes, promoting their interaction with ICAM-1 expressed by endothelial cells, the first step leading to the eventual migration to the extravascular space. The combination of increased vascular permeability and of induction of cell adhesion molecules is believed to be responsible for the accumulation of neutrophils and monocytes in areas where antigens and antibodies have reacted.
6. Once in the tissues, the PMN will recognize the Fc regions of tissue-bound antibodies, as well as C3b bound to the corresponding immune complexes, and will release their enzymatic contents, which include **proteases** and **collagenase**. These enzymes split complement components and generate bioactive fragments, enhancing the inflammatory reaction, and cause tissue damage (i.e., destruction of the basement membrane), which eventually may compromise the function of the affected organ.
7. The pathological sequence of events after the reaction of antibasement membrane antibodies with their corresponding antigens is indistinguishable from the reactions triggered by the deposition of soluble immune complexes or by the reaction of circulating antibodies with antigens passively fixed to a tissue, considered as type III hypersensitivity reactions.

C. **Nephrotoxic (Masugi) Nephritis.** This experimental model of immunologically mediated nephritis, named after the scientist who developed it, is induced by injection of heterologous antibasement membrane antibodies into healthy animals. Those antibodies combine with basement membrane antigens, particularly at the glomerular level, and trigger the development of glomerulonephritis. This experimental model has been extremely useful to demonstrate the pathogenic importance of complement activation and of neutrophil accumulation.

1. If, instead of complete antibodies, one injects Fab or $F(ab')_2$ fragments that do not activate complement, the accumulation of neutrophils in the glomeruli fails to take place, and tissue damage will be minimal to nonexistent.
2. Similar protection against the development of glomerulonephritis is observed when animals are rendered C3 deficient by injection of cobra venom factor prior to the administration of antibasement membrane antibodies, or when those antibodies are administered to animals rendered neutropenic by administration of cytotoxic drugs or of antineutrophil antibodies.

V. IMMUNE COMPLEX-INDUCED HYPERSENSITIVITY REACTIONS (TYPE III HYPERSENSITIVITY)

In the course of acute or chronic infections, or as a consequence of the production of autoantibodies, antigen–antibody complexes (also known as **immune complexes**) are likely to be formed in circulation or in tissues to which the pertinent microbial or self-antigens have been adsorbed. Both scenarios can lead to inflammatory changes that are characteristic of the so-called immune complex diseases (see Chap. 25).

A. **The Fate of Circulating Immune Complexes.** Circulating immune complexes are usually adsorbed to red cells and cleared by the phagocytic system. In most cases, this will be an inconsequential sequence of events, but, in other cases, when the clearance capacity of the phagocytic system is exceeded, inflammatory reactions can be triggered by the deposition of those immune complexes in tissues. A simplified sequence of events leading to immune complex-induced inflammation is shown in Figure 22.1.

B. **In Situ Formation of Immune Complexes.** The adsorption of circulating antigens of microbial origin or released by dying cells to a variety of tissues seems to be a relatively common event. If the same antigens trigger a humoral immune response, immune complex formation may take place in the tissues where the antigens are adsorbed, in which case clearance by the phagocytic system may become impossible. In fact, tissue-bound immune complexes are very strong activators of the complement system and of phagocytic cells, triggering a sequence of events leading to tissue inflammation virtually identical to that observed in cases of in situ immune reactions involving tissue antigens and the corresponding antibodies.

C. **The Arthus Reaction.** This reaction was first described at the turn of the century by Arthus, who observed that the intradermal injection of antigen into an animal previously sensitized results in a local inflammatory reaction. This reaction is edematous in the early stages, but later can become hemorrhagic, and, eventually, necrotic. A human equivalent of this reaction can be observed in some reactions to immunization boosters in individuals who have already reached high levels of immunity.

   1. **Pathogenesis and general characteristics**

      a. The reaction is due to the combination of complement-fixing IgG anti-

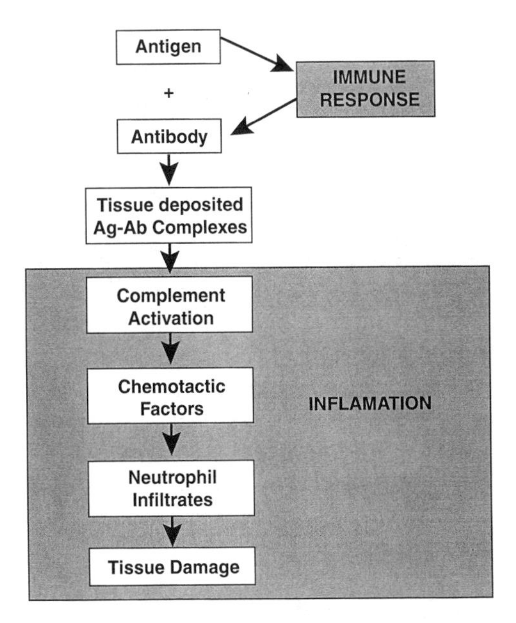

Figure 22.1 Diagrammatic representation of the sequence of events triggered by the deposition of soluble immune complexes that eventually results in inflammation and tissue damage.

bodies (characteristically predominating in hyperimmune states in most species) and tissue-fixed antigens.

b. The lag time between antigen challenge and the reaction is usually 6 hours, which is considerably longer than the lag time of an immediate hypersensitivity reaction, but considerably shorter than that of a delayed hypersensitivity reaction.

c. Although Arthus reactions are typically elicited in the skin, the same pathogenic mechanisms can lead to organ lesions whenever the antigen, although intrinsically soluble, is unable to diffuse freely and remains retained in or around its penetration point (e.g., the perialveolar spaces for inhaled antigens).

2. **Experimental studies.** The Arthus reaction, easy to induce in a variety of laboratory animals, is one of the best-studied models of immune complex disease.

   a. **Pathological changes.** Immunohistological studies have shown that soon after antigen is injected in the skin, IgG antibody and C3 will appear in perivascular deposits at the site of injection. This is followed by massive influx of granulocytes, believed to result from activation of the complement system by the in situ formed immune complexes.

   b. **Role of granulocytes.** If the animals used for induction of an Arthus reaction are first made neutropenic by administration of nitrogen mustard or antineutrophil serum, the inflammatory reaction is prevented.

   c. **Spontaneous healing.** In spite of their pathogenic role, granulocytes will actively engulf and eliminate the tissue-deposited IC, eliminating the trigger for the inflammatory reaction. As the IC are eliminated, the cellular infiltrate changes from a predominance of neutrophils and other granulocytes to a predominance of mononuclear cells, which is usually associated with the healing stage.

D. **Serum Sickness**

1. **Background.** In the preantibiotic era, the treatment of rabies, bacterial pneumonia, diphtheria, and other infections involved the administration of heterologous antisera as a way to transfer immunity to the offending agents. In many instances, serotherapy appeared to be successful, and the patient improved, but a week to 10 days after the injection of heterologous antiserum, the patient developed what was termed as **serum sickness**: a combination of cutaneous rash (often purpuric), fever, arthralgias, mild acute glomerulonephritis, and carditis.
2. **Serum sickness in the 1990s.** Currently, serum sickness as a complication of passive immunotherapy with heterologous antisera is seen after injection of heterologous antisera to snake venoms, after the administration of mouse monoclonal antibodies in cancer immunotherapy, and after the administration of heterologous (monoclonal or polyclonal) antilymphocyte sera in transplanted patients. But, it can also be a side effect of some forms of drug therapy, particularly with penicillin and related drugs.
3. **Experimental models.** Serum sickness is extremely easy to reproduce in experimental animals through the injection of heterologous proteins. Basically two types of experimental serum sickness can be induced: (a) acute, after a single immunization with a large dose of protein; (b) chronic, after repeated daily injections of small doses of protein. While acute serum sickness is reversible, the chronic form, which closely resembles human glomerulonephritis, is usually associated with irreversible damage.
4. **Pathogenesis**
   a. **Induction of serum sickness.** In all types of serum sickness, the initial event is the triggering of a humoral immune response, which explains the lag period of 7 to 10 days between the injection of heterologous protein (or drug) and the beginning of clinical symptoms. The lag period is shorter and the reaction more severe if there has been presensitization to the antigen in question.
   b. **Formation of soluble immune complexes.** As soon as antibodies are secreted, they combine with the antigens (which at that time are still present in relatively large concentrations in the serum of the injected individual or experimental animal). Initially, the antigen–antibody reaction will take place in conditions of great antigen excess, and the resulting complexes are too small to activate complement or to be taken up by phagocytic cells, and will remain in circulation without major consequences.
   c. **Deposition of immune complexes in tissues.** As the immune response progresses and increasing amounts of antibody are produced, the antigen–antibody ratio will be such that intermediate-size immune complexes will be formed, and these are potentially pathogenic since they are

not too large to cross the endothelial barrier (particularly if vascular permeability is increased) but are large enough to activate complement and induce inflammation.

d. **Consequences of immune complex deposition.**
   i. The deposition of immune complexes can take place in different organs, such as the myocardium (causing myocardial inflammation), skin (causing erythematous rashes), joints (causing arthritis), and kidney (causing acute glomerulonephritis).
   ii. Soluble immune complexes can also be absorbed by formed elements of the blood, particularly erythrocytes, neutrophils, and platelets. Hemolytic anemia, thrombocytopenia, and neutropenia can result from the activation of the complement system by cell-associated immune complexes. Purpuric rashes due to thrombocytopenia are frequently seen in serum sickness.

e. **Key role of neutrophils and the complement system in serum sickness.** As in the case of the Arthus reaction, the inflammatory changes associated with serum sickness do not take place or are very mild if complement or neutrophils are depleted.

VI. DELAYED (TYPE IV) HYPERSENSITIVITY REACTIONS

In contrast to the other types of hypersensitivity reactions discussed above, type IV or delayed hypersensitivity is a manifestation of cell-mediated immunity. In other words, this type of hypersensitivity reaction is due to the activation of specifically sensitized T lymphocytes rather than to an antigen–antibody reaction.

A. **The Tuberculin Test as a Prototype Type IV Reaction.** Intradermal injection of tuberculin or PPD into an individual that has been previously sensitized (by exposure to *Mycobacterium tuberculosis* or by BCG vaccination) is followed, 24 hours after the injection, by a skin reaction at the site of injection characterized by **redness and induration**. Histologically, the reaction is characterized by perivenular mononuclear cell infiltration, often described as "perivascular cuffing." Macrophages can be seen infiltrating the dermis. If the reaction is intense, a central necrotic area may develop. The cellular nature of the perivascular infiltrate, which contrasts with the predominantly edematous reaction in a cutaneous type I hypersensitivity reaction, is responsible for the induration.

B. **Experimental Studies**
   1. **Transfer of delayed hypersensitivity.** When guinea pigs are immunized with egg albumin and adjuvant, not only do they become allergic, as discussed earlier, but they develop cell-mediated hypersensitivity to the antigen. This duality can be demonstrated by passively transferring serum and lymphocytes from a sensitized animal to different unsensitized recipients of the same strain and challenging the passively immunized animals with egg albumin. The animals that received serum will develop an anaphylactic response immediately after challenge, while those that received lymphocytes will only show signs of a considerably less severe reaction after at least 24 hours have elapsed from the time of challenge.

2. **Experimental contact hypersensitivity.** Most of our knowledge about the pathogenesis of delayed hypersensitivity reactions derives from experimental studies involving contact hypersensitivity. Experimental sensitization through the skin is relatively easy to induce by percutaneous application of low-molecular-weight substances such as picric acid or dinitrochlorobenzene (DNCB). The initial application leads to sensitization, and a second application will elicit a delayed hypersensitivity reaction in the area where the antigen is applied.
   a. **Induction.** The compounds used to induce contact hypersensitivity are not immunogenic by themselves. It is believed that these compounds couple spontaneously to an endogenous carrier protein, and as a result of this coupling, the small molecule will act as a hapten, while the endogenous protein will play the role of a carrier.
      i. A common denominator of the sensitizing compounds is the expression of reactive groups, such as Cl, F, Br, and SO_3H, which enables them to bind covalently to the carrier protein.
      ii. Spontaneous sensitization to drugs, chemicals, or metals is believed to involve diffusion of the haptenic substance into the dermis, mostly through the sweat glands (hydrophobic substances appear to penetrate the skin more easily than hydrophilic substances). Once in the dermis, the haptenic groups will react spontaneously with "carrier" proteins and trigger an immune reaction.
      iii. By a pathway that has not been defined, the hapten carrier conjugates are taken up by the Langerhans cells of the epidermis, and a sensitizing peptide is presented in association with MHC-II molecules. Since the carrier protein is self, it would be expected that the sensitizing peptide contained the covalently associated sensitizing compound.
      iv. A unique feature of delayed hypersensitivity is that T lymphocytes are mostly involved in the antihapten response, while in most experimentally induced hapten-carrier responses, the hapten is recognized by B lymphocytes. This may be explained, at least in part, by the fact that Langerhans cells migrate to regional lymph nodes, where they become interdigitating cells and predominantly populate the paracortical areas, where they are in optimal conditions to present antigens to CD4+ T lymphocytes (see Chap. 2).
   b. **Effector mechanisms.** The initial sensitization results in the acquisition of immunological memory. Later, when the sensitized individual is challenged with the same chemical, sensitized T cells will be stimulated into functionally active cells, releasing a variety of cytokines, which include chemotactic and activating chemokines for monocyte/macrophages, basophils, eosinophils, and neutrophils.
      i. The release of chemotactic chemokines such as IL-8, RANTES, macrophage chemotactic proteins, and migration inhibitory factor is a key factor in attracting and "fixing" lymphocytes, monocytes, and granulocytes into the area.
      ii. Other cytokines released by activated lymphocytes, particularly

TNF-α and IL-1 up-regulate the expression of cell adhesion molecules (CAMs) in endothelial cells, facilitating the adhesion of leukocytes to the endothelium, a key step in the extravascular migration of inflammatory cells.

iii. As a result of the release of chemokines and of the up-regulation of CAMs, a cellular infiltrate predominantly constituted by mononuclear cells forms in the area where the sensitizing compound has been reintroduced 24 to 48 hours after exposure.

iv. The tissue damage that takes place in this type of reaction is likely to be due to the effects of active oxygen radicals and enzymes (particularly proteases, collagenase, and cathepsins) released by the infiltrating leukocytes, activated by the chemokines and other cytokines.

v. In severe cases, a contact hypersensitivity reaction may take an exudative, edematous, highly inflammatory character. The release of proteases from monocytes and macrophages may trigger the complement-dependent inflammatory pathways by directly splitting C3 and C5; C5a will add its chemotactic effects to those of chemokines released by activated mononuclear cells, and will also cause increased vascular permeability, a constant feature of complement-dependent inflammatory processes. It is not surprising, therefore, that a reaction which at the onset is cell mediated and associated to a mononuclear cell infiltrate may, in time, evolve into a more classical inflammatory process with predominance of neutrophils and a more edematous character, less characteristic of a cell-mediated reaction.

C. **Contact Hypersensitivity in Humans.** Contact hypersensitivity reactions are observed with some frequency in humans due to spontaneous sensitization to a variety of substances.

1. **Plant cathecols** are apparently responsible for the hypersensitivity reactions to poison ivy and poison oak.
2. A variety of **chemicals** can be implicated in hypersensitivity reactions to cosmetics and leather.
3. Topically used **drugs**, particularly sulfonamides, often cause contact hypersensitivity.
4. Metals such as nickel can be involved in reactions triggered by the use of bracelets, earrings, or thimbles.

The diagnosis is usually based on a careful history of exposure to potential sensitizing agents and on the observation of the distribution of lesions that can be very informative about the source of sensitization. Patch tests using small pieces of filter paper impregnated with suspected sensitizing agents that are taped to the back of the patient can be used to identify the sensitizing substance.

D. **The Jones–Mote Reaction.** Following challenge with an intradermal injection of a small dose of a protein to which an individual has been previously sensitized, a delayed reaction (with a lag of 24 hours), somewhat different from a classic delayed hypersensitivity reaction, may be seen. The skin appears more erythematous and less indurated, and the infiltrating cells are mostly lymphocytes and basophils, the last sometimes predominating. The reaction has also been described, for this reason, as cutaneous basophilic hypersensitivity. Experimentally, it has been demonstrated that this reaction is triggered as a consequence of the antigenic stimulation of sensitized T lymphocytes.

E. **Homograft Rejection.** A most striking clinical manifestation of a delayed hypersensitivity reaction is the rejection of a graft. In classic chronic rejection, the graft recipient's immune system is first sensitized to tissue antigens of the donor. After clonal expansion, activated T lymphocytes will reach the target organ, recognize the foreign antigen, and initiate a sequence of events that leads to inflammation and eventual necrosis of the organ. This topic will be discussed in detail in Chapter 27.

F. **Systemic Consequences of Cell-Mediated Hypersensitivity Reactions.** While type IV hypersensitivity reactions with cutaneous expression usually have no systemic repercussions, cell-mediated hypersensitivity reactions localized to internal organs, such as the formation of granulomatous lesions caused by chronic infections with *Mycobacteria*, may be associated with systemic reactions. Cytokines released by activated lymphocytes and inflammatory cells play a major pathogenic role in such reactions.

1. **Proinflammatory cytokines**, particularly IL-1, activate the hypothalamic temperature regulating center and cause fever, thus acting as a central **pyrogenic factor**.
2. TNF-α is also pyrogenic, both directly and by inducing the release of IL-1 by endothelial cells and monocytes. In addition, these cytokines activate the synthesis of acute phase proteins (e.g., C-reactive protein) by the liver.
3. Prolonged release of TNF-α, on the other hand, may have deleterious effects since this factor contributes to the development of **cachexia**. The way in which TNF-α causes cachexia has been recently elucidated: the factor inhibits lipoprotein lipase and, as a consequence, there is an accumulation of triglyceride-rich particles in the serum and a lack of the breakdown of triglycerides into glycerol and free fatty acids. This results in decreased incorporation of triglycerides into the adipose tissue, and, consequently, in a negative metabolic balance. The cells continue to break down stored triglycerides by other pathways to generate energy, and the used triglycerides are not replaced. Cachexia is often a preterminal development in patients with severe chronic infections.

SELF-EVALUATION

Questions

Choose the ONE *best* answer.

22.1 Which of the following is believed to be the triggering step of a cutaneous hypersensitivity reaction to nickel?
- A. Activation of the complement system
- B. Formation of nickel–protein complexes
- C. Induction of IgE antibody synthesis
- D. Migration of inflammatory cells to the site of exposure
- E. Release of proinflammatory cytokines by activated lymphocytes

22.2 Which of the following antigens is LEAST likely to induce a delayed hypersensitivity reaction?
  A. Candida antigen
  B. Mumps antigen
  C. Ragweed pollen
  D. Tetanus toxoid
  E. Tuberculin

22.3 Which of the following is a pathogenic step characteristically associated with an Arthus reaction?
  A. Complement activation
  B. Formation of mononuclear cell infiltrates
  C. High levels of IgM antibodies
  D. Histamine release
  E. Induction of IgE antibodies

22.4 Which of the following histopathological features is most likely to be evident on a skin biopsy of a patient having a delayed hypersensitivity reaction?
  A. Deposits of immunoglobulins and complement in and around the arterial wall
  B. Massive edema of the subcutaneous tissues
  C. Necrosis of the epidermis
  D. Periarteriolar neutrophil infiltrates
  E. Perivenular mononuclear cell infiltrates

22.5 The persistence of an immediate hypersensitivity reaction for several hours is most likely a consequence of the:
  A. Fixation of mononuclear cells at the site of the reaction
  B. Persistent activation of the complement system
  C. Recruitment of basophils as a consequence of the up-regulation of endothelial cell adhesion molecules
  D. Release of leukotrienes C4, D4, and E4
  E. Systemic effects of proinflammatory cytokines

22.6 The pathogenesis of type II (cytotoxic) hypersensitivity reactions typically involves:
  A. Adsorption of antigen–antibody complexes to cell membranes
  B. Cell or tissue damage caused by activated cytotoxic T lymphocytes
  C. Complement activation by IgM or IgG antibodies
  D. Direct cell lysis caused by cytotoxic antibodies
  E. Release of histamine in the early stages

22.7 Progressive weight loss in a patient suffering from a chronic granulomatous infection with predominant mononuclear cell involvement is likely to be due to:
  A. Loss of protein by kidneys damaged as a consequence of immune complex deposition
  B. Massive edema
  C. Massive utilization of glucose by activated mononuclear cells
  D. Persistent high fever
  E. Persistent release of TNF-α

Questions 22.8–22.10

The group of questions below consists of a set of lettered headings followed by a list of numbered words or phrases. For EACH numbered word or phrase, select the ONE lettered heading that is most closely related to it. The same heading can be used once, more than once, or not at all.

A. Cell-mediated (delayed) hypersensitivity
B. Cutaneous basophilic hypersensitivity
C. Cytotoxic hypersensitivity
D. IgE-mediated anaphylaxis
E. Immune complex-mediated hypersensitivity

22.8 Chronic "drug" dermatitis in the upper eyelids of a 24-year-old woman
22.9 Serum sickness-associated glomerulonephritis
22.10 Shock following a penicillin injection

Answers

22.1 (B) The substances involved in contact dermatitis are often small molecules that apparently act as haptens after spontaneously reacting with an endogenous protein that will serve as a carrier. Without this step, cutaneous dermatitis will not develop.

22.2 (C) Pollens are usually involved in type I hypersensitivity, and the skin tests with pollens elicit typical immediate hypersensitivity reactions.

22.3 (A) The Arthus reactions involve primarily complement-fixing antibodies. The cellular infiltrate is mainly composed of neutrophils and other granulocytes.

22.4 (E) In contrast to type II hypersensitivity, the cellular infiltrates in delayed hypersensitivity are perivenular and usually show a predominance of mononuclear cells.

22.5 (D) The initial phase of an immediate hypersensitivity reaction is due to mast cell and basophil degranulation with release of stored histamine. In later stages, the reaction is sustained by the secretion of SRS-A (a mixture of leukotrienes C4, D4, and E4), which are synthesized de novo by stimulated basophils and mast cells.

22.6 (C) Complement activation is the key element and will either lead to direct cytolysis or to phagocytosis.

22.7 (E) TNF-α inhibits lipoprotein lipase and, as a consequence, there is an accumulation of triglyceride-rich particles in the serum and a lack of the breakdown of triglycerides into glycerol and free fatty acids. The incorporation of triglycerides into the adipose tissue is inhibited, resulting in a negative metabolic balance: the cells continue to break down stored triglycerides by other pathways to generate energy, and the used triglycerides are not replaced. The kidneys are usually not affected (IC deposition is a feature of type II hypersensitivity reactions). Massive edema due to capillary breakage is seen in patients receiving large amounts of recombinant IL-2, but not in patients with T-lymphocyte activation. Fever, by itself, does not lead to cachexia.

22.8 (A) Sensitization to cosmetics such as eye shadow is not uncommon and involves a type IV reaction, usually induced by formalin, which is added as a preservative.

22.9 (E) Serum sickness can be associated with glomerulonephritis secondary to the deposition of immune complexes involving heterologous antigens and the corresponding antibodies.

BIBLIOGRAPHY

Brostoff, J., Scadding, G.K., Male, D., and Roitt, I.M., eds. *Clinical Immunology*. Gower Medical Publishers, London, 1991.

Collins, T. Adhesion molecules in leukocyte emigration. *Sci. Am. Med., 2(6)*:28, 1995.

Kaplan, A.P., Kuna, P., and Reddigari, S.R. Chemokines and the allergic response. *Exp. Dermatol., 4*:260, 1995.

Kavanaugh, A. Adhesion molecules as therapeutic targets in the treatment of allergic and immunologically mediated disorders. *Clin. Immunol. Immunopathol., 80(part 2)*:S40, 1996.

Schroeder, J.T., Kagey-Sobotka, A., and Lichtenstein, L.M. The role of the basophil in allergic inflammation. *Allergy, 50*:463, 1995.

Schwiebert, L.A., Beck, L.A., Stellato, C., et al. Glucocorticosteroid inhibition of cytokine production: relevance to antiallergic actions. *J. Allergy Clin. Immunol., 97*:143, 1996.

Sell, S., ed. *Immunology, Immunopathology and Immunity, 5th ed.*, Appleton & Lange, Stamford, CT, 1996.

Tracey, K.J., and Grami, A. Tumor necrosis factor: a pleiotrope cytokine and therapeutic target. *Annu. Rev. Med., 45*:491, 1994.

23
IgE-Mediated (Immediate) Hypersensitivity

Jean-Michel Goust and Albert F. Finn, Jr.

I. INTRODUCTION

The term immediate hypersensitivity describes a key characteristic of IgE-mediated hypersensitivity reactions: the **short time lag** (seconds to minutes) **between antigen exposure and the onset of clinical symptoms**. This is because the initial symptoms of immediate hypersensitivity depend on the release of preformed mediators stored in cytoplasmic granules of basophils and mast cells; the release is triggered by the reaction of membrane-bound **IgE** with the corresponding antigen (also known as **allergen**, by being involved in allergic reactions).

II. MAJOR CLINICAL EXPRESSIONS

Immediate hypersensitivity or allergic reactions can have a variety of clinical expressions, including **anaphylaxis, bronchial asthma, urticaria (hives), and rhinitis (hay fever).** Table 23.1 summarizes the morbidity and mortality data for the two most severe types of allergic reactions, anaphylaxis and asthma.

A. **Anaphylaxis** is an acute life-threatening IgE-mediated reaction usually affecting multiple organs.
   1. The **time of onset of symptoms** depends on the level of hypersensitivity and the amount, diffusibility, and site of exposure to the antigen. In a typical case, manifestations begin within 5 to 10 minutes after antigenic challenge. Reactions that appear more slowly tend to be less severe. Intervals longer than 2 hours leave the diagnosis of anaphylaxis open to question.
   2. Multiple organ systems are usually affected, including the skin (pruritus, flushing, urticaria, angioedema), respiratory tract (bronchospasm and laryngeal edema), and cardiovascular system (hypotension and cardiac arrhythmias).
   3. As a rule, most of the acute manifestations subside within 1 or 2 hours. However, similar symptoms of variable intensity may occur 6–12 hours later. This late-phase reaction results from cytokine release from activated basophils, secondary immune cell activation, and further elaboration and release

Table 23.1 Morbidity and Mortality from Systemic Anaphylaxis and Bronchial Asthma in the United States

	Morbidity	Mortality
Systemic anaphylaxis		
Caused by antibiotics	10–40 : 100,000 injections	1 : 100,000 injections
Caused by insect bites	10 : 100,000 persons/year	10–80/year
Asthma	10 million persons (4–5% of U.S. population)	5–30 years of age: 0.4/100,000/year; >60 years of age: 10/100,000/year

of mediators of inflammation—mechanisms very similar to those responsible for the late phase of asthma.

4. When death occurs, it is usually due to laryngeal edema, intractable bronchospasm, hypotensive shock, or cardiac arrhythmias developing within the first 2 hours.

B. **Atopy** is defined as a genetically determined state of IgE-related disease. Its most common clinical manifestations include: asthma, rhinitis, urticaria, and atopic dermatitis.
   1. Allergic asthma, by its potential severity and frequency, is the most important manifestation of atopy.
   2. Not all cases of asthma are of proven allergic etiology. The differential characteristics of allergic (extrinsic) and nonallergic (intrinsic) asthma are summarized in Table 23.2. The major difference between both is the strong association of allergic asthma with specific IgE antibodies and eosinophils, both of which play important pathophysiological roles.

Table 23.2 Major Characteristics of Allergic and Nonallergic Bronchial Asthma

Symptoms: Dyspnea with prolonged expiratory phase; may be associated with cough and sputum
Chest x-rays: Hyperlucency (reflecting impaired expiratory capacity), bronchial thickening

	Allergic	Nonallergic
Blood	Eosinophilia	± Eosinophil count
Sputum	Eosinophilia	± Eosinophils
Total IgE	Raised	Normal
Antigen-specific IgE	Raised	None
Pathology	Obstruction of airways due to smooth muscle hypertrophy with constriction and mucosal edema; hypertrophy of mucous glands; eosinophil infiltration	Similar to changes seen in allergic asthma, with variable numbers of eosinophils
Frequency:	Children: 80%[a]	20%
	Adults: 60%[a]	40%

[a]% of total number of bronchial asthma cases seen in each age range.

Case 23.1

A 20-year-old female attending college in the southeast presented to the infirmary with wheezing, nocturnal cough, and shortness of breath. She reported that since the fall she had worsening of her asthma as well as nasal congestion and paroxysms of sneezing. During the summer at her parents' residence in Los Alamos, New Mexico, she had minimal nasal congestion in the mornings. She admitted to a long history of mild wheezing as a child. This semester she resides in a home with several other students. There is a cat in the house, although she avoids it as it makes her eyes itch. Presently, she uses her albuterol inhaler frequently during the day and at night.

On physical exam, she was audibly wheezing and mildly tachypneic. Her conjunctivae were mildly injected. The nasal mucosa was edematous and pale with copious watery secretions. Ears were normal, and mouth was unremarkable. Auscultation of the chest revealed a prolonged expiratory phase with diffuse wheezing bilaterally. Extremities were without clubbing, cyanosis, or edema.

Lab data: complete blood count—Hgb = 13.5 g/dL; HCT = 39%; WBC = $9.8 \times 10^6/\mu l$ (normal 4.8–10.8); normal differential. Arterial blood gases: pH = 7.44; pCO_2 = 38 mmHg; pO_2 = 78 mmHg; O_2 saturation = 93%. IgE = 689 IU/mL. Chest X-ray: PA and lateral views of the chest revealed increased AP diameter and flattening of the diaphragm. There were no infiltrates or effusions. Ventilatory studies: Peak flow was 200 L/min with improvement to 320 L/min after nebulized albuterol.

This case raises several questions:

- What possible factors contributed to her respiratory problems?
- Why does she have difficulty exhaling, though it improves with a beta-agonist?
- What histolopathological findings would be expected on a biopsy of the bronchial mucosa?
- Which cytokines are critically involved in the pathogenesis of this patient's allergic disease?
- What therapeutic agents are useful acutely and for long-term therapy?

III. PATHOGENESIS

The pathogenesis of immediate hypersensitivity reactions involves a well-defined sequence of events:

- Synthesis of IgE antibodies.
- Binding of IgE antibodies to FcεI receptors on basophils and mast cells; once bound, IgE acts as an antigen receptor.
- Cross-linking of receptor-bound IgE by a multivalent antigen signals the release of preformed vasoactive compounds and the synthesis and later release of mediators of inflammation.
- The preformed substances released by basophils and mast cells have significant effects on target tissues, such as smooth muscle, vascular endothelium, and mucous glands. They also act as chemoattractant cytokines and may elicit central nervous system-mediated reflexes (e.g., sneezing).

- Furthermore, activated basophils and mast cells can synthesize and express IL-4 and the CD40 ligand, both essential factors to stimulate IgE synthesis (see Chap. 11).

A. Immunoglobulin E Antibodies

1. Historical overview

a. The first demonstration that serum contains a factor capable of mediating specific allergic reactions was published in 1921 by **Prausnitz and Kustner**. The injection of serum from a fish-allergic person (Kustner) into Dr. Prausnitz's skin, and subsequent exposure of Dr. Prausnitz to fish antigen injected in the same site, resulted in an allergic wheal and flare response.

b. In 1967, **Ishizaka** and collaborators isolated a new class of immunoglobulin, designated as IgE, from the serum of ragweed-allergic individuals. Several patients with IgE-producing plasmocytomas were subsequently discovered and provided a source of very large amounts of monoclonal IgE, which greatly facilitated further studies of IgE structure and the production of anti-IgE antibodies.

2. Quantitation of IgG and of IgE antibodies

a. The total IgE concentration, even in allergic individuals, is extremely low, and not detectable by most routine assays used for the quantitation of IgG, IgA, and IgM. The concentration of specific IgE antibody to any given allergen is a very small fraction of the total IgE.

b. Early attempts to measure IgE involved cumbersome and often unreliable bioassays. The availability of anti-IgE antibodies allowed the development of radioimmunoassays sufficiently sensitive to determine serum IgE levels accurately.

   i. The **paper disk radioimmunosorbent test (PRIST)** was one of the first solid-phase radioimmunoassays (see Chap. 15) introduced in diagnostic medicine. This assay, diagrammatically summarized in Figure 23.1, measures total serum IgE.
   - A serum sample is added to a small piece of adsorbent paper to which anti-IgE antibodies are covalently bound. IgE is captured by the immobilized antibody.
   - ^{125}Iodine-labeled anti-IgE antibodies are subsequently added and will bind to the paper-bound IgE. The radioactivity counted in the solid phase is directly related to the IgE level in the serum tested.
   - The results are expressed in nanograms/mL (1 ng = 10^{-6} mg), or in International Units (1 I.U. = 2.5 ng/mL); 180 IU/mL is considered as the upper limit for normal adults. Allergic individuals often have elevated levels of IgE, but a significant number of asymptomatic individuals may also have elevated IgE levels. Therefore, a diagnosis of immediate hypersensitivity cannot be based solely on the determination of abnormally elevated IgE levels.

   ii. The **radioallergosorbent test (RAST),** diagrammatically summarized in Figure 23.2, is a solid-phase radioimmunoassay that deter-

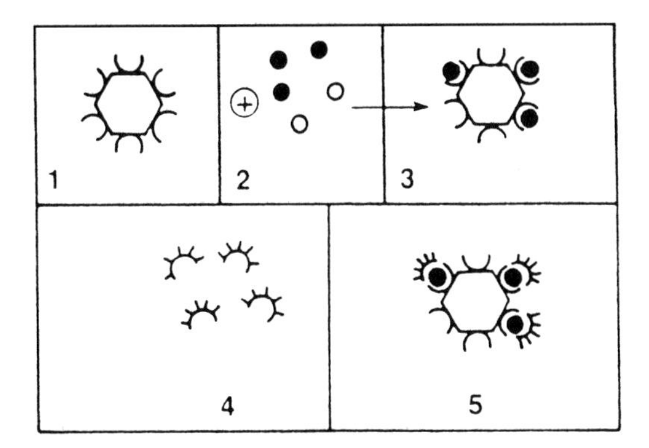

1. Anti-IgE coupled to paper disc
2. Serum IgE
3. Binding of Serum IgE to paper disc
4. Radiolabeled anti-IgE
5. Binding of the Radiolabel to disc-bound IgE

Figure 23.1 Diagrammatic representation of the general principles of the radioimmunosorbent test for IgE quantitation.

mines antigen-specific IgE, which from the diagnostic point of view is considerably more relevant than the measurement of total serum IgE levels.

- A given allergen (ragweed antigen, penicillin, β-lactoglobulin, etc.) is covalently bound to a solid phase (e.g., polydextran beads)
- A patient's serum is added to antigen-coated beads; the antigen-specific IgE, if present, will bind to the immobilized antigen.
- After washing off unbound immunoglobulins, radiolabeled anti-IgE is added. The amount of bead-bound radioactivity counted after washing off unbound labeled antibody is directly related to the concentration of antigen-specific IgE present in the serum.

3. **Skin tests.** Although the RAST assays are highly specific and accurate, they are expensive, lack sensitivity, and the range of antigens for which there are available tests is limited. In addition, some authors have cast doubts about the biological relevance of the RAST assay results. The alternative method for diagnosis of specific allergies is a provocation skin test, which allows the testing of a wider array of antigens and provides results that appear to correlate better with clinical data.
   a. Small amounts of purified allergens are injected percutaneously or intradermally in known patterns and patients are observed for about 30 minutes to 1 hour. Classic IgE-mediated hypersensitivity reactions present as a wheal and flare at the site of the allergen exposure, which develops in a matter of minutes.
   b. In highly sensitized individuals, there is always a risk of anaphylaxis, even after minimal challenge. Because of this risk, these tests should

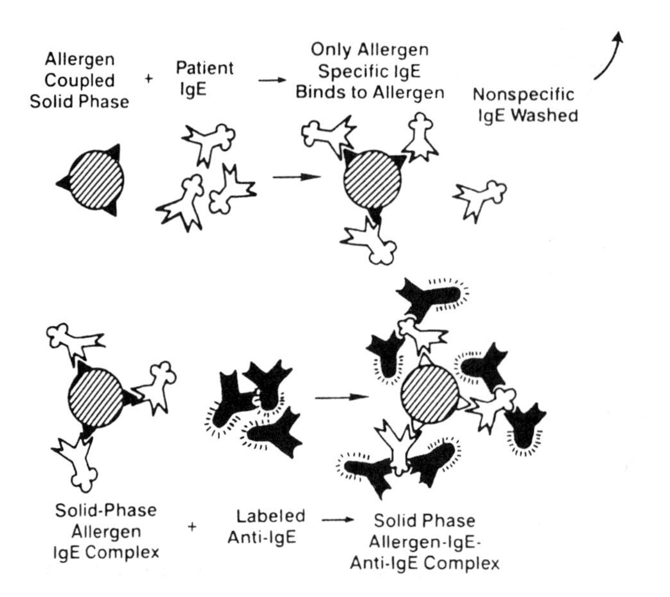

Figure 23.2 Diagrammatic representation of the general principles of the radioallergosorbent test (RAST) for quantitation of specific IgE antibodies.

always be performed by trained professionals in a properly equipped clinical facility.

B. The IgE Antibody Response

1. IgE is predominantly synthesized in lymphoid tissues of the respiratory and gastrointestinal tract. In developing countries, the main antigenic stimulus for IgE synthesis are parasites (particularly nematodes). Levels of circulating IgE considered as normal in a developing country with endemic parasitism are two to three orders of magnitude higher than in the western world. The same perimucosal tissues are stimulated by the vast majority of allergens, which are either ingested or inhaled.
2. In the perimucosal tissues only B lymphocytes with membrane IgE will differentiate into IgE-producing plasma cells. Those IgE-carrying B cells are only a small fraction of the total B-cell population in the submucosa, but are over-represented in the perimucosal lymphoid tissues compared to other lymphoid territories.
3. During the primary immune response to an allergen or a parasite, most of the IgE synthesized appears to be of low affinity. The changes occurring after a second exposure include the synthesis of IgE of progressively higher affinity, probably as a consequence of somatic hypermutations (see Chaps. 7 and 12). This may be the reason why allergic reactions very seldom develop after the first exposure to an allergen.
4. A hypersensitivity reaction may sometimes develop after what appears to be a first exposure. In these instances, one must consider the possibility of cross-reaction between a substance to which the individual was previously sensitized and the substance that elicits the allergic reaction. Such cross-reactions are usually due to molecular mimicry (see Chap. 18) and can be quite unpredictable.

5. Repeated exposures to parasites or allergens will stimulate the differentiation of memory cells and the proportion of circulating high-affinity antigen-specific IgE will also increase with repeated exposures. In patients suffering from severe pollen allergies, antigen-specific IgE may constitute up to 50% of the total IgE.

C. **Genetic Control of IgE Synthesis**

1. The study of total IgE levels in normal nonallergic individuals shows a trimodal distribution: high, intermediate, and low producers (Fig. 23.3). Family studies further suggested that the ability to produce high levels of IgE is a recessive trait, controlled by unknown genes independent of the HLA system. A candidate gene has been localized to chromosome 5(5q31-q33)—in close vicinity to the genes for IL-4, IL-5, and IL-13.
2. The gene in question seems to influence not only the synthesis of IgE, but also determines bronchial hyperresponsiveness to histamine and other mediators.
3. On the other hand, the tendency to develop allergic disorders in response to specific allergens is HLA-linked. For instance, the ability to produce antigen-specific IgE after exposure to the Ra5 antigen of ragweed is observed more often in HLA B7, DR2 individuals than in the general population. Recent studies suggest that various HLA class II antigens are associated with high responses to many different allergens (Fig. 23.4).
4. The biological basis for the association between MHC-II molecules and allergic predisposition is believed to be one of the many expressions of the control exerted by APC on the immune response by virtue of the fact that the MHC-II repertoire will determine what antigen-derived peptides will be most efficiently presented to helper T cells.
5. The genes controlling high IgE levels and high antibody synthesis after exposure to allergens appear to have synergistic effects. For example, an HLA B7, DR2 individual that is also genetically predisposed to hyperproduce IgE is likely to have a more severe allergic disorder than an individual without this genetic combination.

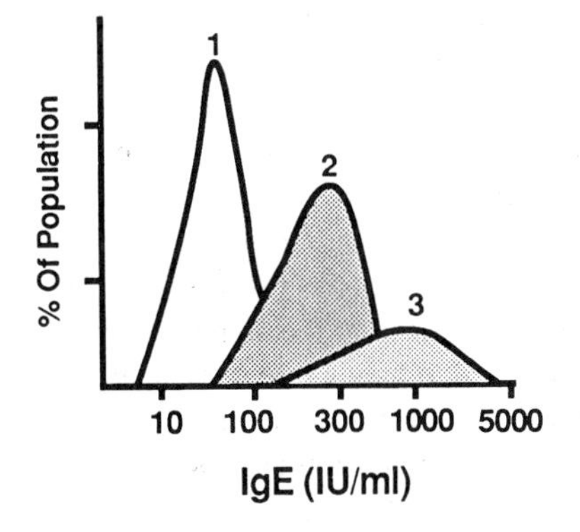

Figure 23.3 Distribution of IgE levels in a population of nonallergic individuals. Three subpopulations appear to exist: one constituted by low-responder individuals (1), one by high-responder individuals (3), and a third population of individuals with intermediate levels of IgE (2).

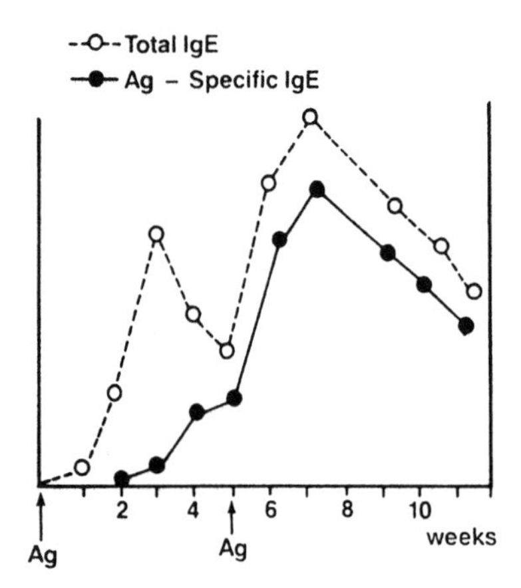

Figure 23.4 Longitudinal evolution of total IgE and antigen-specific IgE levels during an immune-response to an allergen.

D. T-B Cooperation in Antigen-Specific IgE Production

1. The influence of the MHC genes on the response to most allergens strongly indicates that the IgE response must be T-dependent. Support for this view was found in a rat model for high IgE responses. Rats infected with helminths exhibit a marked and persistent primary and secondary IgE response. The height and duration of this response in rats has been demonstrated to result from a predominance of activated helper T cells over relatively low numbers of T cells with suppressor activity.
2. The production of high levels of IgE has been shown to be dependent on the activation of IL-4-producing TH2 cells. IL-4 has been clearly demonstrated to promote IgE synthesis by activated B lymphocytes.
3. Both in humans and animal models, the production of allergen-specific IgE in humans persists long after the second exposure. This may result from the capture of IgE-containing immune complexes by dendritic cells, which express the FcεRII (CD23). Membrane-bound immune complexes are known-to persist for longer periods of time, constantly stimulating B cells and promoting the persistence of secondary immune responses and the differentiation of antibodies with increased affinity (see Chap. 12). At the same time, other immune complexes, perhaps containing IgG antibody, will be internalized, processed, and the source of allergen-derived peptides that will continue to activate the TH2 subpopulation.
4. In addition, IL-4 is both a growth and differentiation factor for mast cells and basophils, causing their number to increase in all tissues where they will be able to bind more IgE. IL-5, which is produced at the same time by the same activated CD4 lymphocytes, is believed to play a role in the late phase of allergic reactions, which is discussed later in this chapter.

E. Interaction of Immunoglobulin E with Cell Surface Receptors

1. Two types of Fc receptors reacting with IgE molecules have been characterized:

a. A unique **high-affinity receptor** designated as **Fcε-RI**, expressed on the surface of basophils and mast cells. Most IgE antibodies interact with this receptor and become cell-associated soon after secretion from plasma cells.
   i. The structure of the Fcσ-RI is unique among the well-characterized lymphoid cell receptors. It is constituted by three subunits: a heterodimer formed by the interaction of two chains (α and β), and a homodimer of a third type of chain (γ chain). The whole molecule is therefore designated as $\alpha\beta\gamma_2$ (Fig. 23.5).
   ii. The external domain of the α chain binds the Fc portion of IgE. The β chain has four membrane-spanning domains and thereby resembles G-protein-linked receptors; thus, it is likely that this chain plays a role in signal transduction, probably with the contribution of the γ chain homodimer, because its amino acid sequence is remarkably homologous to that of the ζ chain of the T-lymphocyte CD3 complex.
   iii. Very little is known concerning the intracellular second messengers involved in transduction of the signal that is generated by the cross-linking of Fcε-RI receptors, but its consequences are well known: release of granule contents into the extracellular space and activation of the synthesis of additional mediators.

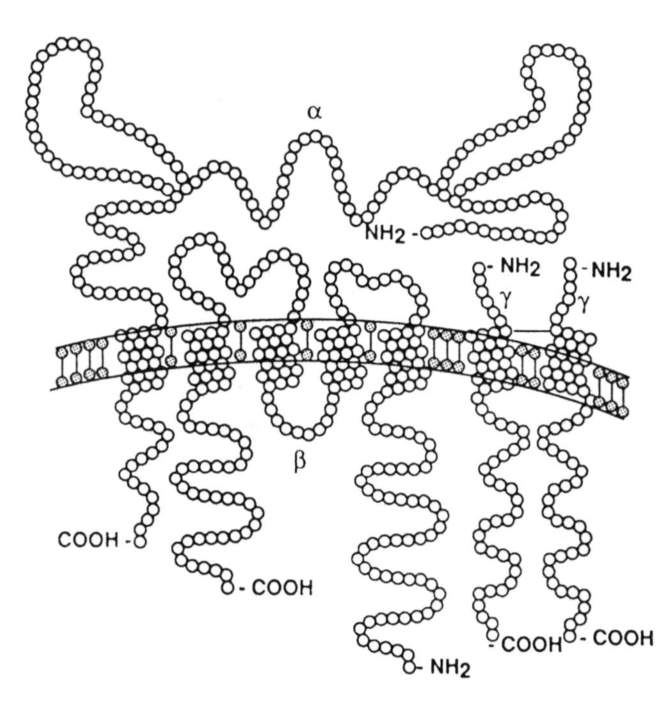

Figure 23.5 Diagrammatic representation of the primary structure of the Fcσ-RI. (Modified from Metzger, H. *Clin. Exp. Allergy*, *21*:1, 1991.)

iv. The interaction between the Fcε-RI and IgE is consistent with a simple bimolecular forward reaction and a first-order reverse reaction:

$$\text{IgE} + \text{receptor} \underset{k^{-1}}{\overset{k^{1}}{\rightleftharpoons}} \text{IgE-receptor complex}$$

The affinity constant of the interaction, $KA = k^1/k^{-1}$ ranges from 10^8 to 10^{10} M/L^{-1}. Because of the high affinity of the interaction between IgE and this Fcε-RI, IgE binds rapidly and very strongly to cells expressing it and is released from these cells very slowly. Passively transferred IgE remains cell bound for several weeks in the skin of normal humans.

v. Because IgE molecules are not produced by the mast cells and basophils, there is no clonal restriction at the mast cell/basophil level. Therefore, if the patient produces IgE antibodies to more than one allergen, IgE antibodies of different specificities may be bound by each basophil or mast cell.

vi. The interaction between IgE and Fcε-RI does not result in cell activation. IgE serves as an antigen receptor for mast cells and basophils. Receptor-bound IgE discriminates among antigens, binding exclusively to those to which the patient has become sensitized.

vii. Receptor-bound IgE must be cross-linked in order for basophils and mast cells to release their intracellular mediators (Fig. 23.6).
- The physiological cross-linking agent is the allergen, which is multivalent.
- Cross-linking of receptor-bound IgE can also be induced with anti-IgE antibodies or with their divalent F(ab′)2 fragments.
- Unoccupied receptors may be cross-linked with aggregated Fcε fragments.

All these types of cross-linking are equally efficient in inducing the release of mediators.

viii. IgE-anti-hapten antibodies are bound by mast cells and basophils, but to stimulate the release of mediators carrier-bound haptens need to be used, because soluble, univalent haptens do not cross-link membrane IgE molecules. Actually, univalent haptens block the binding sites of membrane-bound IgE molecules, and inhibit further reaction with carrier-bound hapten.

ix. Cross-linking of receptor-bound IgE is not the only signal leading to the liberation of mediators from basophils and mast cells; these cells also respond to C3a, C5a, basic lysosomal proteins, and kinins. It is apparent that there are multiple pathways for mast-cell activation and that the participation of cell-bound IgE is not always needed.

b. **Fcε-RII** (CD23) is expressed on the membrane of lymphocytes, platelets, eosinophils, and dendritic cells and binds IgE with lower affinity than Fcε-RI. The role of Fcε-RII on dendritic cells has been previously discussed and it is supposed to be involved in targeting eosinophils to

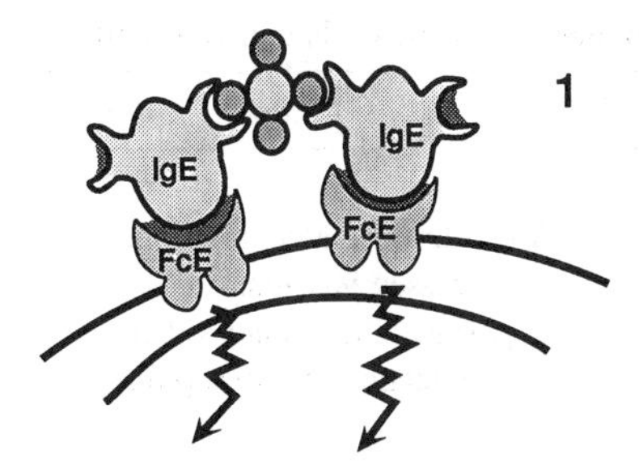

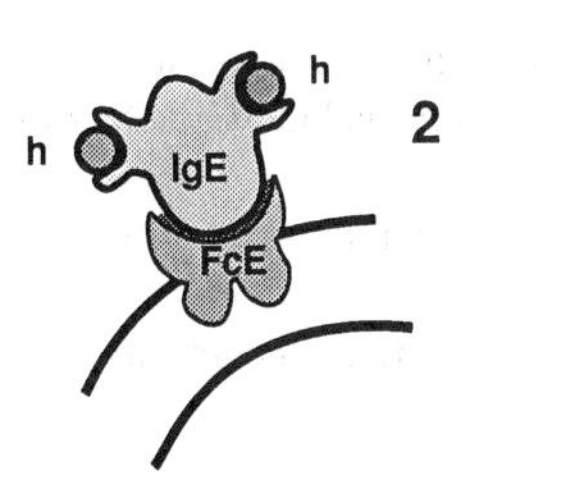

Figure 23.6 Diagrammatic depiction of the conditions required for stimulation of mediator release by mast cells and basophils. In panel 1, the reaction of membrane-bound IgE with a polyvalent antigen, leading to cross-linking of IgE molecules, is represented. This type of reaction leads to mediator release. In panel 2, the reaction of membrane-bound IgE with a monovalent hapten is illustrated. This reaction does not lead to mediator release.

parasites in one of the different variations of ADCC; its role on platelets and lymphocytes is unclear.

D. Early and Late Phases in Type I Hypersensitivity

1. **Early phase.** The metachromatic cytoplasmic granules of basophils and mast cells contain a variety of preformed mediators (Table 23.3). After cross-linking of Fc-receptor-associated IgE, mast cells and basophils undergo a series of biochemical and structural changes.
   a. The first change to be detected is the polymerization of microtubules, which is energy-dependent (inhibited by 2-deoxyglucose), enhanced by the addition of 3-5 GMP, and inhibited by the addition of 3-5 AMP and colchicine.
   b. The polymerization of microtubules allows the transport of the cytoplasmic granules to the cell membrane to which they fuse.
   c. This is followed by the opening of the granules and the release of histamine and other preformed mediators, such as platelet activating factor (PAF) and chemotactic factors for eosinophils (ECFA) into the surrounding medium. In vitro, this sequence of events takes 30 to 60 seconds.
   d. **Histamine** is the mediator responsible for most of the symptoms observed during the early phase of allergic reactions. Since the constricting effect of histamine on the smooth muscle lasts only 1 or 2 hours, this phase stops shortly after most of the granules have been emptied.

Table 23.3 Mediators of Immediate Hypersensitivity Produced by Mast Cells and Basophils

Mediators	Structure	Actions
Stored		
Histamine	5-β-imidazolylethylamine (M.W. 111)	Smooth muscle contraction; increased vascular permeability; many others.
Eosinophil chemotactic factors of anaphylaxis (ECF-A)	Acidic tetrapeptides (M.W. 360–390); Others (M.W. 500–3000)	Chemotactic for eosinophils
Proteolytic enzymes	Trypsin and other enzymes in human mast cells; chymotrypsin in rat mast cells	Actions in vivo unknown, possibly include C′ activation
Heparin	Acidic proteoglycan (M.W. ≈ 750,000)	Anticoagulant; C′ inhibitor
Neutrophil chemotactic factor	Poorly characterized activity with M.W. > 750,000	Chemotactic for neutrophils
Platelet activating factor (PAF)[b]	Phospholipid (M.W. 300–500)[b]	Platelet aggregation and release reaction; increased vascular permeability; eosinophil chemotaxis
Other granuloproteins	Numerous poorly characterized peptides	In vivo significance not yet known
Newly synthesized (upon stimulation of mast cells or basophils)		
Slow-reacting substance of anaphylaxis (SRS-A)	Leukotrienes C4, D4, E4 (derived from arachidonic acid, M.W. 439–625)	Smooth muscle contraction; increased vascular permeability
Prostaglandin D2	Cyclooxygenase product of arachidonic acid[a]	Smooth mucles contraction
Leukotriene B4	Eicosotetraenoate product of arachidonic acid (M.W. 336)	Increased vascular permeability; chemotactic for eosinophils and neutrophils; neutrophil aggregation

[a]Produced exclusively by mast cells.
[b]Also released after denovo synthesis.

2. **Late phase.** The cells remain viable after degranulation and proceed to synthesize other substances that will be released at a later time, causing the late phase of a type I hypersensitivity reaction.
   a. The mediators responsible for the late phase of the response are not detected until several hours after release of histamine and other preformed mediators. The long latency period between cell stimulation and detection of these mediators suggests that they are synthesized by mast cells after stimulation. The following are the main mediators involved in the late phase:

i. **Slow-reacting substance of anaphylaxis (SRS-A)**, which reaches effective concentrations only 5 to 6 hours after challenge and has effects on target cells lasting for several hours. SRS-A is constituted by a mixture of three **leukotrienes** (LT) designated as **LTC4, LTD4, and LTE4**. LTC4 and LTD4 are severalfold more potent than histamine in causing constriction of the peripheral airways and also induce mucous secretion in human bronchi.

ii. **Platelet activating factor (PAF)**, which is also stored in granules, is a phospholipid whose effects include platelet aggregation, chemotaxis of eosinophils, release of vasoactive amines from platelets, eosinophils, and other larger cells, and increased vascular permeability (both due to a direct effect and to the release of vasoactive amines). PAF serves to play an important role in the late phase of asthma.

3. **Role of eosinophils in the late phase**

a. Eosinophils are attracted to the site where an immediate hypersensitivity reaction is taking place by chemotactic factors released by basophils, mast cells, and TH2 lymphocytes.

i. ECF-A and PAF are preformed chemotactic factors released during degranulation.

ii. Leukotriene B4 is synthesized by stimulated basophils/mast cells.

iii. Interleukin-5 is synthesized by activated TH2 lymphocytes.

b. In many cases, the appearance of eosinophils signals the onset of internal negative feedback and control mechanisms that terminate the acute event. This effect is associated with the production and release of enzymes, particularly histaminase (which degrades histamine) and phospholipase D (which degrades PAF). Active oxygen radicals released by stimulated granulocytes, including eosinophils and perhaps neutrophils (which are also attracted by ECF-A and LTB4), cause the breakdown of SRS-A. Histamine itself can contribute to the down-regulation of the allergic reaction by binding to a type III histamine receptor expressed on basophils and mast cells; the occupancy of this receptor leads to an intracellular increase in the level of cAMP that inhibits further release of histamine (negative feedback).

c. In some patients, eosinophil infiltrates are associated with intense inflammation, which causes prolongation of symptoms. For example, asthmatic patients may develop a prolonged crisis during which the symptoms remain severe, and breathing becomes progressively more difficult, leading to a situation of increasing respiratory distress that does not respond to the usual treatment.

i. In these patients very heavy peribronchial cellular infiltrates of the epithelium and lamina propria are found. In these infiltrates eosinophils predominate, but containing also T lymphocytes and plasma cells are also present (chronic eosinophilic bronchitis).

ii. The infiltrating lymphocytes are mostly activated CD4, CD25+ T lymphocytes, whose cytokine mRNA pattern is typical of the TH2 subpopulation.

iii. IL-5 released from the infiltrating activated TH2 lymphocytes seems to attract, retain, and activate eosinophils.
iv. Activated eosinophils release two toxic proteins: **eosinophilic cationic protein** and **major basic protein**. These proteins decrease ciliary movement, are cytotoxic for bronchial epithelium, and cause nonspecific bronchial hyperactivity and cellular denudation.
v. Individuals allergic to multiple environmental allergens suffer chronic inflammation of the airways which, after decades, leads to a chronic obstructive pulmonary disease. This state has been termed "chronic eosinophilic desquamative bronchitis."

d. A similar pathogenic sequence seems responsible for a state of bronchial hyperresponsiveness to many stimuli, which develops in the wake of an asthma crisis.

4. The severity of the clinical symptoms in an immediate hypersensitivity reaction is directly related to the amount of mediators released and produced, which in turn is determined by the number of "sensitized" cells stimulated by the antigen. The expression of early and late phases can be discriminated clinically:
   a. In the case of a positive immediate reaction elicited by a skin test, the early phase resolves in 30 to 60 minutes, the late phase generally peaks at 6 to 8 hours and resolves at 24 hours.
   b. In the case of an asthma crisis, the early phase is characterized by shortness of breath and nonproductive cough and lasts 4 to 5 hours. If the exposure to the responsible airborne allergen is very intense, life-threatening bronchospasm may develop. With less intense or chronic exposure, the initial symptoms of wheezing and dry cough will linger for a few hours.
   c. Around the sixth hour, the cough starts to produce sputum, signaling the onset of the late phase during which the dyspnea may become more severe. In very severe cases, death may occur at this late stage because of persistent bronchoconstriction and peribronchial inflammation associated with increased mucous secretion, both factors contributing to severe airflow obstruction.

III. PREVENTION AND THERAPY

A. Prevention

1. **Environmental control**, trying to prevent exposure to the allergen, is possible in monoallergies; however, it cannot be easily achieved by individuals with multiple allergies.
2. **Hyposensitization** is generally useful in monoallergies and is the standard of care in individuals with insect venom IgE-mediated hypersensitivity. It may also have beneficial results in patients suffering from pollen and perennial allergies (i.e., dust mites, cat dander). Hyposensitization is achieved by subcutaneous injection of very small quantities of the sensitizing antigen, starting at the nanogram level, and increasing the dosage on a weekly basis. This induces the production of IgG **blocking antibodies**; and

an increase in the number of antigen-specific suppressor lymphocytes with concomitant decline of serum IgE levels. Because both effects tend to be simultaneous, they appear to correlate with a decrease of the allergic symptoms (Fig. 23.7).

a. Conceptually, circulating blocking antibodies of the IgG class should have a protective effect by combining with the antigen before it reaches the cell-bound IgE. These blocking antibodies do not interfere with a RAST assay for IgE antibodies. A significant clinical improvement correlates better with an increase in blocking IgG than with a decrease in antigen-specific IgE.

b. The beneficial effect of the competition between IgG and IgE antibodies is easy to understand in cases of insect venom anaphylaxis in which the allergen is injected directly into the circulation where it can be "blocked" by circulating IgG; however, it is more difficult to understand the protective mechanism involved in respiratory allergies, when the allergen has almost direct access to the sensitized cells without entering the systemic circulation.

B. **Drug Therapy.** Various drugs are used to treat or prevent immediate hypersensitivity reactions. Some inhibit or decrease mediator's release by mast cells or basophils; others block or reverse the effect of mediators. The complex interactions of different drugs able to influence mediator release are summarized in Figure 23.8.

1. Localized allergic reactions [seasonal rhinitis (hay fever), perennial rhinitis, urticaria] respond often favorably to antihistaminic compounds, which compete with histamine in the binding to type I receptors at the target cell level. Systemic reactions often require very energetic and urgent measures, particularly the administration of epinephrine (see below).
2. Bronchial asthma presents very complex therapeutic problems. Current therapy is based upon the understanding that an initial acute bronchoconstrictive attack is followed by progressive inflammation in the airways and, later, by increased bronchial responsiveness. Each phase requires a different treatment.

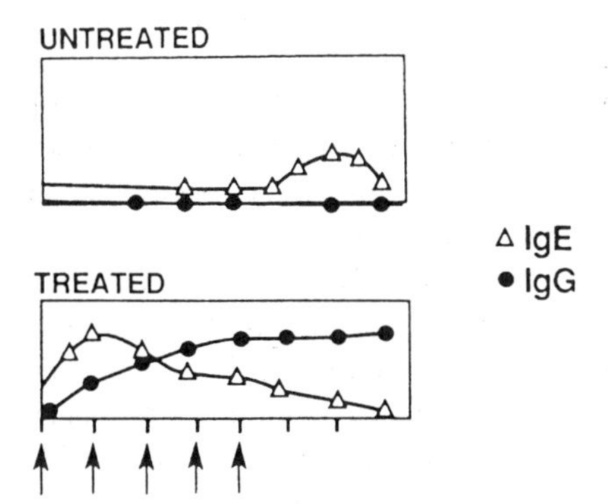

Figure 23.7 Evolution of allergen-specific IgE and IgG levels in patients submitted to hyposensitization (treated) and control patients (untreated).

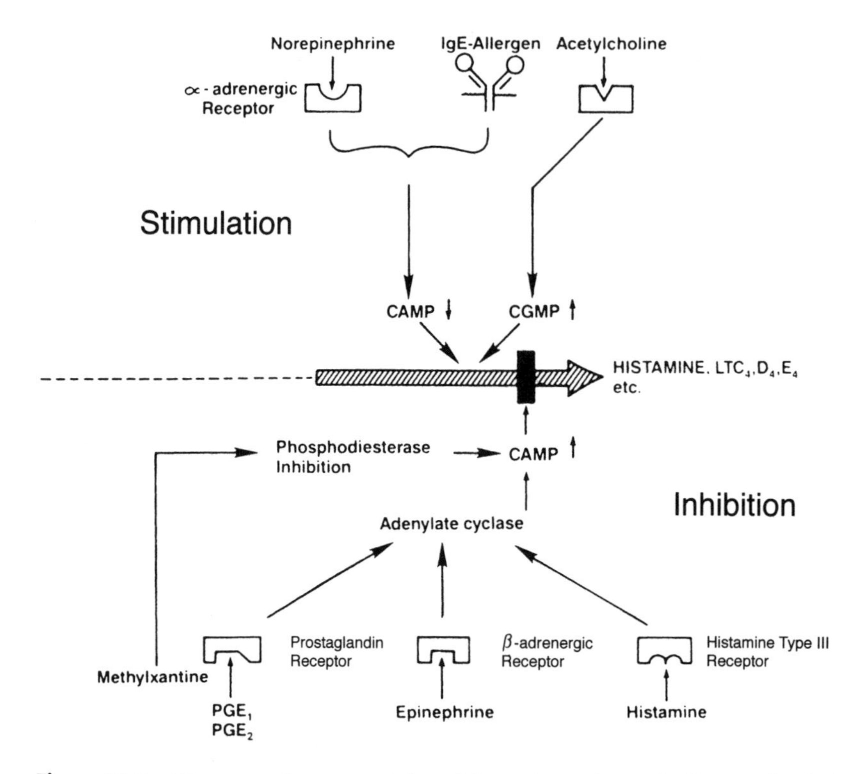

Figure 23.8 Diagrammatic representation of the major pathways leading to stimulation and inhibition of mediator-release by basophils and mast cells. (Modified from David, J., and Rocklin, R.E. Immediate hypersensitivity. In *Scientific American Medicine*. E. Rubinstein, and D.D. Federman, eds. Scientific American, Inc., New York, 1983, Section 6, Chap. IX.)

a. In the early phase, relief of bronchial obstruction is the initial therapeutic goal. β-Adrenergic agonists, methylxanthines, and anticholinergics are the main drugs used.

   i. **β-adrenergic receptor stimulators** (epinephrine, isoproterenol, albuterol) increase cAMP levels by stimulating membrane adenylcyclase directly, inhibiting further degranulation of mast cells and basophils. As stated earlier, epinephrine is the drug of choice for treatment of severe allergic reactions such as anaphylaxis or severe asthma. However, these drugs do not affect eosinophils so that when patients reach the stage of eosinophilic bronchitis, β-agonists will no longer be effective. The patient will have a tendency to increase their use to try to achieve symptomatic relief. But, since eosinophils are unaffected, the bronchitis progresses and can reach a stage where the patient is at risk for death or near death. A patient's increasing need for β-agonists should be considered a sign of worsening lower airway disease.

   ii. **Methylxanthines** (e.g., theophylline) block phosphodiesterases, leading to a persistently high intracellular level of cAMP, which in turn inhibits histamine release. However, recent studies have led to questioning this interpretation because the levels of serum methylxanthine reached during the treatments are much lower than those needed to inhibit phosphodiesterases.

iii. **Anticholinergic drugs.** Most cholinergic agents, raising intracellular levels of cyclic GMP, have an enhancing effect on mediator release: their use must be avoided in asthmatic patients since they aggravate the symptoms. On the opposite side, anticholinergic drugs that block vagal cholinergic tone may be useful but are not as efficient as β-agonists.

b. In the late phase, treatment needs to focus on the eosinophilic bronchitis, an inflammatory reaction no longer responding to the agents useful for treatment in the early phase. Therefore, the treatment's goals are quite different and will include locally administered corticosteroids, cromolyn, and, in the future, leukotriene-modifying drugs.

i. **Glucocorticoids** have no direct action on mast cells in the lung but strongly inhibit eosinophil degranulation, and thus they have an excellent anti-inflammatory effect which inhibits the progression of the late phase, preventing or reducing bronchial hyper-responsiveness.
   - These effects can be achieved safely through the administration of glucocorticoids by aerosol, delivering small intrabronchial concentrations that result in maximal anti-inflammatory effects locally with low daily doses. This avoids the risk of systemic side effects previously incurred with chronic systemic administration. Glucocorticoids administered by aerosol are now recommended for the treatment of chronic asthma, regardless of its degree of severity.
   - Systemic administration of glucocorticoids is reserved for treating severe acute episodes, for preventing the development of the severe late-phase reactions, or for treating severe recalcitrant asthma. Their mild immunosuppressive effects may also be significant by reducing the magnitude of the IgE response.

ii. **Disodium cromoglycate** (cromolyn) and nedocromil sodium, not shown in Figure 23.8, are very efficient prophylactic drugs. Their mechanism of action is still being investigated, but they decrease bronchoconstriction and have proven invaluable in helping to reduce the need for glucocorticoids.

iii. Several **leukotriene-modifying drugs**, currently at various stages of development, appear promising and may become available within the next few years. These agents include lipoxygenase enzyme inhibitors, which down-regulate the production of leukotrienes, and leukotriene receptor antagonists, which block the effects of leukotrienes.

Case 23.1 Revisited

- This 20-year-old college student with rhinitis and asthma represents an individual with atopic disease. She has symptoms consistent with inflammation in the upper and lower airways. Her environment changed with the new semester, and she had worsening of her allergic disease. It is probable that her exposure to cat allergens, as well as increased levels of mold and dust mite allergens, is contributing to her symptomatology.

- When allergens enter the airways during respiration, they induce an allergic response in the respiratory mucosa. In the nasal airways, the inflammation and glandular hypersecretion results in nasal airway obstruction. In the lower airway, the main changes are an eosinophilic inflammation associated with smooth muscle hypertrophy and bronchospasm. These changes in the lower airway result in a reduction of the airways caliber, which is more accentuated during expiration, causing air trapping and difficulty exhaling. The use of a beta-agonist results in smooth muscle relaxation and the ability to exhale against a lessened airway resistance.
- A biopsy of the bronchial mucosa would reveal marked inflammation of the mucosa. There would be a cellular infiltrate with abundant eosinophils. Glandular hyperplasia and denudation of the ciliated epithelial cells would also likely be present. The basement membrane would appear thickened.
- This patient has an elevated IgE and likely has specific IgE for allergens derived from sources such as mold spores, dust mites, and cat dander. IL-4 is a cytokine critical to the elaboration of IgE. In addition, the inflammation is typically eosinophilic, and IL-5 is a necessary chemokine for eosinophils.
- Acutely, this patient needs to have relief of the airway obstruction. Bronchodilators, including inhaled beta-agonists and theophylline, would be used. Long-term, the patient would require anti-inflammatory agents. Initially, systemic glucocorticoids would be employed. However, for long-term therapy, inhaled agents such as aerosolized glucocorticoids, cromolyn, and nedocromil would be preferred.

SELF-EVALUATION

Questions

Choose the ONE *best* answer.

23.1 Which of the following characteristics is associated with the Fcε-RI?
   A. Expression on lymphocytes, monocytes, and eosinophils
   B. High-affinity binding of its ligand
   C. Involvement on antibody-dependent cellular cytotoxicty reactions
   D. Lack of signaling properties
   E. Structural classification as a member of the immunoglobulin superfamily

23.2 Which of the following agents are preferred for treatment of the inflammatory component associated with chronic asthma?
   A. Aerosolized glucocorticoids
   B. Antihistaminics
   C. β-Adrenergic receptor stimulators
   D. Methylxanthines
   E. Systemic glucocorticoids

23.3 Which of the following statements best characterizes the changes observed in IgE as a consequence of successful hyposensitization?
   A. Combines with the injected allergen and forms circulating immune complexes
   B. Decreases in serum
   C. Increases in serum
   D. Is not significantly affected
   E. Is replaced by IgG at the cell membrane level

23.4 Which one of the following pharmacological agents should be immediately administered to a patient with suspected anaphylaxis?
A. Antihistaminics
B. Cholinergic drugs
C. Epinephrine
D. Methylxanthines
E. Sodium cromoglycate

23.5 Which of the following is coupled to the solid phase in a radioallergosorbent test (RAST) assay?
A. A given allergen
B. Anti-IgE
C. Antibodies to a given allergen
D. IgE
E. The patient's serum

23.6 Which of the following newly synthesized mast-cell mediators is responsible for the attraction of eosinophils to the peribronchial tissues in the late stages of an asthma attack?
A. Eosinophil chemotactic factor-A
B. Leukotriene B-4 (LTB-4)
C. Major basic protein
D. Platelet aggregation factor
E. Prostaglandin E2

23.7 Of the following reagents, which one will be able to induce the release of histamine from the mast cells of a ragweed-sensitized individual?
A. A univalent fragment of ragweed
B. F(ab′)2 from an anti-IgE antibody
C. Fab from an anti-IgE antibody
D. IgE antiragweed
E. IgG antiragweed

23.8 A major control mechanism of type I hypersensitivity reactions mediated by eosinophils is the release of:
A. Cationic proteins
B. Histaminase
C. Leukotrienes C4, D4, and E4
D. Major basic protein
E. Platelet activating factor

23.9 Which of the following mediators is NOT involved (directly or indirectly) in negative feedback reactions in immediate hypersensitivity?
A. Eosinophil chemotactic factor-A
B. Histaminase
C. Histamine
D. Phospholipase D
E. Prostaglandin D2

23.10 What is the meaning of the incidental finding of an elevated serum IgE level of 500 IU/mL in a clinically asymptomatic individual?
A. The subject is atopic
B. HLA-B7 is likely to be represented on the individual's phenotype
C. Hyposensitization is not likely to be effective

D. The individual is a high IgE producer
E. The individual is likely to develop allergies

Answers

23.1 (B) The Fcε-RI is the Fc receptor with highest affinity for its ligand (IgE). It is only expressed on basophils and mast cells, which do not mediate ADCC reactions, and is structurally unrelated to the immunoglobulin superfamily.

23.2 (A) Aerosolized glucocorticoids are preferred to systemic glucocorticoids because they are effective in most cases with less risk of development of side effects. Antihistaminics and methylxanthines may have anti-inflammatory properties, but are not sufficiently effective to be useful as primary treatment for chronic asthma.

23.3 (B) Hyposensitization appeared to induce and to increase the activity of antigen-specific suppressor cells that will lead to a decrease in serum IgE levels.

23.4 (C) Epinephrine is the drug of choice for immediate treatment of anaphylaxis.

23.5 (A) The allergen is coupled to the solid phase; if IgE antibodies are present in the patient's serum, they will become bound to the antigen in the solid phase, and their presence can be revealed with a radiolabeled anti-IgE antibody.

23.6 (B) The two other chemotactic factors for eosinophils, platelet activating factor (PAF) and eosinophil chemotactic factor-A (ECF-A), are preformed and released in the early phase.

23.7 (B) The release of histamine requires the cross-linking of membrane IgE which can be induced either by complete anti-IgE antibodies, bivalent $F(ab')_2$ fragments of anti-IgE antibodies, or multivalent antigen of the right specificity.

23.8 (B)

23.9 (E) Histamine, by reacting with type III histamine receptors in basophils and mast cells, will inhibit further histamine release; phospholipase D degrades PAF; histaminase degrades histamine; LTB-4 is chemotactic for eosinophils, which release phospholipase D, histaminase, and other protective factors.

23.10 (D) Many nonallergic individuals may have IgE values above the upper limit of normalcy. It is not possible to conclude that an individual with high IgE levels is more likely to become allergic.

BIBLIOGRAPHY

Bochner, B.S., Undem, B.J., and Lichtenstein, L.M. Immunological aspects of asthma. *Annu. Rev. Immunol.*, *12*:295, 1994.

Bracquet, P., Touqui, L., Shen, T.Y., and Vargaftig, B.B. Perspectives in platelet-activating factor research. *Pharmacol. Rev.*, *39*:97, 1987.

Burrows, B., and Lebowitz, M.D. The β-agonist dilemma. *N. Engl. J. Med.*, *326*: 560, 1992.

Geha, R.S. Regulation of IgE synthesis in humans. *J. Allergy Clin. Immunol.*, *90*:143, 1992.

Goetzl, E.J., Payan, D.G., and Goldman, D.W. Immunopathogenetic role of leukotrienes in human diseases. *J. Clin. Immunol.*, *4*:79, 1984.

International Consensus Report on Diagnosis and Management of Asthma. U.S. Dept. of Health and Human Services Publication 92-3091, 1992.

Postma D., Bleekeer E.R., Amelung P.J., Holroyd, K.J., Jianfeng, X., Panhuysen, C.I.M., Meyers, D., and Levitt, R.C. Genetic susceptibility to asthma-bronchial hyperresponsiveness coinherited with a major gene for atopy. *N. Engl. J. Med.*, *333*: 894, 1995.

Weller P.F. The immunobiology of eosinophils. *N. Engl. J. Med.*, *324*:1110, 1991.

24
Immunohematology

Gabriel Virella and Mary Ann Spivey

I. INTRODUCTION: BLOOD GROUPS

A. **The ABO System.** The first human red-cell antigen system to be characterized was the ABO blood group system. Specificity is determined by the terminal sugar in an oligosaccharide structure. The terminal sugars of the oligosaccharides defining groups A and B are immunogenic. In group O the precursor H oligosaccharide is unaltered. The red cells express either A, B, both A and B, or neither, and antibodies are found in serum to antigens not expressed by the red cells, as shown in Table 24.1.

1. The ABO group of a given individual is determined by testing both cells and serum. The subject's red cells are mixed with serum containing known antibody, and the subject's serum is tested against cells possessing known antigen. For example, the cells of a group A individual are agglutinated by anti-A serum but not by anti-B serum, and his serum agglutinates type B cells but not type A cells. The typing of cells as group O is done by exclusion (a cell not reacting with anti-A or anti-B is considered to be of blood group O).
2. The anti-A and anti-B isoagglutinins are synthesized as a consequence of cross-immunization with enterobacteriaceae that have outer membrane oligosaccharides strikingly similar to those that define the A and B antigens (see Chap. 13). For example, a newborn with group A blood will not have anti-B in his or her serum, since there has been no opportunity to undergo cross-immunization. When the intestine is eventually colonized by the normal microbial flora, the infant will start to develop anti-B, but will not produce anti-A because of tolerance to his or her own blood group antigens (see Table 24.1).
3. The inheritance of the ABO groups follows simple Mendelian rules; with three common allelic genes: A, B, and O (A can be subdivided into A_1 and A_2), of which any individual will carry two, one inherited from the mother, and one from the father.

B. **The Rh System**

1. **Historical overview.** In 1939, **Philip Levine** discovered that the sera of most women who gave birth to infants with hemolytic disease contained an antibody that reacted with the red cells of the infant and with the red cells of 85% of Caucasians. In 1940, **Landsteiner and Wiener** injected blood from the

Table 24.1 The ABO System

Red-cell antigen	Serum isoagglutinins	Blood group
A	Anti-B	A
B	Anti-A	B
A and B	None	AB
None	Anti-A and -B	O

monkey *Macacus rhesus* into rabbits and guinea pigs and discovered that the resulting antibody agglutinated Rhesus red cells and appeared to have the same specificity as the neonatal antibody. The donors whose cells were agglutinated by the antibody to Rhesus red cells were termed Rh positive; those whose cells were not agglutinated were termed Rh negative. It is now known that the antibody obtained by Landsteiner and Wiener reacts with an antigen (LW) that is different but closely related to the one that is recognized in human hemolytic disease, but the Rh nomenclature was retained.

2. **Theories, nomenclatures, and antigens of the Rh system.** The Rh system is now known to have many antigens in addition to the one originally described, and several nomenclature systems are in use.
   a. According to the **Fisher–Race** theory, the Rh gene complex is formed by combinations of three pairs of allelic genes: Cc, Dd, Ee. The possible combinations are: Dce, DCe, DcE, DCE, dce, dCe, dcE, and dCE. The three closely linked gene loci are inherited as a gene complex. Thus a DCe/DcE individual can only pass DCe or DcE to his offspring and no other combination. The original antigen discovered is called **D** and people who possess it are called **Rh positive**. The antigen **d** has never been discovered, and the symbol "d" is used to denote the absence of **D**. All individuals lacking the D antigen are termed **Rh negative**. The most frequent genotype of D-negative individuals is dce/dce. The lack of one

Table 24.2 Comparison of the Fisher–Race and Wiener Notations for the Rh System

Fisher–Race notation		Wiener notation		
Gene complex	Antigens	Genes	Agglutinogens	Factors
Dce	D,c,e	R^0	Rh_o	**Rh_o,hr′,hr″**
DCe	D,C,e	R^1	Rh_1	**Rh_o,rh′,hr″**
DcE	D,c,E	R^2	Rh_2	**Rho,hr′,rh″**
DCE	D,C,E,	R^z	Rh_z	**Rh_o,rh′,rh″**
dce	d,c,e	r	rh	**hr′,hr″**
dCe	d,C,e,	r′	rh′	**rh′,hr″**
dcE	d,c,E	r″	rh″	**hr′,rh″**
dCE	d,C,E	r^y	rh_y	**rh′,rh″**

of the postulated alleles seems to imply that the genetic basis of the Fisher–Race theory and nomenclature are not correct, but the use of this nomenclature has been retained, since it is easier to understand than any other.

b. The second most common nomenclature is that proposed by Wiener, who theorized multiple alleles at a single complex locus, each locus determining its particular agglutinogen comprising multiple factors that were designated by bold-face type. The equivalents of the most common Rh factors in the Fisher–Race and Wiener nomenclature are shown in Table 24.2.

c. Recent studies analyzing DNA from donors of different Rh phenotypes have found two structural genes within the Rh locus of Rh (D) positive individuals and only one present in Rh-negative persons. Therefore, one gene appears to encode the D protein and the other governs the presence of C, c, E, and e.

C. Other Blood Groups. Several other blood group systems with clinical relevance have been characterized. Other than those caused by clerical error, most transfusion reactions are due to sensitization against alloantigens of the Rh, Kell, Duffy, and Kidd systems, of which the Kell system is the most polymorphic. In contrast, most cases of autoimmune hemolytic anemia involve autoantibodies directed to public antigens (antigens common to most, if not all, humans), such as the I antigen or core Rh antigens.

D. Laboratory Determination of Blood Types

1. **Reagents.** Most reagents consist of monoclonal antibodies, usually of mouse origin, used individually or blended, and directed against the different blood group antigens that are used for blood group typing. A major advantage of the use of monoclonals is their specificity, minimizing the possibility of false-positive reactions due to additional contaminating antibodies found in human serum reagents. An important disadvantage derives from the fact that monoclonal antibodies react with a single epitope and the blood group antigens have multiple epitopes. Thus, individuals missing the epitope recognized by the antibody may be typed as negative. This problem is significantly reduced by using a blend of monoclonal antibodies, each one of them recognizing a different epitope of a given antigen.

2. **Tests**

a. **Direct hemagglutination** is the simplest, preferred test. It is easy to perform with typing reagents containing IgM antibodies that directly agglutinate cells expressing the corresponding antigen. Reagents containing IgG antibodies can also be used in a direct hemagglutination test. In one approach, protein is added in relatively high concentration to the reagent for the purpose of dissipating the repulsive forces that keep the red cells apart. As a consequence, the red cells can be directly agglutinated by IgG antibodies. A second approach involves modification of the IgG antibodies by mild reduction of their interchain disulfide bonds to produce "unfolded" molecules, capable of direct agglutination of red cells.

b. **Indirect antiglobulin test.** In general, reagents containing IgG antibodies are used in an indirect antiglobulin test (see below) as a way to induce the agglutination of red cells coated with the corresponding antibodies.

E. **Direct and Indirect Antiglobulin (Coombs) Tests.** In 1945, **Coombs, Mourant,** and **Race** described the use of antihuman globulin serum to detect red cell–bound nonagglutinating antibodies. There are two basic types of antiglobulin or Coombs tests.

1. The **direct antiglobulin test** is performed to detect in vivo sensitization of red cells or, in other words, sensitization that has occurred in the patient (Fig. 24.1). The test is performed by adding antihuman IgG (and/or antihuman complement, to react with complement components bound to the red cells as a consequence of the antigen–antibody reaction) to the patient's washed red cells. If IgG antibody is bound to the red cells, agglutination (positive result) is observed after addition of the antiglobulin reagent and centrifugation. The **direct** antiglobulin test is an aid in diagnosis and investigation of: hemolytic disease of the newborn; autoimmune hemolytic anemia; drug-induced hemolytic anemia; and hemolytic transfusion reactions.
2. The **indirect antiglobulin** test detects in vitro sensitization, which is sensitization that has been allowed to occur in the test tube under optimal conditions (Fig. 24.2). Therefore, the test is used to investigate the presence of nonagglutinating red-cell antibodies in a patient's serum. The test is performed in two steps (hence the designation *indirect*): a serum suspected of containing red-cell antibodies is incubated with normal red blood cells; and after washing unbound antibodies, antihuman IgG (and/or anticomplement) antibodies are added to the red cells as in the direct test.
   The **indirect** antiglobulin test is useful in: detecting and characterizing red-cell antibodies using test cells of known antigenic composition (antibody screening); crossmatching; and phenotyping blood cells for antigens not demonstrable by other techniques

Immunohematology

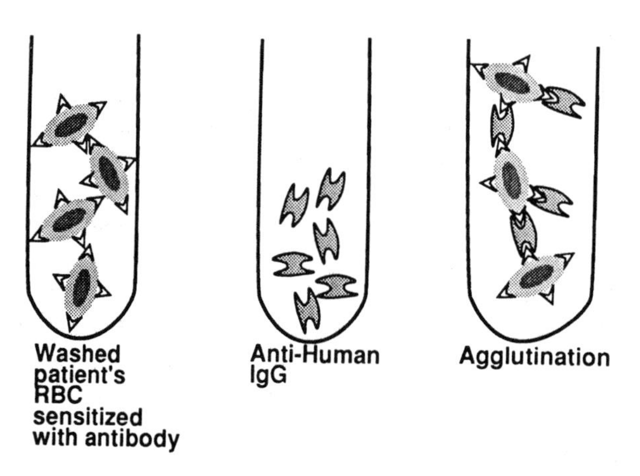

Figure 24.1 Diagrammatic representation of a direct Coombs test using anti-IgG antibodies.

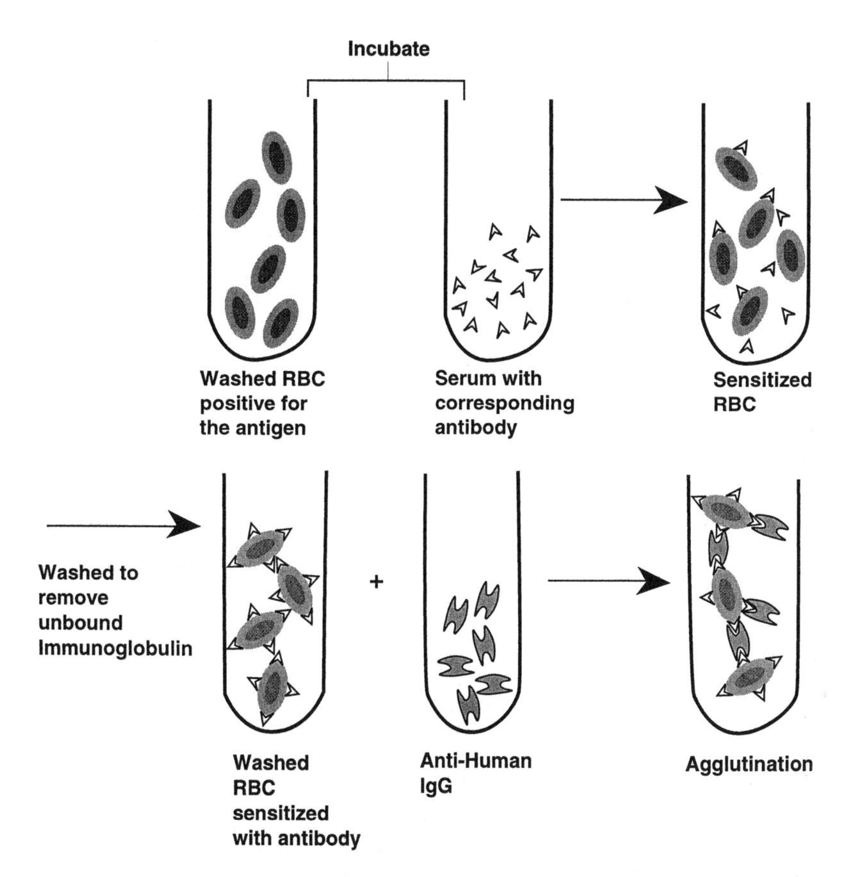

Figure 24.2 Diagrammatic representation of an indirect Coombs' test.

II. BLOOD TRANSFUSION IMMUNOLOGY

A. Blood Testing

1. **Compatibility testing.** Before a blood transfusion, a series of procedures needs to be done to establish the proper selection of blood for the patient. Basically, those procedures try to establish ABO and Rh compatibility between donor and recipient and to rule out the existence of antibodies in the recipient's serum which could react with transfused red cells.
   a. To establish the ABO and Rh compatibility between donor and recipient, both the recipient and the blood to be transfused are **typed**.
   b. To rule out the existence of antibodies (other than anti-A or anti-B), a general **antibody screening test** is performed with **group O red cells of known composition**, which are first incubated with the patient's serum to check for agglutination; if the direct agglutination test is negative, an indirect antiglobulin (Coombs) test is performed.
2. **The cross-match.** The most direct way to detect antibodies in the recipient's serum that could cause hemolysis of the transfused red cells is to test the patient's serum with the donor's cells (**major cross-match**).

a. The complete cross-match also involves the same tests as the antibody screening test described above.
b. An **abbreviated** version of the **cross-match** is often performed in patients with a negative antibody screening test. This consists of immediately centrifuging a mixture of the patient's serum and donor cells to detect agglutination; this primarily checks for ABO incompatibility.

3. The **minor cross-match**, which consists of testing a patient's cells with donor serum, is of little significance and rarely performed, since any donor antibodies would be greatly diluted in the recipient's plasma and rarely cause clinical problems.
4. **Implications of positive antibody screening for transfusion.** Donor blood found to contain antibodies can be safely transfused as packed red cells, containing very little plasma. This is a routine blood bank procedure, and no whole blood units containing clinically significant red-cell antibodies are issued. Such blood is issued as packed red cells and the plasma is discarded. If a patient has a positive antibody screening test due to a clinically significant antibody, the antibody is identified using a panel of cells of known antigenic composition and antigen negative blood is selected for transfusion.

B. Blood Transfusion Reactions. Transfusion reactions may occur due to a wide variety of causes (Table 24.3). Among them, the most severe are those associated with hemolysis, which may be life-threatening. A list of the causes of fatal transfusion reactions reported to the FDA from 1985–1987 is reproduced in Table 24.4.

1. The most frequent cause is an ABO mismatch due to clerical error, resulting in the transfusion of the wrong blood.
2. Transfusion of blood incompatible for other blood groups to a patient previously sensitized during pregnancy or as a consequence of earlier transfusions can also cause a hemolytic reaction.
3. Patients with autoimmune hemolytic anemia often have antibodies reacting with "public" antigens expressed by red cells from virtually all donors as well as their own, and are likely to develop hemolysis whenever a transfusion is administered to them.

C. Hemolytic Reactions

1. **Pathogenesis.** Hemolytic reactions can be classified as intravascular or extravascular.
   a. **Intravascular hemolytic reactions** are triggered by the binding of preformed IgM antibodies to the red cells.
      i. IgM antibodies are very effective in causing the activation of the complement system. Massive complement activation by red-cell

Table 24.3 Classification of Transfusion Reactions

A. Nonimmune
B. Immune
   1. Red-cell incompatibility
   2. Incompatibilities associated with platelets and leukocytes
   3. Incompatibilities due to antiallotypic antibodies (anti-Gm or Am antibodies)

Table 24.4 Summary of Fatal Transfusion Reactions[a]

Causes	No.
Hemolytic reactions	
ABO incompatible transfusions	29
Collection errors	7
Blood bank clerical errors	8
Blood bank technical errors	1
Nursing unit errors	11
Undetermined	2
Non-ABO incompatible transfusions[b]	6
No detectable antibody	3
Glycerol	1
Nonhemolytic reactions	26
Bacterial contamination	11[c]
Acute respiratory distress	9
Anaphylaxis	6

[a]Reported to the Food and Drug Administration from 1985 to 1987.
[b]Including anti-Jk^b, -c, Fy^a, and -K.
[c]In nine cases, the source of contamination was a platelet preparation.
Source: Beig, K., Calhoun, A., and Petz, L.D. ISBT & AABB Joint Congress, Los Angeles, CA, 1990, Abstract S282.

antibodies causes intravascular red-cell lysis, with release of hemoglobin into the circulation. Most of the free hemoglobin forms complexes with haptoglobin.

ii. Due to the massive release of soluble complement fragments (e.g., C3a and C5a) with anaphylotoxic properties, the patient may suffer generalized vasodilatation, hypotension, and shock.

iii. Because of the interrelationships between the complement and clotting systems, disseminated intravascular coagulation may occur during a severe transfusion reaction.

iv. As a consequence of the nephrotoxicity of free hemoglobin, the patient may develop acute renal failure, usually due to acute tubular necrosis. This only happens when the amount of release hemoglobin exceeds the binding capacity of haptoglobin.

b. **Extravascular hemolytic reactions** are caused by the opsonization of red cells with IgG antibodies.

i. IgG red-cell antibodies can activate complement but do not cause spontaneous red-cell lysis.

ii. Red cells opsonized with IgG (often with associated C3b) are efficiently taken up and destroyed by phagocytic cells, particularly splenic and hepatic macrophages.

iii. These reactions are usually less severe than intravascular transfusion reactions. In addition, transfusion reactions may be delayed when an anamnestic response in a patient with undetectable antibody is the precipitating factor.

iv. Typically, a positive direct antiglobulin (Coombs) test will be noted after transfusion in association with a rapidly diminishing red-cell concentration.

2. **Clinical presentation**
   a. The most common initial symptom in a hemolytic transfusion reaction is fever, frequently associated with chills.
   b. Dark urine (due to hemosiderinuria or, rarely, to hemoglobinuria) may be the first symptom noticed by the patient in cases of rapid intravascular hemolysis.
   c. During surgery, the only symptom may be bleeding and/or hypotension.
   d. With progression of the reaction, the patient may experience chest pains, dyspnea, hypotension, and shock.
   e. Renal damage is indicated by back pain, oliguria, and in most severe cases, anuria.
3. **Laboratory investigation**
   a. Immediately after a hemolytic transfusion reaction is suspected, the following procedures must be done:
      i. A clerical check to detect any errors that may have resulted in the administration of a unit of blood to the wrong patient.
      ii. Confirmation of intravascular hemolysis by visual or photometric comparison of pre- and postreaction plasma specimens for free hemoglobin (the prereaction specimen should be light yellow, and the postreaction sample should have a pink/red discoloration).
      iii. Direct antiglobulin (Coombs) test on pre- and postreaction blood samples.
   b. If any of the above procedures gives a positive result supporting a diagnosis of intravascular hemolysis, additional serological investigations are indicated, including the following:
      i. Repeat ABO and Rh typing on patient and donor samples.
      ii. Repeat antibody screening and cross-matching.
      iii. If an anti-red-cell antibody is detected, determine its specificity using a red-cell panel in which group O red cells of varied antigenic composition are incubated with the patient's serum to determine which RBC antigens are recognized by the patient's antibody.
   c. Additionally, one or several of the following confirmatory tests may be performed:
      i. Measurement of serum haptoglobin which decreases due to the uptake of hemoglobin–haptoglobin complexes by the reticuloendothelial system.
      ii. Measurement of unconjugated bilirubin on blood drawn 5 to 7 hours after transfusion (the concentration should rise as the released hemoglobin is processed).
      iii. Determination of free hemoglobin and/or hemosiderin in the urine (neither is normally detected in the urine).

D. Nonhemolytic Immune Transfusion Reactions

1. **Antileukocyte antibodies**
   a. When a patient has antibodies directed to leukocyte antigens, a transfu-

sion of any blood product containing cells expressing those antigens can elicit a febrile transfusion reaction. Leukocyte-reduced blood products should be used for transfusions in patients with recurrent febrile reactions.

b. Special problems are presented by patients requiring platelet concentrates that have developed anti-HLA antibodies or antibodies directed to platelet-specific antigens (HPA antigens). In such cases, it will be necessary to give HLA- or HPA-matched platelets, since platelets will be rapidly destroyed if given to a sensitized individual with circulating antibodies to the antigens expressed by the donor's platelets.

c. Transfusion of blood products containing antibodies to leukocyte antigens expressed by the patient receiving the transfusion can induce intravascular leukocyte aggregation. These aggregates are usually trapped in the pulmonary microcirculation, causing acute respiratory distress, and, in some cases, noncardiogenic pulmonary edema. A similar situation may emerge when granulocyte concentrates are given to a patient with antileukocyte antibodies reactive with the transfused granulocytes.

2. **Anti-IgA antibodies.** The transfusion of any IgA-containing blood product into a patient with high titers of preformed anti-IgA antibodies can cause an anaphylactic transfusion reaction.
   a. Anti-IgA antibodies are mostly detected in immunodeficient individuals, particularly those with IgA deficiency.
   b. It is very important to test for anti-IgA antibodies in any patient with known IgA deficiency who is going to require a transfusion, even if the patient has never been previously transfused. If an anti-IgA antibody is detected in a titer judged to represent a risk for the patient, it is important to administer packed red cells with all traces of plasma removed by extensive washing. If plasma products are needed, they should be obtained from IgA-deficient donors.

III. HEMOLYTIC DISEASE OF THE NEWBORN (ERYTHROBLASTOSIS FETALIS)

Case 1

A 25-year-old gravida 1, para 0, woman who had not received prenatal care appeared at the emergency room just prior to delivering a 3.5-kg baby girl. The mother was found to be group O, Rh negative, and her antibody screen was negative. Twenty hours later, the nurse observed that the neonate was jaundiced. A hemogram with differential showed WBC of 6200/μL, RBC of 4.1×10^6 μL, hemoglobin of 15 g/dL. The differential showed 5% reticulocytes.

This case raises several questions:

- What is the most probable cause of the neonatal jaundice, and what treatment, if any, is usually indicated in such cases?

- What laboratory tests should be ordered to investigate the cause of this newborn's jaundice?
- Can this situation be prevented? How?

A. Pathogenesis

1. Immunological destruction of fetal and/or newborn erythrocytes is likely to occur when **IgG antibodies** are present in the maternal circulation directed against the antigen(s) present on the fetal red blood cells (only IgG antibodies can cross the placenta and reach the fetal circulation).
2. The two types of incompatibility most usually involved in hemolytic disease of the newborn are anti-D and anti-A or anti-B antibodies. Anti-A or anti-B antibodies are usually IgM, but, in some circumstances, IgG antibodies may develop (usually in group O mothers). This can be secondary to immune stimulation (some vaccines contain blood group substances or cross-reactive polysaccharides), or may occur without apparent cause for unknown reasons.
3. **Mechanisms of sensitization**
   a. Although the exchange of red cells between mother and fetus is prevented by the placental barrier during pregnancy, about two-thirds of all women, after delivery (or miscarriage), have fetal red cells in their circulation.
   b. If the mother is Rh-negative and the infant is Rh-positive, the mother may produce antibodies to the D antigen. The immune response is usually initiated at term, when large amounts of fetal red cells reach maternal circulation. In subsequent pregnancies, even the small number of red cells crossing the placenta during pregnancy are significant to elicit a strong secondary response, with production of IgG antibodies.
   c. As IgG antibodies are produced in larger amounts, they will cross the placenta, bind to the Rh-positive cells, and cause their destruction in the spleen through Fc-mediated phagocytosis.
   d. Usually, the first child is not affected, since the red cells that cross the placenta after the 28th week of gestation do so in small numbers and may not elicit a primary immune response.
   e. IgG anti-D antibodies do not appear to activate the complement system, perhaps because the distribution of the D antigen on the red-cell surface is too sparse to allow the formation of IgG doublets with sufficient density of IgG molecules to induce complement activation. Complement, however, is not required for phagocytosis, which is mediated by the Fc receptors in monocytes and macrophages.

B. Epidemiology

1. The frequency of clinically evident hemolytic disease of the newborn was estimated to be about 0.5% of total births, with a mortality rate close to 6% among affected newborns prior to the introduction of immunoprophylaxis. Recent figures are considerably lower: 0.15 to 0.3% incidence of clinically evident disease, and the perinatal mortality rate appears to be declining to about 4% of affected newborns.
2. Ninety-five percent of the cases of hemolytic disease of the newborn requiring therapy were due to Rh incompatibility, involving sensitization against

the D antigen. Due to the introduction of immunoprophylaxis, the proportion of cases due to anti-D antibodies decreased, while the proportion of cases due to other Rh antibodies, and to antibodies to antigens of other systems, increased.

C. **Clinical Presentation.** The usual clinical features of this disease are anemia and jaundice present at birth, or more frequently, in the first 24 hours of life. In severe cases, the infant may die in utero. Other severely affected children who survive until the third day develop signs of central nervous system damage, attributed to the high unconjugated bilirubin concentrations (Kernicterus). The peripheral blood shows reticulocytes and circulating erythroblasts (hence the term "erythroblastosis fetalis").

D. **Immunological Diagnosis.** A positive direct Coombs (antiglobulin) test with cord RBC is invariably found in cases of Rh incompatibility, although 40% of the cases with a positive reaction do not require treatment. In ABO incompatibility, the direct antiglobulin test is usually weakly positive and may be confirmed by eluting antibodies from the infant's red cells and testing the eluate with A and B cells.

E. **Prevention and Treatment**
   1. Rh hemolytic disease of the newborn is rarely seen when mother and infant are incompatible in both Rh and ABO systems. In such cases, the ABO isoagglutinins in the maternal circulation appear to eliminate any fetal red cells before maternal sensitization occurs.
   2. The above observation led to a very effective form for prevention of Rh hemolytic disease of the newborn, achieved by the administration of anti-D IgG antibodies (**Rh immune globulin**) to Rh-negative mothers.
      a. The therapeutic anti-D antibody is prepared from the plasma of previously immunized mothers with persistently high titers, or from male donors immunized against Rh-positive RBC.
      b. The schedule of administration involves two separate doses:
         i. A **postdelivery dose** is administered in the first 72 hours after delivery of the first baby (before sensitization has had time to occur). The passively administered anti-D IgG prevents the emergence of maternal anti-D antibodies, by an unknown mechanism. The rate of success is 98 to 99%.
         ii. **Antepartum administration** of a full dose of Rh immune globulin at the 28th week of pregnancy is also recommended, in addition to the postpartum administration. The rationale for this approach is to avoid sensitization due to prenatal spontaneous or posttraumatic bleeding. Prenatal anti-D prophylaxis is also indicated at the time that an Rh-negative pregnant woman is submitted to amniocentesis and must be continued at 12-week intervals, until delivery, to maintain sufficient protection.
      c. The recommended full dose is 300 μg IM which can be increased if there is laboratory evidence of severe fetomaternal hemorrhage (by tests able to determine the number of fetal red cells in maternal peripheral blood, from which one can calculate the volume of fetomaternal hemorrhage). Smaller doses (50 μg) should be given after therapeutic or spontaneous abortion in the first trimester.

3. To prevent serious hemolytic disease of the newborn in their infants, pregnant Rh-negative women who have a clinically significant antibody in maternal circulation are carefully monitored. Amniocentesis is usually performed if the antibody has an antiglobulin titer greater than 16 or if the woman has a history of a previously affected child. The amniotic fluid is examined for bile pigments at appropriate intervals, and the severity of the disease is assessed according to those levels.
   a. If the pregnancy is over 32 weeks, labor may be induced and, if necessary, the baby can be exchange-transfused after delivery.
   b. If the pregnancy is less than 32 weeks, or fetal lung maturity is inadequate (judged by the lecithin/sphyngomyelin ratio in amniotic fluid), intrauterine transfusion may be performed by transfusing O, Rh-negative red cells to the fetus.

Case 1 Revisited

- Many clinical conditions can cause neonatal jaundice. In a blood group O, Rh-negative mother, hemolytic disease of the newborn secondary to anti-Rh or anti-AB antibodies needs to be considered. In a gravida 1, para 0 female, the disease is unlikely to be due to Rh incompatibility and ABO hemolytic disease of the newborn is usually mild. Treatment is not usually required. If indicated, phototherapy will usually reduce the bilirubin concentration and exchange transfusion is rarely necessary.
- The following tests were ordered on the newborn:

Blood group and Rh type:	A, Rh positive
Characterization of antibodies:	
direct antiglobulin test	Weakly positive
eluted from RBC	Anti-A
Bilirubin, total	7.4 mg/dL
Bilirubin, direct	0.1 mg/dL

  The conclusion from the laboratory tests was that the child had jaundice secondary to a mild hemolytic anemia of immune cause.
- Prevention of hemolytic disease of the newborn is a multistep process. First, this woman should have had a blood typing and antibody screening test ordered in the first trimester. In Rh-negative women, the antibody screening test is repeated at 28 weeks. If a woman has a positive antibody screening test, the antibody must be identified and its clinical significance assessed. This basically means determining whether it is IgG and can cross the placenta and react with incompatible fetal cells at body temperature. Clinically significant antibodies must be monitored closely throughout pregnancy so that treatment such as early delivery or intrauterine transfusions may be given if necessary. In addition, if anti-D antibodies were not detected in this patient, she should have been given a full dose of Rh immune globulin at 28 weeks and again within 72 hours after delivery. The risk of sensitization for an Rh-negative woman delivering her first Rh-positive infant is about 8%. The postpartum dose protects at the time of delivery when the largest number of fetal cells enters the maternal circulation and reduces the risk to about 1%. The antepartum dose prevents a small number of women who have larger than normal amounts of fetal cells entering their circulation during pregnancy from becoming sensitized and decreases the risk. ABO hemolytic disease of the newborn cannot be prevented but it is rarely serious.

IV. IMMUNE HEMOLYTIC ANEMIAS

A. **Introduction.** The designation of hemolytic anemias includes a heterogeneous group of diseases whose common denominator is the exaggerated destruction of red cells (hemolysis). In this chapter we will discuss only the hemolytic anemias in which an abnormal immune response plays the major pathogenic role.

Case 2

A 65-year-old man being treated for essential hypertension with a combination of thiazide and α-methyldopa was seen by his internist. He was complaining of tiredness and shortness of breath. The following laboratory results were obtained:

Hemoglobin	10 g/dL
Hematocrit	31%
Reticulocytes	8%
Bilirubin, direct	1.5 mg/dL
Bilirubin, total	3.6 mg/dL
Direct antiglobulin test	Positive with anti-IgG
Indirect antiglobulin test	Positive

Panels performed on both the serum and an eluate from the patient's red cells revealed positive reactions with all cells tested, indicating the presence of an antibody of broad specificity.

This case raises several questions:

- What are the two most probable causes of this patient's anemia, and why is it important to distinguish between the two?
- What is the pathogenesis of the two types of anemia most likely involved?
- What immediate measure(s) should be instituted?

B. **Autoimmune Hemolytic Anemia (Warm Antibody Type).** This is the most common form of autoimmune hemolytic anemia. It can be idiopathic (often following overt or subclinical viral infection) or secondary, as shown in Table 24.5.

1. **Pathogenesis.** Warm autoimmune hemolytic anemia is due to the spontaneous emergence of IgG antibodies that may have a simple Rh specificity such as anti-e, or uncharacterized specificities common to almost all normal red cells ("public" antigens, thought to be the core of the Rh substance). In many patients, one can find antibodies of more than one specificity. The end result is that the serum from patients with autoimmune hemolytic anemia of the warm type is likely to react with most, if not all, the red cells tested. These antibodies usually cause shortening of red-cell life due to the uptake and destruction by phagocytic cells in the spleen and liver.
2. **Diagnosis.** Diagnosis relies on the demonstration of antibodies coating the red cells or circulating in the serum.
   a. RBC-fixed antibodies are detected by the **direct antiglobulin (Coombs) test.** The test can be done using anti-IgG antiglobulin, anticomplement, or polyspecific antiglobulin serum that has both anti-IgG and anticomplement. The polyspecific or broad-spectrum antiglobulin sera produce positive results in higher numbers of patients, as shown in Table 24.6.

Table 24.5 Immune Hemolytic Anemias

Autoimmune hemolytic anemias (AIHA)
Warm antibody AIHA
Idiopathic (unassociated with another disease)
Secondary (associated with chronic lymphocytic leukemia, lymphomas, systemic lupus erythematosus, etc.)
Cold antibody AIHA
Idiopathic cold hemagglutinin disease
Secondary cold hemagglutinin syndrome
Associated with *M. pneumoniae* infection
Associated with chronic lymphocytic leukemia, lymphomas, etc.
Immune drug-induced hemolytic anemia
Alloantibody-induced immune hemolytic anemia
Hemolytic transfusion reactions
Hemolytic disease of the newborn

Modified from Petz, L.D., and Garraty, G. Laboratory correlations in immune hemolytic anemias. In *Laboratory Diagnosis of Immunologic Disorders*, G.N. Vyas, D.P. Stites, and G. Brechter, eds. Grune & Stratton, New York, 1974.

b. The search for antibodies in serum is carried out by the **indirect antiglobulin test.** Circulating antibodies are only present when the red cells have been maximally coated, and the test is positive in only 40% of the cases tested with untreated red cells. A higher positivity rate (up to 80%) can be achieved by using red cells treated with enzymes such as trypsin, papain, ficin, and bromelin in the agglutination assays. The treatment of red cells with these enzymes increases their agglutinability by either increasing the exposure of antigenic determinants or by reducing the surface charge of the red cells. In the investigation of warm-type AIHA, all tests are carried out at 37°C.

C. **Cold Agglutinin Disease and Cold Agglutinin Syndromes.** These diseases can also be idiopathic or secondary.

1. **Pathogenesis.** The cold agglutinins are classically IgM (very rarely IgA or IgG), and react with red cells at temperatures below normal body temperature.

   a. In chronic, idiopathic, cold agglutinin diseases, 95% or more of the antibodies, which are **IgMκ**, react with the **I antigen**. This is the adult specificity of the I, i system. The fetus expresses the i antigen, common to primates and other mammalians, which is the precursor of the I specificity. The newborn expresses i predominantly over I; in the adult, the situation is reversed.

   b. In postinfectious cold agglutinin syndrome, the antibodies are also predominantly IgM, but contain both κ and λ light chains, suggesting their polyclonal origin. The cold agglutinins that appear in patients with *Mycoplasma pneumoniae* infections are usually reactive with the I antigen, while those that appear in association with **infectious mononucleosis** usually react with the i antigen.

   c. The range of thermal reactivity of cold agglutinins may reach up to

Table 24.6 Typical Results of Serological Investigations in Patients with Autoimmune Hemolytic Anemia

	Cells			Serum	
	Direct Coombs test				
	Antibody to	Positivity rate	Antibody isotype	Serological characteristics	Ab specificity
Warm AIHA	IgG	30%	IgG	Positive indirect Coombs test (40%)	Rh system antigens ("public")
	IgG + C′	50%		Agglutination of enzyme-treated	
	C′	20%		RBC (80%)	
Cold Agglutinin Disease	C′		IgM	Monoclonal IgMκ agglutinates RBC to titers >1024 at 4°C	I antigen

Modified from Petz, L.D., and Garraty, G. Laboratory correlations in immune hemolytic anemias. In *Laboratory Diagnosis of Immunologic Disorders*, G.N. Vyas, D.P. Stites, and G. Brechter, eds. Grune & Stratton, New York, 1974.

35°C. Such temperatures are not difficult to experience in exposed parts of the body during the winter. Cold-induced intravascular agglutination, causing ischemia of cold-exposed areas, and hemolysis are the main pathogenic mechanisms in cold agglutinin disease.

2. **Clinical presentation.** Hemolysis is usually mild, but in some cases may be severe, leading to acute tubular necrosis. But in most cases the clinical picture is dominated by symptoms of cold sensitivity (Raynaud's phenomenon, vascular purpura, and tissue necrosis in exposed extremities).
3. **Laboratory diagnosis.** Testing for cold agglutinins is usually done by incubating a series of dilutions of the patient's serum (obtained by clotting and centrifuging the blood at 37°C immediately after drawing) with normal group O RBC at 4°C.
   a. Titers up to the hundred thousands can be observed in patients with cold agglutinin disease.
   b. Intermediate titers (below 1000) can be found in patients with *Mycoplasma pneumoniae* infections (postinfectious cold agglutinins).
   c. Low titers (less than 64) can be found in normal, asymptomatic individuals.

D. Drug-Induced Hemolytic Anemia. Three different types of immune mechanisms may play a role in drug-induced hemolytic anemias, as summarized in Table 24.7. It is important to differentiate between drug-induced hemolytic anemia and warm autoimmune hemolytic anemia, since cessation of the drug alone will usually halt the drug-induced hemolytic process.

1. **Formation of soluble immune complexes between the drug and the corresponding antibodies**, which is followed by nonspecific adsorption to red cells, and complement activation.
   a. When IgM antibodies are predominantly involved, intravascular hemo-

Table 24.7 Correlation Between Mechanisms of Red-Cell Sensitization and Laboratory Features in Drug-Induced Immunohematological Abnormalities

			Serological evaluation	
Mechanism	Prototype drugs	Clinical findings	Direct Coombs	*In vitro* tests and AB identification
Immune complex formation	Quinidine Phenacetin	Intravascular hemolysis; renal failure; thrombocytopenia	C usually IgG rarely	Drug + serum + RBC; Ab is often IgM
Drug adsorption to RBC	Penicillins	Extravascular hemolysis associated with high doses of penicillin i.v.	Strongly positive with anti-IgG[a]	Drug-coated RBC + serum; antibody is IgG
Membrane modification causing nonimmunological adsorption of proteins	Cephalosporins	Asymptomatic	Positive with a variety of antisera	Drug-coated RBC + serum; no specific antibody involved
Autoimmune	α-methyldopa	Hemolysis in .8% of patients taking this medication	Strongly positive with anti-IgG[a]	Normal RBC + serum. Autoantibody to RBC identical to Ab in warm AIHA

[a]When hemolytic anemia is present.
Modified from Garraty, G., and Petz, L.D. *Am. J. Med.*, *58*:398, 1975.

lysis is frequent and the direct Coombs test is usually positive if anti-complement antibodies are used.

b. IgG antibodies can also form immune complexes with different types of antigens and be adsorbed onto red cells and platelets. In vitro, such adsorption is not followed by hemolysis or by phagocytosis of red cells, but in vivo it has been reported to be associated with intravascular hemolysis.

c. The absorption of IgG-containing immune complexes to platelets is also the cause of **drug-induced thrombocytopenia**. Quinine, quinidine, digitoxin, gold, meprobamate, chlorothiazide, rifampin, and the sulfonamides have been reported to cause this type of drug-induced thrombocytopenia.

2. **Adsorption of the drug onto the red cells.** Adsorption of drugs by red cells may cause hemolytic anemia by several different mechanisms:
   a. The adsorbed drug functions as hapten and the RBC as carrier, and an immune response against the drug ensues. The antibodies, usually IgG, are present in high titers, and may activate complement after binding to the drug adsorbed to the red cells, inducing hemolysis (Fig. 24.3) or phagocytosis. Penicillin (when administered in high doses by the IV route) and cephalosporins can induce this type of hemolytic anemia.
   b. Some cephalosporins (such as cephalothin) have been shown to modify the red-cell membrane which becomes able to adsorb proteins non-specifically, a fact that can lead to a positive direct Coombs test but not to hemolytic anemia.
3. **Induction of a truly autoimmune hemolytic anemia.** The anemia induced by **α-methyldopa** (Aldomet) is the most frequent type of drug-induced hemolytic anemia. It is particularly interesting from the pathogenic point of view in that it is indistinguishable from a true warm autoimmune hemolytic anemia.

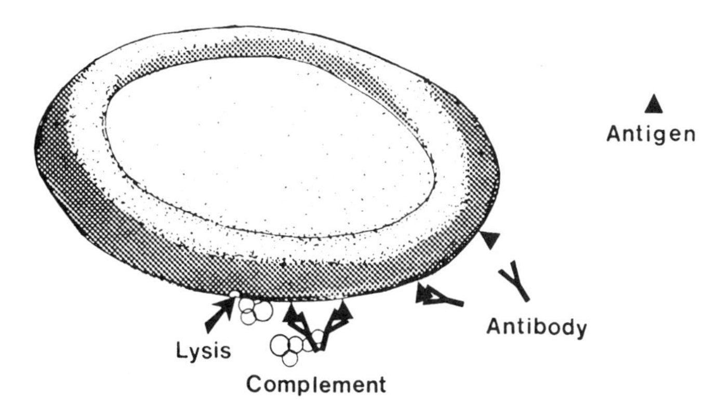

Figure 24.3 Diagrammatic representation of the pathogenesis of drug-induced hemolytic anemia as a consequence of adsorption of a drug to the red-cell membrane.

a. Ten to 15% of the patients receiving the drug will have a positive Coombs test, and 0.8% of the patients develop clinically evident hemolytic anemia.
b. α-Methyldopa is unquestionably the trigger for this type of anemia, but the antibodies are of the IgG1 isotype and react with Rh antigens. It is believed that the drug changes the membrane of red-cell precursors, causing the formation of antibodies reactive with a modified Rh precursor. Once formed, the anti-red-cell antibodies will react in the absence of the drug, as true autoantibodies.
c. L-dopa, a related drug used for treatment of Parkinson's disease, can also cause autoimmune hemolytic anemia. Both α-methyldopa and L-dopa also stimulate the production of antinuclear antibodies.

Case 2 Revisited

- The two most probable causes of this patient's hemolytic anemia were warm autoimmune hemolytic anemia and drug-induced hemolytic anemia. It is important to distinguish the two because the treatment is significantly different. The history of hypertension and treatment with α-methyldopa should alert the physician toward the possibility of a drug-induced hemolytic anemia. Laboratory tests usually do not differentiate between the two conditions because the reactivity of the antibodies is virtually identical.
- Warm-type hemolytic anemia is an autoimmune condition, which can present itself as the only manifestation of autoimmunity or as part of the constellation of a systemic autoimmune disease. The autoantibodies are of broad specificity, reacting with public erythrocyte antigens expressed by almost every individual. Drug-induced hemolytic anemia can be caused by antibodies directed to the drug, due either to previous adsorption of the drug to the red cell or to adsorption of preformed antigen–antibody complexes to the red cell, or (as it is the case in the hemolytic anemia associated with α-methyldopa) by anti-red-cell antibodies identical to those detected in true autoimmune hemolytic anemia. How α-methyldopa causes the production of these antibodies is the subject of speculation. It is believed that the drug may alter the conformation of the Rh complex on the red-cell membrane triggering the synthesis of antibodies that cross-react with unchanged Rh substances. Thus, once induced, the autoantibody recognizes the red cell rather than α-methyldopa, and withdrawal of the drug may not result in immediate improvement.
- In all cases of drug-induced hemolytic anemia, it is important to stop the administration of the drug as soon as the diagnosis is established. In the case of α-methyldopa-induced hemolytic anemia, the improvement will be gradual, because the autoantibodies react with the red cells, rather than with the drug, but the antibody titers will decrease with time and, after a point, their concentration may still be sufficient to cause a positive direct antiglobulin test, but not to cause significant anemia. In contrast, treatment of true autoimmune hemolytic anemia is rather more complex, involving administration of steroids, and in cases not responding to steroids, splenectomy and/or administration of immunosuppressive drugs.

SELF-EVALUATION

Questions

Choose the ONE *best* answer.

24.1 A direct Coombs test using antisera to IgG is virtually always positive in:
- A. Females with circulating anti-D antibodies
- B. Newborns with Rh hemolytic disease
- C. Patients with cold hemagglutinin disease
- D. Patients with α-methyldopa-induced hemolytic anemia
- E. Patients with warm-type autoimmune hemolytic anemia

24.2 The pathogenesis of penicillin-induced hemolytic anemia involves:
- A. Drug adsorption to red cells and reaction with antipenicillin antibodies
- B. Emergence of a neoantigen on the red-cell membrane
- C. Formation of soluble IC, adsorption to red-cell membranes, and complement activation or phagocytosis.
- D. Nonspecific adsorption and activation of complement components
- E. None of the above

24.3 Which of the following drugs induces the production of autoantibodies that react with public red-cell antigens?
- A. α-Methyldopa
- B. Cephalosporin
- C. Penicillin
- D. Phenacetin
- E. Quinidine

24.4 The destruction of Rh-positive erythrocytes after exposure to IgG anti-D antibodies is due to:
- A. Complement activation
- B. Fc-mediated phagocytosis
- C. C3b-mediated phagocytosis
- D. C3d-mediated phagocytosis
- E. A combination of Fc and C3b-mediated phagocytosis

24.5 In a patient with penicillin-induced hemolytic anemia, you should be concerned with the induction of a similar situation if prescribing:
- A. Aminoglycosides
- B. Aspirin
- C. Cephalosporins
- D. Quinidine
- E. Sulfonamides

24.6 An A, Rh-negative female is unlikely to be sensitized by a first Rh-positive baby if:
- A. The baby is B, Rh positive
- B. The baby is A, Rh positive
- C. The baby is O, Rh positive
- D. The father is A, Rh positive
- E. The father is B, Rh positive

24.7 The major cross-match is used to detect antibodies in:
- A. The donor's red cells
- B. The donor's serum

C. The recipient's cells
D. The recipient's serum
E. Both donor's and recipient's sera

24.8 The reason why the Coombs test is NOT used for the diagnosis of cold agglutinin disease (CAD) is that the:
A. Disease is diagnosed by the cold agglutinin test
B. Red cells from patients with CAD autoagglutinate at 37°C
C. Red cells hemolyse spontaneously at room temperature
D. Red-cell-bound IgM does not activate the complement system
E. Results are always negative

24.9 The reason why it is often difficult to transfuse patients with warm AIHA is that the antibodies in the patient's serum:
A. Are complement fixing
B. Are polyspecific and react with many different red-cell antigens
C. Present in very high titers
D. React at body temperature
E. React with "public" red-cell antigens

24.10 A large majority of fatal transfusion reactions result from:
A. Allergic reactions
B. Anti-IgA antibodies
C. Antibodies to HLA antigens
D. Bacterial contamination of transfused blood
E. Human error

Answers

24.1 (B) Rh hemolytic disease is, by definition, due to IgG antibodies that cross the placenta and bind to the newborn's erythrocytes, where they will be easily detected by a direct Coombs test using anti-IgG antibodies.

24.2 (A)

24.3 (A) α-Methyldopa (Aldomet) induces a very unique type of hemolytic anemia in which the antibodies react with red-cell antigens of the Rh complex and not to the drug itself.

24.4 (B) IgG anti-D antibodies do NOT cause complement fixation after binding to the red-cell membrane because the D antigen molecules are too spaced to allow the IgG molecules to form the duplets essential for complement activation.

24.5 (C) There can be cross-reaction between penicillin and the cephalosporins; both groups of antibiotics belong to the β-lactam group and therefore have to share part of their structure.

24.6 (A) The maternal anti-A and anti-B isoagglutinins will function as a natural anti-red-cell antibody, destroying incompatible fetal red cells before there is an opportunity to induce the anti-D immune response. The paternal blood group is irrelevant.

24.7 (D)

24.8 (A) The patient's red cells may autoagglutinate and lyse, but usually at low temperatures, and definitely not at 37°C. The direct Coombs test can be positive, using anti-C3 antibodies, but the cold agglutinin test is preferred because, when positive at a high titer in a symptomatic patient, it established the specific diagnosis of cold hemagglutinin disease.

24.9 (E) The main problem when transfusions are needed for patients with AIHA is the difficulty in finding cells that are not agglutinated by the patient's serum. This is due to the fact that the autoantibodies react with "public" red-cell antigens (i.e., antigens that most every individual carries on his or her erythrocytes).

24.10 (E) Human error is the most frequent cause, and bacterial contamination of transfused blood is the second most frequent cause of fatal transfusion reactions.

BIBLIOGRAPHY

Brostoff, J., Scadding, G.K., Male, D., and Roitt, I.M. *Clinical Immunology.* Gower Medical Publishers, London, 1991.

Jeter, E.K., and Spivey, M.A. *Introduction to Transfusion Medicine: A Case Study Approach.* American Association of Blood Banks, Bethesda, MD, 1996.

Menitove, J.E. Transfusion practices in the 1990s. *Annu. Rev. Med., 42*:297, 1991.

Mollison, P.L. *Blood Transfusion in Clinical Medicine*, 9th ed. Oxford, Blackwell Scientific Publications, 1993.

Petz, L.D., and Garraty, G. *Acquired Immune Hemolytic Anemias.* Churchill Livingstone, New York, 1980.

Petz, L.D., Swisher, S., Kleinman, S., et al., eds. *Clinical Practice of Transfusion Medicine*, 3rd ed. Churchill Livingstone, New York, 1996.

Rudmann, S.V. *Textbook of Blood Banking and Transfusion Medicine.* W.B. Saunders Co., Philadelphia, 1995.

Stangenberg, M., Selbing, A., Lingman, G., and Westgren, M. Rhesus immunization: New perspectives in maternal-fetal medicine. *Obstet. Gynecol. Surv., 46*:189, 1991.

Wells, J.V., and Isbister, J.P. Hematologic diseases. In *Basic and Clinical Immunology*, 7th ed (D.P. Stites, and A.I. Terr, eds.). Lange, Norwalk, CT, 1991, p. 476.

25
Immune Complex Diseases

Gabriel Virella and George C. Tsokos

I. INTRODUCTION

The formation of circulating antigen–antibody (Ag.Ab) complexes is one of the natural events that characterizes the immunological response against soluble antigens. Normally, immune complexes (IC) formed by soluble proteins and their respective antibodies are promptly phagocytosed and eliminated from circulation without any detectable adverse effects on the host.

A. **Serum Sickness.** In the late 1800s and early 1900s, passive immunization with equine antisera was a common therapy for severe bacterial infections. It was often noted that 1 to 2 weeks after administration of the horse antisera, when the symptoms of acute infection had often disappeared, the patient would start to complain of arthralgias and exanthematous rash, and had proteinuria and an abnormal urinary sediment, suggestive of glomerulonephritis. von Pirquet coined the term **serum sickness** to designate this condition.

B. **Experimental Models of Serum Sickness.** Germuth, Dixon, and co-workers carried out detailed studies on rabbits in whom serum sickness was induced by injection of single doses of heterologous proteins.

1. As summarized in Figure 25.1, after the lag time necessary for antibody production, soluble immune complexes were detected in serum, serum complement levels decreased, and the rabbits developed glomerulonephritis, myocarditis, and arthritis.
2. The onset of disease coincided with the disappearance of circulating antigen, while free circulating antibody appeared in circulation soon after the beginning of symptoms.
3. Both the experimental one-shot serum sickness or human serum sickness are usually transient and will leave no permanent sequelae. However, if the organism is chronically exposed to antigen (as in chronic serum sickness), irreversible lesions will develop.

II. PHYSIOPATHOLOGY OF IMMUNE COMPLEX DISEASE

The formation of an immune complex does not have direct pathological consequences. The pathogenic consequences of immune complex formation depend on the ability of those

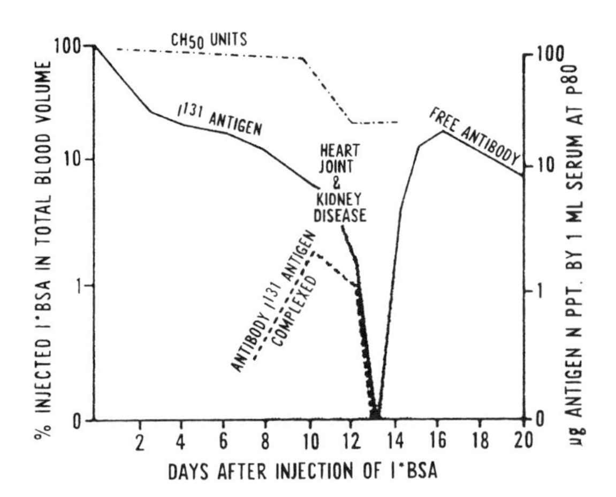

Figure 25.1 Diagrammatic representation of the sequence of events that takes place during the induction of acute serum sickness in rabbits. Six days after injection of radiolabeled BSA, the synthesis anti-BSA antibodies starts being produced and forms complexes with the antigen, which is eliminated rapidly with the circulation. The maximal concentration of immune complexes precedes a decrease in complement levels and the appearance of histological abnormalities on the heart, joints, and kidney. After the antigen is totally eliminated, the antibody becomes detectable and the pathological lesions heal without permanent sequelae. (Reproduced with permission from Cochrane, C.G., and Koffler, D. *Adv. Immunol.*, *16*:185, 1973.)

immune complexes to (1) form a stable bond with a circulating cell or to leave the intravascular compartment and become tissue-fixed; (2) activate the complement system; (3) interact with cells able to release enzymes and mediators involved in inflammation. All these properties are related to the physicochemical characteristics of immune complexes (Table 25.1).

Table 25.1 Antigen–Antibody Characteristics That Affect the Pathogenicity of IC

Antibody
Class
Valence
Affinity for Fc Receptors
Ability to bind and activate complement
Affinity
Charge
Amount
Antigen
Size
Valence
Chemical composition
Charge
Amount

A. **Physicochemical Characteristics of Immune Complexes.** Size, affinity of the Ag.Ab reaction, and class and subclass of antibodies involved in immune complex formation are among the most important determinants of the pathogenic significance of IC.
   1. Very large Ag.Ab aggregates containing IgG1 or IgG3 antibodies will activate complement very effectively, but are usually nonpathogenic. This is due to a combination of facts: very avid ingestion and degradation by phagocytic cells, and difficulty in diffusing across the endothelial barrier.
   2. Very small complexes (Ag1.Ab1-3), even when involving IgG1 and IgG3 antibodies, are able to diffuse easily into the extravascular compartment, but are usually nonpathogenic because of their inability to activate complement.
   3. The most potentially pathogenic IC are those of intermediate size (Ag2-3.Ab2-6), particularly when involving complement-fixing antibodies (IgG1, IgG3) of moderate to high affinity. However, a main question that needs to be answered is how these IC diffuse into the subendothelial space.

B. **Immune Complex Formation and Cell Interactions**
   1. **Circulating immune complexes.** IC that are formed in the circulation may be deposited in various tissues where they cause inflammation and tissue damage. Characteristically, multiple organs and tissues may be affected and the clinical paradigms are diseases such as serum sickness and systemic erythematosus. The interaction of IC with cells able to release mediators of inflammation appears to be considerably enhanced if the IC are surface-bound, rather than soluble. This immobilization of immune complexes along vessel walls is likely to be mediated by:
      a. C3 receptors, such as those located on the glomerular epithelium
      b. Clq receptors, expressed by endothelial cells
      c. Fc receptors, expressed on the renal interstitium and by damaged endothelium
      d. Affinity of some antigen moieties in IC for specific tissues, such as that of DNA for glomerular basement membrane and collagen
   2. **Immune complex formation in the interstitial space.** Direct injection of antigen into a tissue will result in local IC formation with circulating antibody. Examples include the Arthus reaction (antigen injected into the dermis binds circulating antibody) and hypersensitivity pneumonitis (antigen inhaled forms IC with circulating antibody).
   3. **Formation of immune complexes in situ.** Antibody may form an IC with an antigen present on the cell membrane of circulating or tissue-fixed cells.
      a. Immune complexes formed on cell membranes can lead to the destruction of the cell, either by promoting phagocytosis or by causing complement-mediated lysis. This mechanism is responsible for the development of various immune cytopenias.
      b. Autoantibodies may bind to basement membranes by virtue of recognizing an intrinsic membrane antigen, such as in Goodpasture's syndrome, or by reacting with an antigen that has become adsorbed to a basement membrane due to charge interactions, such as seems to be the case of DNA, which binds to basement membrane collagen.
      c. The formation of IC in synovial membranes may also result from a two-

step in situ reaction between monomeric, freely diffusible autoantibodies, and synovial membrane antigens.

4. **Adsorption and transfer of immune complexes.** Circulating IC can bind to platelets and red cells.
   a. **Human platelets** express Fc receptors, specific for all IgG subclasses and CR4, which binds the C3dg fragment of C3.
   b. **Red blood cells** (RBC) express CR1, through which C3b-containing IC can be bound. In addition, IC can bind to RBC through nonspecific interactions of low affinity which do not require the presence of complement.
   c. The binding of IC to peripheral blood cells has both physiological and pathological implications.
      i. Immune complex binding to RBC is believed to be an important mechanism for clearance of soluble IC from the systemic circulation. Experimental work in primates and metabolic studies of labeled IC in humans show that RBC-bound IC are maintained in the intravascular compartment until they reach the liver, where they are presented to phagocytic cells. The phagocytic cells have Fc receptors able to bind the IC with greater affinity than the red cells; as a consequence, the IC are removed from the RBC membrane, while the red cells remain undamaged (Fig. 25.2).
      ii. On the other hand, cell-bound IC can cause cell death either as a consequence of complement-mediated lysis or of phagocytosis.

C. **Tissue Deposition of IC.** How immune complexes induce inflammation and tissue damage is not completely understood, particularly when the tissue damage is believed to result from the extravascular deposition of circulating immune complexes. A major obstacle to such deposition is the endothelial barrier, which is poorly permeable even to intermediate-sized immune complexes.

1. Some of the most frequent localizations of deposited IC are around the small vessels of the skin, particularly in the lower limbs, the kidney glomeruli, the choroid plexus, and the joints. It is likely that regional factors may influence the selectivity of IC deposition.
   a. The preferential involvement of the lower limbs in IC-related skin vasculitis may result from the simple fact that the circulation is slowest and the hydrostatic pressure highest in the lower limbs.
   b. The frequent involvement of the kidney in IC-associated disease may be a consequence of the existence of C3b receptors in the renal epithelial cells, Fc receptors in the renal interstitium, and a collagen-rich structure (the basement membrane), which can also be involved in nonspecific interactions.
2. In some cases, **in situ deposition of IC** may be a two-step process: first the antigen may diffuse across the endothelium and become associated with extravascular structures. Monomeric antibody, equally diffusible, may later become associated with the immobilized antigen. The best example for this sequence is SLE-associated nephritis: DNA becomes associated with the glomerular basement membrane and anti-DNA antibodies cross the endo-

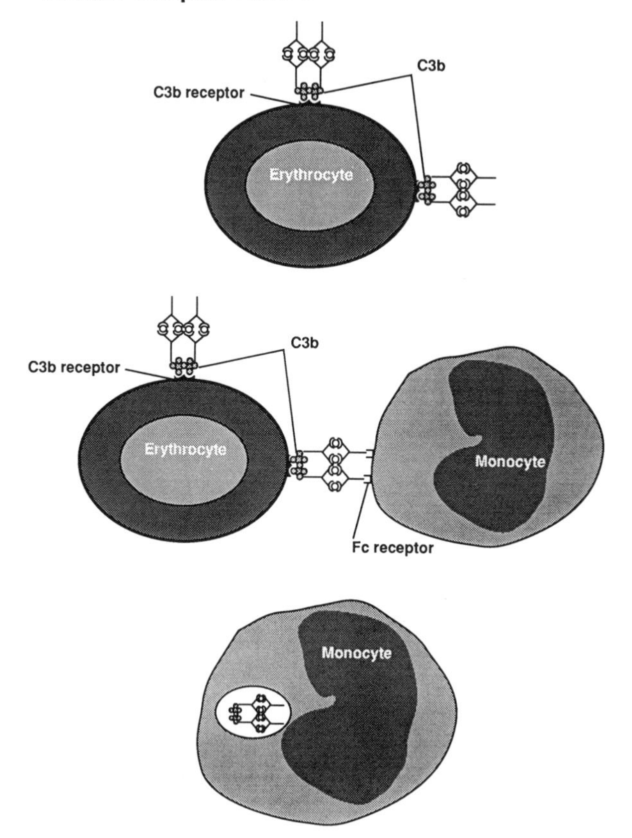

Figure 25.2 Diagrammatic representation of the protective role of erythrocytes against the development of immune complex disease. Erythrocytes can adsorb circulating IC through C3b receptors or through nonspecific interactions. RBC-adsorbed IC persist in circulation until the IC are stripped from the RBC surface by phagocytic cells expressing Fc receptors, which bind the IC with greater avidity. Once taken up by phagocytic cells, the IC are degraded, and this uptake is responsible for their disappearance from circulation.

thelial barrier and become associated with DNA at the basement membrane level.

3. Any pathogenic sequence involving the **deposition of circulating IC** has to account with increased vascular permeability in the microcirculation, allowing the diffusion of small- to medium-sized soluble IC to the subendothelial spaces (Fig. 25.3).
   a. The initial step is likely to be the activation of monocytes or granulocytes by immobilized IC, resulting in the release of vasoactive amines and cytokines. Receptor-mediated interactions involving Fc receptors or complement receptors on endothelial cells could play the initiating role by immobilizing IC at the level of the microvasculature.
   b. The retention of soluble IC diffusing through the endothelium in the

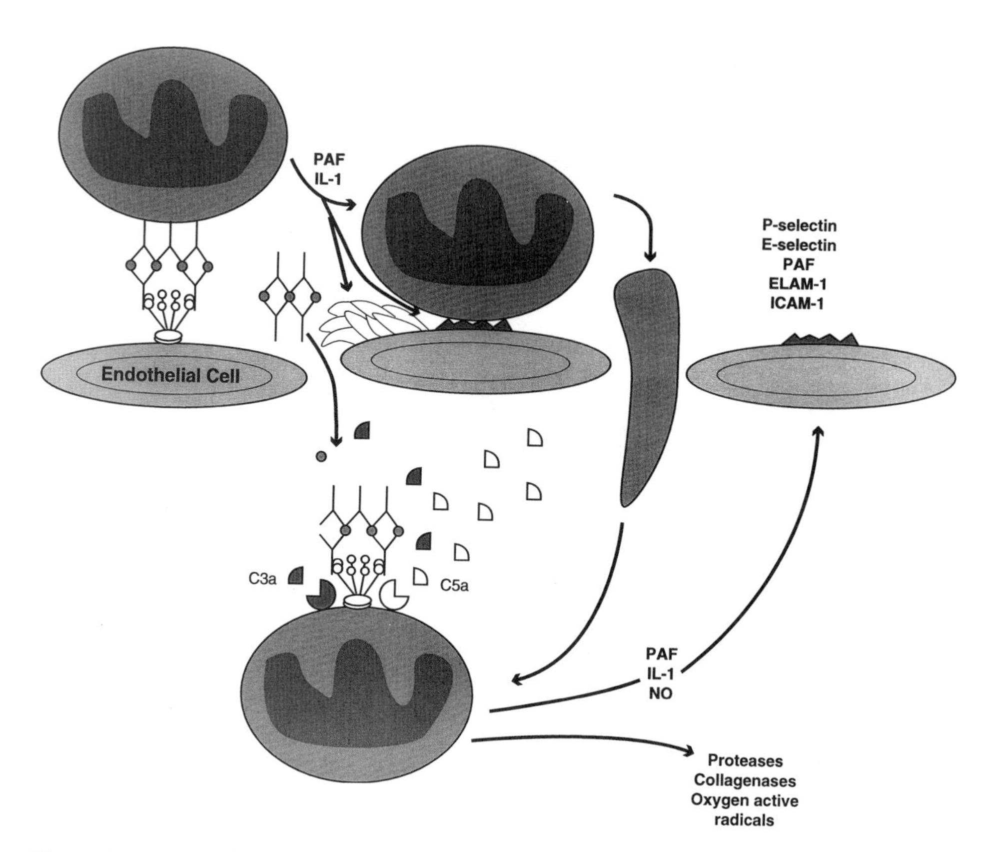

Figure 25.3 The initial stages of circulating immune complex deposition require a sequence of events that enables circulating IC and inflammatory cells to cross the endothelial barrier. In this representation of such a hypothetical sequence, the first event is the binding of circulating IC containing IgG antibodies and associated C1q to C1q receptors on the endothelial cell. The immobilized IC are then able to interact with circulating cells expressing Fcγ receptors, such as PMN leukocytes. Such interaction results in PMN activation and release of mediators, such as platelet-activating factor (PAF) and IL-1. These two mediators have a variety of effects: PAF induces vasodilation and activates platelets, which form aggregates and release vasoactive amines. The resulting increased vascular permeability allows circulating IC to cross the endothelial barrier. IL-1 activates endothelial cells and induces the expression of selectin molecules that interact with glycoproteins and sialoglycoproteins on the leukocyte membrane. This interaction slows down PMNs along the endothelial surface, a phenomenon known as "rolling." As endothelial cells continue to receive activating signals, they start expressing membrane-associated PAF, which interacts with PAF receptors on neutrophils, ICAM-1, which interacts with leukocyte integrins of the CD11/CD18 family, and VCAM-1, which interacts with VLA (very late antigen)-1, up-regulated on PMN leukocytes as a consequence of the occupancy of Fc receptors. This promotes firm adhesion of leukocytes to endothelial cells. At the same time, the IC that diffuse to the subendothelial space activate complement and generate chemotactic factors, such as C5a and C3a. Adherent PMN leukocytes are attracted to the area and insinuate themselves between endothelial cells, reaching the area of IC deposition. Interaction with those extravascular IC with associated C3b delivers additional activation signals to already primed granulocytes, resulting in the release of proteases, collagenases, oxygen active radicals, and nitric oxide. These compounds can cause tissue damage and can further increase vascular permeability, and in so doing contribute to the perpetuation of an inflammatory reaction.

kidney could be mediated by C3b receptors of the renal epithelial cells, or by Fc receptors in the renal interstitium. The mechanisms involved in selective retention of soluble IC in other tissues, such as the skin and choroid plexus, are poorly understood.

D. **Inflammatory Circuits Triggered by Immune Complexes.** The development of inflammatory changes after extravascular formation or deposition of immune complexes is not observed in experimental animals depleted of neutrophils or complement. Complement components play a significant role as opsonins and chemotactic factors, while activated granulocytes can release a wide variety of proinflammatory compounds and proteolytic enzymes, that mediate tissue damage (Table 25.2).

1. **Complement activation.** IC containing IgM, IgG1, and/or IgG3 antibodies, able to activate the complement system by the classical pathway, are believed to have the greatest pathogenic potential.
   a. Activation of the complement cascade results in the generation of chemotactic and proinflammatory fragments, such as **C3a** and **C5a** (see Chap. 9).
      i. C5a and C3a are chemotactic for neutrophils, attracting them to the area of IC deposition and stimulating their respiratory burst and release of granule constituents.
      ii. C5a increases vascular permeability directly as well as indirectly (by causing the release of histamine and vasoactive amines from basophils and mast cells).
      iii. C5a enhances the expression of the CD11/CD18 complex on neutrophil membranes, enhancing their adhesiveness to endothelial cells.
   b. The combination of chemotaxis, increased adherence, and increased vascular permeability plays a crucial role in promoting extravascular emigration of leukocytes, as discussed below.
2. **Emigration of leukocytes into the extravascular compartment.** The in-

Table 25.2 Elements Involved in Immune-Complex-Mediated Immunopathology

A. CELLS

Polymorphonuclear leukocytes—PMN-depleted animals do not develop arthritis or Arthus reaction.

Monocytes/macrophages—monocyte depletion in experimental glomerulonephritis decreases proteinuria.

B. SOLUBLE FACTORS (circulating and released locally)

Complement fragments—complement-depleted animals develop less severe forms of serum sickness.

Lymphokines/cytokines—corticosteroids reduce interleukin release and have beneficial effects in the treatment of immune complex disease.

Lysosomal enzymes—the main mediators of PMN-induced tissue damage.

Prostaglandins—important mediators of the inflammatory reaction; their synthesis is inhibited by aspirin and most nonsteroidal anti-inflammatory agents.

flammatory process triggered by immune complexes is characteristically associated with extravascular granulocyte infiltrates. The emigration of neutrophils and other granulocytes is regulated by a series of interactions with endothelial cells, known as the **adhesion cascade**. This cascade involves the following sequence of events:

a. The initial event involves the up-regulation of selectins (P-selectin and E-selectin) on endothelial cells, which can be caused by a variety of stimuli (e.g., histamine, thrombin, bradykinin, leukotriene C4, free oxygen radicals, or cytokines). The consequence of this up-regulation is the **slowing down** (**rolling**) and loose attachment of leukocytes (which express a third selectin, L-selectin, which binds to membrane oligosaccharides on endothelial cells). These initial interactions are unstable and transient.
b. The endothelial cells, in response to persistent activating signals, express PAF and ICAM-1 on the membrane. Neutrophils express constitutively a PAF receptor, which allows rolling cells to interact with membrane-bound PAF.
c. The interaction of neutrophils with PAF, as well as signals received in the form of chemotactic cytokines such as IL-8 (which can also be released by endothelial cells), activates neutrophils and induces the expression of integrins:
   i. CD11a/CD18, LFA-1, and related molecules
   ii. very late antigen-4 (VLA-4)
d. The interaction between integrins expressed by neutrophils and molecules of the immunoglobulin superfamily expressed by endothelial cells (ICAM-1 and related antigens bind LFA-1 and related molecules; VCAM-1 binds VLA-4) causes **firm adhesion** (**sticking**) of inflammatory cells to the endothelial surface, which is an essential step leading to their extravascular migration. VLA-4 is also expressed on the membrane of lymphocytes and monocytes, and its interaction with endothelial VCAM-1 allows the recruitment of these cells to the site of inflammation.
e. The interactions between integrins and their ligands are important for the development of vasculitic lesions in patients with systemic lupus erythematosus and other systemic autoimmune disorders and of purulent exudates in infection sites. As discussed in Chapter 17, patients with genetic defect in the expression of CD18 and related CAMs fail to form abscesses because their neutrophils do not express these molecules and fail to migrate.
f. The actual transmigration of leukocytes into the subendothelial space seems to involve yet another set of CAMs, particularly one member of the immunoglobulin superfamily known as PECAM (platelet endothelial cell adhesion molecule), which is expressed both at sites of intercellular junction and on the membranes of leukocytes. PECAM-1 interacts with itself and its expression is up-regulated on both endothelial cells and leukocytes by a variety of activating signals.
g. The egression of leukocytes from the vessel wall is directed by chemoattractant molecules released into the extravascular space and involves diapedesis through endothelial cell junctions.

3. **Activation of polymorphonuclear leukocytes and tissue damage**
   a. As leukocytes begin to reach the site of immune complex formation or deposition, they continue to receive activating signals, their activation brings about the release of additional chemotactic factors and continuing up-regulation of CAMs on endothelial cells, the efflux of phagocytic cells to the subendothelial space will intensify, and the conditions needed for self-perpetuation of the inflammatory process are created.
      i. All polymorphonuclear leukocytes express Fcγ receptors and C3b receptors that mediate binding and ingestion of IC. This process is associated with activation of a variety of functions and with the release of a variety of cytokines, enzymes, and other mediators.
      ii. One of the mediators released by activated neutrophils is **platelet activating factor** (**PAF**), which will promote the self-perpetuation of the inflammatory process by:
         - increasing vascular permeability (directly or as a consequence of the activation of platelets, which release vasoactive amines).
         - inducing the up-regulation of the CD11/CD18 complex on neutrophils.
         - inducing monocytes to release IL-1 and TNF-α, which activate endothelial cells and promote the up-regulation of adhesion molecules (E and P selectins) and the synthesis of PAF and IL-8.
   b. Granulocytes trying to engulf large IC aggregates or immobilized IC are activated and release their enzymatic contents, including proteases, collagenases, and oxygen active radicals. These compounds can damage cells, digest basement membranes and collagen-rich structures, and contribute to the perpetuation of the inflammatory reaction by causing direct breakdown of C5, generating additional C5a, and C3, generating C3b, which in turn promotes activation of the alternative pathway. Thus, the inflammatory reaction continues to intensify until it manifests itself clinically. The clinical manifestations of immune complex disease depend on the intensity of the inflammatory reaction and on the tissue(s) predominantly affected by IC deposition.

III. HOST FACTORS THAT INFLUENCE THE DEVELOPMENT OF IMMUNE COMPLEX DISEASE

The development of immune complex disease in experimental animals is clearly dependent on host factors.

A. If several rabbits of the same strain, age, weight, and sex are immunized with identical amounts of a heterologous protein by the same route, only a fraction of the immunized animals will form antibodies, and of those, only some will develop immune complex disease.

B. The magnitude of the response primarily depends on genetic factors, and the extent of tissue involvement is likely to depend on the general characteristics of the antibodies produced (such as affinity, ability to bind complement, and to interact with cell receptors) as well as on the functional state of the RES of the animal.

1. The **affinity and number of available Fcγ receptors** on professional phagocytic cells (PMN leukocytes, monocytes, macrophages) are important in the expression of IC disease. If IC are predominantly taken up by phagocytic cells in tissues where they abound, such as the liver and spleen, the likelihood of developing tissue inflammation is minimal.
   a. Support for the importance of Fc-mediated clearance of IC as a protective mechanism was obtained in experiments in which the Fc receptors were blocked. This resulted in decreased IC clearance and increased pathogenicity.
   b. Patients with systemic lupus erythematosus and rheumatoid arthritis have decreased ability to clear antibody-sensitized red cells, indicating a general inability to clear circulating IC. Recently, a subpopulation of patients with lupus who develop nephritis were found to have distinct FcγRII allele expression. It is speculated that the abnormal allele is functionally deficient, and that the lack of clearance of IC may then favor glomerular deposition.
2. The **ability to interact with complement receptors** may also be an important determinant of pathogenicity. Clq, C3b, C3c, and C3d are readily detected in IC. This may allow IC to bind to cells expressing the corresponding complement receptors.
   a. Binding of IC on the CR1 that is expressed on the surface membrane of red cells facilitates their clearance for the circulation (see above).
   b. The numbers of CR1 on the surface of red cells from patients with lupus and other IC diseases is decreased and this may contribute to decreased IC clearance.
   c. Although it has been claimed that the decreased CR1 expression in lupus patients is genetically determined, it is more widely believed that blocking of the receptors by IC causes a decrease in the number of available receptors.

IV. DETECTION OF SOLUBLE IMMUNE COMPLEXES

Many techniques have been proposed for the detection of soluble immune complexes. In general, these techniques are based either on the *physical properties* [e.g., precipitation with polyethylene glycol (PEG) or cryoprecipitation] or on the *biological characteristics* of the IC. The latter techniques make use of various properties of IC such as their ability to bind Clq or their binding to cells that express CR1 and CR2 (Raji cell assay). Table 25.3 lists those assays that have achieved wider use, some of which will be discussed below.

A. Detection of cryoglobulins. Circulating IC are often formed in antigen excess, with low-affinity antibodies, and remain soluble at room temperature. However, if the serum containing these IC is cooled to 4°C for about 72 hours, the stability of the antigen–antibody reaction increases and eventually there is sufficient cross-linking to result in the formation of large aggregates that precipitate spontaneously. Because antibodies are the main constituents of these cold precipitates, and because antibodies are globulins, the precipitated proteins are designated as **cyroglobulins**.

1. The following are the basic steps followed for detection of a cryoglobulin:

Table 25.3 Most Commonly Used Screening Tests for Soluble IC

A.	**Based on thermosolubility**
	Cryoprecipitation
B.	**Based on differential PEG solubility**
	Measurement of PEG-induced turbidity
	Assay of total protein in PEG precipitates
	Assay of IgG in PEG precipitates
C.	**Based on interactions with complement**
	Solid-phase Clq binding assay
D.	**Based on IC-cell interactions**
	Raji cell assay.

a. Blood is drawn and clotted at 37°C.
b. Serum is immediately separated, without cooling.
c. Immediately after centrifugation, 5–10 mL of serum are placed in a refrigerator at 4-8°C.
d. The serum is examined daily, up to 72 hours, for appearance of a precipitate (Fig. 25.4).

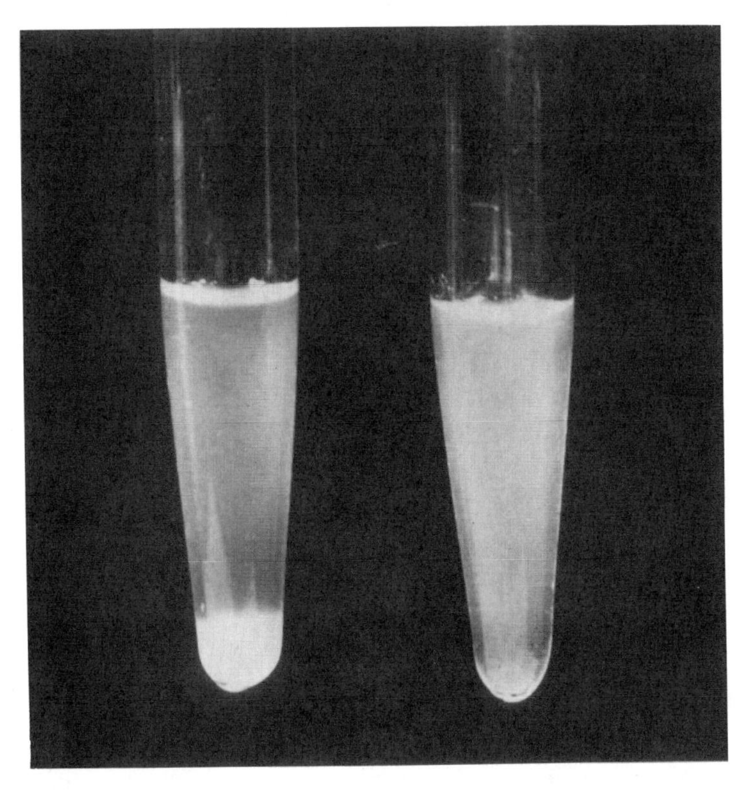

Figure 25.4 Screening of cryoglobulins. Two test tubes were filled with sera from a patient (left) and a healthy volunteer (right). After 48 hours at 4°C, a precipitate is obvious in the patient's serum, but is not present in the control.

e. Characteristically, this precipitate will resuspend if the serum is warmed to 37°C.

2. The formation of a visible precipitate is reported as a positive screening test and should be followed by washing the precipitate (to eliminate contaminant proteins), redissolution, and immunochemical characterization of the proteins contained in the precipitate (Fig. 25.5).
3. Cryoglobulins can be classified into two major types according to their constitution:
   a. **Monoclonal cryoglobulins**, containing immunoglobulin of one single isotype and one single light-chain class.
   b. **Mixed cryoglobulins**, containing two or three immunoglobulin isotypes, one of which (usually IgM) can be a monoclonal component (with one single light-chain type and one single heavy-chain class), while the remaining immunoglobulins are polyclonal. Complement components (C3, C1q) can also be found in the cryoprecipitates containing mixed cryoglobulins.
4. **Monoclonal cryoglobulins** are usually detected in patients with plasma cell malignancies and in some cases of idiopathic cryoglobulinemia. Monoclonal cryoglobulins are essentially monoclonal proteins with abnormal thermal behavior, and their existence has no correlation with immune complex formation or any special diagnostic significance besides the possibility of creating conditions favorable for the development of the hyperviscosity syndrome (see Chap. 29).
5. **Mixed cryoglobulins**, on the contrary, represent cold-precipitable immune complexes. One of the immunoglobulins present in the precipitate (usually the monoclonal component) is an antibody that reacts with the other immunoglobulin(s) that constitute the cryoglobulin.
   a. The most frequent type of mixed cryoglobulin is IgM-IgG, in which IgM is a "*rheumatoid factor.*" It is believed that, at least in some cases,

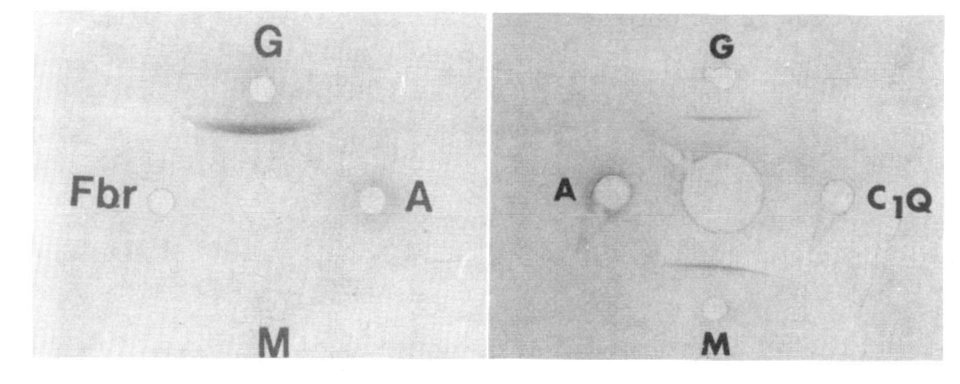

Figure 25.5 Characterization of two washed cryoglobulins. In both studies, the washed and redissolved cryoglobulin was placed on the center well and four different antisera in the surrounding wells. The cryoglobulin studied on the left reacted with anti-IgG (G) only and was classified as a monoclonal cryoglobulin; the one studied on the right reacted with anti-IgG (G) and anti-IgM (M) and was classified as a mixed cryoglobulin.

the IgM antibody is directed to determinants expressed by IgG antibodies bound to their corresponding antigens (Fig. 25.6).

b. Evidence supporting the involvement of infectious agents in the formation of mixed cryoglobulins has been obtained by identifying antigens and/or antibodies in the cryoprecipitates, particularly hepatitis viruses or antigens derived from them.

B. **Techniques Based on the Precipitation of Soluble Immune Complexes with Polyethylene Glycol.** Low concentrations of PEG (3-4%) cause preferential precipitation of IC relative to monomeric immunoglobulins. This is the basis for simple techniques that can be used for screening or purification of soluble IC.

1. For screening purposes, a dilution of patient's serum and PEG are mixed, and the turbidity of the sample is measured after adequate incubation; a direct correlation can be established between the degree of turbidity and the concentration of IC in the serum.
2. Other assays involve the incubation of mixtures of PEG and sera until a precipitate is obtained, and then total protein, IgG, or Clq are measured in the precipitate. This technique is also frequently used as the initial step in immune complex isolation protocols.

C. **Assays Based on Interactions with Complement.** One of the most widely used techniques for general screening of IC is based on the binding of Clq.

1. The original method employed purified ^{125}I-labeled Clq as a tracer for soluble IC precipitated with either a second antibody (antihuman IgG) or polyethylene glycol.

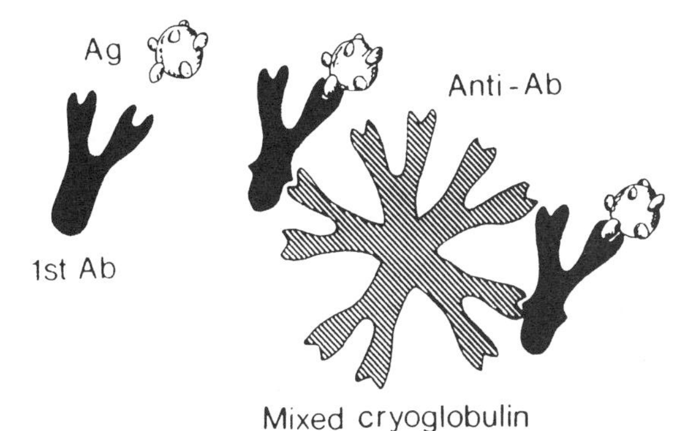

Figure 25.6 Diagrammatic representation of the pathogenesis of mixed cryoglobulins. Initially, an antimicrobial antibody (for example) of the IgG class is produced. This antibody, a consequence of binding to the antigen, exposes a cryptic antigenic determinant, which is recognized by an IgM antiglobulin. The combination of this IgM with the first IgG antibody and the microbial antigen constitutes the mixed cryoglobulin cryoprecipitates. Viral antigens and corresponding antibodies (for example, HBsAg and anti-HBsAg) have been characterized in cryoprecipitates from patients with mixed cryoglobulins, both with and without a history of previous viral hepatitis B infection. It is believed that this mechanism, antiviral IgG combined with an IgM antibody, accounts for over 50% of the cases of **essential** or **idiopathic cryoglobulinemia** (cryoglobulinemia appearing in patients without evidence of any other disease).

2. A solid-phase technique was later introduced, which proved to be equally simple, but considerably more precise and reproducible.
   a. Purified Clq is immobilized in the wells of a microtiter plate
   b. An adequate dilution of an IC-containing sample is added to the Clq-coated well.
   c. The IC contained in the added sample will be bound to the immobilized Clq.
   d. To determine whether IC are bound to Clq, enzyme-labeled antihuman IgG antibodies are added to the wells; their retention on the plate is directly proportional to the IC captured by the immobilized Clq (Fig. 25.7).

D. **The Raji Cell Assay.** The Raji cell assay is another popular assay for soluble IC.
   1. Raji cells (a lymphoblastoid cell line that can be indefinitely grown in culture) have receptors for C3b, Clq, and for the Fc fragment of IgG (these last ones are quantitatively the least important) through which IC can be bound to their surface.
   2. The cells are incubated with IC-containing samples and ^{125}I-labeled anti-immunoglobulin antiserum (reacting with all major immunoglobulin isotypes) is used to detect cell-bound immunoglobulins, which in their majority will correspond to the antibody moiety of complement-binding IC (Fig. 25.8).
   3. The Raji cell assay has not been successfully introduced in clinical diagnostic laboratories for two main reasons:

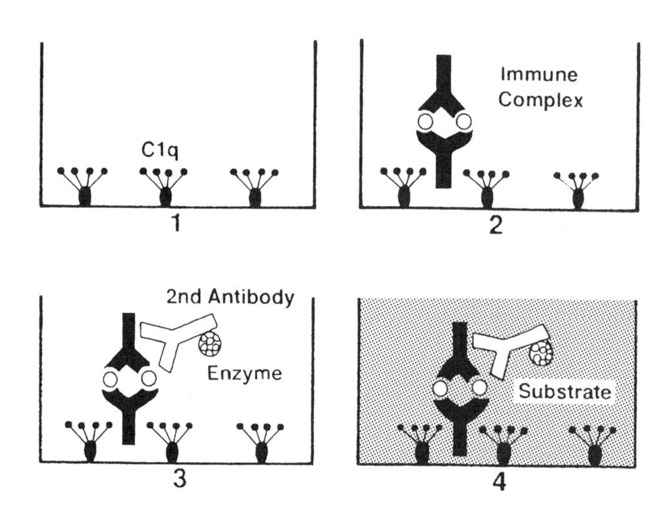

Figure 25.7 Diagrammatic representation of the principle of the solid-phase Clq binding assay. Purified Clq is immobilized by adsorption to a solid phase. IC containing IgG antibodies will bind to the immobilized Clq. After washing off unbound proteins, an enzyme-labeled anti-IgG antibody is added to the solid phase. This antibody will bind to the immobilized immune complexes. After washing off unbound labeled antibody, a colorless substrate will be added to the solid phase. The substrate is broken down into a colored compound, and the intensity of the color is directly proportional to the concentration of labeled antibody bound to the solid phase, which in turn is proportional to the concentration of IC bound to Clq.

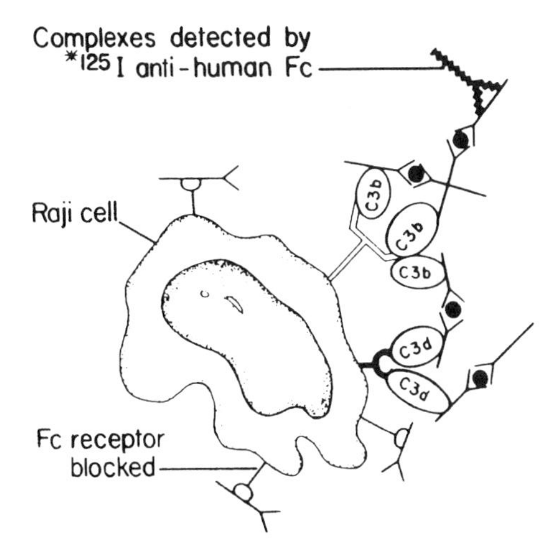

Figure 25.8 Diagrammatic representation of the Raji cell radioimmunoassay. The Raji cell has receptors for the Fc fragment and for C3b and C3d (predominant in number). The Fc receptors account for a background binding and can be blocked with nonhuman IgG. Soluble IC will be bound by the cells through complement components and be detected using a radiolabeled anti-human Fc. (Reproduced with permission from Williams, R.C. *Immune Complexes in Clinical and Experimental Medicine*, Harvard University Press, Cambridge, MA, 1980.)

a. The assay requires the availability of tissue culture facilities and technical expertise to keep a cell line in continuous culture and to check the expression of C3b receptors on the cells.
b. Antilymphocyte antibodies are a recognized cause of false-positive results, and, unfortunately, these antibodies and soluble IC do often coexist in patients with autoimmune disorders.

E. **Specific Immune Complex Screening Tests.** The detection of IC through their antigen or antibody moieties is conceptually very attractive. Indeed, the probability of obtaining false-positive results is considerably reduced when specific tests are used. The main difficulties with this approach are the wide variety of antigen–antibody systems involved and the lack of knowledge concerning the precise antigen–antibody systems involved in a given patient. However, in some cases it has been possible to detect specific antigens and/or antibodies in cryoprecipitates, PEG-precipitates, or other IC-enriched fractions obtained from patient sera, and such studies are very valuable in establishing the etiology of specific types of IC disease.

V. THE ROLE OF IMMUNE COMPLEXES IN HUMAN DISEASE

A. **Classification of Immune Complex Diseases.** Immune complexes have been implicated in human disease either through demonstration in serum or through identification in tissues where lesions are found. Most often, the antibody moiety of the immune complex is detected, and knowledge about the antigens involved is still very fragmentary. However, one of the most common classifications of IC disease, proposed by a WHO-sponsored committee, is based on the nature of the antigens involved (Table 25.4)

B. **Clinical Expression of IC Disease.** The clinical expression of IC disease depends on the target organs where the deposition of IC predominates (see Chap. 24 and 25).
   1. The kidney is very frequently affected (systemic lupus erythematosus, mixed cryoglobulinemia, chronic infections, poststreptococcal glomerulonephritis, purpura hypergammaglobulinemia, serum sickness, etc.), usually with glomerulonephritis as the predominating feature.
   2. The joints are predominantly affected in rheumatoid arthritis.
   3. The skin is affected in cases of serum sickness, mixed cryoglobulinemia, and vasculitis.
   4. The lungs are affected in extrinsic alveolitis.
   5. The reasons why the target organs can vary from disease to disease include:
      a. In some cases, such as in extrinsic alveolitis, the route of exposure to the antigen is a major determinant for the involvement of the lungs.
      b. The kidneys, due to their physiological role and to the existence of C3 and Fc receptors in different anatomical structures, appear to be an ideal organ for IC trapping.
      c. In rheumatoid arthritis, IC appear to be present not only in the circulation, but also to be present (and probably formed) in and around the joints, and although they do not appear to be the initiating factor for

Table 25.4 A Classification of Immune Complex Diseases According to the Antigen Involved

1. **ICD involving endogenous antigens**
   - **Immunoglobulin antigens** (e.g., rheumatoid arthritis, hypergammaglobulinemic purpura)
   - **Nuclear antigens** (e.g., systemic lupus erythematosus)
   - **Specific cellular antigens** (e.g., tumors, autoimmune diseases)
   - **Modified lipoproteins** (e.g., atherosclerosis)
2. **ICD involving exogenous antigens**
   - **Medicinal antigens** (e.g., serum sickness, drug allergy)
   - **Environmental antigens**
     - Inhaled (e.g., extrinsic alveolitis)
     - Ingested (e.g., dermatitis herpetiformes)
   - **Antigens from infectious organisms**
     - Viral (e.g., hepatitis B, hepatitis C, HIV infection)
     - Bacterial (e.g., poststreptococcal glomerulonephritis, subacute endocarditis, leprosy, syphilis)
     - Protozoan (e.g., malaria, trypanosomiasis)
     - Helminthic (e.g., schistosomiasis, onchocerciasis)
3. **ICD involving unknown antigens**

   This category includes most forms of chronic immune complex glomerulonephritis, vasculitis with or without eosinophilia, and many cases of mixed cryoglobulinemia.

the disease, their potential for perpetuating the inflammatory lesions is unquestionable.

d. The reasons for some IC to be trapped in the skin and lead to vasculitis and other dermatological manifestations are unknown.

VI. THERAPEUTIC APPROACHES TO IMMUNE COMPLEX DISEASE

The most common types of therapy in immune complex disease are based on four main approaches: (1) eradication of the source of persistent antigen production (e.g., infections, tumors); (2) turning off the inflammatory reaction (using corticosteroids and nonsteroidal agents); (3) suppression of antibody production (using immunosuppressive drugs such as cyclophosphamide, azathioprine, or cyclosporine); and (4) removal of soluble IC from the circulation by plasmapheresis.

A. **Administration of Anti-Inflammatory and Immunosuppressive Drugs** is the mainstay of IC disease treatment, particularly when autoimmune reactions are its basis.

B. **Plasmapheresis** consists of the removal of blood (up to 5 L each time), separation and reinfusion of red cells, and replacement of the patient's plasma by normal plasma or plasma-replacing solutions. Plasmapheresis appears most beneficial when associated with the administration of immunosuppressive drugs; by itself, it can even induce severe clinical deterioration, perhaps related to changes in the immunoregulatory circuits. The main drawbacks of plasmapheresis are its high cost (derived from the sophisticated equipment used and from the cost of plasma and plasma-replacing products) and the fact that it can only be adequately performed in a well-equipped medical center.

SELF-EVALUATION

Questions

Choose the ONE *best* answer.

25.1 A patient injected with horse antirattlesnake venom serum complains of general weakness, headaches, muscular and joint pains, and notices that his urine is darker in color 10 days after the injection. A urine test shows increased elimination of proteins. Laboratory tests show normal immunoglobulin levels and low serum C4 and C3. The most likely cause for this clinical situation is:

A. Delayed hypersensitivity to horse proteins
B. Deposition of antigen–antibody complexes made of horse proteins and human immunoglobulins
C. Deposition of antigen–antibody complexes made of snake venom proteins and horse antibody
D. Immediate hypersensitivity to snake venom
E. Systemic reaction to snake venom released after the effects of the antitoxin have disappeared

25.2 The binding of IC to RBC is believed to play a physiological role by:

A. Causing cold agglutination
B. Causing hemolysis

C. Delivering IC to the RES
D. Promoting tissue deposition of IC
E. Shortening red-cell half-life

25.3 Which one of the following cytokines and soluble factors is released by activated PMN and contributes to the migration of neutrophils toward an inflamed tissue?
A. C5a
B. Elastase
C. IL-1
D. Platelet-activating factor
E. TNF-α

25.4 The effects of C5a include all of the following **EXCEPT**:
A. Chemotaxis
B. Degranulation of basophils and mast cells
C. Induction of PAF synthesis by endothelial cells
D. PMN aggregation
E. Vasodilation

25.5 Which one of the following procedures or measures is useful for the treatment of immune complex disease?
A. Administration of antihistaminic drugs
B. Blocking phagocyte Fc receptors with intravenous gammaglobulin
C. Inducing neutropenia with nitrogen mustard
D. Lowering complement levels with cobra venom
E. Plasmapheresis

25.6 All of the following are considered as possible manifestations of IC disease **EXCEPT**:
A. Extrinsic alveolitis
B. Glomerulonephritis
C. Immune hemolytic anemia
D. Type I cryoglobulinemia
E. Vasculitis

25.7 An example of IC disease involving exogenous antigens is:
A. Hypergammaglobulinemic purpura
B. Malaria-associated glomerulonephritis
C. Rheumatoid arthritis
D. Sjögren's syndrome
E. Systemic lupus erythematosus-associated glomerulonephritis

25.8 Factors associated with increased risk of development of immune complex disease include all of the following **EXCEPT**:
A. Blockade of the RES
B. Formation of immune complexes at antigen excess
C. Involvement of IgG1 and IgG3 in the formation of IC
D. Release of vasoactive substances
E. Strong immune response against the offending antigen

Questions 25.9 and 25.10 refer to the following case:

A 36-year-old woman with a history of drug abuse, hepatitis C, and treatment with IFN-γ (6 months prior) presents with recent proven streptococcal throat infection, low-grade fever, malaise, migratory joint pains affecting predominantly the knees and elbows, and a non-blanching maculopapular rash distributed over her lower extremities and back.

25.9 The MOST likely diagnosis for this patient is:
A. Infectious arthritis
B. Post-infectious vasculitis
C. Rheumatic fever
D. Rheumatoid arthritis
E. Thrombocytopenic purpura

25.10 Which of the following tests would give the most valuable information about the pathogenesis of this disease?
A. ASO titer
B. Cryoglobulin assay
C. Hepatitis C virus serologies
D. Liver biopsy
E. Liver enzymes

Answers

25.1 (B) This is a classic scenario for serum sickness. The horse antisnake venom serum will initially neutralize the poison and promote its elimination, but enough horse serum proteins will be left in circulation to induce an immune response. As soon as antihorse protein antibodies are secreted, antigen–antibody complexes will be formed, and their deposition will cause pathological symptoms, namely, glomerulonephritis, which is reflected by hematuria and proteinuria.

25.2 (C) RBC-bound IC are kept in the intravascular compartment until taken up by the phagocytic cells of the RES. This is believed to be a physiological protective function.

25.3 (D) Of the five listed mediators, only PAF and elastase are released in large concentrations by neutrophils. Of the two, PAF is the only one involved in promotion of neutrophil-EC interactions, which is the essential first step for the migration of neutrophils toward areas of tissue inflammation.

25.4 (C) The synthesis of platelet-activating factor (PAF) by endothelial cells is triggered by IL-1 and TNF-α.

25.5 (E) Plasmapheresis has been used with some success in the treatment of very severe IC disease, as a way to induce a quick reduction of circulating IC levels. The induction of neutropenia or complement depletion reduces the severity of IC disease in experimental animals, but would be rather risky in humans. Blocking phagocyte Fc receptors is likely to have adverse effects by prolonging the persistence of IC in circulation.

25.6 (D) Type I (monoclonal) cryoglobulinemia is not an expression of IC disease, but rather the laboratory expression of a peculiar tendency for some rare monoclonal proteins to self-aggregate and precipitate at cold temperatures.

25.7 (B) Malaria, such as many other chronic infections, can be associated with glomerulonephritis secondary to the deposition of IC formed by *Plasmodium* antigens and the corresponding antibodies. Thus, it can be considered an example of IC disease involving exogenous antigens.

25.8 (E) A strong immune response is usually associated with formation of medium to large complement-fixing IC, which are rapidly taken up by phagocytic cells, promoting rapid elimination of the antigen without apparent consequences for the host.

25.9 (B) The development of a nonblanching maculopapular rash in a patient with hepatitis C is highly suggestive of postinfectious vasculitis, secondary to the deposition of antigen–antibody complexes. Neither rheumatic fever nor infectious arthritis is usually associated with vasculitis; rheumatoid arthritis has no relationship either with viral hepatitis or with streptococcal infections, and affects predominantly the smaller joints on the hands.

25.10 (B) Around 5% of patients with hepatitis C have mixed cryoglobulinemia that manifests with malaise, arthritis, and skin vasculitis, secondary to the deposition of immune complexes that contain viral antigens, rheumatoid factor, and antivirus C antibodies. Thus, the finding of a mixed cryoglobulin, indicating the presence of circulating immune complexes, has direct relevance to the understanding of the pathogenesis of the pathological process that results in vasculitis and inflammatory arthritis.

BIBLIOGRAPHY

Albelda, S.M., Smith, C.W., and Ward, P.A. Adhesion molecules and inflammatory injury. *FASEB J.*, *8*: 504–512, 1994.

Bielory, L., Gascon, P., Lawley, T.J., Young, N.S., and Frank, M.M. Human serum sickness: A prospective analysis of 35 patients treated with equine anti-thymocyte globulin for bone marrow failure. *Medicine*, *67*:40, 1988.

Brown, E.J. Complement receptors and phagocytosis. *Curr. Opin. Immunol.* *3*:76–82, 1991.

Camussi, G., Tetta, C., and Baglioni, C. The role of platelet-activating factor in inflammation. *Clin. Immunol. Immunopathol.*, *57*:331, 1990.

Collins, T. Adhesion molecules and leukocyte emigration. *Sci. Med.*, *2*(6):28–37, 1995.

Frank, M.M., Hamburger, M.I., Laeley, T.J., Kimberly, R.P., and Plotz, P.H. Defective reticuloendothelial system Fc-receptor function in systemic lupus erythematosus. *N. Engl. J. Med.*, *300*:518–523, 1979.

Hebert, L.A. The clearance of immune complexes from the circulation of man and other primates. *Am. J. Kidney Dis.*, *XVII*(3):352, 1991.

Tsokos G.C. Lymphocytes, cytokines, inflammation, and immune trafficking. *Curr. Opin. Rheumatol.*, *7*:376, 1995.

Virella, G., and Glassman, A.B. Apheresis, exchange, adsorption and filtration of plasma: Four approaches to the removal of undesirable circulating substances. *Biomed. Pharmacotherapy*, *40*:286, 1986.

Wilson, C.B. Immune aspects of renal diseases. *J. Am. Med. Assoc.*, *258*:2957, 1987.

Zimmerman, G.A., Prescott, S.M., and McIntyre, T.M. Endothelial cell interactions with granulocytes: Tethering and signaling molecules. *Immunol. Today*, *13*:93, 1992.

26
Immune System Modulators

Jean-Michel Goust, Henry C. Stevenson-Perez, and Gabriel Virella

I. INTRODUCTION

A. **Immune System Modulators** are agents, principally drugs, which adjust the activity of a patient's immune response, either up or down, until a desired level of immunity is reached. The principal targets of immune modulation are the specific components of the immune response, T- and B-lymphocyte clones, which can hopefully be selectively "fine-tuned" in their function to promote the better health of the patient. Three general clinical scenarios dominate the immunomodulation landscape:
   1. **Immunosuppressive therapies.** These are utilized when specific T and B lymphocytes of the patient's immune system have become activated against the patient's own body organs, such as in autoimmune diseases (see Chap. 18) or in organ transplantation (see Chap. 27).
   2. In patients with IgE-mediated hypersensitivity (see Chap. 23), the immunomodulation goal is to **desensitize** the patient (i.e., to induce a state of hyporesponsiveness of the B lymphocytes that produces IgE in response to a particular environmental antigen ("allergen").
   3. A third modulator option is to attempt to boost the overall B- and T-lymphocyte function of the patient (**immunopotentiation**): this can be accomplished either by actively stimulating the patient's own immune system to higher performance levels through immunization techniques or by passively introducing protective immune system components from outside sources, such as gamma globulin, into the patient's body (see Chap. 30).

B. **Anti-Inflammatory Drugs.** The effects of these drugs do not focus upon the function of T and B lymphocytes, but are rather directed toward changing the function of the nonspecific inflammatory components of the immune system, mononuclear phagocytes, polymorphonuclear granulocytes, natural killer cells, and mast cells. These cells play an essential role in the elimination of infectious agents (Chap. 13) and are also the key to inflammatory processes associated with anti-infectious responses, autoimmunity, or hypersensitivity, which can have devastating effects for the patient.

II. IMMUNOSUPPRESSION

A. **Introduction.** Suppression of the immune response is at present the only efficient therapy in most autoimmune diseases, in the control of transplant rejection, and in other situations in which the immune system plays a significant pathogenic role.

1. Most of the currently used immunosuppressive drugs have a generalized, nonspecific suppressive effect. Some immunosuppressants have effects practically limited to either humoral or cell-mediated immunity, but they still lead to generalized immunosuppression.
2. More recently, a variety of new biological agents have been tried in different immunosuppressive regimens, including monoclonal antibodies to T cells and their subsets, immunotoxins, IL-2-toxin conjugates, anti-idiotypic antibodies, etc., with the goal of developing more specific and effective therapies (see Chap. 27). In many cases, these agents are still in the early stages of evaluation, and it is too soon to issue definite judgments about their usefulness. It is, however, unquestionable that they are the prototypes of approaches that will be more and more used in the near future.

B. **Immunosuppressive Drugs: Pharmacological and Immunological Aspects.** A variety of drugs, ranging from glucocorticoids to cytotoxic drugs, has been used for the purpose of suppressing undesirable immune responses. While many of these drugs are loosely termed "immunosuppressive," they differ widely in their mechanisms of action, toxicity, and efficacy. The exact mechanisms of action of immunosuppressive drugs are difficult to determine, partly because the physiology of the immune response has not yet been completely elucidated. Possible target sites are: phagocytosis and antigen-processing by macrophages; antigen recognition by lymphocytes; differentiation of lymphocytes; proliferation of T and B lymphocytes; production of interleukins and cytokines; and the activation of immune effector mechanisms, including the production and release of cytotoxic leukocytes, antibodies, and/or delayed hypersensitivity mediators.

1. **Hormones.** The major agents in this group of substances are the **glucocorticoid** hormones of the adrenal complex cortex and their synthetic analogs (**glucocorticoids** or **corticosteroids**, such as prednisone and prednisolone). The mechanisms of action of glucocorticoids are still being defined, but they can be divided into three major effects:
   a. **Induction of apoptosis.** At certain dosage levels, treatment with glucocorticoids may produce a rapid and profound lymphopenia. This is particularly true in cases of lymphocytic leukemia and it is a consequence of the induction of apoptosis.
      i. The molecular mechanism of glucocorticoid-induced apoptosis hinges on the activation of an endonuclease, which is normally inactive due to its association to a protein.
      ii. All cellular effects of glucocorticoids, including the induction of apoptosis, requires association with a **glucocorticoid cytoplasmic receptor**. The glucocorticoid-receptor complex is translocated to the nucleus, where it binds to regulatory DNA sequences (**glucocorticoid-responsive elements**).
      iii. The mechanism of activation of the endonuclease involved in

apoptosis by glucocorticoids is uncertain. Two possibilities have been supported by experimental evidence:

- down-regulation of the synthesis of the protein which inactivates the endonuclease
- activation of an ICE-like protease which degrades the inactivating protein.

b. **Down-regulation of cytokine synthesis**. The administration of glucocorticoids is followed by a general down-regulation of cytokine synthesis. This effect is secondary to the inhibition of nuclear binding proteins that activate the expression of cytokine genes. Two mechanisms (not mutually exclusive) have been proposed (Fig. 26.1).

i. After combining with the glucocorticoid cytoplasmic receptor, the glucocorticoid-receptor complex is translocated to the nucleus, where it blocks the association of AP-1 with promoter sequences controlling the expression of cytokine genes.

ii. The translocated glucocorticoid-receptor complex binds to the promoter of the inhibitory protein that regulates the activity of NFκB

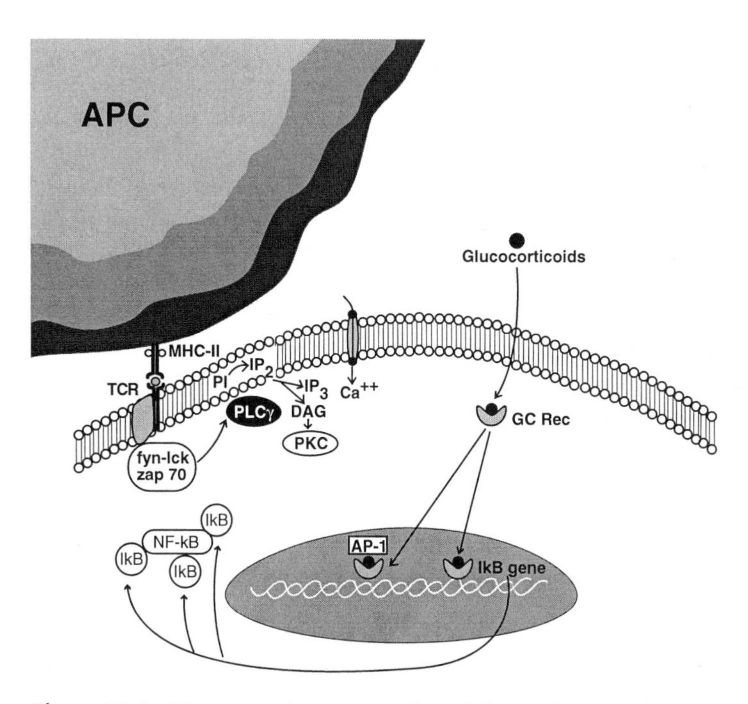

Figure 26.1 Diagrammatic representation of the mechanism of action responsible for the down-regulation of cytokine synthesis by glucocorticoids. Once internalized in the cytoplasm, glucocorticoids combine with their receptor and the complex is translocated to the nucleus, where it interacts with the nuclear binding protein AP-1, preventing its interaction with cytokine gene enhancers, and with a glucocorticoid-responsive element that up-regulates the expression of the IκB gene. The excess of IκB protein prevents the activation of NF-κB, another nuclear binding protein that normally activates the expression of cytokine genes (see Chaps. 4 and 11).

(IκB; see Chap. 4 and 11) and induces its expression. The synthesis of abnormally high levels of IκB results in the inactivation of NFκB, thus neutralizing a second nuclear binding protein that enhances the expression of cytokine genes.

c. **Anti-inflammatory effects.** The anti-inflammatory effect of glucocorticoids is probably the most significant from the pharmacological point of view. Several actions of glucocorticoids combine to induce this anti-inflammatory effect.
   i. Down-regulation of the synthesis of proinflammatory cytokines.
   ii. Reduced expression of CAMs on the vessel wall, partly as a consequence of the down-regulation of cytokine synthesis (proinflammatory cytokines up-regulate CAM expression), partly as a consequence of a direct down-regulation of the expression of those molecules. The modulation of CAM expression markedly alters the trafficking of human immune system cells, particularly neutrophils and T lymphocytes, which remain sequestered on the bone marrow and lymph nodes, impairing their ability to generate both specific immune responses and nonspecific inflammatory responses.
   iii. The synthesis of phospholipase A_2 is down-regulated due to the binding of the glucocorticoid-receptor complex to a DNA glucocorticoid-responsive element that has a negative effect on the expression of the phospholipase A_2 gene. Consequently, the synthesis of leukotrienes, prostaglandins, and platelet activating factor is down-regulated.
   iv. The nitric oxide synthetase gene is also down-regulated; thus, the release of nitric oxide is reduced, eliminating its vasodilator effect.
   v. In contrast, glucocorticoids up-regulate the expression of lipocortin-1, a protein that has anti-inflammatory effects in part due to its ability to inactivate preformed phospholipase A_2.

d. **Nonsteroidal anti-inflammatory drugs (NSAID)** are a cluster of diverse acid molecules, totally unrelated to the corticosteroid family, which share certain anti-inflammatory properties.
   i. Examples of NSAID in clinical use today include aspirin, ibuprofen, and indomethacin.
   ii. The anti-inflammatory potency of each member of the NSAID family appears to correlate directly with its ability to inhibit prostaglandin synthesis, chiefly through inhibition of cellular cyclooxygenase activity.
   iii. The NSAID do not appear to modulate cytokine release or to alter T- and B-lymphocyte cell trafficking, and thus have virtually no immunosuppressive effects.

2. **Alkylating agents and antimetabolites.** Several commonly used immunosuppressants fit into this category. They can be generically classified into two large groups: *cycle-active* and *noncycle-active*.
   a. **Cycle-active agents** may kill by acting at different parts of the cycle; antimetabolites such as **methotrexate, azathioprine,** and **6-mercaptopurine (6-MP)** appear to act only on cells in the S-phase, when DNA is actively synthesized. In a nonstimulated lymphocyte population, many

of the cells are not in cycle, but rather in a prolonged G_1 or "G_0" (totally resting) state. These cells are not affected by exposure to cycle-active agents.

b. **Alkylating agents** such as **cyclophosphamide** or **x-rays**, although able to kill cells *in cycle* to a greater degree than cells not in cycle, can also kill nondividing cells.

c. **Immunosuppressive effects.** The three most commonly used cytotoxic drugs—cyclophosphamide, azathioprine, and methotrexate—all suppress primary and secondary humoral immune responses, delayed hypersensitivity, skin graft rejection, and autoimmune disease in animals. However, some striking differences in the mechanism of action of these three agents have become apparent (see Table 26.1).

   i. In studies of the effects of cyclophosphamide, methotrexate, and 6-MP on antibody production in mice, one can compare the dose that kills 5% of the animals within 1 week (LD5) with the dose required to reduce the antibody response of the mice by a factor of 2 (inhibitory dose; ID2); a therapeutic index (TI) can be calculated which is defined as the ratio of the two doses (LD5/ID2). Cyclophosphamide has the highest therapeutic index followed by methotrexate and 6-MP (Table 26.2).

   ii. Sharply different effects of azathioprine and cyclophosphamide on humoral antibody production have been demonstrated in patients, using flagellin as a test antigen. There was a significant suppression of antibody response to flagellin in cyclophosphamide-treated patients, while the responses of azathioprine-treated patients did not differ significantly from those of nontreated control patients.

   iii. Several investigators have also shown that cyclophosphamide can decrease the production of anti-DNA antibodies, both in NZB mice and humans. This suggests that cyclophosphamide can inhibit an ongoing immune response, whereas azathioprine and 6-MP cannot. This, of course, is the situation that one faces in the treatment of

Table 26.1 Summary of Effects of Drugs with Alkylating and Antimetabolite Activity[a]

Effect	Cyclophosphamide	Azathioprine and 6-MP	Methotrexate
Reduced primary immune response	++	++	++
Reduced secondary immune response	++	±	+
Reduced immune complexes	++	0	0
Anti-inflammatory effect	+	++	+
Mitostatic effect	++	++	++
Reduced delayed hypersensitivity	++	+	+
Suppression of passive transfer of cellular immunity	++	±	±
Lymphopenia	++	±	±
Facilitation of tolerance induction	++	+	+

[a]On the basis of a combination of experimental and clinical data.

Table 26.2 Therapeutic Indices of Cytotoxic Agents Inhibiting Antibody Production

Agent	LD5[a]	ID2[b]	TI[c]
Cyclophosphamide	300.0	50.0	6.0
Methotrexate	6.3	1.25	5.0
6-Mercaptopurine	240.0	100.0	2.4

[a]LD5 dose (in mg/kg) killing 5% of animals within 1 week.
[b]ID2 dose (in mg/kg) lowering antibody titer to $1/2^2$.
[c]Therapeutic index = LD/ID.

patients with autoimmune disease, since the relevant immune responses are already established by the time they are recognized and treated. In patients with systemic lupus erythematosus (SLE), cyclophosphamide treatment can reverse the deposition of immune complexes in the dermoepidermal junction (which correlates with renal disease), whereas steroid therapy alone does not.

iv. Studies of the effects of these drugs on cellular immunity have shown that all three depress cellular immunity. However, comparative studies show a greater effect with cyclophosphamide. While both cyclophosphamide and methotrexate are more effective than 6-MP in suppressing a PPD skin test in experimental animals, only cyclophosphamide depletes thymus-dependent areas of the lymph nodes. The in vitro response of lymphocytes to PHA and other mitogens is likewise inhibited only by cyclophosphamide. In addition, tolerance induction is much easier to achieve in mice treated with cyclophosphamide than with azathioprine or methotrexate.

3. **Inhibitors of signal transduction: Cyclosporine A (CsA), tacrolimus (Prograf, FK506), and rapamycin (Sirolimus).** These compounds are fungal metabolites with immunosuppressive properties. Structurally they are macrocyclic compounds; cyclosporine A (CsA) is a macrocyclic peptide while FK506 and rapamycin are macrocyclic lactones (macrolides). All are virtually devoid of toxicity for leukocyte precursors, and, hence, do not cause leukopenia or lymphopenia. Their molecular mechanism of action depends on their binding to cytoplasmic proteins, which appear to be involved in the process of signal transduction essential for lymphocyte activation and/or proliferation (Fig. 26.2). Because of their ability to bind immunosuppressive compounds, these proteins are collectively known as **immunophilins**.

   a. **Cyclosporine A** is cyclic undecapeptide obtained from *Tolypocladium inflatum*. It has a uniquely selective effect on T lymphocytes, suppressing humoral T-dependent responses and cell-mediated immune responses.

      i. The **mechanism of action of CsA** involves its high-affinity binding to **cyclophilin**. The CsA-cyclophilin complex binds and inactivates **calcineurin**. This protein, upon activation, acquires phosphatase proteins and, in turn, activates NFAT by dephosphorylation. NFAT is a nuclear binding protein involved in the control of the expression of the IL-2 gene and other cytokine genes (see Chap. 11). As a consequence of the inactivation of calcineurin, Ca^{2+}-associated T-cell activation pathways, such as those triggered by anti-CD3 antibodies or the occupancy of the TcR, are inhibited.

ii. As a consequence of the block of a critical step cell in the activation pathway, there is a general **down-regulation of the production of IL-2, interferon-γ, IL-3, IL-4, GM-CSF, and TNF-α**. The expression of the CD40 ligand is also down-regulated. Helper (CD4+) T cells are the chief cellular target for CsA; T cells with suppressor activity, on the other hand, appear to proliferate at higher rates. This differential effect is reflected in humans by a reversal of the CD4/CD8 ratio rates and by a relative increase in suppressor lymphocyte function. However, the activation of cytotoxic T lymphocytes is also inhibited, apparently due to both the lack of stimulatory signals provided by IL-2 and interferon-γ and to a direct inhibitory effect on cytotoxic T-cell precursors.

iii. CsA has a remarkable ability to **prolong graft survival**. In experimental animals, even short courses of CsA can result in significant prolongation of kidney graft survival, suggesting that the drug facilitates the development of low-dose tolerance. In humans, used alone or in association with steroids, it reduces the number of rejection episodes in renal transplantation, even in patients with cytotoxic antibodies and receiving poorly matched organs. It also induces a substantially longer survival of kidney, liver, and especially heart transplants, and reduces the incidence and severity of graft-versus-host disease in bone marrow transplantation.

iv. The main advantages of CsA are its **selective effect on helper T lymphocytes** and its excellent **steroid-sparing effect**. The dosages of steroids necessary to achieve effective immunosuppression are considerably lower when steroids are associated to CsA than when they are associated to other immunosuppressive drugs. As a result, the incidence of infection is substantially reduced, although cytomegalovirus infections are relatively common in CsA-treated patients.

v. CsA has many serious **side effects**, including tremors, hypercholesterolemia, increased hairiness, headaches, swelling of the gums, etc.
   - The most serious side effect is **nephrotoxicity**, a major concern in patients with kidney grafts. The renal toxicity is frequently associated with hypertension, which, in turn, has a negative impact in all patients, but especially in those receiving a heart transplant.
   - The de novo appearance of **lymphoproliferative syndromes** after CsA therapy, by itself or in combination with other immunosuppressive agents (e.g., antilymphocyte globulin), is another significant problem.
   - **Accelerated atherosclerosis** has been observed in heart trans-

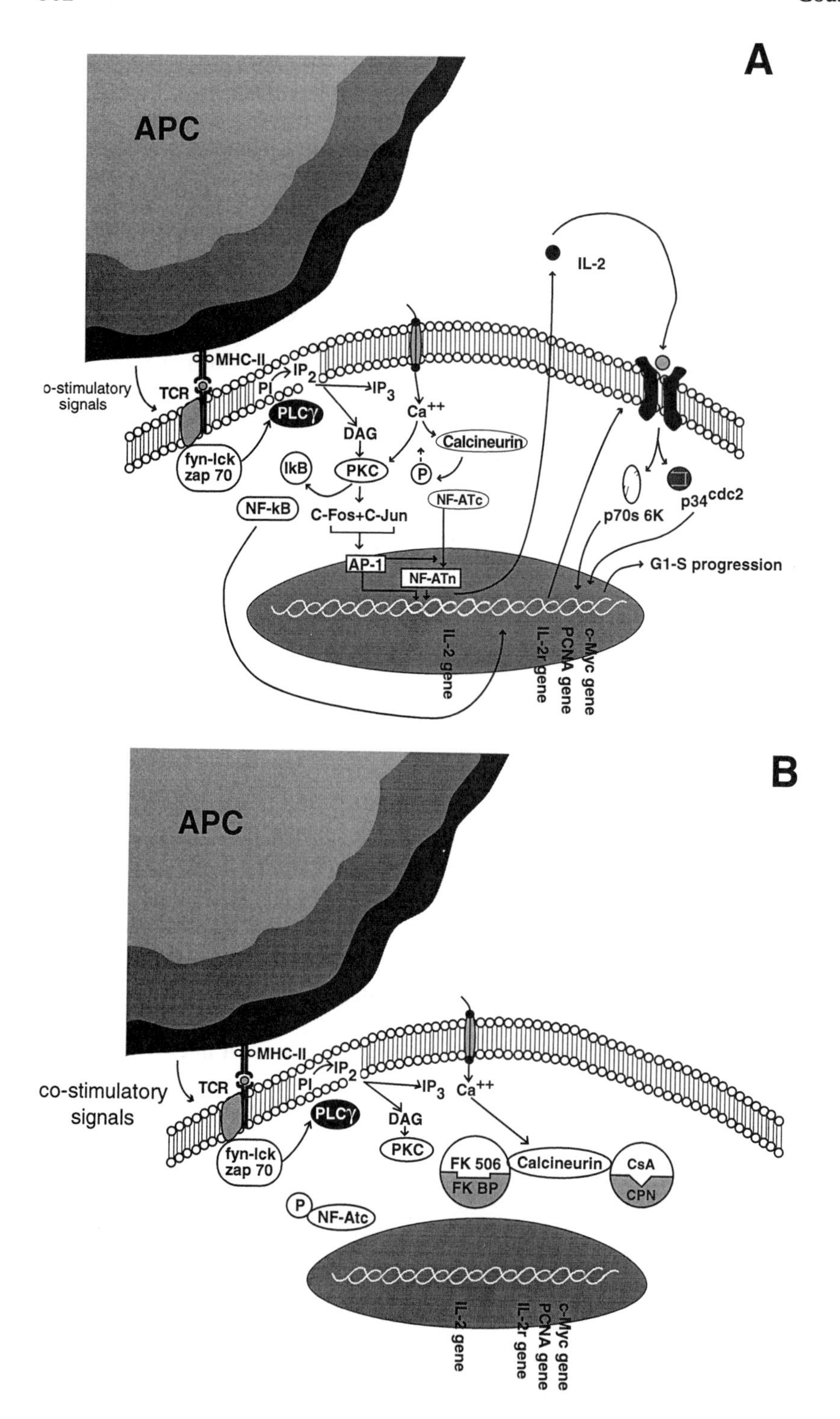
A
APC
IL-2
MHC-II
o-stimulatory signals
TCR
PI
IP2
IP3
PLCγ
Ca++
DAG
Calcineurin
fyn-lck zap 70
IkB
PKC
P
NF-ATc
NF-kB
C-Fos+C-Jun
p70s 6K
p34cdc2
AP-1
NF-ATn
G1-S progression
IL-2 gene
IL-2r gene
PCNA gene
c-Myc gene
B
APC
MHC-II
co-stimulatory signals
TCR
PI
IP2
IP3
Ca++
PLCγ
DAG
PKC
fyn-lck zap 70
FK 506
FK BP
Calcineurin
CsA
CPN
P
NF-Atc
IL-2 gene
IL-2r gene
PCNA gene
c-Myc gene

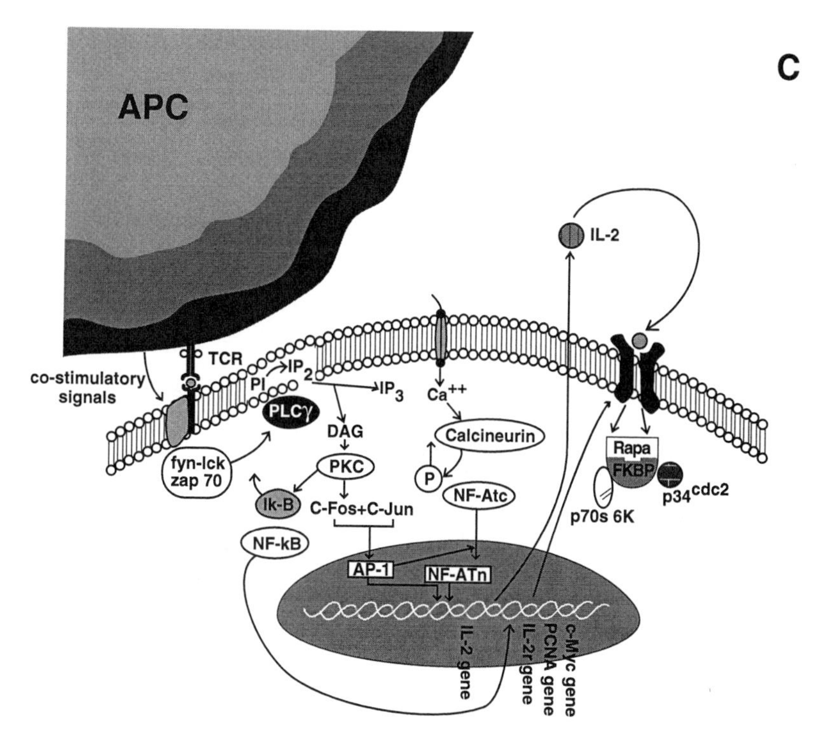

Figure 26.2 Diagrammatic representation of the mechanism of action of the inhibitors of signal transduction. Panel A illustrates the main steps of the calcineurin activation pathway and the autocrine activation pathway mediated by IL-2, discussed in detail in Chapter 11. Panel B illustrates the mechanism of action of cyclosporine A and of tacrolimus (FK506). Cyclosporine A associates with cyclophilin (CPN) and the resulting complex binds and inactivates calcineurin. As a consequence, NFAT remains phosphorylated and inactive. Tacrolimus achieves the same results, but it associates to a different cytoplasmic protein—FKBP. Panel C illustrates the effect of rapamycin. This macrolide combines with the same cytoplasmic protein as tacrolimus (FKBP), and the complex inactivates two protein kinases, p70 s6 and $p34^{cdc2}$, activated as a consequence of the interaction of IL-2 and IL-4 with their respective receptors, and are required for cells to progress from G1 to S.

plant recipients surviving for over 2 years, but the role played by CsA in this complication is not clear.

vi. Recent observations suggest that the long-term immunosuppression achieved with associations of low-dose CsA, steroids, and azathioprine may be preferable to long-term administration of high doses of CsA, at least in patients with kidney transplants.

b. **Tacrolimus (Prograf, FK506)** is produced by a different fungus (*Streptomyces tsukubaensis*) with a similar mechanism of action to CsA but is 10 to 100 times more active.

i. **Mechanism of action.** The cytoplasmic target of tacrolimus is a different protein, known as **FKBP** (**FK**506 **B**inding **P**rotein). The tacrolimus-**FKBP** complex has a very similar effect to the cyclophilin-CsA complex, inhibiting calcineurin and preventing the activation of NF-AT. Not surprisingly, the cellular effects of FK506 and CsA are almost identical.

ii. Clinically, tacrolimus has been mainly used to date in **liver transplantation**. It was first used to reverse rejection in patients unresponsive to other immunosuppressive agents. Its later use as primary immunosuppressant followed and resulted in improved patient survival.

iii. Its side effects are similar to those of CsA. Neurotoxicity (including rare cases of severe irreversible encephalopathy), gastrointestinal intolerance, and infections (particularly by cytomegalovirus) are the most prominent complications of its use.

c. **Rapamycin (Sirolimus)**, produced by *Streptomyces hygroscopicus*, is structurally similar to tacrolimus, but has a different intracellular target and different pharmacological properties.

i. **Mechanism of action.** Rapamycin binds to the same cyclophilin (FKBP) as tacrolimus, but the complex formed by cyclophilin and FKBP (and probably other proteins) does not interact with calcineurin, but rather with other cellular targets. The best characterized targets are two kinases, p70 s6 and $p34^{cdc2}$, whose activation appears to be mediated by the interaction of IL-2 and IL-4 with their respective receptors, and are required for cells to progress through the replication cycle.

ii. While CsA and FK506 inhibit the transition of lymphocytes from G_0 to G_1, rapamycin inhibits cell division later in G_1, prior to the S phase. Also, rapamycin inhibits both Ca^{2+} -dependent and independent activation pathways, does not inhibit IL-2 synthesis, but inhibits the response of IL-2- and IL-4-sensitive cell lines to exogenous IL-2 or IL-4.

iii. Because of the different mechanism of action, rapamycin is effective even when added 12 hours after in vitro mitogenic stimulation of T cells, while CsA and FK506 are only effective when added to the cultures not later than 3 hours after the mitogen.

iv. Rapamycin has been successfully used for prevention of graft rejection in experimental animals, and is currently the object of clinical trials in human transplantation.

C. Use of Immunosuppressive Drugs in Hypersensitivity and Autoimmune Diseases

1. **Glucocorticoids.** Glucocorticoid administration can be life saving in certain acute disorders, such as bronchial asthma and autoimmune thrombocytopenic purpura and can induce significant improvement in chronic warm autoantibody hemolytic anemia, autoimmune chronic active hepatitis, autoimmune thrombocytopenic purpura, systemic lupus erythematosus, and a variety of chronic hypersensitivity conditions. Steroids are also part of most immunosuppressive regimens used for preventing the rejection of transplanted organs.

2. **Cytotoxic agents and cyclosporine A**

a. Many non-neoplastic diseases that are either proven or presumed to be immunologically mediated have been treated with cytotoxic drugs. Results of controlled trials of **azathioprine** and **cyclophosphamide** suggest that both drugs, when given in sufficient quantity, may be capable of suppressing disease activity and eliminate the need for long-term therapy with steroids.

b. **Cyclophosphamide** has been demonstrated to be the only effective means of achieving immunosuppression (and sometimes clinical cure) in certain steroid-resistant diseases, such as Wegener's granulomatosis. Cyclophosphamide is also the drug of choice for the treatment of lupus glomerulonephritis and other vasculitides (see Chap. 24).

c. **Azathioprine** has also been used in the treatment of patients with SLE. Controlled studies demonstrated a number of beneficial effects (i.e., an increase in creatinine clearance, a decrease in proteinuria, and a decrease in mortality). However, a decrease in glomerular cell proliferation has been noted in renal biopsies of SLE patients receiving azathioprine and upon discontinuation of treatment severe exacerbations of the disease have been reported.

d. **Cyclosporine A** has not been as widely used in the treatment of autoimmune disorders as azathioprine and cyclophosphamide, with the exception of type I (insulin-dependent) diabetes and myasthenia gravis. In these conditions, considerable clinical improvement may be seen while the drug is being administered, but relapses occur as soon as CsA is suspended.

e. **Combinations of glucocorticoids and cytotoxic agents** have been used in most diseases that were classically treated with glucocorticoids alone, and although controlled trials are still required to assess overall benefit in many of these diseases, it should be stated that perhaps their major advantage is the possibility of reducing the dose of steroids when such drugs are added to corticosteroid therapy—the previously mentioned "steroid-sparing" effect.

D. Adverse Consequences of Prolonged Immunosuppression

1. **Bone marrow suppression** is the most common type of toxicity. It is due to the effects on the bone marrow and hemopoiesis, particularly when cytotoxic drugs are used to suppress the immune response.

   a. The degree of bone marrow suppression observed with cytotoxic drugs is usually dose related and can be modulated by dose changes, although in rare cases the bone marrow failure may become irreversible.

   b. When **neutropenia** develops, severe pyogenic and fungal infections are likely to develop; these infections are extremely difficult to treat, often being the cause of death. For this reason, **neutropenia is considered as the most serious side effect of immunosuppression**, and continuous monitoring of white-cell count is essential in patients treated with these drugs. The availability of recombinant CSF and GM-CSF provides the means to considerably shorten the period during which a patient remains neutropenic (see below).

2. **Infections** are another common side effect in patients treated with all types of immunosuppressive drugs.

   a. The most frequent manifestation of global immunosuppression due to cytotoxic or immunosuppressive drugs is the inability to mount a primary immune response after adequate immunization.

   b. Two main features characterize the infections of immunosuppressed patients.

      i. Usually involve low-grade pathogens or opportunistic microorganisms not usually associated with clinical disease. For example, infections by *Pneumocystis carinii*, *Candida albicans*, and viruses of the herpes group are unusually frequent in severely immunosuppressed patients

ii. The extent and distribution of the infection are unusual, differing from those commonly observed in noncompromised hosts. Systemic candidiasis, measles encephalitis, measles retinitis, and cerebral toxoplasmosis are but a few examples of atypical infections almost exclusive to immunocompromised patients.

c. Because the depression of cellular immunity is the goal pursued when these drugs are used, it makes patients more vulnerable to viral infections, such as herpes simplex and varicella, which may disseminate with a fatal outcome. The incidence of herpes zoster (shingles) is also increased, but the course of the disease is similar to that seen in otherwise normal individuals. The impairment of cell-mediated immunity is also probably responsible for the severity of mycobacterial and fungal infections (which are much more likely to disseminate during immunosuppressive treatment), and for opportunistic infections with cytomegalovirus, *Pneumocystis carinii*, and fungi such as *Candida* and *Aspergillus*.

3. **Increased incidence of neoplasms** is a major concern in patients chronically immunosuppressed. Although the precise role of the immune system in eliminating neoplastic clones in a normal individual is not clear, the incidence of malignancies is clearly elevated in patients receiving immunosuppressive drugs. The most frequently seen malignancies in immunosuppressed patients are skin cancers and lymphoreticular neoplasms. Also, the location and pattern of spread of those malignancies is unusual. For example, intracerebral lymphoma is virtually restricted to immunosuppressed patients, where it occurs in frequencies as high as 50% in certain immunosuppressed populations. The intracerebral lymphomas have a highly malignant evolution and are always B-cell lymphomas expressing Epstein-Barr virus. It has been postulated that they emerge in individuals carrying the virus in a latent stage as a consequence of a deficient immunosurveillance that normally would control the proliferation of viral-infected B lymphocytes.

4. **Other side effects** include hair loss or alopecia and major constitutional symptoms (e.g., nausea, vomiting, anorexia, malaise, etc.), chromosomal changes and teratogenic effects.

III. IMMUNOPOTENTIATION

A. **Introduction.** Many compounds and biological substances have been used in attempts to restore normal immune system function in clinical conditions in which it is believed to be functionally altered. All of these types of therapeutic interventions fall under the general designation of **immunopotentiation**, which can be defined as any type of therapeutic intervention aimed at restoring the normal function of the immune system.

B. **Biological Response Modifiers (BRM)** is a term used to designate a variety of soluble compounds that allow the various elements of the immune system to

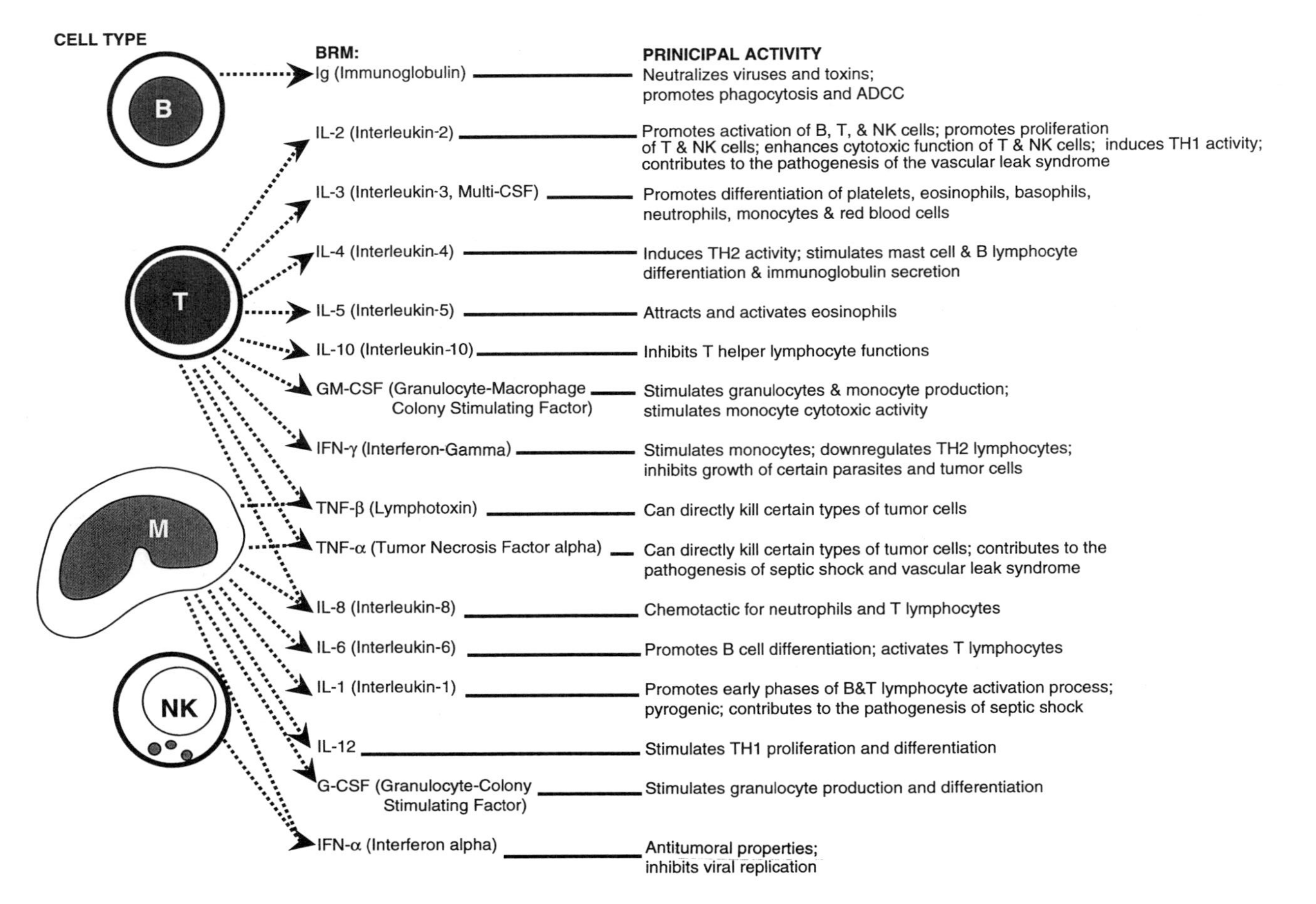

Figure 26.3 Diagrammatic representation of the source and principal mechanisms of action of the most important biological response modifiers secreted by mononuclear leukocytes.

communicate with one another. This communication network allows for "up-regulation" and coordination of immune responses when needed, and "down-regulation" of immune responses when no longer needed by the host. Figure 26.3 lists the most important of the BRM compounds which have been studied and which appear to have some clinical applicability.

1. **Structure.** All BRM have a surprising degree of structural similarity to one another; this is felt to be reflective of an early gene reduplication event that occurred as the immune system of mammals evolved.
2. **Function.** The BRM appear to form a very delicate network of communication signals between the four principal mononuclear leukocyte subsets that participate in the immune response. We are still without a complete understanding as to how these interactions occur; however, the release of these signals appears to be different from hormones. Hormones tend to act at a site far distant from the original cell that secreted the signal; in contrast, BRM appear to work in the immediate vicinity of the cell type that secretes them. The timing of the release of the BRM coupled with the intensity of their secretion appears to provide the overall balance of signals required to orchestrate the immune response.
3. **Cellular sources**
   a. **T lymphocytes** produce a wide range of BRM.
      i. **Interleukin-2** (**IL-2**) is a factor that induces the proliferation of T lymphocytes, B lymphocytes, and NK cells. This factor is also capable of up-regulating the tumor-killing capabilities of T lymphocytes and NK cells.
      ii. **Interleukin-3** (**IL-3**) is a growth factor for stem cells in the bone marrow that stimulates the production of many types of leukocytes from the bone marrow when needed.
      iii. **Interleukin-4** (**IL-4**) is the main determinant of TH2 differentiation; it also stimulates B lymphocytes to proliferate and induces the synthesis of IgE antibodies.
      iv. **Interleukin-5** (**IL-5**) has chemotactic and activating properties for eosinophils.
      v. **Interleukin-10** (**IL-10**) down-regulates the synthesis of other interleukins and turns off both TH1 and TH2 activities.
      vi. **Granulocyte–monocyte colony simulating factor** (**GM-CSF**) stimulates the bone marrow to produce both granulocytes and monocytes. It also appears capable of up-regulating the spontaneous killing capability of monocytes.
      vii. **Interferon-γ** (**IFN-γ**) is a protein with moderate antiviral activity, which is the main mediator responsible for the activation of monocytes and macrophages.
      viii. **Tumor necrosis factor β** (**TNF-β**, **lymphotoxin**) is a molecule secreted by cytotoxic T lymphocytes, which appears capable of killing tumor cells directly, perhaps by making their cell membranes overly permeable.

   b. **Natural killer cells** also secrete a series of BRM, particularly **interferon-α** (**IFN-α**), an antiviral agent that promotes the activation of NK lymphocytes in vitro. Its precise physiological significance may include autocrine or paracrine stimulation of NK cells and, in addition, interferon-α appears capable of directly inhibiting the growth of certain types of tumor cells, such as the cells that proliferate in **hairy cell leukemia** (see Chap. 28).

c. **Mononuclear phagocytes (monocytes and macrophages)** share with NK lymphocytes the ability to secrete **interferon-α**. In addition, they secrete several types of interleukins and cytokines. Some of them (IL-6, IL-8, and TNF-α) are also produced by T lymphocytes.
   i. **Interleukin-1 (IL-1)**, also produced by other antigen-presenting cells, including B lymphocytes, promotes the early phases of the B- and T-lymphocyte activation processes.
   ii. **Interleukin-6 (IL-6)** promotes B-lymphocyte differentiation and immunoglobulin secretion.
   iii. **Interleukin-8 (IL-8)** is a chemotactic factor for lymphocytes and neutrophils.
   iv. **Interleukin-12** is responsible for the activation and differentiation of TH1 lymphocytes.
   v. **Tumor necrosis factor-α (TNF-α)** has a variety of effects, particularly causing the death of certain types of tumor cells.
   vi. Two types of **colony stimulating factors**, **G-CSF** and **M-CSF**. **M-CSF** promotes the production and activation of monocytes, and **G-CSF** stimulates the production of granulocytes from stem cells in the bone marrow.

4. **Clinical applications**
   a. G-CSF and GM-CSF, which are produced by recombinant DNA technology, have been approved by the FDA for use in patients with bone marrow depression, such as cancer patients and patients who receive bone marrow transplants. In these patients, bone marrow depression is usually associated with intensive chemotherapy or irradiation, and the use of colony stimulating factors avoids the profound and prolonged neutropenia seen in controls not receiving these compounds, with a corresponding decrease in morbidity and mortality. On the other hand, administration of G-CSF or GM-CSF is expensive and can be associated with severe side effects.
   b. Also, encouraging results have been obtained with several BRM as primary agents in the treatment of a variety of malignancies (see Chap. 28).
   c. Efforts to enhance cellular immunity in immunodeficient patients often meet with negative or conflicting results. For example, efforts to correct the profound immunological dysfunction characteristic of the Acquired Immunodeficiency Syndrome (AIDS) have met with limited success.
      i. As discussed in detail in Chapter 30, the profound immunodepression that develops in AIDS patients is believed to be secondary to the infection of helper T lymphocytes and monocytes by the causative retrovirus, HIV-1. Nonspecific therapies aimed at restoring the depressed immune function of AIDS patients have included the administration of:
         - Colony-stimulating factors (e.g., **GM-CSF**) to boost the pro-

duction and release of noninfected leukocytes from the bone marrow.

- IL-12 to expand the numbers and augment the function of TH1 lymphocytes.

ii. In vitro data suggest that immunostimulation should benefit the patients if viral replication could be simultaneously controlled and several groups of investigators are carrying out clinical trials with different combinations of antiviral drugs and BRM (particularly IL-2), hoping to find the optimal combination that will benefit the patients.

C. **Active Immunization as an Immunomodulating Intervention.** Active immunization can be used for prevention of infectious diseases, as discussed in detail in Chapter 12, or to enhance the immunological defenses in patients already infected. The immunotherapeutic use of vaccines has been the object of experimental protocols in two areas:

1. **HIV infection.** The administration of low doses of killed HIV vaccine is currently being tried in HIV-positive individuals, with the goal of stimulating the activity of TH1 helper lymphocytes, believed to be essential for the differentiation of cytotoxic T lymphocytes with antiviral activity (see Chap. 30). So far these attempts do not seem to be successful.
2. **Cancer.** Anticancer vaccines have been the object of considerable interest and are also the object of ongoing trials (see Chap. 28).

Vaccines to stimulate resistance against antibiotic-resistant bacteria are also under development.

D. **Dialyzable Leukocyte Extracts and Transfer Factor**

1. In a series of classic experiments, Lawrence showed that the injection of an extract of lymphocytes from a tuberculin-positive donor to a tuberculin-negative recipient resulted in acquisition of tuberculin reactivity by the latter. Lawrence coined the term *transfer factor* to designate the unknown agent responsible for transfer of tuberculin sensitivity. Nowadays, it is known that this activity is contained in the dialyzable (low-molecular-weight) fraction of a leukocyte extract, so the term **dialyzable leukocyte extract** (DLE) is preferred. The term **transfer factor** is reserved for DLE components with antigen-specific activity of which we have limited knowledge.
2. In normal human subjects, injection of DLE has been shown to transfer both skin test reactivity (delayed hypersensitivity) and the ability to produce various lymphokines in vitro in the presence of the same antigen(s) to which the donor of the leukocytes responded. Crude DLE preparations have been shown to contain approximately 160 separate moieties. The mode(s) of action of DLE and transfer factor remain undetermined. The chemical nature of the mediator(s) remains poorly characterized, and the gene(s) responsible for its synthesis has/have not been identified. The lack of chemically defined products and the fact that most reports of the clinical use of DLE and transfer factor are based on isolated case observations or on small, uncontrolled series have impacted negatively on their general acceptance as therapeutically useful BRMs.
3. Transfer factor has been tried in HIV-infected patients as a way to induce adoptive immunity.

E. **Thymic Hormones.** Immunotherapeutic applications of thymic hormones have received increasing attention in recent years. Many peptides with thymic hormonelike activity have been isolated and described, including thymosin, *facteur thymique serique* (FTS, serum thymic factor), and thymopoietin.

1. **Thymosin** is a mixture of different peptides (thymosin α_1, thymosin β_3, thymosin β_4, etc.) with different biological activities.
   a. Thymosin α_1, the most widely studied of the thymic hormones, promotes T-cell differentiation and also may have other effects on cell-mediated immunity.
   b. β_3 and β_4 thymosins promote early stem-cell differentiation to prothymocytes.
   c. A number of laboratories have isolated these different thymic hormones, and some of the fractions have been purified, sequenced, and chemically synthesized. Thymosin α_1 has been biologically synthesized with the use of recombinant DNA.
   d. Although animal experiments have shown little or no toxicity, reactions to thymosin α_1 have been reported in a few patients, presumably to bovine-specific antigens since thymosin of bovine origin is used generally for immunotherapy.
   e. Thymic hormones have been used most extensively in patients with acquired or congenital T-cell defects, and dramatic improvements have been reported in some instances. In some patients with congenital T-cell deficiency, combined therapy using both thymosin and DLE appears to be beneficial, even though thymosin or DLE alone is without effect.
   f. Thymus extracts also are being tried in some "autoimmune diseases" such as systemic lupus erythematosus, since it appears that they can induce the proliferation and differentiation of suppressor T cells in patients with apparent deficiencies in suppressor activity and, conversely, can cause a reduction in suppressor activity (or an increase in helper activity) in some hypogammaglobulinemic patients.

F. **Bacterial and Chemical Immunomodulators**

1. **Bacterial immunomodulators.** Killed bacteria and several substances of bacterial origin are capable of activating the immunological function of mononuclear leukocyte subset cells.
   a. **Bacillus Calmette-Guérin (BCG)** and ***Corynebacterium parvum*** have been extensively used for their adjuvant therapy in therapeutic protocols aimed at stimulating antitumoral immunological mechanisms. At the cellular level, these bacteria appear mainly to activate macrophages.
   b. **Muramyl dipeptide (MDP)**, a dipeptide moiety extracted from the cell walls of *Mycobacterium tuberculosis*, stimulates both monocytes and T lymphocytes. It has been tried in HIV-infected patients as a nonspecific immunostimulant.
2. **Chemical immunomodulators**
   a. **Levamisole** is an antihelminthic drug used in veterinary medicine that has been found to have immunostimulantory properties. It immunopotentiates the graft-versus-host reaction in rats, and in some animal diseases causes an apparent increase in host resistance to tumor cells. It

acts on the cellular limb of the immune system, and can restore impaired cell-mediated immune responses to normal levels but fails to hyperstimulate the normal functioning immune system. Thus, it shows true immunomodulatory activity.

i. The primary **mechanism of action** of levamisole may be to facilitate the participation of monocytes in the cellular immune response, apparently by enhancing monocyte chemotaxis. In addition, it has been reported to increase DNA synthesis of T lymphocytes and to augment their proliferative responses to mitogens, as well as their production of mediators of cellular immunity in vitro.

ii. **Clinical applications.** In humans, levamisole has been reported to restore delayed hypersensitivity reactions in anergic cancer patients and to be of some benefit in the treatment of aphthous stomatitis, rheumatoid arthritis, systemic lupus erythematosus, viral diseases, and chronic staphylococcal infections. Its most promising application appears to be in patients with resected colorectal carcinomas. It increases the duration of disease-free intervals.

iii. **Side effects.** Levamisole has a variety of side effects; patients may complain of nausea, flu-like malaise, and cutaneous rashes disappearing after cessation of therapy. The most serious side effect is agranulocytopenia, which is reversed upon therapy termination, but white cell counts should be monitored in patients taking the drug for prolonged periods.

b. **Isoprinosine** (**Inosine Prabonex**, **ISO**) is a synthetic immunomodulatory drug recently approved for clinical use in the United States. It appears to be effective in a wide variety of viral diseases.

i. **Mechanisms of action**
- In vitro experiments have shown that ISO inhibits the replication of both DNA and RNA viruses in tissue culture, including herpes simplex, adenovirus, and vaccinia (DNA viruses), poliovirus, influenzae types A and B, rhinovirus, ECHO, and Eastern equine encephalitis (RNA viruses).
- Isoprinosine potentiates cell-mediated immune responsiveness in vivo, and a major factor in its effectiveness against viral infections appears to be its ability to prevent the depression of cell-mediated immunity that has been shown to occur during viral infection and to persist for 4 to 6 weeks thereafter.

ii. **Clinical use.** Toxicological, teratogenic, and carcinogenic studies have demonstrated that ISO is safe, well tolerated, and remarkably free of side effects, even upon prolonged administration. The clinical efficacy of ISO has been well documented in double-blind trials:
- ISO produces striking decreases in both the duration of infection and severity of symptoms in a whole host of viral diseases, including influenza virus infections, rhinovirus infections, herpes labialis and herpes progenitalis, herpes zoster, viral hepatitis, rubella, and viral otitis.

- Of particular interest are the results of ISO therapy in subacute sclerosing panencephalitis (SSPE), a progressive disease due to a chronic measles virus infection, which results in complete debilitation and eventual death of the patient. ISO has been reported to halt the progression of SSPE when given in stages I and II of the disease in 80% of the patients, provided it is administered for at least 6 months. Indeed, ISO is the only agent to date with documented beneficial effects in SSPE patients.
- Isoprinosine has been considered as potentially useful in AIDS, where the combination of its immunomodulating activity with reported anti-HIV activity would appear to be ideal. However, ISO does not appear to induce consistent clinical improvement in the patients to which it has been administered.

SELF-EVALUATION

Questions

Choose the ONE *best* answer.

26.1 The immunosuppressive effect of cyclosporine A is best explained by its:
   A. Ability to block calcineurin activation
   B. Ability to induce lymphocyte apoptosis
   C. Cytotoxic effect on CD8+ cells
   D. Inhibition of signal transduction from the IL-2 receptor
   E. Suppression of bone marrow lymphocytopoiesis

26.2 Which of the following side effects of immunosuppressive drugs has the greatest clinical impact?
   A. Anorexia
   B. Alopecia
   C. Chromosomal changes
   D. Nausea and vomiting
   E. Neutropenia

26.3 The lymphocytopenia seen a few hours after administration of a large dose of prednisone to a patient with lymphocytic leukemia is due to:
   A. Activation of cytotoxic cells
   B. Bone marrow depression
   C. Massive lymphocyte apoptosis
   D. Redistribution of peripheral blood lymphocytes
   E. Stimulation of NK cell activity

26.4 Which one of the following types of malignancies is characteristic of chronically immunosuppressed patients?
   A. Colon carcinoma
   B. Disseminated melanoma
   C. Hodgkin's lymphoma
   D. Lymphocytic leukemia
   E. Lymphoproliferative syndromes

26.5 The clinical effects of isoprinosine in patients with viral infections are explained by the following effect(s):
   A. Inhibition of viral replication

B. Potentiation of cell-mediated immunity
C. Potentiation of humoral immunity
D. Potentiation of humoral immunity and inhibition of viral replication
E. Potentiation of cellular immunity and inhibition of viral replication

26.6 Which one of the listed infectious agents is LEAST likely to be unusually frequent as a cause of opportunistic infections in patients with depression of cell-mediated immunity caused by chronic immunosuppressive therapy?
A. *Pneumocystis carinii*
B. *Aspergillus fumigatus*
C. *Staphylococcus aureus*
D. *Mycobacterium tuberculosis*
E. Herpes zoster virus

26.7 Which of the following immunomodulatory drugs down-regulates the expression of cell adhesion molecules on endothelial surfaces proximal to an inflammatory site?
A. Aspirin
B. Cyclophosphamide
C. Ibuprofen
D. Indomethacin
E. Prednisolone

26.8 Which of the following therapies is indicated in a child with humoral immunodeficiency?
A. Dialyzable leukocyte extract
B. Thymosin
C. Interleukin-2
D. Intravenous gammaglobulin
E. Interferon-γ

26.9 Which one of the following biological response modifiers would be indicated in a patient with agranulocytosis?
A. GM-CSF
B. Interferon-γ
C. Interleukin-3
D. Interleukin-8
E. Transfer factor

26.10 The pursuit of chemotherapy regimens with a steroid-sparing effect is motivated by the:
A. Need to increase the doses of steroids for more complete immunosuppression
B. Need to avoid side effects associated to high dosages of steroids
C. The synergism between steroids and cytotoxic drugs
D. Protective effect that steroids have relative to the side effects of cytotoxic drugs
E. Protective effect that cytotoxic drugs have relative to the side effects of steroids

Answers

26.1 (A) Cyclosporine A binds to cyclophilin and the complex prevents the activation of calcineurin. Because of the inactivation of calcineurin, NFκB remains phosphorylated, does not translocate to the nucleus, and the activation pathway leading to IL-2 synthesis is interrupted.

26.2 (E) All the listed side effects can be seen during administration of cytotoxic drugs, but neutropenia is the most serious because the infections secondary to neutropenia are very difficult to treat.

26.3 (C) A large dose of steroids leads to a pronounced decrease in the lymphocyte counts of patients with T-lymphocytic leukemia secondary to the induction of apoptosis in the proliferating cells.

26.4 (E) Two types of tumors predominate above all in immunosuppressed patients: epithelial carcinomas and lymphoproliferative syndromes, particularly B-cell lymphomas. The lymphomas often have atypical locations, such as intracerebral.

26.5 (E) Isoprinosine combines two potentially beneficial effects in viral infections: it inhibits viral replication, while at the same time it seems to prevent the depression of cell-mediated immunity that is often associated with viral infections.

26.6 (C) The depression of cell-mediated immunity is associated mainly with viral and opportunistic infections, as well as with increased severity of infections by intracellular parasites such as *M. tuberculosis*. Pyogenic infections, such as those caused by *S. aureus*, will be a serious problem in the neutropenic patient, but not in patients with depressed cell-mediated immunity.

26.7 (E) Of the listed drugs, only four have anti-inflammatory properties (cyclophosphamide is strictly immunosuppressive), and of the four, only prednisolone, a glucocorticoid, is able to down-regulate the expression of cell adhesion molecules.

26.8 (D) The preferred immunotherapy in a child with humoral immunodeficiency is intravenous injection of human gamma globulin, which will passively transfer antibodies to most common pathogens to the deficient child.

26.9 (A) GM-CSF will stimulate production of all granulocytes and monocytes. It has been used with some success in patients with neutropenia or with agranulocytosis (lack of all types of granulocytes).

26.10 (B) Although steroids can effectively induce a suppression of the immune response, this requires large doses for prolonged periods of time, and the side effects can be life threatening. Thus, the association of steroids with other immunosuppressants allows the use of lower doses of the combined drugs, reducing the side effects of all. Steroid-sparing effects is a term that specifically refers to the decrease of steroid doses due to their association to other immunosuppressive drugs.

BIBLIOGRAPHY

Auphan, N., DiDonato, J.A., Rosette, C. et al. Immunosuppression by glucocorticoids: inhibition of NF-kB activity through induction of IkB synthesis. *Science*, *270*:286, 1995.

Barnes, P.J. and Adcock, I. Anti-inflammatory actions of steroids: molecular mechanisms. *TIPS*, *14*:436, 1993.

Bierer, B. Mechanisms of action of immunosuppressive agents: cyclosporin A, FK506, and rapamycin. *Proc. Assoc. Am. Phys.*, *107*:28, 1995.

Braun, W., Kallen, J., Mikol, V., et al. Three dimensional structure and actions of immunosuppressants and their immunophilins. *FASEB J.*, *9*:63, 1995.

Brooks, P.M. and Day, D.O. Nonsteroidal antiinflammatory drugs—differences and similarities. *N. Engl. J. Med.*, *324*:1716, 1991.

Clipstone, N.A. and Crabtree, G. R. Calcineurin is a key signaling enzyme in T lymphocyte activation and the target of the immunosuppressive drugs cyclosporin A and FK506. *Ann. N.Y. Acad. Sci.*, *696*:20, 1993.

Ehrke, M.J., Mihich, E., Berd, D., and Mastrangelo, M.J. Effects of anticancer drugs on the immune system in humans. *Sem. Oncol.*, *16*:230, 1989.

Hassner, A., and Adelman, D.C. Biologic response modifiers in primary immunodeficiency disorders. *Ann. Intern. Med.*, *115*:294, 1991.

Kerman, R.H. Effects of cyclosporine immunosuppression in humans. *Transplant. Proc.*, *20 (S2)*:143, 1988.

Metcalf, D. Control of granulocytes and macrophages: Molecular, cellular, and clinical aspects. *Science*, *254*:529, 1991.

Montague, J.W. and Cidlowski, J.A. Glucocorticoid-induced death of immune cells: Mechanism of action. *Curr. Top. Microbiol. Immunol.*, *200*:51, 1995.

Pedersen, C., Sandström, E., Peterson, C.S., and Norkrans, G., et al. The efficacy of inosine prabonex in preventing the acquired immunodeficiency syndrome in patients with human immunodeficiency virus infection. *N. Engl. J. Med.*, *322*:1757, 1990.

Stevenson, H.C., Green, I., Hamilton, J.M., Calabro, B.A., and Parkinson, D.R. Levamisole: Known effects on the immune system, clinical results, and future applications to the treatment of cancer. *J. Clin. Oncol.*, *9*:2052, 1991.

27
Transplantation Immunology

Gabriel Virella and Jonathan S. Bromberg

I. INTRODUCTION

The replacement of defective organs with transplants was one of the impossible dreams of medicine for many centuries. Its realization required a multitude of important steps: surgical asepsis; development of surgical techniques of vascular anastomosis; understanding of the cellular basis of rejection phenomena; and introduction of drugs and antisera effective in the control of rejection.

A. By the early 1970s tissue and organ transplantation emerged as a major area of interest for surgeons and physicians. Kidney and bone marrow transplants have become routine in most industrialized countries and lead in frequency, followed by heart, liver, pancreas, bone, and lung transplants, in order of decreasing frequency. Transplantation of intestine, trachea, and bladder is still in experimental development. Other tissues and organs will certainly follow.

B. The success of an organ transplant is a function of several variables. However, **the major determinant of acceptance or nonacceptance (rejection) of a technically perfect graft is the magnitude of the immunologically mediated response against the graft**. The likelihood of acceptance or rejection is closely related to the extent of genetic differences between the donor and recipient of the graft.

1. Transplantation of organs between animals of the same inbred strain or between homozygous twins is successful and does not elicit an immune rejection response.
2. Transplants between distantly related individuals or across species barriers (xenogeneic) are always rapidly rejected.

In humans, genetic diversity between individuals is currently the main obstacle to successful transplantation.

II. DONOR-RECIPIENT MATCHING

Prevention of rejection is more desirable than trying to treat established rejection and is achieved by careful matching of donor and recipient and by manipulation of the recipient's immune response. Avoidance of antigenic differences between the donor and recipient is a crucial factor for the success of a transplant. Although many different antigenic systems show allotypic variation, in transplantation practice only the ABO blood groups and the HLA system are routinely typed.

A. **ABO Incompatibility is** generally considered an insurmountable obstacle to transplantation since it leads to an accelerated rejection response, called hyperacute rejection (see below), probably because A and B antigens are expressed on vascular endothelium. However, some groups have reported successful grafting of HLA-compatible but ABO-incompatible organs after removing anti-A and/or anti-B isohemagglutinins by plasmapheresis or by extracorporeal immunoadsorption. In extremely urgent cases of liver transplantation, ABO matching is sometimes ignored, and reasonable graft function and survival can be seen.

B. **HLA Matching.** The practical significance of HLA matching varies depending on the organ to be transplanted.

   1. **Kidney transplantation.** HLA matching is considered important for kidney transplantation, since there is a positive correlation between the number of HLA antigens common to the donor and recipient and the survival of the transplanted kidney.

      a. In the case of grafts donated from living relatives, HLA identical sibling grafts have the best outcome, followed by haploidentical grafts, which in turn do better than two haplotype incompatible grafts (Fig. 27.1).

      b. Cadaveric transplants matched for HLA-A, B, and DR achieve survivals similar to those obtained with transplants between two haplotype-matched living related individuals. The significance of MHC-II typing is supported by European centers, suggesting that kidney transplants matched only for HLA-B and HLA-DR have an excellent outcome (83% graft survival at 12 months).

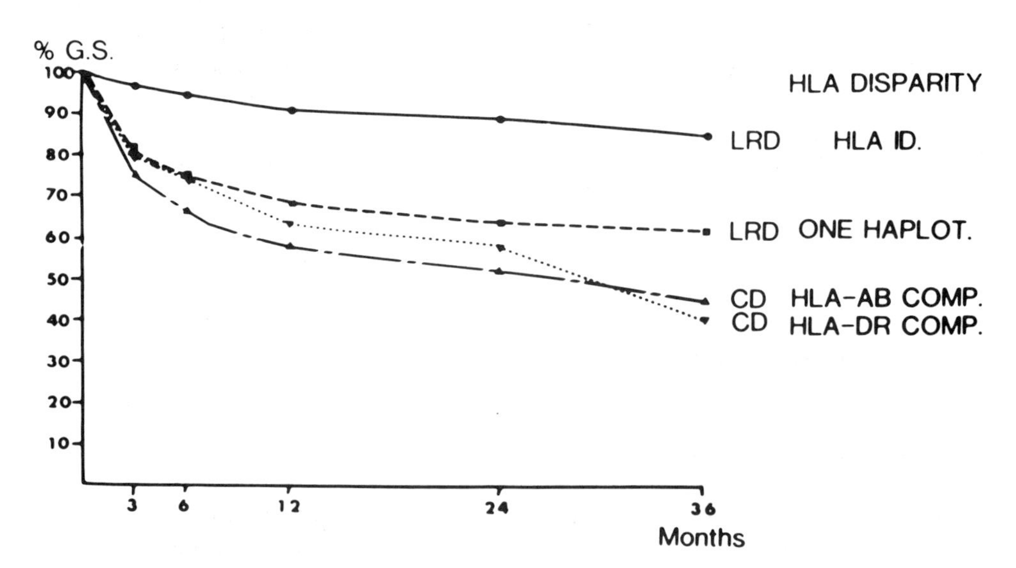

Figure 27.1 Cumulative graft survival of living related donor (LRD) and cadaveric donor (CD) transplants. (Reproduced with permission from Flatmark and Thorsby, *Transplant. Proc. 14*:61, 1982.)

2. **Bone marrow transplantation.** HLA typing is also very important in the case of bone marrow transplantation.
   a. This type of transplant presents a special problem in that it is necessary to avoid both the rejection of the grafted tissue by the host, and the damage of host tissue by the transplanted lymphocytes, a phenomenon called graft-versus-host disease (see later in this chapter). A living relative of the recipient is therefore usually the preferred donor and by order of preference an identical twin, an HLA identical sibling, or a haplotype-identical relative are selected.
   b. It is important to check for cytotoxic antibodies in the recipient's serum directed against the donor's lymphocytes as this could be a sign that the recipient is already immune to a potential donor. This is achieved by means of a test called a **cross-match**. The recipient's serum is tested against lymphocytes from the potential donor(s) and against a cell panel of known phenotypes. This test is useful to prevent rapid rejection of the grafted tissue or organ.
   c. **Mixed-lymphocyte cultures** are used to try to avoid the graft-versus-host reaction, which could emerge even in HLA-identical siblings due to incompatibilities in as yet undefined antigenic systems. Cultures are set up by mixing the recipient's lymphocytes with lymphocytes from potential donors. A well-matched donor–recipient pair should react minimally to one another in these cultures.
3. **Other organs**
   a. In the case of **liver transplantation**, HLA matching is *not* important for graft acceptance and survival. This unexpected finding may be due to the remarkable regenerative capacity of the liver.
   b. For **pancreas transplantation**, HLA matching, particularly MHC class II, promotes graft survival.
   c. For **heart and lung transplantation**, HLA matching is not generally performed because of the scarcity of available donor organs.

III. GRAFT REJECTION

A. **General Concepts.** Graft rejection is the consequence of an immune response mounted by the recipient against the graft as a consequence of the incompatibility between tissue antigens of the donor and recipient.
   1. Cells that express class II MHC antigens (such as passenger leukocytes in the case of solid organ transplants) play a major role in sensitizing the immune system of the recipient.
   2. The sensitization of alloreactive helper T lymphocytes from the recipient is followed by their clonal expansion, which in turn is the cause of multiple immunological and inflammatory phenomena, some mediated by activated T lymphocytes and others mediated by antibodies, which eventually result in graft rejection.
   3. Rejection episodes are traditionally classified as **hyperacute, acute, and chronic,** based primarily on the time elapsed between transplantation and the rejection episode.

B. Hyperacute (Early) Rejection

1. Hyperacute rejection occurs usually within the first few hours post-transplantation and is mediated by preformed antibodies against ABO or MHC antigens of the graft. It is also possible that antibodies directed against other alloantigens such as vascular endothelial antigens may also play a role in this type of rejection.
2. Once the antibodies bind to the transplanted tissues, rejection can be caused either by activation of the complement system, which results in the chemotactic attraction of granulocytes and the triggering of inflammatory circuits, and/or by antibody-mediated cellular cytotoxicity (ADCC).
3. A major pathological feature of hyperacute rejection is the formation of massive intravascular platelet aggregates leading to thrombosis, ischemia, and necrosis. The formation of platelet thrombi probably results from several factors, including release of platelet activating factor (PAF) from immunologically damaged endothelial cells and/or from activated neutrophils.
4. Hyperacute rejection episodes are untreatable and result in graft loss. With proper cross-matching techniques, this type of rejection should be almost 100% avoidable.
5. The major limitation to xenogeneic transplantation (e.g., pig to human) is hyperacute rejection by antibodies to cellular antigens that all humans make, even prior to any known exposure to xenogeneic tissues (natural antibodies).

C. Acute Rejection. Acute rejection occurs mostly in the first few days or weeks after transplantation. Up to 70% of graft recipients experience one or more acute rejection episodes.

1. When taking place in the first few days after grafting, it may correspond to a **secondary (second set) immune response**, implying that the patient had been previously sensitized to the HLA antigens present in the organ donor (as a consequence of a previous transplant, pregnancy, or blood transfusions).
2. When occurring past the first week after grafting, it usually corresponds to a **first set (primary) response**.
3. Acute rejection is predominantly mediated by **T lymphocytes**, and controversy has arisen concerning the relative importance of CD8+ cytotoxic lymphocytes versus helper CD4+ lymphocytes. Most likely, both subsets play important roles.
   a. In rejected organs, the cellular infiltrates contain mostly monocytes and T lymphocytes of both helper and cytotoxic phenotypes, and lesser frequencies of B lymphocytes, NK cells, neutrophils, and eosinophils. All these cells have the potential to play significant roles in the rejection process.
   b. CD4+ helper T lymphocytes are believed to play the key role, because of their release growth factors:
      i. IL-2 and IL-4 promote the expansion of CD8+ lymphocytes and B cells
      ii. Interferon-γ enhances the expression of MHC class II antigens in the graft

iii. Chemotactic interleukins, such as IL-8 (also released by activated monocytes and macrophages), attract lymphocytes and granulocytes to the transplanted organ.

c. In most cases, acute rejection, if detected early, can be reversed by increasing the dose of immunosuppressive agents or by briefly administering additional immunosuppressants. However, this simple approach is complicated by the uncertainties that often surround the diagnosis of rejection.

3. **Diagnosis of acute rejection**
   a. The initial diagnosis is usually based on **clinical suspicion**. Functional deterioration of the grafted organ is the main basis for considering the diagnosis of acute rejection.
   b. Confirmation usually requires a **biopsy** of the grafted organ. There are established histological criteria for the identification of an acute rejection reaction in transplanted organs. A hallmark finding in a rejecting graft is **mononuclear cell infiltration**, as would be expected in a typical delayed hypersensitivity reaction. However, mononuclear cell infiltrates (not as intense as in cases of acute rejection) can also sometimes be observed in transplanted organs that are apparently functioning normally.
   c. **Noninvasive diagnosis of rejection.** Since biopsy is an invasive procedure with potential complications and pitfalls, several approaches to the noninvasive diagnosis of rejection have been attempted. Particular attention has been directed to the measurement of cytokines released by activated T lymphocytes, such as IL-2, in serum and in urine (in the case of renal transplants). However, these tests have been found to be lacking in sensitivity and specificity.

D. Delayed or Chronic Rejection is characterized by an insidiously progressive loss of function of the grafted organ. Recent data show a positive correlation between the number of HLA incompatibilities and the progression of chronic rejection, which is difficult to control by any type of therapy.

1. It is not certain if chronic rejection is a unique process or if it represents the final common pathway of multiple injuries occurring over a protracted period of time, including acute rejection episodes, infection, and atherosclerosis.
2. The functional deterioration associated with chronic rejection seems to be due to both immune and nonimmune processes.
   a. The immune component of chronic rejection is believed to cause vascular endothelial injury. A variety of cells, such as granulocytes, monocytes, and platelets have an increased tendency to adhere to injured vascular endothelium, and the expression of PAF on the membrane of endothelial cells may be one of the major factors determining the adherence of neutrophils and platelets, both types of cells having PAF receptors on their membranes.
   b. A variety of interleukins and soluble factors are released by activated leukocytes at the level of the damaged vessel walls, including IL-1 and platelet-derived growth factor (PDGF). The damaged endothelium is covered by a layer of platelets and fibrin, and eventually by proliferat-

ing fibroblasts and smooth muscle cells. The end result is a proliferative lesion in the vessels as a consequence of the inflammatory nature of the process, which progresses toward fibrosis and occlusion.

Case

A 45-year-old white male with glomerulonephritis underwent a cadaveric renal transplant. The renal allograft was functional within 24 hours of transplantation and over the next week postoperatively the patient's creatinine dropped from 10.2 to 1.7 mg/dL. The patient recuperated from the operation well and was discharged from the hospital on the ninth postoperative day. He returned to the hospital 4 days later complaining of fever, malaise, decreased urine output, weight gain, and increased blood pressure. Laboratory investigation revealed a creatinine level that had risen to 3.2 mg/dL. Percutaneous renal biopsy was obtained and pathological examination of the tissue revealed a prominent interstitial lymphocytic infiltrate with tubular necrosis. The patient was treated with high-dose intravenous glucocorticoids for 6 days. During this time the fever abated, urine output rose, and creatinine decreased to 2.2. The patient was discharged home in good clinical condition with stable and acceptable renal function. He continued to do well with maintenance doses of prednisone and cyclosporine until approximately 8 weeks after transplant, when he presented with fever, malaise, diffuse abdominal pain, bloody stools, and shortness of breath. Chest x-ray revealed a fine reticular interstitial infiltrate of the lungs and blood gas analysis showed hypoxia while breathing room air. Physical examination revealed rhonchi and diminished transmission of breath sounds in both sides of the thorax, more accentuated on the bases, minimal hepatosplenomegaly, and a diffusely tender abdomen. Occult blood was detected in the stools.

- What was the most likely cause of functional deterioration soon after the transplant?
- Were the symptoms 8 weeks after the transplant related to the same process that affected the patient 6 weeks earlier?
- Should any tests be ordered to clarify the patient's later complaints?
- What therapeutic options would you consider?

III. IMMUNOSUPPRESSION

A. **General Concepts.** The ideal transplantation should take place among genetically identical individuals. This is only possible in the rare event of transplantation between identical twins. Thus, the success of clinical transplantation depends heavily on the use of nonspecific immunosuppressive agents which, by decreasing the magnitude of immunological rejection responses, prolong graft survival. Current immunosuppression in transplanted patients is achieved by the use of cytotoxic/immunosuppressant drugs and biological response modifiers, such as antilymphocyte antibodies.

B. **Chemical Immunosuppression.** Several drugs are currently used to induce immunosuppression, including glucocorticoid, antimetabolites, cyclosporine A, and tacrolimus (FK507).

1. **Glucocorticoids** are used to treat and prevent rejection. They have multiple effects on the immune system, including lymphocyte apoptosis, inhibition of antigen-driven T-lymphocyte proliferation, inhibition of IL-1 and IL-2

release, and inhibition of chemotaxis. Because of the side effects associated with the use of glucocorticoids in relatively large doses for long periods of time (as required in transplantation) they are usually administered together with other immunosuppressant drugs, allowing the reduction of steroid doses below levels causing major side effects.

2. **Antimetabolites** are mostly used in the prevention of rejection episodes.
   a. **Azathioprine (Imuran)** undergoes metabolic conversion into 6-mercaptopurine, which inhibits purine nucleotide synthesis and prevents lymphocyte proliferation (both T and B).
   b. **Mycophenolate mofetil (CellCept)** is converted to mycophenolic acid, which is an inhibitor of inosine monophosphate dehydrogenase, a participant in guanosine nucleotide synthesis.
   c. **Cyclophosphamide (Cytoxan)** is an alkylating agent that modifies DNA and prevents lymphocyte replication. All these agents inhibit DNA replication, lymphocyte proliferation, and the expansion of antigen-reactive clones of lymphocytes.
3. **Signal transduction inhibitors** (see Chap. 26)
   a. **Cyclosporine A (Sandimmune, CsA)** is used in the prevention and treatment of rejection. Its introduction in 1983 had a marked impact on the survival of transplanted organs, which increased by at least 20–30% in the case of kidney and heart grafts. The revival of interest in heart transplants in the last decade was a direct consequence of the availability of CsA, and the success of liver transplants is also directly related to the use of this drug.
      i. The effects of CsA are mainly related to the activity of transcriptional activators controlling the expression of IL-2 and other lymphokine genes in helper T cells, thus curtailing the onset of both cellular and humoral immune responses.
      ii. CsA is particularly helpful in the prevention of rejection, usually administered in association with glucocorticoids, because of its **steroid-sparing effect.**
      iii. CsA itself has marked toxicity. It is nephrotoxic (and that raises considerable problems in patients receiving kidney transplants, in which it will be necessary to differentiate between acute rejection and CsA toxicity) and causes hypertension. Monitoring of circulating cyclosporine levels is essential to minimize the toxic effects of this drug.
   b. **Tacrolimus (Prograf, FK506)** has a mechanism of action similar to CsA but it is able to reverse rejection episodes in patients unresponsive to other immunosuppressive agents.
      i. Its use as primary immunosuppressant in liver transplant was reported to result in improved patient survival, better than could be achieved with other immunosuppressant agents or any combination of immunosuppressants.
      ii. Tacrolimus has numerous toxic effects, including nephrotoxicity and neurotoxicity.
      iii. Tacrolimus and CsA are used similarly in combination with glucocorticoids and antimetabolites.

c. **Rapamycin (Sirolimus)**, is currently undergoing phase III clinic trials and should soon be added to the list of available agents in clinical transplantation.

B. **Biological Response Modifiers**

1. **Antithymocyte** and **antilymphocyte globulins** were among the earliest successful therapeutic agents used in the prevention of graft rejection.
   a. These reagents are gamma globulin fractions separated from the sera of animals (usually horses) injected with human thymic lymphocytes or human peripheral blood lymphocytes.
   b. They are very effective in the prevention and reversal of rejection episodes, and their mechanism of action is related to the destruction or inhibition of recipient lymphocytes.
   c. Their main drawbacks have been related to difficulty in obtaining standardized preparations, reactivity with other cell types, and frequent sensitization of the patients, which often leads to serum sickness.
2. **Anti-T-cell monoclonal antibodies** derived from mouse B cells and directed against human T cells, particularly those reacting with the **CD3** marker (**OKT3**), have been extensively used in the management of transplanted patients.
   a. Their mechanism of action is not entirely clear. OKT3 has been reported to cause depletion of CD3+ T lymphocytes, and it is likely that the depletion is due to complement-mediated lysis, opsonization, ingestion by phagocytic cells, and ADCC. OKT3 also causes down-modulation of CD3 on the cell surface of otherwise viable T cells.
   b. These antibodies are predominantly used for the treatment of acute rejection. In addition, some groups use the monoclonal antibody as "induction" treatment at the time of transplantation to prevent rejection.
   c. The possibility of using monoclonal antibodies to treat repeated episodes of rejection is limited by the sensitization of the patients receiving the antibody, which often causes **serum sickness**. It must be noted that in spite of concomitant immunosuppression, up to 30% of patients become sensitized to various antibody preparations. However, changing to a different monoclonal antibody can meet with success in hypersensitive patients.
   d. Besides serum sickness, monoclonal and polyclonal antibodies can cause what is known as the **cytokine syndrome**, which presents with fever, chills, headaches, vomiting, diarrhea, muscle cramps, and vascular leakage and transudation. Experiments and data suggest that the syndrome is caused by massive interleukin release from T cells activated as a consequence of the binding of these antibodies to the lymphocyte membrane.
   e. The problem of sensitization can be minimized when genetically engineered monoclonal antibodies containing the binding site of a murine monoclonal antibody and the constant regions of a human immunoglobulin are used. These "humanized" monoclonals have been used on a limited scale, but the current results suggest that prolonged administration may become possible.

f. Both monoclonal and polyclonal antilymphocyte antibody preparations are strongly immunosuppressive, so the risk of developing life-threatening infections or non-Hodgkin's lymphomas is markedly higher in patients treated with them (see below). As a result, treatment with these agents usually does not exceed 14 days, and repeated courses of antibody treatment are usually contraindicated.

C. **Total Lymphoid Irradiation.** Irradiation of those areas of the body where the lymphoid tissues are concentrated is almost exclusively used to prepare leukemic patients for bone marrow transplantation.

1. Irradiation combines two potential benefits in this circumstance: the elimination of malignant cells and the ablation of the immune system.
2. Another immunosuppressive effect of irradiation is due to the greater radiosensitivity of helper T cells, resulting in the survival and predominance of suppressor/cytotoxic T cells among the residual lymphocyte population after irradiation.
3. Transplantation preconditioning protocols with total lymphoid irradiation have met with some impressive success: patients are reported to require very low doses of maintenance immunosuppressive drugs, and, in a few cases, it has been possible to withdraw the immunosuppressive drugs completely.

D. **The Transfusion Effect.** Although blood transfusions were generally avoided in potential transplant recipients due to the fear of sensitization to HLA, blood group, and other antigens, several groups reported in the early 1980s that kidney graft survival was longer in patients who had received blood prior to transplantation (Fig. 27.2). This led to attempts to precondition transplant recipients with multiple pretransplant transfusions. Several interesting observations were recorded during these attempts.

1. The effect is more pronounced if MHC-II-expressing cells are included in the transfusion. Thus, the administration of whole blood, packed cells, or buffy coat are more efficient than the administration of washed red cells in improving graft survival.
2. The protection can be induced with a few donor-specific transfusions but usually requires multiple random transfusions. This probably reflects an MHC-specific effect, which is obviously easier to achieve when the transfused cells have the same MHC as the graft.
3. Third, the transfusion effect is delayed, usually seen about 2 weeks after a donor-specific transfusion. Following transfusion, there is a depression of cellular immunity which, according to some studies, seems to become more accentuated and long lasting with repeated transfusions.
4. Based on those observations, several theories were formulated concerning the mechanisms underlying the transfusion effect:
   a. **Stimulation of suppressor T cells**
   b. **Clonal deletion.** According to the proponents of this theory, pretransplant transfusion would sensitize the patient against MHC antigens of the transplanted tissue. When the patient receives the graft, he or she also receives high doses of immunosuppressive agents, and the memory T lymphocytes responding to the grafted tissue do not receive adequate help and undergo apoptosis, thus achieving a situation of clonal deletion.

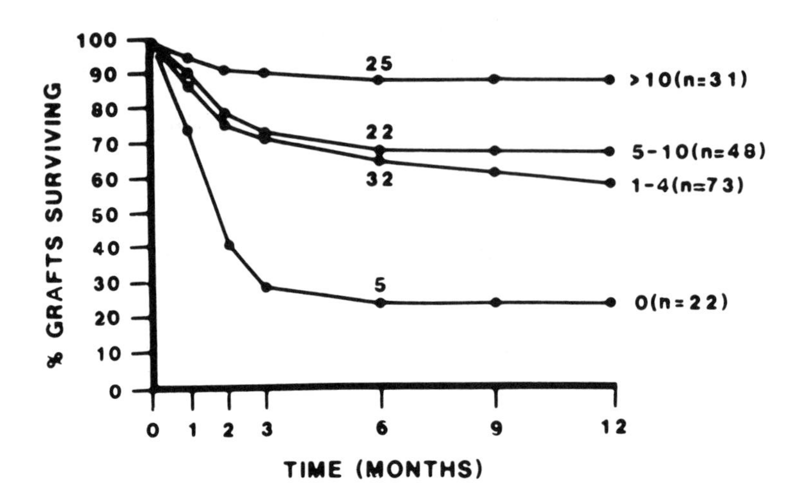

Figure 27.2 Actuarial kidney graft survival rates according to the number of transfusions received before transplantation. Numbers of transfusions are indicated at the end of each curve and numbers of patients are given in parentheses. Numbers of graft survivals for each group at 6 months are as indicated. Number of patients at risk at 6 months are indicated. p (weighted regression analysis) was <0.0001 at 3, 6, and 12 months, indicating that the improvement in graft outcome was dependent upon an increased number of pretransplant transfusions. (Reproduced with permission from Opelz, Graver, and Terasaki. *Lancet*, *I*:1223, 1981.)

c. **Induction of "enhancing" antibodies** has also been suggested as an explanation for the transfusion effect. This could be achieved through three possible pathways:
   i. Antibodies directed against MHC-II molecules could block the initiation of the rejection reaction.
   ii. Anti-MHC-II antibodies would bind to complexes of cells expressing the non-self alloantigen and host cells recognizing alloantigen, promoting the phagocytosis of the cellular complex via Fc-mediated opsonization.
   iii. Anti-idiotypic antibodies, reacting with the T-cell receptors of effector T cells, would inhibit rejection by either physically blocking subsequent interactions between cytotoxic T cells and grafted tissues, or by up-regulating suppressor T cells which prevent the expansion of helper and cytotoxic T-cell populations.

5. About 30–40% of patients given pretransplant transfusions develop **cytotoxic antibodies** to multiple HLA specificities in high titers which, in effect, delay or preclude transplantation. This observation led to the suggestion that pretransplant transfusions actually helped separate responsive patients who develop cytotoxic antibodies and would tend to reject the graft from nonresponsive individuals, who would be more likely to accept the graft. Most authors, however, believe that this selection cannot totally explain the transfusion effect.
6. Recent evidence suggests that the administration of blood transfusions may also transfer donor-derived hematopoietic stem-cell precursors. This could result in the establishment of a low level of donor-derived cells within the

recipient bone marrow and peripheral blood and result in a state of **microchimerism**. The microchimeric state could then potentially result in tolerance and improved graft survival.

7. While the debate has raged about the mechanism, the transfusion effect has become less and less evident, apparently as a consequence of better outcomes in nontransfused patients, due to improvements in matching and immunosuppression protocols. Thus, deliberate pretransplant transfusions have ceased to be used without ever having determined their mechanism of immunosuppression.

E. **Immunosuppression Side Effects.** Effective long-term immunosuppression is inevitably associated with a state of immunoincompetence. Two major types of complications may result from this:

1. **Opportunistic infections.** The immunosuppressed patient is susceptible to a wide variety of infections, particularly caused by infectious agents that are not often seen as pathogens in immunocompetent individuals, such as cytomegalovirus, herpes viruses (EBV, HSV, VZV), *Pneumocystis carinii*, *Toxoplasma gondii*, and fungi (e.g., *Candida albicans*). Cytomegalovirus infections are particularly ominous because this virus can further interfere with the host's immune competence, and may also trigger rejection in a nonspecific way.
   a. The incidence of infections in transplant patients can be reduced by prophylactic therapy with intravenous gamma globulin, which is part of most post-bone marrow transplant protocols, since those patients are probably the most profoundly immunosuppressed.
   b. Bone marrow and solid organ transplant patients also usually receive prophylactic antibiotics such as trimethoprim-sulfamethoxazole (for *Pneumocystis*, urinary tract infections, pneumonia, and cholangitis prophylaxis), acyclovir (for herpes and CMV prophylaxis), and nystatin (for *Candida* prophylaxis).
2. **Malignancies.** Either as a consequence of the oncogenic properties of some immunosuppressive agents or as a consequence of disturbed immunosurveillance, the incidence of **malignancies** is significantly increased in transplant patients.
   a. In those patients with survival times following transplantation of 10 years or longer, the frequency of **skin cancer** (squamous cell carcinoma) may be up to 40%, although the lesions are no more invasive than in normal individuals.
   b. An additional 10% may develop other types of malignancies, including **EBV-associated, post-transplant lymphoproliferative disorder** (PTLD).
      i. The reasons for the predominance of skin cancer and PTLD among transplant patients may relate to the inability of the immune system to respond to papilloma viruses and EBV, which are etiological agents for these malignancies, respectively.
      ii. Some of these PTLD are reversible with interruption or reduction of immunosuppressive therapy, while others are true malignant lymphomas that may spread to areas usually spared in nontransplanted patients, such as the brain.
   c. Common cancers such as colon, lung, and breast are *no* more common

in transplant patients than in the normal population. This suggests that immune surveillance is *not* important for most common cancers.

3. **Other side effects.** Each immunosuppressive drug is also associated with specific drug-induced side effects.
   a. Glucocorticoids can cause obesity, insulin-resistant diabetes, cataracts, avascular necrosis of the femoral heads, and thinning of the skin.
   b. Antimetabolites are associated with decreased blood counts and bone marrow depression.
   c. Cyclosporine and tacrolimus cause hypertension, nephrotoxicity, and neurotoxicity.

IV. GRAFT-VERSUS-HOST REACTION

A. General Concepts. Whenever a patient with a profound immunodeficiency (primary, secondary, or iatrogenic) receives a graft of an organ rich in immunocompetent cells, there is a considerable risk that a **graft-versus-host (GVH) reaction** may develop. The probability of developing a GVH reaction is greatest in the 2-month period immediately following transplantation.

B. Incidence

1. GVH reactions are a significant problem in infants and children with primary immunodeficiencies in whom a **bone marrow** or **thymus** transplant is performed with the goal of reconstituting the immune system, and in adults receiving a bone marrow transplant.
2. Small bowel, heart-lung, and even liver transplantation rank second in risk of causing graft-versus-host reactions, since these organs have a substantial amount of lymphoid tissue.
3. Transplantation of organs such as the heart and kidneys, poor in endogenous lymphoid tissue, very rarely results in a graft-versus-host reaction.

C. Pathogenesis

1. Patients receiving bone marrow transplants receive cytotoxic and immunosuppressive therapy, and their immune system is completely or partially destroyed.
2. When a graft containing immunocompetent cells is placed into an immunoincompetent host, the transplanted cells can recognize as non-self the host antigens. In response to these antigenic differences, the donor T lymphocytes become activated, proliferate, and differentiate into helper and effector cells which attack the host cells and tissues, producing the signs and symptoms of GVH disease.
3. At the peak of a GVH reaction the majority of the proliferating cells are of host origin and include T and B lymphocytes and monocytes/macrophages. The proliferation of host cells is probably a result of the release of nonspecific mitogenic and differentiation factors by activated donor T lymphocytes.
4. The crucial role played by the donor T cells is demonstrated by the fact that their elimination from a bone marrow graft avoids GVH reactions.

D. Pathology

1. The initial proliferation of donor T cells takes place in lymphoid tissues, particularly in the liver and spleen (leading to hepatomegaly and splenomegaly).

2. Later, at the peak of the proliferative reaction, the skin and intestinal walls are heavily infiltrated leading to severe skin rashes or exfoliative dermatitis and severe diarrhea.

E. **Treatment**

1. All immunosuppressive drugs used in the prevention and treatment of rejection have been used for treatment of the GVH reaction.
2. **Thalidomide**, the tranquilizer drug that achieved notoriety due to its teratogenic effects, has been used successfully for the control of chronic GVH unresponsive to traditional immunosuppressants, and it may become an extremely useful drug in the near future.

F. **Prevention.** Once a GVH reaction is initiated, its control may be extremely difficult. Thus, great emphasis is placed on prevention.

1. All immunosuppressive drugs discussed earlier in this chapter have been used by themselves or in different combinations in attempts to prevent GVH, with variable success.
2. **T-cell depletion** on the graft is a more successful approach and can be achieved by pretreatment of the bone marrow with anti-lymphocyte/thymocyte immunoglobulin, or with monoclonal antibodies reacting with T cells (e.g., anti-CD3).
3. The major problem with this approach is that the transplantation of T-cell-depleted bone marrow into immunosuppressed adults may result in a persistent state of severe immunodeficiency. These data suggest that the T cells actually facilitate the engraftment of the donor cells within the host bone marrow, although the mechanism for this is not understood.
4. In addition, a low-grade, controllable GVHD is often associated with better outcomes in leukemic patients, perhaps as a result of the elimination of leukemic cells by the grafted lymphocytes (graft-versus-leukemia effect).

Case Revisited

- The initial deterioration of kidney function seen 2 weeks after the graft would most likely correspond to a first-set acute rejection, but the possibility that cyclosporine toxicity was causing the symptoms could not be immediately ruled out. The findings on the biopsy, however, were typical of acute rejection and treatment with large doses of glucocorticoids was instituted.
- The clinical deterioration seen 8 weeks after the transplant was obviously systemic in nature. GVH was not a very likely possibility because of its very infrequent association with kidney grafts. A systemic CMV infection was suspected and confirmed by bronchoscopic biopsy, which revealed cells with intranuclear inclusion bodies on interstitial cells, by colonoscopy, with colonic biopsies of superficial ulcers of the right colon, which also revealed cells with intranuclear inclusions, and by blood cultures that were positive for CMV virus.
- A systemic CMV infection in an immunosuppressed patient requires energetic antiviral therapy. The patient was started on intravenous gancyclovir and over the course of the next 2 weeks gradually improved and resolved his lung and colon pathology.

SELF-EVALUATION

Questions

Choose the ONE *best* answer.

27.1 Which one of the following organ or tissue transplants is *least* likely to cause a graft-verus-host reaction?
  A. Bone marrow
  B. Heart-lungs
  C. Kidney
  D. Liver
  E. Small intestine

27.2 A 25-year-old female with chronic renal failure received a renal transplant. She was given cyclosporine A, steroids, and azathioprine, and appeared to be recovering uneventfully for the first 2 weeks. At day 18 post-transplant, the level of serum creatinine jumped from 2.1 to 3.5 mg/dL. A renal biopsy was obtained and showed heavy peritubular mononuclear cell infiltrates. Which one of the following events is most likely to have triggered this cellular reaction in the transplanted kidney?
  A. Complement activation by antigraft antibodies
  B. Recognition of cell-bound IgG antibodies by large granular lymphocytes
  C. Release of interleukin-8 from activated T lymphocytes
  D. Release of LTB-4 from activated monocytes
  E. Release of PAF from activated PMN

27.3 A major complication resulting from the use of monoclonal anti-CD3 antibodies in the treatment of graft rejection is:
  A. Graft-versus-host reaction
  B. Neutropenia
  C. Non-Hodgkin's lymphomas
  D. Urinary tract infections
  E. Serum sickness

27.4 The major pathogenic factor(s) in hyperacute graft rejection is(are):
  A. Anti-Rh antibodies
  B. Killer (K) lymphocytes
  C. Natural killer cells
  D. Predifferentiated cytotoxic T lymphocytes
  E. Preformed antibodies

27.5 Athymic nude mice are transplanted with bone marrow (B.M.) from genetically unrelated and immunocompetent Balb/c mice. Identify in the following chart the most likely combination of results seen in the grafted mice.

Nude recipient/Balb/c donor

	B.M. graft	Systemic effects
A	Rejected	None
B	Rejected	Splenomegaly, diarrhea, wasting
C	Accepted	None
D	Accepted	Splenomegaly, diarrhea, wasting
E	Accepted	Lymphomas, infections

27.6 Which of the following is the main risk associated with bone marrow transplant to an infant with lack of functional T-cell differentiation?
A. B-cell replacement without T-cell replacement
B. Development of autoimmune hemolytic anemia
C. Lack of correction of the defect due to rejection of the graft
D. T-cell replacement without B-cell replacement
E. Uncontrolled proliferation of alloreactive donor T lymphocytes

27.7 The main problem facing surgeons trying to perform xenogeneic transplants is:
A. Scarcity of suitable sources for the grafts
B. Technical difficulties associated with the size of the organs
C. Development of chronic rejection after a few months
D. Severe hyperacute rejection
E. Ethical problems

Each question below consists of a set of lettered headings followed by numbered phrases. For each numbered phrase, choose the ONE lettered heading that is best. In each group, the lettered heading may be used once, more than once, or not at all.
A. Bone marrow transplant
B. Heart transplant
C. Kidney transplant
D. Liver transplant
E. Pancreas transplant

27.8 Highest risk of GVH reaction
27.9 Patient survival time has significantly increased with the use of cyclosporine
27.10 Graft survival is inversely related to the number of mismatches for HLA-A, B, and DR.

Answers

27.1 (C) The likelihood of developing a GVH reaction is minimal when solid organs with minimal endogenous lymphoid tissue are grafted, such as the kidney and the heart. However, the likelihood increases in a heart-lung transplant due to the lung-associated lymphoid tissues, as well as in a small intestine graft, due to the peri-intestinal lymphoid tissues.

27.2 (C) A reaction appearing 18 days after graft could represent a first-set acute rejection or a manifestation of drug toxicity. The massive mononuclear cell infiltrate observed in the kidney biopsy is indicative of a cell-mediated reaction, such as it is classically associated with first-set acute rejection. The main cause for the accumulation of mononuclear cells is the release of IL-8 by activated lymphocytes involved in the rejection of the grafted organ.

27.3 (C) Although many complications may result from the use of mouse monoclonal anti-CD3 antibodies in the treatment and prevention of rejection episodes due to the profound immunosuppression associated with their use and to their heterologous nature, a major life-threatening complication is the induction of EBV-related non-Hodgkin's B-cell lymphomas and post-transplant lymphoproliferative disorders. Serum sickness is usually reversible and can be avoided by suspending the administration of anti-CD3, but malignant lymphomas continue to develop.

27.4 (E)

27.5 (D) Athymic mice will lack T cells and will not be able to reject the graft. However, the grafted T cells will be able to mount a graft-versus-host response, and the three major manifestations are splenomegaly, diarrhea, and wasting. Lymphomas and infections are more prevalent in immunocompromised animals, but not as a consequence of the bone marrow transplant.

27.6 (E) While a successful bone marrow graft should be curative for a patient with ADA deficiency, the main problem is the possible development of an uncontrolled graft-versus-host reaction due to the proliferation of grafted T cells recognizing alloantigens in the donor's tissues as nonself.

27.7 (D) Humans have preformed antibodies reactive with tissues from a variety of species, including those most often considered as possible organ sources. Those preformed antibodies cause hyperacute rejection, thus limiting the therapeutic value of the graft, which in most cases would be tried as a temporary palliative measure.

27.8 (A)

27.9 (B) Cyclosporine has also been successfully used for the treatment of other types of transplants, but in no case has the impact on survival been so dramatic as in the case of heart transplants. Liver transplants were not tried on a large scale until cyclosporine was available, and the immunosuppressant preferred for liver transplants is tacrolimus (FK506).

27.10 (C) Of the listed transplants, the one in which more data have accumulated concerning the relationship between graft survival and HLA typing is kidney transplant. Available data indicate that perfectly matched cadaveric allografts can have survivals identical to living related donor transplants among two haplotype-identical siblings. In the case of bone marrow grafting, the main concern is not graft survival, facilitated by the profound state of immunodeficiency of the recipient, but rather ensuring that a surviving graft does not induce a GVH reaction.

BIBLIOGRAPHY

Bromberg, J.S., and Grossman, R.A. Care of the organ transplant recipient. *J. Am. Board Fam. Pract.*, *6*:563, 1993.

Cicciarelli, J., and Terasaki, P. Matching cadaver transplants achieve graft survivals comparable to living related transplants. *Transplant Proc.*, *23*:1284, 1991.

Cobbold, S.P. Monoclonal antibody therapy for the induction of transplantation tolerance. *Immunol. Lett.*, *29*:117, 1991.

Ettenger, R., and Ferstenberg, L.B. Basic immunology of transplantation. *Perspect. Pediatr. Pathol.*, *14*:9, 1991.

Fugle, S. Immunophenotypic analysis of leukocyte infiltration in the renal transplant. *Immunol. Lett.*, *29*:143, 1991.

Hisanga, M., Hundrieser, J., Boker, K., et al. Development, stability, and clinical correlations of allogeneic microchimerism after solid organ transplantation. *Transplantation*, *61*:40, 1996.

Hooks, M.A., Wade, C.S., and Millikan, W.J., Jr. Muromonab CD3: A review of its pharmacology, pharmacokinetics, and clinical use in transplantation. *Pharmacotherapy*, *11*:26, 1991.

McDiarmid, S.V., Farmer, D.A., Goldstein, L.I., et al. A randomized prospective trial of steroid withdrawal after liver transplantation. *Transplantation*, *60*:1443, 1995.

Parr, M.D., Messino, M.J., and McIntyre, W. Allogeneic bone marrow transplantation: Procedures and complications. *Am. J. Hosp. Pharm.*, *48*:127, 1991.

Sollinger, H.W. Mycophenolate mofetil for the prevention of acute rejection in primary cadaveric renal allograft recipients. *Transplantation*, *60*:225, 1995.

Thompson, A.W. The immunosuppressive macrolides FK-506 and rapamycin. *Immunol. Lett.*, *29*:105, 1991.

Wood, K.J. Alternative approaches for the induction of transplantation tolerance. *Immunol. Lett.*, *29*:133, 1991.

28
Tumor Immunology

Henry C. Stevenson-Perez and Kwong-Y. Tsang

I. INTRODUCTION

One of the most intellectually attractive concepts in clinical immunology has been the postulated role of the immune system as an antitumor surveillance system. It is becoming clear that such a postulate may have serious flaws, but nevertheless it has been the major impetus for decades of medical research in this area. For the past 100 years, efforts at numerous levels have been mounted in the hopes of developing an immunological cure for cancer. Unfortunately, we are still faced with many agonizing dilemmas, including an incomplete understanding of the nature of the malignant process, as well as of the mechanisms for human immune system responses to malignant cells.

II. MALIGNANT TRANSFORMATION

Cancer is a term used to encompass a wide range of clinical disease states that involve virtually every tissue type of the body and every stage of differentiation of these tissues. The common denominator of all types of malignancies is the uncontrolled proliferation of cancer cells, said to have undergone "malignant transformation" as they demonstrate a loss of functions and characteristics associated with their nontransformed counterparts.

A. General Characteristics of Transformed Cells

1. Transformed cells are considered to have become **de-differentiated**. As a consequence, they cease participating in the overall mission of supporting the survival of the organism as a whole and begin competing with normal cells for both space and the limited resources that the organism has available.
2. Because transformed cells are capable of successfully competing with normal cells, they tend to **proliferate more rapidly than normal cells**.
3. As a consequence of their loss of differentiation, tumor cells may **express developmental antigens** that are usually seen only in the prenatal period. Examples of these antigens are alpha-fetoprotein and carcinoembryonic antigen.
4. Tumor cells express other types of unique antigens as well.
   a. One type is the so-called "**tumor-specific antigens**," membrane mole-

cules found on tumor cells from the same tissue type; upon occasion, tumor-specific antigens may be uniquely different from one cancer patient to another, even though both patients have the same type of tumor.

b. Another group of molecules found on malignant cells are the **common membrane markers of malignancy** that are shared on virtually all tumor cell types. These common tumor markers include the membrane glycoproteins and/or glycolipids recognized by NK cells (see Chap. 11).

5. In addition, highly malignant cells seem to lose their "homing instinct" and start invading the vasculature or lymphatic system, thus spreading to sites of the body far removed from the original tissue site, a process known as metastatic spread.

B. Genetic Basis of Malignant Transformation

1. **Tumor cells and programmed cell death.** Most mammalian cells, including tumor cells appear to be genetically programmed to spontaneously die at specified time intervals and/or under specified environmental conditions, a process known as **apoptosis**. A number of factors, including immunological influences, are designed to activate apoptosis, which is an essential part of regulatory and effector processes (see Chap. 11).
   a. One of the final common pathways to apoptosis activation is a group of serine proteases, including interleukin-1 converting enzyme (ICE) and other related "ICE-like" proteins. These apoptosis-promoting enzymes are in turn influenced by a group of apoptosis-controlling genes, including the p53 protein.
   b. Normal p53 is a phosphoprotein that arrests cell division in those cells that contain damaged DNA. If such cellular DNA cannot be repaired, p53 activates the apoptosis pathway. Thus, a normally functioning p53 gene protects the body from cells that contain DNA damage and mutations. A number of human cancers, including colon and lung carcinomas, as well as osteosarcomas, are associated with either a missing or mutated p53 gene.
2. **Mutations affecting proto-oncogenes.** Proto-oncogenes (such as "myc" and "ras") are cellular homologs of genes found in tumor viruses that appear to code for protein products that are an integral part of the normal pathways of cellular activation. Overexpression of mutated proto-oncogenes has been identified in a variety of malignancies, suggesting that the overproduction of hyperactive analogs of normal cellular activation and regulatory proteins is intimately associated with malignant transformation of certain tissues.
3. Malignant growth seems to result from several additive factors, including loss of tumor-protecting genes and overexpression of normal or mutated proto-oncogenes.
   a. A variety of cellular insults, including certain viruses, certain chemicals, and certain types of radiation injury are apparently able to cause the multiple sequential changes necessary for transformation.
   b. The unifying hypothesis regarding these potential causes of cancer focus on the ability of these agents to interact with and alter the structure and/or function of the normal host DNA, including proto-oncogenes and tumor-suppressing genes.

III. TUMOR ANTIGENS

Tumor cells possess many antigens not found in differentiated cells. These antigens vary in their biological role, cell location, and specificity. From the broad public health point of view, those antigens that are unique to the tumors of a given individual are currently of relatively little utility, except for the possibility of using those antigens as the basis for individualized tumor vaccines, as discussed later in this chapter.

A. **Tumor-Associated Antigens** (TAA) are not unique to transformed cells. They can be expressed by normal cells at very low levels or only during differentiation, but their expression is de-repressed or considerably enhanced as a result of malignant transformation.
   1. **Silent tumor-associated genes** are not expressed in normal cells but are actively transcribed in tumor cells.
      a. In mice, an example of a product of a silent gene is the **thymic leukemia antigen**, a nonpolymorphic MHC-I-like molecule. This TAA is expressed by leukemic cells of all strains of mice.
      b. No examples of this type of TAA have yet been characterized in human malignancies.
   2. **Tissue-specific genes** or **differentiation genes** are present in the surface of normal cells or may be shed to the circulation, but the levels of expression are usually very low in normal cells.
      a. **Malignant B-cell lymphomas** express the **CD10** membrane marker, which is expressed at very low levels in normal B cells.
      b. The **prostate-specific antigen** (PSA) is a kallikrein-like serine protease produced exclusively by the epithelial cells in the prostate gland, and is detectable at relatively high levels in seminal plasma and at very low levels in the serum of healthy men. The assay of serum PSA levels is a very useful marker of **prostate carcinoma**, perhaps the most meaningful serum marker for neoplasia. In healthy men, the levels of PSA vary between 0.65 ± 0.66 ng/mL at ages 21–30 to 1.15 ± 0.68 ng/mL at ages 61–70. Significantly elevated levels are assayed in 63 to 86% of patients with prostatic carcinoma, depending of the stage.
   3. **Oncofetal antigens** are cell-surface antigens that are normally expressed only in embryos or in fetuses, but not in adult tissues. By mechanisms unknown, the expression of these genes can be de-repressed in conditions of intense cellular proliferation, including malignancies. Two of the best-known tumor antigens are carcinoembryonic antigen and alpha-fetoprotein.
      a. **Carcinoembryonic antigen (CEA)** is a glycoprotein with an approximate molecular weight of 200,000 daltons, initially discovered in extracts of human colon tumors.
         i. High plasma CEA levels were initially reported in 35 of 36 **colorectal cancers** and in 3 of 32 other cancers of the digestive tract. However, subsequent studies have shown not only significantly lower positivity among colorectal cancer patients (i.e., an average of 30–40% significant elevations), but also higher positivity rates among patients with nondigestive tract cancer (approximately 50%).
         ii. High CEA levels have been detected in patients with cancers of

the pancreas, stomach, breast, lung and many other neoplasms. It can be released in large concentrations by noncancerous, inflammatory lesions of the colon—such as ulcerative colitis, sigmoiditis, and polyposis. Positivity has also been quite frequently noted in cases of alcoholic cirrhosis, alcoholic pancreatitis, peptic ulcer, and in heavy smokers.

iii. At present, therefore, the assay of CEA should not be considered as a diagnostic test. However, the CEA assay appears to be of much greater value for the follow-up of previously CEA-positive cancer patients who have already undergone potentially curative therapy. A successful surgical resection of a colorectal tumor is associated with a rapid postoperative decline of serum CEA to a normal value, and no recurrence has been detected among those patients whose CEA values remained negative for a certain period of time. Positive CEA reappearance generally precedes clinical detection of CEA-positive colon cancer recurrence.

iv. By constantly monitoring patients' plasma CEA levels postsurgery, chemotherapy or radiotherapy can be initiated as soon as CEA levels start rising, well before clinical signs of recurrence are detectable.

b. **α-fetoprotein (AFP)**, a glycoprotein with an approximate molecular weight of 70,000 daltons, is found in the fetal sera of all mammalian species, and is synthesized mainly in the yolk sac endoderm and liver parenchymal cells. AFP levels in fetal sera begin to decline in the last months of gestation and, although still present at birth, AFP totally disappears within a very short period thereafter.

i. AFP is produced by certain tumor cells, namely **hepatocellular carcinoma** and **germ-cell teratocarcinoma**. Elevated serum AFP levels (i.e., 500–1000 ng/mL) may be utilized as a specific marker for the detection of these tumors (the positive correlation of elevated AFP levels with hepatocellular carcinoma is about 80% in Africa, but only 50% in Europe and the United States).

ii. Some colorectal cancer patients with liver metastases and some acute hepatitis patients have been described to show elevated AFP levels. Low levels of AFP can be detected in normal adult sera, in the sera of acute hepatitis patients, and in the sera of pregnant women.

iii. As with the use of CEA to monitor the recurrence of CEA-positive colon tumors, clinicians constantly monitor the AFP levels of patients with resected germ cell tumors as an early marker of tumor recurrence.

B. **Tumor Antigens Encoded by Mutant Cellular Genes**. Some human malignancies are the result of mutations caused by physical or chemical carcinogens in cellular genes that control cell growth. Such mutations are reflected by the synthesis of abnormal products.

1. **Ras proto-oncogene products** include the p21 Ras proteins and other related gene products. Ras proteins bind guanine nucleotides (GTP and GDP) and possess intrinsic GTPase activity. The mutations associated with

Ras genes in malignant cells seem to lead to single amino acid substitutions at specific positions (12, 13, or 61), which result in increased enzymatic activity of the gene product. As a consequence, the cells acquire transforming capacity.

2. **Tumor antigens encoded by genomes of oncogenic viruses**. There is considerable evidence for the involvement of viruses in the pathogenesis of animal and human malignancies. Integrated proviral genomes are usually detected in virus-induced tumors, which may also express viral genome-encoded proteins.
    a. At least four different viruses have been proved to be associated with human tumors: three DNA viruses (Epstein-Barr virus or EBV, papilloma virus, and hepatitis B virus) and one RNA virus (human T-cell leukemia virus-I, HTLV-1).
    b. EBV is associated with three major types of tumors: endemic Burkitt's lymphoma, non-Hodgkin's lymphomas, and nasopharyngeal carcinoma. A variety of viral genes is expressed in those malignancies, including some that code for antigens that have shown to be the target for cytotoxic T lymphocytes.
    c. HTLV-1 encoded proteins are also detected in HTLV-1-associated leukemias and lymphomas, some of them expressed in the membrane of malignant cells. However, whether the immune system recognizes those proteins is not known.

IV. IMMUNOLOGICAL DEFENSE MECHANISMS AGAINST TUMORS

Tumor-associated antigens can activate a complete set of specific and nonspecific defense mechanisms—both humoral and cellular. Host antitumor immunity involves a complex series of events that result from the participation of the various "arms" of the immune system. These "arms" include T lymphocytes, B lymphocytes, monocytes/macrophages, NK cells, and a variety of cytokines which regulate the interactions of the different cells involved in the process.

A. **Specific Immune Reactions Against Tumor Cells** involve T and B lymphocytes.
    1. **T lymphocytes** play an important dual role, as cytotoxic effector cells and as the central modulating cells that control the specific cell-mediated antitumor immune responses and up-regulate nonspecific killing mechanisms.
    2. **B lymphocytes** produce tumor-specific antibodies (specific humoral immunity), which may induce complement-dependent cytotoxicity of tumor cells or may mediate antibody-dependent cell-mediated cytotoxicity (ADCC).

B. **Antibody-Dependent Cellular Cytotoxicity** can be mediated by a variety of cells expressing Fcγ receptors (NK cells, monocytes/macrophages, and granulocytes) recognizing and destroying IgG-coated tumor cells.

C. **Nonspecific Killing** can involve the same cell populations involved in ADCC. In this case the cells recognize a common membrane marker of cancerous cells that appears to be shared by many different types of malignant cells. In the case of NK cells, activation of the killing function also requires the lack of down-regulating signals due to modified expression of MHC proteins in the malignant cell (see Chap. 11).

D. Modulation of Nonspecific Killing

1. The activity of the cells involved in nonspecific killing can be augmented in vitro by interferon-γ, interleukin-2, and adjuvants such as BCG and *Corynebacterium parvum.*
2. The activation of T lymphocytes by tumor cell products as a consequence of antigen recognition may result in the secretion of nonspecific immunoregulatory factors (see below) that are capable of "up-regulating" the tumor killing function of mononuclear phagocytes, NK cells, and granulocytes, as well as enhancing the ability of NK cells and monocytes to participate in antibody-dependent cellular cytotoxicity (ADCC) against tumor cells.

E. Interactions Between Transformed Cells and the Immune System. From an immunological perspective, there are four aspects of cancer immunobiology that have commanded most attention recently.

1. The expression of tumor-specific antigens and common membrane markers of malignancy may allow the immune system to distinguish malignant cells from normal cell types and promote their destruction.
2. Since the immune system is also a potent source of growth factors (some of which may enhance cancer cell growth), immune-system-mediated tumor promotion may occur.
3. Mediators released by the immune system may limit transformed cell growth, not by directly killing those cells, but by promoting their differentiation into cells with more normal functions.
4. The emergence of cancer cells within the body may not be a rare or unusual event at all. Of the trillions of normal cells found in the body, several hundred per day may be undergoing malignant degeneration in response to the cancer-promoting stimuli cited above; the immune system may possibly play a significant role in halting the growth of these cells and preventing the development of overt malignancy.

F. Immune Surveillance

1. The most popular attempt to conceptualize the role of the immune system in protection against tumors is the immunological surveillance theory, as postulated by Thomas and Burnet. According to this theory:
   a. Frequent spontaneous mutations naturally occur in somatic cells, some of which may result in malignant transformation.
   b. Such transformations are often accompanied by the production of abnormal cell-surface protein antigens and, consequently, many neoplastically transformed cells are sufficiently antigenic in the host to cause their elimination by cytotoxic T lymphocytes and other immune system components.
   c. In rare instances, weakly immunogenic cells or mutant cells that fail to express tumor marker antigens can escape this defense mechanism and develop into overt malignant tumors.
   d. Alternatively, certain transformed cells may release factors that shield them from the immune system attack or otherwise actively down-regulate local protective immunity.
2. The increased incidence of tumors in individuals whose immune competence is impaired or depressed, whether by genetic factors or by treatment

with immunosuppressive drugs or irradiation, appears to be the corollary supporting the immunosurveillance theory. In those individuals, even highly immunogenic cells would survive and proliferate. A major problem with this corollary, however, is the fact that the tumors appearing in these individuals do not represent a random cross-section of those seen in the population at large (see Chaps. 26, 27, and 30).

3. Although the concept of immunological surveillance was widely accepted, several studies seriously question its validity; to this day it remains hypothetical. In general, it may be reasonable to assume that there are immunological mechanisms that can be triggered as a defense against neoplastic outgrowth, and the resultant biotherapeutic attempts to enhance the functions of such immune mechanisms are valid; however, the assumption that malignant neoplastic outgrowths occur only if the tumor is not antigenic or if the host immune recognition and/or defense capability is somewhat impaired is not supported by most investigators. Moreover, there is evidence for tumor growth enhancement as a consequence of the activation of the immune system, as discussed below.

V. ESCAPE FROM HOST IMMUNITY

The postulated host defense mechanisms against cancer are obviously ineffective whenever tumor growth occurs. Several mechanisms have been proposed to explain how tumor cells escape or evade immunity and grow unchecked.

A. **Immunoselection**. Most tumor cell populations appear to be heterogeneous mixtures of different subclones, some of which may have decreased antigenicity. These poorly antigenic variants may have an advantage under the selective pressure exerted on the tumor cells by the host immunity and their proliferation will increase the general immunoresistance of the tumor.

B. **Enhancement**. Immunological enhancement described the paradoxical overgrowth of tumors observed in animal experiments when animals are immunized with inactivated tumor cells before they are infected with live cells in an attempt to increase the immune response of the host against the tumor. Several mechanisms have been proposed to explain the enhancement in tumor growth caused by an antitumoral response.

    1. Production of noncytotoxic antitumor antibodies (**blocking antibodies**) which bind inconsequentially to the antigenic sites of the tumor and yet prevent the recognition of the same sites by potentially effective cytotoxic T lymphocytes.
    2. Free tumor antigen or antigen–antibody complexes present in the serum of the tumor-bearing animals (**blocking factors**) may block the immune response by either blocking the antigen receptors of the cytotoxic immune cells or, in the case of immune complexes, stimulating suppressor cell activity.
    3. Activated leukocytes may promote malignant cell proliferation by releasing growth factors active on transformed cells. Such factors include interleukin-2 (IL-2), epidermal growth factor (EGF), and transforming growth factor-β (TGF-β).

VI. CANCER BIOTHERAPY

The object of biotherapeutic manipulations in cancer patients is to enhance the ability of the immune system to control or eliminate the abnormal cells. As summarized in Figure 28.1, inhibition or promotion of tumor growth by the immune system involves a rather complex set of interactions:

- At the hypothesized earliest level (the single malignant cell), immunosurveillance mechanisms may prevent the emergence of a tumor.
- In established cancer, immunosurveillance has obviously been ineffective. Any immunological inhibitory effect on tumor growth has to be induced by biological therapeutic interventions (cancer biotherapy).
- On the other hand, there is at least the theoretical possibility for immune system-related tumorigenesis (i.e., stimulation of cells already committed to malignant transformation by mediators released by immune cells, such as leukocyte-derived growth factor).
- Established tumor progression in certain instances may also be related to the release of tumor-promoting growth factors by activated immune system cells. Obviously, any attempts to enhance the antitumoral response have the risk of causing immunological enhancement.

A. **General overview of cancer biotherapy**. Cancer biotherapy approaches can be classified into two broad groups.

1. **Active biotherapies**, which require an intact patient immune system capable of actively identifying and eliminating malignant cells in response to a

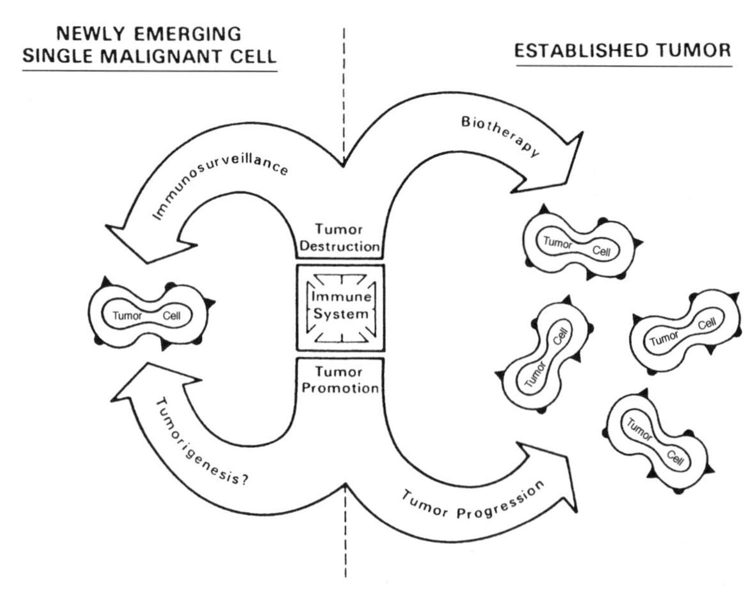

Figure 28.1 Diagram representing the interrelations that may exist between the immune system and tumor cells. The immune system may either prevent tumor growth (spontaneously or after proper activation) or promote tumor growth.

biotherapeutic stimulus. A good example of this type of biotherapy are tumor vaccines.

2. **Passive biotherapies** somehow "short-circuit" active immunity; the patient's immune system need not be capable of all tumor recognition/destruction functions; biotherapeutic replacement(s) for certain critical steps are administered to the patient. Adoptive cellular immunotherapy and administration of immunotoxins are representative of passive cancer biotherapies.

   Cancer biotherapy research activity is concentrated into four areas that will be discussed in the following sections: tumor vaccines; gene therapy approaches; administration of biological compounds and/or immunostimulatory chemicals (biological response modifiers; BRM); adoptive cellular immunotherapy; and affinity column apheresis.

B. **Tumor Vaccines**. The vaccination of cancer patients with tumor cell preparations is the earliest form of cancer biotherapy research developed, in an attempt to replicate the vaccination strategies that have been shown to be effective in controlling a variety of infectious diseases.

1. **Types of tumor vaccines**. As shown in Figure 28.2, tumor vaccines may be prepared from a patient's own autologous tumor or from similar allogeneic tumor cell lines.
   a. Some vaccines are made with **whole tumor cells** treated with a variety of enzymatic, chemical, and/or radiation procedures to enhance the expression of tumor antigens and prevent cell division.
   b. **Cell-free extracts** of tumor cell antigen(s) have also been used in immunization protocols.
   c. **Recombinant vaccines** are being evaluated.
      i. Recombinant vaccines for colon carcinoma and prostate carcinoma are being tested. Both use genetically engineered vaccinia virus as the vector, adding either the CEA or the PSA genes to the viral genome. Preliminary data suggest that these vaccines elicit T-lymphocyte-mediated immunity in humans and nonhuman primates.
      ii. A second example is the inoculation of cancer patients with their own tumor cells transfected in vitro with genes for specific immunomodulatory cytokines (such as IL-2 and TNF-α) in the hope of recruiting immune cells to the vaccination site, followed by elicitation of specific T-lymphocyte-mediated immunity.
   d. Biological adjuvants, such as BCG, may be coadministered with both cellular and cell-free vaccines in the hope of enhancing their efficiency.

2. **Limitations of tumor vaccines**. There is a lack of universal effectiveness of tumor vaccines to date. Several factors may account for the disappointing results:
   a. Tumor vaccines have traditionally been considered as examples of active biotherapies; they require that the patient have both intact recognition and cell-mediated and humoral immune effector mechanisms. Many vaccinated patients only demonstrate enhanced cell-mediated immunity upon laboratory testing while other tumor vaccines may

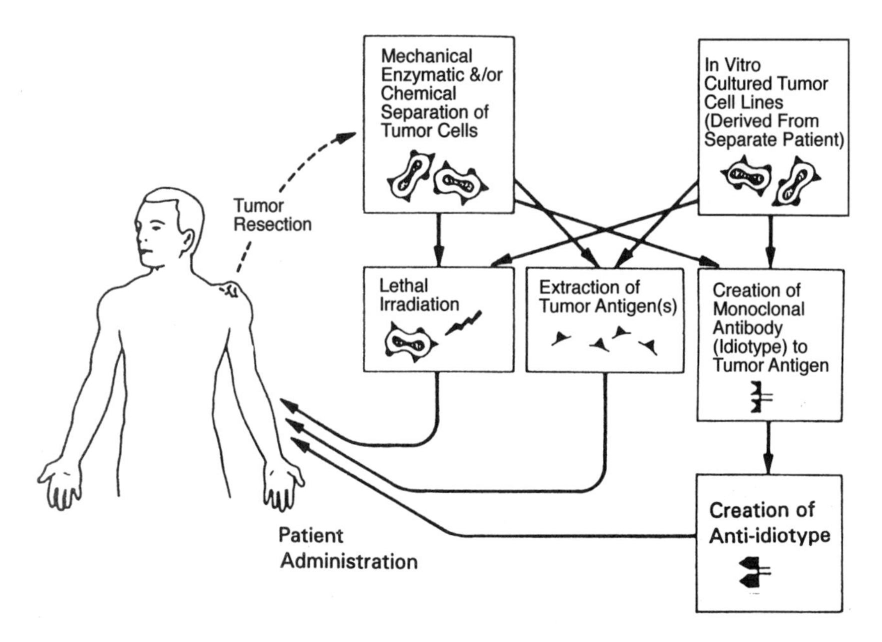

Figure 28.2 Diagrammatic representation of the protocols used for the development of tumor vaccines and anti-idiotypic vaccines for use in tumor immunotherapy.

mainly stimulate a modest humoral response. In either case, established tumor burden is generally not affected.

b. It is possible that other required (but less well-defined) immune system effector mechanisms are defective in certain cancer patients, limiting tumor cell elimination even in the face of antitumor immune response.

C. **Gene Therapy of Cancer**. Gene therapy is defined as a therapeutic technique in which a functioning gene is transferred into a patient's somatic cells for the purpose of correcting an inborn genetic error, or to provide a new genetic property to those cells.

1. In cancer treatment, such gene therapy could:
   a. Target emerging cancer cells and impart them with new molecular immune recognition molecules
   b. Restore defective apoptosis genes
   c. Target immune system cells and provide enhanced tumor recognition, tumor homing, and/or tumor destruction properties to these immune system cells.
2. A variety of DNA gene delivery systems (known as "vectors") are being tested, including retroviruses, pox viruses, adenoviruses, and herpes viruses. Cancer patient trials are currently being developed to insert the genes for IL-2, TNF-α, and IL-1β in tumor cells, hoping that the expression of these cytokine genes will contribute to the local activation of the immune system.

D. **Biological Response Modifiers**. The major categories of BRMs that are employed as biotherapies have been cited in Chapter 26; they are generally administered systemically (into the circulatory system) of the cancer patient. Since most BRMs have been successfully produced by genetic engineering techniques, abundant quantities of the agents are available for use in clinical trials; they are substantially pure and free of toxins.

1. **Interferon-α**
   a. **Clinical applications**
      i. **Interferon-α** has been successfully adopted in the treatment of **hairy cell leukemia**, a chronically progressive leukemia for which there was no effective therapy. Modest doses of interferon-α are capable of inducing prolonged durable remissions in those patients.
      ii. Interferon-α is also approved for use in the treatment of Kaposi's sarcoma in its early stages and it also been shown to induce remissions in certain types of lymphoma and chronic myelogenous leukemia and in certain solid tumors (such as melanoma and renal cell carcinoma).
   b. **Mechanism of action**. The mechanism(s) of action of interferon-α in patients who experience clinical remissions of their tumors are not entirely understood; however, much of the evidence favors a direct antiproliferative effect on the tumor cells rather than an up-regulation of the patient's immune system. This antiproliferative effect may be related to the ability that interferon-α seems to have to promote redifferentiation of malignant cells.
2. **Transretinoic acid** is a vitamin A analog that has been found to induce the rapid remission of acute promyelocytic leukemia. The mechanism of action appears to be similar to that of interferon-α (i.e., to promote the redifferentiation of malignant cells, perhaps by inhibiting growth factor release by activated immune cells). One major advantage of this compound is that it is effective after modest doses administered orally.
3. **Custom-made fusion cytokines**, manufactured by recombinant technology, may incorporate on a single fusion molecule both IL-3 and GM-CSF activities. This hybrid cytokine is far more effective in stimulating white-cell production by the bone marrow than either cytokine by itself. Fusion molecules aimed at stimulating antitumoral immune responses are in the early stages of development.
4. **Antitumor monoclonal antibodies**. Infusion of IgG or IgM monoclonal antibodies with tumor antigen specificity should induce tumor cell cytotoxicity by complement activation following binding to tumor cells. However, membrane attack complex activation has not been found in most cancer patients treated with murine monoclonal antibodies (MoAb), perhaps because most murine MoAb activate human complement poorly and mammalian cells have anticomplementary structures (see Chap. 9). Two important limitations have been found with the therapeutic use of murine monoclonal antibodies:
   a. Murine hybridoma-derived antibodies are antigenic to humans. With repeated injections of murine monoclonal antibodies, the host will eventually develop antimouse immunoglobulin antibodies; once antimouse antibodies appear in the circulation, any further injections of monoclonal antibodies will result in the rapid formation of immune complexes. The formation of immune complexes has two main conse-

quences: rapid elimination of the monoclonal antibody, with loss of therapeutic efficiency, and the potential development of serum sickness (see Chap. 21).

b. An additional limitation to the use of monoclonal antibodies is the fact that the tumor cells can escape their effects by at least two different mechanisms:
   i. Antigenic modulation (i.e., shedding or internalization of the surface antigen recognized by the monoclonal antibody), which in some cases seems to be followed by a complete loss of expression of the tumor antigen in question.
   ii. Emergence of antigenically different mutants, which are selected by the monoclonal antibody itself.
c. The immunogenicity question has been addressed by the development of **chimeric (humanized) monoclonal antibodies** produced by immortalized cell lines to which hybrid immunoglobulin genes carry the variable region genes of a murine hybridoma of desired specificity, and the constant region genes of human immunoglobulins carry the antigen specificity of the murine antibody and the constant regions of a human antibody, thereby being considerably less immunogenic than a murine monoclonal antibody.
d. The antigenic modulation/variation question could be solved by strategies in which monoclonal antibodies are used as targeting devices for other BRM or toxic compounds that can promote cytotoxicity more effectively than the monoclonal antibody by itself.
   i. **Conjugate monoclonal antibodies** could potentially focus tumor-killing BRM to appropriate targets in the cancer patient [e.g., by coupling of lymphotoxin (TNF-β) to a tumor-reactive MoAb] or amplify the immune response at the tumor site (e.g., by coupling interferon-γ to a Fab region obtained from an antitumor antibody).
   ii. **Immunotoxins**, obtained by coupling toxic compounds (including radioisotopes, chemotherapeutic agents, or biological toxins) to monoclonal antibodies, with preservation of the bioactivity of both, should target the toxic compound to malignant cells. Such monoclonal antibody "immunotoxins" have been used with success in experimental models but have not yet been shown to be reproducibly effective in the cancer patient setting. This could reflect our current inability to focus enough immunotoxin exclusively at the sites of tumor burden.

5. **Limitations of biological response modifiers**. There are dozens of biological compounds and biologically active chemicals that are undergoing testing as BRMs. None of these agents (either singly or in combination) has of yet been shown to be generally useful; in most series, even in "biotherapy responsive" malignancies, response rates are less than 20%. This is perhaps not surprising given the complexity of the "immune system symphony." The delicate checks and balances that are required for the successful operation of this system would theoretically preclude duplication by the administration of just a single BRM component. Thus, it is possible that the use of

single BRMs should be focused on those malignancies that are known to be directly down-regulated by these agents. Alternatively, whenever a better understanding of the immunological defects in the immune system of cancer patients is obtained, it may then be possible to administer the appropriate combination of BRM (at the proper timing and dosage) required to reproduce the normal anticancer "immunological symphony."

E. **Adoptive Cellular Immunotherapy**. The initial adoptive cellular immunotherapy trials were based on the observation that immune cells isolated from cancer patients could be up-regulated in the laboratory to the point that they would become cytotoxic to tumor cell lines in vitro. Thus, various trials began in which patient leukocyte subsets were removed, up-regulated in the laboratory, and then readministered to the patient.
   1. **Tumor-infiltrating lymphocyte (TIL) or tumor-derived activated cell (TDAC)**. As shown in Figure 28.3, the extraction of T lymphocytes from tumor specimens and their expansion and activation in vitro in interleukin-2 produces the so-called tumor-infiltrating lymphocyte (TIL) or tumor-derived activated cell (TDAC). TIL/TDAC (once generated in the laboratory) can be readministered to the patient, usually with supplementary interleukin-2 to maintain their cytotoxic activity and cyclophosphamide to limit the generation of suppressor T lymphocytes.
   2. **Lymphokine-activated killer (LAK) cells** are obtained by removal of NK cells from the peripheral blood by apheresis, followed by activation and expansion in vitro by incubation with interleukin-2. The resulting LAK cells are then readministered to the patient with supplementary interleukin-2 to maintain their cytotoxic activity.
   3. **Activated killer monocytes (AKM)** are obtained from blood monocytes extracted by apheresis from the cancer patient and activated in vitro to enhance cytotoxic activity with interferon-γ and other BRM. The resultant AKM can then be readministered to the patient with supplementary interferon-γ to maintain the up-regulation of their cytotoxic activity.
   4. **Limitations of adoptive cell immunotherapies**. Therapeutic trials of adoptive cell immunotherapies are truly monumental clinical research efforts. The clinical responses seen in some of the trials (e.g., in selected patients with advanced melanoma and renal cell carcinoma, approximately 20% of patients respond to LAK cell therapy, sometimes with prolonged, complete remissions) represent the best scientific information to date that indicates that the immune system can eradicate an established tumor. However, each trial requires a tremendous effort on the part of clinicians and laboratory and support personnel. It has been estimated that certain of these therapies may cost as much as $100,000 per patient. The time involved, the expense, and the toxicity of certain of these interventions require that much additional research be done to simplify and improve these forms of cancer biotherapy if they are to become clinically useful.

F. **Affinity Column Apheresis** requires the passage of the plasma of cancer patients over specially formulated affinity columns in order to either (1) remove blocking factors of appropriate immune responses or (2) activate tumor cytotoxic factors in the plasma as shown in Figure 28.4.
   1. The original rationale for affinity column apheresis were research observa-

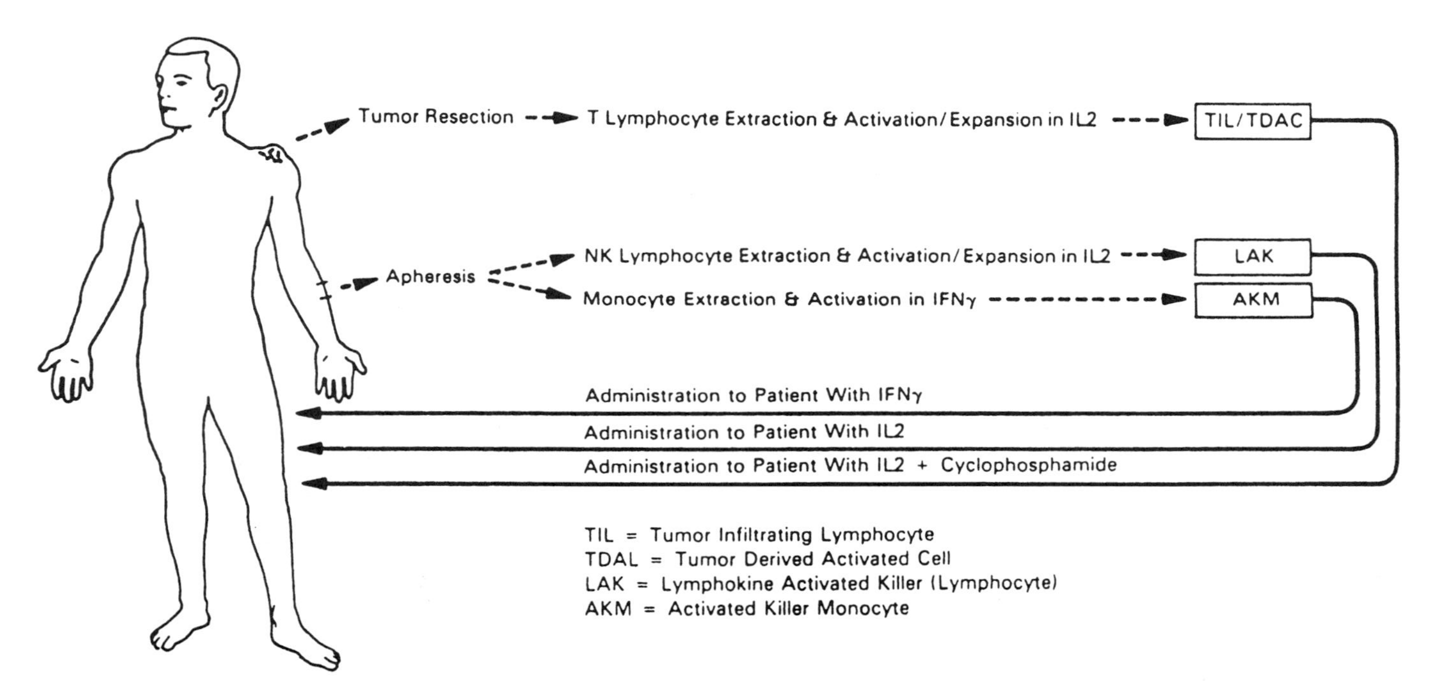

Figure 28.3 Diagrammatic representation of the main protocols developed for adoptive cellular immunotherapy.

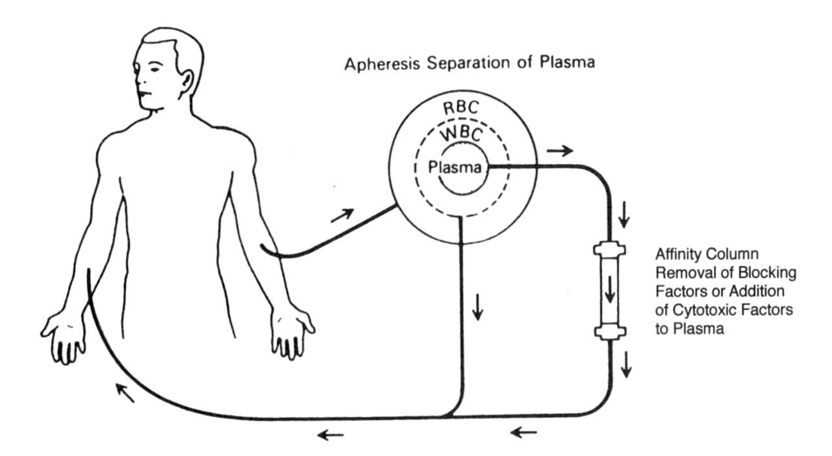

Figure 28.4 Schematic representation of the general protocol for the use of affinity column apheresis in cancer immunotherapy.

tions indicating that the plasma of cancer patients contained "blocking factors" of the immune response. Such blocking factors could include shed tumor antigens or antigen–antibody complexes.

2. Many of the affinity column apheresis cancer therapy trials consist of passing patient plasma over IgG-binding substrates, such as Staphylococcus protein A columns, to remove IgG-containing immune complexes. However, many other reactions affecting complement and other proteins take place in these columns and the precise mechanism of action that may result in a reduction of tumor mass is not known.
3. The results of affinity column apheresis trials have been inconsistent to date.

VII. TUMOR IMMUNOLOGY AND THE FUTURE OF CANCER BIOTHERAPY

The initial excitement regarding cancer biotherapy in the 1980s has not yet resulted in any dramatic overall net population benefit with regard to cancer treatment or prevention. Considering our incomplete understanding of the nature of the malignancy process and the nature of the immune response to malignant cells, this may not be entirely surprising. There is a need to continue the research into the interrelations between malignant cells and the immune system in vitro and animal models. As our basic knowledge improves, new biotherapeutic approaches will continue to emerge and old approaches will be refined. Carefully devised clinical trials are then essential to determine the toxicity and the efficacy rates of these approaches. Positive results with a biotherapeutic approach will stimulate new research to better define their mechanism of action and increase its efficiency. Ultimately, one hopes that strategies will emerge that will not only be useful to fight established malignancies, but will allow us to define strategies that will maintain enhanced immunosurveillance in normal individuals, thus preventing the emergence of clinical cancer. Time will tell if those goals can be achieved.

SELF-EVALUATION

Questions

Choose the ONE *best* answer.

28.1 In vitro activation and expansion of NK cells to be used in adoptive cellular immunotherapy can be best achieved with:
- A. GM-CSF
- B. Interferon-γ
- C. Interleukin-2
- D. Interleukin-6
- E. Transfer factor

28.2 High serum levels of α-fetoprotein (AFP) are detected in:
- A. Heavy smokers
- B. Patients with alcoholic cirrhosis
- C. Patients with germ-cell teratocarcinoma
- D. Patients with localized cancer of the colon
- E. Pregnant women

28.3 Which of the following is not considered a passive form of biotherapy?
- A. Injection of lethally irradiated autologous tumor cells plus BCG
- B. In vivo administration of monoclonal antitumor antibody
- C. In vivo administration of IL-2
- D. Readministration of activated killer monocytes (AKM)
- E. Readministration of activated tumor-infiltrating lymphocytes (TIL)

28.4 Activated mononuclear phagocytes:
- A. Are a part of the nonspecific component of cell-mediated immunity
- B. Are not suitable for adoptive cellular immunotherapy
- C. Destroy normal tissues
- D. Express antigen-specific receptors
- E. Kill melanoma cells better than colon cancer cells

28.5 The antitumoral effect of interferon-α seems related to its ability to:
- A. Enhance the cytotoxicity of peripheral blood monocytes
- B. Prevent viral replication
- C. Potentiate the activity of helper T cells
- D. Promote redifferentiation of malignant cells
- E. Stimulate cytotoxic T cells

Each set of questions below consists of a set of lettered headings followed by numbered phrases. For each numbered phrase, choose the ONE lettered heading that is best. In each group, the lettered heading may be used once, more than once, or not at all.
- A. α-Fetoprotein (AFP)
- B. Carcinoembryonic antigen
- C. Both
- D. Neither

28.6 Pathognomonic when positive

28.7 Elevated serum levels are found in patients with localized colon cancer

28.8 Useful in the management of patients with cancer

For Questions 28.9–28.10, match the pathological condition with the BRM which has been used with greater success for its treatment:

A. Interferon-α
B. Anti-idiotypic antibody
C. LAK cells
D. Retinoic acid
E. Tumor vaccine

28.9 Hairy cell leukemia
28.10 Malignant melanoma

Answers

28.1 (C) IL-2 promotes the expansion and activation of NK cell in vitro. The cells resulting from this process are designated as lymphokine-activated killer (LAK) cells. Other immunotherapy protocols have been tried in which patient cells have been expanded in vitro and reinfused into the patient, including monocytes (expanded with interferon-γ) and tumor-infiltrating lymphocytes (expanded and activated with IL-2).

28.2 (C) Germ-cell teratocarcinoma and hepatocellular carcinoma are the situations associated with the highest concentrations of circulating α-fetoprotein.

28.3 (A) Injection of killed tumor cells and BCG is a form of tumor vaccination, whose objective is to actively induce specific antitumor immunity.

28.4 (A) Activated monocytes seem to kill tumor cells by recognizing a "common membrane marker" of malignant cells. Neither the monocyte recognition system nor the recognized marker is specific for a given type of tumor.

28.5 (D) Interferon-α seems to induce redifferentiation of the malignant cells characteristic of hairy cell leukemia. Retinoic acid has been found to have a similar effect on acute promyelocytic leukemia.

28.6 (D) Both types of antigens can be increased in conditions other than hepatoma or colon carcinoma; therefore, they are NOT pathognomonic for a given malignancy.

28.7 (B) α-Fetoprotein can also be increased in patients with disseminated colon carcinoma, particularly when metastases are formed in the liver.

28.8 (C) Both CEA and α-fetoprotein levels can be used to monitor progression after surgical excision of a tumor producing either one of these oncofetal antigens.

28.9 (A) The two most successful clinical applications of interferon-α as a biotherapeutic anticancer agent have been in hairy cell leukemia and Kaposi's sarcoma.

28.10 (C) The best documented successes of LAK therapy have been in patients with advanced malignant melanoma.

REFERENCES

Chen, Z.X., Xue, Y.Q., Zhang, R., Tao, R.F., et al. A clinical and experimental study of all-trans retinoic acid-treated acute promyelocytic leukemia patients. *Blood*, *78*:1413, 1991.

Chomienne, C., Ballerini, P. Balitrand, N., Daniel, M.T., et al. All-trans retinoic acid in acute promyelocytic leukemias. II. In vitro studies: Structure-function relationship. *Blood*, *76*:1710, 1990.

Gressler, V.H., Weinkauff, R.E., Franklin, W.A., Golomb, H.M. Is there a direct differentiation-inducing effect of human recombinant interferon on hairy cell leukemia in vitro? *Cancer*, *64*:374, 1989.

Oesterling, J.E. Prostate-specific antigen: a critical assessment of the most useful tumor marker for adenocarcinoma of the prostate. *J. Urol.*, *145*:907, 1991.

Oldham, R.K. Immunoconjugates: Drugs and toxins. In: *Principles of Cancer Biotherapy*, R.K. Oldham, ed. Raven Press, New York, 1987.

Ravikumar, T.S., and Steele, G.D., Jr. Modern immunotherapy of cancer. *Adv. Surg.*, *24*:41, 1991.

Restifo, N.R. Recombinant anticancer vaccines. *Sci. Am. Cancer* 2:16, 1996.

Rosenberg, S.A. Adoptive cellular therapy in patients with advanced cancer. *Biol. Ther. Cancer Updates*, *1(1)*:1, 1991.

Stevenson, H.C., ed. *Adoptive Cellular Immunotherapy of Cancer*. Marcel Dekker, New York, 1989.

Tsang, K.Y., Zaremba, S., Nieroda, C.A., et al. Generation of human cytotoxic T cells specific for human carcinoembryonic antigen epitopes from patients immunized with recombinant vaccinia-CEA vaccine. *J. Natl. Cancer Inst.*, *87*:982, 1995.

Vaux, D.L. and Strasser, A. The molecular biology of apoptosis. *Proc. Natl. Acad. Sci.*, *93*:2239, 1996.

29
Malignancies of the Immune System

Gabriel Virella and Jean-Michel Goust

I. INTRODUCTION

A. **Lymphocytes** are frequently affected by neoplastic mutations, perhaps as a consequence of their intense mitotic activity. Lymphocyte malignancies can be broadly classified into B-cell and T-cell malignancies.
   1. B-cell malignancies can be identified by the production of abnormal amounts of homogeneous immunoglobulins (or fragments thereof) resulting from the monoclonal proliferation of immunoglobulin-secreting B cells or plasma cells or by specific cell markers.
   2. T-cell malignancies are usually identified through specific cell markers.

II. B-CELL DYSCRASIAS

A. **Definitions**
   1. The abnormally homogeneous immunoglobulins characteristic of malignant proliferation of immunoglobulin-producing cells are designated as **monoclonal proteins** because they represent the activity of a single mutant clone. However, monoclonal proteins may be detected in patients without overt signals of malignancy (some mutations may lead to clonal expansion without uncontrolled cell proliferation).
   2. The conditions associated with detection of monoclonal proteins are generically designated as **monoclonal gammopathies**, **plasma cell dyscrasias**, or **B-cell dyscrasias** (from the Greek *dyskrasis*, meaning "bad mixture," often used to designate hematological disorders affecting one particular cell line).
   3. **Monoclonal proteins** or **paraproteins**, in practical terms, are defined by the fact that they are constituted by large amounts of identical molecules, carrying one single heavy-chain class and one single light-chain type, or, in some cases, by isolated heavy or light chains of a single type (Fig. 29.1).

B. **Diagnosis of B-Cell Dyscrasias**
   1. In general, the diagnosis of a B-cell dyscrasia relies on the demonstration of a monoclonal protein. Secreted paraproteins are detected by a combination of methods.
      a. Initial screening usually involves the **electrophoretic separation of**

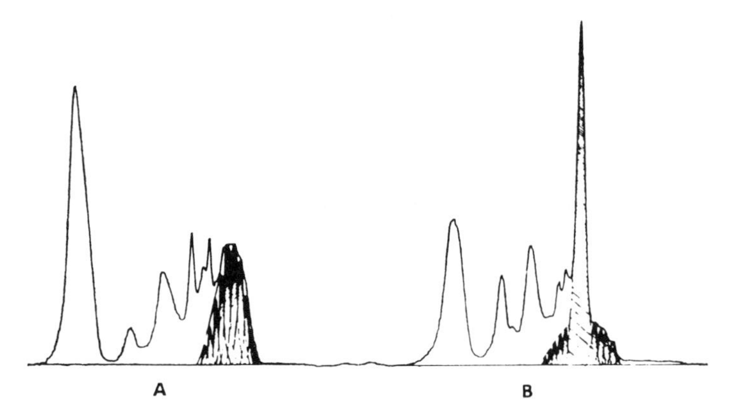

Figure 29.1 Concept of monoclonal gammopathy: In normal sera or reactive plasmacytosis the gamma globulin fraction is made up of the sum of a large number of different antibodies, each one of them produced by a different plasma cell clone (A); if a B-cell clone escapes normal proliferation control and expands, the product of this clone, made up of millions of structurally identical molecules, will predominate over all other clonal products and appear on the electrophoretic separation as a narrow-based, homogeneous peak in the gamma globulin fraction (B).

serum and urine from the suspected case (Fig. 29.2). To be sure that urinary proteins are not overlooked because of their low concentration, the urine sample must be concentrated.

b. Electrophoretic studies must usually be supplemented by **immunoelectrophoresis** or by **immunofixation**. These studies are essential in order to characterize the paraproteins as containing one of the five possible classes of immunoglobulins, and one of the two possible types of light chains (Fig. 29.3), necessary criteria to confirm the monoclonal nature of a suspected electrophoretic spike.

c. The diagnosis of some specific B-cell dyscrasias, such as **light-chain disease** (a variant of multiple myeloma characterized by the synthesis of large amounts of homogeneous light chains), **Waldenstrom's macro-**

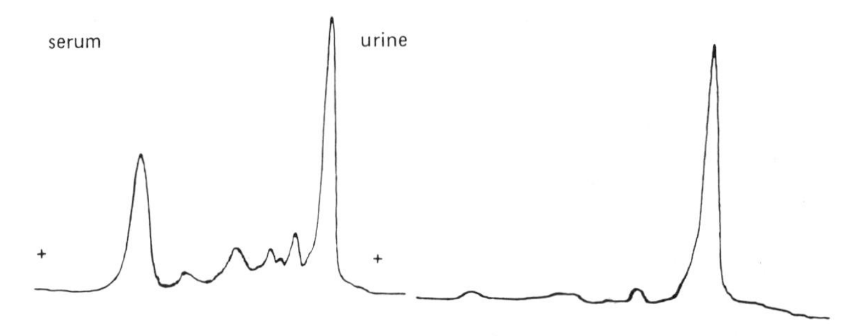

Figure 29.2 Electrophoresis of serum and urine proteins from a patient with multiple myeloma. The serum, shown on the left, depicts a very sharp peak in the gamma globulin region, with a base narrower than that of albumin, corresponding to an IgG monoclonal component. The urine, shown on the right, indicates a sharp fraction in the gamma region with only traces of albumin, meaning that the monoclonal peak is constituted by proteins smaller than albumin, able to cross the glomerular filter. This monoclonal protein in the urine was constituted by free κ-type light chains (κ-type Bence-Jones protein).

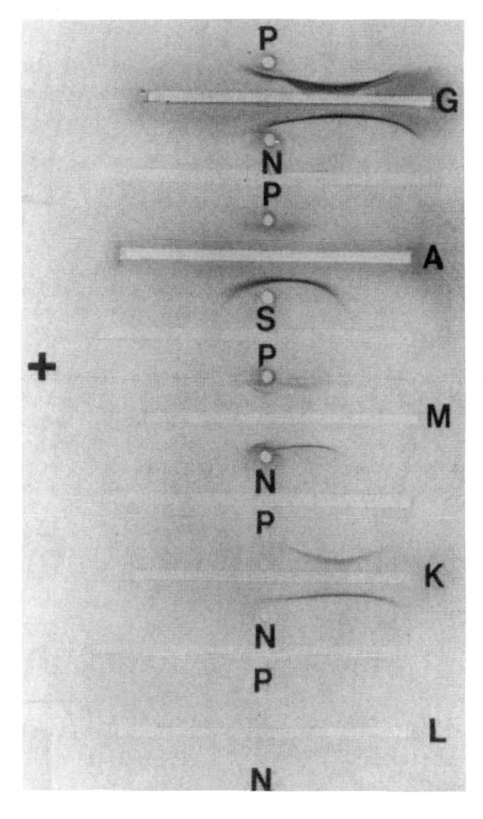

Figure 29.3 Immunoelectrophoretic analysis of the serum protein of a patient with multiple myeloma (P), compared to serum from a normal control (N). The following antisera were used: anti-IgG (G), anti-IgA (A), anti-IgM (M), anti-kappa chains (K), and anti-lambda chains (L). Notice the abnormally homogeneous shape of the precipitation arcs corresponding to the patient's serum when studied with anti-IgG and anti-κ chains, allowing the identification of an IgG κ monoclonal component.

globulinemia, or the **heavy-chain diseases**, depends on the immunochemical characterization of the paraprotein.

d. In some instances, plasma cell dyscrasias do not result in the secretion of paraproteins. In rare cases of **multiple myeloma**, for example, the neoplastic mutation alters the synthetic process so profoundly that no paraproteins are produced (**nonsecretory myeloma**).
e. B-lymphocyte malignancies also may not be associated with the synthesis of monoclonal proteins.
   i. **Chronic lymphocytic leukemias** are B-cell malignancies in more than 90% of the cases, but only one-third show paraproteins; the remainder have monoclonal cell surface immunoglobulins only.
   ii. **B-cell acute lymphocytic leukemias** show rearrangements of their immunoglobulin heavy-chain genes in chromosome 14 and the malignant cells may synthesize μ chains, but those remain intracytoplasmic and there is no detectable monoclonal protein in serum or urine.

2. In the majority of cases, the finding of a monoclonal protein in a laboratory

test does not give a very precise diagnostic indication. For example, the isolated finding of homogeneous free light chains (**Bence-Jones protein**) in the urine may correspond to one of the following B-cell dyscrasias: (1) light-chain disease; (2) chronic lymphocytic leukemia; (3) lymphocytic lymphoma; or (4) "benign" or "idiopathic" monoclonal gammopathy. The precise diagnosis depends on a combination of clinical and laboratory data, as discussed in detail later in this chapter.

Case 29.1

A 55-year-old woman was admitted with progressive weakness, malaise, fatigue, and loss of vision. She also felt short of breath after climbing one flight of stairs. She claimed that for the last 3 months she had experienced a progressive loss of bilateral vision, which she described as progressive blurring, to the point that she could hardly walk around unassisted. She had also lost 5 pounds since she started feeling sick and had noticed easy gum bleeding after brushing her teeth. She had also experienced three episodes of nose bleeding in the last few weeks.

Her past medical history included two normal pregnancies, diffuse rheumatic pains for the last year, and an episode of pneumonia 6 months ago.

Physical examination showed the patient to be thin, with normal vital signs. Examination of the head showed pale conjunctival mucosae. Fundoscopy showed irregularly dilated, tortuous veins, with flame-shaped retinal hemorrhages and "cotton wool" exudates in both eyes. Ear, nose, and throat appeared normal. Neck examination revealed no abnormalities. No lymph nodes were palpable. Chest and abdominal examination were within normal limits. Pelvic and neurological examination also showed no abnormalities.

Serum protein electrophoresis—abnormal spike in the gamma region, which was typed by immunofixation as IgA kappa. Serum immunoglobulin levels—IgG: 340 mg/dL; IgA: 5800 mg/dL; IgM: 47 mg/dL. Complete blood count—RBC: $3.8 \times 10^6/\mu L$ (normal 4.2–5.4); WBC: $5.9 \times 10^3/\mu L$ (normal 4.8–10.8); normal differential; erythrocyte hemoglobin: 8.9 g/dL (normal 12.0–16.0). X-ray survey of the skeleton: multiple osteolytic lesions in the skull, femur, humerus, pelvis, spine, and ribs.

This case raises several questions:

- Why did this patient present with weakness, easy fatigue, and loss of vision?
- What is the significance of the fundoscopic abnormalities and exaggerated mucosal bleeding?
- What is the nature of the osteolytic lesions seen in the bone?
- What is the diagnosis of this condition?
- What is the best treatment for this patient?

C. Physiopathology of B-Cell Dyscrasias

1. **Direct pathological consequences of malignant B-cell proliferation** include:
   a. Enlargement of lymph nodes, spleen, and liver, as seen in lymphomas and some leukemias.
   b. Leukemic invasion of peripheral blood, characteristic of B-cell leukemias.
   c. Compressive and obstructive symptoms can result from the proliferation of plasma cells in soft tissues. Oropharyngeal plasmocytomas often lead

to obstructive symptoms. Heavy-chain-producing intestinal lymphomas, when grossly nodular, can lead to intestinal obstruction.

d. Intestinal malabsorption is typical of α-chain disease. It results from extensive infiltration of the intestinal submucosa by malignant B cells, causing total disruption of the normal submucosal architecture.

2. **General metabolic disturbances** are responsible for some major pathological manifestations of B-cell dyscrasia, including:
   a. **Bone destruction**, which does not result directly from B-cell proliferation but rather from **osteoclast hyperactivity** secondary to the release of several soluble osteoclast-stimulating mediators by the malignant B cells and/or by activated T cells, including **IL-6, lymphotoxin (TNF-β)**, **macrophage colony-stimulating factor** (**M-CSF**), and **IL-1β**. It has been proposed that the initial event in the evolution toward multiple myeloma is the exaggerated synthesis of IL-1β and/or TNF-β by transformed plasma cells. These cytokines activate osteoblasts and/or stromal cells and induce the secretion of IL-6, which acts as a growth factor for plasma cells and as an activating factor for osteoclasts. Osteoclasts are also activated by M-CSF, also produced by neoplastic plasma cells.
   b. **Renal insufficiency** can result from a diversity of factors, such as hypercalcemia (secondary to bone reabsorption), hyperuricemia, deposition of amyloid substance in the kidney, clogging of glomeruli or tubuli with paraprotein (favored by dehydration), and plasmocytic infiltration of the kidney.
   c. **Anemia** (normochromic, normocytic) is frequent and is basically due to decreased production of red cells. A moderate shortening of red-cell survival is also common.
3. **Serum hyperviscosity**
   a. The viscosity of serum relative to water increases with protein concentration. **IgM** and **polymeric IgA**, due to their molecular complexity and high intrinsic viscosity, lead to disproportionate increases of serum viscosity (Fig. 29.4).
   b. The **hyperviscosity syndrome** is a frequent manifestation of **Waldenstrom's macroglobulinemia**; however, it is also observed in multiple myeloma patients, mainly in those with IgA paraproteins, and occasionally in IgG myeloma.
   c. The symptoms of serum hyperviscosity are related to high protein concentration, expanded plasma volume, and sluggishness of circulation. Table 29.1 lists the main signs and symptoms of the syndrome. Typical fundoscopic changes are shown in Figure 29.5.
4. **Immunodeficiency**
   a. One of the classic clinical features of malignant B-cell dyscrasias, particularly of multiple myeloma, is the increased tendency for pyogenic infections. This is paralleled by decreased levels of normal immunoglobulins and decreased antibody production after active immunization.
   b. The depression of the immune response in patients with multiple myeloma appears to be multifactorial.
      i. In IgG myeloma, the large amounts of circulatory IgG present are likely to have a feedback effect, depressing normal IgG synthesis.

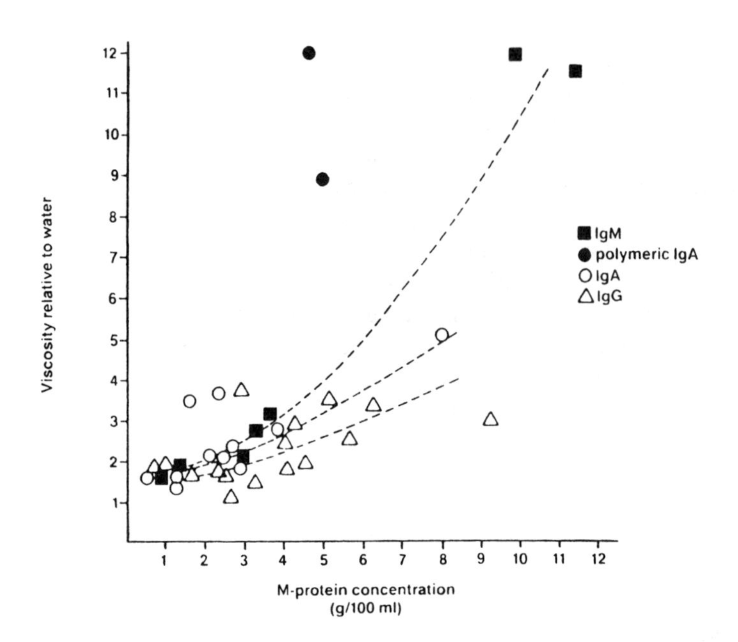

Figure 29.4 Plot of relative serum viscosity versus monoclonal protein concentration in sera containing IgG, IgA, and IgM monoclonal proteins. The highest viscosity values were determined in sera containing IgM or polymeric IgA monoclonal proteins.

Table 29.1 Clinical Manifestations of the Hyperviscosity Syndrome[a]

Ocular
Variable degrees of vision impairment
Fundoscopic changes
Dilation and tortuosity of retinal veins ("string-of-sausage" appearance)
Retinal hemorrhage and "cotton-wool" exudates
Papilledema
Hematological
Mucosal bleeding (oral cavity, nose, gastrointestinal tract, urinary tract)
Prolonged bleeding after trauma or surgery
Neurological
Headaches, somnolence, coma
Dizziness, vertigo
Seizures, EEG changes
Hearing loss
Renal
Renal insufficiency (acute or chronic) due to (a) clogging of the glomerular vessels with paraprotein and (b) diminished concentrating and diluting abilities
Cardiovascular
Congestive heart failure secondary to expanded plasma volume

[a]Modified from Bloch, K.J., and Maki, D.G. *Sem. Hematol.*, *10*:113, 1974.

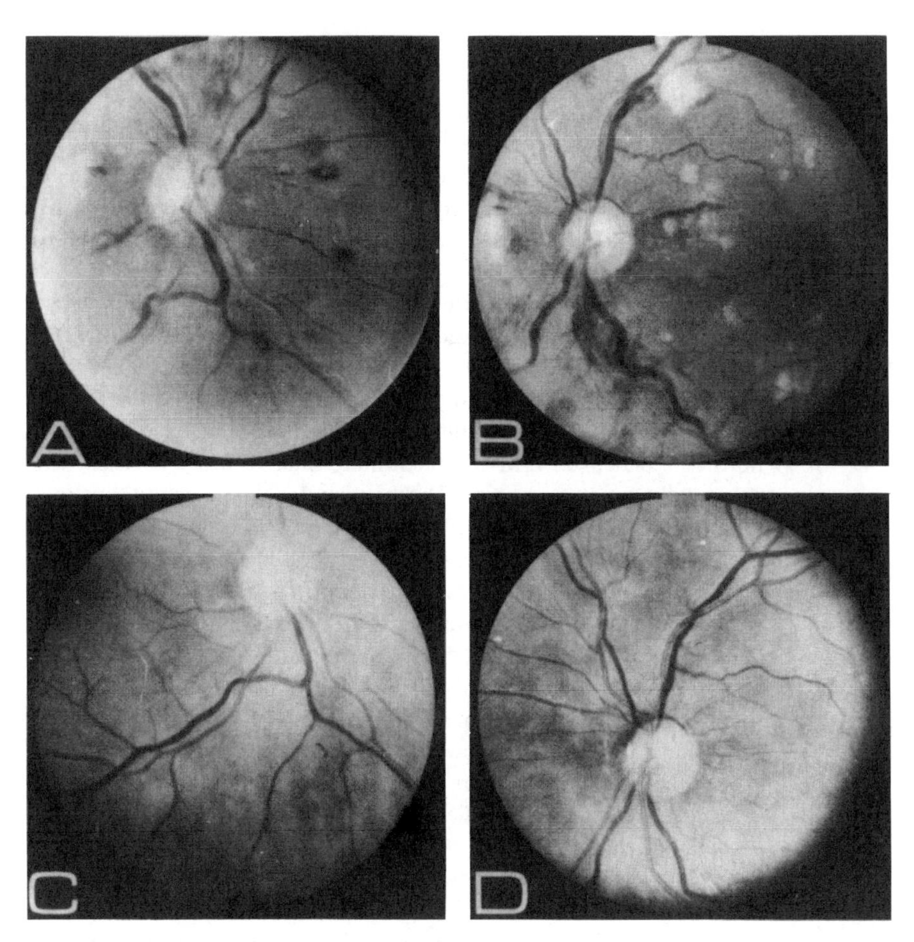

Figure 29.5 Fundoscopic examination of a patient with hyperviscosity syndrome. (A) and (B) were obtained from the right and left eyes, respectively, at the time of admission. Flame-shaped hemorrhages, "cotton-wool" exudates, and irregular dilation of retinal veins are evident. (C) and (D) were obtained from the same eyes after 5 months of therapy, showing total normalization. (From Virella, G., et al. Polymerized monoclonal IgA in two patients with myelomatosis and hyperviscosity syndrome. *Br. J. Haematol.*, *30*:479, 1975. (Reproduced with permission of the publisher.)

ii. A more general mechanism of suppression of the humoral response seems to be mediated by suppressor cells. Phagocytic monocytes (and to a lesser extent T cells) appear to mediate this suppression, which can be demonstrated by cocultures of peripheral mononuclear leukocytes from normal donors and myeloma patients, resulting in impairment of the function of the normal B lymphocytes.

iii. Other abnormalities that may contribute to the predisposition to infections are defects in neutrophil responses and impairment of Fcγ receptor functions, which are more likely to be present in patients with renal failure. Anemia also seems to predispose to infections, for unknown reasons.

c. In **chronic lymphocytic leukemia**, in addition to a depression of humoral immunity (milder than that seen in multiple myeloma), there is a depression of T-cell counts and function. Viral and fungal infections, as well as cases of disseminated infection after administration of live attenuated viral vaccines, have been reported in patients with this type of leukemia.

4. **Pathological consequences of the immunological activity of a paraprotein**. Most paraproteins have unknown and inconsequential antibody activities, but in some exceptional cases the reactivity of a monoclonal protein may be directly responsible for some of the manifestations of the disease.

a. **Cold agglutinin disease** results from the synthesis of large concentrations of monoclonal IgM (IgMκ in more than 90% of the cases) with cold agglutinating properties. Those monoclonal cold agglutinins react with the **I antigen** expressed by the erythrocytes of all normal adults, and the reactivity is only evident at temperatures below normal body temperature. Usually, the sera suspected to contain them are tested at 4°C on a direct hemagglutination test using O-positive red cells as antigens. For these reasons, these autoantibodies are known as cold agglutinins. Monoclonal cold agglutinins can be detected:

i. In cases of IgM-producing B-cell malignancy (Waldenstrom's macroglobulinemia, see below), when the IgM paraprotein behaves as a cold agglutinin. The titers of cold agglutinins in such cases are very high, and usually the patient will have symptoms attributable to the cold agglutinin.

ii. In patients with symptomatic cold agglutinin disease associated to high titers of a monoclonal cold agglutinin in which there is no evidence of B-cell dyscrasia other than the presence of a monoclonal anti-I antibody and an increase in the numbers of lymphoplasmocytic cells in the bone marrow.

The clinical manifestations of cold agglutinin disease fall into two categories:

i. **Cold-induced hemolytic anemia**, which is usually mild but in some severe cases can be intense enough to lead to acute renal failure.

ii. **Cold-induced ischemia**, due to massive intracapillary agglutination in cold-exposed areas.

b. **Hyperlipidemia**. A pronounced increase in serum lipid and lipoprotein levels can often be detected in patients with monoclonal gammopathies and, in some cases, the monoclonal protein has antibody activity to lipoproteins. It has been demonstrated that the binding of antibodies to the lipoprotein molecules alters the uptake and intracellular processing of the lipoprotein, resulting in hyperlipidemia and in increased accumulation of cholesterol in macrophages.

D. Clinical Presentations of B-Cell Dyscrasias

1. Multiple myeloma

a. **Clinical features**. The most frequent clinical symptoms of multiple myeloma are (1) **bone pain** and "spontaneous" or "pathological" **fractures**; (2) malaise, headaches, or other symptoms related to **hyperviscosity**; (3) weakness and **anemia**; (4) repeated **infections**; and (5) **renal failure**.

i. The presentation of multiple myeloma can vary considerably. In some cases, **anemia** is the leading feature, and the diagnosis is established when the cause of anemia is investigated. Hemoglobin levels below 7.5 g/dL are usually associated with poor prognosis.

ii. Other cases are first seen in a rheumatology out-patient clinic, due to "**bone pains**."

iii. Cases with advanced bone destruction may reveal themselves by a fracture after minimal trauma (known as **"pathological" fractures**).

iv. Symptoms related to **hyperviscosity** (see Table 29.1) may also lead to hospitalization.

v. **Repeated infections** and **renal failure**, which usually occur in advanced stages of the disease, are among the most frequent causes of death, but rarely constitute the presenting symptoms. Infection is associated with an increased risk of death and the prognosis of a multiple myeloma patient with renal failure, particularly when his blood urea nitrogen exceeds 80 mg/dL, is also poor.

vi. Several parameters have been found to be associated with rapidly growing myelomas and poor prognosis. In general, these parameters reflect the magnitude of the malignant cell population and its degree of dedifferentiation and include: heavy plasma cell infiltration of the bone marrow (>30%); numerous lytic bone lesions; Bence-Jones proteinuria; low hemoglobin (<11 g/L); expression of myeloid cell markers and the CALLA antigen on the membrane of the malignant plasma cells; expression of the *c-myc* gene; and high serum levels of β_2-microglobulin, thymidine kinase, and neopterin.

b. **Laboratory diagnosis**. A typical case of multiple myeloma will present at least two elements of the following diagnostic triad: (1) **bone lesions**; (2) **monoclonal protein** in serum and/or urine; and (3) bone marrow **plasmacytosis**.

i. **Bone lesions** are typically osteolytic (appear in the x-ray as punched-out areas without peripheral osteosclerosis) and multiple (several punched-out areas appear in the same bone and can be seen in a number of bones in the same patient) (Fig. 29.6). Practically all bones can be affected.

- In advanced cases, **pathological fractures** can occur in the long bones, skull, or spinal column.
- Rarely, a single bone lesion may be detected in one patient; however, such a "**solitary bone plasmacytoma**" is in fact rarely solitary, and bone marrow aspiration will reveal diffuse plasmacytosis in most cases.
- Exceptionally, a patient with monoclonal gammopathy and diffuse plasmacytosis can present with no evident bone lesions, or with generalized osteoporosis.

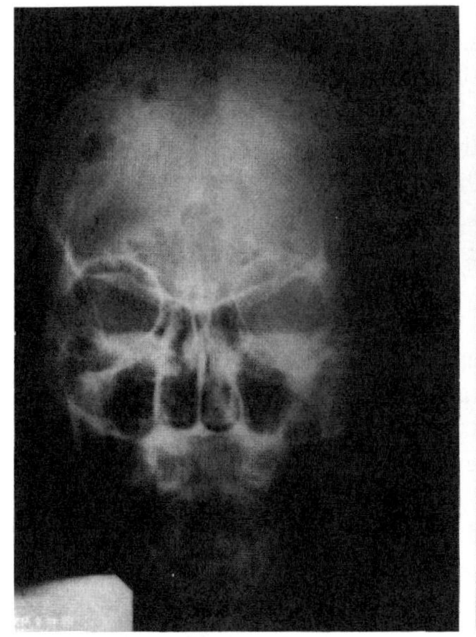
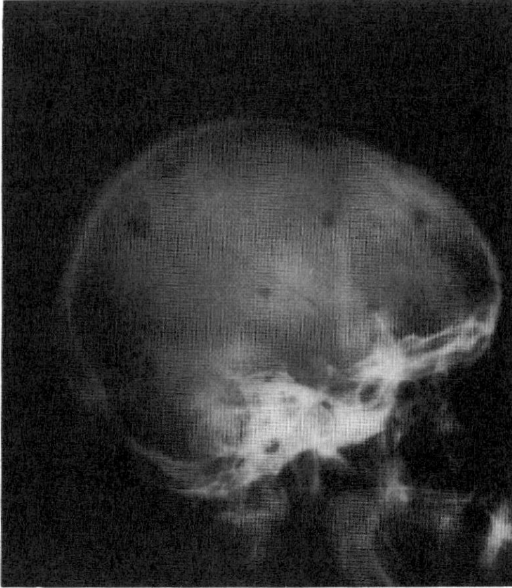

Figure 29.6 X-ray of the skull of a patient with multiple myeloma showing typical osteolytic lesions. (Courtesy of Dr. S. Richardson.)

ii. **Monoclonal protein in serum and/or urine**. In 98% of the cases of multiple myeloma a monoclonal protein can be detected by adequate studies. The distribution of monoclonal proteins among the different immunoglobulin classes closely parallels the relative proportions of those immunoglobulins in normal serum: 60–70% of the proteins are typed as IgG, 20–30% as IgA, 1–2% as IgD, and, very rarely one monoclonal protein can be typed as IgE. A single light-chain type is found in these paraproteins. For example, IgG paraproteins can be *either* κ or λ.

- The finding of a heterogeneous increase of IgG or any other immunoglobulin (i.e., and increase of both IgGκ and IgGλ molecules) is not compatible with a diagnosis of multiple myeloma.
- In the urine, the most frequent finding is the elimination of **free light chains**, κ or λ (**Bence-Jones proteins**). These light chains are usually found in addition to a monoclonal immunoglobulin detectable in serum, but in about 20–30% of patients with multiple myeloma, the only abnormal proteins to be found are the free monoclonal light chains in the urine. Some authors give the designation of **light-chain disease** to the form of multiple myeloma in which the paraprotein consists of free light chains.
- Very rarely (about 2% of cases) no monoclonal paraprotein is detected in the serum or urine of a patient with a typical clinical picture or multiple myeloma. This situation is designated as **nonsecretory myeloma**. In most nonsecretory myelomas, immunofluorescence studies have demonstrated intracellular mono-

clonal proteins that are not secreted into the extracellular spaces. Nonsecretory myelomas have a very poor prognosis.

iii. **Bone marrow plasmacytosis**. In multiple myeloma bone marrow, aspirates show increased numbers of plasma cells with a more or less mature appearance (Fig. 29.7). The plasma cell infiltration can be massive, with sheets of plasma cells occupying the bone marrow.

- An increase in the number of plasma cells in the bone marrow, even when associated with morphological aberrations, is not sufficient to differentiate between malignant and reactive plasma cell proliferations.
- The differential diagnosis between malignant and reactive plasma cell proliferations should be based on the immunochemical characteristics of the patient's immunoglobulins. While a patient with a malignant B-cell dyscrasia will show either a monoclonal protein or low immunoglobulin levels (it if is a case of nonsecretory myeloma), reactive plasmacytosis is invariably associated with a polyclonal increase of immunoglobulins.

c. **Management**

i. **Staging**. The management of a patient with a monoclonal gammopathy rests on a decision about whether the malignancy is dormant or aggressive, and the latter ones can be staged by a variety of parameters indicating early stages versus full-blown or terminal stages. Several laboratory parameters have been suggested to be helpful in the decision-making process:

- A high labeling index, usually determined by counting the number of plasma cells that show bright nuclear staining with a fluorescent monoclonal antibody to 5-bromo-2-deoxyuridine, an enzyme that becomes associated with the DNA of actively proliferating cells, is considered indicative of an aggressive tumor.

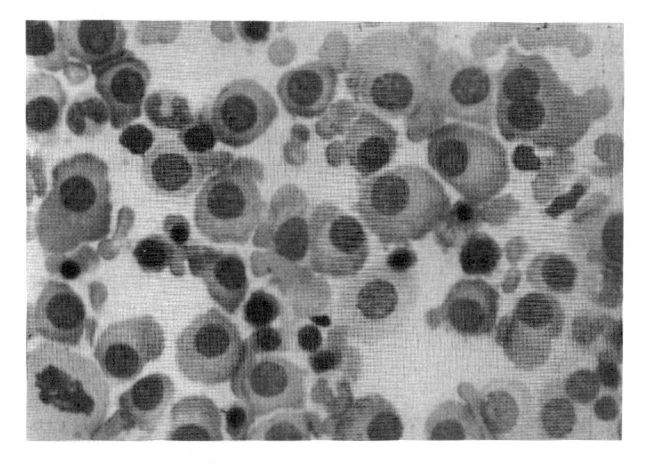

Figure 29.7 Plasma cell infiltration of the bone marrow in a patient with multiple myeloma. Note the binucleated plasma cell in the upper right corner of this figure.

- Other markers associated with aggressive tumors include high levels of: β_2 microglobulin (produced and shed by actively proliferatingplasmacells);thymidinekinase(anotherenzyme involved in DNA synthesis which is released by dividing cells); IL-6 (the main cytokine proposed to mediate plasmocytoma cell growth); C-reactive protein (a direct reflection of the levels of IL-6, since this interleukin stimulates C-reactive protein synthesis); and a combination of assays for the labeling index and β_2 microglobulin levels is believed to give the best indication about the degree of aggressiveness of a multiple myeloma.

ii. **Chemotherapy**
- The standard treatment for multiple myeloma is the administrationofacombinationofmelphalanandprednisoneinintermittent high-dosage cycles
- Combination chemotherapy regimens (including these two drugs plus several others, such as vincristine, nitrosourea, cyclophosphamide, and doxorubocin) have gained favor with more complete and durable responses.
- Even higher rates of remission can be obtained with administration of several chemotherapeutic agents followed by autologous bone marrow or stem-cell transplantation. The preferred approach is to obtain stem cells (CD34+) from the peripheral blood after an initial administration of large doses of cytotoxic drugs (particularly cyclophosphamide) plus GM-CSF, and use those cells to reconstitute the immune system after more extensive ablation of the bone marrow.

iii. **Biological response modifiers**
- **Administration of interferon-α** as a form of maintenance therapy after a favorable response is obtained with chemotherapy has been reported to extend the duration of relatively symptom-free periods.
- In vitro experiments suggest that **interferon-γ** inhibits the IL-6-induced growth of some myeloma cell lines, and that **retinoic acid** (an agent known to induce the redifferentiation of malignant cells) depresses the expression of IL-6 receptors on other myeloma cell lines.

iv. **Plasmapheresis**, which consists of replacing the patient's plasma by normal plasma or a plasma-replacing solution is indicated in cases with hyperviscosity, since the reduction of viscosity secondary to the reduction in the circulating levels of monoclonal protein caused by chemotherapy takes time and the rapid correction of hyperviscosity may be essential for proper management. The rapidly beneficial effects of plasmapheresis are illustrated in Figure 29.4, which shows the normalization of retinal changes observed after plasmapheresis.

v. Other **supportive measures** include hemodialysis or peritoneal dialysis in cases with renal insufficiency, and antibiotic therapy and prophylactic administration of gamma globulins in patients with recurrent infections.

2. **Plasma cell leukemia** is the designation applied to cases in which large numbers of plasma cells can be detected in the peripheral blood (exceeding 5–10% of the total white blood cell count). Besides the leukemic picture in the peripheral blood, the remaining clinical and laboratory features of plasma cell leukemia are usually indistinguishable from those of multiple myeloma. The prognosis is generally poor; this may reflect a higher degree of dedifferentiation on the part of malignant plasma cells that abandon their normal territory.
3. **"Benign" or "idiopathic" monoclonal gammopathies**. The designation of benign or idiopathic monoclonal gammopathy is used when a monoclonal protein is found in an asymptomatic individual or in a patient with a disease totally unrelated to B-lymphocyte or plasma cell proliferation (solid tumors, chronic hepatobiliary disease, different forms of non-B-cell leukemia, rheumatoid arthritis, etc.).
   a. **Frequency**. Scandinavian authors conducting extensive population studies have given an average figure for the incidence of idiopathic monoclonal gammopathies of about 1%, by far the most common form of B-cell dyscrasia. The incidence seems to increase in the elderly, up to about 20% in 90-year-old and older individuals.
   b. **Clinical significance**. The clinical significance of the finding of a benign or idiopathic monoclonal gammopathy lies in the need to make a differential diagnosis with a malignant B-cell dyscrasia in its early stages. Several criteria for the differential diagnosis between benign and malignant plasma cell dyscrasias have been proposed (Table 29.2).
   c. **Management**. A good practical rule is to assume that any monoclonal gammopathy detected unexpectedly during the investigation of a condition not clearly related to a B-cell malignancy, or during screening of a normal population (particularly in individuals of advanced age), should be considered as benign until proven otherwise. The best attitude in such cases is to withhold cytotoxic therapy and follow the patients closely, every 3 to 6 months, measuring the amount of paraprotein; malignant

Table 29.2 Laboratory Features That Have Been Proposed for the Differentiation Between Malignant and Idiopathic Monoclonal Gammopathies

Feature	Benign	Malignant
Paraprotein	Complete molecule; little or no Bence-Jones protein	Bence-Jones proteinuria > 0.6 g/day
Normal immunoglobulins	Conserved	Depressed
Serum paraprotein	< 1 g/100 mL	> 1 g/100 mL
Serum albumin	> 3 g/100 mL	< 3 g/100 mL
Hemoglobin	> 10 g/100 mL	< 10 g/100 mL
Serum urea	< 80 mg/100 mL	> 80 mg/100 mL
Serum β_2-microglobulin	< 700 μg/100 mL	> 700 μg/100 mL
Nonspecific proteinuria	Absent	Present
Numbers of B cells in peripheral blood	Normal	Decreased
Labeling index	Normal	Increased

cases show a progressive increase, whereas in benign cases the levels remain stable. Patients with benign gammopathy have to be observed at least yearly after the first 2 years of follow-up, since there are documented cases of malignant evolution after 5 or more years of "benign" behavior.

4. **Waldenstrom's macroglobulinemia** was first reported by a Swedish internist, Dr. **Jan Waldenstrom**, as a malignancy of lymphoplasmacytoid cells associated with increased levels of serum macroglobulins. After IgM was structurally and antigenically characterized, this B-cell dyscrasia was redefined to include the existence of a monoclonal IgM protein in the serum as its diagnostic hallmark.
   a. **Clinical features**. Waldenstrom's macroglobulinemia is clinically characterized by weakness and anemia; hyperviscosity-related symptoms; hepatomegaly, splenomegaly, and lymphadenopathy; and diffuse osteoporosis. Symptoms typical of multiple myeloma, such as bone pain or "spontaneous" fractures, are rare. The immunosuppression is also milder than in multiple myeloma. Hypercalcemia, leukopenia, thrombocytopenia, and azotemia are rarely seen. Renal insufficiency, when present, is usually a manifestation of serum hyperviscosity and can be reversed by plasmapheresis and/or peritoneal dialysis.
   b. **Diagnosis**. The two main diagnostic features of Waldenstrom's macroglobulinemia are presence of an **IgM monoclonal protein**; and **pleomorphic infiltration of the bone marrow** with plasma cells, lymphocytes, and lymphoplasmacytic cells.
   c. **Management**. Waldenstrom's macroglobulinemia is a disease of old age and frequently follows a benign course. The most common life-threatening complications result from serum hyperviscosity. In such cases, repeated plasmapheresis is often sufficient to keep the patient asymptomatic, avoiding the use of cytotoxic drugs and their side effects. Cytotoxic therapy may be required, due to the severity of the symptoms or the impossibility of keeping the patient on repeated plasmapheresis. Chlorambucil (leukeran) is the drug of choice, usually given in a continuous low dosage.
5. **The heavy-chain diseases**. Some B-cell dyscrasias are associated with the exclusive production of heavy chains (or fragments thereof) or with the synthesis of abnormal heavy chains that are not assembled as complete immunoglobulin molecules and are excreted as free heavy chains. Both types of abnormality can be on the basis of a heavy-chain disease. The heavy-chain diseases are classified according to the isotype of the abnormal heavy chain as γ, α, μ, and δ (one single case of δ chain disease has been reported, and ϵ-chain disease has yet to be described).
   a. **α-Chain disease**. This is the most common and best defined heavy-chain disease. It affects patients in all age groups, even children, and is more frequent in the Mediterranean countries, particularly affecting individuals of Jewish or Arab ancestry.
      i. Clinically, it is indistinguishable from the so-called **Mediterranean-type abdominal lymphoma**, characterized by diarrhea and malabsorption unresponsive to gluten withdrawal, with progressive wasting and death. Intestinal x-ray changes, suggestive of diffuse

infiltration of the small intestine such as thickened mucosal folds. Intestinal biopsy reveals diffuse infiltration of the submucosa by reticulolymphocytic cells.

ii. Diagnosis relies on the demonstration of free alpha chains, usually in serum. Routine electrophoresis usually fails to show a monoclonal component, but immunoelectrophoresis shows an abnormal IgA that does not react with antisera specific for light chains.

b. **γ-Chain disease**. This was the first form of heavy-chain disease discovered. Clinically, it appears as a lymphoma with lymphadenopathy, splenomegaly, and hepatomegaly. Bone marrow and lymph node biopsies show lymphoplasmocytic proliferation. The diagnosis is based on the immunochemical demonstration of free γ chains in the serum and/or urine.

c. **μ-Chain disease**. This variant of heavy-chain disease is less frequent than either γ- or α-heavy-chain diseases, and clinically it is indistinguishable from chronic lymphocytic leukemia or lymphocytic lymphoma, with marked Bence-Jones proteinuria and small amounts of free μ chains detectable in the serum and sometimes also in the urine.

Case 29.1 Revisited

- The history of progressive weakness, malaise, fatigue, with shortness of breath after climbing one flight of stairs in a 55-year-old woman is suggestive of anemia, supported by the observation of pale conjunctival mucosae. The anemia of multiple myeloma has an obscure etiology and is neither hemolytic nor secondary to iron deficiency.
- Progressive loss of vision with bilateral retinal changes (irregularly dilated, tortuous veins, with flame-shaped retinal hemorrhages and "cotton wool" exudates) is very suggestive of serum hyperviscosity, which is also associated with increased bleeding. Serum hyperviscosity causes venous stasis with exudation and bleeding in the retinal capillaries, which leads to progressive vision impairment.
- In a patient with serum hyperviscosity, the main differential diagnosis is between multiple myeloma and Waldenstrom's macroglobulinemia, since this last disease is frequently associated with the hyperviscosity syndrome. The finding of osteolytic lesions, a very high level of IgA and low IgM, strongly indicates multiple myeloma as the most likely diagnosis. Additional laboratory tests in this patient revealed a monoclonal protein of gamma mobility, characterized as predominantly polymeric IgA, normal urine proteins, a serum viscosity of 13.2 relative to distilled water (normal <2), and extensive plasma cell infiltrates in the bone marrow. All these findings were compatible with the diagnosis of multiple myeloma and hyperviscosity syndrome.
- The patient was treated with plasmapheresis (two to three sessions a week in each of which 300–500 mL of plasma were exchanged) and cyclophosphamide. A marked clinical improvement of all symptoms related to serum hyperviscosity was seen soon after plasmapheresis was initiated and repeated measurements of serum viscosity showed normalization after 1 month of combined therapy. After one and a half months of combined therapy, plasmapheresis was stopped and the patient continued to be treated with cyclophosphamide. Fundoscopy revealed normalization of the retina after 5 months of treatment.

III. LEUKEMIAS AND LYMPHOMAS

A. **Nomenclature**. The malignant proliferations of leukocytes can be classified by a variety of criteria. One first important distinction is made between leukemia and lymphoma.

1. **Leukemia** refers to any malignant proliferation of leukocytes in which the abnormal cell population can be easily detected in the peripheral blood and in the bone marrow. Leukemias may involve any type of hemopoietic cell, including granulocytes, red cells, and platelets.
2. **Lymphoma** refers to localized lymphocyte malignancies, often forming solid tumors, predominantly affecting the lymph nodes and other lymphoid organs. Lymphomas are always lymphocytic malignancies.
3. Leukemias are often classified as **acute** or **chronic**, based on their clinical evolution and morphological characteristics which are closely related.
   a. **Acute leukemias** follow a very rapid progression toward death if left untreated. Many immature and **atypical cells** can be seen in the peripheral blood of patients with acute leukemias.
   b. **Chronic leukemias** have a more protracted evolution; differentiated cells predominate in the peripheral blood of patients with chronic leukemia.
   c. Leukemic states may evolve from a chronic form to an acute disease, and the type of cell that is proliferating may also change during the course of the disease. For example, transition from a chronic granulocytic stage to an acute and very often fatal lymphoblastic leukemia is characteristic of chronic myelocytic leukemia.
   d. All malignant proliferations of cells identifiable as lymphocytes are classified as either T- or B-cell malignancies, based on a variety of characteristics:
      i. Identification of the malignant cells as immunoglobulin-producing cells allows their classification as B-cell malignancies.
      ii. Cell membrane markers are widely used to classify malignant lymphocyte proliferations.
      iii. Molecular genetic procedures may be used to determine whether the heavy-chain genes or the T-cell receptor genes are rearranged in a malignant lymphocyte population.

Case 29.2

A 48-year-old black male who worked as a graphic designer had emigrated from Jamaica 33 years ago. He was referred to the Dermatology clinic of the University Hospital for investigation of an atypical dermatitis, fever, and nonproductive cough. His main complaint was of a progressive skin rash which his family doctor did not know how to manage. The patient claimed that he noticed the first skin lesions three months earlier and started coughing more than usual 2 weeks prior to the time he sought medical attention. He also referred to a weight loss of 19 lb. in 2 months. The erythematous lesions initially were very small and barely noticeable but had been spreading very fast during the last 2 weeks. Physical examination showed an underweight male in no acute distress. Blood pressure was 136/65, pulse 98/min., respiration 30/min., temperature 101.9°F (38.3°C). A generalized skin rash sparing very few areas of the body was seen.

The skin was red, thickened and infiltrated, feeling like cardboard at the touch. In addition, two skin ulcers were seen. One, on the forearm, had an approximate diameter of 1 in. and the second, on the lateral aspect of his right thigh, was larger, with a diameter of 2.5 in. During physical examination, it was noted that pressure of the sixth and seventh ribs, on the right side, caused severe pain. On questioning about the chest pain, the patient referred that it started suddenly 2 days earlier, when he was trying to lift a suitcase. He also noticed that, after that, coughing caused pain. Two lymph nodes the size of a cherry were felt on the right side of the neck, four lymph nodes were felt in the left axilla (the largest about 1 in. in diameter), and three smaller nodes were felt in the right axilla. All nodes were firm, smooth, nontender, and mobile under the skin. Diffuse rhonchi were audible in both lungs. The liver was nontender and palpated 3 in. below the costal margin. The rest of the examination was normal. A chest x-ray showed bilateral diffuse interstitial infiltrates, severe osteoporosis of the ribs and vertebrae, and a fracture of the sixth and seventh ribs on the right. A CBC and differential revealed (normal values in parentheses) RBC = $2.9 \times 10^6/\mu L$ ($4.8 \pm 0.6 \times 10^6/\mu L$), hemoglobin of 7.3 g/dL (14 ± 2 g/dL), WBC of 23,000 μL (4.0–10.5 $10^3/\mu L$) with 37% lymphocytes (20–45%). Lymphocyte subpopulations were as follows: CD3+: 86% ($60 \pm 10\%$); CD4+: 88% ($40 \pm 10\%$); CD8+: 12% ($15 \pm 10\%$); CD1+: 0% (0%); Tdt^+: 0% (0%); CD25+: 28% (<1%); CD4, CD25+; 26% (<1%). Serum calcium was 12.2 mg/dL (8.5–10.6 mg/dL). Serum immunoglobulins: IgG: 500 mg/dL (600 to 1300); IgA: 48 mg/dL (60 to 300); IgM: 21 mg/dL (30 to 150).

This case raises several questions:

- What is the most likely diagnosis?
- What test should be done to confirm the most likely diagnosis?
- What is the nature of the patient's skin rash?
- Why did the patient develop rib fractures with minimal trauma?
- What is the meaning of the large percentage of CD25+ cells?
- What is the meaning of the bilateral diffuse interstitial infiltrates seen on the chest x-ray?

B. **Chronic Lymphocytic Leukemia** is a B-cell malignancy and has many features in common with Waldenstrom's macroglobulinemia: it is a disease of old age, often with a relatively benign course. Central to its pathogenesis seems to be an overexpression of the *bcl-2* gene, which inhibits apoptosis.

1. **Clinical symptoms** are often absent or very mild. Malaise, fatigue, or enlargement of the lymphoid tissues felt by the patient are the most frequent presenting complaints. Physical diagnosis shows enlargement of the lymph nodes, spleen, and liver. Viral infections, such as herpes and herpes zoster, and fungal infections are frequent in these patients, pointing to a T-cell deficiency that is confirmed by the finding of reduced numbers of T cells and reduced responses to T-cell mitogens. The prognosis is determined by the frequency of severe opportunistic infections.
2. **Diagnosis**
   a. **Serum and urinary immunoglobulins**. Most patients are hypogammaglobulinemic. Rarely, IgM monoclonal proteins may be detected. Bence-Jones proteins can be detected in the concentrated urine of approximately one-third of the patients.
   b. **Cell-associated immunoglobulins**. About 98% of the patients carry monoclonal IgM on the membrane of the leukemic cells. In some cases, cytoplasmic retention of immunoglobulins can also be demonstrated.

c. **Membrane markers**. The membrane markers of the leukemic cells are identical to those of mature B cells, with the following exceptions: CD11a/18 and CD22 are not expressed, while CD5 (a T-cell marker also expressed by a B-cell subset) is expressed in over 85% of the cases.

C. **Hairy Cell Leukemia** is a malignancy of uncertain classification, predominantly affecting elderly males.

1. **Clinical presentation**. Patients present with a nonspecific clinical picture of malaise, fatigue, and frequent infectious episodes. The physical examination usually shows splenomegaly and sometimes generalized lymphadenopathy.
2. **Diagnosis**. The finding of atypical lymphocytes with numerous fingerlike (or hairy) projections in the peripheral blood defined the disease (which derives its name from the morphological characteristics of the abnormal lymphocytes). The abnormal cells have mixed characteristics.
   a. Express membrane immunoglobulins, often with several isotypes present and synthesize monotypic heavy and light chains, suggesting a B-lymphocyte origin.
   b. Have monocyte/macrophage functions and markers including phagocytic properties, ability to produce and release lysozyme and peroxidase, presence of intracellular tartarate-resistant acid phosphatase.
   c. After mitogenic stimulation, the abnormal lymphocytes may express the CD2 and CD3 membrane markers characteristic of T lymphocytes. These findings have been interpreted as indicating the proliferation of: B-cell precursor cells sharing monocytic and T-cell markers and functions; malignant chimaeric cells with multiple lineages; and malignant lymphocytes with aberrant gene expression.

   Without additional data it is impossible to decide which of these possibilities is more likely to reflect accurately the nature of the malignant proliferation in this type of leukemia.
3. **Management**. **Interferon-α** is therapeutically useful (sometimes inducing permanent remissions) in hairy cell leukemia. This seems to result from a **direct antiproliferative effect** that is attributed to the ability of interferon-α to promote redifferentiation of malignant cells, stopping their uncontrolled multiplication.

D. **Acute Lymphocytic Leukemias** are those acute leukemias in which the malignant cells seen in the peripheral blood are immature lymphocytes (lymphoblasts). These leukemias usually have a very poor prognosis. Death usually occurs as a consequence of the massive lymphocytic proliferation in the bone marrow, where the proliferating cells overwhelm and smother the normal hemopoietic cells.

1. **Classification**. With the introduction of monoclonal antibodies directed against T- and B-cell markers, it was determined that the large majority (about 95%) of acute lymphocytic leukemias are B-cell-derived because the proliferating cells express the CD19 and CD20 B-cell markers. The remaining 5% of these leukemias are of the T-cell type.
2. **Enzymatic markers of acute lymphocytic leukemia**. The expression of enzymes of the purine salvage pathway is altered in acute lymphocytic leukemia.

a. **Adenosine deaminase** (**ADA**) is often overexpressed. In patients with increased ADA, 2-deoxycorfomycin, a drug that specifically inhibits ADA, has remarkable therapeutic effects.
b. **Terminal deoxynucleotidyl transferase** (Tdt) is not expressed by adult lymphocytes but is reexpressed by about 80% of all cases of this type of leukemia; it constitutes a useful marker because its levels fall during remission and increase again before a clinically apparent relapse.

3. **B-cell acute lymphocytic leukemia**. In most cases, the malignant cells do not express membrane immunoglobulins, do not have intracellular immunoglobulins, have rearranged heavy-chain genes, and express the common acute lymphocytic leukemia antigen (CALLA).
   a. The **common acute lymphocytic leukemia antigen** (**CALLA**) is present in the majority of non-T, non-B acute lymphocytic leukemia lymphocytes, almost always expressed in association with a B cell marker such as the CD19 or CD20 antigens. CALLA positivity identifies patients with a more favorable prognosis.
   b. CALLA is also expressed by the lymphoblasts seen during the blastic crisis of patients with chronic myelocytic leukemia (CML); these lymphoblasts also express B-cell markers such as the CD20 antigen, establishing their identity as B-cell precursors. When the blast cells seen in the blastic crisis of a patient with CML are $CALLA^+$ and CD20+, the crisis responds well to chemotherapy; when none of these markers is expressed, survival is limited to a few days.
   c. Acute lymphocytic leukemia in which the leukemic cells express membrane or cytoplasmic immunoglobulins has very poor prognosis and survive less than a year unless very aggressively treated.
   d. **Monoclonal anti-CALLA antibodies** have been used therapeutically in acute lymphocytic leukemia with disappointing results. A sharp decrease in leukemic cell counts is observed after administration of antibody, but this effect is usually of short duration since the $CALLA^+$ lymphocytic population is soon replaced by a $CALLA^-$ population (antigenic modulation) not affected by further administration of antibody. Also, the prolonged administration of monoclonal anti-CALLA of murine origin leads to the development of antimouse immunoglobulin antibodies, which cause rapid elimination of anti-CALLA antibodies from the patient's circulation and may also cause serum sickness.
4. **T-cell acute lymphocytic leukemia** usually has a worse prognosis than B-cell acute lymphocytic leukemia (patients with T-cell acute lymphocytic leukemia have less than a 20% probability to remain in remission for more than 2 years). Chromosomal abnormalities involving the T-cell receptor genes have been observed in at least 40% of the T-cell leukemias. One of the most frequent is a translocation of the area of chromosome 14 which carries the α gene of the T-cell receptor to the area of chromosome 8 which has the *c-myc* gene. Equally frequent is a translocation of the area of chromosome 7, which contains the β chain of the T-cell receptor to chromosome 11.

   Three distinct subgroups of leukemic processes can be defined in the group of T-cell acute lymphocytic leukemia depending on the expression of T-cell membrane markers (Table 29.3).

a. The first and largest group includes cases in which the proliferating T cells express the CD5 and CD2 markers. The malignant cell is, therefore, an early T-cell precursor which mutated before the full rearrangement of the T-cell receptor genes so that the CD3 molecule is not expressed.
b. The second group is constituted by cases in which the proliferating cells have reached a later stage of T-cell differentiation: They express CD3, co-express both CD4 and CD8 markers, and are also positive for the CD1 marker, indicating an aberrant reversal of a partially differentiated T lymphocyte to an earlier ontogenic stage.
c. The third group is constituted by cases with proliferating mature T cells, sharing markers (CD4 or CD8, in association with both CD2 and CD3) with the lymphocytes normally found in the peripheral blood and lymphoid organs.

Because of the relative rarity of T-cell acute lymphocytic leukemia, it has not yet been possible to establish whether any of the subgroups of this disease has a worse prognosis than that of T-cell acute lymphocytic leukemia in general.

5. **T-cell acute lymphocytic leukemia with thymic mass**. This type of leukemia affects males predominantly, and the patients are somewhat older than the children who are affected with B-cell acute lymphocytic leukemia. A remarkable feature of this group is the frequent finding of a thymic mass, suggesting that the malignant process arose from this organ.

E. **T-cell leukemia associated to HTLV-1**. This type of T-cell leukemia has a very unique geographic distribution, closely associated to the first identified human retrovirus (**human T-cell lymphotropic virus-I** or **HTLV-I**) which is very prevalent in Japan and the Caribbean basin (where the rates of infection reach endemic proportions); the virus has also been reported, although with lower frequency, in the southern U.S.

1. **Pathogenesis**
   a. **HTLV-1** is an exogenous retrovirus, fully able to replicate and to be transmitted horizontally. Its genome contains a transforming gene, *tax*,

Table 29.3 Classification of T-Cell Acute Lymphocytic Leukemia According to Membrane-Associated Markers Recognized by Monoclonal Antibodies

Patient	Marker					
Group	CD5	CD2	CD3	CD4	CD8	CD1
1	+	+	−	−	−	−
2	+	+	−	+[a]	+[a]	+
3	+	+	+	+[b]	+[b]	−

[a]CD4 and CD8 are coexpressed by the malignant cells.
[b]The malignant cells express *either* CD4 *or* CD8, but not both.

whose gene product modifies the nuclear binding protein NFkB, leading to the permanent overexpression of IL-2 receptors (CD25) in the infected cells. This has several consequences:

i. This type of T-cell leukemia is easily distinguishable by the fact that the proliferating cells are easily labeled with anti-CD25 monoclonal antibodies.
ii. IL-2 stimulates the growth of the leukemic T cells in long-term culture.
iii. Some patients have malignant T cells that not only express CD25, but release high concentrations of IL-2, leading to an autocrine circuit of T-cell proliferation.
iv. Other interleukin-coding genes are also activated, including the one coding for IL1-β, which indirectly causes osteoclast activation. As a consequence of osteoclast activation, bone resorption and hypercalcemia are prominent in these patients.

b. **Other factors**. The HTLV-I-associated T-cell leukemia develops 10 to 20 years after infection with the virus. This very long latency period and the fact that T-acute lymphocytic leukemia is only seen in a fraction of the HTLV-I-infected individuals (4 to 5% of the seropositive individuals) suggests that malignant transformation must not result exclusively from the viral infection, but the nature of the additional cofactors leading to leukemic transformation is unknown.

c. **Secondary immunosuppression** may develop in patients with HTLV-1 leukemia. Several factors contribute to the state of immunosuppression:

i. IL-2 receptors are shed from the membrane of the leukemic cells and can be detected in high concentrations in the circulation. Soluble IL-2 receptors diffuse into the extracellular spaces and adsorb the IL-2 that is necessary for the activation of normal T cells, causing a functional deficiency of this interleukin.
ii. The proliferating CD4+ cells function as suppressor–inducers and turn on cells with suppressor activity.

2. **Clinical presentations**

a. **Erythroderma and skin ulceration** which are associated with a dense lymphocytic infiltration of the dermis and epidermis. It is believed that increased venous permeability, probably caused by an increased local concentration of IL-2 and other interleukins, is responsible for the formation of cellular infiltrates, which, in turn, interfere with proper oxygenation of tissues, leading to localized ischemia and necrosis.

b. **Hypercalcemia and spontaneous fractures** resulting from excessive bone resorption.

c. **Opportunistic infections**, such as *Pneumocystis carinii* pneumonia. Hypogammaglobulinemia is common in these patients.

F. **Sézary Syndrome and Mycoses Fungoides** are cutaneous T-cell lymphomas that have also been related to HTLV-I infection.

1. **Sézary syndrome** is an exfoliative erythroderma with generalized lymphadenopathy and circulating atypical cells with a characteristic multilobulated nucleus (Sézary cells). The skin is the original site of malignant cell proliferation and the phase of cutaneous lymphoma can last many years

with little evidence of extracutaneous dissemination. The leukemic evolution is associated with the invasion of the peripheral by malignant cells. The malignant cells infiltrating the skin or circulating in the blood are CD4+ and behave functionally as helper T cells when mixed in vitro with T-cell-depleted lymphocytes from a normal donor and antigenically stimulated.

2. **Mycoses fungoides** is clinically similar to the cutaneous phase of the Sézary syndrome, and the infiltrating cells in the skin are also CD4+. No leukemic stage seems to develop in patients afflicted with the disease. However, the lymphocytes from patients with mycoses fungoides suppress the response of normal allogeneic T and B cells.

G. **Epstein-Barr Virus (EBV)-Associated B-Cell Lymphomas**

1. **Burkitt's lymphoma**. Burkitt's lymphoma, endemic in certain areas of Africa and sporadic in the U.S., has been characterized as a B-cell lymphoma expressing monotypic surface IgM. Burkitt's lymphoma is epidemiologically linked to infection of the B lymphocytes with the Epstein-Barr virus (EBV). The malignant B cells in BL usually express a single EBV gene product, the nuclear antigen EBNA-1, which is essential for establishment of latency, but has no known transforming properties. It is possible that the EBV infection has as its main role the promotion of a state of active B-cell proliferation that may favor the occurrence of the translocations involving the region of chromosome 8 coding for *c-myc*.
2. **B-cell lymphomas in immunocompromised patients**. B-cell lymphomas are frequently detected in immunodeficient or iatrogenically immunosuppressed patients, and almost in all cases there is evidence of association with EBV. In those cases, a variety of viral-coded proteins are expressed on the malignant cells, including six different nuclear antigens and three different membrane proteins. Of the proteins coded by nuclear antigens, EBNA-2 protein has immortalizing properties, transactivating the cyclin-2 gene and others, EBNA-LP impairs the function of the products of two tumor suppressor genes, p53 and the retinoblastoma gene product, and the latent membrane protein 1 (LMP-1) is considered as a transforming gene whose activity seems to be mediated by the activation of a Ca^{2+}/calmodulin-dependent protein kinase.
3. **Hodgkin's disease**. EBV genomes and gene products can be detected in a significant number of Hodgkin's disease lymph node biopsies. More significant is the fact that LMP-1 is among the expressed proteins.

Case 29.2 Revisited

- The presentation of a native of the Caribbean basin with an erythemato-ulcerative skin rash, pneumonia, and leukocytosis with a marked increased of the CD4 population and of double staining CD4-CD25 lymphocytes, with osteoporosis and hypercalcemia, is typical for an HTLV-1-associated T-cell leukemia.
- The most informative test from the diagnostic point of view would be a serological assay for anti-HTLV-1 antibodies, which was positive in this patient. In addition, biopsies of the ulcerative skin lesions and enlarged lymph nodes were compatible with lymphoma, the cells infiltrating the biopsied lymph node were identified as CD4+, CD25+, peripheral blood lymphocytes showed a vigorous mitogenic re-

sponse to IL-2, circulating levels of soluble IL-2 receptors were high (1000 IU/mL, normal < 277 U/mL), and the blood levels of paratohormone were normal.

- The skin rash was due to a dense lymphocytic infiltration of the dermis and epidermis, secondary to an increased venous permeability, probably caused by an increased local concentration of IL-2 and other interleukins. The intense cellular infiltrate interferes with proper oxygenation of tissues, and the resulting ischemia leads to localized necrosis.
- Transformed CD4+ cells produce IL-1β and other less well-defined mediators that activate osteoclasts and induce bone resorption and hypercalcemia. Because of the loss of calcium the bones become fragile and may break with minimal trauma.
- ATLL is a malignancy of mature CD3+, CD4+ T cells caused by the HTLV-1 virus. The virus has a transforming gene (*tax*) that becomes overexpressed, and the protein coded by that gene modifies cellular transactivating proteins, such as NF-B, increasing their activity. As a consequence, the transformed cells overexpress IL-2 receptors and are able to proliferate spontaneously by using their own IL-2 (whose synthesis is also enhanced by the *tax* gene product) as a growth factor.
- The bilateral interstitial infiltrates suggest a pneumonitic process, which could be due to viruses or fungi, such as *Pneumocystis carinii*. *P. carinii*, is one of the most prevalent causes of opportunistic pneumonia in immunocompromised individuals. This patient was immunocompromised, as shown by his very low levels of circulating immunoglobulins. In addition, lymphocyte mitogenic responses to PHA and ConA were depressed, revealing a functional impairment of cell-mediated immunity. Examination of a broncho-alveolar lavage sample was positive for the typical silver-staining cysts of *P. carinii*.

SELF-EVALUATION

Questions

Choose the ONE *best* answer.

29.1 HTLV-1 induces a T-cell leukemia in which the proliferating cells permanently overexpress:
- A. *c-myc* mRNA
- B. CALLA antigen
- C. CD5
- D. CD25
- E. Terminal deoxynucleotidyl transferase (Tdt)

29.2 In the common type of chronic lymphocytic leukemia:
- A. Bence-Jones proteinuria is rarely detected (<5% of patients)
- B. Most lymphocytes will stain with labeled anti-CD19
- C. The levels of serum immunoglobulins are usually normal
- D. There is an expansion of CD3+/CD5+ T lymphocytes
- E. There is no significant compromise of cell-mediated immunity

29.3 Of the following lymphocytic malignancies, which one is usually due to the uncontrolled expansion of transformed T lymphocytes?
- A. Acute lymphocytic leukemia
- B. Burkitt's lymphoma
- C. Chronic lymphocytic leukemia
- D. Hairy cell leukemia
- E. Sézary syndrome

29.4 The finding of Bence-Jones protein in the urine:
- A. Allows a diagnosis of multiple myeloma
- B. Has no definite diagnostic implications
- C. Is characteristic of nonsecretory forms of multiple myeloma
- D. Is proof of the existence of a B-cell dyscrasia
- E. Rules out a diagnosis of benign gammopathy

29.5 The finding of a polyclonal increase of serum immunoglobulins:
- A. Has been reported in some cases of Waldenstrom's macroglobulinemia
- B. Is a good prognosis indicator in multiple myeloma
- C. Is characteristic of all heavy-chain diseases
- D. Is not compatible with a diagnosis of B-cell dyscrasia
- E. Rules out a diagnosis of multiple myeloma

29.6 Which of the following is associated with poor prognosis in a patient with acute lymphocytic leukemia?
- A. 8:14 chromosomal translocation
- B. Expression of the CALLA antigen
- C. Expression of the CD5 marker
- D. Expression of the CD19 marker
- E. Positive staining for intracytoplasmic immunoglobulins

29.7 The distinction between multiple myeloma and reactive plasmacytosis can be readily based on the:
- A. Characterization of the gammopathy as monoclonal or polyclonal
- B. Levels of serum immunoglobulins
- C. Number of plasma cells in the bone marrow
- D. Quantitation of light chains in the urine
- E. Results of a skeletal x-ray survey

29.8 The most reliable criterion for differentiation between benign and malignant B-cell dyscrasias is:
- A. Age of the patient
- B. Levels of normal immunoglobulins
- C. Levels of serum albumin, hemoglobin, and urea
- D. Presence of Bence-Jones proteinuria
- E. Progressive increase of paraprotein levels

29.9 Which of the following would be an unexpected finding in Waldenstrom's macroglobulinemia:
- A. Anemia
- B. Hypercalcemia
- C. Increased numbers of plasma cells, lymphocytes, and lymphoplasmacytic blastic forms in the bone marrow
- D. Increased serum viscosity
- E. Normal or near-normal levels of IgG

29.10 A 59-year-old man has been complaining of weakness, repeated pulmonary infections, and "rheumatic" pains for 2 years. He has been hospitalized because he broke his right humerus falling from a chair. Serum immunoglobulins are: IgG: 600 mg/100 mL; IgA: 450 mg/100 mL; IgM: 200 mg/100 mL. The patient eliminates 2 g of protein daily in the urine. Which of the following tests will be most useful for diagnosis:

A. Determination of serum viscosity
B. Quantitation of urinary light chains
C. Serum calcium levels
D. Serum electrophoresis and/or immunoelectrophoresis
E. Urine electrophoresis and/or immunoelectrophoresis

Answers

29.1 (D) In individuals with HTLV-1, the leukemic process is characterized by the permanent expression of the IL-2 receptor (CD25).

29.2 (B) Bence-Jones proteinuria is detected in about one-third of the patients; the transformed lymphocytes are of the B lineage and express CD19 and CD5, but not CD3; most patients are hypogammaglobulinemic and cell-mediated immunity is often compromised.

29.3 (E) The Sézary syndrome is a malignancy of helper T cells.

29.4 (D) Bence-Jones proteinuria is most often associated with multiple myeloma but can also be seen in patients with other types of B-cell malignancies and even in patients without evidence of malignant B-cell proliferation. But it can always be considered as proof of the existence of a B-cell dyscrasia.

29.5 (E) Although in some cases of lymphocytic lymphoma, a malignant, monoclonal, B-cell proliferation may exist in association with a reactive plasmacytosis leading to the simultaneous presence of a monoclonal protein and polyclonal hypergammaglobulinemia; this is not seen in multiple myeloma, in which the residual nonmonoclonal immunoglobulins are always normal or reduced in their levels.

29.6 (E)

29.7 (A) Reactive plasmacytosis is associated with polyclonal gammopathy while multiple myeloma is characterized by a monoclonal gammopathy. The numbers of plasma cells in the bone marrow can be identical in both cases, and even atypical forms can be seen in patients with reactive plasmacytosis. A skeletal x-ray survey may be normal or show nonspecific osteoporosis in patients with multiple myeloma, and the finding of osteolytic bone lesions is not unique to multiple myeloma.

29.8 (E) No other parameter is as reliable.

29.9 (B) Hypercalcemia is usually seen in multiple myeloma, as a consequence of disseminated osteolysis; in Waldenstrom's macroglobulinemia, there is no appreciable increase of serum calcium, as a rule.

29.10 (E) Since the serum appears not to contain a monoclonal protein (normal to low-normal immunoglobulin levels), but proteinuria is definitely increased, the best approach would be to characterize the urinary proteins, looking for a Bence-Jones protein. The quantitation of light chains in the urine is not as reliable as the characterization of the light chains as monoclonal by a combination of electrophoretic and immunochemical techniques.

REFERENCES

Barlogie, B., Jagannath, S., Vesole, D., et al. Autologous and allogeneic transplants for multiple myeloma. *Semin. Hematol.*, *32*:31, 1995.

Bartl, R., Frisch, B., Diem, H., et al. Histologic, biochemical, and clinical parameters for monitoring multiple myeloma. *Cancer*, *68*:2241, 1991.

Bataille, R., Chappard, D., Marcelli, C., et al. Recruitment of new osteoblasts and osteoclasts is the earliest critical event in the pathogenesis of human multiple myeloma. *J. Clin. Invest.*, *88*:62, 1991.

Borden, E.C. Innovative treatment strategies for non-Hodgkin's lymphoma and multiple myeloma. *Semin. Oncol.*, *21* (Suppl. 14):14, 1994.

Camilleri-Broet, S., Davi, F., Feuillard, J., et al. High expression of latent membrane protein 1 of Epstein-Barr virus and BCL-2 oncoprotein in acquired immunodeficiency syndrome-related primary brain lymphomas. *Blood*, *86*:432, 1995.

Dimopoulos, M.A. and Alexanian, R. Waldenstrom's macroglobulinemia. *Blood*, *83*:1452, 1994.

Fermand, J.P. and Brouet, J.C. Marrow transplantation for myeloma. *Annu. Rev. Med.*, *46*:299, 1995.

Jarrett, R.F. Viruses and Hodgkin's disease. *Leukemia*, *7* (Suppl. 2):S78, 1993.

Joshua, D.E., Brown, R.D., and Gibson, J. Prognostic factors in myeloma: what they tell us about the pathophysiology of the disease. *Leuk. Lymphoma*, *15*:375, 1994.

Klein, B. Cytokine, cytokine receptors, transduction signals, and oncogenes in human multiple myeloma. *Semin. Hematol.*, *32*:4, 1995.

Kyle, R.A. Newer approaches to the management of multiple myeloma. *Cancer*, *72* (Suppl. 11):3489, 1993.

Kyle, R.A. The monoclonal gammopathies. *Clin. Chem.*, *40*:2154, 1994.

Mundy, G.R. Mechanisms of osteolytic bone destruction. *Bone*, *12* (suppl. 1):S1, 1991.

Niedobitek, G., Agathanggelou, A., Rowe, M., et al. Heterogeneous expression of Epstein-Barr virus latent proteins in endemic Burkitt's lymphoma. *Blood*, *86*:659, 1995.

Oken, M.M. Standard treatment for multiple myeloma. *Mayo Clin. Proc.*, *69*:781, 1994.

Rayner, H.C., Haynes, A.P., Thompson, H.R., et al. Perspectives in multiple myeloma: Survival, prognostic factors and disease complications in a single centre between 1975 and 1986. *Q. J. Med.*, *290*:517, 1991.

Rozman, C. and Montserrat, E. Chronic lymphocytic leukemia. *N. Engl. J. Med.*, *333*:1052, 1995.

30
Immunodeficiency Diseases

Gabriel Virella

I. INTRODUCTION

A. Immunodeficiency diseases and syndromes are the cause of significant mortality and morbidity, as well as a source of extremely valuable information about the physiology of the human immune system. Most immunodeficient patients have secondary forms of immunodeficiency, caused by either pathological conditions that affect the immune system or the administration of therapeutic compounds with immunosuppressive effects.

B. A functional defect of the immune system is suspected when a patient has: unusual frequency of infections with common or opportunistic microorganisms; unusually severe infections; and failure to eradicate infections with antibiotics to which the microorganisms are sensitive.

C. A good history, a careful workup of the infectious episodes, and a thorough physical examination are essential for the initial evaluation of a suspected immunodeficiency, providing useful clues about the type of immunodeficiency (Table 30.1). A good family history is very important. Early death of older siblings suffering from repeated infectious episodes is often the only way to document the hereditary character of a congenital immunodeficiency.

D. Once an immunodeficiency is suspected, investigations need to be undertaken with the purpose of documenting and characterizing the immunodeficiency state. Known causes of secondary immunodeficiency need to be ruled out and, if present, therapy will be directed at eliminating them. If a diagnosis of primary immunodeficiency is made, it will be important to define the degree of compromise of the different mechanisms of immunological defense—cellular immunity, humoral immunity, phagocytosis, and complement—in order to select the most effective type of therapy.

Case 1

A 23-year-old female was admitted with a clinical diagnosis of bacterial pneumonia. She complained of rigors and chills, productive cough, and shortness of breath. Personal history was remarkable for repeated episodes of pneumonia since 8 years of age, with hospitalizations at 8, 16, 21, 22 (at 5 months of an otherwise uneventful pregnancy) and 24 years of age for pulmonary infections, which were treated with antibiotics. At 9 years

Table 30.1 Clues About the Nature of an Immunodeficiency Disease Derived from History, Physical Examination, and General Diagnostic Procedures

Predominant infections	Physical examination findings	Type of immunodeficiency
Repeated pyogenic infections (tonsillitis, otitis, pneumonia, disseminated impetigo)	Peripheral lymph nodes and tonsils are atrophic	B-lymphocyte deficiency
Severe mycotic infection; development of active infection after administration of a live virus vaccine		T-lymphocyte deficiency
Neonatal tetany with cardiac malformations, mongoloid facies, severe viral and mycotic infections	Lack of thymic shade on x-ray	DiGeorge syndrome
Abscess-forming infection with low-grade pathogens (*Staphylococcus epidermidis*, *Serratia marcescens*, *Aspergillus* sp).	Scars from previous bone, liver, or soft tissue abscesses, draining lymphadenitis; hepatosplenomegaly	Neutrophil deficiency
Repeated infections with *Neisseria* sp.		Complement deficiency
Persistent infection of the lungs, diffuse mucosal moniliasis, chronic diarrhea, and wasting early in life	Growth retardation; other associated congenital abnormalities	Severe combined immunodeficiency

of age, she was successfully treated for pulmonary tuberculosis. At 12, she had an episode of generalized pyoderma. She received the usual childhood immunizations without complications. She had chickenpox at 3 years of age, with normal evolution. Physical examination revealed a sick-looking female, with a temperature of 102.5°F (39.2°C), pulse of 85/min., respirations of 26/min., BP 120/80. Chest percussion revealed signs of consolidation on the lower right hemithorax. Auscultation of the same area revealed rhonchi and rales, with dulled breath sounds. A chest x-ray showed consolidation of the right lower lobe. CBC and differential showed RBC of 3,600,000/μL, Hgb of 12 g/L, WBC of 12,800/μL with 60% neutrophils, ESR of 122 mm/h. Lymphocyte subpopulations were: CD3+ lymphocytes: 1200/μL; CD4+ lymphocytes: 620/μL; CD8+ lymphocytes: 575/μL; CD20+ lymphocytes: 390/μL. Serum immunoglobulin concentrations were: IgG 80 mg/dL; IgA, IgM, and IgD: not detectable. A sputum culture was positive for *Haemophilus influenzae*.

This case raises several questions:

- What arm of the immune system appears to be most severely involved in this patient?
- Is the patient's child at risk for developing this disease?
- Should any other tests be run to clarify the pathogenesis of this condition?
- What are the most likely complications to develop in this patient?
- What is the best therapy for this case?

II. DIAGNOSTIC STUDIES

Once a patient with suspected immunodeficiency has been identified, it is necessary to attempt to confirm the diagnosis. This can be done in many different ways, which may vary from patient to patient. The main approaches for the diagnosis of deficiencies of humoral and cellular immunity, phagocytic function, and complement have been discussed in earlier chapters. It is necessary to stress at this point that investigation of an immunodeficiency is a step-by-step procedure in which tests are ordered to confirm clinical impressions or the results of previous tests.

A. An **initial evaluation** may involve simple assays such as a white blood cell count with leukocyte differential, immunoglobulin assay, isohemagglutinin titers, lymphocyte subpopulation counts, an NBT test for phagocytic function, and a CH50 test for complement activity.

B. Depending on the results of this initial evaluation and on the clinical impression as to the nature of the immunodeficiency, **further tests** should be ordered so as to obtain the best possible definition of the nature and degree of immunodeficiency, information essential for rational therapeutic choices.

C. Immunodeficiencies can be classified by a variety of criteria, such as the spectrum of infections, their primary or secondary nature, the main limb of the immune system affected, and, in the case of hereditary primary immunodeficiencies, their mechanism of genetic transmission (Table 30.2).

III. PRIMARY IMMUNODEFICIENCY DISEASES

A simplified classification of the most important primary immunodeficiency diseases is given in Table 30.3. In the following pages, a brief outline of the main features of representative primary immunodeficiency diseases (with the exception of phagocytic deficiencies, discussed in Chap. 17) is presented.

A. Humoral Immunodeficiencies are those in which antibody synthesis is pre-

Table 30.2 Criteria for Classification of Immunodeficiency States

By their range
Broad spectrum
Restricted ("antigen-selective")
By their etiology
Primary
Secondary
By the limb of the immune system predominantly affected
Humoral immune deficiencies
Cellular immune deficiencies
Combined immune deficiencies
Phagocyte dysfunction syndromes
Complement deficiencies
By the mechanism of transmission
Genetically transmitted
X-linked
Autosomal recessive
Autosomal dominant
Sporadic

Table 30.3 Classification of Primary Immunodeficiency Diseases

Humoral immunodeficiencies
1. Lack of B-lymphocyte development
    - Infantile hypogammaglobulinemia (Bruton-Janeway syndrome)
2. Abnormal immunoregulation
    - Transient hypogammaglobulinemia of infancy
3. Variable or undetermined pathogenesis
    - Common variable, unclassifiable immunodeficiency
    - Selective IgA deficiency
    - Hyper-IgM syndrome
    - Antigen-specific deficiencies

Cellular (T-cell) immunodeficiencies
1. Lack of thymic development
    - Congenital thymic aplasia (DiGeorge syndrome)
2. Undetermined pathogenesis
    - Chronic mucocutaneous candidiasis

Combined immunodeficiencies
1. Lack of stem-cell development
    - Severe combined immunodeficiency (Swiss-type agammaglobulinemia)
    - Nezelof's syndrome
2. Enzymatic deficiency
    - ADA deficiency
3. Deficient DNA repair
    - Immunodeficiency with ataxia-telangiectasia
4. Impaired antigen presentation
    - Bare lymphocyte syndrome
    - MHC-II deficiency syndrome
5. Helper T-cell deficiency
    - Primary
    - Secondary to IL-2 deficiency
6. Undetermined pathogenesis
    - Immunodeficiency with eczema and thrombocytopenia (Wiskott-Aldrich syndrome)

Complement deficiencies
1. Early component deficiencies
2. C3 deficiency
3. Factor H and factor I deficiencies
4. Late component deficiencies

Phagocytic deficiencies
1. Chronic granulomatous disease
2. Myeloperoxidase deficiency
3. Chediak-Higashi syndrome
4. Job's syndrome

dominantly impaired. The general characteristics of the most important primary immunodeficiencies included in this group are summarized in Table 30.4.

1. **Infantile agammaglobulinemia (Bruton-Janeway syndrome).** Infantile agammaglobulinemia is the prototype of "pure" B-cell deficiency.
    a. **Genetic transmission and molecular basis.**
        i. In the majority of cases, the disease is transmitted as a sex-linked

Table 30.4 Summary of the Main Characteristics of Primary Humoral Immune Deficiencies

Characteristic	Infantile agammaglobulinemia	Common variable immunodeficiency	Hyper-IgM syndrome	Transient hypogammaglobulinemia of infancy	IgA deficiency
Genetics	Usually X-linked	Variable	Usually X-linked	?	?
Molecular basis	Lack of Bruton's tyrosine kinase (BTK)	Variable, ill defined	Lack of CD40 ligand on T cells (gp39)	Unknown	Unknown
Lymphoid tissues	Lack of development of B-cell territories (follicles)	Follicular necrobiosis, reticulum cell hyperplasia	Normal	Normal	Normal
B lymphocytes	Very low to absent	Normal numbers, abnormal differentiation or function	Normal numbers	Normal numbers	Normal numbers
Serum immunoglobulins	Very low	Low to very low levels	Low IgG and IgA, high IgM	Low for age	Low to undetectable IgA
Infections	Bacterial (pyogenic)	Bacterial, parasitic (Giardia)	Bacterial	Bacterial	Bacterial (particularly when associated to IgG2 deficiency)
Treatment	Gamma globulin IV	Gamma globulin IV	Gamma globulin IV	Gamma globulin IV (if necessary)	Gamma globulin IV (when associated to IgG2 deficiency)

trait. The defective gene is located on Xq21.2-22, the locus coding for the B-cell progenitor kinase or **Bruton's tyrosine kinase (Btk)**. Agammaglobulinemic patients have mutations at different sites, which result either in the lack of synthesis of the kinase or in the synthesis of an inactive kinase.

ii. **Btk** plays an important role in B-cell differentiation and maturation, and is also part of the group of tyrosine kinases involved in B-cell signaling in adult life. Most cases of infantile agammaglobulinemia are associated with mutations affecting Btk, but some patients with similar mutations have very mild forms of immunodeficiency with variable levels of immunoglobulins, suggesting that B-cell differentiation may depend on additional co-factors, not yet identified.

b. **Clinical presentation**. Infectious symptoms usually begin early in infancy (8 months to 3 years).

i. Patients suffer from **repeated infections caused by common pyogenic organisms** (*S. pneumoniae, N. meningitidis, H. influenzae, S. aureus*)—pyoderma, purulent conjunctivitis, pharyngitis, otitis media, sinusitis, bronchitis, pneumonia, empyema, purulent arthritis, meningitis, and septicemia.

ii. Chronic obstructive lung disease and bronchiectasis develop as a consequence of repeated bronchopulmonary infections.

iii. Infections with *Giardia lamblia* are diagnosed with increased frequency in these patients and may lead to chronic diarrhea and malabsorption.

iv. Agammaglobulinemic patients are at risk of developing paralytic polio after vaccination with the attenuated virus; they also are at risk of developing chronic viral meningoencephalitis, usually caused by an echovirus.

v. Arthritis of the large joints develops in about 30–35% of the cases and is believed to be infectious, caused by *Ureaplasma urealyticum*.

c. **Laboratory studies**

i. **Very low immunoglobulin levels** (usually less than 100 mg/dL for the sum of the three major isotypes)

ii. Undetectable isohemagglutinins

iii. Failure to produce antibodies in response to active immunization with toxoids, polysaccharides, and bacteriophage øX174.

iv. Peripheral blood lymphocyte counts are usually normal, T-lymphocyte counts are normal or elevated, T-lymphocyte subsets are normal, and T-lymphocyte function is also normal. **B lymphocytes**, on the contrary, are **absent or greatly reduced in the peripheral blood**.

v. Histological examination of a peripheral lymph node draining the site of an antigenic challenge (often difficult to localize) shows **lack of germinal centers and secondary follicles**. Peri-intestinal lymphoid tissues are also abnormal, showing lack of development of germinal centers. **Plasma cells are absent** both from peripheral lymphoid tissues and from bone marrow. Adenoids, tonsils, and peripheral lymph nodes are hypoplastic. In contrast, the thymus has normal structure, and the T-cell-dependent areas in peripheral lymphoid organs are normally populated. Normal numbers of B-cell precursors can be demonstrated in the bone marrow suggesting that the basic defect is a maturation block.

d. **Therapy**. This condition is best treated with replacement therapy using **gamma globulin** (a plasma fraction containing predominantly IgG, obtained from normal healthy donors) administered intravenously.

2. **Transient hypogammaglobulinemia of infancy**. As a consequence of a delay in the infant's B-cell functional maturation, the hypogammaglobulinemia normally occurring during the second and third months of life, because of progressive catabolism of maternal IgG, may persist until 2–3 years of age and become progressively more accentuated (relative to age-matched controls).
   a. **Clinical presentation**. Most patients are referred because of an increased frequency and/or severity of bacterial infections.
   b. **Laboratory findings**. Low-for-age circulating immunoglobulin levels is the diagnostic hallmark. Differentiation with more severe forms of humoral immunodeficiencies is usually based on functional tests and enumeration of B cells.
      i. Lymphocyte mitogenic responses and antibody response to challenge with toxoids are usually normal.
      ii. Peripheral blood B lymphocytes are usually normal in number; in most cases, a deficiency of helper T-cell function appears to be responsible for the delay in immunoglobulin synthesis.
   c. **Therapy**. Intravenous **gamma globulin** is indicated until the child's immunoglobulin levels normalize. With time, most children will develop normal immune function.

3. **Common, variable, unclassified immunodeficiency ("acquired" hypogammaglobulinemia)**. This designation includes a large number of cases of primary immunodeficiency, heterogeneous in presentation, with variable age of onset and patterns of inheritance, whose clinical picture is similar to X-linked agammaglobulinemia, but usually with a less severe course.
   a. **Physiopathology**. Several variants of common variable immunodeficiency were recognized by a panel of experts who met under the auspices of the W.H.O. in 1983.
      i. Most variants of "acquired hypogammaglobulinemia" have normal or increased numbers of B lymphocytes in peripheral blood, but the B cells remain immature and do not respond adequately to in vivo stimulation.
      ii. T-cell function appears deficient in most cases, with abnormally low proliferative responses to T-cell mitogens. T-cell receptor stimulation is followed by reduced release of interleukins and reduced expression of gp39. Thus, lack of proper T-cell help seems responsible for the lack of B-lymphocyte responses.
      iii. In some patients the defect seems to result from excessive suppressor T-lymphocyte activity.
   b. **Clinical presentation**. Sinusitis and bacterial pneumonia are the predominant infections. Intestinal *giardiasis* is common, and in some patients can lead to malabsorption. Opportunistic infections involving *P. carinii*, mycobacteria, viruses, and other fungi are also more frequent in these patients.

c. **Laboratory findings**
   i. Serum **immunoglobulin levels are variably depressed**, and, as a rule, the patients fail to respond to produce antibodies after proper antigenic stimulation.
   ii. Normal or increased numbers of B lymphocytes in peripheral blood, which can be stimulated in vitro to produce immunoglobulins.
   iii. Tonsils, lymph nodes, and spleen may be enlarged. Lymph node biopsies show morphological changes including necrobiosis of the follicles (also seen in the spleen) and/or reticulum cell hyperplasia (which may be the major contributing factor for the development of lymphadenopathy and splenomegaly, and, in some patients seem to evolve into lymphoreticular malignancies).

d. **Treatment** usually involves administration of intravenous **gamma globulin**.

4. **Immunoglobulin A deficiency**. IgA deficiency is the most common immunodeficiency (detected in 1 out of 500–800 normal Caucasian individuals).
   a. **Physiopathology**
      i. Phenotypic studies of circulating B cells show patterns similar to those of cord blood B lymphocytes, suggesting a **differentiation abnormality**, sometimes reflected by a defect in secretion of intracytoplasmic IgA.
      ii. In other cases, there is evidence for **immunoregulatory defects**:
         - Predominant synthesis of IgGl and IgG3 antibodies to pneumococcal polysaccharides, even when their serum levels of IgG2 are normal (IgG2 is usually the immunoglobulin isotype of antipolysaccharide antibodies).
         - **Longitudinal variations in IgA levels**. In children, a delayed increase of IgA to normal levels is the most frequently observed variation; in adults, IgA levels may fluctuate widely, from very low to normal and back to very low.
      iii. **Anti-IgA antibodies** reacting with isotypic or allotypic determinants of IgA can be detected in about one-third of the patients, usually in low titers.
         - When present in high titers, anti-IgA antibodies can cause hypersensitivity reactions (which may be fatal) upon transfusion of IgA-containing blood products.
         - Anti-IgA antibodies may contribute to accentuate and perpetuate the state of IgA deficiency. The administration of radiolabeled IgA to patients with anti-IgA antibodies is followed by its rapid elimination from the circulation (in a matter of hours). More significantly, a comparison of the levels of residual IgA in patients with and without anti-IgA antibodies demonstrated that those with antibodies have the lowest levels.
   b. **Clinical presentation**
      i. Most cases of IgA deficiency are asymptomatic.
      ii. Patients with **combined IgA and IgG2 deficiency** have frequent infections caused by bacteria with polysaccharidic capsules.

iii. Many IgA-deficient individuals have antibodies to food proteins, which, in most cases, appear to be of no consequence.
iv. Infections with *Giardia lamblia* are more frequent in patients with IgA deficiency than in individuals with normal IgA levels. As in patients with agammaglobulinemia this parasitic infection may lead to chronic diarrhea and malabsorption.
v. IgA deficiency can be associated with "autoimmune" disorders (especially pernicious anemia) and with a complex syndrome of lymphoid hyperplasia of the intestine, diarrhea, and malabsorption (**Crabbé's syndrome**).

c. **Therapy**
i. Treatment is usually symptomatic, using antibiotics as needed if patients are infected.
ii. Replacement therapy for IgA deficiency is questionable, because the IgA content of commercial gamma globulins is variable, and the short half-life of IgA (5–6 days) would require very frequent administration of replacement IgA. To complicate matters further, there is always the possibility that administration of IgA-containing gamma globulin may trigger a hypersensitivity reaction in a patient with high levels of anti-IgA antibodies.
iii. Administration of intravenous gamma globulin is indicated in patients with combined IgA and IgG2 deficiency, or in IgA-deficient patients who fail to produce antibodies to bacterial polysaccharides.

d. **Prevention of hypersensitivity reactions due to anti-IgA antibodies**. Anti-IgA antibodies should be assayed in any known IgA-deficient patient considered for transfusion, gamma globulin administration, or elective surgery. If high titers of such antibodies are found, the blood bank needs to be notified so that steps can be taken to make sure that any blood transfused to the patient is IgA-depleted or a gamma globulin preparation lacking IgA should be selected. Transfusion of IgA-depleted blood can be achieved by obtaining compatible blood from a healthy IgA-deficient donor, or by using extensively washed red cells.

5. **Hyper IgM syndrome**. This syndrome is characterized by the assay of low to very low levels of IgG, IgA, and IgD in association to a marked elevation of IgM, and sometimes IgD. In 70% of the cases, the disease is X-linked.

a. **Genetics and physiopathology**
i. **X-linked hyper IgM syndrome** usually involves a mutation of the CD40 ligand (gp39) gene, located on Xq26-27. As a consequence, gp39 is not expressed by T cells and the signals mediated by gp39-CD40 interactions, essential for B-lymphocyte differentiation and switching from IgM synthesis to the synthesis of other immunoglobulin classes, are not delivered. Thus, these patients fail to switch from IgM to IgG (IgA, IgE) synthesis during an immune response and germinal centers do not develop in lymphoid tissues, which contain normal numbers of T and B lymphocytes.
ii. Other patients with hyper-IgM syndrome do express gp39. In those patients, the molecular defect is believed to involve the second

message systems that transduce activation signals after gp39-CD40 interaction.

b. **Clinical presentation**. Increased frequency of pyogenic infections, similar to those affecting patients with infantile agammaglobulinemia. Infections with *Pneumocystis carinii* are also frequently diagnosed in these patients.

c. **Therapy**. Intravenous administration of gamma globulin.

6. **Antigen-selective immune deficiencies**. These are immunodeficiencies in which the affected patients fail to respond to a given antigen, while exhibiting normal immune responses to most other antigens.

a. **Physiopathology**. At least theoretically, two basic mechanisms can underlie antigen-specific immune deficiencies.

i. "Holes" in the repertoire of the responding cells, implying that no binding sites for a given antigen are available either at the T- or the B-cell level. Considering that immunogenic proteins are complex molecules with a variety of different epitopes, it is difficult to envisage how this mechanism could be involved. In the case of polysaccharides, simpler in structure and presenting a limited number of epitopes to the immune system, the hypothesis is more plausible.

ii. Inefficient antigen presentation to helper T cells, implying that the nonresponse is a consequence of the lack of MHC-II molecules with adequate sites for binding of key peptides derived from antigen processing. This mechanism would apply only to T-dependent responses.

b. **Clinical presentation**

i. Antigen-selective immune deficiencies are often undiagnosed, and often **asymptomatic**. For example, using tetanus toxoid as immunogen, one can detect about 1 in 100 individuals whose humoral response is consistently low or undetectable, and the same is probably true with any other immunogen.

ii. When the patient suffers from frequent infections, a diagnosis of antigen-selective immune response may be missed because most of the tests used for general evaluation of the immune system are usually within normal limits.

iii. In some symptomatic cases, one or more of the IgG subclasses may be deficient. IgG2 deficiency, as stated above, can be associated with infections by bacteria whose polysaccharide capsules are a major virulence factor (such as *Streptococcus pneumoniae* and *Haemophilus influenzae*).

c. **Diagnosis**. The best way to diagnose this type of deficiency when symptomatic is to isolate and identify the organisms infecting the patient and investigate whether the patient can generate specific immunity to the identified infectious agent(s).

d. **Therapy**. Intravenous gamma globulin administration.

B. Cellular Immunodeficiencies are those in which cell-mediated immunity is predominantly impaired. The general characteristics of the most important primary immunodeficiencies included in this group are summarized in Table 30.5.

Table 30.5 Summary of the Main Characteristics of Primary Cellular and Combined Immune Deficiences

Characteristic	Thymic aplasia (DiGeorge syndrome)	Severe combined immune deficiency	MHC deficiencies	IL-2 synthesis deficiency
Genetics	Not inherited	Variable	?	?
Molecular basis	Chromosomal deletion (22q11)	1. Deficient IL-2 (4,7) receptor γ chain 2. Deficient ZAP-70 kinase 3. ADA deficiency 4. RAG mutations	Lack of expression of MHC-I (bare lymphocyte syndrome) or MHC-II	CD4 T-cell deficiency; defects in signal transduction
Lymphoid tissues	Thymic aplasia; depletion of T-cell areas	Thymic asplasia; general atrophy of lymphoid organs	Normal	Normal
T lymphocytes	Low to very low	Very low	Normal to low (low CD4 counts in MHC-II deficiency)	Low CD4 count in some cases
B lymphoctes	Normal numbers, deficient function	Very low or undetectable	Normal numbers, deficient function	Normal numbers
Serum immunoglobulins	Variable levels	Low levels	Low levels	Low levels
Infections	Viral, bacterial	All types, with chronic or persistent evolution	All types	Bacterial (opportunistic and pyogenic)
Treatment	Fetal thymus transplant	Bone marrow graft; PEG-ADA, gene therapy	Gamma globulin IV (if pyogenic infections predominate)	Gamma globulin IV (if pyogenic infections predominate)

1. **Congenital thymic aplasia (DiGeorge Syndrome)**. The DiGeorge syndrome can be considered as the paradigm of a pure T-cell deficiency.
   a. **Etiology and pathogenesis**
      i. Although the DiGeorge syndrome is a **congenital** immunodeficiency, it is not hereditarily transmitted. It is believed to be caused by an intrauterine infection prior to the eighth week of life, possibly of viral etiology.
      ii. It is associated with microdeletions of chromosomal region 22q11.
      iii. From the immunological point of view, it results from defective embryogenesis of the third and fourth pharyngeal clefts at 6 to 8 weeks of fetal life, leading to deficient development of the thymus and parathyroids. Other congenital malformations are common in these patients.
   b. **Clinical presentation**. The main clinical features include:
      i. **Neonatal tetany**, which results from hypocalcemia secondary to hypoparathyroidism
      ii. Abnormalities of the heart and large vessels
      iii. Facial dysmorphism
      iv. Mental subnormality
      v. Frequent infectious episodes
   c. **Laboratory findings**
      i. The **thymic image** on a chest x-ray is absent or extremely reduced in size.
      ii. Variable degree of T-lymphocyte deficiency. In some cases, there are residual T lymphocytes and/or partial thymus function (partial DiGeorge syndrome). In these cases, if the patient can be kept alive for a number of years, a slow development of immune functions may take place.
      iii. The numbers of B lymphocytes in peripheral blood are normal in patients with DiGeorge syndrome and humoral immunity is not severely impaired, as a rule. Low levels of immunoglobulins and increased frequency of viral and bacterial infections have been reported in some cases, probably due to lack of T-cell help.
   d. **Therapy**
      i. The best treatment for complete DiGeorge syndrome is the **transplantation of a fetal thymus**.
      ii. If residual T lymphocytes can be detected, the administration of immunomodulating agents (e.g., thymosin or transfer factor) may be of benefit and allow the patient to remain relatively infection-free.
2. **Chronic mucocutaneous candidiasis**. Some patients with chronic infection of skin and mucosae with *Candida albicans* have been shown to have a selective deficiency of cell-mediated immunity.
   a. **Physiopathology**. Skin tests with Candida antigens and in vitro lymphocyte proliferation responses to *C. albicans* reveal a selective lack of reactivity. T-lymphocyte functions are normal when tested with other antigens and mitogens. The humoral response to *C. albicans* is also normal.

b. **Therapy**. Symptomatic therapy with antimycotic agents is often unsuccessful, and efforts to correct the deficiency with immunomodulators (particularly using transfer factor prepared from the lymphocytes of donors with strong CMI against *C. albicans*) have met with some success, although the improvement is usually temporary.

C. **Combined Immunodeficiencies**

1. **Severe, combined immunodeficiency (SCID, Swiss-type agammaglobulinemia)** is believed to result from the lack of differentiation of stem cells.
   a. **Genetics and physiopathology**. It is an inheritable disorder that can be inherited as an X-linked recessive form or as an autosomal recessive form (also known as *Swiss-type agammaglobulinemia*).
      i. The sex-linked form is associated with a defect of the gene that codes for a polypeptide chain (γc) common to several interleukin receptors (IL-2, IL-4, IL-7, IL-11, and IL-15). This chain is involved in signaling of second messages, and in its absence T-cell precursors fail to receive the signals necessary for their proliferation and differentiation.
      ii. The autosomic recessive forms are associated with two types of abnormalities: deficiency of ZAP-70, a protein kinase implicated in signaling through the TcR (see Chaps. 4 and 11); and deficiencies of purine salvage enzymes, such as adenosine deaminase (ADA) and nucleoside phosphorylase.
      iii. The pathogenesis of SCID associated with ADA deficiency has been studied in detail.
         - ADA catabolizes the deamination of adenosine and 2′-deoxyadenosine. Therefore, the lack of ADA causes the intracellular accumulation of these two compounds.
         - 2′-deoxyadenosine is phosphorylated intracellularly and the activity of the phosphorylating enzyme is greater than the activity of the dephosphorylating enzyme. Consequently, there is a marked accumulation of deoxyadenosine triphosphate.
         - DeoxyATP is a feedback inhibitor of ribonucleotide reductase, an enzyme required for normal DNA synthesis. As a consequence, DNA synthesis will be greatly impaired, and no cell proliferation will be observed after any type of stimulation.
         - In addition, 2′-deoxyadenosine is reported to cause chromosome breakage, and this mechanism could be the basis for the severe lymphopenia observed in these patients.
         - The reason why lymphocytes are predominantly affected over other cells that also produce ADA in normal individuals is that immature T cells are among those cells with higher ADA levels (together with brain and gastrointestinal tract cells).
   b. **Clinical presentation**. Symptoms start very early in life, usually by 3 months of age. Survival beyond the first year of life is rare.
      i. Persistent infections of the lungs, often caused by opportunistic agents such as *Pneumocystic carinii*.
      ii. Severe mucocutaneous candidiasis.

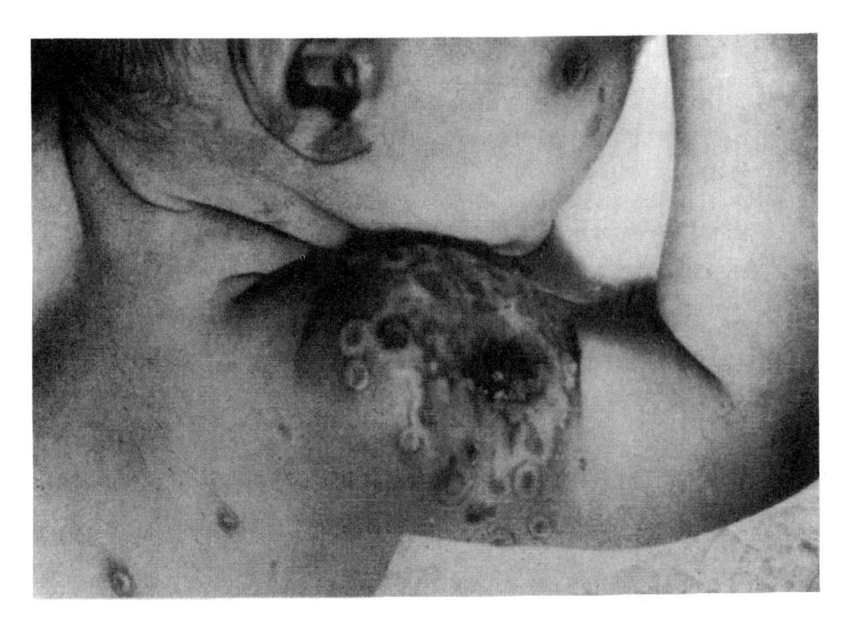

Figure 30.1 Vaccinia gangrenosa reaction in a 6-month-old child with combined immunodeficiency. Note the extensive gangrenous and necrotic lesions with satellite poxes around the gangrenous lesion that eventually spread to the face and buttocks. (Reproduced with permission from Good, R.A. et al., *Progr. Allergy*, *6*:187, 1962.)

iii. Severe infections after immunization with live, attenuated organisms (Fig. 30.1).
iv. Chronic, untractable diarrhea, wasting, and runting.
v. Physical examination shows **absence of all lymphoid tissues**: atrophic tonsils, very small or undetectable lymph nodes, absent thymic shadow on x-ray of the thorax plus signs of pulmonary infection, evidence of poor physical development, and oral thrush.

c. **Laboratory findings**
i. The lack of differentiation of stem cells results in **profound lymphopenia**, with deficiency of both T and B cells. Neutropenia can also be seen in some patients. In cases of ZAP kinase deficiency, lymphocyte counts may be normal or close to normal, but the T lymphocytes do not respond to stimulation.
ii. The **deficiency in cell-mediated immunity** is reflected by negative skin tests, delayed rejection of allogeneic skin grafts, and lack of response of cultured mononuclear cells to T-cell mitogens and anti-CD3 monoclonal antibodies.
iii. **Immunoglobulins are usually low**, but in some cases can be normal or irregularly affected. B cells and plasma cells are usually low or undetectable, but can be normal or increased in numbers in some patients. Even in these cases, however, antibody responses are very low to absent.
iv. **ADA deficiency** is diagnosed by lack of detection of the enzyme.

Red cells, lymphocytes, fibroblasts, amniotic cells, fetal blood, chorionic villous samples are all adequate for testing ADA activity.

d. **Therapy**

i. All forms can be corrected with a **bone marrow graft** from HLA-DR matched siblings. The graft is usually successful, but there is a great risk for the development of **graft-versus-host disease** (GVHD). GVHD can develop also after transfusion of any fresh blood components contaminated with viable T lymphocytes and is characterized by fever, maculopapular rash involving the volar surfaces, diarrhea and protein-losing enteropathy, Coombs' positive hemolytic anemia, thrombocytopenia, and splenomegaly. In full-blown cases, the outcome is generally poor, with death occurring within 10–14 days from the onset of symptomatology. The reaction may be prevented in the case of transfusion by using frozen or irradiated blood products. Current attempts at eliminating all cells except stem cells from bone marrow grafts appear promising, and the successful grafting of parental haploidentical bone marrow has been reported.

ii. **ADA deficiency** was first treated by **transfusion of frozen, irradiated red cells** (to kill any contaminating lymphocytes), as a source of ADA. Later, it was found that administration of bovine **ADA plus polyethylene glycol** (PEG) was more effective. The addition of PEG resulted in decreased immunogenicity and increased half-life of the bovine ADA. Finally, ADA was the first human disease to be successfully treated by **gene therapy**. The protocol involves harvesting peripheral blood T lymphocytes from the patients, transfecting the ADA gene using a retrovirus as vector, expanding the transfected T lymphocytes in culture, and readministering them to the patient. In the first treated patient, normal peripheral blood T-lymphocyte counts and clinical improvement were seen after three infusions of 1.3×10^{10} genetically corrected cells. The infusions need to be periodically repeated, since the ADA^+ T-lymphocyte population will eventually decline. The normalization of T-cell counts probably reflects the fact that the transfected ADA^+ cells will produce excess ADA, which will diffuse into genetically deficient cells unable to synthesize it.

2. **Combined immunodeficiency with abnormal immunoglobulin synthesis (Nezelof's syndrome)**. Clinically, this situation is very similar to those cases of SCID in which variable numbers of B lymphocytes and variable levels of immunoglobulins can be assayed. There is no well-defined pattern of inheritance.

3. **Immunodeficiency with ataxia-telangiectasia**

a. **Genetics and physiopathology**. Ataxia-telangiectasia is genetically transmitted following an autosomal recessive pattern of inheritance. It is believed that the disease may result from a deficiency of DNA repair enzymes, as suggested by the high frequency of lymphoreticular malignancies. In addition, the enzyme defect seems to result in a generalized defect in tissue maturation, affecting many tissues, but with particular significance in the brain capillary vessels. Persistently increased levels of serum α-fetoprotein and carcinoembryonic antigen in many patients with this disease support this last postulate.

b. **Clinical presentation**
   i. The initial symptoms are of progressive cerebellar **ataxia** beginning in early childhood associated with insidiously developing **telangiectasia** (first appearing as a dilation of the conjunctival vessels). The capillary abnormalities are systemically distributed and involve the cerebellum, causing the motor difficulties characteristic of ataxia. In late childhood, recurrent sinobronchial **infections** begin, leading to bronchiectasia.
   ii. The immunodeficiency is characterized by associations of thymic hypoplasia, T-cell deficiency, and low immunoglobulin levels, particularly IgA which is low or absent in 80% of the patients.
   iii. The prognosis is poor and there is no effective therapy. Death usually occurs before puberty, most frequently as a consequence of lymphoreticular malignancies or of the rupture of telangiectatic brain vessels.

4. **IL-2 synthesis deficiency**. Combined immunodeficiency associated with a deficiency in IL-2 synthesis and IL-2 receptor can be seen in patients with **congenital deficiency of CD4+ cells** as well as in patients with normal numbers of CD4+ cells.
   a. **Genetics and physiopathology**
      i. In patients with congenital deficiency of CD4+ cells, the defect of IL-2 production and IL-2 receptor expression is a direct consequence of the lack of differentiation of this lymphocyte subpopulation. At least in one case of CD4+ deficiency, the CD4 gene was identified in the patient's cells, although no transcription products could be detected. Thus, the defect may result either from minor gene alterations, undetectable by our current methodologies, or from lack of transcriptional activation of a normal gene.
      ii. In patients with normal numbers of helper T lymphocytes, the defect can be found in a mutated IL-2 gene or in the system of second messenger molecules and transacting proteins (particularly NFAT) which mediate the activation of cellular genes as a result of antigenic or mitogenic stimulation.
   b. **Clinical presentation**. In general, the affected children have a very early onset of symptoms and suffer both from opportunistic and bacterial infections. Immunoglobulin levels tend to be decreased, and antibody responses after active immunization are subnormal.
   c. **Laboratory findings**. The number of CD4+ cells may be very low, absent, or normal. In either case, the peripheral blood lymphocytes fail to proliferate and to release IL-2 after stimulation with T-cell mitogens, antigens, or with anti-CD2, anti-TCR, and anti-CD3 monoclonal antibodies.

5. **Deficient expression of MHC molecules**. The lack of expression of either MHC-I or MHC-II molecules is associated with combined immunodeficiency.

a. The **bare lymphocyte syndrome** is characterized by a deficient expression of HLA-A, B, and C markers and absence of β_2-microglobulin on lymphocyte membranes. In one family, the defect has been localized to the a mutation of one of the genes coding for the transporter proteins (TAP-2) essential for proper intracellular assembly of the MHC-I molecules (see Chap. 4).
   i. Although some patients with this syndrome may be asymptomatic, most suffer from infections. In some cases, the infection pattern, involving *Pneumocystis carinii* and other fungi, is suggestive of combined immunodeficiency; in other patients the symptoms are mainly due to infections with pyogenic bacteria. The link between lack of expression of MHC class I markers and humoral immunodeficiency is unclear.
   ii. Laboratory findings include lymphopenia, poor mitogenic responses, low immunoglobulin levels, and lack of antibody responses. B cells are usually detected, but plasma cells are absent.

b. **MHC class II deficiency** is inherited as an autosomal recessive trait, apparently resulting in abnormal transcription regulation of the MHC genes. It is associated with a severe form of combined immunodeficiency, with absent cellular and humoral immune responses after immunization. These patients have a **low number of helper T lymphocytes**, which results in lack of differentiation of B lymphocytes into antibody-producing cells. This syndrome provides strong support for the theory that suggests that the interaction between double positive CD4+, CD8+ thymocytes and MHC-II molecules is the essential stimulus for the differentiation of CD4+ helper T lymphocytes.
   i. Patients present with protracted diarrhea, secondary to infections with candidiasis or *Cryptosporidium*, leading to malabsorption and failure to thrive. Pulmonary infections are also frequent. Residual cytotoxic T-cell function is reflected by the ability of these children to reject grafted cells and tissues.
   ii. Laboratory findings include normal counts of CD3+ lymphocytes associated to low numbers of CD4+ cells.
   iii. The prognosis is very poor, death tends to occur before the second decade of life.
   iv. Bone marrow transplantation, if successful, can correct the deficiency.

6. **Immunodeficiency with thrombocytopenia and eczema (Wiskott-Aldrich syndrome)**

a. **Genetics and pathogenesis.** It is a genetically transmitted disease, with a X-linked recessive pattern of inheritance. The responsible gene has been located to Xp11.23, which encoded a protein of unknown function. Platelets and T lymphocytes are predominantly affected. T lymphocytes show disorganization of the cytoskeleton and loss of microvilli.

b. **Clinical presentation.** The Wiskott-Aldrich syndrome is characterized by **eczema, thrombocytopenia**, and frequent **infections**.
   i. Early in life the infections tend to be caused by encapsulated pyogenic bacteria, such as *Streptococcus pneumoniae*, *Neisseria*

meningitidis, and *Haemophilus*, which correlates with the inability to respond to bacterial polysaccharides.

ii. Later in life, the patients can suffer from a variety of opportunistic infections—viral, bacterial, mycotic, and parasitic—reflecting a deterioration of both cell-mediated and humoral immune functions.

c. **Laboratory findings**

i. The finding of profound **thrombocytopenia** with **small-sized platelets** very early in life is considered diagnostic.

ii. **Low IgM** levels with normal or high levels of other immunoglobulins.

iii. **Lack of antibody response to polysaccharides**.

iv. Lymphocyte count and function are normal in early infancy, but as the infant grows, lymphopenia develops, evident in the peripheral blood, thymus, and all other lymphoid tissues.

v. Infection is the most frequent cause of death, but some children develop lymphoreticular malignancies which can have a fatal evolution.

d. **Therapy**

i. The immunological defects can be corrected by **bone marrow transplantation**.

ii. Given the potential risk of this type of therapy, attempts at immunostimulation with transfer factor or at replacement therapy with intravenous gamma globulin have been made with mixed success.

iii. Thrombocytopenia usually responds to splenectomy, but this is a risky intervention in a patient whose immune and clotting systems are compromised.

D. Complement Deficiencies. Deficiencies of virtually all components of the complement cascade have been reported in the literature. The deficiencies can be grouped as early component (Cl, C2, C4) deficiencies, late component (C5 to C9) deficiencies, and C3 deficiency (Table 30.6).

1. **Early component deficiencies**. Deficiencies of Cl, C2, and C4 can be asymptomatic, associated with predisposition to infections (particularly when the deficiencies of C2 and C4 are complete or associated with factor B deficiency), or, most frequently, associated with clinical symptoms suggestive of **autoimmune disease**. C2 and C4 deficiencies are often associated with a syndrome mimicking systemic lupus erythematosus (SLE), although the clinical evolution is more benign, and the kidneys are usually spared. Persistent levels of circulating immune complexes are often found in these patients, probably as a result of altered dynamics of IC clearance, since these patients will not be generating normal amounts of C3b fragments due to the interruption in the activation sequence due to the lack of C2. The prolonged persistence of IC in circulation is likely to be an important pathogenic factor.

2. **C3 deficiency**

a. Primary C3 deficiency is a rare condition, transmitted as an autosomal recessive trait. Patients with C3 deficiency have an inability to opsonize antigens and suffer from **recurrent pyogenic infections** from early in life, with a clinical picture similar to that of X-linked infantile agammaglobulinemia, in spite of normal B- and T-cell function. This is not

Table 30.6 Summary of the Main Characteristics of Primary Complement Deficiencies, Using Common Variable Immunodeficiency as a Term of Reference

Characteristic	Common variable immune deficiency	Early complement component deficiencies	C3 deficiency	Late complement component deficiencies
Genetics	Variable	Undefined	Autosomal recessive	Variable
Molecular basis	Variable, ill-defined	Lack of synthesis	Lack of synthesis	Lack of synthesis
Lymphoid tissues	Follicular necrobiosis, reticulum cell hyperplasia	Normal	Normal	Normal
B lymphocytes	Normal numbers, abnormal differentiation or function	Normal	Normal	Normal
Serum immunoglobulins	Low to very low levels	Normal to high	Normal to high	Normal to high
Infections	Bacterial, parasitic (Giardia)	Pyogenic bacteria	Pyogenic bacteria	Capsulated bacteria, esp. *N. meningitidis*
Autoimmunity	Rheumatoid arthritis	SLE-like syndrome	Vasculitis, glomerulonephritis	None

surprising, given the pivotal role played by C3 in complement activation and in opsonization of microbial agents. Furthermore, C3-deficient patients may present manifestations of immune complex disease, such as glomerulonephritis and vasculitis.

b. **Factor I (C3b inactivator) and factor H deficiencies** result in a deficiency of C3 secondary to exaggerated catabolism of this complement component (it has been estimated that C3 is catabolized at four times the normal rate). Patients present with recurrent **pyogenic infections**, particularly with encapsulated bacteria (such as *S. pneumoniae* and *N. meningitidis*), and immune complex disease. "Anaphylactoid" reactions secondary to the spontaneous generation of C3a are also frequent in these patients.

3. **Late component deficiencies**. Deficiencies of C5, C6, C7, C8, and C9 have been reported to be associated with **increased frequency of infections** most frequently involving bacteria with polysaccharide-rich capsules (particularly *Neisseria* species).

Case 1 Revisited

- The prevalence of bacterial infections (pneumonia, pyoderma) caused by classic pyogenic organisms (e.g., *Haemophilus influenzae*) is strongly suggestive of humoral immunodeficiency. The lack of adverse reactions after the usual childhood immunizations, which include several attenuated viruses, and the uneventful course of chickenpox strongly suggest that cell-mediated immunity is normal.

- The most likely diagnosis, due to the humoral immunity deficiency, apparent integrity of cell-mediated immunity, normal numbers of B lymphocytes, and age of onset (8 years) is common, variable immunodeficiency. The pattern of inheritance is variable, but in this case is most likely autosomal recessive. A male child would not likely be affected, but a female child should be considered as potentially carrying the same disease. Diagnosis would be impossible in the neonatal and early infancy period, since any child born to this female would be severely hypogammaglobulinemic and likely to suffer from bacterial infections. Also, the onset of disease could be delayed by several years.
- Two types of tests could be valuable in this patient: the assay of anti-*H. influenzae* antibodies (was negative) and a lymphocyte transformation assay, measuring IL-2 release after stimulation with PHA and anti-CD3 monoclonal antibodies. Patients with common variable immunodeficiency often show abnormal mitogenic responses, low CD40 ligand expression, and subnormal IL-2 release after T-cell mitogenic stimulation, suggesting that the basis of the disease is lack of T-cell help.
- As in infantile agammaglobulinemia, chronic obstructive lung disease and bronchiectasis are likely to develop as a consequence of repeated bronchopulmonary infections.
- The best treatment for this condition is the administration of intravenous gamma globulin. The periodicity of administration needs to be established for each individual patient, based on the longitudinal variation in serum immunoglobulin levels, and on the duration of the symptom-free period after gamma globulin administration. In most cases, administration of IV gamma globulin needs to be repeated every 3–4 weeks.

V. SECONDARY IMMUNE DEFICIENCIES

A. **General Considerations**. Many factors influencing the function of the immune system can lead to variable degrees of immunoincompetence. Infections, exposure to toxic environmental factors, physical trauma, and therapeutic interventions can all be associated with immune dysfunction. In some cases, the primary disease that causes the immunodeficiency is very obvious, while in others, a high degree of suspicion is necessary for its detection. The pathogenic mechanisms are very clear in some cases, and totally obscure in others. The following is a brief summary of the major types of secondary immunodeficiencies, followed by a more detailed discussion of the acquired immunodeficiency syndrome (AIDS).

Case 2

A 30-year-old male is seen by his family physician with a 3-week history of headaches, dry cough, and shortness of breath with exertion that was increasingly severe, and a 40-lb. weight loss over 6 months. He was observed to have a fever of 100.5°F, clear lungs. He became more dyspneic and was seen 3 days later with cough productive of white sputum and severe dyspnea with exertion. Arterial blood gases showed a PO_2 of 38 mmHg. He started noticing difficulty swallowing solid food at that time. On physical exam he was a thin male in no apparent distress sitting in bed. Vital signs were: Temp

100°F, pulse 100/min., respiration 28/min., BP 116/84, weight 110 lb. There were white plaques on the tongue and buccal mucosa that could be removed with a tongue depressor. There was an occasional wheeze but otherwise clear lung fields. He became dyspneic with conservation or walking across the room.
The WBC count was 5,000/μL with 86% neutrophils and 5% lymphocytes. Hemoglobin was 11.7 g/dL and hematocrit was 34%. Chest x-ray showed diffuse granular opacities over both lung fields. Sputum Gram stain and culture were negative. A bronchoalveolar lavage examination was positive for *Pneumocystis carinii.*

This case raises several questions:

- What type of immunodeficiency could be affecting this patient?
- Is there evidence for another infection besides pneumonia?
- What test(s) should be ordered to investigate the status of the immune system?
- Is the patient at risk for any other type of infection?
- What therapeutic and prophylactic measures should be taken?

B. Immunodeficiency Associated with Malnutrition

1. **Severe protein-calorie malnutrition** is primarily associated with a depression of cell-mediated immunity. Anergy, low T-lymphocyte counts, depressed lymphocyte reactivity to PHA, and depressed release of cytokines have been reported in malnourished populations by different groups.
2. In **kwashiorkor**, which is due to a combination of protein-calorie malnutrition and deficiency in trace elements and vitamins, the degree of immunodeficiency seems to be more profound. Affected children seem to have a delayed maturation of the B-cell system and often have low levels of mucosal IgA, without apparent clinical reflection. Efforts to study the humoral immune response to active immunization have yielded variable results. The complement system and neutrophil functions have been reported as depressed, but the phagocytic impairment is mild and hypocomplementemia seems to be due in major part of consumption as a consequence of infections.
3. Several causes for the immune deficiency associated with malnutrition have been suggested, including general metabolic depression, thymic atrophy with low levels of thymic factors, depressed numbers of helper T lymphocytes (which could account for the variable compromise of humoral immunity), and impaired cytokine release.
4. Malnourished children should not be vaccinated with live, attenuated vaccines, which are generally contraindicated in immunodeficient patients.

C. Immunodeficiency Associated with Zinc Deficiency

1. **Acrodermatitis enteropathica** is a rare congenital disease in which diarrhea and malabsorption (affecting zinc, among other nutrients) play a key pathogenic role. Affected patients often present with epidermolysis bullosa and generalized candidiasis, associated with combined immunodeficiency that can be corrected with zinc supplementation.
2. **Secondary zinc deficiency** can develop due to low meat consumption, high-fiber diet, chronic diarrhea, chronic kidney insufficiency, anorexia nervosa and bulimia, alcoholism, diabetes, psoriasis, hemodialysis, parenteral alimentation, etc. By itself it does not appear to lead to a depletion severe enough to result in immunodeficiency, but may be one of several factors adversely affecting the immune system.

3. The basis for the depression of cell-mediated immunity in zinc deficiency has not been established, but it has been proposed that zinc may be essential for the normal activity of cellular protein kinases involved in signal transduction during lymphocyte activation.

D. **Immunodeficiency Associated with Vitamin Deficiencies**. Several vitamin deficiencies are associated with and presumably the cause of abnormalities of the immune response, particularly when associated with protein-calorie malnutrition. The molecular mechanisms underlying these deficiencies have not been defined.
   1. **Deficiencies of pyridoxine, folic acid, and vitamin A** are usually associated with cellular immunodeficiency.
   2. **Pantothenic acid deficiency** is usually associated with a depression of the primary and secondary humoral immune responses.
   3. **Vitamin E deficiency** is associated with a combined immunodeficiency.

E. **Immunodeficiency Associated with Renal Failure**. Patients with renal failure have depressed cell-mediated immunity, as reflected in cutaneous anergy, delayed skin graft rejection, lymphopenia, and poor T-lymphocyte responses to mitogenic stimulation. Humoral immunity can also be affected, particularly in patients with the nephrotic syndrome, who may lose significant amounts of IgG in their urine.
   1. Several factors seem to contribute to the **depression of cell-mediated immunity**:
      a. Release of a soluble suppressor factor, as shown by experiments demonstrating that plasma or serum from uremic patients suppresses the mitogenic responses of normal lymphocytes in vitro. The responsible factors have a molecular weight less than 20,000 kDa, and it has been suggested that methylguanidine and "middle molecules" (molecular weight 1200) are responsible. These molecules can be isolated from uremic sera and have been shown to suppress in vitro mitogenic responses of normal T lymphocytes.
      b. In dialyzed patients there is a paradoxical activation of the immune system, which results in excessive and dysregulated release and consumption of IL-2, resulting in decreased bioavailability of this cytokine.
   2. Fc-mediated phagocytosis is depressed in patients with severe renal failure, perhaps secondary to increased levels of endogenous glucocorticoid levels. There is also evidence of a compromise of the capacity of monocytes to function as antigen-presenting cells. These abnormalities are reproducible when normal monocytes are incubated with uremic serum. Dialysis may accentuate the problem as a consequence of complement activation in the dialysis membranes, which causes a poorly understood down-regulation of the expression of CAMs by phagocytic cells.
   3. Patients with chronic renal failure are often treated with immunosuppressive drugs and have semipermanent intravenous catheters inserted. These factors contribute to an increased susceptibility to opportunistic infections.

F. **Burn-Associated Immunodeficiency**. Bacterial infections are a frequent and severe complication in burn patients, often leading to death. There are several factors that may contribute to the incidence of infections in burned patients, including the presence of open and infected wounds, a general metabolic disequilibrium, and a wide spectrum of immunological abnormalities.

1. **Depressed neutrophil function** is a major factor contributing to the lowered resistance to infection. Defective chemotaxis and reduced respiratory burst are the most prominent abnormalities. Several factors may contribute to this depression:
   a. Exaggerated complement activation (mostly by proteases released in injured tissues) cause the release of large concentrations of C5a that may disturb proper chemotactic responses and cause massive activation of granulocytes. When the already activated granulocytes reach the infected tissues, they may no longer be responsive to additional stimulation.
   b. Bacterial endotoxins, prostaglandins, and β-endorphins have been suggested as additional factors that adversely affect phagocytic cell functions. The involvement of prostaglandins has been supported by studies in experimental animals, in which administration of **cyclooxygenase blockers** normalizes phagocytic cell functions.
   c. Another contributing factor seems to be the low opsonic power of the burn blister fluid, which has very low levels of both complement and immunoglobulins.
2. **Impairment of cell-mediated immunity**, suggested by a prolongation of skin homograft survival and depressed delayed hypersensitivity responses, and supported by depressed responses to mitogenic stimuli, depressed mixed lymphocyte culture reactions, has also been well documented.
   a. A major functional abnormality of T lymphocytes isolated from burned patients is their **depressed release of IL-2** after mitogenic stimulation. This depression may be secondary to the release of immunosuppressive factors by the burned tissues.
      i. A 10-kDa glycopeptide inhibits IL-2 release by mitogenically stimulated normal T lymphocytes.
      ii. A 1000-kDa lipid-protein complex released by damaged epithelial cells inhibits the cellular response to IL-2.
      iii. **PGE_2**, released by overactive monocytes, causes an increase of intracellular cAMP in T cells, resulting in an inhibition of cell proliferation.
   b. The levels of GM-CSF are also reduced after thermal injury. In thermally injured experimental animals, administration of this cytokine restores T-cell proliferation and IL-2 release.

G. Iatrogenically-Induced Immune Deficiencies. A wide range of therapeutic interventions has been shown to cause functional depression of the immune system. On top of the list is the administration of cytotoxic/immunosuppressive drugs (see Chap. 26), but many other medical procedures have unexpected effects on the immune system.

1. **Postsurgery immunodeficiency**. Any type of surgery represents an acute trauma that can reduce the anti-infectious defenses in a variety of ways (Fig. 30.2).
   a. Surgery always involves trauma.
      i. The surgical incision disrupts the integrity of the skin, a very important barrier against infection.
      ii. Intestinal surgery promotes spreading of bacteria from a highly contaminated organ into surrounding tissues.

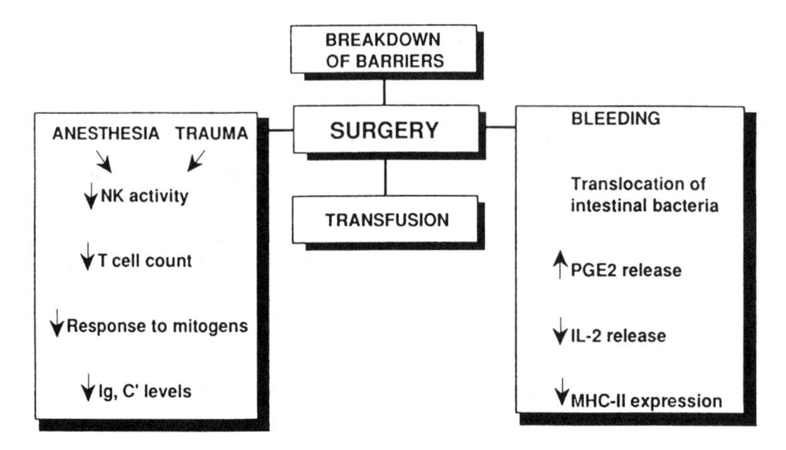

Figure 30.2 Diagrammatic representation of the different factors contributing to the immune suppression associated with surgery.

iii. The introduction of intravenous lines and catheters opens new routes for the penetration of opportunistic agents into the skin.
iv. Severe blood loss can cause massive entrance of intestinal bacteria into the portal circulation, and subsequently, into the systemic circulation (phenomenon known as *bacterial translocation*).

b. Surgery and general anesthesia are associated with **transient depression of immune functions**, affecting the mitogenic responses of PBL, cutaneous hypersensitivity, and antibody synthesis. Multiple factors seem to contribute to the depression of the immune system:
   i. A transient severe **lymphopenia** can occur in the immediate postoperative period.
   ii. Exaggerated release of **PGE_2**, due to the posttraumatic activation of inflammatory cells, depressed T-lymphocyte and accessory cell functions.
   iii. **Blood loss** can be associated with a reduction of IL-2 release by activated T lymphocytes and MHC-II expression by accessory cells, and with reduced B-lymphocyte responses to antigenic stimulation.
   iv. **Transfusions** have a poorly understood immunosuppressive effect (see Chap. 27).
   v. **Anesthesia** and administration of opiates (as pain killers) can lead to a depression of phagocytic cell functions and to a reduced activity of NK cells.

c. Complete normalization of immune function may take 10 days (for mitogenic responses) to a month (for delayed hypersensitivity reactions and the humoral immune response).

d. **Splenectomy** is a particularly important cause of immune depression.
   i. The removal of the spleen represents the loss of a major filtration organ, very important for the removal of circulating bacteria. In

addition, the spleen plays a significant role in recruiting immunocytes in the initial phases of the immune responses.

ii. Splenectomized patients are weakly responsive to polysaccharides and their circulating B cells respond poorly to mitogenic stimulation.

iii. As a consequence, splenectomized patients are prone to develop bacteremia and sepsis; the offending organisms are common pyogenic bacteria, particularly those with antiphagocytic polysaccharide capsules, such as *Streptococcus pneumoniae* (50% of the cases), *Haemophilus influenzae*, and *Neisseria meningitidis* as well as other common bacteria with unrelated virulence factors, such as *Staphylococcus aureus* and group A *Streptococcus*. It has also been demonstrated that about one-third of the cases of human infection by *Babesia*, an intracellular sporozoan, have happened in splenectomized patients.

iv. Similar defects are noticed in patients with sickle-cell anemia, who develop splenic atrophy as a consequence of repeated infections and fibrosis (autosplenectomy).

e. **Thymectomy** was frequently done in neonates with congenital heart disease to ensure proper surgical access. It is generally believed that thymectomy after birth has few (if any) effects on the development of the immune system of humans and there is no conclusive evidence suggesting otherwise.

H. **Immunosuppression Associated with Drug Abuse**. There is considerable interest in defining the effects of drug abuse on the immune system. Unfortunately, most data concerning the immunological effects of drugs of abuse is based on in vitro experiments or on studies carried out with laboratory animals, which may or may not reflect the in vivo effects of these compounds in humans.

1. **Alcoholism**. Chronic alcoholism is associated with a depression of CMI, but it could be argued that factors other than ethanol consumption, such as malnutrition and vitamin deficiencies, could be the major determinants of the impairment of the immune system. A direct effect of ethanol is supported by animal experiments, in which both T-lymphocyte functions and B-lymphocyte responses to T-dependent antigens are compromised after 8 days of ethanol administration.

2. **Cannabinoids**. There is little concrete evidence for an immunosuppressive effect of cannabinoids in humans, except for depressed results in in vitro T-lymphocyte function tests and for an increased incidence of genital herpes among young adults who use cannabinoids. In laboratory animals, cannabinoid administration predominantly affects T- and B-lymphocyte functions, increases the sensitivity to endotoxin, and increases the frequency of infections by intracellular agents, such as *Listeria monocytogenes*. However, the conditions of administration of these compounds to laboratory animals are rather different than the conditions surrounding their use as recreational drugs.

3. **Opiates**

a. **Cocaine** has been shown to have direct effects over human T lymphocytes in vitro, but the required concentrations greatly exceed the plasma levels measured in addicts. The results of studies carried out in addicts have been contradictory.

b. I.V. **heroin** use is associated with a high frequency of infections. In many instances, the infection (thrombophlebitis, soft tissue abscesses, osteomyelitis, septic arthritis, hepatitis B and D, and HIV infection) seems clearly related to the use of infected needles, but in other cases (bacterial pneumonias, tuberculosis), the infection could result from a depression of the immune system. However, to date no conclusive evidence supporting a depressive effect of heroin over the immune system has been published.

I. Immunosuppression Associated with Infections

1. **Bacterial infections**
   a. Disseminated mycobacterial infections are often associated with a state of **anergy**. Patients fail to respond to the intradermal inoculation of tuberculin and other antigens, and their in vitro lymphocyte responses to PHA and the mycobacterial antigens are depressed. The mechanisms leading to anergy are poorly understood and probably involve more than a single factor:
      i. An increased production of IL-10 and IL-4 could reduce the activity of TH1 helper lymphocytes, thus depressing cell-mediated immunity.
      ii. Mycobacteria infect phagocytic monocytes, and intracellular infection is associated with a depression of both the antigen-presentation capacities and the ability to deliver costimulatory signals to T lymphocytes (for example, the expression of the B7 molecule is depressed). In addition, infected monocytes/macrophages may release nitric oxide, which inactivates lymphocytes in the proximity of the infected cells.

   b. The **release of soluble immunosuppressive compounds** has been demonstrated for several bacteria. Several different substances, including enzymes (ribonuclease and asparaginase), exotoxins (such as staphylococcal enterotoxins), and other proteins have been shown to have immunosuppressive properties, although probably their effects are limited to reducing the specific anti-infectious immune response. The staphylococcal enterotoxins are part of a group of bacterial proteins known as **superantigens** (see Chap. 13). In vitro, most superantigens have stimulatory properties, but when administered in vivo induce generalized immunosuppression (perhaps as a consequence of indiscriminate, nonspecific T-cell activation).

2. **Parasitic infections**. Parasitic infections due to protozoa often seem to be associated with suppression of the immune response to the parasite itself. In some cases, however, there is evidence of the induction of a more generalized state of immunosuppression.
   a. Acute infections with *Trypanosoma cruzi* are associated with CMI depression that can be easily reproduced in laboratory animals. Both in humans and experimental animals, there is a reduced expression of IL-2 receptors, which can be interpreted as resulting from a down-regulation of TH1 cells, either by cytokines or by suppressor compounds released by the parasite.

b. Similar mechanisms seem to account for the generalized immunosuppression observed in experimental animals infected with *Toxoplasma*, *Schistosoma*, *Leishmania*, and *Plasmodia*.

3. **Viral infections**. The AIDS epidemic has certainly focused our attention in the interplay between viruses and immunity. However, HIV is certainly not the only virus able to interfere with the immune system.
   a. The development of a transitory state of anergy during the acute stage of **measles** was first reported by Von Pirquet in 1908.
      i. During the 3 to 4 weeks following the acute phase of measles, patients show lymphopenia and the residual population of peripheral blood lymphocytes shows poor responses to mitogens and antigens such as PHA and *Candida albicans*.
      ii. The cause for this state of anergy is the infection of both T and B lymphocytes by the measles virus, which prevents the infected cells from proliferating in response to antigenic or mitogenic stimuli. The infected cells stop progressing on their division cycle at the terminal G1 phase, and, while interferon-γ and IL-2 are normally secreted, RNA synthesis is globally decreased and the activated lymphocytes fail to develop into fully mature and functional helper or cytotoxic T lymphocytes or into immunoglobulin-secreting plasmablasts.
   b. Other viruses, such as **cytomegalovirus (CMV)** and the **rubella virus**, can cause immunosuppression. CMV mainly depresses the specific response to the virus, while the rubella virus induces a generalized immunosuppression, similar to that caused by the measles virus.
   c. Viruses can also release suppressor factors (*Herpes simplex* virus secretes a protein similar to IL-10, which can down-regulate cytokine release by activated T lymphocytes) and interfere with antigen presentation (adenovirus infection is associated with a depressed expression of MHC-I molecules). However, patients infected with these viruses do not develop generalized immunosuppression, so it seems likely that the significance of these mechanisms is mostly related to promoting conditions favorable for the persistence of the infection.

VI. THE ACQUIRED IMMUNODEFICIENCY SYNDROME (AIDS)

A. **Etiology and Definitions**. The designation of AIDS is given to the profound immune deficiency that develops in the vast majority of individuals infected by the **human immunodeficiency virus (HIV)**.

1. **HIV** belongs to the *Lentiviridae* family of **retrovirus**. Two major variants of the virus have been identified: HIV-1, the original virus characterized by Montaigner and Gallo; and HIV-2, prevalent in West Africa. HIV-2 is associated with clinical disease identical to that caused by HIV-1, but appears less virulent, and it is felt that it is not spreading so widely and rapidly as HIV-1.
2. The major **structural components of HIV** are:
   a. Nucleic acid, constituted by two identical (+)RNA strands. The genome includes the usual retroviral genes:

i. **gag**, which codes for structural proteins.

ii. **pol**, which codes for the **reverse transcriptase**. In its intact form or after fragmentation, the pol gene product has several different enzymatic activities: polymerase; ribonuclease; and endonuclease (integrase, ligase).

iii. **env**, which codes for envelope glycoproteins

iv. **regulatory genes**, including:

- **tat** (transactivator of transcription); codes for a protein (p16) which binds to a region near the 5′ end of a nascent viral RNA strand known as TAR (transactivator response sequence) and promotes full and effective transcription of that strand. Soluble Tat protein (p16) is released by infected cells and taken up by both infected and noninfected cells. When taken up by infected cells it promotes viral genome expression; in noninfected cells it mainly induces transcription of cellular genes, creating ideal conditions for infection of the cell.
- **rev** (regulator of expression of viral proteins); codes for a second protein p19, which promotes the expression of HIV-1 structural proteins.

b. A nucleocapsid whose major component is protein (p) **p24**. The nucleocapsid encloses the two (+)RNA strands and the enzymes listed above: reverse transcriptase, ribonuclease, endonuclease, and protease.

c. A protein matrix underlying the envelope, formed by p17.

d. The envelope, in which glycoprotein spikes (**gp160**) are inserted. The **envelope glycoproteins** are composed of a transmembrane segment (**gp41**) that is noncovalently linked to the external major glycoprotein (**gp120**). gp120 is responsible for the interaction with CD4, the main cellular receptor for HIV; gp41 has fusogenic properties, and promotes the fusion of the viral envelope with the cell membrane, which results in injection of the nucleocapsid into the cytoplasm.

3. The **replication** of HIV requires reverse transcription of the viral RNA into double-stranded DNA. The insertion of viral DNA into cellular DNA is mediated by the viral endonuclease. Once integrated, the virus may remain dormant or may actively replicate. The factors controlling the activation of viral replication are under intense scrutiny.

4. The designation of acquired immunodeficiency syndrome (AIDS) is applied when an HIV-positive patient presents one or more of the following features:

a. **Opportunistic infections** (see Table 30.7).

b. **Progressive wasting syndrome** (adults) or **failure to thrive** (infants).

c. **Unusually frequent or severe infections** not considered as opportunistic, such as recurrent bacterial pneumonia or pulmonary tuberculosis. Recurrent bacterial infections are the most common infectious presentation of AIDS in infants and children.

d. Specific neoplastic diseases—**Kaposi's sarcoma**, **non-Hodgkin's lymphoma**, **invasive cervical carcinoma**.

e. Neuropsychiatric diseases such as **encephalopathy** (dementia) and **progressive multifocal leukoencephalopathy** (due to reactivation of

Table 30.7 Opportunistic Infections Characteristically Associated with AIDS

Pneumocystis carinii pneumonia
Chronic cryptosporidiosis or isosporiasis causing untractable diarrhea
Toxoplasmosis
Extraintestinal strongyloidosis
Candidiasis (oral candidiasis is common as a prodromal manifestation and is considered as a marker of progression towards AIDS; esophageal, bronchial, and pulmonary candidiasis are pathognomonic)
Cryptococcosis
Histoplasmosis
Infections caused by atypical Mycobacteria, such as *M. avium intracellulare*
Pulmonary and extrapulmonary tuberculosis (often resistant to therapy)
Disseminated cytomegalovirus infection (may affect the retina and cause blindness)
Disseminated herpes simplex infection
Multidermatomal herpes zoster
Recurrent *Salmonella* bacteremia
Progressive multifocal leukoencephalopathy
Invasive nocardiosis

an infection with the JC virus) or significant developmental delays or deterioration in children.

f. **Lymphocytic interstitial pneumonitis** in infants and children.

g. A **CD4+ cell count below 200/mm^3**.

5. The diagnosis of AIDS needs to be supported by evidence of HIV infection, such as:
   a. Isolation of the virus or detection of viral genomic material by PCR.
   b. Positive serological tests (as discussed later in this chapter).

B. Epidemiology. The main modes of transmission of HIV and the population groups affected by HIV infection in the U.S. are summarized in Tables 30.8 and 30.9. In the U.S., as well as in most western industrialized countries, the type of sexual contact with greatest risk is male-to-male, followed by male-to-female and female-to-female. Heterosexual transmission is considerably more common in third world countries, but is also on the rise in the U.S. Factors associated with

Table 30.8 Main Modes of HIV Transmission

1.	Sexual contact (by order of greater to smaller risk: male-to-male; male-to-female; female-to-female); receptive anal intercourse is considered as most risky.
2.	Sharing of needles and syringes among intravenous drug users
3.	Mother to child, transplacental or perinatal
4.	Blood and blood products[a]
5.	Transplantation of infected organs
6.	Artificial insemination and by maternal milk (rarely)

[a]Transmission by blood, blood products, and organ transplantation is currently highly unlikely because of the screening of blood in blood banks and the requirement that organ donors must be HIV-negative. Furthermore, all potential donors who engage in high-risk activities are asked not to donate or to self defer.

Table 30.9 Exposure Category Groups for AIDS

1. Homosexual or bisexual men (56%)
2. Heterosexual intravenous drug users (23%)
3. Homosexual or bisexual IV drug users (6%)
4. Heterosexual contact cases (6%)
5. Children (2%)[a]
6. Recipients of blood, blood components, or organ tansplants (2%)
7. Hemophiliacs (1%)

[a]Children infected by HIV in most cases (82%) have a mother with or at risk for HIV infection; those cases are usually found in minority populations. 10% of the children are infected through blood transfusions, and 5% have been infected because of hemophiliac disorders and administration of contaminated clotting factors.

increased risk of venereal transmission include receptive anal intercourse, IV drug-using partner, presence of genital ulcers, and multiple partners.

C. **HIV and the Immune System**

1. **Infected cell populations**. The most common routes of infection for HIV are mucosal abrasions or direct injection of contaminated blood or blood products.
   a. When penetration occurs through a mucosal surface, the virus is taken up by submucosal Langerhans cells, which transport it to the regional lymph nodes, where it is transmitted to **CD4+ T cells**, particularly those that coexpress the CD45RO marker, considered as activated or memory helper T lymphocytes. Only activated CD4+ T cells are believed to be susceptible to infection.
   b. When the virus is introduced directly into the blood stream, it will most likely be filtered in the spleen, adsorbed by **monocytes, macrophages**, and related cells, which express CD4-like molecules on the membrane (at much lower levels than the CD4 molecule on helper T lymphocytes), and from those passed to CD4+ T cells.
   c. The preferential infection of macrophages and related cells vs. CD4 lymphocytes depends on the affinity of HIV strains for co-receptors. The infection of macrophages involves interaction with β-chemokine receptors (CCR-5 or CKR-5; see Chap. 11) while the infection of CD4+ T cells involves the CXCR-4 molecule (fusin). Some strains which can infect both CD4 T lymphocytes and macrophages use CCR5 to infect macrophages, and a third chemokine receptor, CCR2b, to infect lymphocytes.
   d. The infection of **monocytes, macrophages**, and related cells is productive but not cytotoxic, and the infected cells become a source of persistent viral infection. Thus, the lymphoid tissues (lymph nodes and peri-intestinal lymphoid tissue) are the main reservoirs for virus burden in the infected individual. The virus replicates continuously, even in the early stages of asymptomatic disease.
2. **Early viremic stage**
   a. In the early stages of the infection, the virus appears to replicate at a

very low level and both a transient decrease of total CD4+ cells and a rise in circulating HIV-infected CD4+ T cells may be detected. Soluble p24 protein may be detected in circulation as early as 5–10 days after infection and circulating infectious virus is detectable for variable times, usually starting about the same time as p24, peaking 10–20 days after infection, and persisting in circulation until free anti-HIV antibody becomes detectable (**seroconversion**).

b. After seroconversion, integrated and soluble viral genomes continue to be detectable by PCR and other gene amplification techniques. Whether some individuals may completely eliminate the virus from their organisms is not known, although data collected on long-term survivors suggest that this may be the case in very rare instances. Understanding why some individuals are long-term survivors while others develop AIDS rather swiftly is a major priority in AIDS research. At this point it appears as if both host and microbial factors are involved:
   i. The genetic constitution of the individual may be critical. Differences in MHC repertoire and transport-associated proteins are emerging as related to the evolution of HIV infection.
   ii. The mode of exposure to HIV may play a significant role. Mucosal exposure to low virus loads seems to induce protective CMI at the mucosal level, and individuals may remain seronegative in spite of repeated exposures.
   iii. At least in some cases, long-term survivors seem to be infected by strains of reduced pathogenicity, which replicate less effectively, and are associated with lower viral loads. Indeed, there is an inverse correlation between the number of HIV-1 RNA copies in plasma and the duration of the asymptomatic period. It has been reported that only 8% of HIV-infected patients with less than 4350 copies of viral RNA/mL of plasma at the time of diagnosis developed AIDS after 5 years of follow up.
   iv. Regardless of what factors determine the size of the viral load in a given patient, **low viral loads are associated with prolonged survival**.

3. **Asymptomatic stage: humoral and cellular immune responses**. HIV-positive patients remain asymptomatic for variable periods of time, often exceeding 10–15 years (the average length of the asymptomatic period is currently 14 years). During that period of time, the virus replicates actively, but a steady state is reached in which the number of dying CD4+ T cells is roughly equivalent to the number of CD4+ T cells released from the bone marrow. It is likely that a vigorous anti-HIV immune response may be important to maintain this steady state.
   a. A strong **humoral immune response** against HIV can be detected in most patients. **Neutralizing antibodies**, which inhibit the infectivity of free HIV in vitro, directed against epitopes of gp120 and gp41, can be demonstrated. Also potentially protective are **ADCC-promoting antibodies**, which react with gp160 expressed on the membrane of infected cells. The rate of progression to AIDS and the mortality rate are considerably higher in individuals lacking neutralizing antibodies.

However, neutralizing antibodies do not prevent infected individuals from eventually developing AIDS, in part due to the high frequency of mutations in gp120, which result in the developing of mutants not neutralizable by previously existing antibodies.

b. On the negative side, **enhancing antibodies**, which react with gp41 antibodies and enhance HIV infectivity by an unknown mechanism, have also been demonstrated. In some studies, the presence of **HIV-enhancing antibodies** appears to be correlated with progression toward AIDS.

c. The general consensus is that **the humoral response elicited by HIV does not eliminate the infection and does not prevent evolution to AIDS**.

d. **Cell-mediated immune responses** involve MHC-I restricted CD8+ T lymphocytes, which recognize a variety of epitopes in *gag*, *env*, *nef*, and *pol* HIV proteins, presumably captured as short peptides during protein synthesis and expressed in association with MHC-I proteins. Cell-mediated cytotoxic reactions seem to be especially prominent in HIV-positive individuals who remain asymptomatic for prolonged periods of time. CD8 lymphocytes are also able, at least in vitro, to release cytokines (particularly RANTES, MIPI-α and β, and a recently discovered interleukin, IL-16) which appear to act by blocking the chemokine coreceptors used by HIV to penetrate uninfected CD4+ cells as well as compounds that inhibit HIV replication. Thus, **cell-mediated immunity seems able either to block infection or to reduce viral replication to levels tolerated by the immune system**.

4. **Decline of immune functions**

a. The evolution toward AIDS is associated with a progressive loss of immunological protective mechanisms, particularly the **progressive depletion of absolute numbers of CD4+ T lymphocytes**. This decline is associated with increased HIV replication, and several lines of evidence suggest that **T-cell activation is essential for the replication of integrated HIV**.

i. In vitro, the replication of integrated HIV is activated by mitogenic or antigenic stimulation of infected T cells as well as by coinfection with viruses of the herpes family.

ii. Several theories have been advanced about the cause of T-cell activation leading to enhanced replication of integrated HIV.

- Activation of infected macrophages and CD4+ lymphocytes by **concurrent infections** (venereal or not).
- Some viral components may act as **superantigens**, interacting directly with the Vβ regions of specific types of T-cell receptors, consequently activating those cells.
- Dendritic cells in the submucosa and lymphoid tissues appear to bind HIV to their surface without becoming infected. Thus, the **activation of HIV-carrying dendritic cells** could lead to clustering with noninfected T cells, which would receive activating signals and HIV virus from the dendritic cells.
- **Tumor necrosis factor-α (TNF-α)** and **interleukin-6 (IL-6)**

have the capacity to activate HIV replication in monocytes and T lymphocytes. In T lymphocytes, TNF-α induces the synthesis of a DNA-binding protein that binds to a nuclear factor kB (NF-kB) site on the HIV-LTR, whose occupancy results in activation of the expression of the integrated genome. There is evidence suggesting that HIV-infected macrophages release increased amounts of TNF-α and IL-6.

b. As a consequence of this enhanced replication, the level of plasma viral RNA increases and a **second wave of viremia** is detectable preceding clinical evolution to AIDS by as much as 14 months. Coinciding with the second wave of viremia, there is a decline in antibody levels, particularly to p24. In terminal stages, both viremia and antibody levels may drop again, perhaps reflecting the total exhaustion of the CD4+ cell population.
c. **T-cell depletion** is the direct cause of the profound immunodepression seen in AIDS patients. Several factors have been suggested to account for the progressive decrease of CD4 cells:
   i. **Direct cytotoxicity** caused by virus replication is probably the most important cause of T-cell death, particularly when viral replication is active. **Accumulation of unintegrated DNA** in the cytoplasm of infected cells is associated with vigorous HIV replication and cell death.
   ii. The cross-linking of CD4 molecules by gp120 is believed to prime T lymphocytes for apoptosis. In this case, active infection may not be essential, but as the apoptosis-primed T cell is activated by some other stimulus it will undergo apoptosis.
   iii. The expression of gp120 with unique sequences of the V_1 to V_2 and of the V_3 regions of gp120 on the membrane of infected T cells promotes the **formation of syncytia** by interaction and fusion with the membranes of noninfected cells expressing CD4. The formation of syncytia allows direct cell–cell transmission of the virus and eventually leads to a reduction in the number of viable T cells. The emergence of strains with the syncytia-inducing sequences in infected patients is usually a late event in the course of HIV infection, and associated with a faster progression to AIDS (median of 23 months).
   iv. The **immune response against viral infected T cells** (mediated both by cytotoxic T cells and by ADCC mechanisms) may also contribute to CD4+ T-cell depletion.
   v. **Coinfection of HIV-infected T cells** with other microorganisms, such as cytomegalovirus or *Mycoplasma fermentants* has synergistic effects in the induction of viral replication and cell death.
   vi. **Precursor cells in the thymus and bone marrow** are also infected by HIV, and this may account for the lack of regeneration of the declining CD4+ lymphocyte pool.
d. Several other factors beyond the depletion of CD4+ cells seem to contribute to the state of marked **immunodepression** associated with full-blown AIDS:

i. **An imbalance of the TH1 and TH2 subsets** seems to precede the evolution toward AIDS. While evidence for predominant TH2 activity in patients evolving toward symptomatic HIV infection has not been observed by several groups, reduction of TH1 activity is inevitable when the absolute CD4 count starts to fall. Patients become anergic, and cytotoxic T-cell and NK-cell activity decrease due to lack of help. Even B-cell responses eventually become depressed (as the TH2 subpopulation also becomes numerically depleted).

ii. **Soluble gp120** is released from infected cells, binds to CD4, and may block the interaction of this molecule with MHC-II antigens, therefore preventing the proper stimulation of helper T cells by antigen-presenting cells.

iii. **Immune complexes** involving viral antigens and the corresponding antibodies may also play a role in depressing immune responses. For example, binding of complexes of gp120 and anti-gp120 to CD4+ molecules of normal lymphocytes results in blocking of activation via the T-cell receptor.

iv. **Infected monocytes** are functionally abnormal, unable to perform chemotaxis, synthesize cytokines, and present antigens to helper T cells.

e. The **humoral immune responses** become suppressed, while at the same time autoantibodies are being synthesized.

i. **Humoral responses** to toxoids **are impaired**; this may be, in part, the result of lack of efficient help, but the B-cell system is affected by HIV infection insofar as it seems to be in a state of permanent polyclonal activation, probably a result of increased release of IL-6 by activated APC and T cells.

ii. As a consequence of polyclonal activation and loss of immunoregulatory mechanisms, a variety of unrelated antibodies start being produced, including antinuclear **autoantibodies** and autoantibodies directed against platelets and lymphocytes. The later autoantibodies may result from cross reactions involving antiviral antibodies—some areas of the gp120 and the MHC molecule share structural similarities (both interact with CD4), and an immunodominant region of gp41 is homologous to the β1 region of MHC-II.

5. **HIV escape from the immune response**. In spite of all the factors that lead to the depletion of immunocompetent cells and subsequent immunodepression, most patients mount an anti-HIV humoral response, which is still obvious when the patient starts to evolve into full-blown AIDS. Several factors may contribute to the "escape" of HIV from the immune response mounted by the patient:

a. HIV can avoid the immune response by integrating its genome in host cell's DNA with minimal expression of the integrated genome.

b. There are many subtypes of HIV, and HIV mutates at a much faster rate than most other viruses. This is due to the fact that the reverse transcriptase is error-prone and lacks copyediting capabilities. The mutations affecting the epitopes of gp120 against which neutralizing antibodies are directed represent a selective advantage to the mutant, able to avoid recognition by the preformed antibodies.

c. Humoral immune responses are relatively inefficient in eliminating virus-infected cells. ADCC and lysis of virus-infected cells after exposure to antibody and complement have been observed in vitro, but it is questionable that these may be significant defense mechanisms in vivo.

D. **Serological Diagnosis**

1. The initial **screening** of anti-HIV antibodies is done by an enzyme-linked immunoassay test (ELISA) using HIV antigens obtained either from infected cells or by recombinant technology. Since this is a screening test, its cutoff (particularly when used to screen blood in blood banks) is set for maximal sensitivity, since it is preferable to discard some blood units that test false-positive than to transfuse contaminated units with low antibody titers.
2. Any positive result on ELISA needs to be confirmed, first by repeating the ELISA to rule out errors or technical problems. If the repeat test is positive, the result should be confirmed primarily by **Western blot** (**immunoblot**). A Western blot is considered positive if antibodies to structural proteins (e.g., p24), enzymes (gp41), and envelope glycoproteins (gp41 or gp120) are simultaneously detected. The accuracy of the combined tests (ELISA and Western blot) is better than 99.5%.

E. **Therapy**

1. **Antiretroviral agents**. Administration of antiretroviral agents is the mainstay of treatment for HIV-infected individuals. The antiretroviral agents currently in use can be divided into several groups:

   a. **Inhibitors of reverse transcription**, which basically can be subdivided into two subgroups:

      i. **Nucleoside analogs**, such as:

      - **Zidovudine** (**Azidodideoxythymidine, ZDV, AZT**), the most widely used antiviral agent for treatment of HIV infections. This compound binds to the reverse transcriptase and blocks subsequent binding of nucleotides, thus inhibiting the polymerase. In addition, zidovudine is phosphorylated by cellular thymidine kinases, is taken up preferentially by the HIV polymerase, and causes termination of DNA transcription.
      - **Zalcitabine (2′,3′-dideoxycytidine, ddC)**
      - **Didanozine (2′,3′-dideoxyinosine, ddI)**
      - **Stavudine (2′3′-didehydro-3′deoxythymidine, d4T)**
      - **Lamivudine (2′-deoxy-3′-thiacytidine, 3TC)**.

      ii. **Non-nucleoside reverse transcriptase blockers**

      - **Nevirapine** is a non-nucleoside chain terminator that is bound to a hydrophobic pocket of the reverse transcriptase at a site different from the site(s) to which other reverse transcriptase inhibitors bind. Thus, strains of HIV resistant to several of the other reverse transcriptase inhibitors are not cross-resistant to nevirapine.

b. **Protease inhibitors**. Several synthetic, nonhydrolyzable synthetic peptides that compete as substrates for the HIV protease have recently been developed. Three drugs are currently in use—saquinavir, ritonavir, and indinavir. HIV-infected cells exposed to these compounds accumulate *gag* polyprotein precursors that are not cleaved due to the inhibition of protease activity. This results in cell death and inhibition of viral replication.

c. **Combination modality therapy**. Because of the onset of viral resistance to AZT seen in many patients after a year or more of treatment, the possible benefits of drug associations are under active scrutiny:

i. Associations of two reverse transcriptase inhibitors with different binding sites in the polymerase (the probability of a double mutant polymerase retaining functional activity becomes infinitesimally low).

ii. Association of three drugs acting at different points of the viral replication cycle (e.g., the association of two reverse transcriptase inhibitors and a protease inhibitor) have induced remarkable reductions in viral load and normalization of immune parameters. The question that remains is whether total viral eradication can be achieved.

2. **Ancillary therapy**. Besides antiretroviral agents, many other therapies have been proven necessary for the proper management of HIV-infected patients.

a. **Treatment and prophylaxis of life-threatening opportunistic infections** has proven to be perhaps the single most effective step for long-term management of HIV-infected patients. Chemoprophylaxis often includes the administration of antimycobacterial drugs, antifungal agents to prevent infections by *Candida albicans*, *Pneumocystis carinii*, and *Cryptococcus neoformans*, antiparasitic drugs to prevent disseminated toxoplasmosis, antiviral agents to prevent disseminated CMV infections, and intravenous gamma globulin to prevent pyogenic infections (mainly in infantile HIV infection).

b. **Biological response modifiers** have been used with a variety of goals:

i. **Erythropoietin** and **granulocyte colony-stimulating factor (G-CSF)** have been administered to patients with neutropenia or red-cell aplasia secondary to the administration of ZDV to promote the proliferation of red-cell and neutrophil precursors.

ii. **Interferon-α** has been approved for administration to patients with Kaposi's sarcoma.

iii. **Immunomodulating agents**, such as thymosin, interleukin-2, and isoprinosine, have been administered hoping to restore the immune functions in AIDS patients or to prevent the evolution toward AIDS in patients still asymptomatic or with pre-AIDS symptoms.

F. Immunoprophylaxis. A great effort is under way to develop a vaccine against HIV. A variety of approaches is being explored by different groups, although up to the present time there has yet to be truly significant progress in the development of a safe and effective HIV vaccine.

1. **Types of vaccines**
   a. **Attenuated vaccines**, based on creating genetically engineered strains lacking some crucial genes, so that the resulting virus causes a harmless infection. This approach has been tried successfully with SIV and an attenuated vaccine may be field-tried soon.
   b. **Killed vaccines** have not been shown to induce protective antiviral immunity in animal trials using SIV. However, there has been considerable interest in using killed vaccines to prevent the emergence of clinical disease. It has been proposed that vaccination with low doses of killed HIV boosts TH1 responses and favors the development of cell-mediated cytotoxicity. The evaluation of the efficacy of this approach is complicated by the fact that the end point for evaluation is the disease-free interval, rather long and variable.
   c. **Recombinant viral particles** made by inserting HIV glycoprotein genes in vaccinia virus genomes, for example, have been shown to induce neutralizing antibodies in animals. In humans, recombinant vaccinia virus vaccines have been tried but their overall effectiveness has not been established.
   d. **Component vaccines** have been prepared using **isolated gp120**, polymerized gp120, or gp120 peptides representing more conserved regions (such as the CD4-binding domain). Recently, Tat protein vaccines have been proposed, with the rationale that antibodies to this protein will prevent the intercellular transactivation of HIV replication mediated by soluble Tat protein.
2. **Vaccine efficacy**. The evaluation of HIV vaccine efficacy is made difficult by the fact that there are no adequate animal models (HIV can be transmitted to chimpanzees, but does not cause clinical disease in these animals) and by the lack of adequate indices of protection in humans.
   a. Most commonly, the assessment of the efficacy of a vaccine is based on the assay of protective antibodies. However, antibodies are not truly protective in the case of HIV.
   b. The lack of protection by antibodies is in great part a result of the variability of the gp120 regions against which neutralizing antibodies are directed. Different strains of HIV diverge by as much as 20% in the structure of gp120 and antibodies elicited with one strain do not cross-neutralize other strains. However, there are some strains more common than others in the infected population (e.g., the MN strain accounts for about 30% of the HIV isolated in the U.S.), and it may be possible to prepare vaccines with mixtures of the most prevalent strains.
   c. It is believed that an efficient vaccine should stimulate ADCC and/or cell-mediated immunity (CMI), which may be the only way to eliminate viral-infected cells that apparently can be involved in the transmission of HIV infection. Cytotoxic lymphocytes reacting with HIV-infected CD4+ lymphocytes have recently been identified, and it is known that certain conserved epitopes of gp120, gp41, and of the Gag protein appear more effective in inducing T-cell-mediated immunity. A trial with a recombinant vaccinia virus expressing a *gag* epitope demonstrated that it effectively induces CD8+ cells with cytotoxic activity specifically

directed to it. The problem is that the assessment of cell-mediated cytotoxicity is much more laborious and expensive than the assessment of humoral immunity.

3. **Vaccine safety**. A variety of safety questions have been raised:
   a. The remote possibility of spontaneous reversion of genetically engineered, defective HIV strains into infectious, pathogenic strains.
   b. The possibility that an immune response directed against epitopes of gp120 would induce cytotoxicity of noninfected CD4+ cells, which are known to adsorb soluble gp120 to their CD4 molecule.
   c. The possible induction of enhancing antibodies that may accelerate the progression of disease by a variety of mechanisms, such as:
      i. Antibodies may promote viral infection of macrophages by forming immune complexes that can be taken up through Fc receptors.
      ii. Antibodies may help to "select" mutant viral strains.
      iii. An ineffective humoral response may shield viral particles and infected cells against the more effective cell-mediated immune mechanisms.
   d. Recombinant vaccines using vaccinia virus as carrier have the problem of being able to cause clinical disease in an asymptomatic but already immunocompromised patient. This becomes a significant risk if these recombinant viruses are used for mass vaccination of populations with a high frequency of HIV-positive individuals whose immunocompetence may not be fully normal.

Case 2 Revisited

- The onset of *Pneumocystis carinii* in a previously healthy young adult with low T-cell count and evidence suggestive of mucocutaneous candidiasis should raise the possibility of the diagnosis of AIDS.
- Two important tests should be immediately ordered in this patient: (a) because of the lymphopenia during an acute infection, a lymphocyte subpopulation profile should be ordered; (b) because of the suspected diagnosis of AIDS, HIV serologies should also be ordered. This patient had a profound CD4+ lymphocyte deficiency (4/μL) and was HIV-positive both by EIA and by Western blot.
- A patient with profound depression of the CD4+ lymphocyte count is at risk for all types of infections by pathogenic and opportunistic agents, including bacteria, viruses, fungi, and parasites. This patient has mucosal candidiasis at the time of diagnosis and developed a systemic infection with *Mycobacterium avium-intracellulare* soon thereafter.
- At the time of diagnosis the most pressing issue was the *Pneumocystis carinii* pneumonia, which was treated with IV sulfamethoxazole-trimethoprim (SMZ-TMP). At the same time, oral fluconazole was started to control the mucosal candidiasis. Antiretroviral therapy was delayed until the patient could be placed on a maintenance dose of SMZ-TMP, because of the combined risk of bone marrow depression that is associated with combinations of ZDV and SMZ-TMP, particularly when high doses of the latter are administered. The diagnosis of disseminated infection with *Mycobacterium avium-intracellulare* was followed by administration of Clarithromycin and ethambutol. After the resolution of the acute infections that affected this patient,

he was placed on chemoprophylaxis with a combination of SMZ-TMP, clarithromycin, and fluconazole, was started with a combination of three antiretroviral drugs, and was instructed to receive periodical *S. pneumoniae* and influenza immunizations. Prevention of infections has resulted in prolonged survival for patients with AIDS.

SELF-EVALUATION

Questions

Choose the ONE *best* answer.

30.1 A large majority of individuals with serum IgA deficiency:
- A. Are asymptomatic
- B. Are equally deficient in IgG2
- C. Have a marked predisposition for upper respiratory infections
- D. Have anti-IgA antibodies
- E. Suffer from diarrhea and malabsorption

30.2 Gamma globulin administration is **not** indicated in cases of:
- A. Combined IgA/IgG2 deficiency
- B. Common, variable immunodeficiency ("acquired" agammaglobulinemia)
- C. Infantile agammaglobulinemia
- D. Isolated IgA deficiency
- E. Transient hypogammaglobulinemia of infancy

30.3 A pediatrician asks for an immunological workup of cellular immunity in a 3-year-old child who has been acutely ill with measles in the past few days. Skin tests with Candidin, SK-SD, PPD, and mumps antigen are negative. Mitogenic responses to PHA, conA, monoclonal anti-CD3 antibody, measles antigen, and tetanus toxoid are depressed. IL-2 release after stimulation of mononuclear cells with PHA is undetectable. In your report to the referring physician, you will state that:
- A. No conclusion is possible
- B. The patient has no immune abnormality
- C. The patient has primary cell-mediated immunodeficiency
- D. The results are difficult to interpret; blood should be collected as soon as possible to repeat in vitro studies
- E. There is a depression of cell-mediated immunity that could be secondary to the viral infection; the studies should be repeated in 4 weeks

Questions 30.4 and 30.5 refer to the following case history:

A previously healthy 6-month-old boy suddenly fell ill with a life-threatening *Pneumocystis carinii* pneumonia. WBC were 5,200/μL (15% neutrophils, 70% lymphocytes). Serum immunoglobulin levels were: IgG: 120 mg/dL; IgA: undetectable; IgM: 1100 mg/dL; Isoagglutinin A titer: 16; CD3+ lymphocytes in peripheral blood: 1100/μL; CD19+ lymphocytes in peripheral blood: 80/μL; mitogenic responses of T lymphocytes to stimulation with PHA and monoclonal antibody to CD3 were within normal limits. PWM stimulation of mononuclear cells was followed by the release of 2 μg of IgM/10^6 cells at day 7; no IgG was detected.

30.4 The most likely diagnosis in this case is:
- A. Common, variable immunodeficiency
- B. Hyper-IgM syndrome

C. IgA deficiency
D. Infantile agammaglobulinemia (Bruton's disease)
E. Neutropenia

30.5 The molecular basis of the immunodeficiency affecting this patient is:
A. Abnormal differentiation of granulocytes
B. Deficiency of the ZAP tyrosine kinase
C. Deficient release of IL-2 by activated T cells
D. Lack of Bruton's tyrosine kinase
E. Lack of interaction between CD40 and gp39 (CD40 ligand)

30.6 Which of the following mechanisms is unlikely to contribute to the immunodepression associated with symptomatic HIV infection?
A. Formation of soluble immune complexes involving gp120 and anti-gp120 antibodies
B. Formation of syncytia involving infected and uninfected CD4+ T lymphocytes
C. Hyperactivity of CD8+ cells with suppressor activity
D. Immunological elimination of HIV-infected T lymphocytes
E. Release of soluble gp120 from infected cells

In **Questions 30.7–30.10**, match EACH numbered word or phrase with the ONE lettered heading that is most closely related to it. Each lettered heading may be selected once, more than once, or not at all.

A. Adenosine deaminase deficiency
B. C6 deficiency
C. Common, variable immunodeficiency
D. DiGeorge syndrome
E. Infantile agammaglobulinemia

30.7 Absence of CD19+ lymphocytes
30.8 Associated with frequent infections with *Neisseria sp.*
30.9 Defective differentiation of the third and fourth pharyngeal pouches
30.10 Deficient helper T-cell function

Answers

30.1 (A) Most IgA-deficient patients are symptomatic; associated IgG2 deficiency is seen most frequently in symptomatic cases; anti-IgA antibodies are relatively rare.

30.2 (D) In isolated IgA-deficiency gamma globulin, administration is not likely to be beneficial because the commercial gamma globulin preparations contain very little IgA, and even if IgA was administered, its relative short half-life would result in its rapid disappearance from circulation.

30.3 (E) A compromise of cell-mediated immunity is frequently seen during the acute phase of measles. The defect is reversible, and the immunological parameters should normalize at the end of the convalescence period.

30.4 (B) The presentation of opportunistic infections in a young boy with low to undetectable levels of IgG, IgA, IgD, and IgE and high levels of IgM is diagnostic of the X-linked form of the hyper-IgM syndrome. T-cell mitogenic responses are normal, but B lymphocytes do not switch from IgM to IgG synthesis.

30.5 (E) The X-linked form of the hyper-IgM syndrome is caused by the lack of expression of CD40. This prevents the signaling of B lymphocytes normally mediated by CD40-gp39 interactions and the mutual stimulation of T and B cells is impaired. As a consequence, the release of cytokines involved in B-cell differentiation is also impaired, and the patient's B cells fail to differentiate into IgG-producing plasma cells.

30.6 (C) CD8+ cells are not directly infected by the HIV virus. Their differentiation into functional cytotoxic cells is compromised by the lack of help from CD4+ cells. There is no evidence for the increased activity of any cell population with suppressor functions in HIV-infected patients.

30.7 (E) In contrast to what is seen in patients with common, variable, immunodeficiency, infants with infantile agammaglobulinemia have a marked depression or absence of B cells in the peripheral blood.

30.8 (B) *Neisseria* infections are characteristically frequent in patients with deficiency of the late complement components (C5 to C9).

30.9 (D) Infants with the DiGeorge syndrome have combined aplasia of the thymus and parathyroids, and other congenital abnormalities of the facial bones, heart, and large vessels.

30.10 (C) The cause of most cases of common, variable immunodeficiency is not lack of B cells (they are present in normal concentrations), but their deficient response to antigenic stimulation, which is secondary to a T helper cell deficiency.

BIBLIOGRAPHY

Advances in Primary Immunodeficiency (The Jeffrey Modell Immunodeficiency Symposium). *Clin. Immunol. Immunopathol.*, *75* (no. 3, part 2):S145, 1995.

Bloom, B.R. A perspective on AIDS vaccines. *Science*, *272*:1888, 1996.

Chandra, R.K. Cellular and molecular basis of nutrition-immunity interactions. *Adv. Exp. Med. Biol.*, *262*:13, 1990.

Cunningham-Rundles, C. Clinical and immunologic analysis of 103 patients with common variable immunodeficiency. *J. Clin. Immunol.*, *9*:22, 1989.

Fauci, A.S., Pantaleo, G., Stanley, S., and Weissman, D. Immunopathogenic mechanisms of HIV infection. *Ann. Intern. Med.*, *124*:654, 1996.

Goldstein, G. HIV-1 Tat protein as a potential AIDS vaccine. *Nature Med.*, *1*:960, 1966.

Grieco, M. and Virella, G. Acquired immunodeficiency syndrome. In *Principles and Practice of Medical Therapy in Pregnancy*, 3rd ed., 1997.

Harris, B.H. and Gelfand, J.A. The immune response to trauma. *Semin. Pediatr. Surg.*, *4*:77, 1995.

Haynes, B.F., Pantaleo, G., and Fauci, A.S. Toward an understanding of the correlates of protective immunity to HIV infection. *Science*, *271*:324, 1996.

Ho, D.D. Viral counts count in HIV infection. *Science*, *272*:1124, 1996.

Richman, D.D. HIV therapeutics. *Science*, *272*:1886, 1996.

Rosen, F.S., Cooper, M.D., and Wedgwood, R.J.P. The primary immunodeficiencies. *N. Engl. J. Med.*, *333*:431, 1995.

Virella, G. Humoral immunity and complement. In *Infections in Immunocompromised Infants and Children*, C.C. Patrick, ed. Churchill-Livingstone, New York, 1992.

Weiss, R. HIV receptors and the pathogenesis of AIDS. *Science*, *272*:1885, 1996.

Yocum, M.W. and Kelso, J.M. Common variable immunodeficiency: The disorder and the treatment. *Mayo Clin. Proc.*, *66*:83, 1991.

Index